CRITICAL ISSUES
IN THE HISTORY OF
SPACEFLIGHT

CRITICAL ISSUES CONFERENCE SPEAKERS

Left to right: James Hansen, John Krige, Asif Siddiqi, David DeVorkin, Howard McCurdy, Woody Kay, Andrew Butrica, Margaret Weitekamp, Phil Scranton, Steven Dick, Stephen Pyne, Alexander Brown, Slava Gerovitch, Stephen Johnson, and David Mindell. Not shown: Roger Launius, Peter Hays, John Logsdon, Todd La Porte, and Diane Vaughan. The image was taken in front of the Space Shuttle *Enterprise* at the Steven F. Udvar-Hazy Center of the National Air and Space Museum.

CRITICAL ISSUES
IN THE HISTORY OF
SPACEFLIGHT

Steven J. Dick and Roger D. Launius
Editors

National Aeronautics and Space Administration
Office of External Relations
History Division
Washington, DC

2006

NASA SP-2006-4702

Library of Congress Cataloging-in-Publication Data

Critical issues in the history of spaceflight / Steven J. Dick and Roger D. Launius, editors.
    p. cm. -- (The NASA history series)
Includes index.
1. Astronautics--History. I. Dick, Steven J. II. Launius, Roger D. III. Series.
TL788.5.C66 2005
629.409--dc22
                        2005035416

Contents

INTRODUCTION

At a May 1981 "Proseminar in Space History" held at the Smithsonian Institution's National Air and Space Museum (NASM) in Washington, DC, historians came together to consider the state of the discipline of space history. It was an historic occasion.[1] The community of scholars interested in the history of spaceflight was not large; previously, well-meaning but untrained aficionados consumed with artifacts had dominated the field, to the exclusion of the larger context.[2] At a fundamental level, this proseminar represented a "declaration of independence" for what might be called the "new aerospace history." In retrospect, it may be interpreted as marking the rise of space history as a recognizable subdiscipline within the field of U.S. history. Bringing together a diverse collection of scholars to review the state of the art in space history, this proseminar helped in a fundamental manner to define the field and to chart a course for future research. Its participants set about the task of charting a course for collecting, preserving, and disseminating the history of space exploration within a larger context of space policy and technology.

In large measure, the course charted by the participants in this 1981 proseminar aided in advancing a very successful agenda of historical research, writing, and understanding of space history. Not every research project has yielded acceptable results, nor can it be expected to do so, but the sum of the effort since 1981 has been impressive. The opportunities for both the exploration of space and for recording its history have been significant. Both endeavors are noble and aimed at the enhancement of humanity. Whither the history of spaceflight? Only time will tell. But there has been an emergent "new aerospace history" of which space history is a central part that moves beyond an overriding concern for the details of the artifact to emphasize the broader role of the spacecraft. More importantly, it emphasizes the whole technological system, including not just the vehicle but also the other components that make up the aerospace climate, as an integral part of the human experience. It suggests that many unanswered questions spur the development of flight and that inquisitive individuals seek to know that which they do not understand.

---

1. Richard F. Hirsh, "Proseminar on Space History, 22 May 1981," *Isis* 73, no. 266 (1982): 96–97. There had been previous gatherings of historians interested in the subject, but these had mostly been oriented toward specific subdisciplines such as space science. See Paul A. Hanle and Del Chamberlain, eds., *Space Science Comes of Age: Perspectives in the History of Space Sciences* (Washington, DC: Smithsonian Institution Press, 1981).

2. At that time, only the several volumes published as part of the NASA History Series, all written by credible scholars, and John M. Logsdon's *The Decision to Go to the Moon: Project Apollo and the National Interest* (Cambridge, MA: MIT Press, 1970) were accepted as works of serious scholarship by the larger historian community.

This assumption arises within historians and is based on their understanding of humans, for technological systems are constructions of the human mind.[3]

This "new aerospace history," therefore, emphasizes research in aerospace topics that are no longer limited to the vehicle-centered, project-focused, scientific internalist style of space history. Many of the recommendations that historian James R. Hansen suggested in an important historiographical article in *Technology and Culture* are beginning to come to fruition.[4] Taken altogether, these tentative explorations of themes build on what has gone before. At the same time, they represent a departure from the simplistic works that preceded them, notably the argumentative volumes and essays that either espouse or ridicule space exploration.

Twenty-four years after the 1981 proseminar, the National Aeronautics and Space Administration (NASA) Headquarters History Division and NASM's Division of Space History brought together another group of scholars—including historians, political scientists, sociologists, public administration scholars, and engineers—to reconsider the state of the discipline. This volume is a collection of essays based on this workshop on "Critical Issues in the History of Spaceflight," held at the Steven F. Udvar-Hazy Center of the National Air and Space Museum on 15–16 March 2005. The meeting was especially timely because it took place at a time of extraordinary transformation for NASA, stemming from the new Space Exploration Vision, announced by President George W. Bush in January 2004, to go to the Moon, Mars, and beyond. This Vision in turn stemmed from a deep reevaluation of NASA's goals in the wake of the Space Shuttle *Columbia* accident and the recommendations of the Columbia Accident Investigation Board. By June 2004, a nine-member Presidential Commission on Implementation of United States Space Exploration Policy, led by former Secretary of the Air Force Edward "Pete" Aldridge, had produced a report on "A Journey to Inspire, Innovate, and Discover." In February 2005, NASA's strategic objectives were released in a report called "The New Age of Exploration." All these documents placed the new vision in the context of the importance of exploration and discovery to the American experience.[5]

---

3. Roger D. Launius discusses the richness of what has been accomplished thus far in "The Historical Dimension of Space Exploration: Reflections and Possibilities," *Space Policy* 16 (2000): 23–38.

4. James R. Hansen, "Aviation History in the Wider Context," *Technology and Culture* 30 (fall 1989): 643–649.

5. Columbia Accident Investigation Board, *Report*, (Washington, DC, 2003), 6 volumes. The President's program for NASA as announced on 14 January 2004 was entitled "A Renewed Spirit of Discovery." It was followed in February by a more detailed "Vision for Space Exploration." The Aldridge Commission report was *A Journey to Inspire, Innovate and Discover*. Events leading up to the Vision are detailed in Frank Sietzen, Jr., and Keith L. Cowing, *New Moon Rising: The Making of America's New Space Vision and the Remaking of NASA* (Burlington, Ontario: Apogee Books, 2004), as well as in the Aldridge report.

As the meeting took place, NASA had not flown a Space Shuttle since the *Columbia* accident on 1 February 2003 and was looking forward to returning to flight in mid-2005. At the same time, the space agency was in the midst of a reorganization and a change in programs of truly historic proportions. The transformation potentially heralded the beginning of a new era, as the Agency's human spaceflight program sought to leave the Space Shuttle behind and depart Earth orbit for the Moon and Mars—something that humans had not done since the end of the Apollo era more than three decades earlier. Because the new Vision was to be achieved with little or no addition to NASA's $16-billion annual budget, attempts to develop an implementation plan set off a debate on the relative merits of other areas of NASA's portfolio. Funding for aeronautics was under severe pressure, with serious implications for NASA's aeronautics research centers at Glenn, Langley, and Ames. In the wake of renewed emphasis on human spaceflight, the space science community was quick to argue that its activities were also exploration, an integral part of the "Moon, Mars, and beyond" vision, and therefore should not be subject to cuts. Earth science—which had been administratively combined with space science as part of the recent transformation—could not so easily make that argument, but it had Congress largely on its side because of the practical implications of the Earth Observing System. Also in the mix was the extraordinary and sustained controversy over a servicing mission for the Hubble Space Telescope, in which the public, Congress, and the science community had strong opinions, mostly favoring a servicing mission. Finally, it was also a time of transition between Administrators: after three years of heading the Agency, Sean O'Keefe departed in February, and on 11 March, the President nominated a new Administrator, Michael Griffin, who was confirmed by the Senate and became the 11th NASA Administrator on 14 April.

As these issues swirled, March 2005 thus proved a particularly appropriate time to assess some of the perennial challenges and concerns of spaceflight, with the primary goal of providing perspective on current issues. Six critical issues were chosen for analysis. The first session examined motivations—the persistent question of why we should go into space at a time when there are so many problems on Earth. The second session provided background on another often-asked question, why should so much be spent on human spaceflight if robotic spacecraft were cheaper and more efficient? The controversy then raging over servicing the Hubble Space Telescope with the Space Shuttle demonstrated that this dichotomy was not quite so simple; without human spaceflight and four servicing missions, the myopic Hubble would never have functioned properly and certainly would not have reached its 15th anniversary on 25 April 2005. The third session could provide only a sampling of case studies of NASA's relations with external groups, in this case with the Department of Defense (DOD), international relations, and a portion of the aerospace industry. The fourth session shed light on another persistent issue: why there has been no replacement for the aging Space Shuttle. The fifth session, on NASA cultures, reflected the preoccupation with safety and risk in the wake of the *Columbia* accident. A concluding session addressed specific questions relating to the historiography of spaceflight and suggested possibilities for future research. After the assessment of distinct issues, it particularly considered the second goal of the meeting: to assess the state of the field of space history.

Two decades had passed since serious attempts had been made to assess the state of the field. In addition to the 1981 proseminar, NASA and NASM joined forces once again to hold a broader meeting in the spring of 1987, published as *A Spacefaring Nation: Perspectives on American Space History and Policy.*[6] In its treatment of issues, *Critical Issues in the History of Spaceflight* is broader in some respects but narrower in others. The title and spirit of the current volume hearkens back to Marshall Clagett's book *Critical Problems in the History of Science*, a collection of essays from a meeting at the beginning of the Space Age that had a considerable influence on the evolution of the history of science.[7] Space history was no part of that volume, but the 50 intervening years have given rise to a new kind of history with links to scientific, technological, political, cultural, and social history.

Although the subject of the meeting was "Critical Issues in the History of Spaceflight," this did not imply that history was the only mode of analysis that could be applied. Experts with a variety of backgrounds brought a variety of approaches to the chosen critical issues, including history, cultural studies, political science, and sociology. The reader will therefore find a range of approaches reflecting these backgrounds.

Certainly not all subjects could be covered at this meeting. NASA's first *A*, aeronautics, was not represented at all—not for a lack of issues, but precisely because an entire conference could be devoted to the subject. In addition, the focus was naturally on NASA and American space history, despite papers on international relations, and comparisons of the U.S. and Soviet space programs. The space sciences also received short shrift in this workshop and in this resulting volume. Again, there is more than enough in this arena to fill an entire volume. The issues encompassed by space history, along with its interconnections with the broader world and with other forms of analysis in

---

6. Martin J. Collins and Sylvia D. Fries, eds., *A Spacefaring Nation: Perspectives on American Space History and Policy* (Washington, DC: Smithsonian Institution Press, 1991). A similar conference hosted by Yale University in 1981 was published as Alex Roland, ed., *A Spacefaring People: Perspectives on Early Space Flight* (Washington, DC: NASA, 1985).

7. Marshall Clagett, *Critical Problems in the History of Science* (Madison: University of Wisconsin Press, 1959).

history and the social sciences, compose a field now grown so large—in scope if not in practitioners—that only a fair sampling can be given here. If this volume serves to stimulate more research in these areas, which we believe are of vital importance to the nation and the world, it will have served its purpose.

The meeting was a small workshop with 18 presentations and several dozen audience members who contributed substantially to the discussions. Even a small workshop, however, engendered numerous logistics. We would like to thank General John R. Dailey, Director of the National Air and Space Museum, for allowing us to use the beautiful Udvar-Hazy Center, just opened in December 2003 and adjacent to Dulles International Airport. It was a pleasure to contemplate space history in the midst of the Concorde, the SR-71 Blackbird, and the Space Shuttle *Enterprise*, among other aviation and space icons, all part of "the cathedral of the artifact," as it was termed during the meeting. For essential logistical help, we thank Nadine Andreassen, Giny Cheong, and Annette Lin, all of the NASA History Division. We are grateful to Chris Brunner and Tim Smith of SAIC for recording the proceedings on videotape. A copy of the video, along with transcripts of the discussions, may be accessed in the NASA Historical Reference Collection at NASA Headquarters.

At the NASA Headquarters Printing and Design Office, our thanks to Lisa Jirousek for copyediting and Shelley Kilmer-Gaul for design and layout. Finally, we wish to acknowledge the many contributions of those who participated in the workshop, both as presenters and from the audience. This book represents a final report on the activities of the workshop, and we hope that it will stimulate additional contemplation, research, and presentation of the history of spaceflight.

Steven J. Dick, NASA Chief Historian
Roger D. Launius, Chair, NASM Department of Space History

SECTION I

MOTIVATIONS FOR SPACEFLIGHT

INTRODUCTION

The first section of this volume examines what is perhaps the most basic question that can be asked of the Space Age: Why do nations undertake spaceflight, and why should they? It is a question equally important for understanding the history of spaceflight and for divining its future. And it is a question that history is in a unique position to illuminate. From its inception in 1957 to "The New Age of Exploration" that NASA proclaimed in 2005 in the wake of the Vision for Space Exploration, the Space Age has inevitably been linked with the idea of exploration as a motivating force. In the opening paper of the conference, Stephen Pyne argues that the idea of exploration and its links to the past need to be examined in more detail and in the context of the cultures in which it is embedded. Many writers, especially journalists, have seen space activities as part of an unbroken line of exploration going back at least to the Renaissance Age of Discovery and even earlier. Richard S. Lewis's *From Vinland to Mars: A Thousand Years of Exploration* is a prime example of this view. By contrast, Pulitzer Prize–winning historian William H. Goetzmann distinguishes a "Second Great Age of Discovery," beginning with 18th-century explorers such as Captain James Cook and Alexander von Humboldt—an age characterized by further geographic exploration, now driven by the scientific revolution and still in progress. Goetzmann sees this fissioning of ages as important to understanding the differences between the two.[1]

While examining the characteristics and lessons of the first two ages, Pyne now proposes a Third Age of Discovery, which segregates space exploration from the motivators of the Second Age and places it with the exploration of the Antarctic and the deep oceans.[2] This distinction, he argues, is important to understanding the unique character of the current age. Just as for the Second Age, science replaced God, commerce replaced gold, and national prestige trumped individual glory, the motivators for the Space Age have changed in part. Most strikingly, at least so far, and perhaps happily, since such encounters in the past have left more than one civilization decimated, explorers of the Space Age have not had to worry about encounters with indigenous inhabitants of the lands they explore.[3]

---

1. William H. Goetzmann, *New Lands, New Men: America and the Second Great Age of Discovery* (New York: Penguin Books, 1987).

2. He also made this case in his article, "Space: A Third Great Age of Discovery," *Space Policy* 4 (August 1988): 187–199.

3. For a discussion of this problem, see Jane M. Young, "'Pity the Indians of Outer Space:' Native American Views of the Space Program," *Western Folklore* 46 (October 1987) 269–279.

Pyne also argues that it is "cultural conditions that prompt and sustain discovery" and that exploration is an invention of particular societies. This is no academic distinction, but one with real-life consequences: if exploration is a cultural invention, then it may pass away as have other cultural inventions and, indeed, as exploration itself has withered in some societies throughout history. This is no less true in the American context than it is in other societies, now or in the past: Carl Sagan, Ray Bradbury, Robert Zubrin, and others have argued that exploration is a societal imperative with unique valences to American history and the American character.[4] Exploration means many things to many people, and historians need to analyze these meanings and understand the myriad ways in which culture imbues exploration with meaning, or with no meaning at all.

Pyne's essay is full of provocative suggestions: that the idea of exploration needs to be decoupled from the idea of colonization; that the Second Age collapsed not only from closed frontiers, but also from a weariness with the Enlightenment enterprise; that geopolitical rivals may divert some of their energies from the battlefield to exploration; that Voyager's Grand Tour may be for the Third Age what Humboldt was for the Second and Magellan for the First; that the Third Age may already be in decline; that cyberspace may be more important in historical terms than outer space; and that although encounters with other cultures were essential to creative individuals and societies in the first two Ages, that possibility is unlikely for the Third Age, at least in the near future, unless by remote radio communication.

In the second paper, Roger Launius takes a broader view of the motivations for spaceflight and enumerates five, and only five, rationales operating over the last 50 years: human destiny and survival of the species; geopolitics, national pride, and prestige; national security and military applications; economic competitiveness and satellite applications; and scientific discovery and understanding. Launius argues that some of these rationales rest on a fundamental desire to become a multiplanetary species and, in particular, to found utopian societies beyond Earth.

In the context of the human destiny argument, Launius finds that the "frontier thesis"—the idea that the existence of a frontier has given Americans their most distinctive characteristics and that space exploration is important for that reason alone—is counterproductive for a postmodern, multiculturalist society. Yet "the final frontier" continues to be a rallying cry for space enthusiasts. Is this inappropriate, or can the frontier thesis be separated from the charges of excessive ethnocentrism?

---

4. See in particular Robert Zubrin, "Epilogue: The Significance of the Martian Frontier," in *The Case for Mars* (New York: Free Press, 1996).

In the area of national security and military applications, Launius emphasizes a fact little known outside the space community: that since 1982, military spending on space has outpaced civilian spending. By 2003, the Department of Defense was spending $19 billion on space, compared to NASA's $14 billion. Obviously, the military is motivated to use space as "high ground." Launius finds that the economic competitiveness argument, though emphasized by the conservative agenda since the 1980s, remains mixed: although communications satellites have proven a commercial success since COMSAT and Intelsat in the early 1960s, other efforts such as Landsat and the Global Positioning System (GPS), while great technical successes, have not yet proven commercially viable. Space tourism and private investment for access to space are barely at the beginning of their potential. Whether these activities become economically viable, thereby causing the commercial motivator to become increasingly important, is one of the great open questions of the Space Age.

Launius discusses science as a motivator at some length; however, in the context of Pyne's paper, it is notable that he does not explicitly include exploration as one of his five motivations, instead viewing it as a means to an end rather than an end in itself. He briefly discusses it in the context of the human destiny argument and the frontier thesis, and he later uses it again in the context of the science motivator, noting that a National Research Council (NRC) study in 2005 proclaimed that "exploration done properly is a form of science."[5] It should be noted that the NRC did so in the context of threatened cuts to space science—money that would go to the new human exploration program—and therefore had a vested interest in relating science to exploration. This raises the interesting question of the differences between science and exploration in principle and in practice. While it is clear that, as Launius argues, there are synergies between science and exploration, one could clearly argue that they are not one and the same. After all, Magellan was an explorer, not a scientist; conversely, many scientists undertake routine science that can hardly be called exploration. One might argue a relationship as follows: when exploration is undertaken, it may lead to discoveries, which then are explained by science and in turn add to the body of scientific knowledge. Alternatively, one might also argue that when exploration is undertaken, it is usually done with an economic, military, or nationalistic purpose in mind, but that exploration, viewed as benign while the true objective may be less so, serves as the rationale. As Pyne puts it in his article, historically "society needed science, science needed exploration, exploration to far countries [or outer space] needed support," at the national level.

---

5. National Research Council, *Science in NASA's Vision for Space Exploration* (NRC: Washington, DC, 2005).

These distinctions are more than semantic in nature—they become an issue of public policy when decisions must be made about the balance between human and robotic exploration (see section II). Although Apollo clearly produced important science, as Launius points out, it was criticized for not generating enough science relative to its high cost. Yet one could argue that the explorations of Apollo represented something beyond science that will be remembered as one of humanity's greatest triumphs. At least some space scientists have come to this realization, despite the high costs and the risks involved in human spaceflight. At a NASA meeting on risk and exploration, Steve Squyres, the science principal investigator for the Mars Exploration Rovers, allowed that he loved his machines, which are still active after 16 months. But, he added, "when I hear people point to Spirit and Opportunity and say that these are examples of why we don't need to send humans to Mars, I get very upset. Because that's not even the right discussion to be having. We must send humans to Mars. We can't do it soon enough for me."[6] Squyres's words reflect a deep truth: even though science may be a motivation for exploration and a product of it, human exploration is more than the sum of all science gained from it. If exploration is a primordial human urge, and in a larger sense the mark of a creative society, to what extent should a society support it in the midst of many other priorities? In a democratic society, that is a question with which the public, and public policy-makers, must grapple.

---

6. Steven J. Dick and Keith Cowing, eds., *Risk and Exploration: Earth, Sea and the Stars* (Washington, DC: NASA SP-2005-4701, 2005), p. 179.

CHAPTER 1

SEEKING NEWER WORLDS:
AN HISTORICAL CONTEXT FOR SPACE EXPLORATION

Stephen J. Pyne

Come, my friends,
Tis not too late to seek a newer world
—Tennyson, "Ulysses," 1842

Nearly 40 years ago, William Goetzmann, in his Pulitzer-winning *Exploration and Empire*, argued that explorers were "programmed" by their sponsoring societies. They saw what they were conditioned to see, and even novelty fell within a range of expected "curiosities" and "marvels." What is true for explorers has been no less true for exploration's philosophers, historians, and enthusiasts. Pundit and public, commentator and scholar, all have become accustomed, if not programmed outright, to see exploration and space as inseparable. Space has become the new frontier; exploration, if it is to thrive, must push to the stars; the solar system is where, in our time, exploration is happening.[1]

Since Sputnik, no survey of exploration has not looked heavenward, and no advocate for space adventuring has failed to trace its pedigree through the lengthening genealogy of the Earth's explorers. But in the particulars they differ; this field, too, has its "lumpers" and "splitters." The lumpers consider the long saga of geographic exploration by Western civilization as continuous and thematically indivisible. The Viking landers on Mars are but an iteration of the long ships that colonized Greenland. The *Eagle*, the Command Module orbiter, and Saturn V rocket that propelled the Apollo XI mission to the Moon are avatars of Columbus's *Niña*, *Pinta*, and *Santa Maria*. The "new ocean" of planetary space is simply extending the bounds of the old. The ur-lumpers would go further. The historic eruption of European exploration was but the most recent device to carry humanity's expansive hopes and ambitions;

---

1. William H. Goetzmann, *Exploration and Empire* (New York: Knopf, 1966).

its origins reside in the genetic code of humanity's inextinguishable curiosity. Even more, space exploration shares an evolutionary impulse. Through humanity, life will clamber out of its home planet much as pioneering species crawled out of the salty seas onto land. The impulse to explore is providential; the chain of discovery, unbroken; the drivers behind it, as full of evolutionary inevitability as the linkage between DNA and proteins.

The most prominent have generally boosted space exploration as necessary, desirable, and inevitable. The argument assumes the form of a syllogism: The urge to explore is a fundamental human trait. Space travel is exploration. Therefore, sending people into space is a fundamental characteristic of our species—what more is there to say? The only impediment to the past serving as prologue to the future is imagination, as translated into political will, expressed as money. From Carl Sagan to Ray Bradbury, such advocates have self-admittedly been fantasists, whether they argued that the motivating vision is embedded in our genes or our souls. But the urge, the motivating imperative, they place within the broad pale of *Homo sapiens sapiens*.[2]

Yet humanity doesn't launch rockets; nations do. So there exist also among the spacefaring folk special themes that place interplanetary exploration within the peculiar frame of a national experience. In particular, there exist groups for whom extra-Earthly exploration is a means to perpetuate or recreate what they regard as the fundamental drivers of American civilization. Space exploration offers the chance to discover another New World and to erect a New America, a technological New Jerusalem, beyond the tug of the Earth's gravitational field and the burdens of its past. Only a New Earth can save the Old. Space colonization would remake William Bradford's vision of Plymouth Plantation into a very high-tech city and transplant it to a very distant hill.[3]

Still, a countercase exists. What expands can also collapse. Ming China launched seven dazzling voyages of discovery and then shut down all foreign travel and prohibited multimasted boats. Medieval Islam sponsored great travelers before shrinking into the ritual pilgrimage of the hajj. The Norse spanned the Atlantic, then withered on the fjords of a new world. Moreover, plenty of peoples stayed where they were: they lacked the technological means, the fiery incentives and desperate insecurities, or the compelling circumstances to push them to explore beyond their homeland. Like Australia's Aborigines, they were content to cycle through their ancestral Dreamtime and felt little

---

2. Examples among the celebrity celebrants might include Carl Sagan, *Cosmos* (New York: Random House, 1980); Ray Bradbury et al., *Mars and the Mind of Mind* (New York: Harper and Row, 1973); and Arthur C. Clarke, *The Exploration of Space* (New York: Harper, 1959).

3. See, as an extreme example, Robert Zubrin, *The Case for Mars* (New York: Touchstone, 1996).

urgency to search beyond the daunting seas or looming peaks. A walkabout was world enough.[4]

In this perspective, what matters are the particulars—the cultural conditions that prompt and sustain discovery. What is commonly called "geographic exploration" has been, in truth, a highly ethnocentric enterprise. It will thrive or shrivel as particular peoples choose. There is nothing predestined about geographic discovery, any more than there is about a Renaissance, a tradition of Gothic cathedrals, or the invention of the electric lightbulb. Exploration as a cultural expression is something peculiar to times, places, and peoples. General historians might site exploration within dramas of human mobility, of empires, of Europe's astonishing millennial-long expansion, and its equally astounding almost-instantaneous implosion. They would grant exploration little intrinsic motivation; explorers would derive their inspiration, no less than their characteristics, from a sustaining society. They view contemporary arguments for space trekking as not grounded in historic reality but inspirational rhetoric.

From such a perspective, the exuberant era of exploration that has dominated the past five centuries, bonded to European expansion, is simply another in a constellation of cultural inventions that have shaped how peoples have encountered a world beyond themselves. It will, in time, pass away as readily as the others; European-based exploration may yet expire, even after 500 years, perhaps exhausted like the cod fisheries of the Grand Banks. The history of exploration bears little similarity to the simplistic narrative of triumphalists. Historians, litterateurs, humanists, and a significant fraction of ordinary citizens may wonder why a chronicle of past contacts, particularly when burdened by imperialism and inflated by tired clichés, should argue for doing more. The record suggests that future worlds will be corrupted as old ones were. The much-abused Earth is world enough. Space exploration may prove to be a defiant last hurrah rather than a daring new departure.

To date, the lumpers have commanded the high ground of historical interpretation and historiography. Dissenters are few, and even they accept space travel's exploring pedigree. Scholarship has hardly begun to parse exploration's long chronicle to understand what features might apply or not apply to the Space

---

4. For a good discussion of the Norse traverse across the Atlantic, see Carl Sauer, *Northern Mists* (Berkeley: University of California Press, 1968). Studies of the Chinese voyages have become a minor cottage industry; see, for example, Gavin Menzies, *1421: The Year China Discovered the World* (New York: Bantam, 2002). On the Islamic eucumene, see Richard Hall, *Empires of the Monsoon: A History of the Indian Ocean and Its Invaders* (London: HarperCollins, 1996). The Polynesian voyages are the subject of endless retellings; an early, defining work is Peter Buck, *Vikings of the Pacific* (Chicago: University of Chicago Press, 1959).

Age. Does the Apollo program resemble Columbus landing in the Bahamas, or the abortive Greenlanders at Vinland? Does Voyager mimic Captain James Cook's circumnavigating sallies, or Roald Amundsen's small-craft threading of the Northwest Passage? Is Mariner a robotic version of Lewis and Clark, leading America to its new westward destiny, or a Zebulon Pike, whose expeditions south led him to a Mexican jail and whose forays north left him dead outside a Canadian fort? The history of exploration is so complex that one can find whatever anecdote and analogy one wants.

How one identifies the exploratory character of space depends on how one interprets the enterprise—whether space travel is primarily about technology, science, adventure, geopolitics, or inspiration. Each theme can lead to its own history. For space as exploration, however, two clans dominate the discourse: space enthusiasts eager to trace the genealogy of exploration from ancient times to contemporary launch sites, and historians anxious to push their erstwhile narratives into today's news. The two display a kind of symmetry, a yin and yang of emphasis. Space enthusiasts tend to condense exploration prior to the mid-20th century into a lengthy prelude, while historians of exploration—there aren't that many—update their chronicles to include space endeavors into a kind of coda. The common assumption is that space is of course exploration, so there is little need to explain how and why. One only needs enough of the past to boost the narrative into orbit, or enough contemporary events to predict the narrative splashdown. Instead of analysis, the ur-lumper rhetoric tends to conflate a cascade of themes: intellection with exploration, exploration with contact, contact with colonization, colonization with human settlement.

Of course there are exceptions; the best scholarship usually is. A good example of exploration considered from the perspective of space is Richard S. Lewis's *From Vinland to Mars: A Thousand Years of Exploration*. The Space Age, he concludes, "can be defined reasonably as the modern extension of a process of exploration that began a thousand years ago with the Norse voyages to Greenland and North America." The common motivator was "intraspecific competition," the deadly contest "among men and families for land, among nations for power and wealth." This persistent trait could yet "carry future generations to the stars." Lewis devotes 100 pages to exploration prior to 1957 and 300 from the International Geophysical Year (IGY) to the Viking landings on Mars. Like most lumper historians, he came to space themes by way of journalism.[5]

---

5. Richard S. Lewis, *From Vinland to Mars: A Thousand Years of Exploration* (New York: Quadrandle, 1976), pp. xi, xii.

Scholar historians, by contrast, are likely to carve up the long chronicle into more manageable units. Unquestionably, the outstanding practitioner is William H. Goetzmann. Building on J. C. Beaglehole's scholarship on Captain Cook and the exploration of the Pacific generally, in which Beaglehole argued that the voyages constituted something new, a renewal of global exploration, Goetzmann has elaborated the concept of a Second Great Age of Discovery, which he believes has not yet ended, which extends unbrokenly into space, and which has fundamental valences with America. "Just as in the Renaissance," he writes, "a New Age of Discovery began—born of competition between men and nations, dependent alike on abstract theory, applied science, now called 'engineering,' visionary imagination and the faith of whole cultures who invested billions of dollars or rubles in the great adventure out into the frontier that President John F. Kennedy called 'this new ocean.'" The undertaking has special resonance for the United States, for "America is the product of an Age of Discovery that never really ended. From the Viking voyages in the 10th century to the lunar voyages of the twentieth, much that is held to be American derives from a sense of the ongoing and complex process of exploration that has made up so much of its history." The explorer, Goetzmann concludes, "stands as a kind of archetypal American."[6]

Even so, Goetzmann, ever the scholar, concluded that the ultimate payoff lay in the realm of knowledge, particularly the peculiar moral understanding that helps us understand who we are and how we should behave. In explicating that understanding, Goetzmann, always the historian, chronicles exploration against the "constant imaginative redefinition of America." In that sense, "America has been almost anything its explorers or their 'programmers' wanted it to be at the time. And yet constant discoveries and rediscoveries have continually changed the meaning of the country for its citizens." Thus, "to many," by implication himself included, the analogy of Apollo to Columbus "seems false." Rather, "what Armstrong and Aldrin and all their heroic space predecessors have revealed is not a series of new worlds for escape and habitation, but a profounder knowledge of the earth's true place in the universe. They have changed once again the entire perspective of the globe and man's place on it." Yet Goetzmann, author of a trilogy of books on American exploration, never included space exploration in those volumes, save allusively in a preface.[7]

---

6. J. C. Beaglehole, *The Exploration of the Pacific*, 2nd ed. (Stanford: Stanford University Press, 1966); William H. Goetzmann, "Exploration's Nation. The Role of Discovery in American History," in *American Civilization: A Portrait from the Twentieth Century*, ed. Daniel J. Boorstin (New York: McGraw-Hill Book Co., 1972), pp. 36, 25.

7. Goetzmann, "Exploration's Nation," pp. 33, 36.

And then there are those for whom space is continuous not merely with exploration but with evolution, for whom the Space Age represents a quantum leap in human existence. The nuances of geographic discovery's changing technologies, beliefs, lore, institutions, and personalities become mere background noise, the junk genes of history. Most practitioners come from literature or natural science, an odd couple joined by conviction and pulp fiction rather than formal scholarship. History is a loose jumble of anecdotes, like oft-told family stories or the sagas of the clan. For them, the future is what matters. What preceded contact is only preamble. What follows will be, in Arthur C. Clarke's words, childhood's end.

Regardless, no one questions the linkage of space with exploration. Their analysis of what that bond is, and what benefits the country might derive, vary. Exploration remains a means to other ends. The recent report of the President's Commission on Implementation of United States Space Exploration Policy described the goal of the "vision" as "to advance U.S. scientific, security, and economic interests . . . ," and not least national prestige. More realistically, at the time Mariner orbited Mars, Bruce Murray observed simply that "we are exploring," that the "very act of exploration is one of the more positive achievements open to a modern industrial society," that space exploration is "as important as music, art, as literature," that it is "one of the most important long-term endeavors of this generation, one upon which our grandchildren and great-grandchildren will look back and say, 'That was good.'" But if space exploration is a cultural enterprise, then it should be examined as such, subject to the same tangible criteria.[8]

All this suggests ample opportunity for future research. There is, first, a place for dedicated analysis beyond the selective anecdote, heroic narrative, and flimsy analogizing. There is little empirical heft, even less quantitative data, and sparse scrutiny of what, exactly, exploration has meant in terms of economics, politics, ethics, knowledge, fiction, and the like. Serious scholarship has not tracked exploration to the extent it has related fields such as the history of physics, military history, government institutions, or even the literature of the western hero. The founding saga of the Great Age of Discovery, Luis de Camões's *Lusiadas*, is, after all, a tragedy, brooded over by the Old Man of Belem, who declaims its debased origins: the enterprise to the Indies will turn out badly, though it cannot be stopped. By the time we arrive at the Space Age, "literature" has come to mean Edgar Rice Burroughs and the imagination of nine-year-old boys. A similar declension seems to affect the scholarship.

---

8. Report of the President's Commission on Implementation of United States Space Exploration Policy, *A Journey to Inspire, Innovate, and Discover* (Washington, DC: Government Printing Office [GPO], June 2004), p. 11; Bruce Murray, quoted in Ray Bradbury et al., *Mars and the Mind of Man*, pp. 24–25.

A part of this general task is comparative study. We know things in the context of other things. We will understand contemporary exploration better if we arrange suites of comparisons with past efforts, understanding their common elements, isolating what features make them similar or different. But this exercise should not be restricted to exploration, comparing itself in various eras. If expeditions are a cultural creation, then they should be compared as well with other undertakings, perhaps with opera, baseball, publishing, art museums, extreme sports. Exploration is not the only way by which a society can express its restlessness or exercise its curiosity. A society must choose to explore. What is the basis for that choice and how does the outcome compare with other choices?

Of particular value to Americans is the need to segregate exploration from colonization. American accounts of the Space Age almost invariably begin with the discovery of North America, preferably by the Norse. This is teleological history: the point, the conclusion, of exploration was to find a New World and, subsequently, to found the United States. The epic of America is its expansion westward. When exploration completed its survey of America, it had to continue elsewhere, to the Poles, for example, and then to planetary space, or else that epic would end. It makes a wonderful national creation story. It works less well as scholarship. The exploration of America was part of a global project, rising and falling with those same geopolitical tides. So it is proving to be with space.

I confess to being a splitter. This is a minority viewpoint without much of a clientele; it may be a singularity. My premises are these: that exploration as an institution is an invention of particular societies; that it derives much of its power because it bonds geographic travel to cultural movements, because it taps into deep rivalries, and because its narrative conveys a moral message; that, while unbroken, the trajectory of a half millennium of exploration by Western civilization can be understood best by parsing its long sweep into smaller increments; and that the future of exploration may become a reversed mirror image of its past. In particular, my splitter history would partition the past half millennium of European exploration from humanity's various migrations, and it would then fraction that grand chronicle into three great ages of discovery, fissioning William Goeztmann's Second Great Age of Discovery into two, adding a Third Age as distinctive from the Second as the Second was from the First.

This is not a commonly held analysis, not least because it compels us to examine differences. It demands that we identify what segregates space exploration from its progenitors beyond exalted claims that, in leaving the Earth's gravitational pull, humanity is, at last, leaving its nest. It places space exploration with the exploration of Antarctica and the deep oceans. It suggests a future that will less resemble the near past than the deep past. The Space Age is different; the Space Age is the same. A splitter history asks, how? and what does it mean? My version of a sample such history follows.

ISLANDS IN THE MIST: THE CASE FOR AGES OF DISCOVERY

Why *three* eras? Why not four, or eight? Why any at all? History is messy, and exploration history, with its perpetual disputes over prediscoveries and rediscoveries, messier still. So consider, as an index of exploration, the case of Pacific islands. None were known empirically to Europe prior to the Great Voyages. While some discoveries, particularly by the Portuguese, were no doubt hoarded as state secrets, the dates of discovery for most are reliably known. Plotting those discoveries by 50-year increments yields three fairly distinct periods (see figure below).

The first coincides with the classic voyages of discovery, led by Portugal from the west and Spain from the east. Every island is new: discovery is rapid and relatively easy (if anything done by ship in those days can be considered easy). Between 1500 and 1550, mariners discovered some 32 islands. They found fewer in the next 50 years, and half as many again in the next 50. By the mid-17th century, the long wave has exhausted itself. Some 75 percent of the discoveries occurred over roughly 75 years.

Discovery of Pacific Islands[9]

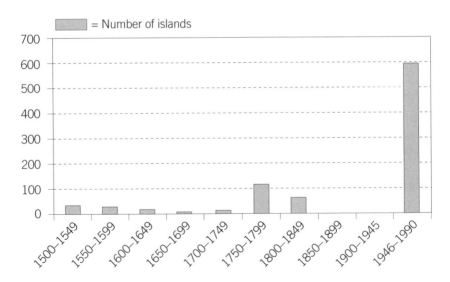

---

9. Data from Henry Menard, *Islands* (New York: Scientific American, 1986).

An explanation for the odd curve, a peak followed by a rapid decay, is simple. Mariners were not searching for islands, but for routes to the great entrepôts of the East. They found islands along the way, and once they plotted out the best paths, they had scant incentive to keep at sea. The latter discoveries happened from miscalculations or accidents—storms, for example, that blew ships off course—or, as the Dutch became more expansive in their plans to outflank the Portuguese, a scattering of islands that they chanced upon. There was no reason to randomly rove the seas. Explorers had completed their task, had hewed routes to the riches, and the discovered islands had been, as it were, the chips that scattered to the side.

Then, in the mid-18th century, after nearly 150 moribund years, the process rekindled: Europeans begin encountering new islands. But these are new Europeans—British and French, mostly—entangled in fresh rivalries, and they have novel purposes. They come as emissaries of the Enlightenment; they are keen to explore nature's economy for its exotic wealth and commercial wonders; they carry naturalists eager to catalog the Great Chain of Being, trace the contours of the world ocean, and draft a new *mappa mundae*; and they haul artists and litterateurs avid for lush tropical utopias. They search out the blank spots of the Pacific. They seek unknown islands as ends in themselves. A great age of circumnavigators commences, of which the three voyages of Captain James Cook are a prime exhibit.

The number of known islands explodes. More islands are discovered in 70 years than in the previous three centuries. But this, too, quickly expires. They reach the last island, Midway, in 1859. Then nothing, and it is a nothing all the more profound because the voyagers have revealed all that exist. By the onset of the 20th century, not only have explorers exhausted the dominion of Pacific islands, the Enlightenment itself has begun to crumble before the intellectual tremors and metaphysical termites of Modernism. For the Pacific Ocean, a second age of exploration ends with traffic in guano and copra; excursions by tourists, adventurers, and anthropologists; and color prints by Gauguin.

How, then, might there be another era? Because mariners went below the deep swells; they traveled by submarine and surveyed the hard-rock topography of the deep ocean by remote sensing devices. They discovered a vast realm of volcanic islands—guyots—that had eroded and subsided beneath the surface. In a few brief decades, an exploring science mapped 596 new Pacific isles. These were more Pacific islands than Western civilization had discovered since Vasco da Gama first landed Portugal at the gates of the Indies. More powerfully, the context of discovery revived with another global rivalry, this one begun in World War II and accelerated during the Cold War; with another intellectual syndrome, the curious culture of Modernism; with another revolution in technology.

The Cold War competition beneath the waves complemented almost perfectly the better-known competition for the high ground beyond the Earth's atmosphere. Space exploration was part and parcel of this same Third Great Age of Discovery. If islands are a reliable index, three eras might equally characterize the vast sweep of Western exploration since the natal times of Henry the Navigator. If space, however, becomes a powerful enough presence, then the Eurocentric frame itself might need resetting and another index might in time be necessary. Perhaps discovered moons might replace encountered isles, although it's hard to imagine a future artist rendering Titan or Europa with the lavish cultural colorations that William Hodge, traveling with Cook, brought to Tahiti.[10]

GREAT VOYAGES: THE FIRST AGE

The Great Age of Discovery opened with centuries of false dawns. Part of the difficulty is disentangling exploration from other forms of travel: from migration, walkabout, exile, wars of conquest, trading expeditions, reconnaissance, long hunts, great treks, missionizing, pilgrimage, tourism, and just plan wanderlust. Roman merchants had contact with the Canaries and Cathay. European pilgrims trekked from Hibernia to the Holy Land. Franciscan scholars trudged to the court of the Great Khan. Each age of expansion, every expansionist people, experienced a burst of discovery about a wider world. What made events of the 15th century special was that these exploring contacts did not end in a rapid contraction. They became welded to a revived expansion of Europe that would stretch over half a millennium; they bonded with revolutionary epochs of learning and political reform. They became institutionalized. Exploration became the outward projection of internal unrest that would not let the momentum long languish.[11]

The Great Voyages began cautiously enough. That Portugal pioneered the practice should alert us to the process's uncertain origins and its often desperate character. There was little in Portuguese history from which someone might predict, in 1450, that it would leap across whole seas and over unknown continents, establish the world's first global empire, and create the raw template

---

10. To match the discovery of Pacific islands with the general swarm of exploration, consult standard references. A sprawl of atlases exist that trace the general contours of geographic exploration, for example, and there is the flawed but indispensable *A History of Geographical Discovery and Exploration*, by J. N. L. Baker (New York: Cooper Square Publishers, 1967).

11. The doyen of the founding Age of Discovery is J. H. Parry. Among his many works, three are especially informative as syntheses: *The Establishment of the European Hegemony, 1415–1715*, 3rd ed., rev. (New York: Harper and Row, 1966); *The Discovery of the Sea* (Berkeley: University of California Press, 1981); and *The Age of Reconnaissance: Discovery, Exploration, and Settlement, 1450–1650* (New York: Praeger, 1969).

for European expansion. Yet that is precisely what happened. For several hundred years, exploring nations sought to emulate the Portuguese paradigm, whose outposts survived until the 21st century. Within a generation, it came to be said that it was the fate of a Portuguese to be born in a small land but to have the whole world to die in.

What happened was that exploration became—directly, or indirectly through charters—an organ of the state, and because no single state dominated Europe, many joined the rush. Geographical exploration became a means of knowing; of creating commercial empires; of outmaneuvering political, economic, religious, and military competitors—it was war, diplomacy, proselytizing, scholarship, and trade by other means. For this reason, it could not cease. For every champion, there existed a handful of challengers. This competitive dynamic—embedded in a squabbling Europe's very fabric—helps explain why European exploration did not crumble as quickly as it congealed. On the contrary, many Europeans absorbed discovery into their culture, even, in some cases, writing explorers into a founding mythology, a cultural creation story. In short, where exploring became a force, something beyond buccaneering, it interbred with the rest of its sustaining society. The broader those cultural kinship ties, the deeper the commitment. Societies dispatched explorers; explorers reshaped society. Exploration became an institution. The explorer became a role.

~ ~ ~

The fabled Great Voyages announced a First Age of Discovery. Its particular domain was the exploration of the world ocean, the discovery that all the world's seas were one, that it was possible to sail from any shore and reach any other. Of course, there were some grand *entradas* in the Americas, and missionaries, Jesuits especially, penetrated into the vast interiors of the Americas, Africa, and Asia. But as J. H. Parry observes, it was the world sea that defined the scope and achievements of the First Age. The mapping its littoral was the era's finest cartographic triumph.[12]

The map reminds us that the First Age coincided with a Renaissance. The era unveiled two new worlds: one of geography, another of learning. Francis Bacon conveyed this sense perfectly when he used as a frontispiece to his *Great Instauration* the image of a sailing ship pushing beyond the Pillars of Hercules. The voyage of discovery became a metaphor for an age of inquiry that would venture far beyond the dominion of the Mediterranean and the inherited wisdom of the ancients. The discoveries overwhelmed a text-based

12. Parry, *Discovery of the Sea.*

scholarship: scholasticism, that arid discourse that resulted from too many scholars and not enough texts, collapsed as new information poured into Europe like New World bullion into Spain and, like it, inspired an inflationary spiral of knowledge.[13]

An age of discovery, however, demands more than curiosity and craft. It has to speak to deeper longings and fears and folk identities. The ships must voyage into a moral universe that explains who a people are and how they should behave, that criticizes and justifies both the sustaining society and those it encounters. The Great Voyages provided that moral shock: they forced Europe to confront beliefs and mores far beyond the common understanding of Western civilization. The Renaissance expansion of Europe profoundly altered Europe's understanding of itself and its place in the world. There was plenty of hollow triumphalism, of course, but those contacts also inspired Montaigne's celebrated preference for the cannibalism of Brazil's noble savages to that of Versailles's courtiers, and Las Casas's excoriating denunciation of the *conquistadores*. They compelled a reexamination of the political and ethical principles underlying Christendom and its secular principalities. Exploration could upset the discovering society as well as the discovered. It often found things it didn't like, not least things about itself. The dark regions held horrors as well as marvels.

For all this, the Portuguese were the originators. If exploration became, as Goetzmann argues, programmed, then the Portuguese paradigm was the template, the default setting for exploration's software. The degree of inter-penetration between exploration and society was astonishing, of which the suite of exploring ships was only a down payment. Consider the founding explorers: Henry the Navigator, late-medieval prince, blurry-eyed speculator, and wastrel, who began the fusion of discovery with state policy; Vasco da Gama, merchant and administrator, representing the bonding of commerce with the state; Afonso de Albuquerque, soldier and strategist, seizing at gun-point the critical nodes of traffic throughout the Indian and South China seas; St. Francis Xavier, tempering the sword with the cross, missionizing in India, the East Indies, and especially Japan, with plans to proselytize in China; and Luis de Camões, adventurer turned litterateur, author of the epic *Os Lusiadas* (1572), which cast contemporary explorers into the mode of classical heroes. Together they embodied, literally, the swirl of Renaissance ambitions—God, gold, glory—while wrapping it in an enduring saga. "Had there been more

---

13. This has long been a common theme. A somewhat eccentric but insightful (and lively) recreation of what it meant can be found in William Manchester's *A World Lit Only by Fire: The Medieval Mind and the Renaissance: Portrait of an Age* (Boston: Little, Brown, 1992), which tracks the imaginative impact of Magellan's voyage.

of the world," Camões wrote, his bold mariners "would have discovered it." Revealingly, all the founders died overseas.[14]

When this tidal bore of discovery passed, it left an institutional berm throughout the strands on every continent save Australia and Antarctica. Portuguese explorers and fishermen plied the Grand Banks and probed the shorelines of North America and the North Atlantic isles. They established colonies in Brazil. They held trading fortresses along the coast of Africa and India, in the Spice Isles, Cape Verde, St. Helena, Tristan da Cunha, and at such major trading entrepôts as Malacca, Macau, and Nagasaki. Probably they had reached Australia, though they found nothing to hold them. Those who followed were interlopers, seeking to poach parts of an empire too vast for tiny Portugal to hold. Or they sought to outflank the Portuguese. That was surely the intention of Christopher Columbus, who after all had learned his mariner's craft sailing on the Portuguese Atlantic circuit. And that was the prospect held by Magellan, who had already been to the East Indies in service to Portugal before, on Columbus's example, he offered fealty to Spain.[15]

The Portuguese paradigm pointed as well to the enormous liabilities inherent in geographic discovery. The overseas posts, never fully staffed, nevertheless siphoned off perhaps a tenth of the Portuguese population. They drained the homeland without demographically overwhelming the colonies. The rapid infusion of knowledge failed to spark a Portuguese renaissance; much of the data was hoarded as a state secret, and the rest demanded an infrastructure of scholarship that did not exist. Worse, the sudden inundation of wealth proved destabilizing. It tempted rulers to indulge personal and geopolitical fantasies, typically expressed as foreign wars. The unwisely sainted Henry was here the prototype. What wealth he gleaned, he sank in futile fighting on Moroccan sands. Exploration could led to profitable colonization where the discovered place was uninhabited, as at Madeira. Where lands were already occupied, colonization led to extravagant wars and bottomless expenses. The paradigm thus had its paradox: exploration required money as well as will, but beyond sacked towns and coastal trade, there was little wealth to get from it. Once permanent, the colonies became not sources of sustainable wealth, but economic placers, quickly

---

14. Even in English, the Portuguese experience looms large. In addition to Parry, see C. R. Boxer, *The Portuguese Seaborne, 1415–1825* (London, Hutchinson, 1969), and *Four Centuries of Portuguese Expansion, 1415–1825: A Succinct Survey* (Johannesburg: Witwatersrand University Press, 1965), as well as Bailey W. Diffie and George D. Winius, *Foundations of the Portuguese Empire, 1415–1580* (Minneapolis: University of Minnesota Press, 1977). Luis de Camões's epic, *The Lusiads,* is translated by Leonard Bacon with an introduction and notes (New York, Hispanic Society of America, 1950).

15. On the Dutch strategy, see C. R. Boxer, *The Dutch Seaborne Empire, 1600–1800* (New York, Knopf, 1965).

plundered, before plummeting into fiscal sinks. An exploring imperium proved easier to grab than to hold.[16]

The Portuguese paradigm should remind us how much geographic exploration has morphed over the centuries. By the late 18th century, as motivating forces, science had replaced God; commerce, gold; and national prestige, individual glory. The issues are even more serious for space exploration, although America's spacefaring traverse through the solar system may be the closest geographic romp comparable in scale to Portugal's. But there the similarities cease. Pioneer did not have to force access to the outer planets by the sword, Mariner did not have to proselytize, and Voyager did not have to wrestle with restless indigenes and obstreperous crews. Instead of Camões, American letters had Norman Mailer and Ray Bradbury, neither of whom had been in space, and instead of classic heroes, Renaissance versions of Odysseus, we had Tom Wolfe's test pilots, forever fretting about drinking and screwing and their ranking on the ziggurat. No one wrote about the vessels themselves, any more than the 16th century did about the *Victoria*. Mostly, Portugal's voyages were a prelude to imperium, an extension of ancient empire-building by new means. America's probes were valenced to the limited conflict of the Cold War. If Portugal faltered, someone else would move in. If America stalled, the void might widen.

Corps of Discovery: The Coming of the Second Age

The inflection to what William Goetzmann has termed a Second Great Age of Discovery was messier than the paradigm of Pacific islands suggests. Yet the same basics apply. By the early 18th century, exploration had become moribund; mariners did more poaching and piracy than original probing, like William Daumpier more buccaneer than naturalist; the explorer blurred into the fantasist and fraud, the promoters of the Mississippi and South Seas bubbles, the Lemuel Gulliver of Jonathan Swift's savage satire, or with the forlorn adventures of Daniel Defoe's Robinson Crusoe, who curses a woeful addiction to adventuring that repeatedly brings him to grief. Exploring expeditions persisted largely because interlopers tried to outflank established competitors, but little new was added. Exploration seemed destined to be left marooned on the shore of a fast-ebbing historical tide.[17]

Then the cultural dynamics changed. The long rivalry between Britain and France, the penetration of high culture by the Enlightenment, a hunger for

---

16. Peter Russell, *Prince Henry "the Navigator": A Life* (New Haven: Yale University Press, 2000). In fact, all the standard accounts of the Portuguese eruption, even the most celebratory, relate the same sad decline.

17. See William H. Goetzmann, *New Lands, New Men: The United States and the Second Great Age of Discovery* (New York: Viking, 1986).

new markets, all combined to move Europe again out of dry dock and onto the high seas of exploration and empire. Naturalists lengthened their excursions; artists painted natural scenes; *philosophes* looked to pure nature for guidance. The Grand Tour became a global excursion around the Earth. Perhaps most extraordinarily, the missionary emerged out of a secularizing chrysalis into the naturalist. Increasingly, scientists replaced priests as the chroniclers and observers of expeditions, and scientific inquiry substituted for the proselytizing that had helped justify an often violent and tragic collision of cultures.[18]

From Linnaeus's apostles gathering the fruits of nature from the Americas to Antarctica, to expeditions measuring the arc of the meridian and the transit of Venus, explorers swarmed across the Earth and often sailed around it. Over the next century, every aspiring great power dispatched fleets to seek out new wealth and knowledge, to loudly go where others had not yet staked claims. Cook, Vancouver, Bougainville, LePerouse, Wilkes, Bellingshausen, Malaspina— these became the Magellans of the Enlightenment. They placed the competition intrinsic to science into the service of geopolitical strife. Once again, the rivalries among the Europeans were as great as anything between Europeans and other peoples. A civilization's internal conflicts drove its outward expressions.

In the process, the old motivations became secularized and updated. In petitioning the Lords Commissioners of the Treasury to support the 1761 transit, the Royal Society of London laid out the new rationales for systematic discovery:

> The Memorial itself plainly shews, that the Motives on which it is founded are the Improvement of Astronomy and the Honour of this Nation [an Englishman, Edmund Halley, had proposed the transit as a means of measuring the astronomical unit] . . . . And it might afford too just ground to Foreigners for reproaching this Nation in general (not inferior to any other in every branch of Learning and more especially in Astronome); if, while the French King is sending observers . . . not only to Pondicherie and the Cape of Good Hope, but also to the Northern Parts of Siberia; and the Court of Russia are doing the same to the most Eastern Confines of the Greater Tartary; not to mention the several Observers who are going to various

---

18. The classic figure, of course, is James Cook, so see J. C. Beaglehole's classic (if exhaustive) biography, *The Life of Captain James Cook* (Stanford, CA: Stanford University Press, 1974). But see also the impact of Linnaeus in *The Compleat Naturalist; a Life of Linnaeus* (New York: Viking Press, 1971), by Wilfrid Blunt, with the assistance of William T. Stearn, and see the impact of Banks, a critical catalyst for whom Patrick O'Brian offers a popular biography, *Joseph Banks: A Life* (London: Collins Harvill, 1987). The literature on all these men and their apostles and imitators is almost oceanic in its extent.

Places, on the same errand from different parts of Europe; England should neglect to send Observers to such places . . . subject to the Crown of Great Britain.

This is by foreign Countries in general expected from us; Because the use that may be derived from this Phaenomenon, will be proportionate to the numbers of distant places where . . . observations . . . shall be made of it; And the Royal Society, being desirous of satisfying the universal Expectations of the World in this respect have thought it incumbent upon them . . . to request your effectual intercession with His Majesty . . . to enable them . . . to accomplish this their desire . . . which . . . would be attended with expense disproportionate to the narrow Circumstances of the Society.

But were the Royal Society in a more affluent State; it would surely tend more to the honour of his Majesty and of the Nation in general, that an Expense of this sort, designed to promote Science and to answer the general Expectation of the World, should not be born by any particular Set of Private Persons.[19]

Here, in a nutshell, were the formal reasons for state sponsorship: society needed science; science needed exploration; exploration to far countries needed support beyond what individuals could contribute; international scholarship and national honor demanded participation. Unsaid, but indispensable, were the rising popular enthusiasms for geographic discovery, bonded not to reason but to sentiment. The Lacondamine expedition to Ecuador commanded public attention not for Lacondamine's meticulous mapping of the Amazon's latitude and longitude, but for Isabella Godin's heart-wrenching journey down it to find her husband. Public interest widened. By the latter part of the 18th century, as select colonies moved from the littoral inland, wider populations found in explorers a Moses-like leader of the people to promised lands. Daniel Boone, not George Washington, for example, would become America's folk-epic hero. From high culture to pop cult, the explorer claimed cultural standing.[20]

Those grand circumnavigations revived geographic exploration, but they mostly proved a means to reposition explorers, who promptly moved inland. The world's continents replaced the world sea as an arena for discovery: the

---

19. Harry Wolfe, *The Transits of Venus* (Princeton: Princeton University Press, 1959), quoted on p. 83.

20. The Lacondamine expedition is not as well known among English speakers as it should be. A good introduction, leading to the successors in South America, is available in Victor von Hagen, *South America Called Them. Explorations of the Great Naturalists: La Condamine, Humboldt, Darwin, Spruce* (New York: A. A. Knopf, 1945). A popular version has recently been published: Robert Whitaker, *The Mapmaker's Wife: A True Tale of Love, Murder, and Survival in the Amazon* (New York: Basic Books, 2004).

cross-continental traverse superseded a circumnavigation as the grand exploring gesture of the age.

The pivotal figure was Alexander von Humboldt, whose five years in Latin America redefined exploration for the era. Humboldt rewove the loose strands into a taut fabric. He projected Linnaeus's natural-history excursion into a cross section of continents. He carried the artist's Grand Tour to the New World. He put legs under Cook's tours and let them trek from the shoreline over vast landscapes. He gave empirical heft to the misty musings of *Naturphilosophie*. He empowered geographic science with a global reach. In the words of Ralph Waldo Emerson, he was one of "those Universal men, like Aristotle." While he was not the first European to paddle up the Orinoco or climb in the Andes, Humboldt was the first of a new kind of European, such that even when explorers of the Second Age revisited sites known to the First, they did so with original eyes and to novel ends. Symbolically, upon his return to Europe, he dined with Thomas Jefferson the same month that Lewis and Clark's Corps of Discovery departed St. Louis. In the person of Humboldt, the explorer embodied the Romantic hero.[21]

The transition matters because, as the 19th century ripened, Europe was no longer content to remain as a trader on the beaches of the world sea. Like its exploring emissaries, it shoved and swarmed inland. Trading ventures became imperial institutions; coastal colonies became continental nations; and the politics of commerce gave way to outright conquest. Thus commenced a grand era of exploring naturalism. New scholarship, particularly sciences, bubbled up out of the slush of specimens shipped home. The returns from the earliest explorers to a particular place were often phenomenal—the scholarly equivalent to placer mining. A revolution in geographic discovery again accompanied a revolution in learning, aptly symbolized by the simultaneous recognition by two exploring naturalists, Charles Darwin and Alfred Wallace, of evolution by natural selection.

The moral drama changed accordingly. Secularization and science translated Vasco da Gama's famous declaration that he had come to the Indies for "Christians and spices" into a cry for civilization and commerce. The deeper drama concerned that fraction of Europe's imperium colonized by European emigrants. Most of what Europe nominally ruled was densely inhabited by long-residing peoples, often in numbers far vaster than that of the rulers. But in some lands, the indigenes were swept away, and into that demographic vacuum European émigrés poured in. These settler societies tended to look

---

21. Probably the best biography of Humboldt is still Helmut de Terra, *Humboldt: The Life and Times of Alexander von Humboldt, 1769–1859* (New York: Knopf, 1955). For a fascinating insight into his cultural impact, however, see Halina Nelken, *Alexander von Humboldt: His Portraits and Their Artists: Documentary Iconography* (Berlin: Dietrich Reimer Verlag, 1980). On his impact in America, see Goetzmann, *New Lands, New Men.*

upon discovery as part of a national epic and to honor explorers as vital protagonists—a Moses, an Aeneas—of those founding events. With Lewis and Clark, for example, the frontiersman morphed into a naturalist, the scout into a scholar, and the adventurer into an Aeneas on his way to the founding of a new civilization. Their subsequent folk expansions proceeded hand in glove with formal exploration. These were new worlds, premised on the prospects for a new order of society. America truly was, in William Goetzmann's words, "exploration's nation," but so were Russia, Australia, Canada, and others.[22]

Discovery metastasized. As measured by the number of exploring expeditions, a slight increase appears in the latter 18th century and then erupts into a supernova of discovery that spans the globe. By the 1870s, explorers had managed continental traverses—cross sections of natural history—for every continent save Antarctica. With the partition of Africa, expeditions proliferated to assess what the lines drawn on maps in a Berlin library actually meant on the ground. Exploration had become an index of national prestige and power. The first International Polar Year (1882) turned attention to the Arctic. An announcement by the Sixth International Geographical Congress in 1896 that Antarctica remained the last continent for untrammeled geographic discovery inspired a stampede to its icy shores; even Belgium and Japan sponsored expeditions. (America's attention remained fixated on the North Pole and that other stampede to the Klondike.) Ernest Shackleton's celebrated Trans-Antarctic Expedition was, after all, an attempt to complete for that continent the grand gesture that had crowned every other.[23]

But Antarctica was the last: there were no more unvisited lands to traverse other than such backwaters as, for example, the Red Centre of Australia, the crenulated valleys and highlands of New Guinea, and the windswept Gobi. The number of exploring expeditions began to decline. Plotting them reveals the Second Age as a kind of historical monadnock, rising like a chronological volcano above a level terrain (see figure on opposite page). The peak crests in the last decades of the 19th century, as exploration crossed the summit of the Second Age. Then it began a descent down the other side.

---

22. Goetzmann, "Exploration's Nation."

23. Accounts from the Heroic Age of Antarctic Exploration are legion. The entire literature is contained—incredibly—within the *Antarctic Bibliography* published by the Library of Congress. A surprisingly good compilation, wonderfully illustrated, is available in Reader's Digest, *Antarctica: Great Stories from the Frozen Continent* (Surrey Hills, New South Wales: Reader's Digest, 1985). For an interpretive summary that places the experience within the Three Ages of Discovery schema, see Stephen J. Pyne, *The Ice: A Journey to Antarctica* (Iowa City, IA: University of Iowa Press, 1986).

Lost Worlds: The Waning of the Second Age

The reasons for the slow bursting of this exploration bubble are many. The simplest is that Europe had completed its swarm over the (to it) unknown surfaces of the planet. There was nowhere else for the Humboldtean explorer to go. Equally, there were no more lands to meaningfully colonize. Instead, Europe turned upon itself in near self-immolation, with two world wars, a depression, and the sudden shedding of its old imperium. The enthusiasm for boundary surveys and natural-history excursions—for imperialism itself—waned with the slaughter of the Great War.

The critical players were exhausted, especially Great Britain. The Second Age had kindled with a rivalry between Britain and France, much as the contest between Portugal and Spain had powered the First Age. Thereafter, virtually every competition featured Britain, which is why its explorers so dominate the age. Britain and France clashed in India, the Pacific, and Africa; Britain and the U.S., in North America; Britain and Russia, the Great Game, across central Asia; Britain and all comers in Antarctica. After the Great War, Britain and France

Great Ages of Discovery[24]

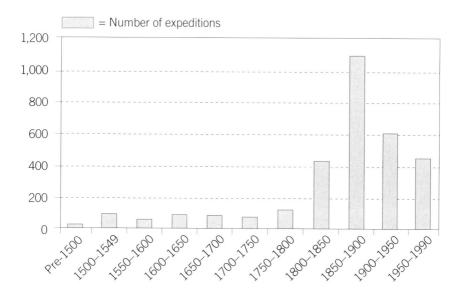

24. Data from J. N. L. Baker, *A History of Geographical Discovery and Exploration*, rev. ed. (G. G. Harrap and Co., 1937); Alex Roland, ed., *A Spacefaring People: Perspectives on Early Spaceflight* (Washington, DC: GPO, 1985); and J. H. Parry, *The Discovery of the Sea* (New York: Dial, 1974).

could no longer afford the enterprise; Russia turned inward with revolution; the U.S. had few places other than Antarctica in which discovery had geopolitical meaning. The Second Great Age of Discovery, like the First before it, deflated.

By the middle 20th century, Kipling's "Recessional" had become prophetic: Europe was rapidly disengaging itself from its imperial past and thus from the exploring energies that had, like lampreys, attached themselves to the institutions of an expansionist era. Decolonization accompanied an implosion of exploration; Europe turned inward, quelling the ancient quarrels that had restlessly and violently propelled it around the globe, pulling itself together rather than projecting itself outward. Antarctica, the deep oceans, interplanetary space—these arenas for geographic discovery might be claimed, but they would not be colonized. No one was willing to wage war over the asteroid belt or Io.

Other reasons were cultural. The Second Age had served as the exploring instrument of the Enlightenment. Geographic discovery had bonded with modern science: no serious expedition could claim public interest without a complement of naturalists, while some of the most robust new sciences like geology and biology relied on exploration to cart back the data that fueled them. Science, particularly natural history, had shown itself as implacably aggressive as politics, full of national rivalries and conceptual competitions, and through exploration, it appeared to answer, or at least could address, questions of keen interest to the culture. It could exhume the age of the Earth, reveal the evolution of life, celebrate natural monuments to nationalism and Nature's God. Artists like Thomas Baines and Thomas Moran joined expeditions or, like John James Audubon, mounted their own surveys; general intellectuals eagerly studied narratives of discovery (even Henry David Thoreau, nestled into his Walden Pond cabin, read the entire five volumes of the Wilkes Expedition). Exploring accounts and traveler narratives were best sellers; explorers became cultural heroes; exploration was part and parcel of national epics; exploration was a means to fame and sometimes fortune. The Second Age, in brief, braided together many of the dominant cultural strands of its times.

By the early 20th century, however, this splendid tapestry was unraveling. The Enlightenment found itself challenged by Modernism's avant-garde: in field after field, intellectuals turned to subjects that no longer lent themselves to explication by exploration. Modernism spread like an intellectual infection, a fever that turned the attention of high culture away from a tangible, commonsense world to an interior realm full of paradoxes. The vital truths no longer lay in the domain of geographic discovery. Art looked to art, mathematics to mathematics, literature to literature. Natural scientists scrutinized the very large and the very small, to red-shifting nebulae and subatomic particles or molecular genes. Artists turned inward, probing themselves and the foundations of art, not outward to representational landscapes. High culture was more inclined to follow Sigmund Freud into the symbol-laden depths of the unconscious or Joseph Conrad into a heart of imperial darkness than to ascend Chimborazo with Humboldt or to row with John Wesley Powell through the gorges of the Grand Canyon. The Second Age sagged not simply from the exhaustion of closed frontiers, but from a more profound weariness with the entire Enlightenment enterprise.

In the early 19th century, an intellectual could claim international acclaim by exploring new lands. By the early 20th, he could not, if he could even find suitable lands. There were a few spectacular exceptions: the gold-prospecting Leahy brothers trooping into the unknown Highlands of New Guinea; Richard Byrd, wistfully erecting Little America on the Ross Ice Shelf; Roy Chapman Andrews, with carbine and Model T, whisking across the Gobi in search of dinosaur eggs, the very model of a Hollywood action hero (and inspiration for Indiana Jones). But there was, overall, a rueful, forlorn quality to the striving, aptly expressed when the American Museum of Natural History, with Andrews at the helm, dispatched an expedition to Shiva Temple, an isolated butte within the Grand Canyon, to look for exotic creatures. Sixty years before, the Canyon had claimed center stage, not only for geographic discovery, but for what it said to fundamental questions about the Earth's age and organic evolution. Now the press boosted a minor foray into a search for lost worlds and possibly living relics from the age of dinosaurs. Lost world, indeed.[25]

Boldly Going Where No One Is: The Third Age

The fascinating question is why the bubble did not burst more catastrophically. One reason is that Western civilization *did* discover new lands to explore. There were the ice sheets (and sub-ice terrains) of Greenland and especially Antarctica; there were the deep oceans; and, of course, a solar system beckoned, full of wonders beyond the vision of Earth-bound observatories. As powerful instruments and remote sensing technologies emerged, as manned vehicles and unmanned probes plummeted to the depths and beyond the atmosphere, the prospects for a revival of exploration became possible.

Yet dazzling technologies and a rekindled curiosity are not enough to sustain an era of exploration: cultural engagement also demands a sharp rivalry. Those competitive energies flourished with the Cold War. In retrospect, the Great Game between the United States and the Soviet Union lasted far less than

25. Contrast, for example, the classic explorers of the Second Age with the career of Roy Chapman Andrews, as described in Charles Gallenkamp, *Dragon Hunter: Roy Chapman Andrews and the Central Asiatic Expeditions* (New York: Viking, 2001). Andrews set out to be an explorer in the classic mode but found that the times had changed. For the story of a contemporary who did manage to make the transition in part, see Carol Gould, *The Remarkable Life of William Beebe: Explorer and Naturalist* (Washington, DC: Island Press, 2004). Beebe famously plunged into the Atlantic in a bathysphere. An attempt to trace the contours of the Second Age's rise and fall, using the Grand Canyon as a test site, can be found in Stephen J. Pyne, *How the Canyon Became Grand: A Short History* (New York: Viking, 1998), which recounts the Shiva Temple saga.

those between Spain and Portugal, or Britain and France, but the era is young, and if it does in fact mark a Third Age, some other competitors, keen to secure national advantage or prestige through sponsored discovery, may emerge. China has announced its intention to land a tikonaut on the Moon; India and Japan have launch capabilities and may choose to compete. Without the Cold War, however, there would have been scant incentive to erect bases on the Antarctic ice, scour the oceans for seamounts and trenches, or launch spacecraft. The Cold War allowed a controlled deceleration of exploring energies, a reversed complement to the British-French competition that helped accelerate the Second Age. Two geopolitical rivals, both with active exploring traditions, chose to divert some of their contest away from battlefields and into untrodden landscapes.

But perhaps more profoundly, exploration did not wither away because the culture, the popular culture, did not wish it to. Exploration had become not only institutionalized, but internalized. This was a civilization that could hardly imagine itself as other than exploring. Explorers flourished, if only in pulp fiction, movies, and adolescent fantasies. Quickly, it forged new institutions, of which the International Geophysical Year is an apt annunciation, and in the Voyager missions, it found what is likely to endure as the great gesture of the Third Age, a traverse through the solar system. Voyager's Grand Tour may serve for this era as Magellan's voyage did for the First and Humboldt's travels did for the Second. Voyager demonstrated both the power and peculiarities of the era.[26]

What has not happened is a new knitting together of exploration and high culture. Instead, popular culture has filled that void, but in ways that resuscitate the images and narrative templates of previous eras. *Star Trek*, for example, is the voyage of the *Beagle* with warp drive. Enthusiasts show Conestoga wagons

---

26. The literature on IGY is large but mostly technical. A good popular survey is J. Tuzo Wilson, *IGY: The Year of the New Moons* (New York: Knopf, 1961). The Third Age has not been the object of a comprehensive survey since space seems to command its own literature and, to put the matter bluntly, the concept is not widely known. Useful starting points for works about space travel are Alex Roland, ed., *A Spacefaring People: Perspectives on Early Space Flight* (Washington, DC: GPO, 1985); William Burrows, *This New Ocean: The Story of the First Space Age* (New York: Random House, 1998); Roger Launius, *Frontiers of Space Exploration,* 2nd ed. (Westport, CT: Greenwood Press, 2004); Roger D. Launius et al., *Reconsidering Sputnik: Forty Years Since the Soviet Satellite* (Amsterdam: Overseas Publishers Association, 2000); and, for the political context of the Cold War, Walter A. McDougall, . . . *The Heavens and the Earth: A Political History of the Space Age* (New York: Basic Books, 1985). The deep-ocean story has been much less described, although declassification of military documents is beginning to change the record. See William Broad, *The Universe Below: Discovering the Secrets of the Deep Ocean* (New York: Simon and Schuster, 1997); Robert D. Ballard with Will Hively, *The Eternal Darkness: A Personal History of Deep-Sea Exploration* (Princeton, NJ: Princeton University Press, 2000); Henry Menard, *The Ocean of Truth: A Personal History of Global Tectonics* (Princeton, NJ: Princeton University Press, 1986); and *Anatomy of an Expedition* (New York: McGraw-Hill, 1969). To measure the contrast with the supreme oceanic expedition the Second Age, see Richard Corfield, *The Silent Landscape: The Scientific Voyage of HMS Challenger* (Washington, DC: Joseph Henry Press, 2003).

trekking to Mars, prairie schooners propelled by solar wind. But popular culture can be fickle and selective. The first *Star Trek* movie, for example, imagined a Voyager spacecraft returning to Earth, stuffed with a universe of wonders, reporting to its "creator." Ten years later, *Star Trek V* opened with a bored Klingon commander blasting a Voyager probe as space junk. Exploring the galaxies needed a story—a deep narrative of moral and imaginative power—as much as dilithium crystals. With neither a rambunctious imperialism nor an eager Enlightenment, the Third Age must, for now, continue its downward declension.

~ ~ ~

There are good reasons, then, for considering the Third Age—our age—as continuous with its predecessors. Yet it is also different, and those differences matter. Most intrinsically, the Third Age is going where no one is or ever has been.

The geographic realms of the Third Age are places where people cannot live off the land. In Antarctica, they can at least breathe. In the deep oceans, beneath the ice sheets, or in space, they can survive only if encased in artificial life-support systems. These are environs that offer no sustaining biota. There is little reason to believe that much more thrives beyond Earth. These geographies remain, for all practical purposes, abiotic worlds. They propel exploration beyond the ethnocentric realm of Western discovery, but also beyond the sphere of the human and perhaps beyond the provenance of life.

This is a cultural barrier to exploration, in comparison to which the limiting velocity of light may prove a mere technological inconvenience. The reason goes to the heart of exploration: that it is not merely an expression of curiosity and wanderlust but involves the encounter with a world beyond our ken that challenges our sense of who we are. It is a moral act, one often tragic, a strong nuclear force that bonds discovery to society. It means that exploration is more than adventuring, more than entertainment, more than inquisitiveness. It means it asks, if indirectly, core questions about what the exploring people are like.

This was unavoidable in the past because almost all previous encounters had involved people. Exploration meant the meeting of one people with another, the transfer of knowledge and experience from one group to another. Most of the world Europe did not discover, except to itself. Almost every place that could have people did have them, and those indigenes proved indispensable. They served as interpreters, translators, native guides, hunters, and collectors. Explorers often succeeded to the extent that they borrowed from or emulated the peoples who already resided in these (for Europe) far and foreign realms. What Europe did was to stitch these separate someones together into a vast cosmological quilt: its voyages of discovery were needles and threads that joined geographic patches into new collective patterns.

The Third Age has no such option. No one will live off the land on Deimos, go native on Titan, absorb the art of Venus, the mythology of Uranus, the religious precepts of Mars, or the literature of Ceres. There will be no one to talk to except ourselves. Discovery will become a colossal exercise in self-reference. Consider some of the iconic images of the American space program. There is the image of Earthrise, which is a view of ourselves from the Moon. And there is Buzz Aldrin, encased like a high-tech Michelin Man, staring into a camera on the lunar surface. His visor, however, reflects back the image of the photographer. In a classic image, *Wanderer Above a Sea of Fog*, Caspar David Friederick could position his painting's observer peering over the shoulder of a Humboldtean traveler, in turn overlooking a valley of mist. In a comparable classic, Neil Armstrong could photograph Aldrin, looking at Armstrong, showing the photographer taking the photograph. That shift in perspective captures exactly the shift from Enlightenment to Modernism and from Second Age to Third. Add to the survey the curious plaques affixed to Pioneer and Voyager, surely indecipherable to any entity that might find them. They are a message in a bottle dispatched to ourselves.

Yet there is promise amid the paradox. For a century, Modernism has grappled precisely with how to reconcile observer with observed, with somehow putting ourselves into the scene. Russell's paradox, Godel's proof, Heisenberg's indeterminacy principle—all struggled with self-reflexivity. They addressed precisely, if abstractly, the conundrum of exploring without an Other. As a result, Modernist art, literature, and philosophy can outfit exploration with the intellectual kit it will need to survive such alien scenes and self-encounters. They can provision it to move beyond the landscapes of earlier eras of discovery.

The other good news is that the coruscating ethical dilemmas of so much earlier exploring and empire-building will disappear. No group need expand at the expense of another. Ethnocentricity will vanish: there is only one culture, that of the explorer. There is no exoecosystem to foul. With no distinctively *human* encounter possible, there is no compelling reason for humans to even serve as explorers. As long as other life or cultures are not present, there is no ethical or political crisis except whatever we choose to impose on ourselves. Beyond the Earth there may well be no morality as traditionally understood, that is, as a means of shaping behavior between peoples. The morality at issue is one of the self, not between the Self and an Other.

The bad news is that exploration's moral power—the tensions, awful and enlightening both, that are involved in a clash of cultures—also vanishes. The price of ethically sanitizing exploration is to strip it of compelling *human* drama. Planetary probes become technical challenges, to make machines to withstand the rigors of space travel, a technological equivalent to extreme sports, like white-water kayaking in Borneo or NASCAR's Daytona 500. The intellectual challenge has telescoped, more or less, into a search for life, notably on Mars. Whether this can command the kind of cultural attention that earlier exploration did is unclear. What is inescapable is yet another paradox: we are safe as long as we don't find life. If we do, then the old morality returns. (Here is the real intergalactic ghoul.) If we decline to revive those concerns, and withdraw, then the primary justification for continued discovery vanishes and the space program becomes a kind of national hobby, a jobs program, or a daytime TV soap opera. But the matter gets even worse.

In past ages, discovery *had* to be done by people. There was no other option by which to learn the languages, to record data and impressions, to gather specimens, to meet other societies and translate their accumulated wisdom. It is impossible to imagine the great expeditions of the past without considering the personality of individual explorers who inspired, collected, witnessed, fought, wrote, sketched, exulted, feared, suffered, and otherwise expressed the aspirations and alarms of their civilization. But it is entirely possible to do so now. Not only is there no encounter between people, there need not even be a human encounterer. People do not have to be physically present at the discoveries of the Third Age, and there are sound reasons for arguing that they should not be.

Nor is the case for planetary colonization truly compelling, not at present, any more than it was for Magellan at the Marianas or Peary at the Pole, or those fatally premature experiments from promoters like Walter Raleigh. The theses advanced to promote outright settlement are historical, culturally bound, and selectively anecdotal: that we need to pioneer to be what we are, that new colonies are a means of renewing civilization, that the Second Age can have a Second Coming. America, in particular, could not survive the closing of the final frontier (although the American Century flourished only after the old frontier nominally shut down).

There is little to justify this assertion. Even considered on economic grounds, Europe's imperial nations boomed only after they shed their foreign colonies. Moreover, advocates for exploration as a prelude to colonization conveniently ignore such fiascos as the Darien debacle—the scheme boosted by William Paterson in the 1690s to establish a Scottish settlement in Panama. The isthmus would be critical to global trade, he insisted; Scotland's economic future and national identity depended on it seizing control of that geopolitical chokepoint; destiny demanded colonization. The outcome was a crushing failure that, not incidentally, bankrupted Scotland and drove it into union with England. Paterson was a visionary: in 200 years, a canal would join the two oceans across Panama. He was also a lethal crank who cost hundreds of lives and ruined a national economy. Successful settlements followed a long gestation period of reconnaissance and aid from indigenes. Examples abound of societies that chose to withdraw into themselves and suffered. There are, equally, examples of societies that chose to push outward and suffered. Portugal, as the founding paradigm, is a good case. Within a generation, it had sunk into collapse, even absorbed by Spain, only emerging fully as a modern state when it finally shed its colonies. The issue is not whether to explore or not, but how to engage the wider world: where, with what means, how much. More likely is an era of space tourism or historical reenactment—Plymouth Colony on the Moon, Golden Goa on Venus, Magellan Tours Takes You to Phobos.[27]

This is precisely what the closest Earth analogues do. No one lives in the Marianas Trench or the Laurentian Abyss. No one homesteads in Antarctica. There are permanent settlements, but not permanent residents. There are no schools because there are no children. There are no families. There is no indigenous society. These enclaves are the scientific equivalent of the commercial and military posts that characterized the early centuries of European expansion, only a fraction of which ever evolved into full-blown colonies, and most of those in defiance of the wishes of the commercial joint-stock companies or royal monopolies that oversaw them. In some ways, the contemporary colonies of the Third Age on Earth offer even less because there are no indigenes with whom to co-inhabit, interbreed, or coerce into labor. (The historic outposts of Europe's exploring imperium tended to be populated by indentured servants, slaves, serfs, soldiers, convicts, religious refugees, or company employees, most of whom survived thanks to the largesse or forced conscription of native peoples; all in all not a formula for the demographic renewal of Earth.)

Within the realm of the solar system—the dominion of the Third Age— the likelihood is that posts, if established and staffed by humans, would involve short tours of duty and high turnovers. The infrastructure would remain; the people would not. Exploration could thrive; outright colonization would not.

BACK TO THE FUTURE: BEYOND THE THIRD AGE

The Third Age encompasses more than space exploration, but the Antarctic has not enough undiscovered terrain to sustain a whole era, and oceanographic exploration has not yet gripped the public imagination, although it might. The future of exploration will depend on the exploration of the solar system. What might it look like?

It will look like what its sustaining society wants it to look like. The possibility exists that political contests will boil over into space, perhaps if China declares a colony on the Moon as essential to its prestige and the European Union joins the

---

27. For a thumbnail of the Darien fiasco, see Arthur Herman, *How the Scots Invented the Modern World* (New York: Three Rivers Press, 2001), pp. 15–37. Interestingly, several nations tried to gain a foothold in Panama and failed, including the United States in the 1850s.

fray, using space as an alternative to military might in the search for a multipolar world. There is a prospect that the search for life will take on an imaginative, even a theological cast, sufficient that a large fraction of the culture wants to pursue it among the planets. It may happen that extreme arts, brash new sciences, an as-yet-undeveloped commerce, an astropolitics, and some critical personalities will combine to kindle a Third Age echo of the Second Age. In some form or another, a virtuous cycle is possible. But it is not likely. For the American economy, the world's greatest debtor, cyberspace is far more significant than outer space. Like Spain before it, the United States squandered its windfall. Something might reverse that slide, but as Damon Runyon advised, the race is not always to the swift nor victory to the strong, but that's where you place your money.

The most plausible prognosis is that the future will resemble the past, that the Second Age's monadnock will mark an axis around which the evolving contours will unfold with rough historical symmetry. The Third Age will resemble the early Second, though in reverse, eventually mimicking with high-tech hardware the tempo of the First. (Even the attrition of spacecraft resembles that of far-sailing mariners.) Expeditions will slide to a new steady state, perhaps on the order of one or two a year (see figure below). These will be complicated probes, requiring years of preparation, similar to the expeditions launched during the Great Voyages and

A Prospective Future for Exploration

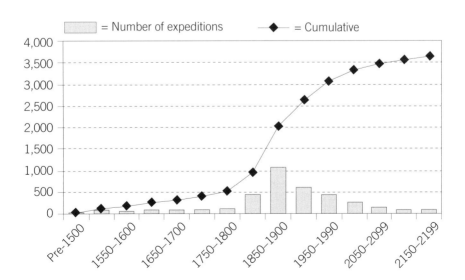

quite unlike the brawling swarm that so inflated the Second Age. The motor for the past half millennium of exploration—Europe's internal quarrels—is now directed inward, to the European Union, or in absorbing rivalries that once drove exploration outward. Such expeditions will be targeted to some particular purpose—commercial, scientific, technological, national prowess, and prestige. They are unlikely to spill out from colonization: they will rather resemble those expeditions that early established trading factories on islands or episodically visited coastlines for barter or sought out new routes. If the process thrives, there will be several rivals, not some collective United Earth Space Agency, and that institutional unrest is what will keep the pot simmering. Steadily, more and more of the solar system will be visited, cataloged, mapped, assessed. Perhaps, here and there, an outpost will appear, staffed for a few years. Reversing this trend would require an immense, global commitment that could only come from some dark necessity or irresistible rivalry, say, the discovery amid the asteroids of some mineral absolutely vital to national existence—the equivalent of the Potosi mines of Mexico, perhaps—or from Venusians announcing that they intend to colonize Mars and the moons of Saturn, and defying Earthlings to stop them.

What might all this portend for NASA? It likely means that exploration will continue to command popular interest, that scientific pursuits may well be sufficient to justify the further exploration of the solar system, with curiosity replacing commerce as a motive force and modern science substituting its own fierce competitive for geopolitical rivalry, and that the cultural continuities inherent in the long trajectory of geographic discovery by Western civilization will persist. For the near future, exploration's own inertia will propel more exploration. But the interpretation also argues that NASA would do well to attend to the differences. It suggests that trekking among the planets will not be the same as crossing a continent or sailing the seas. The distinction is not simply one of technology and vessels, but of psychology and the meaning, ultimately the morality, of what occurs. It suggests that in the future, expeditions will be complex, public commitment modest, and the vigor of the program measured not by the number of expeditions so much as the impact of their novelty. It suggests, as so much other evidence does, that the Apollo program was an aberration and the attempt to institutionalize a successor through the Space Shuttle, an anomaly. It suggests that the chief novelty unveiled by space travel will be the character of exploration itself, that the explorer may be—ought to be—robotic and virtual. It is, in truth, more than a little odd that an enterprise premised on the discovery of the new should be so obsessed with retaining the old, especially cultural archaisms. The vital requirement for future exploration is less a new propulsion mechanism than a new appreciation for how geographic discovery must proceed in a context beyond Earth. Eponymously named spacecraft and planetary rovers may be the future's prosthetic explorers. An obsession with colonization will be a burden rather than a boon.

There may even be a deeper symmetry in the narrative arc of the Great Ages of Discovery. The Grand Ages may themselves end. They were created; they can expire. The conditions that sustained them may cease altogether; they may no longer inspire interest as a tradition worthy of institutional support. One can even imagine a robotic Columbus, ceremoniously announcing an end to the enterprise. If the late 19th century marks a bilateral middle in this saga, that passing may happen some 400 years later, the early 23rd century, where *Star Trek* now resides in the popular imagination. Exploration, even of space, may then exist only in literature, history, film, and popular imagination, and in a past where no one, boldly or otherwise, wishes any longer to go.

CHAPTER 2

COMPELLING RATIONALES FOR SPACEFLIGHT? HISTORY AND THE SEARCH FOR RELEVANCE

Roger D. Launius

Are there compelling reasons to travel into space? Assuming that there are, when did they emerge in the consciousness of the space community, opinion leaders, politicos, larger public? How have those compelling reasons for spaceflight been articulated and adjusted over time? With all of the changes in the larger society during the last half century, do those rationales remain persuasive at the dawn of the 21st century? Finding answers to these questions are probably *the* most critical issues currently facing the space policy community. Of course, these issues may be considered without the use of historical analysis, and many do so, but the debate is immeasurably enriched by an understanding and explication of the historical evolution of the rationales that have been offered for why humanity seeks to fly in space. This essay begins with a discussion of the motivations for spaceflight—ultimately resting on the deep-seated desire to become a multiplanetary species and a quest for utopia beyond this realm—before moving into a sustained discussion of the five rationales for spaceflight that have been advanced over time: national pride/prestige/geopolitics, human destiny/survival of the species, commercial and other applications, national security, and science and technology. All of these have been used over time to support the concept of spaceflight. But are they compelling rationales today? Were they ever? The conclusion of this essay explores the long-term consequences of these rationales.

A QUESTION OF MOTIVATION

Of course, one must ask the question, why did spaceflight advocates believe so thoroughly in the necessity of moving beyond "Mother Earth?" Certainly, they viewed it as a thrilling adventure, one that would test the best that humanity had to offer. Was it simply a problem to be solved, or did they envision something more? Ultimately, what was the point of sending people into space? Is not the expansion of a human presence throughout the cosmos the real, long-term agenda? I am convinced that there was much more to it than just trying to solve an engineering problem, although few of the spaceflight enthusiasts systematically expressed their long-term objectives. In essence, the advocates have long believed that it is human destiny to become a multiplanetary species, not just as an end in itself, but because of the desire to create a utopian society free from the constraints of cultures on Earth.

De Witt Douglas Kilgore has recently noted that this motivation may be characterized as "astrofuturism," the application in the American tradition of technological utopianism responding to the political upheavals of the 20th century. Kilgore asserts that the pro-space utopian impulse was founded in the imperial politics and utopian schemes of the 19th century but envisions outer space as an endless frontier that offers solutions to the economic and political problems that dominate the modern world. Its advocates used the conventions of technological and scientific conquest to express the ideals and contradictions endemic to American culture. Astrofuturists, according to Kilgore, imagined space frontiers that could extend the reach of the human species and heal its historical wounds. Their efforts both replicated dominant social presuppositions and supplied the technologies necessary for the critical utopian projects that emerged in the latter 20th century.[1]

One critical astrofuturist, the American rocket pioneer Robert H. Goddard, wrote effectively about breaking the bonds of Earth to achieve the full potential of the human spirit. A native of Worcester, Massachusetts, Goddard had a surprisingly metaphysical perspective on the cause of human spaceflight. As a boy, while his family was staying at the suburban home of friends in Worcester on 19 October 1899, he climbed into an old cherry tree to prune its dead branches. Instead, he began daydreaming. As he wrote later, "It was one of the quiet, colorful afternoons of sheer beauty which we have in October in New England, and as I looked toward the fields at the east, I imagined how wonderful it would be to make some device which had even the possibility of ascending to Mars, and how it would look on a small scale, if sent up from the meadow at my feet." From that point on, Goddard enthusiastically pursued the idea of spaceflight as a necessary part of human destiny. He wrote in his diary, "Existence at last seemed very purposive." In addition, 19 October became "Anniversary Day," noted in his diary as his personal holiday. He went on to tie space exploration to a surprisingly utopian vision of the future. At his high school oration in 1904, he summarized his life's perspective: "It is difficult to say what is impossible, for the dream of yesterday is the hope of today and the reality of tomorrow."[2] Later he added, "Every vision is a joke, until the first man accomplishes it."

1. See De Witt Douglas Kilgore, *Astrofuturism: Science, Race, and Visions of Utopia in Space* (Philadelphia: University of Pennsylvania Press, 2003).

2. Two solid biographies of Goddard are Milton Lehman, *This High Man: The Life of Robert H. Goddard* (New York: Farrar, Straus, 1963), which is outdated, and David A. Clary, *Rocket Man: Robert H. Goddard and the Birth of the Space Age* (New York: Hyperion, 2003). The quotations are from Esther C. Goddard, ed., and G. Edward Pendray, assoc. ed., *The Papers of Robert H. Goddard*, 3 vols. (New York: McGraw-Hill Book Co., 1970), 1:10, 1:63–66.

The most powerful justification Goddard ever offered for humanity's movement into space was an essay called "The Great Migration," written on 14 January 1918 but not made public until much later. He scrawled across the envelope: "To be given to the Smithsonian Institution, after the owner has finished with it, there to be preserved on file, and used at the discretion of the Institution. The notes should be read thoroughly only by an optimist." In this essay, Goddard postulated a time in the distant future when the Sun had cooled and life on Earth could no longer be sustained. He envisioned gigantic, intergalactic arks taking the essence of the creatures and knowledge of this planet to new homes throughout the vastness of the Milky Way. "It has long been known," he wrote, "that protoplasm can remain inanimate for great periods of time, and can also withstand great cold, if in the granular state." There, amidst the stars, human society would replicate the best of what it had to offer.[3] While the issue of utopianism is implicit, it is still present and offered for Goddard a reason to dedicate his life to building the technology necessary to achieve multiplanetary migration.

These ideas of human destiny and perfect societies on new and perfect worlds have been expanded upon and extended far beyond Goddard's basic vision in numerous subsequent works. Wernher von Braun, the single most important promoter of America's space effort in the 1950s and 1960s, captured the essence of American utopian idealism and used it to justify an aggressive space exploration program.[4] Although a German immigrant to the United States after World War II, or perhaps because of it, he was remarkable in his grasp of what made Americans tick. He spoke often of "The Challenge of the Century" as a continuation of American exploration and settlement and the creation of a perfect society in a new land. "For more than 400 years the history of this nation has been crammed with adventure and excitement and marked by expansion," he said. "Compared with Europe, Africa, and Asia, America was the New World. Its pioneer settlers were daring, energetic, and self-reliant. They were challenged by the promise of unexplored and unsettled territory, and stimulated by the urge to conquer these vast new frontiers." Americans need the space frontier both physically and spiritually, von Braun insisted, and suggested that greater efforts in moving beyond the Earth would lead to a society in which "right relationships" prevailed.[5]

---

3. Robert H. Goddard, "The Great Migration," in *Papers of Robert H. Goddard*, ed. Goddard and Pendray, 3:1611–1612.

4. It is important to understand that this effort to colonize the cosmos was not limited to Goddard. Hermann Oberth wrote, "This is the goal: To make available for life every place where life is possible. To make inhabitable all worlds as yet uninhabitable, and all life purposeful" (Hermann Oberth, *Man into Space* [New York: Harper and Brothers, 1957], p. 167).

5. Wernher von Braun, "The Challenge of the Century," 3 April 1965, Wernher von Braun Biographical File, NASA Historical Reference Collection, Washington, DC.

Von Braun never wavered in his commitment to creating a perfect society in space. In a 1976 speech to the National Space Institute, he pointed to a bright future for humanity if it embarked on the high frontier of space. He said space would "offer new places to live—a chance to organize a new interplanetary society, and make fresh beginnings."[6] He believed this was "as inevitable as the rising of the sun; man has already poked his nose into space and he is not likely to pull it back . . . . There can be no thought of finishing, for aiming at the stars—both literally and figuratively—is the work of generations, and no matter how much progress one makes, there is always the thrill of just beginning."[7]

Gerard K. O'Neill, an experimental physicist at Princeton University, emerged during the 1970s to emphasize the possibilities of human settlement in space colonies. He left an indelible mark on the utopia-in-space movement by advocating the development of gigantic cylinders or spheres of roughly one-half by a few miles in size that would hold a breathable atmosphere, all the ingredients necessary for sustaining crops and life, and include rotating habitats to provide artificial gravity for thousands of inhabitants. While the human race might eventually build millions of these space colonies, each settlement would of necessity be an independent biosphere. Animals and plants endangered on Earth would thrive on these cosmic arks; insect pests would be left behind. Solar power, directed into each colony by huge mirrors, would provide a constant source of nonpolluting energy. Positioned at a specific point between the Earth and the Moon where the gravitation fields are equalized, known as LaGrange Point 5 (L-5), these O'Neill colonies could pursue the perfect society absent the problems of the parent society.[8]

This bold vision catapulted O'Neill into the spotlight of the space community and prompted a collective swoon from the thousands attracted to his ideas. They formed the L-5 Society in 1975 and adopted the slogan "L-5 in 1995." A particularly attractive group of space activists, one of their members wittily opined that they intended to "disband the Society in a mass meeting at L-5."[9] The space settlement mission also received a major boost from numerous science fiction and science fact writers, among them Arthur

6. "For Space Buffs—National Space Institute, You Can Join," *Popular Science* (May 1976): 73.

7. Wernher von Braun, "Crossing the Last Frontier," *Collier's* (22 March 1952): 24–29, 72–73. See also Ernst Stuhlinger and Frederick I. Ordway III, *Wernher von Braun, Crusader for Space: A Biographical Memoir* (Malabar, FL: Robert E. Krieger Company, 1994).

8. Gerard K. O'Neill, "The Colonization of Space," *Physics Today* 27 (September 1974): 32–40; Gerard K. O'Neill, *The High Frontier: Human Colonies in Space* (New York: William Morrow, 1976); Peter E. Glaser, "Energy from the Sun—Its Future," *Science* 162 (1968): 857–860; Peter E. Glaser, "Solar Power via Satellite," *Astronautics & Aeronautics* (August 1973): 60–68; Peter E. Glaser, "An Orbiting Solar Power Station," *Sky and Telescope* (April 1975): 224–228.

9. Michael A. G. Michaud, *Reaching for the High Frontier: The American Pro-Space Movement, 1972–84* (New York: Praeger, 1986), pp. 57–102.

C. Clarke, who popularized O'Neill's concept for colonies in space.[10] The strongly utopian impulse present in the O'Neill movement found voice in the words of aerospace writer T. A. Heppenheimer. "On Earth it is difficult for . . . people to form new nations or region[s] for themselves. But in space it will become easy for ethnic or religious groups, and for many others as well to set up their own colonies," Heppenheimer wrote. "Those who wish to found experimental communities, to try new social forms and practices, will have the opportunity to strike out into the wilderness and establish their ideals in cities in space."[11]

O'Neill's vision of practical and profitable colonies in space found an audience in many quarters of NASA even as it did in the larger pro-space movement. He received funding from NASA's Advanced Programs Office— but only $25,000—to develop his ideas more fully. Senior NASA officials such as Administrator James C. Fletcher and Ames Research Center Director Hans Mark encouraged his efforts. At the same time, some discredited his vision of colonies in space as hopelessly utopian.[12]

In the summer of 1975, NASA officials took O'Neill's ideas seriously enough to convene a study group of scientists, engineers, economists, and sociologists at the Ames Research Center, near San Francisco, to review the idea of space colonization, and followed it up with a study the next summer. Surprisingly, they found enough in the scheme to recommend it. Although budget estimates of $100 billion in then-year dollars accompanied the colonization project, the authors of this study concluded, "in contrast to Apollo, it appears that space colonization may be a paying proposition." For them, it offered "a way out from the sense of closure and of limits which is now oppressive to many people on Earth." The study recommended an international project led by the United States that would result in the establishment of a space colony at L-5. Most importantly, and decidedly utopian in expression, the study concluded:

> The possibility of cooperation among nations, in an enterprise which can yield new wealth for all rather than a conflict over the remaining resources of the Earth, may be far more important in the long run than the immediate return of energy to the Earth. So, too, may be the sense of hope and of

---

10. Arthur C. Clarke, *Rendezvous with Rama* (New York: Bantam Books, 1973).

11. T. A. Heppenheimer, *Colonies in Space* (Harrisburg, PA: Stackpole Books, 1977), pp. 279–280.

12. This would be completely consistent with their ideology. See Roger D. Launius, "A Western Mormon in Washington, D.C.: James C. Fletcher, NASA, and the Final Frontier," *Pacific Historical Review* 64 (May 1995): 217–241; Hans Mark, *The Space Station: A Personal Journey* (Durham, NC: Duke University Press, 1987); "Colonies in Space," *Newsweek* (27 November 1978): 95–101.

new options and opportunities which space colonization can bring to a world which has lost its frontiers.[13]

O'Neill publicized these findings exhaustively, but with political will for an aggressive space effort at low tide in the latter 1970s, nothing came of it.[14]

The utopian impulse has been strong in the history of the pro-space community since that time and has manifested itself in numerous quarters and by various advocates. The libertarian viewpoint of Rick Tumlinson and the Space Frontier Foundation clearly evokes a utopian mindset.[15] The commitment of Lyndon LaRouche to space colonization also bespeaks a utopian vision for the future modeled on his unique political and social ideals.[16] At some level, the rise of a conservative space agenda in the last two decades of the 20th century represented a utopian impulse as well, oriented as it is toward a celebration of the ideology of progress. The placement of the history of the Strategic Defense Initiative/"single stage to orbit" (SSTO)/ space colonization effort in the context of the United States' well-documented political "right turn" may represent the central thrust of space policy since the 1980s. The foundation and growth of this conservative space policy agenda has been well-documented in several historical works. Its linkage to various space advocacy groups, conservative futurists such as Gerry Pournelle, and space-power advocates such as Pete Worden ensured that conservative space advocates were able to manipulate the political system to achieve funding for their technological goals. At sum, they were intent on remaking both this world and outer space into a utopia of their own design.[17]

---

13. Richard D. Johnson and Charles Holbrow, eds., *Space Settlements: A Design Study in Colonization* (Washington, DC: NASA SP-413, 1977), pp. 27–28, a study sponsored by NASA Ames, American Society for Electrical Engineering (ASEE), and Stanford University in the summer of 1975 to look at all aspects of sustained life in space. See also John Billingham, William Gilbreath, Gerard K. O'Neill, and Brian O'Leary, eds., *Space Resources and Space Settlements* (Washington, DC: NASA SP-428, 1979).

14. The latter half of the 1970s might best be viewed as a nadir in human space exploration, with the Apollo program gone and the Shuttle not yet flying. See Louis J. Halle, "A Hopeful Future for Mankind," *Foreign Affairs* 59 (summer 1980): 1129–1136.

15. See Rick N. Tumlinson, "Why Space? Personal Freedom," Message 6 of the Frontier Files, 1995, *http://www.space-frontier.org/frontierfiles.html* (accessed 11 April 2001); Rick N. Tumlinson, "The Foundation Credo—Our View of the Frontier," Part 4 of 4, Frontier Files, 1995, *http://www.space-frontier.org/frontierfiles.html* (accessed 11 April 2001).

16. See Lyndon H. LaRouche, *The LaRouche-Bevel Program to Save the Nation: Reversing 30 Years of Post-Industrial Suicide* (Leesburg, VA: Independents for Economic Recovery, 1992). See especially chap. 11, "Frontier in Space," pp. 88–100.

17. This subject has been discussed in Andrew J. Butrica, *Single Stage to Orbit: Politics, Space Technology, and the Quest for Reusable Rocketry* (Baltimore, MD: Johns Hopkins, 2003), and W. D. Kay, "Space Policy Redefined: The Reagan Administration and the Commercialization of Space," *Business and Economic History* 27 (fall 1998): 237–247.

While the quest for utopia in space has been implicit rather than explicit, there has never been any question but that the long-term objective of spaceflight is human colonization of the cosmos. Virtually all models for the future of spaceflight have at their core human expansion beyond Earth. This model for human colonization of the cosmos was first developed in the 1950s, honed to a fine edge in later years, and carried to its logical conclusion by many in the more recent past. Promises in space of a bountiful future, in which all have enough resources to live a rewarding life, where there is unlimited economic potential, where peace and justice reign for all, and where the perfectibility of humankind is expected are all utopian sentiments. In addition, allusions to spaceflight as an attribute of human destiny and the hearkening back to a positive American frontier experience also stimulate visions of idyllic, perfect places.[18]

There is also a basic belief, utopian at its base, that spaceflight offers the only hope for the continuation of the human race. Asteroids or nuclear holocaust or environmental degradation or even a supernova all spell eventual doom for this planet and all who reside here. Astronaut John Young—veteran of Gemini, Apollo, and Space Shuttle missions—believes that the truly endangered species on Earth are humans. The only way to escape is to leave. The idea of a series of arks containing the living creatures of Earth is especially appealing since Americans so often conceptualize of themselves as called apart to "redeem" the world. Time is short, and every day brings humankind closer to destruction.[19]

Because of spaceflight's critical role in human colonization beyond Earth, it was logical that the early enthusiasts would always envision space exploration with humans at the center. For them, it made no sense to send robots as surrogates. We had to go ourselves because our ultimate purpose was to move outward. And, of course, humans did so with resounding success, landing on the Moon only 12 years after the launch of the first Earth-orbital satellite. Having reached the conclusion that human destiny requires movement outward from Earth and colonization of the solar system and, ultimately, the cosmos, the next question revolves around how it is advocated before the larger public. What rationales have been advanced in support of the grand design of human spaceflight? How effective have they been in garnering support for this great adventure?

---

18. While I do not want to overstate this case, I believe it is a very real aspect of the current spaceflight agenda of a cadre of "true believers." I have emphasized this part of the story in the period since the 1970s in "Perfect Worlds, Perfect Societies: The Persistent Goal of Utopia in Human Spaceflight," *Journal of the British Interplanetary Society* 56 (September/October 2003): 338–349.

19. John W. Young to Steve Hawley et al., "Why the Human Exploration of the Moon and Mars Must Be Accelerated," 9 March 2001, John Young File, folder 18552, NASA Historical Reference Collection, Washington, DC.

FIVE RATIONALES FOR SPACEFLIGHT

From the defining event of Sputnik in 1957, five major themes have been used to justify a large-scale space exploration agenda. None of them explicitly advocate the human colonization of space—although that theme is implicit throughout—and none even hint at the larger utopian objective, despite its fundamental presence within the spaceflight community. The five themes are as follows:

1) Human destiny/survival of the species.
2) Geopolitics/national pride and prestige.
3) National security and military applications.
4) Economic competitiveness and satellite applications.
5) Scientific discovery and understanding.

Those themes have continued to motivate American space policy from the very beginning of the Space Age to the present. Specific aspects of these five rationales have fluctuated over time but remain the primary reasons for the endeavor. Indeed, there are no more nor no less than these five basic rationales.

Human Destiny/Survival of the Species

The first and most common rationale for spaceflight is that an integral part of human nature is a desire for discovery and understanding. In essence, it is human destiny to explore, to learn, and to absorb new knowledge and new territories into the human experience. With the Earth so well known, space exploration advocates argue, exploration and settlement of the Moon and Mars is the next logical step in human exploration. Humans must question and explore and discover or die, advocates for this position insist. It is the "final frontier," and Americans have always responded well to their frontiers.

When speaking and writing of these possibilities, many space advocates explicitly use the language of the "Frontier Thesis," described for America in Frederick Jackson Turner's famous 1893 essay. Turner's "Frontier Thesis" is perhaps the most influential essay ever read at the American Historical Association's annual conference. It has exerted a powerful force in the historiography of the United States, in no small measure because of its powerful statement of American exceptionalism and its justification of conquest. Turner took as his cue an observation in the 1890 U.S. census that the American frontier had, for the first time, closed. He noted, "Up to our own day American history has been in a large degree the history of the colonization of the Great West. The existence of an area of free land, its continuous recession, and the advance of American settlement westward explain American development." He insisted that the frontier made Americans American, gave the nation its democratic character, and ensured the virtues of self-reliance, community, and the promise of justice. He noted that cheap or even free land provided a "safety valve" that protected the nation against uprisings of the poverty-stricken and malcontented. The frontier also produced a people with "coarseness and strength . . . acuteness and inquisitiveness, that practical and inventive turn of mind . . . [full of] restless and nervous energy . . . that buoyancy and exuberance which comes with freedom." It gave the people of the United States, in essence, virtually every positive quality they have ever possessed.[20]

Repeated use of the frontier analogy for spaceflight, with its vision of a new land and a new and better society, has given the American public a distinctive perspective on spacefaring. It always tapped a vein of rich ideological power. The symbolism of the frontier has been critical to understanding how Americans have viewed themselves since at least the end of the 19th century, and perhaps much longer. It conjured up an image of self-reliant Americans moving westward in sweeping waves of discovery, exploration, conquest, and settlement of an untamed wilderness. And in the process of movement, the Europeans who settled North America became an indigenous American people. The frontier concept has always carried with it the ideals of optimism, democracy, and right relationships.

It also summoned in the popular mind a wide range of vivid and memorable tales of heroism, each a morally justified step toward the modern democratic state. While the frontier ideal reduced the complexity of events to a relatively static morality play, avoided matters that challenged or contradicted the myth, viewed Americans moving westward as inherently good and their opponents as evil, and ignored the cultural context of westward migration, it served a critical unifying purpose for spaceflight advocates. Those persuaded by this metaphor (and many have been) recognize that it summons them not only to recall past glories, but also to undertake—or at least to acquiesce in—a heroic engagement under the ideal with the forces of social, political, and economic injustice.[21]

Turner's image of the American frontier has been an especially evocative and somewhat romantic popular theme for proponents of an aggressive space program. The popular conception of "westering" and the settlement of the American continent by Europeans has been a powerful metaphor for the propriety of space exploration and has enjoyed wide usage by supporters of space exploration. It hearkens back to the American West and the frontier in speaking of what might be gained in the unknown of space. But more important, it calls

---

20. Frederick Jackson Turner, "The Significance of the Frontier in American History," *The Frontier in American History* (New York: Holt, Rinehart, and Winston, 1920), pp. 1–38.

21. See Richard Slotkin, *Gunfighter Nation: The Myth of the Frontier in Twentieth-Century America* (New York: Atheneum, 1992).

upon the adventurousness of the American people and offers the promise of a utopian change in society as it moves to a new, untainted place where it could remake society. Such has always been the siren call of the frontier myth.

From Captain James T. Kirk's soliloquy—"Space, the final frontier"—at the beginning of each *Star Trek* episode to President John F. Kennedy's 1962 speech about setting sail on "this new ocean" of space, the exploration and colonization allusion has been a critical component of space program promotion. Astronaut, then Senator, John Glenn captured some of this tenor in 1983 when he summoned images of the American heritage of pioneering and argued that the next great frontier challenge was in space. "It represents the modern frontier for national adventure. Our spirit as a nation is reflected in our willingness to explore the unknown for the benefit of all humanity, and space is a prime medium in which to test our mettle."[22]

Quintessential American novelist James A. Michener also applied this frontier analogy to the space program. In two articles in *Omni* magazine in the early 1980s, he explicitly compared the space program to the Anglo-American westward movement of the 19th century. He described the American sense of pioneering and argued that the next great challenge in this arena is space. "A nation that loses its forward thrust is in danger," he commented; "the way to retain it is exploration." In an eloquent and moving way, he argued for the American space program as the logical means of carrying out exploration. One of these articles had the ironic title of "Manifest Destiny," a blatant hearkening to the ideology of continental expansion that gained preeminence in the 1840s. Michener argued that it is the American destiny to explore and colonize, and space is the next logical place to do this. His statement presents an eloquent and moving defense of America's human space program in all its permutations.[23]

NASA Administrator for nine years in the 1970s and 1980s, James C. Fletcher was especially attracted by the analogy of the American frontier. A Caltech Ph.D., he guided NASA during the critical period of redefining the space program at the conclusion of Apollo and for three years after the *Challenger* accident. But for all his hardheaded practicality, for all his understanding of science, he was enthralled with the frontier allusion and made specific connections to his pioneering ancestors in Utah. He commented:

> History teaches us that the process of pushing back frontiers on
> Earth begins with exploration and discovery and is followed by

---

22. John Glenn, Jr., "The Next 25: Agenda for the U.S.," *IEEE Spectrum* (September 1983): 91.

23. James A. Michener, "Looking toward Space," *Omni* (May 1980): 58. See also James A. Michener, "Manifest Destiny," *Omni* (April 1981): 48–50, 102–104.

> permanent settlements and economic development. Space will be no different . . . . Americans have always moved toward new frontiers because we are, above all, a nation of pioneers with an insatiable urge to know the unknown. Space is no exception to that pioneering spirit.[24]

The frontier myth's accessibility, coming with its utopian imagery, has served the pro-space movement well. Casting decisions on projects as facilitating the opening of this frontier has enormous appeal and has been used repeatedly since the launch of Sputnik.

But invoking Frederick Jackson Turner has become increasingly counterproductive for anyone who appreciates postmodern multicultural society. Historians appropriately criticize Turner's approach as excessively ethnocentric, nationalistic, and somewhat jingoistic. His rhetoric excludes more than it covers, failing to do justice to diverse western people and events. Yale historian Howard R. Lamar believes the Frontier Thesis emphasizes an inappropriate discontinuity between a mythical rural past and an urban-industrial future. Thus, it is unsuitable as a guide for understanding the present or projecting the future. Some scholars also discount its central safety-valve proposition. It may have applied in antebellum America, when many did "go West," they suggest, but failed to hold after the Civil War as the prospect of migration moved beyond the reach of urban slumdwellers and others because of a lack of funds for farming and transportation. In fact, later settlers, mostly the children of farmers, arrived from the fringes of existing settlements. Despite the criticism, the Frontier Thesis has had lasting appeal, in no small measure because it tells Americans how perfect they could become and offers an easily understandable if simplistic explanation for why that is the case. It is a small wonder that the Frontier Thesis would find service among those advocating an aggressive space exploration program![25]

President George W. Bush also supported space exploration as a human-destiny program in his 14 January 2004 announcement of a new vision for NASA. He stated that NASA would return to the Moon and eventually send astronauts to Mars. Doing so, as stated in the White House release on this subject, was human destiny:

---

24. James C. Fletcher, "Our Space Program Is Already Back on Track," *USA Today* (28 July 1987); R. Scott Lloyd, "NASA Head Is Veteran Teacher," *Salt Lake City (UT) Church News* (25 May 1986).

25. See John Mack Faragher, *Rereading Frederick Jackson Turner: The Significance of the Frontier in American History, and Other Essays* (New York: Henry Holt, 1994); Allan G. Bogue, *Frederick Jackson Turner: Strange Roads Going Down* (Norman: University of Oklahoma Press, 1998); Ray Allen Billington, *America's Frontier Heritage* (Albuquerque: University of New Mexico Press, 1974).

> America's history is built on a desire to open new frontiers and to seek new discoveries. Exploration, like investments in other Federal science and technology activities, is an investment in our future. President Bush is committed to a long-term space exploration program benefiting not only scientific research, but also the lives of all Americans. The exploration vision also has the potential to drive innovation, development, and advancement in the aerospace and other high-technology industries.[26]

In explicitly raising the issue of the space frontier, the President followed a long succession of advocates who invoked the happy metaphor of America's westward expansion to support his idea of human destiny.

If human destiny is a positive attribute that generally finds resonance among spaceflight advocates and the general public, there is also a terrifying aspect to this rationale. The flip side of the human-destiny argument is that humanity will not survive if it does not become multiplanetary. Carl Sagan wrote eloquently about the last perfect day on Earth, before the Sun would fundamentally change and end our ability to survive on this planet.[27] In their astrobiology book, *The Life and Death of Planet Earth*, Peter Ward and Donald Brownlee describe the natural life cycle of stars such as our Sun and the planets that circle them. They describe several possible scenarios for the end of life on Earth. Life on Earth will definitely end when the Sun, having used up too much of its hydrogen will become a red giant star and heat the Earth until every living thing, no matter how deep underground, is dead.[28]

While this will happen billions of years in the future, any number of catastrophes could end life on Earth beforehand. A much earlier and quite likely way for life (or at least life as we know it) to end is the way life almost ended 65 million years ago when either an asteroid or a comet crashed into the Earth. The consequences of this collision caused the extinction of the dinosaurs and probably two-thirds of all life on Earth at that time. Enough life survived the harsh environmental aftermath and gave rise to mammals, a highly adaptable species that even survived the last Ice Age.

In 1992, a noted scientist spoke to the American Astronautical Society on the subject "Chicken Little Was Right." The scientist claimed that humans had a greater chance of being killed by a comet or asteroid falling from the sky than dying in an airplane crash. This is true; mathematical calculations confirm

---

26. White House Press Release, "President Bush Announces New Vision for Space Exploration Program," 14 January 2004, *http://www.whitehouse.gov/infocus/space/#* (accessed 30 December 2004).

27. Carl Sagan, *Cosmos* (New York: Random House, 1980), pp. 231–232.

28. See Peter D. Ward and Donald Brownlee, *The Life and Death of Planet Earth: How the New Science of Astrobiology Charts the Ultimate Fate of Our World* (New York: Henry Holt and Co., 2002).

that every individual faces a 1-in-5,000 chance of being killed by some type of extraterrestrial impact. Throughout history, asteroids and comets have struck Earth, and a great galactic asteroid probably killed the dinosaurs. An object probably only 6 to 9 miles wide left a crater 186 miles wide in Mexico's Yucatan Peninsula. This reality entered most people's consciousness in July 1994, when humans for the first time witnessed the devastating impact of a large Near-Earth Object (NEO) into one of the planets in the solar system when Comet Shoemaker-Levy 9 crashed into Jupiter with spectacular results.[29]

With time, a comet or meteoroid will again hit Earth with disastrous consequences. Efforts to catalogue all Earth-crossing asteroids, track their trajectories, and develop countermeasures to destroy or deflect objects on a collision course with Earth are important, but to ensure the survival of the species, humanity must build outposts elsewhere. Astronaut John Young said it best, to paraphrase Pogo, "I have met an endangered species, and it is us."[30]

Geopolitics/National Pride and Prestige

In addition, geopolitics and national prestige have dominated so many of the spaceflight decisions that it sometimes seems trite to suggest that it has been an impressive rationale over the years. Yet there is more to it than that, for while all recognize that prestige sparked and sustained the space race of the 1960s, they fail to recognize that it continues to motivate many politicians to support NASA's programs. John F. Kennedy responded to the challenge of the Soviet Union by announcing the Apollo decision in 1961, and that rivalry sustained the effort. Kennedy put the world on notice that the U.S. would not take a back seat to its superpower rival. As John M. Logsdon commented, "By entering the race with such a visible and dramatic commitment, the United States effectively undercut Soviet space spectaculars without doing much except announcing its intention to join the contest."[31] Kennedy said in 1962 that "we mean to be a part of it [spaceflight]—we mean to lead it. For the eyes of the world now look into space, to the moon and to the planets beyond, and we have vowed that we shall

29. K. Zahnle and M. M. Mac Low, "The Collision of Jupiter and Comet Shoemaker-Levy 9," *Icarus* 108 (1994): 1–17; Paul W. Chodas and Donald K. Yeomans, "The Orbital Motion and Impact Circumstances of Comet Shoemaker-Levy 9," in *International Astronomical Union (IAU) Colloquium 156: Proceedings of the Space Telescope Science Institute Workshop* (held in Baltimore, MD, 9–12 May 1995), ed. Keith S. Noll, Harold A. Weaver, and Paul D. Feldman (New York: Cambridge University Press, 1996), pp. 1–30. On the mass extinction of the dinosaurs, see Walter Alvarez, *T. Rex and the Crater of Doom* (Princeton, NJ: Princeton University Press, 1997).

30. John W. Young, "The Big Picture: Ways to Mitigate or Prevent Very Bad Planet Earth Events," *Space Times: Magazine of the American Astronautical Society* 42 (November/December 2003): 22–23.

31. John M. Logsdon, "An Apollo Perspective," *Astronautics & Aeronautics* (December 1979): 112–117, quotation from p. 115. See also John M. Logsdon, *The Decision to Go to the Moon: Project Apollo and the National Interest* (Cambridge, MA: Massachusetts Institute of Technology [MIT] Press, 1970).

not see it governed by a hostile flag of conquest, but by a banner of freedom and peace. We have vowed that we shall not see space filled with weapons of mass destruction, but with instruments of knowledge and understanding."[32] Apollo was a contest of wills, of political systems, of superpowers. And the United States had to win it. Lyndon Johnson summed this up well with his assertion, "Failure to master space means being second best in every aspect, in the crucial area of our Cold War world. In the eyes of the world first in space means first, period; second in space is second in everything."[33]

Through the decade of the 1960s, prestige dominated much of the discussion of Apollo, even penetrating to the popular culture. Actor Carroll O'Connor perhaps said it best in an episode of *All in the Family* in 1971. Portraying the character of Archie Bunker, the bigoted working-class American whose perspectives were more common in our society than many observers were comfortable with, O'Connor summarized well how most Americans responded to the culture of competence that Apollo engendered. He observed that he had "a genuine facsimile of the Apollo 14 insignia. That's the thing that sets the US of A apart from . . . all them other losers."[34] In very specific terms, Archie Bunker encapsulated for everyone what set the United States apart from every other nation in the world: success in spaceflight. At a basic level, Apollo provided the impetus for the perception of NASA as a culture of competence, one of the great myths emerging from the lunar landing program.

The United States went to Moon for prestige purposes, but it also built the Space Shuttle and embarked on the space station for prestige purposes as well. The turning point for Richard Nixon's decision to proceed with the Space Shuttle for post-Apollo spaceflight came in August 1971 when Caspar Weinberger wrote an impassioned memorandum to the President that not to do so "would be confirming in some respects, a belief that I fear is gaining credence at home and abroad: That our best years are behind us, that we are turning inward, reducing our defense commitments, and voluntarily starting to give up our super-power status, and our desire to maintain world superiority." Weinberger appealed directly to the prestige argument by concluding, "America should be able to afford something besides increased welfare, programs to repair our cities, or Appalachian relief and the like." In a handwritten scrawl on

---

32. President John F. Kennedy, "Address at Rice University on the Nation's Space Effort," 12 September 1962, Houston, TX, John Fitzgerald Kennedy Library, Boston, MA, available online at *http://www.cs.umb.edu/jfklibrary/j091262.htm* (accessed 27 October 2002).

33. Lyndon B. Johnson, quoted in Walter A. McDougall, "Technocracy and Statecraft in the Space Age: Toward the History of a Saltation," *American Historical Review* 87 (October 1982): 1010–1040, quotation from p. 1025.

34. "Carroll O'Connor Obituary," on *Morning Edition* (National Public Radio program), 22 June 2001. This report by Andy Bowers is available online at *http://www.npr.org* (accessed 2 July 2001).

Weinberger's memo, Richard Nixon indicated, "I agree with Cap."[35] Prestige also entered into the decision in one other way. Nixon was also unwilling to go down in history as the President who gave away the nation's leadership in the exploration of space and ended the practice of flying astronauts, and a decision against the Shuttle, in his mind, would have done both.[36]

An object lesson in the role of humans in space exploration. Charles Conrad, Jr., Apollo 12 commander, examines the robotic Surveyor III spacecraft during the second extravehicular activity (EVA-2) in 1969. The Lunar Module (LM) *Intrepid* is in the right background. This picture was taken by astronaut Alan L. Bean, Lunar Module pilot. The *Intrepid* landed on the Moon's Ocean of Storms only 600 feet from Surveyor III. The television camera and several other components were taken from Surveyor III and brought back to Earth for scientific analysis. Surveyor III soft-landed on the Moon on 19 April 1967. Interestingly, microbes from Earth on the Surveyor spacecraft survived in hibernation during a three-year stay on the lunar surface and revived upon return to Earth. This suggests the resilience of life in the harsh environment of space. *(NASA JSC photo no. AS12-48-7136)*

---

35. Caspar W. Weinberger to President Richard M. Nixon, via George Shultz, "Future of NASA," 12 August 1971, White House, Richard M. Nixon, President, 1968–1971 File, NASA Historical Reference Collection, Washington, DC.

36. John Erlichman interview, Washington, DC, by John M. Logsdon, 6 May 1983, NASA Historical Reference Collection, Washington, DC. See also George M. Low, NASA Deputy Administrator, to James C. Fletcher, NASA Administrator, "Items of Interest," 12 August 1971, and James C. Fletcher, NASA Administrator, to Jonathan Rose, Special Assistant to the President, 22 November 1971, both in Fletcher Correspondence, folder 4247, NASA Historical Reference Collection, Washington, DC.

Prestige also played a key role in the decision to build a space station. At a 1 December 1983 meeting in the White House, NASA Administrator James M. Beggs asked President Ronald Reagan to approve his agency's space station plans. Beggs stressed the space station's potential contribution to the leadership of the United States on the world's stage. He knew that Ronald Reagan had long been concerned with a perceived withering of American prestige vis-à-vis the Soviet Union. The station, he argued, would help to quell that declension. But as the punch line for the briefing, Beggs hit Reagan between the eyes with a photo of a Salyut space station overflying the United States. He emphasized that the Soviet Union already had this modest space station and was planning a larger orbital facility. Should not the United States have one as well? Reagan agreed it should.[37]

Prestige will ensure that no matter how difficult the challenges and overbearing the obstacles, the United States will continue to fly humans in space indefinitely. In the aftermath of the *Columbia* accident on 1 February 2003 that took the lives of seven astronauts, when it appeared that all reason for human spaceflight should be questioned, no one seriously considered ending the program. Instead, support for the effort came from all quarters. Even President George W. Bush, who had always been silent on spaceflight before, stepped forward on the day of the accident to say that "the cause in which they died will continue. Mankind is led into the darkness beyond our world by the inspiration of discovery and the longing to understand. Our journey into space will go on."[38]

Of course, there is a positive aspect to this prestige that is very present throughout the age of spaceflight. One might call this pride, which aims to make Americans feel good about what they are doing. There is a genuine excitement and interest in space science/technology that the human spaceflight program produces. This is not new, and it remains critical to understanding this rationale for spaceflight. One might ask, as John Krige recently did, "How many people would come to the National Air and Space Museum [NASM] if it was just the NAM, and why are the human in space objects the ones that attract the most attention?"[39] As Krige asserts, the prestige factor disguises a critical foreign policy component in all of these human spaceflight programs. National leaders supported Apollo, the Space Shuttle, and the space station efforts not on their merits, but on the image they projected. Their initial

37. "Revised Talking Points for the Space Station Presentation to the President and the Cabinet Council," 30 November 1983, with attached "Presentation on Space Station," 1 December 1983, Reagan/NASA Correspondence, folder 12766, NASA Historical Reference Collection, Washington, DC.

38. Statement by President George W. Bush, The Cabinet Room, 2:04 p.m. eastern standard time (EST), 1 February 2003, in Bush, George W., folder 18262, NASA Historical Reference Collection, Washington, DC.

39. John Krige to author, e-mail message, "Space Rationales," 2 February 2005.

and continued support rested on the value they offered not as instruments of science, military prowess, economics, or the like, but on their usefulness as icons mobilized to buttress America's position in the world. Accordingly, despite some truly significant accomplishments, they have been in no small measure symbolic for the majority of those observing them. That is certainly not all bad, and one might say essentially the same thing about the United States' nuclear arsenal during the Cold War. The missiles and strategic bombers served to deter the Soviet Union, offering a symbolic threat more than one in reality since the doomsday scenarios their use would unleash were too devastating to contemplate.[40] Might this be a powerful enough motivation to continue human spaceflight indefinitely?

The fundamental importance of human spaceflight as an instrument of U.S. foreign policy—which is not necessarily identical with national prestige and geopolitics but is closely allied—should not be mislaid in this discussion. It served, and continues to do so, as an instrument for projecting the image of a positive, open, dynamic American society abroad. What of the good will generated by the United States in opening spaceflight to foreign astronauts during the Shuttle era? What about the significance of binding allies more closely to the United States through numerous international efforts ranging from robotic missions to the International Space Station? The foreign policy dimension of international human spaceflight should not be underestimated.

National Security and Military Applications

Another rationale for spaceflight has involved national defense and military space activity. From the beginning, national leaders sought to use space to ensure U.S. security from nuclear holocaust. In October 1951, Wernher von Braun proposed in the pages of *Popular Science* the building of a space station because "the nation which first owns such a bomb-dropping space station might be in a position virtually to control the earth."[41] In 1952, a popular conception of the U.S.-occupied space station showed it as a platform from which to observe the Soviet Union and the rest of the globe in the interest of national security. As the editors of *Collier's* magazine editorialized, "The U.S. must immediately embark on a long-range development program to secure for the West 'space superiority.' If we do not, somebody else will . . . . A ruthless foe established on a space station could actually subjugate the peoples of the world."[42]

---

40. There has been an enormous amount of historical literature on this subject. See especially Fred Kaplan, *The Wizards of Armageddon* (Stanford, CA: Stanford University Press, 1991); Herman Kahn, *Thinking about the Unthinkable* (New York: Touchstone Books, 1985); Paul S. Boyer, *By the Bomb's Early Light: American Thought and Culture at the Dawn of the Atomic Age* (New York: Pantheon Books, 1985).

41. "Giant Doughnut is Proposed as Space Station," *Popular Science* (October 1951): 120–121.

42. "What Are We Waiting For?" *Collier's* 129 (22 March 1952): 23.

Early in the 1950s, the U.S. military recognized that space represented the new high ground and that they had to control it. Numerous defense officials referred to space as the high seas of the future. The nation that could exploit the potential benefits of this ultimate strategic high ground for military purposes would dominate the rest of the world. The nation's goals for space dominance have revolved since that time around four interrelated strategic issues:

1) Space is a geographic location like air, land, and sea. Any national security capabilities for these other regions must be replicated in space. The Department of Defense, therefore, must control the use of space and defend its military and civil assets from foreign attack.

2) A strong national security presence in space is vital, even during times of peace. Military strategists long have maintained that those nations most successful at controlling the seas are the same nations that tend most to succeed politically and economically. Space is becoming the seas of the future.

3) Space must be dominated during wartime. That requires that the U.S. be prepared to protect U.S. access to space while denying its enemies' access to space. It also means that the U.S. must be capable of exploiting the space regime, especially preferred orbits and missile lanes.

4) National security requires that the United States enhance space resources for a variety of Earth-oriented missions: command, control, communications, and intelligence (C^3I); early warning; weather forecasting; navigation; antisatellite; space-to-ground attack; and missile defense.[43]

The U.S. military also argued for a human capability to fly in space for rapid deployment of troops to hot spots anywhere around the Earth, but they never managed to convince the political leadership of the nation and, despite periodic attempts, never gained a human military mission. The human spaceflight enterprise also gained energy from Cold War rivalries in the 1950s and 1960s as international prestige, translated into American support from nonaligned nations, found an important place in the space policy agenda. Human spaceflight also had a strong military nature during the 1980s, when astronauts from the military services deployed reconnaissance satellites into Earth orbit from the Space Shuttle. A human military presence in space promises to remain a prospect for national security well into the 21st century.[44]

---

43. Bryan Johnson, "Political Economy—The Military Use of Space," 14 May 1999, *http://www. suite101.com/article.cfm/political_economy/19993* (accessed 31 December 2004).

44. See Roger D. Launius, *Space Stations: Base Camps to the Stars* (Washington, DC: Smithsonian Books, 2003), pp. 26–35, 114–121.

As it stands, the military has employed space-based and space-transiting resources for more than 40 years. The major systems include the following:

- Ballistic missiles.
- Reconnaissance satellites, both imagery and signals intelligence.
- Navigational satellites, the Global Positioning System.
- Weather and communications satellites.
- Early-warning satellites.
- Ballistic missile defense.

Collectively, these resources have been enormously important in winning the Cold War and ensuring American preeminence at the dawn of the 21st century.[45] No one questions the legitimate role of space resources in the security of the United States. Indeed, the national defense space budget of the United States exceeded NASA's space budget in 1982 and has far outdistanced its spending since that time. In fiscal year 2003, for example, the Department of Defense's spending on space was $19.39 billion, while NASA's space budget was $14.36 billion.[46]

Economic Competitiveness and Satellite Applications

The fourth rationale of economic competitiveness and commercial applications has provided another reason for engaging in spaceflight. Satellite communications is still the only truly commercial space technology to be developed in the more than 45 years since the beginning of the Space Age in 1957. It generates billions of dollars annually in sales of products and services. The first inkling of what this business might look like appeared in the fall of 1945 when a then-obscure RAF electronics officer and member of the British Interplanetary Society, Arthur C. Clarke, wrote a short article in *Wireless World* that described the use of satellites in 24-hour "geosynchronous" orbits some 26,000 miles above the Earth to distribute television programs.[47]

Perhaps the first person to evaluate both the technical and financial possibilities of satellite communications was John R. Pierce of AT&T's Bell Labs. In the mid-1950s, he argued that a communications "mirror" in space would be worth as much as a billion dollars. His estimate was conservative. Following

45. See Everett Carl Dolman, *Astropolitik: Classical Geopolitics in the Space Age* (Portland, OR: Frank Cass Publishers, 2001); M. Mowthorp, "U.S. Military Space Policy, 1945–1992," *Space Policy* 18 (February 2002): 25–36; Roger Handberg, "Review Article: Military Space Policy: Debating the Future," *Astropolitics* 2 (spring 2004): 79–89.

46. *Aeronautics and Space Report of the President, Fiscal Year 2003 Activities* (Washington, DC: NASA NP-2004-17-389-HQ, 2004), p. 139.

47. Arthur C. Clarke, "Extra-Terrestrial Relays: Can Rocket Stations Give World-Wide Radio Coverage?" *Wireless World* (October 1945): 305–308.

Pierce's leadership, in 1960 AT&T filed with the Federal Communications Commission (FCC) for permission to launch a communications satellite as an experiment.[48] This shocked the Kennedy administration, many of whose senior officials believed that AT&T was seeking to extend its telephone monopoly into the "new high ground" of space. They did not approve, and the U.S. government scrambled to implement a new regulatory environment, something that cheered AT&T's telecommunications rivals if not AT&T itself. Accordingly, NASA was directed to enter the fray in developing this new technology, and in 1961, it awarded contracts to RCA and Hughes Aircraft to build communication satellites, Relay and Syncom. Both, government officials believed, would help offset AT&T's technological lead in the field. This policy succeeded. By 1964, two AT&T Telstars, two Relays, and two Syncoms had operated successfully in space and technological "know-how" had been transferred to companies other than AT&T.

At the same time and largely for similar reasons, the Kennedy administration sponsored the Communications Satellite Act of 1962. This law created the Communications Satellite Corporation (COMSAT), with ownership divided 50-50 between the general public and the various telecommunications corporations. Later, COMSAT became the American manager of an emerging global system known as the International Telecommunications Satellite Consortium (INTELSAT) formed on 20 August 1964. On 6 April 1965, COMSAT's first satellite, Early Bird, was launched from Cape Canaveral. Global satellite communications had begun.[49] From a few hundred telephone circuits in 1965, the INTELSAT system rapidly grew to become a massive organization providing millions of telephone circuits. And the costs persistently declined, making the backers of this technology appear geniuses. Whereas customers had paid as much as $10 per minute using older, cable-based technology, the new satellites reduced costs to less than $1 per minute.[50] Even before this time, government officials realized they had a "winner" on their hands. In 1964, NASA Administrator James E. Webb asked his staff, "How did we get so much communication satellite technology for so little money?"[51] His question was not satisfactorily answered by his NASA lieutenants, but space commerce

48. This story is well told in David J. Whalen, *The Origins of Satellite Communications, 1945–1965* (Washington, DC: Smithsonian Institution Press, 2002).

49. See Hugh R. Slotten, "Satellite Communications, Globalization, and the Cold War," *Technology and Culture* 43 (April 2002): 315–360; Whalen, *The Origins of Satellite Communications*; Andrew J. Butrica, ed., *Beyond the Ionosphere: Fifty Years of Satellite Communication* (Washington, DC: NASA SP-4217, 1997); Heather E. Hudson, *Communications Satellites: Their Development and Impact* (New York: Free Press, 1990).

50. David J. Whalen, "Communications Satellites: Making the Global Village Possible," *http://www.hq.nasa.gov/office/pao/History/satcomhistory.html* (accessed 1 January 2005).

51. Paraphrase of Webb's comments at the 22 September 1964 Program Review, referred to in W. A. Radius to ADA/Shapley, 10 December 1965, Thompson papers, NASA Historical Reference Collection, Washington, DC.

has been dominated by satellite communications, and Webb and his successors have ballyhooed it ever since. The sale of all components associated with satellite communications—development, launch, operations—surpassed $100 billion a year in the first part of the 21st century.

There may be other commercially viable space-based industries that will prove lucrative, but they do not yet exist. Many believed that the Landsat Earth remote sensing efforts of the 1970s and since would turn into a commercial activity, but it failed to gain a market despite its significance as a scientific effort. More recently, remote sensing of various types and for a multitude of activities may be on the verge of takeoff, but this remains to be seen. Many observers point to the growth of space-based navigation as another economically viable activity, but they tend to omit the fact that the constellation of satellites—the Global Positioning System (GPS)—is provided gratis by the Department of Defense, and without this critical infrastructure, it is problematic that much commercial activity would be forthcoming.[52]

In recent years, the economic rationale has become stronger and even more explicit as space applications become increasingly central for maintaining United States global economic competitiveness. Ronald Reagan's presidential administration especially emphasized enlarging the role of the private sector, and its priorities have remained in place thereafter. For instance, in the context of space access, the American political right argued an ideology of progress aimed at private development of space-access technology. This led to changes in the government environment, especially regulations that eased authorizations for launch services, and in the encouragement of private rocket-development projects. Such success stories as the Pegasus air-launched booster for small payloads built by Orbital Sciences, Inc., emerged from this cauldron of entrepreneurship. Even such projects as the X-33/VentureStar™, begun in 1995, used a public-private partnership model between NASA and Lockheed Martin, with each contributing to the development of a small suborbital vehicle that could demonstrate the technologies required for an operational SSTO launcher. The X-33 project had an ambitious timetable to fly by 2001, but instead, NASA canceled the program without flying any hardware.[53]

---

52. Ray A. Williamson and John C. Baker, "Current U.S. Remote Sensing Policies: Opportunities and Challenges," *Space Policy* 20 (May 2004): 109–116; Sameer Kumar and Kevin B. Moore, "The Evolution of Global Positioning System (GPS) Technology," *Journal of Science Education and Technology* 11 (March 2002): 59–80; Irene A. Miller, "GPS and Beyond: How the Aviation Industry is Advancing the Usefulness of GPS," *Quest: The History of Spaceflight Quarterly* 7 (winter 1998): 41–45.

53. Frank Sietzen, "VentureStar Will Need Public Funding," *Space Daily Express*, 16 February 1998, NASA Historical Reference Collection, Washington, DC; Butrica, *Single Stage to Orbit*, pp. 13–28; Leonard David, "NASA Shuts Down X-33, X-34 Programs," *Space.com*, 1 March 2001, *http://www.space.com/missionlaunches/missions/x33_cancel_010301.html* (accessed 28 March 2003); Andrew J. Butrica, "The Quest for Reusability," in *To Reach the High Frontier: A History of U.S. Launch Vehicles*, ed. Roger D. Launius and Dennis R. Jenkins (Lexington: University Press of Kentucky, 2001), pp. 443–469.

One of the key initiatives in this effort for human spaceflight is tourism, a major aspect that envisages hotels in Earth orbit and lunar vacation packages. In 1995, Patrick Collins, Richard Stockmans, and M. Maita undertook a market study on the potential demand for space tourism for the National Aerospace Laboratory in Tokyo, Japan. In the first actual market research of its type, they suggested that space tourism services would be very popular both in North America and Japan, the two leading economies in the world. Overall, 60 percent of the people surveyed "want to visit space for themselves" and were interested in traveling to space for a vacation. Accordingly, the authors found that a market of 1 million passengers per year paying $10,000 per person would generate revenues of $10 billion per year. Thus the market potential of space tourism is somewhat similar to that of the Concorde.[54] Adding fuel to this belief, NASA engineer Barbara Stone opined at a 1996 conference that "studies and surveys world-wide suggest that space tourism has the potential to be the next major space business."[55]

Several futurists believe that by the year 2030, there will be space tourists taking their vacations, albeit exceptionally expensive ones, in low-Earth orbit. Market studies suggest that there are more than 1,000 people per year willing to spend $1 million each for a weekend in space. Even at multimillion-dollar prices, it could become a billion-dollar-per-year business, space economist Patrick Collins believes, and could grow significantly in the future. If the cost of a space vacation dropped to about $25,000 per person, the number of people making the flight would rise to about 700,000 each year, he predicts. This represents a revenue stream of $17.5 billion per year.[56]

The industry is already beginning to see the first space tourists, as Dennis Tito pioneered the way by spending a week in April 2001 on the International Space Station (ISS). In so doing, advocates of space tourism believed that he has challenged and overturned the dominant paradigm of human spaceflight: national control of who flies in space overseen with a heavy hand by NASA and the Russian Space Agency. Dennis Tito's saga began in June 2000 when

54. Patrick Collins, Richard Stockman, and M. Maita, "Demand for Space Tourism in America and Japan, and its Implications for Future Space Activities," AAS (American Astronautical Society) paper no. AAS 95-605, AAS vol. 91, 1995, pp. 601–610. Available online at *http://www.spacefuture.com/archive/ demand_for_space_tourism_in_america_and_japan.shtml* (accessed 2 January 2005).

55. Barbara A. Stone, "Space Tourism: The Making of a New Industry" (paper presented at the International Symposium on Space Technology and Science [ISTS], 1996), copy available in NASA Historical Reference Collection, Washington, DC.

56. Patrick Collins, "The Space Tourism Industry in 2030," in *Space 2000: Proceedings of the Seventh International Conference and Exposition on Engineering, Construction, Operations, and Business in Space*, ed. Stewart W. Johnson, Koon Meng Chua, Rodney G. Galloway, and Philip J. Richter (Reston, VA: American Society of Civil Engineers, 2000), pp. 594–603; Roy W. Estess interview, Johnson Space Center Director, Stennis Space Center, MS, 25 June 2002.

he signed a deal with MirCorp to fly aboard a Soyuz rocket to the Russian space station *Mir*. MirCorp acted as Tito's broker with the Russian space firm Energia, which owned both Mir and the rocket that would get Tito into space. While MirCorp had grandiose plans for operating a space station supporting tourists and commercial activities, they failed to obtain the venture capital necessary to make it a reality. Despite these efforts, MirCorp failed to raise enough money to keep Mir in orbit, and the Russians announced in December 2000 that they would deorbit the space station.

This forced Tito to look elsewhere for a trip into space, and he negotiated a deal with the Russians fly aboard a Soyuz rocket to the International Space Station. While the cash-starved Russian Space Agency was happy to make this deal, no one bothered to discuss it with any of the international partners building ISS. A meltdown in public relations ensued, and NASA led the other partners in a rebellion that reached high into the political systems of the United States and Russia. NASA tried to persuade Tito to postpone his flight in February 2001, ostensibly to undergo two months of additional training before flying in October, but really to win time to convince the Russians not to allow Tito to fly to ISS. NASA and the other international partners building ISS argued that this slippage was paramount because of safety considerations on orbit. Ever a cagey gamester, Tito saw the trap and refused. He forced a confrontation with NASA at the gates of Johnson Space Center in March, where he planned to undergo training in preparation for an April 2001 flight. NASA lost that argument and was crucified by space enthusiasts for trying to block access to space for ordinary tourists. The Johnson Space Center acting Director at the time, Roy W. Estess, reflected a year later that he and his staff did not handle the Tito episode well and would have been better off to embrace the effort, as always ensuring the safety of the mission.[57]

With that one incident in Houston, Tito became a cause célèbre among space activists and NASA haters, who viewed him as the vanguard of a new age of space for everyone. Space psychologist Albert A. Harrison summarized the beliefs of many when he opined that "tourism is one of the world's largest industries and Russia's sale of a twenty million-dollar space station ticket to Dennis Tito represents but the first attempt to pry open the door for civilians in space. (Is there an irony that the Russians are the entrepreneurs prying open the door for space tourism while the Americans try to preserve a government monopoly?)"[58] A *Space.com* Web site visitors poll taken in early May 2001—which did not represent a random sample by any means but suggested where

---

57. Estess interview, 25 June 2002.

58. Albert A. Harrison, "Our Future Beyond Earth," *Space Times: Magazine of the American Astronautical Society* 40 (July–August 2001): 12.

the space enthusiasts came down on the issue—showed that 75 percent of respondents supported Tito's flight, 24 percent believed he should not have flown, and 1 percent were undecided.[59]

Tito would not allow anything or anyone to stand in his way, and many space activists cheered as he thumbed his nose at "big, bad NASA" to take his week-long vacation on the ISS at the end of April 2001. In making his way over the objections of NASA, Tito may have paved the way for other millionaires to follow. South African millionaire Mark Shuttleworth also flew aboard ISS in the fall of 2001, without the rancor of the Tito mission. Others may make similar excursions in the future, either paying their own way or obtaining corporate sponsorships. Space policy analyst Dwayne A. Day does not believe this is the best way to open the space frontier. He wrote, "Now that Tito has flown, it will not be the Earth-shattering precedent that space enthusiasts hoped for . . . . [I]s it any easier for the average citizen to raise $20 million in cash and buy a seat on a Soyuz than it is to get a Ph.D. in engineering and join the astronaut corps? No. Far from opening a frontier, Tito's flight symbolizes just how out of reach space remains for the common person."[60]

The flight of Dennis Tito offers an ambivalent precedent for the opening of spaceflight to the average person. Space tourism seems only a little closer today, even with the ISS, than it did in earlier eras. If there is a way to bring down the cost of access to space, then this dynamic may change, but until then, it does not much matter how many space stations are in orbit. Without a convenient, safe, reliable, and less costly means to reach them, little will change.[61] Once less expensive access to space is attained, an opening of the space frontier may take place in much the same way as the American continental frontier emerged in the 19th century, through a linkage of courage and curiosity with capitalism. As it does so, the role of the government should become less dominant in space. NASA will continue research and development for space systems and carry out far-reaching space science activities. But widespread human spaceflight should become the province of the commercial sector in the first half of the 21st century.

---

59. "SPACE.com Survey Reveals Strong Public Support for Dennis Tito's Flight," *Space.com*, 7 May 2001, *http://www.space.com/news/tito_poll_010507.html* (accessed 14 August 2002).

60. Ibid.; Dwayne A. Day, "From Astropower to Everyman to Rich Man: The Changing Human Face of Spaceflight," *Space Times: Magazine of the American Astronautical Society* 40 (July–August 2001): 22–23.

61. The issue of space access is critical to opening any part of space to broad usage. See Roger D. Launius and Lori B. Garver, "Between a Rocket and a Hard Place: Episodes in the Evolution of Launch Vehicle Technology," IAA-00-IAA.2.2.02 (paper presented at the 51st International Astronautical Congress, Rio de Janeiro, Brazil, 2–6 October 2000); Roger D. Launius and Dennis R. Jenkins, eds., *To Reach the High Frontier: A History of U.S. Launch Vehicles* (Lexington: University Press of Kentucky, 2002); Howard E. McCurdy, "The Cost of Space Flight," *Space Policy* 10 (November 1994): 277–289; Craig R. Reed, "Factors Affecting U.S. Commercial Space Launch Industry Competitiveness," *Business and Economic History* 27 (fall 1998): 222–236.

In addition to the ISS efforts of Tito and Shuttleworth, to help make space tourism a reality, Peter Diamandis publicly announced the "X Prize" project at a gala dinner in St. Louis, Missouri, on 18 May 1996. Designed to encourage private space investment, the X Prize offered $10 million to the first team that could launch a privately funded space vehicle into a suborbital trajectory twice within a two-week period. It had to be capable of carrying a pilot and two passengers more than 100 kilometers above the Earth. At the kickoff, numerous commentators linked the X Prize to the prospects for space tourism. NASA Administrator Daniel S. Goldin attended this event and said, "I hope my grandson who is 2 years old will be able to go on a trip to a lunar hotel." Of course, in October 2004, Burt Rutan's entry into the X Prize competition, *SpaceShipOne*, successfully claimed the prize. He and his benefactor, Microsoft billionaire Paul Allen, received numerous accolades for this accomplishment, including *Time* magazine's award for "coolest invention" of the year.[62]

Does the success of *SpaceShipOne* signal an opening of a new commercial space market? Brian Berger, writing for *Space.com*, made this observation on 29 December 2004:

> The dream of opening space to the general public was given a tremendous boost in 2004 with SpaceShipOne's prize-winning suborbital jaunt and congressional legislation to help establish a space travel industry in the United States. But even the biggest champions of commercial spaceflight acknowledge that a vital space tourism market is still years from becoming reality.[63]

It remains to be seen whether these efforts signal a new and exciting possibility of future space tourism. There are many questions yet to be answered, ranging from safety to economic viability to legal restrictions. While there have been some interesting developments in the last few years, much has yet to happen before space tourism finds realization; it remains a tantalizing possibility for the first half of the 21st century.

Scientific Discovery and Understanding

Finally, there exists the ideal of the pursuit of abstract scientific knowledge—learning more about the universe to expand the human mind—and pure science and exploration of the unknown will remain an important aspect of spaceflight

---

62. Chris Taylor and Kristina Dell, "The Sky's the Limit," *Time* (29 November 2004): 62.
63. Brian Berger, "2004: The Year Space Tourism Finally Took Off," *Space.com*, 29 December 2004, *http://www.space.com/spacenews/business_tourism2004_041229.html* (accessed 2 January 2005).

well into the foreseeable future. This goal clearly motivates the scientific probes sent to all of the planets of the solar system save Pluto. It propels a wide range of efforts to explore Mars, Jupiter, and Saturn projected for the early part of the 21st century.[64] It energizes such efforts as the James Webb Space Telescope, which promises to revolutionize our knowledge of the universe through, among other possibilities, the imaging of Earth-like planets around other stars.

And from the beginning, science has been a critical goal in spaceflight. The National Aeronautics and Space Act of 1958 that created the National Aeronautics and Space Administration (NASA) stated that its mandate included "the expansion of human knowledge of phenomena in the atmosphere and space." This idea has continually drawn verbal and fiscal support, but knowledge for its own sake has proven less important than the pursuit of knowledge that enables some practical social or economic payoff.[65]

Even the Apollo missions to the Moon, certainly inaugurated as a Cold War effort to best the Soviet Union and establish the United States as the preeminent world power, succeeded in enhancing scientific understanding.[66] The scientific experiments placed on the Moon and the lunar soil samples returned through Project Apollo have provided grist for scientists' investigations of the solar system ever since. The scientific return was significant, even though the Apollo program did not answer conclusively the age-old questions of lunar origins and evolution. For example, the origin of the Moon is still a subject of considerable scientific debate, but because of the harvest from lunar exploration during the Apollo era, currently the most accepted theory is that the Moon was formed by debris from a massive collision with the young Earth about 4.6 billion years ago. Prior to the study of the Apollo lunar rock and soil samples in the 1970s, however, confusion ruled among scientists about lunar origins as competing schools battled among themselves for dominance of their particular viewpoint in the textbooks. Indeed, determining the Moon's origins became the single most significant scientific objective of Project Apollo.[67]

---

64. An excellent discussion of all space probes launched to date may be found in Asif A. Siddiqi, *Deep Space Chronicle: A Chronology of Deep Space and Planetary Probes, 1958–2000* (Washington, DC: NASA SP-2002-4524, 2002).

65. John M. Logsdon, moderator, *The Legislative Origins of the National Aeronautics and Space Act of 1958: Proceedings of an Oral History Workshop* (Washington, DC: Monographs in Aerospace History, No. 8, 1998).

66. See W. David Compton, *Where No Man Has Gone Before: A History of Apollo Lunar Exploration Missions* (Washington, DC: NASA SP-4214, 1989); David M. Harland, *Exploring the Moon: The Apollo Expeditions* (Chichester, U.K.: Springer Praxis, 1999); Don E. Wilhelms, *To a Rocky Moon: A Geologist's History of Lunar Exploration* (Tucson: University of Arizona Press, 1993); Paul D. Spudis, *The Once and Future Moon* (Washington, DC: Smithsonian Institution Press, 1996); Donald A. Beattie, *Taking Science to the Moon: Lunar Experiments and the Apollo Program* (Baltimore, MD: Johns Hopkins, 2001).

67. Stephen G. Brush, "Early History of Selenogony," in *Origin of the Moon*, ed. William K. Hartman, Roger J. Phillips, and G. Jeffrey Taylor (Houston, TX: Lunar & Planetary Institute, 1986), pp. 3–15;

continued on the next page

Through a laborious polling of lunar scientists in the mid-1990s, the staff of the Curator for Planetary Materials Office at the Johnson Space Center, Houston, Texas, compiled a list of the top 10 scientific discoveries made as a result of the Apollo expeditions to the Moon. Collectively, they describe the current state of knowledge about this fascinating astronomical artifact.[68] The quest for knowledge about the Moon continues. In the 1990s, more than 60 research laboratories throughout the world continued studies of the Apollo lunar samples. Many analytical technologies, including some that did not exist in 1969–1972, when the Apollo missions returned the lunar samples, were being applied by a new generation of scientists.[69]

In the case of Apollo, and many others both before and since, a linkage between the spirit and need of scientific inquiry and the spirit and need for exploration served as strong synergetic forces for human spaceflight. This synergy arose explicitly in the National Research Council's 2005 study, *Science in NASA's Vision for Space Exploration*. It asserted: "Exploration is a key step in the search for fundamental and systematic understanding of the universe around us. Exploration done properly is a form of science."[70] As commentator David West Reynolds has noted, "Space probes like Voyager, Hubble, and Sojourner can accomplish *space exploration* as well as *space science* when they send back compelling images that can be appreciated by the public. Space science is at its best when it offers new vista along with its valuable data."[71]

The performance of scientific experiments on the Space Shuttle and the science program envisioned for the ISS demonstrate the same positive linkages at the beginning of the 21st century. Without question, the Space Shuttle has served as a significant test bed for scientific inquiry. While the program was not conceptualized as a science effort—rather it was a technology demonstrator and workhorse for space access—it has been used as a platform for all manner

---

continued from the previous page

Stephen G. Brush, "From Bump to Clump: Theories of the Origins of the Solar System, 1900–1960," in *Space Science Comes of Age: Perspectives in the History of the Space Sciences*, ed. Paul A. Hanle and Von D. Chamberlain (Washington, DC: Smithsonian Institution Press, 1981), pp. 78–100; Stephen G. Brush, "A History of Modern Selenogony: Theoretical Origins of the Moon from Capture to Crash, 1955–1984," *Space Science Reviews* 47 (1988): 211–273; Stephen G. Brush, "Nickel for Your Thoughts: Urey and the Origin of the Moon," *Science* 217 (3 September 1982): 891–898.

68. Curator for Planetary Materials, Johnson Space Center, "Top Ten Scientific Discoveries Made During Apollo Exploration of the Moon," 28 October 1996, NASA Historical Reference Collection, Washington, DC.

69. See G. Ryder, "Apollo's Gift: The Moon," *Astronomy* 22 (July 1994): 40–45; G. Jeffrey Taylor, "The Scientific Legacy of Apollo," *Scientific American* 271 (July 1994): 26–33.

70. National Research Council, *Science in NASA's Vision for Space Exploration* (Washington, DC: National Academies Press, 2005), p. 1.

71. David West Reynolds, "Astronauts: Only Exploration Justifies the Cost and Risk," *Space Times: The Magazine of the American Astronautical Society* 42 (November/December 2003): 4–7, quotation from p. 6.

In an instance of irony of the first order, astronaut Dale A. Gardner, having just completed the major portion of his second EVA in three days, holds up a "For Sale" sign during STS-51A in 1984. While he was probably referring to the two satellites, Palapa B-2 and Westar 6, that they retrieved from orbit, the sign speaks volumes about the lack of a compelling rationale for human spaceflight. On-orbit services provided a reason to send humans into space, but it was very much an approach that was not economically viable, as each Shuttle mission was estimated to cost at least $400 million, whereas a normal satellite and launch services cost less than half of that. *(NASA JSC photo no. 51A-104-049)*

of microgravity and space science enterprises. President Nixon, announcing the decision to build the Space Shuttle in 1972, minimized its scientific role. Instead, he argued that it was "the right step for America to take, in moving out from our present beach-head in the sky to achieve a real working presence in space—because the Space Shuttle will give us routine access to space."[72]

Even so, the Space Shuttle has been a useful instrument in the hands of scientists. Each of its more than 100 flights has undertaken some scientific experiments, ranging from the deployment of important space probes to other planets, through the periodic flight of the European-built "Spacelab" science

---

72. As an example of the scientific activities undertaken on the Shuttle see Kenneth Souza, Guy Etheridge, and Paul X. Callahan, *Life into Space: Space Life Sciences Experiments, Ames Research Center, Kennedy Space Center, 1991–1998* (Washington, DC: NASA SP-2000-534, 2000). On Nixon and the Shuttle decision, see T. A. Heppenheimer, *Space Shuttle Decision, 1965–1972,* vol. 1, *History of the Space Shuttle* (Washington, DC: Smithsonian Institution Press, 2002).

module, to a dramatic set of Earth observations over a 20-year period.[73] One example of a momentous science experiment, among others that might be offered, is the flight of the Italian Tethered Satellite System, designed to investigate new sources of spacecraft power and ways to study Earth's upper atmosphere, on STS-75 in 1996. It demonstrated that tethered systems might be used to generate thrust to compensate for atmospheric drag on orbiting platforms such as the International Space Station. Deploying a tether towards Earth could place movable science platforms in hard-to-study atmospheric zones. Tethers also could be used as antennas to transmit extremely low-frequency signals able to penetrate land and seawater, providing for communications not possible with standard radio. In addition, nonelectrical tethers may be used to generate artificial gravity and to boost payloads to higher orbits.[74]

Of course, some astoundingly significant scientific discoveries have resulted from robotic missions. But, if the purpose of spaceflight is to create a perfect society elsewhere, this necessitates human migration as its core activity. There would be very little reason to limit spaceflight to robotic explorers in this context. Robots might be useful servants—even the modern equivalent of slaves making our lives luxurious—but scientific understanding that might be gained by satellites remote from Earth would be decidedly less important than human spaceflight since the goal is migration. Second, while we seek to migrate into space as a method of ensuring human survival, such a goal is essentially a utopian dream based on expedition myths, and the popular culture treatment of robotics wholly failed to anticipate the degree to which we could send surrogates to do our work. This situation led to specific policy decisions and programs that focused on human spaceflight as the core function of the endeavor.

Many scientists believe that robotic spaceflight is the sine qua non of the Space Age, to the exclusion of a human presence. This is a dichotomy that began with the launch of the first missions into space and has been a perennial debate ever since. If anything, it has grown even more heated as robotic spacecraft have advanced in capability over time. Homer E. Newell, who directed NASA's space science program between 1958 and 1973, commented on this problem during the Apollo program:

---

73. David Shapland and Michael Rycroft, *Spacelab: Research in Earth Orbit* (Cambridge, England: Cambridge University Press, 1984); Douglas R. Lord, *Spacelab: An International Success Story* (Washington, DC: NASA, 1987); *Science in Orbit: The Shuttle and Spacelab Experience, 1981–1986* (Washington, DC: NASA, 1988); Arturo Russo, *Big Technology, Little Science: The European Use of Spacelab* (Netherlands: European Space Agency [ESA] Publications Division, ESA HSR-19, August 1997); Lorenza Sebesta, *Spacelab in Context* (Netherlands: ESA Publications Division, ESA HSR-21, October 1997).

74. Dennis Papadopoulos, Adam T. Drobot, and Nobie Stone, "The Flight of the Tethered Satellite System," *EOS* 73 (July 1992): 321–323; L. R. O. Storey, "The Shuttle Electrodynamic Tether Mission," *Environmental and Space Electromagnetics* (1991): 37–41.

The most spectacular aspects of space exploration in that last 30 years have been accomplished by robotic probes to other planets of the solar system. Here in the Spacecraft Assembly and Encapsulation Facility-2 (SAEF-2), Jet Propulsion Laboratory workers are closing up the metal "petals" of the Mars Pathfinder lander in 1996. The small Sojourner rover is visible on one of the three petals. On 4 July 1997, Pathfinder soft-landed on the Martian surface and provided spectacular imagery and important scientific data about the red planet's past. Among other findings, scientists learned that Mars had once been a watery planet. *(NASA JPL photo no. 96PC-1130)*

For space science one of the most difficult problems of leadership, both inside and outside NASA, concerned the manned spaceflight program. Underlying the prevailing discontent in the scientific community regarding this program was a rather general conviction that virtually everything that men could do in the investigation of space, including the moon and planets, automated spacecraft could also do and at much lower cost. This conviction was reinforced by the Apollo program's being primarily engineering in character. Indeed, until after the success of *Apollo 11,* science was the least of Apollo engineers' concerns. Further, the manned project appeared to devour huge sums, only small fractions of which could have greatly enhanced the unmanned space science program.[75]

The scientists viewed the debate over human versus robotic space missions in part as a zero–sum game. The expansive costs of human spaceflight might be

---

75. Newell, *Beyond the Atmosphere*, p. 290.

more effectively utilized for scientific purposes by sending only robots. They perceived inefficiency, redundancy, and enormous costs to keep astronauts alive as waste, and with only a small percentage of that funding, they believed they could accomplish so much.

The internecine warfare between advocates of human exploration and colonization of regions beyond Earth and the supporters of spaceflight for scientific purposes grew more heated as time passed. Space science leaders such as Homer Newell, Lloyd Berkner, and John E. Naugle established the science element of spaceflight during the 1960s and achieved stunning success in gaining a significant percentage of the NASA budget each year for those activities, usually about 25 to 30 percent. Using that funding, throughout the 1960s they created meaningful missions yielding useful scientific data and, in the process, established a community of scientists dedicated both to NASA and to robotic missions.[76] For example, by 1967, NASA had 942 scientists from 297 institutions involved as investigators in its various science projects. In 1996 alone, it flew 121 experiments on spacecraft and 99 sounding rockets. As Homer Newell reported, "In 1966 we evaluated 366 proposals for flight experiments, 248 of which were selected for flight. An additional 1,329 unsolicited proposals for SR&T work were evaluated."[77] At some level, as these statistics suggest, NASA co-opted some of the opposition to human spaceflight by, in effect, placing scientists on its payroll. Indeed, some NASA officials have expressed anger at University of Iowa astrophysicist James A. Van Allen's persistent criticism of human spaceflight as ungratefulness for all of the space agency's support over the years. One told a group of NASA public affairs officers in 1996 that "NASA made Van Allen, and now all he does is condemn us."[78]

Space science missions remain one of the most visible and popular aspects of the spaceflight agenda. While some of the work requires a human presence, usually to undertake scientific experiments aimed at understanding the bio-medical aspects of long-duration spaceflight, most of it is done exceptionally well by robotic explorers. The stunning success of a succession of missions to Mars, as well as to other places, demonstrates this beyond all doubt. No one questions the value of scientific space missions, but many question the necessity of placing humans aboard spacecraft undertaking those scientific efforts.

---

76. This story is well told in Joseph N. Tatarewicz, *Space Technology & Planetary Astronomy* (Bloomington: Indiana University Press, 1990) and Ronald E. Doel, *Solar System Astronomy in America: Communities, Patronage, and Interdisciplinary Research, 1920–1960* (New York: Cambridge University Press, 1996).

77. NASA Office of Space Science and Applications, "Program Review: Science and Applications Management," 22 June 1967, Space Science and Applications Files, NASA Historical Reference Collection, Washington, DC.

78. Roger Launius witnessed this discussion on 17 October 1998, during a meeting discussing the upcoming flight of John Glenn on STS-95.

HISTORY AND THE SEARCH FOR RELEVANCE

Of the five rationales that may be advanced in support of spaceflight, the human dimension is the only part that is seriously questioned. Military, economic, and scientific efforts in space, many observers have concluded, do not require human missions beyond the Earth. Even though the possibility of a human presence might be desirable in the future—such as in the case of space tourism, certain types of scientific inquiry, and a possible human military presence—thus far, reasons for humans in space to support these activities have remained elusive. Only the human destiny/survival of the species and the national prestige and geopolitics agendas require humans to fly in space. Not all are persuaded by these rationales to expend the considerable resources necessary to continue them. This especially occurred in the aftermath of the *Columbia* accident of 1 February 2003, with the grounding of the Shuttle fleet while an investigation of the cause of the accident took place and the fleet could be retrofitted to overcome the cause of the accident. Initially, NASA leaders promised to return to flight in the fall of 2003. Most observers believed that was unrealistic and possibly motivated by a "can-do" agency's optimism and bravado. Then it slipped into 2004 and finally to the middle 2005 as the Agency's engineers found more and more that needed to be fixed in the aging fleet of orbiters. At the same time, the price tag associated with the Shuttle's return to flight climbed.[79]

The accident called into question long-term assured human access to space. After more than four decades of human spaceflight, this problem is now thornier than ever because of the Shuttle's grounding and compounded every day that the fleet remains inactive. Is the United States as a nation willing to endure a period of several years when humans do not fly in space like we did between the time of the Apollo-Soyuz Test Project in 1975 and the first Shuttle mission in 1981? Are American citizens willing to end human spaceflight altogether? The answer to both of these questions for most Americans is probably "no," but while the support for human spaceflight is broad, it does not seem to be very deep.

Many Americans hold seemingly contradictory attitudes on human space exploration. Most are in favor of the human exploration and development of space and view it as important but also believe that federal money could be well spent on other programs. This relates closely to empirical research on other aspects of public policy. The American public is notorious for its willingness to support programs in principle but to oppose their funding

---

79. Richard O. Covey et al., "Interim Report: Return to Flight Task Group," 20 January 2004, NASA Historical Reference Collection, Washington, DC.

at levels appropriate to sustain them. Most are also in favor of NASA as an organization but are relatively unfamiliar with the majority of its activities and objectives and sometimes question individual projects. It is a little like how the overlanders traveling to Oregon in the 19th century described the Platte River on the Great Plains: "a mile wide and an inch deep." Americans appreciate and support—in principle—human spaceflight and recognize the astronauts as heroes but believe it is overly expensive. So what do we do for the future? It seems uncertain at present.[80]

Are these sufficient rationales to sustain human spaceflight indefinitely? Only time will tell. The first three rationales have not up to now required a human presence to be effective, but the last two have been used repeatedly to justify an aggressive human spaceflight agenda. The last two rationales— the human destiny/survival of the species and national prestige/geopolitics arguments—have been salient from the beginning of the Space Age. As John M. Logsdon, the dean of space policy, recently wrote:

> Most public justifications for accepting the costs and risks of putting humans in orbit and then sending them away from Earth have stressed motivations such as delivering scientific payoffs, generating economic benefits, developing new technology, motivating students to study science and engineering, and trumpeting the frontier character of the U.S. society. No doubt space exploration does provide these benefits, but even combined, they have added up to a less-than-decisive argument for a sustained commitment to the exploratory enterprise. The United States has committed to keeping humans in space, but since 1972 they have been circling the planet in low-Earth orbit, not exploring the solar system. The principal rationales that have supported the U.S. human spaceflight effort to date have seldom been publicly articulated. And those rationales were developed in the context of the U.S.-Soviet Cold War and may no longer be relevant.[81]

---

80. Howard E. McCurdy to author, 12 December 2002, copy in possession of author; Roger D. Launius, "Public Opinion Polls and Perceptions of U.S. Human Spaceflight," *Space Policy* 19 (August 2003): 163–175. The best book on the overland migration to Oregon and California remains John D. Unruh, *The Plains Across: The Overland Emigrants and the Trans-Mississippi West, 1840–60* (Urbana: University of Illinois Press, 1979). In it, Unruh describes the overlanders' view of the Platte River: it was something they recognized as necessary but did not enjoy.

81. John M. Logsdon, "A Sustainable Rationale for Human Spaceflight," *Issues in Science and Technology* (winter 2003), available online at *http://www.issues.org/issues/20.2/p_logsdon.html* (accessed 3 August 2004).

Indeed, over time, the traditional arguments have become less powerful as drivers of support for the space program. Since the age of Apollo in the early 1970s, most Americans have taken human spaceflight as a reality that is unchanging but treated the NASA efforts to fly the Space Shuttle and build a space station as necessary rather than desirable. No national commitment to a multibillion-dollar investment for this effort ever took place. Instead, the effort proceeded on inertia not unlike that seen in many other public policy sectors where there is no perceived crisis.

HUMAN AND ROBOTIC EXPLORATION

INTRODUCTION

No one realized it initially, but the intricate linking of humans and machines in spaceflight has been one of the most significant aspects of the endeavor. While humans have always been viewed as preeminent in spaceflight, the technology they employed—either in piloted spacecraft or in semiautonomous robots—proved critical to space exploration. This section explores the relationship between humans and machines in the evolution of spaceflight. The three essays consider strikingly different approaches to analyzing the human-machine interface in space exploration.

The chapter by Howard E. McCurdy, a senior space policy historian, addresses the classic debate over the primacy of human versus robotic spaceflight. He finds that the development of spaceflight technology always outstripped the slow evolution of human spaceflight, despite the overwhelming excitement associated with the human element. Virtually no one in history succeeded in making meaningful predictions about this discrepancy. For example, when Arthur C. Clarke envisioned geosynchronous telecommunications satellites in 1945, he believed that they would require humans working on board to keep the satellite operational. In such a situation, it is easy to conceive of the motivation that led people like Clarke and Wernher von Braun to imagine the necessity to station large human crews in space. Some of the most forward-thinking spaceflight advocates, in this instance, utterly failed to anticipate the electronics/digital revolution then just beginning. Humans, spaceflight visionaries always argued, were a critical element in the exploration of the solar system and, ultimately, beyond.[1]

With the rapid advance of electronics in the 1960s, however, some began to question the role of humans in space exploration. It is much less expensive and risky to send robot explorers than to go ourselves. This debate reached saliency early on and became an important part of the space policy debate by the latter 20th century. This has led many scientists and not a few others to question its merits. In the summer of 2004, esteemed space scientist James A. Van Allen asked the poignant question, "Is human spaceflight obsolete?" He commented:

---

1. Arthur C. Clarke, "Extra-Terrestrial Relays: Can Rocket Stations Give World-Wide Radio Coverage?" *Wireless World* (October 1945): 305–308; Wernher von Braun with Cornelius Ryan, "Can We Get to Mars?" *Collier's* (30 April 1954): 22–28; Randy Liebermann, "The *Collier's* and Disney Series," in *Blueprint for Space: From Science Fiction to Science Fact*, ed. Frederick I. Ordway III and Randy Liebermann (Washington, DC: Smithsonian Institution Press, 1991), pp. 135–144; "Giant Doughnut is Proposed as Space Station," *Popular Science* (October 1951): 120–121.

My position is that it is high time for a calm debate on more fundamental questions. Does human spaceflight continue to serve a compelling cultural purpose and/or our national interest? Or does human spaceflight simply have a life of its own, without a realistic objective that is remotely commensurate with its costs? Or, indeed, is human spaceflight now obsolete? . . . Risk is high, cost is enormous, science is insignificant. Does anyone have a good rationale for sending humans into space?[2]

The response offered is one that emphasizes human colonization on other planets, moons, and asteroids. As one observer who went by the pseudonym Hans L. D. G. Starlife noted on an Internet discussion list where Van Allen's arguments arose:

Sure, if it's all about science, you can always raise these questions. But it's not, and it never has been—whatever the scientists themselves try to make us believe. The human expansion into space is about totally different things—although like many times before, it isn't fully apparent until we can see it in the light of history . . . .

In a very long-range perspective, it's easy to see that these ventures, simply make up the path of evolution for Human civilization, not much different from how biological evolution works. Indeed, Human spaceflight is precisely what Van Allen argues it's not: it does and should have a life of its own. Now is the time to once and for all to SEPARATE the case for Human spaceflight with the case for science. These are two different agendas—both worthwhile—and sometimes crossing their paths, but having their own sets of motives and rationales![3]

Indeed, for people of this persuasion, spaceflight is all about making human civilization anew, making it in the mold of the best ideas of those who are founding settlements beyond Earth. It is, and in reality always has been, about creating a technological utopia.

---

2. James A. Van Allen, "Is Human Spaceflight Obsolete?" *Issues in Science and Technology* 20 (summer 2004), *http://www.issues.org/20.4/p_van_allen.html* (accessed 3 August 2004).

3. Hans L. D. G. Starlife, "On to Mars," *Quark Soup*, 27 July 2004, *http://davidappell.com/archives/00000202.htm* (accessed 3 August 2004).

McCurdy finds that while this debate over primacy in space missions has intensified with time, it does not really consider the core issues at play in space policy. As he notes, the human-robotic debate leaves unaddressed the manner in which humans and machines might become even more tightly linked in future spaceflight activities. McCurdy comments that "the classical visions of human and robotic spaceflight as presented in the popular culture contain instabilities likely to lessen the future influence of these visions. The emerging alternatives are quite exotic and beyond the mainstream of current thinking, yet interesting to contemplate. They may or may not occur. Their consideration, nonetheless, helps to enlarge the contemplation of the directions that future space exploration might take."

In essence, McCurdy suggests that the old paradigm for human exploration—ultimately becoming an interstellar species—is outmoded and ready for replacement. He specifically looks to the future of humans and robots in space and suggests that a posthuman cyborg species may realize a dramatic future in an extraterrestrial environment. This form of speculative futurism in a postbiological universe in which humans may become more robotlike may seem inappropriate for some historians. A question that might be considered is whether or not McCurdy has abandoned traditional modes of argumentation and analysis in favor of political commentary. A related question might focus on whether there even is a traditional mode of argumentation. Regardless of the answers to these questions (and those answers are highly idiosyncratic), there is no question but that McCurdy's essay is highly stimulating and provocative.

Alternatively, Slava Gerovitch's essay on "Human-Machine Issues in the Soviet Space Program" takes a much more traditional historical approach of narrating the evolution of relationships in the Soviet space program between humans and machines. He finds that from the early days of human spaceflight in the Soviet Union, a debate raged between the pilots/cosmonauts and the aerospace engineers over the degree of control held by each group in human-rated spacecraft. The engineers placed much greater emphasis on automatic control systems and sought to reduce drastically the role of astronauts on board a spacecraft. These space engineers often viewed the astronaut as a "weak link" in the spacecraft control system. Of course, the question of whether machines could perform control functions better than people became the subject of a considerable internal controversy. The cybernetics movement attempted to undermine the existing hierarchies of knowledge and power by introducing computer-based models and decision-making mechanisms into a wide range of scientific disciplines. By focusing on the debate over the nature and extent of on-board automation in Soviet spacecraft, Gerovitch illuminates a fascinating world of divergent professional groups within the Soviet space community and how they negotiated their place and their priorities in the system.

Finally, "Human and Machine in the History of Spaceflight," by David A. Mindell, argues for a new research agenda in the history of human spaceflight that moves beyond the virtual catechism of retelling of a specific myth and in that retelling performing a specific purpose. Much of this work has been not so much history as it has been "tribal rituals, meant to comfort the old and indoctrinate the young."[4] He notes that "a series of questions about human/machine interaction in the history of spaceflight can open up new research avenues into what some might think is a well-worn historical topic . . . . The human/machine relationship, as a meeting point for the social and technical aspects of a system, provides access to a variety of other aspects of space history that are otherwise difficult to integrate."

Collectively, these three essays provide a window into a unique area for consideration in the history of spaceflight. All are intellectually, artistically, and historically sound. All make important contributions to the history of human spaceflight and its relationship to robotics and space technology. All offer stimulating conclusions to be pondered, accepted, rejected, or revised as appropriate.

---

4. Alex Roland, "How We Won the Moon," *New York Times Book Review* (17 July 1994): 1, 25.

CHAPTER 3

OBSERVATIONS ON
THE ROBOTIC VERSUS HUMAN ISSUE
IN SPACEFLIGHT

Howard E. McCurdy

Since the beginning of the Space Age, people have debated the merits of human versus robotic flight. Some have argued for automated activities or what many—without apparent reference to the presence of women in space—term "unmanned" flight. Astrophysicist James A. Van Allen, designer of the experiment package for the first U.S. orbital satellite, insists that the whole history of spaceflight provides "overwhelming evidence that space science is best served by unmanned, automated, commandable spacecraft."[1] Historian Alex Roland maintains that "for virtually any specific mission that can be identified in space, an unmanned spacecraft can be built to conduct it more cheaply and reliably."[2]

To supporters of human spaceflight, such arguments are misplaced. The relative effectiveness of humans and robots seems irrelevant to people whose primary objective remains the movement of humankind into space. When asked to justify his upcoming lunar voyage, astronaut Neil Armstrong explained that "the objective of this flight is precisely to take man to the moon, make a landing there, and return."[3] From that point of view, human spaceflight provides its own justification. Robots serve as precursors to human flight, not as substitutes for it. Even if robots were more effective, advocates of human flight would not rely entirely upon them. The whole purpose of spaceflight is to prepare humankind to migrate off of the Earth and into the cosmos.

This essay presents a series of observations regarding the relative merits of the longstanding historical debate over human and robotic flight; it is speculative in nature and suggestive of future scholarship. It is also provocative and tentative. And it is an important debate. In many ways, the human and robotic

---

1. James Van Allen, "Space Station and Manned Flights Raise NASA Program Balance Issues," *Aviation Week & Space Technology* (25 January 1988): 153.

2. Alex Roland, "NASA's Manned Space Nonsense," *New York Times* (4 October 1987): sec. 4, p. 23.

3. Apollo 11 crew premission press conference, 5 July 1969, 2:00 p.m., Apollo 11 mission file, NASA Historical Reference Collection, Washington, DC.

perspectives present the two principal visions that motivate space exploration. The first anticipates the widespread migration of humans off the Earth's surface, while the latter emphasizes the advantages of scientific discovery.

In its speculative sections, the essay anticipates the manner in which the human versus robotic issue might change as space exploration matures. If cosmic exploration continues over the timespans anticipated by its advocates, changes in the dominant visions are probably inevitable. For many years, the robotic vision has stood as the sole alternative to the dominant vision of human spaceflight articulated by early advocates such as Wernher von Braun. This essay suggests that the classical visions of human and robotic spaceflight as presented in the popular culture contain instabilities likely to lessen their future influence. Two emerging alternatives are quite exotic and beyond the mainstream of current thinking, yet interesting to contemplate. They may or may not occur. Their consideration, nonetheless, helps to enlarge the contemplation of the directions that future space exploration might take.

CLASSICAL APPROACHES TO SPACE EXPLORATION

One of the most influential visions of spaceflight, prepared before humans entered space, appeared in the 22 March 1952 issue of *Collier's* magazine. Accompanying an article by Wernher von Braun, a two-page panorama prepared by Chesley Bonestell artistically illustrates human activity in low-Earth orbit. From a point of view well above the Isthmus of Panama, the viewer receives an enticing vision of small space tugs transporting astronauts between a winged space shuttle and a large, rotating space station.[4]

Visions of space exploration, often initiated in science fiction and articulated in popular outlets, shape public policy. They generate public interest, help place exploration on the governmental agenda, and prepare the citizenry for concrete proposals. Especially in the United States, the popular culture of space exploration has played a significant role in determining the types of activities public officials have sought to accomplish.[5] Not by accident did the members of the 1986 National Commission on Space choose to begin their report with a reproduction of the famous Bonestell diorama, juxtaposed with a Robert McCall painting of the actual facilities.[6]

Less well recalled is an object in the painting that Bonestell placed between the winged shuttle and the 250-foot-wide space station. The cylin-

---

4. Wernher von Braun, "Crossing the Last Frontier," *Collier's* (22 March 1952): 24–25.

5. See Howard E. McCurdy, *Space and the American Imagination* (Washington: Smithsonian Institution Press, 1997).

6. National Commission on Space (Thomas O. Paine, chair), *Pioneering the Space Frontier* (New York: Bantam Books, 1986).

drical object, surrounded by three astronauts, is an orbiting space telescope. Von Braun explained that the telescope would operate in a robotic fashion, without humans on board, since "the movements of an operator would disturb the alignment."[7] The panorama contains both human and robotic elements, yet the presence of a facility that anticipated the Hubble Space Telescope is not well remembered.

As is typical of images transmitted through popular culture, people selectively emphasize elements of the motivating material. The elements that emerge typically resonate with traditions and ideas popular at that time, being so familiar as to require little explanation. The early use of frontier analogies to explain space exploration is a preeminent example of this tendency. The editors at *Collier's* titled the accompanying article "Crossing the Last Frontier." Building transportation systems to transport people to the equivalent of frontier stations resonated well with the pioneering experience from which Americans had only recently emerged.

The inclusion of an orbiting telescope helped von Braun justify the presence of humans in this new frontier. What are astronauts doing to the remotely controlled observatory, and why is it orbiting near the space station? Given the existing state of technology for collecting images from space, von Braun explained, humans would be needed to retrieve and change the film.

As is typical of motivating visions, the expectations made powerful by reference to analogies from the past can be made weak by their encounter with the future. It is a familiar pattern. A vision of the future emerges and becomes part of the popular culture when it resonates so well with the experience of people contemplating a common past. To the extent that the vision is rooted in old and inapplicable analogies, or fails to account for developments yet to fully emerge, it acquires instabilities likely to plague its accomplishment.

The people who popularized the dominant vision of human spaceflight failed to anticipate technical developments that would make the conduct of robotic activities much easier than anticipated. Von Braun believed that astronauts would be needed to change the film in space telescopes. Arthur C. Clarke thought that astronauts would be needed to operate communication satellites. Producers of the classic 1950 film *Rocketship X-M* reinforced a popular misconception when they announced that radio waves from control stations on Earth would not be able to reach a spacecraft bound for Mars, thereby requiring a thinking presence on all missions into the celestial realm.[8]

---

7. Von Braun, "Crossing the Last Frontier," p. 72.
8. Arthur C. Clarke, "Extra-Terrestrial Relays: Can Rocket Stations Give World-wide Radio Coverage?" *Wireless World* (October 1945): 305–308; Kurt Neumann, *Rocketship X-M* (Kippert, 1950).

Generated at the midpoint of the 20th century, the dominant visions helping to define the impending Space Age failed to anticipate the manner in which electronic technology would expand robotic capabilities. The anticipated difficulties of operating remotely controlled telescopes and satellites provided a major justification for the presence of human crews. Real advances in remote sensing, solid-state transistors, and deep space communications allowed robotic flight to advance well beyond initial expectations and more rapidly than human flight.

What appears to be a failure of anticipation may in large measure arise from a failure of vision, a subtle but important distinction. A failure of anticipation implies an inability to foresee (one could say imagine) future events. Vision, as the term is commonly employed, represents a process in which imagination is joined with forces that motivate people to accept the dream.[9]

It is my contention that both the human and robotic space visions contain elements that make them attractive when viewed as continuations of past traditions. The visions do not fare as well when contemplated from the perspective of emerging trends. In essence, the dominant human and robotic visions account for the past more effectively than they address the future. This explanation requires an historical survey of the human and robotic visions, especially as they appear in popular culture, and some speculation about future developments.

HISTORY AND THE HUMAN SPACEFLIGHT VISION

The vision of human spaceflight is a familiar one. It begins with brave souls venturing in small ships through difficult substance to distant lands. Voyages of discovery produce scientific insights, including the identification of new species. Scientific gain, however, did not provide the ultimate motivation for new voyages. Settlers and entrepreneurs followed the early expeditions, extending technological civilization into new realms and distant lands.

Familiar analogies for the spacefaring vision are easy to find. Rocket ships are the equivalent of sailing vessels that cross terrestrial seas and flying machines that plow through the air. Space stations and extraterrestrial bases serve as the 21st-century equivalent of forts on the outer edges of settlement, providing sanctuaries from hostile forces as well as departure points for places beyond. The expectation of extraterrestrial life grows out of the manner in which the leaders of terrestrial expeditions returned with samples of strange life-forms from the lands they explored. Extraterrestrial colonies are portrayed as pioneer settlements, with their promise of fresh starts and the abandonment of old ways.

---

9. See John P. Kotter, *Leading Change* (Boston: Harvard Business School Press, 1996).

The power of the human spaceflight visions rests on a set of mutually compatible images, drawn from the recent (and frequently romanticized) memory of terrestrial events. Space offers a realm in which humans can continue the centuries-old tradition of terrestrial exploration. It allows nations to demonstrate their technological prowess and provides new lands for settlement and exploitation. It satisfies the apparent human need for human migration. It promotes the utopian belief that life will be better in newly created settlements beyond the reach of the "old world." These are familiar images, not hard to

An iconic image seen everywhere, this photograph shows Gemini astronaut Edward H. White II on 3 June 1965, when he became the first American to step outside his spacecraft for a "spacewalk." For 23 minutes, White floated and maneuvered himself around the Gemini spacecraft while logging 6,500 miles during his orbital stroll. The astronaut as central figure in space exploration has dominated imagery since before the beginning of the Space Age, but is it an accurate depiction of the future? *(NASA JSC photo no. S65-30433)*

explain to an often inattentive public. It is not hard for the average person to understand what is meant by space as "this new ocean" or new initiatives as "pioneering the space frontier."[10]

The human spaceflight vision arose during the first half of the 20th century, at a time when the opportunities for terrestrial exploration of the traditional sort seemed to be winding down. The rise of the human spacefaring vision with the nearly simultaneous decline of the heroic age of terrestrial exploration was not coincidental. The spacefaring vision offered an opportunity to continue the virtues thought to accompany terrestrial exploration and settlement in a new realm. Few developments had more influence on the popular acceptance of space exploration in the mid-20th century than the recent memory of terrestrial expeditions crossing Earthly lands and seas.

Intensive promotion of space exploration began just as the heroic era of terrestrial exploration came to a close. The latter is generally marked by the 1929 expedition of Richard E. Byrd to Antarctica, the first such incursion to substitute fully modern technology for dependence upon human skills. Byrd's expedition followed a series of polar expeditions that depended heavily upon the personal qualities of their human leaders. Among these were the efforts of separate parties led by Roald Amundsen and Robert Scott to reach the South Pole during the Antarctic summer of 1911–12 and the survival of the Trans-Antarctic Expedition of 1914 led by Ernest Shackleton. Both Amundsen and Scott reached the South Pole, but Scott and his four companions perished on the return voyage. Trapped in the polar ice, Shackleton led the crew of the *Endurance* on a 17-month odyssey that remains one of history's greatest stories of human triumph over extreme adversity. The polar expeditions followed a century marked by similarly heroic expeditions such as those led by Meriwether Lewis and William Clark and John Wesley Powell in the American West, Henry Morton Stanley in Africa, and the astonishingly influential voyage of Charles Darwin as the ship's naturalist on the HMS *Beagle*.

Expeditions in the heroic mold followed a well-established formula. Expedition leaders operated autonomously, without the technology necessary to maintain regular contact with their sponsors or home base. Typically, the public did not learn of their expeditions' achievements until the leaders emerged from isolation and reported their findings through lectures and

---

10. Loyd S. Swenson, James M. Grimwood, and Charles C. Alexander, *This New Ocean: A History of Project Mercury* (Washington, DC: NASA SP-4201, 1966), about connecting two distant points within the universe; National Commission on Space, *Pioneering the Space Frontier*. The term "this new ocean" is derived from "this new sea," a phrase employed by John F. Kennedy's "Address at Rice University in Houston on the Nation's Space Effort," 12 September 1962, in U.S. President (1961–1963 Kennedy), *Public Papers of the Presidents of the United States: John F. Kennedy, 1962* (Washington, DC: Government Printing Office, 1963), p. 373.

publications. In nearly all cases, the public did not know whether the members of the expedition under way were dead or alive. Cut off from their sponsors and home port, members of terrestrial expeditions were obliged to rely on their own skills to repair equipment and gain sustenance from local resources. Given the conditions they faced, expedition leaders depended upon human ingenuity rather than machine technology to survive and complete their discoveries. Terrestrial expeditions in the heroic tradition served as an expression of the power of humans to overcome natural obstacles without resorting to the conveniences of the industrializing world.

Such traditions provided the inspiration for the vision of human space-flight that gained popular acceptance during the middle years of the 20th century. Between 1950 and 1954, Wernher von Braun prepared a series of plans for the exploration of the Moon and Mars that recounted the heroic expeditions of preceding centuries. His proposal for a Mars mission was especially impressive. It called for a flotilla of 10 ships, guided by a 70-person crew, departing on a 30-month voyage. To prepare their landing site, pilots would descend in one of the ships to the polar ice cap of Mars—the only surface thought to be sufficiently smooth to permit a skid-assisted landing. From there, the crew would commence a 4,000-mile trek in pressurized trac-tors over unfamiliar terrain to the Martian equator, where they would bulldoze a landing strip for additional craft. Commenting on the attractive power of such schemes, von Braun remarked, "I knew how Columbus had felt."[11]

Von Braun's vision dominated popular presentations of the spacefaring vision during the mid-20th century. The image of winged spaceships, orbiting space stations, lunar expeditions, and voyages to Mars reappeared in the earliest long-range plans of the National Aeronautics and Space Administration. The vision remained the dominant paradigm for human spaceflight from the 1961 decision to go to the Moon through the 2004 presidential call for a return to the lunar surface and expeditions to Mars.[12] Yet this vision was already outdated in terrestrial terms when it first appeared.

Beginning with the Byrd expedition to Antarctica in 1929, expedition leaders came to rely much more on machines than on human heroics to

---

11. Quoted in Daniel Lang, "A Reporter at Large: A Romantic Urge," *New Yorker* 27 (21 April 1951): 74. See also Wernher von Braun, *The Mars Project* (Champaign: University of Illinois Press, 1991); von Braun with Cornelius Ryan, "Can We Get to Mars?" *Collier's* (30 April 1954): 22–28; von Braun, "Man on the Moon: The Journey," *Collier's* (18 October 1952): 52–60; Fred L. Whipple and von Braun, "The Exploration," *Collier's* (25 October 1952): 38–48.

12. NASA Office of Program Planning and Evaluation, "The Long Range Plan of the National Aeronautics and Space Administration," 16 December 1959; Space Task Group, *The Post-Apollo Space Program: Directions for the Future* (Washington, DC: Executive Office of the President, 1969); NASA, "President Bush Delivers Remarks on U.S. Space Policy," news release, 14 January 2004.

accomplish their goals. Byrd and his compatriots brought three airplanes and an aerial camera to Antarctica, which they flew over the South Pole. They brought 24 radio transmitters, 31 receivers, and 5 radio engineers, which they used to maintain communication with the outside world. The Byrd expedition, like others that followed, replaced the need for exceptional heroics with a dependence upon machines.

Basic plans for human spaceflight embodied language that recounted the spirit of heroic exploration. This occurred in spite of the program's obvious dependence upon machines of the sort that had caused the heroic tradition to disappear on Earth. The earliest astronauts were portrayed as heroic explorers even though they were selected to be mostly passive passengers on spacecraft treated more like guided missiles than ships at sea. Winged spaceships and large space stations proved much harder to construct than airplanes and frontier forts, notwithstanding the relative simplicity of their terrestrial analogies. Human space missions were controlled extensively from the ground, thereby forgoing the heroic tradition established by ship captains at sea.

Hence, the vision of human spaceflight was outmoded in terrestrial terms 30 years before it began. Yet spaceflight advocates clung to it, a testament to its motivating power. Much of its persistence arose from a supporting feature—the belief in American exceptionalism and the ability of space activities to maintain it.

The doctrine of American exceptionalism has appeared in a number of forms. Alexis de Tocqueville noted how conditions in New World settlements promoted innovation and a spirit of cooperation. This insight reappeared in the writings of 20th-century social scientists such as the historian Louis Hartz and the political scientist Aaron Wildavsky. Hartz traced American exceptionalism to the absence of rigid class distinctions such as those that dominated feudal arrangements in Europe. The doctrine achieved its most influential form in the frontier thesis promulgated by Frederick Jackson Turner in 1893. Jackson traced what he saw as the distinctive characteristics of American society to the presence of open land on a continental frontier. From this perspective, inquisitiveness, inventiveness, individualism, democracy, and equality grew out of the experience of founding new settlements free from the persistence of old arrangements.[13]

Turner's thesis has been dismissed by academic historians, yet it continues to possess special appeal to people unschooled in the nuances of historical

13. Alexis de Tocqueville, *Democracy in America* (New York: Random House, 1994); Louis Hartz, *The Liberal Tradition in America: an Interpretation of American Political Thought Since the Revolution* (New York: Harcourt, Brace, 1955); Aaron B. Wildavsky, *The Rise of Radical Egalitarianism* (Washington, DC: American University Press, 1991); Frederick Jackson Turner, "The Significance of the Frontier in American History," in *Rereading Frederick Jackson Turner*, ed. John M. Faragher (New York: Henry Holt, 1994).

research. The gap between academic intellectualism and popular opinion is in few places more pronounced than in the advocacy of human spaceflight. Human spaceflight advocates repeatedly cite the importance of "new frontiers" in sustaining the values of American exceptionalism.

At its heart, American exceptionalism is a utopian doctrine closely associated with the belief that people can improve the human condition by moving to new lands. Much of the interest in transforming Mars into an inhabitable sphere and establishing other space colonies arises from the utopian belief that life gets better when humans are allowed to start anew. The settlement schemes of space advocates such as Gerard O'Neill and Robert Zubrin embrace utopian themes, as does the work of science fiction writers such as Ray Bradbury.[14]

Academic historians point out that distinctive characteristics such as those valued by space advocates can arise from a number of cultural conditions and that the association of frontier life with values such as equality and individualism ignores actual events. Such criticism has had little effect on the popular promotion of human spaceflight. Its advocates continue to emphasize American exceptionalism and its linkage to the opportunities provided by the space frontier. Given the cultural history of the United States, this is a particularly appealing doctrine to the descendants of European settlers. The thought that the United States is becoming more like countries of the "old" world simply increases the interest in recreating conditions thought to make America unique.

The theory of American exceptionalism and its association with frontier life is dubious history. Whatever controversy it engenders as a historical doctrine, however, is overshadowed by the biological issues involved. Humans are a remarkably well-suited species for terrestrial migration. In fact, the ability of humans to adapt to a very wide range of terrestrial conditions through their tool-making capabilities may be the most distinguishing characteristic of the species as an earthly life-form. That adaptation has taken place on a terrestrial surface marked by a specific gravity condition, a protective atmosphere, and a magnetic field that shields earthly life-forms from cosmic violence. None of those conditions exist in outer space. Nearly all of the biological advantages that humans possess for Earthly migration disappear as they move away from the Earth. One pair of authors likens the use of human tool-making capabilities to overcome cosmic conditions to the thought that a fish might be able to survive on land if it had the ability to surround itself with a bubble of water.[15]

---

14. Gerard K. O'Neill, *The High Frontier: Human Colonies in Space* (New York: William Morrow, 1976); Robert Zubrin, *Entering Space: Creating a Spacefaring Civilization* (New York: Jeremy P. Tarcher/Putnam, 1999); Ray Bradbury, *The Martian Chronicles* (New York: Bantam Books, 1950).

15. Manfred E. Clynes and Nathan S. Kline, "Cyborgs and Space," *Astronautics* (September 1960): 29–33.

Early experience suggests that the ability of humans to transport conditions favorable to the maintenance of life in outer space is severely limited.

Accomplishments during the first half century of spaceflight have not favored human spaceflight. The human space endeavor has not kept pace with expectations. The inspirational value of elaborate visions such as those contained in the 1969 report of the Space Task Group or the popular film *2001: A Space Odyssey* far exceeded the capacity of humans to achieve them. The relatively uninspiring tasks of constructing near-Earth space stations and reusable spacecraft have taken far longer and cost far more than anticipated. With the exception of the landings on the Moon, human spaceflight has turned out to be much harder than people standing at the beginning of the Space Age envisioned it to be.

In practical terms, humans will probably return to the Moon and visit Mars. By necessity, they may rendezvous with nearby asteroids. They may establish Martian bases of the sort found at the Earth's South Pole, for reasons of scientific inquiry and national prestige. Their ability to populate Mars or other local spheres is debatable, and the idea that humans in large numbers may undertake interstellar journeys using conventional spacecraft is more doubtful still.

The human spaceflight vision is likely to end at Mars or some nearby place in the inner solar system. Ultimately, the human spaceflight vision will disappear because it is an old vision, tied to past events that become more distant with each succeeding generation. The spacefaring vision helped people standing at the midpoint of the 20th century express their loss at the passing of the heroic age of terrestrial exploration. Such nostalgia is likely to hold less appeal as new generations and developments emerge.

ROBOTS IN SPACE

While attractive in a number of respects, the robotic spaceflight alternative suffers from many of the same difficulties as the human flight paradigm. On the surface, as its advocates insist, robots may seem better suited to spaceflight than human beings. Yet as cultural phenomena, the robotic perspective similarly draws its motive force from social movements located in a rapidly receding past. The image of robotics contained in those movements fails to account for many new developments in technology.

The term "robot" is taken from the Czech word *robota*. In its purest form, it refers to statute labor or compulsory service of the type demanded of European peasants. In feudal Europe, aristocrats required peasants to work

This image represents the epitome of the NASA perspective that humans and robots will explore the solar system together. Here Sojourner, the Mars Pathfinder rover of 1997 named after former slave and famous abolitionist Sojourner Truth, is visited many years after its mission by a descendant of its namesake in this artist's rendering by Pat Rawlings. Sojourner the rover paved the way for those that followed. *(NASA image no. S99-04192)*

without remuneration for limited periods of time in the fields of noblemen. The Czech playwright Karel Capek used the term in a 1921 play, *R.U.R.* (Rossum's Universal Robots) to characterize mandatory factory work that was tedious and unrewarding. In Capek's play, factory work is performed not by people but by biologically produced human substitutes who are engineered to complete their work more efficiently than human counterparts.[16]

Therein lies the fundamental difficulty with robotics as a social phenomenon. Robots are viewed as machine-age products designed to serve as human substitutes. To anyone vaguely familiar with industrial-age technology, the implications are obvious. At the least, robots serve in the master-servant relationship characteristic of Edwardian times. At the worst, they are slaves.

The concept of slavery or involuntary servitude was well understood during the early stages of the industrial revolution. The practice of slavery existed scarcely a generation before the advent of wide-scale industrialization in America, and social commentators criticized the practices that tended to create "wage slavery" in industrial plants. Nineteenth-century law treated slaves as property without the rights accorded citizens of the United States, while factory practices treated workers as elements of production interchangeable with machines.

As servants or slaves, robots are not expected to possess human or sentient qualities. Even where robots take the physical form of human beings, they remain machines. The ultimate trust in the ability of humans to control robots forms the basis for Isaac Asimov's three laws of robotics, first elucidated in a 1942 story titled "Runaround":

> A robot may not injure a human being, or, through inaction, allow a human being to come to harm . . . . A robot must obey the orders given it by human beings except where such orders would conflict with the First Law . . . . A robot must protect its own existence as long as such protection does not conflict with the First or Second Laws.[17]

In the dominant fictional depiction of their relationships in space, robots commonly serve as companions to humans engaged in various extraterrestrial activities. This approach is well represented by robots such as Asimov's QT-1 from his early short story "Reason," Lieutenant Commander Data from *Star Trek: the Next Generation*, and the high-strung C3PO and the astromech

---

16. Peter Kussi, ed., *Toward the Radical Center: A Karel Capek Reader* (Highland Park, NJ: Catbird Press, 1990).

17. Isaac Asimov, *I, Robot* (New York: Random House, 1950), p. 37.

R2D2 of *Star Wars* fame. In the realm of fiction, Space Age robots exist to extend the capabilities of humans who travel alongside them. This creates a fundamental contradiction in the use of robots for space activities. If robots are merely machines, they can be treated as subhuman objects. They can be sent on perilous missions and programmed to perform their duties without the opportunity for earthly return, requirements that would never be permitted for expeditions with humans on board. At the same time, developments in robotics promise ever-increasing levels of sophistication—even to the level that they become sentient beings.

In the fictional setting, exploitive treatment of robots is rarely regarded as ethical. Even if robots are machines, humans treat them in considerate ways. Thoughtfulness for the "feelings" of robots grows directly out of misgivings regarding the treatment of factory workers, servants, and slaves. In a direct retelling of the Dred Scott case, writers for the *Star Trek* episode "The Measure of a Man" question whether the android Data should be treated as property or a human being. Data is a machine, albeit one that resembles a human being, and as such can be reassigned by a commander under the regulations governing the disposal of Federation property. Dred Scott was a 19th-century slave who sued in U.S. courts to maintain his freedom on the grounds that he was being reassigned from a state in which slavery was illegal into one which still permitted its practice. The Supreme Court ruled in 1857 that the provisions of the U.S. Constitution applicable to Scott were the ones that dealt with the property rights of owners rather than the personal rights of citizens, thereby helping to precipitate the Civil War. The Judge Advocate General in the *Star Trek* episode issues a contrary opinion. Data may be a machine, the jurist rules, but he has the right to be treated like a person.[18]

Social commentators find themselves caught between their insistence that robots are merely machines and the necessity of treating them with respect. In his classic work *Do Androids Dream of Electric Sheep?*, Philip K. Dick contemplates the morality of locating and shutting down wayward robots. (The story formed the basis for the classic 1982 science fiction film *Blade Runner.*) In a retelling of the fugitive slave law, the novel deals with android servants who escape from their masters on Mars and attempt to hide on Earth. To encourage emigration to Mars, the government grants each settler a personal android servant which becomes the emigrant's private property. The androids attempt to escape and sometimes murder their masters. The circumstances posed by the novel, Dick admits, duplicate the conditions of the Nat Turner rebellion in the pre–Civil War American South.

---

18. Robert Scheerer, "The Measure of a Man," *Star Trek: The Next Generation*, 13 February 1989, production 135, Paramount Pictures.

Dick eventually concludes that the androids are merely machines. They are worthy of careful treatment, as would be the case with any piece of expensive equipment, but are not persons in the conventional use of that term. Answering the title of his book, Dick concludes that androids would not dream of electric sheep unless they were programmed to do so, nor would they assign any particular value to the experience unless so instructed.[19]

Isaac Asimov wrestled with the same conundrum throughout his literary career. On the one hand, he railed against what he termed the "Frankenstein complex"—the tendency of writers to produce stories about robots gone bad. Nearly every robot story Asimov read as a young person presented "hordes of clanking murderous robots." The basic story, he observed, was "as old as the human imagination."[20] Humans who attempted to improve their condition through invention, like Icarus who flew too close to the Sun, were penalized by the gods. In a similar manner, humans who invented exceptional machines would be punished by their creations. Asimov absolutely rejected that point of view. All technologies, from fire to the automobile, possess dangers when misused. To Asimov, that did not justify their abandonment.

Robots were merely machines, Asimov insisted. Some aspects of their operation might prove faulty but were always subject to improvement. Said Asimov of his robotic creations: "I saw them as machines—advanced machines—but machines. They might be dangerous but surely safety factors would be built in."[21]

At the same time, Asimov could not resist the temptation to treat his creations anthropomorphically. He gave them human faces and human emotions and human needs. In one of his most famous robot stories, "Bicentennial Man," Asimov describes a robot that wants to become a person. Originally programmed to work as a household servant, the robot acquires artistic sensitivity through an error in the plotting of what Asimov terms its positronic pathways. Over a period of nearly 200 years, the robot replaces its machine parts with human prosthetics and wins its freedom. Yet it does not possess a human brain, a distinction that Asimov characterizes as "a steel wall a mile high and a mile thick."[22] A human brain is subject to irreplaceable decay. The price for becoming human, Asimov declares, is eventual death. It is a price that the robot is willing to pay.

The conceptual challenges of resolving the treatment of robots in practice are not as farfetched as they may seem. Throughout the early stages of

---

19. Philip K. Dick, *Do Androids Dream of Electric Sheep?* (New York: Ballantine Books, 1968).

20. Isaac Asimov, *Gold: The Final Science Fiction Collection* (New York: Eos, 2003), pp. 192, 193, 196.

21. Ibid., p. 195.

22. Isaac Asimov, *Robot Visions* (New York: Penguin Putnam, 1991), p. 287. The story first appeared in 1976.

the space program, humans allowed robots little autonomy. Robots operated under tight constraints and remote control. With the advent of planetary rovers, robots were allowed higher degrees of freedom. Should robots ever be used for interstellar investigation, they will require autonomous operating capability. They will need the capability to repair themselves without human intervention and possibly the ability to reproduce their parts.

The extent to which this will require the treatment of robots as sentient beings is as yet unknown. From a strictly industrial-age point of view, they will remain machines. Industrial thinkers like Frederick Taylor treated humans like machinery with interchangeable parts. Why would someone who adopts industrial-age thinking assign a higher status to intelligent equipment? A necessary requirement of space exploration, however, is the disappearance of organizational doctrines rooted in a pure mechanistic point of view. Space exploration requires organizational techniques that promote exceptionally high levels of creativity, reliability, and interactive complexity. It requires electronic equipment, most notably computers, whose basic conception rests more in the postindustrial age than the industrial. The traditional, assembly-line mentality that characterized the early industrial revolution is no longer appropriate for space travel, neither from an organizational nor a technological point of view.

Yet this is the very point of view around which the doctrine of robotics revolves. As a cultural phenomenon, robotics is rooted in an industrial-era vision of machinery and the period of human servitude from which it emerged. Whatever one may think about the technical advantages of unmanned spaceflight, its origins as a cultural doctrine are as traditional as those associated with human cosmic travel. The latter draws its force from romantic images of terrestrial exploration and frontier settlements; the former finds its potency in the fascination with machines that characterized the early industrial revolution and an idealized image of master-servant relationships.

The limitations of the robotic perspective are apparent in the seeming inability of its advocates to imagine such machines operating without direct human control. Very few of the robot stories prepared by Isaac Asimov present robots working alone. One notable exception is "Victory Unintentional," in which three incredibly hardy robots visit an invidiously hostile civilization on the planet Jupiter preparing for space travel.[23] The Jovians mistakenly identify the robots as human emissaries from Earth and, convinced that the Earthlings are indestructible, decide to abandon their spacefaring plans. The story departs so radically from Asimov's standard robot fare that he excluded it from his collection of *I, Robot* tales.

---

23. "Victory Unintentional" was published in the August 1942 issue of *Super Science Stories*.

The standard robot story involves machines working alongside human beings. The television series *Lost in Space* that ran from 1965 to 1968 featured a large robot that one critic characterized as a metal version of the canine *Lassie*, another popular television show from that period.[24] The Robinson family treated the robot as a member of the family, much like an intelligent pet. In *The Day the Earth Stood Still*, the alien portrayed by Michael Rennie travels with a robot named Gort who serves as the ship's chief medical officer and a ruthless enforcer of the extraterrestrial doctrine of arms control.

The official NASA policy for the use of robots in space exploration remains one of complementary capability. When pressed to comment on the virtues of manned and unmanned spaceflight, NASA's leaders repeat the dominant vision that it will be "robots and humans together."[25]

The treatment of robots in fiction is not unlike that accorded animals in space. The first animal to orbit the Earth, a Russian dog named Laika, was allowed to die in space. In a 1953 proposal for the use of monkeys to test living conditions on board a "baby space station," Wernher von Braun suggested that the animals be euthanized before reentry using "a quick-acting lethal gas."[26] To a certain extent, this recalled the polar practice wherein expedition members ate their dogs as the animals' usefulness for transport declined. Such treatment was not enforced upon the chimpanzees that tested conditions in NASA's Mercury space capsules before humans climbed in. The chimpanzees returned home, as did most of the subsequent Russian dogs to fly in space. In spite of their lower status as flight subjects, these animals were accorded appropriate respect. They came to be treated more like sentient beings.

Visionaries like Asimov predicted the widespread use of robots as personal servants by the end of the 20th century. His initial robot story, titled "Robbie," is set in New York City in the year 1998, a time by which Asimov anticipated the mass production of robotic servants for service on Earth and in space. People like Asimov anticipated a new machine age dominated by intelligent robots. In fact, the machine age departed. In its place, the postindustrial era appeared. In spite of his abiding interest in the workings of his robots' "positronic brains," Asimov wholly failed to anticipate the advent of personal computers and information networks that have come to characterize the postindustrial era.

Early images of computers in popular space literature are similar to those accorded mechanical robots. Sophisticated computers acquire a sense of their own existence and often behave in a roguish fashion. In the classic film and

---

24. "Robot B9 from Lost in Space," *http://www.jeffbots.com/b9robot.html* (accessed 10 July 2004).
25. See, for example, NASA, "Humans, Robots Work Together to Test 'Spacewalk Squad' Concept," news release 03-227, 2 July 2003.
26. Wernher von Braun with Cornelius Ryan, "Baby Space Station," *Collier's* (27 June 1953): 40.

novel *2001: A Space Odyssey*, the HAL-9000 computer attempts to seize control of the ship and kills all but one member of the crew. It resists the efforts of the remaining astronaut to disconnect it. The notion that humans might construct computers so advanced that they acquire self-awareness appears frequently in fictional and popular treatments of the subject.

Robots have already been used to explore the solar system. They have returned samples from the Moon, and they will likely return samples from Mars. They will closely inspect other planets and their moons. They will rove, dig, possibly swim, and explore. They have and will continue to reach the outer limits of the local solar system.

As a philosophy of exploration, nonetheless, robotics is full of contradictions and outdated metaphors. It remains a machine-age concept in a cybernetic world. Machine-age philosophies are fundamentally concerned with control, both in large organizations and the design of processes such as the assembly line. As with Asimov's three laws, the means of control are rooted in jurisprudence. Rules remain the primary means of control under the machine philosophy. Yet rules are largely inappropriate to the cybernetic models associated with postindustrial processes and information networks. The dominant metaphor for the cybernetic world is the brain, with its qualities of redundancy and creative problem solving.

Robots will surely continue to explore the local solar system. They may develop sufficient capacities to explore regions beyond. Such capabilities, as in the field of artificial intelligence, may lead to sentient qualities of the sort currently found in science fiction. Developing levels of self-consciousness, they might even come to think of themselves as superior beings. This is not guaranteed, but one cannot rule out the possibility. If this occurs, such robots would probably be treated with ever-increasing degrees of respect and kindness. This is the Asimov vision—sophisticated machinery with sentient characteristics operating under human control treated in a humane manner. The scenario is farfetched, but one that would pose no basic difficulty to the expanded use of robots for space exploration.[27]

A darker alternative exists. It is the vision presented in fictional devices such as *Blade Runner* and the behavior of the HAL-9000 in *2001*. Humans might treat such creations inhumanly. In *Blade Runner*, biologically manufactured robots are programmed to die after four years of operation. Having achieved self-consciousness, they understandably object to this policy. The HAL-9000 computer does not want to be shut off either. This scenario, while entertaining, seems flawed in a number of ways. It requires humans to treat intelligent

---

27. Some theorists believe that this is a given. See Ray Kurzweil, *The Age of Spiritual Machines: When Computers Exceed Human Intelligence* (New York: Penguin, 2000).

robots like slaves, a philosophy not too compatible with the guiding moral doctrines of the postindustrial world. It also suggests that humans would use advanced technology to build robots. As will be seen in a following section of this paper, a more likely scenario is that humans would use such technology to improve themselves. If humans ever develop the technology to construct biologically derived androids, they will by necessity acquire the technology to recreate themselves. That is a more profoundly interesting possibility.

Nonetheless, the image of intelligent but angry robots is not an impossibility. Humans are capable of great kindness toward their creations, but also great cruelty. The image of the mad robot attracts great interest because it says something cogent about human behavior. The concept of machines as slaves may be outmoded, but the worldwide traffic in humans pressed into forms of slavery continues.

In practical terms, the robotic vision will be weighed against the advantages and disadvantages of alternative schemes. This is inevitable. In that respect, the robotic vision, with its traditional quality, may have difficulty competing with approaches that better fit modern technological and cultural developments. One of the most challenging alternatives arises out of the developments in the increasingly strange world of astrophysics.

ASTROPHYSICS AND
THE ELECTROMAGNETIC SPACE PROGRAM

Recount for a moment the framework for the observations presented in this essay. To a substantial degree, the vision of space travel is a blank tablet onto which its advocates project images drawn from their own hopes for the culture at large. By necessity, those images change as actual ventures encounter reality. They also change as new generations of people project fresh hopes and cultural beliefs onto the space tableau. As reality intrudes and old cultural fascinations fade, so may old visions. This often encourages advocates to draw selectively what appear to be new ideas from old images—statements and visions not fully recognized until the new visions begin to take form.

One of the most pervasive expectations of the early 20th century held that Mars and Venus would turn out to be habitable planets not far different from the Earth. This expectation, presented in works both scientific and fictional, fueled much of the public interest in human spaceflight. Spaceflight enthusiasts hoped to fly to Mars and Venus and discover new life. Revelation of their inhospitable nature did not destroy that expectation so much as redirect it. Beginning in the last decade of the 20th century, much of the interest in habitable objects began to shift toward extrasolar planets.

The variance between the proximity of the inner planets of the local solar system and the challenges of reaching extrasolar spheres is extreme.

One can speculate on the manner by which this reality, joined with the continuing search for habitable objects, may affect the spaceflight vision. One commentator, proceeding from the mathematics of probabilities, estimates the average distance between life-supporting planets within the Milky Way galaxy to be about 50 light-years. (This is the estimated distance to planets on which life as we understand it might live. The average distance between planets possessing complex or intelligent life-forms may be substantially more.) Fifty light-years is merely an estimate—the actual number is unknown at this time. Nonetheless, it does illustrate the nature of the reality.

A typical voyage from Earth to Mars, using a fast-transit approach, covers about 500 million kilometers (300 million miles). This is the route followed by the robots Spirit and Opportunity that arrived at Mars in 2004. The difference between a fast-transit voyage to Mars and a journey of 50 light-years is a factor of 1 million. The two robots took seven months to reach Mars; a similar journey to a planet really capable of supporting human life might take 500,000 years. Regardless of the accuracy of the underlying estimate (it could be wrong by a factor of 10), the resulting distances pose a substantial barrier to people embracing the traditional vision of space exploration.

The energy requirements for crossing such distances are prodigious in the extreme. Fictional space captains may zip around the galaxy at warp speed, but serious proposals for interstellar flight have been confined to fractions in the 10 to 20 percent of light-speed range. Accelerating spacecraft to such velocities would require energy sources as yet undeveloped, such as fusion power or antimatter drives. For human flight, it would also require very large, multigenerational spacecraft. The people who began any such a voyage would not live to see its completion.[28]

The substantial engineering challenges involved in interstellar transit have forced its most serious advocates to emphasize robotic payloads. Even so, robotic expeditions suffer severe restrictions. A proposal by members of the British Interplanetary Society for a 50-year expedition to Barnard's Star promised a scientific payload with the impressive mass of 500 tons. The energy requirements needed to accelerate the robotic payload to one-eighth light speed proved so prodigious, however, that no fuel remained to help the spacecraft slow down. The expedition plan, named Project Daedalus, called for the spacecraft to zip past its destination at interstellar speeds. NASA executive George Mueller attempted to resolve this difficulty in his proposal for a 25-year voyage to Alpha Centauri 3, powered again by antimatter drive and achieving a peak velocity of two-tenths light speed. Assuming sufficient

---

28. On the technologies of this type of spaceflight, see Yoji Kondo, ed., *Interstellar Travel & Multi-Generational Space Ships* (Burlington, Ontario: Apogee Books, 2003).

fuel for deceleration, the resulting calculations left room for a robotic payload that weighed just 1 ton.[29]

The practical challenges of traveling to nearby solar systems, whether with human or robotic payloads, well exceed those of local flight. Concurrently, popular interest in the machine-age social issues that helped to spawn robotic dreams has declined. Might some other approach prove more compatible with the personal experiences of postindustrial people, while at the same time offering a better solution to practical difficulties of interstellar contact?

Such an approach exists—and if the combination of personal imagination and practical reality affecting previous spacefaring visions continues to foster new ones, it could create a significant variation in the classic human versus robotic debate. The new vision could arise from that pervasive symbol of postindustrial life, the computer. As noted in the previous section, the use of personal computers is as widespread as people in the early 20th century believed the employment of robots would be. The computer is as compatible with the electronic thinking that dominates the postindustrial age as the fascination with rockets and other machines was with the industrial.

A method for achieving light-speed velocities with very low energy requirements exists within the world of electronics. In 1974, astronomers Frank Drake and Carl Sagan aimed the Arecibo Radio Telescope at the globular star cluster M-13 and dispatched a binary code message at light speed. When properly deciphered, the message contained diagrams depicting a human being, the chemical makeup of Earth life, and the position of the home planet in the solar system. Sagan estimated that the chances of communicating with a civilization residing in the 100,000-star cluster were 50-50. Since the star cluster resides outside of the Milky Way galaxy, however, any return message traveling at light speed will not arrive for 48,000 years.

Civilizations capable of communicating in the electromagnetic spectrum may exist much closer to the Earth. During the 1970s, space advocates proposed a $20-billion government-funded listening system called Project Cyclops. In support of the initiative, NASA Administrator James Fletcher told a gathering of engineers that the Milky Way galaxy "must be full of voices, calling from star to star in a myriad of tongues." Fletcher was a lay minister in the Church of Jesus Christ of Latter-day Saints, which subscribes to the theological doctrine that God has created a plurality of worlds populated with human beings.[30]

---

29. Alan Bond and Anthony R. Martin, "Project Daedalus: The Mission Profile," in "Project Daedalus—The Final Report of the BIS Starship Study," ed. A. R. Martin, *JBIS: Journal of the British Interplanetary Society* (Supplement, 1978): S37–S42; George Mueller, "Antimatter & Distant Space Flight," *Spaceflight* 25 (May 1983): 104–107.

30. James C. Fletcher, "NASA and the 'Now' Syndrome," NASA brochure, text from an address to the National Academy of Engineering, Washington, DC, November 1975, p. 7, NASA Historical
continued on the next page

The prospect of spending billions of dollars on an approach to space exploration departing so radically from the traditional human and robotic vision sunk the initiative. Bereft of public funding, advocates sought private contributions for what became known as the Search for Extraterrestrial Intelligence (SETI).[31]

Technical developments of a practical nature may cause future lawmakers to fund extrasolar investigations. Propelled by widespread interest in the discovery of extrasolar planets, NASA officials have recommended the creation of space telescopes capable of recording light waves reflected from such objects. Beginning in the last decade of the 20th century, astronomers began confirming the presence of planets orbiting nearby stars using indirect means, such as variations in the positions of central stars as would be produced by orbiting spheres. More than 100 planets were discovered in the first decade of observation. Space-based telescopes utilizing the technology of interferometry could capture images of such bodies. This would require a large number of telescopes, flying in formation, assembling light waves from nearby solar systems in such a manner that the electromagnetic waves from the central star nullify each other. The bright glare from the central object would disappear, revealing the reflected light from objects orbiting the central star.

NASA officials created a hint of what such a technology might produce in 2003 when they aimed the Mars Global Surveyor toward the inner solar system and captured an image of Earth some 86 million miles away. The image shows Earth half lit. Cloud cover is clearly visible. With small adjustments in technology, the color of the seas appeared. Spectral studies of such an image would reveal water vapor, free oxygen, and trace amounts of methane and carbon dioxide—signatures of a planet populated with living beings.

Space scientists would like to know how many such spheres occupy the stellar neighborhood and the fraction of such bodies that might support complex life. Inspection through the electromagnetic spectrum is a far more efficient means of locating such bodies than the random dispatch of very large spacecraft with extraordinarily large energy requirements. Given 21st-century technologies, the electromagnetic spectrum would prove superior to human and robotic flight for investigations outside of the local solar system.

Where this may lead is as yet unknown. It is a history that has not yet occurred. Nonetheless, the confluence of social interest and practical reality suggests that it might form the basis for an alternative vision of considerable

continued from the previous page

Reference Collection, Washington, DC. On Fletcher's religious background, see Roger D. Launius, "A Western Mormon in Washington, D.C.: James C. Fletcher, NASA, and the Final Frontier," *Pacific Historical Review* 64 (May 1995): 217–241.

31. See Steven J. Dick and James E. Strick, *The Living Universe: NASA and the Development of Astrobiology* (New Brunswick, NJ: Rutgers University Press, 2004).

power. At the present time, it is relatively undeveloped—but no more so than the conventional reality of spaceflight remained until its popularization during the mid-20th century.

The electromagnetic space program anticipates possible communication at or even exceeding light speeds. The possibility of such developments has caused some people to contemplate the manner in which electromagnetic communication might be combined with traditional interest in human spaceflight. In 1985, one of the principal proponents of the Search for Extraterrestrial Intelligence, Carl Sagan, presented a draft of a science fiction novel to physicist Kip Thorne. Sagan suggested that Earthlings searching through the electromagnetic spectrum might discover devices that would cause objects to evade the cosmological limits imposed by conventional space and time. In the novel and film, titled *Contact*, the plans for such a device are supplied through a radio message received from outer space. The device, in Sagan's original draft, allowed humans to create a black hole. Thorne, who was completing a book on black holes and hyperspace, suggested that Sagan instead employ a series of wormholes.[32]

The laws of quantum gravity, Thorne observes, require that nature produce "exceedingly small wormholes."[33] A wormhole is a short tunnel connecting two distant points within the universe, moving outside the four dimensions that humans conventionally experience. Theory suggests that wormholes disappear as soon as they appear, but Thorne speculates that a technologically advanced civilization might employ the laws of quantum gravity to hold a wormhole open long enough to travel through it.[34] In this respect, fantastic tales in which children drop into rabbit holes or step through wardrobes and emerge in other worlds might provide the cultural inspiration for 21st-century space travel.

In Sagan's novel, engineers construct a device that creates an access point to an exit located in the vicinity of Vega some 26 light-years away. This cosmological tunnel provides access to additional passageways leading throughout the galaxy. Raised on the conventional image of space exploration, Sagan cannot resist the temptation to dispatch a human crew through the transit device. In the book, five individuals travel in a dodecahedron to Vega and beyond. Movie producers simplified the narrative to a single passenger, the central character played by actress Jodie Foster.

---

32. Kip S. Thorne, *Black Holes & Time Warps* (New York: W. W. Norton, 1994), pp. 483–484; Carl Sagan, *Contact: A Novel* (New York: Simon and Schuster, 1985). Sagan continued to refer to the tunnel as "the black hole, if that was what it really was," p. 335.

33. Thorne, *Black Holes & Time Warps*, p. 55.

34. Michael S. Morris and Kip S. Thorne, "Wormholes in spacetime and their use for interstellar travel: A tool for teaching general relativity," *American Journal of Physics* 56 (May 1988): 395–412.

In his book and a series of accompanying articles, Thorne explores whether a wormhole might be used for communication or transport of the conventional sort. Unlike a black hole, whose force would stretch and destroy any conventionally arranged object or message that entered it, a wormhole provides some possibility of transit. "We do not understand the laws of quantum gravity well enough to deduce . . . whether the quantum construction of wormholes is possible," Thorne observes. Nonetheless, physicists do understand how such a wormhole, if one were constructed, might be held open "by threading it with exotic material."[35]

Viewed from the perspective of conventional spaceflight, visions of electromagnetic communication and shortcuts through space and time are certainly strange. So far, no significant public funds have been provided for such activities. Yet the possibility of studying extrasolar planets is no more fantastic today than space travel seemed to a public raised on images of Martian canals and Buck Rogers in the early 20th century, and advances in modern physics continue to produce startlingly strange results. No one can predict with certainty where such developments might lead. The history of space travel does suggest, however, that prevailing visions depend considerably upon public interests and technological reality.

A POSTBIOLOGICAL PERSPECTIVE

The other alternative perspective on space travel is so strange that it makes the discussion of wormholes and extraterrestrial communications appear commonplace by comparison. For many years, NASA leaders have insisted that humans and robots will explore space together. The other alternative suggests that humans and machines will do more than travel together. As a result of space travel, they might merge into what Steven Dick has characterized as a "postbiological universe."[36]

A curious discussion surrounding the Search for Extraterrestrial Intelligence provided the reality check helping to motivate this perspective. In assessing the possibility of contacting extraterrestrial beings, Frank Drake prepared a formula that famously calculated the number of communicative civilizations that might exist within the Milky Way galaxy at the present time. The final parameter in the equation measures the average length of time that a communicative civilization survives. The parameter, labeled L, imposes a paradox raised by the physicist Enrico Fermi. If the value of L is small—on the

35. Thorne, *Black Holes & Time Warps*, p. 498.

36. Steven J. Dick, "They Aren't Who You Think," *Mercury* (November–December 2003): 18–26; Dick, "Cultural Evolution, the Postbiological Universe and SETI," *International Journal of Astrobiology* 2, no. 1 (2003): 65–74.

order of a few hundred years—then the predicted number of civilizations capable of communicating with one another in the Milky Way at any time rapidly approaches "one." In other words, humans are alone—and destined soon to revert to some pretechnological state.

Conversely, suppose that the value of L is very large. Given the age of the universe and the history of stars, the first technological civilizations could have emerged 3 billion years ago. Those that managed to survive infancy could have endured for hundreds of millions of years. The potential age of technological civilizations existing at the present time might range from 1 to 3 billion years.[37] Therein arises the paradox. Given the amount of time required for interstellar travel relative to the parameter L, intelligent extraterrestrials should already be here. Since this does not appear to be the case, it follows on the basis of the Drake formula that the longevity of technological civilizations must be very small. This is very disappointing to people anticipating a lengthy lifespan for human culture.

The fault, however, may lie in the formula. Drake's formula contains no parameter for the probability that the beings creating a technological civilization may evolve into something else. Yet this possibility has been raised repeatedly by science fiction writers. Many have foreseen the arrival of mutated life-forms, often as a result of horrible wars. H. G. Wells described a world full of Morlocks and Eloi in *The Time Machine*, while Pierre Boulle predicted the rise of intelligent chimpanzees in *Planet of the Apes*. In his early science novel *Orphans of the Sky*, Robert Heinlein allows the alterations to occur on an intergenerational spaceship bound for the Alpha Centauri star system. Succeeding crew members become mutants that dwell in the ship's core and simple farmers who, blissfully unaware that they live on a giant spaceship, occupy the outer shell.[38]

In the works of Arthur C. Clarke, similar transformations occur. Unlike other authors, Clarke presents such transformations in a uniformly positive way. To Clarke, space travel provides access to technologies that transform biological creatures into more immortal, spiritual beings. This optimistic vision forms the principal theme in Clarke's fictional work. It appears in *Childhood's End*, one of his first novels, in which alien beings oversee the total transformation of the human race. It reappears in *Rendezvous with Rama*, in which the extraterrestrial creators of a gigantic starship have long since

37. Mario Livio, "How Rare Are Extraterrestrial Civilizations, and When Did They Emerge?" *Astrophysical Journal* 511 (20 January 1999): 429–431; N. S. Kardashev, "Cosmology and Civilizations," *Astrophysics and Space Science* 252 (1997): 25–40.

38. H. G. Wells, *The Time Machine* (London: Everyman, 1995); Pierre Boulle, *La plane'te des singes*, translated by Xan Fielding as *Planet of the Apes* (New York: Vanguard Press, 1963); Robert A. Heinlein, *Orphans of the Sky: A Novel* (New York: Putnam, 1963).

evolved into a higher spiritual form. Most significantly, it dominates the central narrative in Clarke's classic novel and screenplay, *2001: A Space Odyssey*. In that story, an alien monolith provides a passageway for the transformation of the sole surviving astronaut on a deep space mission. The astronaut enters a passageway generated by the monolith and reappears as a "star child" with supernatural powers.[39]

From a cultural perspective, the transformations Clarke presents contain a message quite familiar to human beings. Clarke's characters achieve forms of immortality through space travel. Practically every human culture and nearly all religions contain messages about resurrection, typically achieved through some sort of physical dying and rebirth. Most space advocates are reluctant to discuss the possibility of physical transformation through space travel, perhaps out of a desire to appear scientifically sober. To the extent that visions of space travel rest upon a foundation of cultural expectations, however, few expectations are more widespread than those concerning the desire for immortality through some sort of physical transformation.

The existence of those expectations has provided the cultural foundation for a modern movement known as "transhumanism." This rather strange philosophy is a product of conversations taking place largely on the Internet. Transhumanism is "a radical new approach to future-oriented thinking" that utilizes advances in science and technology "to eliminate aging and greatly enhance human intellectual, physical, and psychological capacities."[40] Transhumanists believe that advances in computer capacity and nanotechnology will allow genetic change to occur very soon—possibly within the 21st century. The result, they believe, will be a "posthuman" species as superior to *homo sapiens* as humans are to the primates. The new species will survive for very long periods of time, perhaps approaching immortality.

Transhumanism is not a movement focused on space travel, although its applications to that endeavor are readily apparent. If humans or the species they produce are able to live under the severe conditions and extraordinarily long periods of time required for interstellar travel, many of the barriers to extended journeys would disappear. Physical modifications beneficial for space travel might include induced hibernation, a staple element in science fiction stories.[41] It could extend to physical alterations experienced by humans born on worlds with different gravities. Extraordinary lifespans would change the human perspective

---

39. Arthur C. Clarke, *Childhood's End* (New York: Harcourt, Brace & World, 1953); Clarke, *Rendezvous with Rama* (New York: Harcourt, Brace, Jovanovich, 1973); Clarke, *2001: A Space Odyssey* (New York: New American Library, 1968).

40. Nick Bostrom, "The Transhumanist FAQ," October 2003, *http://www.transhumanism.org/resources/faq.html* (accessed 5 January 2005).

41. See Stanley Kubrick, *2001: A Space Odyssey* (Metro-Goldwyn-Mayer, 1968); Gordon Carroll, David Giler, and Walter Hill, *Alien* (20th Century Fox, 1979).

of time and might allow the completion of lengthy interstellar voyages within a single generation. Combined with new insights into the structure of the universe, it might allow reconstructed beings to move through space in ways that humans could not survive. Given sufficient time, posthumans or their descendants might fulfill the science fiction dream of space travel by experiencing near immortality.

As is typical of such movements, the new approach has motivated current generations to rediscover words and works not previously emphasized under conventional visions. A leading approach within the transhumanist movement envisions the merging of human and machine parts. The resulting creatures are known as cyborgs, a term originally presented in a 1960 paper by Manfred Clynes and Nathan Kline on the challenges of space travel. Clynes and Kline suggested a number of modifications to the human body that would allow some of the basic requirements of extraterrestrial survival to take place automatically. They proposed induced hypothermia as a means of reducing energy requirements, drugs that might combat weightlessness, and an inverse fuel cell that would take the place of lungs.[42]

Cyborgs appear frequently in science fiction stories. The concept received popular attention in a 1972 novel by Martin Caidin that formed the basis for the television series *The Six Million Dollar Man*. A number of *Star Trek* episodes feature cyborgs, and the 1996 *Star Trek* movie *First Contact* presents an extraterrestrial life-form known as "the Borg." Part organic, part machine, the Borg are insectlike creatures that share a single mind.[43]

A person no less notable than Robert Goddard contemplated methods for transporting creatures through space in something other than their current bodily form. To assure the continuation of Earthly life, he recommended that distant spheres be seeded with what he termed protoplasm, dispatched on one-way journeys from Earth to distant spheres. Over time, the material would evolve into Earthly life-forms. Goddard suggested that the spacecraft also transport the accumulated knowledge of humankind "in as light, condensed, and indestructible a form as possible."[44] Goddard's proposal anticipated the development of microtechnologies and discovery of human DNA, which were unknown at the time. Lying so far from conventional visions of space travel, Goddard's speculations on interstellar flight received much less attention than his work on rocketry, but they could be selectively rediscovered if interest in transhumanistic space travel appears.

---

42. Clynes and Kline, "Cyborgs and Space."

43. Martin Caidin, *Cyborg: A Novel* (New York: Arbor House, 1972); Glen A. Larson et al., *The Six Million Dollar Man* (Silverton and Universal Production for the American Broadcasting Company [ABC], 1973–78); Rick Berman, *Star Trek: First Contact* (Paramount Pictures, 1996).

44. Esther C. Goddard, ed., and G. Edward Pendray, assoc. ed., *The Papers of Robert H. Goddard*, vol. 3 (New York: McGraw-Hill Book Co., 1970), p. 1612.

In his discussion of postbiological civilizations, Steven Dick refers to the work of the British philosopher Olaf Stapledon, who wrote science fiction novels and essays during the first half of the 20th century. Speaking of *homo sapiens*, Stapledon, in a classic 1948 address to the British Interplanetary Society, insisted that maintenance of the human physical form need not provide the ultimate justification for space travel. Rather, he emphasized the preservation of what he called the "spiritual experience" of being human. Stapledon surmised that the process of adapting humans to fit alien environments might prove easier given sufficient time than carrying Earthly conditions and unaltered humans to distant objects. Stapledonian thinking, as Dick describes it, takes into account "the evolution of biology and culture" alongside the process of space travel over very long time periods.[45] The works of Stapledon and those of the early-20th-century philosopher J. D. Bernal, on which he drew, are considered "classics" in a modern movement that did not exist when the works first appeared.[46]

In a half-serious sort of way, Steven Dick uses the postbiological perspective to solve the Fermi paradox. People searching for extraterrestrial civilizations listen for radio transmissions of the sort produced by human technology. Radio and television signals from Earth, however, are hardly 100 years old. As noted above, an extraterrestrial civilization mastering advanced technologies might have survived for billions of years. Over those time periods, such creatures would have evolved either naturally or through self-imposed means. As Dick notes, the transformation could have produced beings that no longer communicate through the electromagnetic spectrum. Fulfilling one of the ultimate spacefaring dreams, they might have attained a form of spiritual or electronic immortality.

The ultimate result of many such evolutionary sequences is hard to imagine. It might result in the modification of biological creatures into forms more suitable for living under conditions beyond their home planet. It might result in species that prefer not to be confined to wet, rocky spheres. Perhaps such species prefer to communicate over vast distances at speeds that seem sluggish to *homo sapiens* with traditionally short lifespans. Over lengthy periods of time, the iterations might produce creatures with little resemblance to species from which they emerged. Referring to such creatures on other planets, Dick observes, "It is entirely possible that the differences between our minds and theirs is so great that communication is impossible." His comments are equally applicable to new forms that might someday arise from Earthly life.[47]

---

45. Olaf Stapledon, "Interplanetary Man?" in *An Olaf Stapledon Reader*, ed. Robert Crossley (Syracuse, NY: Syracuse University Press, 1997), p. 234; Dick, "Cultural Evolution, the Postbiological Universe and SETI," p. 65.

46. J. D. Bernal, *The World, the Flesh, and the Devil* (Bloomington: Indiana University Press, 1969; original publication, 1929).

47. Dick, "Cultural Evolution, the Postbiological Universe and SETI," p. 72.

CONCLUSION

The original vision that helped to motivate the first phase of space travel favored human over robotic flight. Completion of the human spaceflight vision, with its winged spaceships, orbiting space stations, lunar bases, and planetary expeditions, proved more difficult than anticipated. During the same period, robotic activities overcame many of the technical obstacles expected to retard that approach. In spite of its rapid development, however, robotic technology did not supplant human activities. On balance, the two approaches achieved a state of approximate parity after one-half century of cosmic flight.

Scientists and engineers provided a vivid demonstration of the relative status of robotic and human flight during the 2004 debate over the repair of the Hubble Space Telescope. Rarely does a single flight activity permit a direct, head-to-head comparison between human and robotic approaches. More often, the debate arises in the context of different missions, such as the choice between the replacement of an aging Space Shuttle and the desire to launch another robotic probe to the outer planets.[48] In 2004, however, a special group from U.S. National Academy of Sciences reported on the relative merits of robotic and human spaceflight approaches to the task of servicing the batteries and gyroscopes on the 14-year-old Hubble Space Telescope. The group concluded that a robotic mission was not inherently superior but would probably involve more time and risk than an astronaut-guided repair. Further analysis suggested that the robotic mission would cost as much a Shuttle flight for the same purpose.[49]

Exploring the relative advantages of human and robotic flight in a manner similar to the calculations performed for the Hubble rescue mission is a productive avenue for future research. So is a reexamination of the underlying visions. As the generation of space advocates raised on the pioneering paradigm of human flight is replaced by young people raised in the computer age, the underlying cultural interests in space exploration may shift. Few people have attempted to study the manner in which a generation shift could affect the supporting visions of spaceflight possessed by the public at large.

So far, neither the human nor the robotic approach has achieved a commanding advantage over the other. Both continue to receive substantial support. Human space travel has fallen well short of the original vision, and robotic

---

48. See James A. Van Allen, "Space Science, Space Technology and the Space Station," *Scientific American* 254 (January 1986): 32–39.

49. Committee on the Assessment of Options for Extending the Life of the Hubble Space Telescope, Space Studies Board, Aeronautics and Space Engineering Board, National Research Council, *Assessment of Options for Extending the Life of the Hubble Space Telescope: Final Report* (Washington, DC: National Academies Press, 2005). See also U.S. House of Representatives Committee on Science, Hearing Charter, "Options for Hubble Science," 2 February 2005.

flight has exceeded initial expectations. Such observations, however, rest on a remarkably short base of practical experience and public perspectives, especially when viewed in cosmic terms. Over much longer timespans, the situation as it presently exists will probably change and do so in fundamental ways. These changes are not well represented in the current human versus robotic debate.

Ultimately, the classic human versus robotic debate fails to capture the full scope of the space endeavor because it fails to account for time. Time will present new opportunities, new visions, and new generations with different dreams to fulfill. The traditional human versus robotic controversy will suffer as time passes because it is essentially rooted in the past. Whatever technical merits guide the two points of view, the cultural context of both perspectives draws upon social movements that no longer play a dominant role in terrestrial affairs.

Seen from the perspective of the past, the human spaceflight movement resides in a utopian vision of Earthly activities that romanticizes events such as the settlement of the North American continent by Europeans and the "golden age" of terrestrial exploration. Even if the motivating events did occur as described by advocates of space travel—which is doubtful—they are not easily transferred to the reality of space.

Robotic flight does not fare much better. An analysis of the social commentary on robotics sets that movement squarely in the context of the industrial revolution and the disappearance of involuntary servitude. Support for robotics, especially as it appears in science fiction, arises from the utopian belief that industrial-age machines can be engineered to work like obedient servants, toiling alongside humans and relieving them of the need to perform dangerous or tedious space activities. This outlook is well expressed by the early belief that space robots would take the form of androids—machines in human form performing human work. In general, however, robotic spacecraft have not adopted the human form. When urged to propose a robot for the Hubble repair, NASA officials eschewed plans for an androidlike Robonaut in favor of a mechanism that looked like a Transformers toy. A concept under development at NASA's Johnson Space Center, Robonaut is an automated device with the arms, torso, and head of an astronaut. It looks like a human being. NASA officials instead suggested a design based on the Canadian-built Special Purpose Dexterous Manipulator (Dextre) designed for the International Space Station.[50]

The industrial age, with its emphasis upon machines that perform human functions like lifting and digging, encouraged the contemplation of robots that did the work of human beings. The industrial age, however, has been

---

50. NASA, "Robonaut," *http://robonaut.jsc.nasa.gov* (accessed 5 January 2005); Francis Reedy, "Hubble: Robot to the Rescue?" *Astronomy*, 12 August 2004, *http://www.astronomy.com/asy/default. aspx?c=a&id=2377* (accessed 5 January 2005).

supplanted by the postindustrial, with its emphasis upon electronic networks and computers. This may encourage popular interest to move away from space robots as human substitutes toward machines of a different sort. In the future, such machines might take the form of elaborate space telescopes that rely upon electromagnetic techniques to investigate extraterrestrial phenomena, cosmic listening posts, or even devices built to evade the conventional notions of space and time.

At the highest level, the human versus robot debate fails to account for changes in the species who frame it. People who envision the ultimate purpose of space activity anticipate its continuation over extraordinarily long periods. Commenting on the necessity of spaceflight, Robert Goddard noted that *homo sapiens* would need to move once "the sun grows colder," an event not likely to occur for billions of years. Setting a shorter but nonetheless epochal timeframe, astronomer Carl Sagan predicted that the galactic collisions that destroy species every 10 to 30 million years would force human migration. "Such a discussion may seem academic in the extreme," Goddard remarked, noting the very long time periods involved. Yet people who investigate space tend to think in cosmological terms. The ultimate choice, concluded Sagan, "is spaceflight or extinction."[51]

The introduction of very long periods of time creates a dynamic situation not extensively analyzed in the traditional human versus robot debate. A species that survives long enough to overcome solar destruction would certainly undergo genetic modification. This could occur gradually, or the species might acquire the means to reengineer lifeforms, including its own, in ways that make space travel more accessible. Either way, changes will occur over the periods of time during which space enthusiasts hope to prosper and survive.

Under such conditions, reconsideration of original expectations is inevitable. The human and robotic visions that motivated the first half century of spaceflight may continue to play a powerful role, especially for the exploration of the solar system. Yet it would be foolish to assume that they will be the only visions to ever inspire public policy and captivate public attention.

Rather than view the progress of space exploration as a two-sided contest between humans and robots, it is probably wise to consider what other visions might emerge. The history of space exploration suggests that motivating visions arise from social outlooks and the tempering influence of physical reality. This chapter has reviewed the human and robotic spaceflight visions and, from this perspective, speculated on the type of visions that might motivate future space activities. What arises is something more than the conventional two-sided debate—a future with perhaps four points of view.

---

51. Goddard and Pendray, eds., *Papers of Robert H. Goddard*, vol. 3, p. 1612; Carl Sagan, *Pale Blue Dot: A Vision of the Human Future in Space* (New York: Random House, 1994), p. 327.

CHAPTER 4

HUMAN-MACHINE ISSUES IN
THE SOVIET SPACE PROGRAM[1]

Slava Gerovitch

In December 1968, Lieutenant General Nikolai Kamanin, the Deputy Chief of the Air Force's General Staff in charge of cosmonaut selection and training, wrote an article for the *Red Star*, the Soviet Armed Forces newspaper, about the forthcoming launch of Apollo 8. He entitled his article "Unjustified Risk" and said all the right things that Soviet propaganda norms prescribed in this case. But he also kept a private diary. In that diary, he confessed what he could not say in an open publication. "Why do the Americans attempt a circumlunar flight before we do?" he asked. Part of his private answer was that Soviet spacecraft designers "over-automated" their spacecraft and relegated the cosmonaut to the role of a monitor, if not a mere passenger. The attempts to create a fully automatic control system for the Soyuz spacecraft, he believed, critically delayed its development. "We have fallen behind the United States for two or three years," he wrote in the diary. "We could have been first on the Moon."[2]

Kamanin's criticism was shared by many in the cosmonaut corps who described the Soviet approach to the division of function between human and machine as "the domination of automata."[3] Yet among the spacecraft designers,

1. I wish to thank David Mindell, whose work on human-machine issues in the U.S. space program provided an important reference point for my own study of a parallel Soviet story. Many ideas for this paper emerged out of discussions with David in the course of our collaboration on a project on the history of the Apollo Guidance Computer between 2001 and 2003, and later during our work on a joint paper for the 2004 annual meeting of the Society for the History of Technology in Amsterdam. I wish to express my gratitude to Asif Siddiqi and Valentina Ponomareva for sharing their insights into the history of the Soviet space program, as well as copies of relevant archival documents. I am also indebted to Stanislav Marchenko, Georgii Priss, Viktor Przhiyalkovsky, Irina Solov'eva, Vladimir Syromiatnikov, Iurii Tiapchenko, Iurii Zybin, and the staff of the Archive for Scientific and Technical Documentation in Moscow for providing invaluable help with my research. I am especially thankful to John L. Goodman for his detailed comments on early versions of this paper.

2. Nikolai Kamanin, *Skrytyi kosmos*, vol. 3, *1967–1968* (Moscow: Novosti kosmonavtiki, 1999), p. 335 (12 December 1968).

3. Georgii Beregovoi, as quoted in Valentina Ponomareva, "Nachalo vtorogo etapa razvitiia pilotiruemoi kosmonavtiki (1965–1970 gg.)," in *Issledovaniia po istorii i teorii razvitiia aviatsionnoi i raketno-kosmicheskoi tekhniki,* vyp. 8-10, ed. Boris Raushenbakh (Moscow: Nauka, 2001), p. 166.

a different point of view prevailed. They regarded the high degree of auto-mation on Soviet spacecraft as a remarkable achievement. The leading control system designer Boris Chertok, for example, praised the implementation of fully automatic docking on Soyuz, in contrast to the human-mediated rendezvous procedure on Apollo. "We did not copy the American approach," he argued, "and that proved to be one of the strengths of Soviet cosmonautics."[4]

The historiography of the Soviet space program has devoted little attention to on-board automation, treating it largely as a narrow technical issue. Yet the intensity of debates within the Soviet space program over the division of control functions between human and machine, both in the design phase and during spaceflights, indicates that the issue has fundamental importance. The success or failure of specific missions often depended on crucial control decisions made by the crew, the on-board automatics, or the ground control. The correctness and timeliness of such decisions critically depended on the integration of human decision-makers into a large, complex, technological system.

The problem of on-board automation, which tied together the interests of different professional groups, provides a window into the internal politics of the Soviet space program. Recent scholarship on the Soviet space program has largely been devoted to biographies, organizational history, and policy analysis, emphasizing the competition among different design bureaus and the lack of a coherent government policy.[5] While most accounts focus on only one of the relevant groups—the cosmonauts, the engineers, or the policy-making community—a study of human-machine issues illuminates the roles of all major professional groups within the Soviet space program. Aviation designers, rocket engineers, human engineering specialists, and cosmonauts had very different assumptions about the role of the human on board a spacecraft. A study of the actual division of function between human and machine on board would help us understand the role of these groups in shaping the Soviet space program.

The issue of on-board automation is also closely linked to the definition of the cosmonaut profession. Debates on the relative importance of cosmonauts' skills as pilots, engineers, or researchers reveal the connections between technological choices, professional identity, and the social status of cosmonauts. The seemingly

---

4. Boris E. Chertok, *Rakety i liudi*, vol. 3, *Goriachie dni kholodnoi voiny*, 3rd ed. (Moscow: Mashinostroenie, 2002), p. 393.

5. For recent biographies, see Iaroslav Golovanov, *Korolev: Fakty i mify* (Moscow: Nauka, 1994); James Harford, *Korolev: How One Man Masterminded the Soviet Drive to Beat America to the Moon* (New York: John Wiley & Sons, 1997). For analysis of the inner workings of space policy-making and institutional conflicts, see William P. Barry, "The Missile Design Bureaus and Soviet Piloted Space Policy, 1953–1974" (Ph.D. diss., Oxford University, 1995); Roger D. Launius, John M. Logsdon, and Robert W. Smith, eds., *Reconsidering Sputnik: Forty Years Since the Soviet Satellite* (Amsterdam, Netherlands: Harwood Academic Publishers, 2000); Asif A. Siddiqi, *Challenge to Apollo: The Soviet Union and the Space Race, 1945–1974* (Washington, DC: NASA SP-4408, 2000).

The original 1960 group of cosmonauts is shown in May 1961 at the seaside port of Sochi. The names of many of these men were considered state secrets for more than 25 years. Sitting in front, from left to right: Pavel Popovich, Viktor Gorbatko, Yevgeniy Khrunov, Yuri Gagarin, Chief Designer Sergey Korolev, his wife Nina Koroleva with Popovich's daughter Natasha, Cosmonaut Training Center Director Yevgeniy Karpov, parachute trainer Nikolay Nikitin, and physician Yevgeniy Fedorov. Standing the second row, from left to right: Aleksey Leonov, Andrian Nikolayev, Mars Rafikov, Dmitriy Zaykin, Boris Volynov, German Titov, Grigoriy Nelyubov, Valeriy Bykovskiy, and Georgiy Shonin. In the back, from left to right: Valentin Filatyev, Ivan Anikeyev, and Pavel Belyayeu. Four cosmonauts were missing from the photograph: Anatoliy Kartashov and Valentin Varlamov had both been dropped from training because of injuries; Valentin Bondarenko died in a training accident a few months before; and Vladimir Komarov was indisposed. I. Snegirev took the original photo. *(NASA photo no. cosmonauts01)*

technical problem of on-board automation raises larger questions of the nature and purpose of human spaceflight. An examination of different approaches to human-machine issues uncovers competing visions of spaceflight as a piloting mission, an engineering task, or a research enterprise.

Comparative studies of the American and Soviet aerospace industries have addressed the role of the national context in space engineering.[6] Soviet space program participants often regarded the U.S. as the paragon of a "human-centered" approach to spacecraft design. A leading spacecraft designer, for exam-

---

6. See Stephen J. Garber, "Birds of a Feather? How Politics and Culture Affected the Designs of the U.S. Space Shuttle and the Soviet Buran" (master's thesis, Virginia Institute of Technology, 2002); Leon Trilling, "Styles of Military Technical Development: Soviet and U.S. Jet Fighters, 1945–1960," in *Science, Technology, and the Military*, ed. E. Mendelsohn, M. R. Smith, and P. Weingart (Dordrecht, Netherlands: Kluwer, 1988), pp. 155–185.

ple, remarked: "Americans rely on the human being, while we are installing heavy trunks of triple-redundancy automatics."[7] A closer look at both American and Soviet space programs through the prism of on-board automation reveals a more complex picture. By exploring the arguments of internal debates, the diversity of engineering cultures, and the negotiations among various groups favoring different approaches to automation, one could critically reexamine the stereotype of fixed "national styles" in space engineering.

In this essay, I shall review a number of human-machine issues raised at different phases in the Soviet space program from the early 1960s to the late 1970s. From my perspective, the problem of on-board automation was not a purely technical issue, but also a political issue—not in terms of big politics, but in terms of "small" politics, local politics. My approach is to examine how technological choices were shaped by power relations, institutional cultures, and informal decision-making mechanisms, and how these choices, in turn, had significant ramifications for the direction of the Soviet space program and ultimately defined not only the functions of machines, but also the roles of human beings.

I will argue that the Soviet approach to the problem of on-board automation was neither fixed nor predetermined; it evolved over time and diversified across different institutions and projects. Instead of a single, dominating approach, we find a series of debates, negotiations, and compromises. In my view, the division of function between human and machine on board had much to do with the division of power on the ground among different groups involved in the debates over automation. I will illustrate how these episodes can be taken as entry points into larger historical issues about politics, organization, and culture of the Soviet space enterprise. Finally, I will suggest directions for further research into this subject.

AUTOMATION ON VOSTOK:
TECHNOLOGICAL, DISCIPLINARY, AND MEDICAL FACTORS

The first spacecraft—the Soviet Vostok and the American Mercury—were both fully automated and were flight-tested first in the unpiloted mode. Yet there was one important difference: the astronaut on board had a wider range of manual control functions than the cosmonaut. This can be illustrated by a simple comparison of the control panels of Vostok and Mercury. The Vostok panel had only 4 switches and 35 indicators, while the Mercury instrument panel had 56 switches and 76 indicators.[8] There were only two manual control

---

7. Chertok, *Rakety i liudi*, vol. 3, p. 257.

8. For a comparison of the technical parameters of manual control panels on American and Soviet spacecraft, see Georgii T. Beregovoi et al., *Eksperimentalno-psikhologicheskie issledovaniia v aviatsii i kosmonavtike* (Moscow: Nauka, 1978), pp. 62–63.

functions that a cosmonaut could perform in case of emergency: orientation of the spacecraft into correct attitude and firing of the retrorocket for descent.[9]

The range of manual control functions available to and actually performed by American astronauts was much wider. They could override the automatic system in such essential tasks as separating the spacecraft from the booster, activating the emergency rescue system, parachute release, dropping the main parachute in case of failure and activating the second parachute, correcting the on-board control system, and many other functions not available to Soviet cosmonauts.[10]

Different authors have offered a number of explanations for the Soviet reliance on automation in the case of Vostok:

1) *High reliability of automatic control:* Soviet rockets could lift greater weights, and therefore the Soviets could install redundant sets of automatic equipment to ensure its reliability.

2) *Disciplinary bias of rocket engineers:* Unlike American space engineers, who came from the aviation industry, Soviet spacecraft designers drew on specific engineering traditions in rocketry, and they were not accustomed to assign humans a significant role on board.

3) *Health and safety concerns:* There existed doubts about the cosmonaut's mental and physical capacity to operate the spacecraft in orbit.

Some of these explanations do have a grain of truth. Yet they mostly reflect partisan positions in internal Soviet debates over the proper division of control functions between human and machine.

The first, "technological" explanation is most favored by spacecraft designers, who view it as an "objective" basis for automation. Indeed, the Vostok rocket could lift to the orbit a 4.5-ton spacecraft, while the Americans could launch only 1.3 to 1.8 tons. Using this extra weight, the argument goes, the Soviets could afford to build redundant, more reliable systems and to construct a fully automatic spacecraft, while the Americans were forced to delegate some of the functions to the astronaut on board. The space journalist Iaroslav Golovanov wrote: "The American astronaut had to work more than the Soviet cosmonaut because the weight of Vostok was more than twice

---

9. Valentina Ponomareva, "Osobennosti razvitiia pilotiruemoi kosmonavtiki na nachal'nom etape," in *Iz istorii raketno-kosmicheskoi nauki i tekhniki*, vyp. 3, ed. V. S. Avduevskii et al. (Moscow: IIET RAN, 1999), pp. 132–167; Siddiqi, *Challenge to Apollo*, p. 196.

10. Robert B. Voas, "A Description of the Astronaut's Task in Project Mercury," *Human Factors* (July 1961): 149–165.

the weight of Mercury, and this made it possible to relieve [the cosmonaut] of many in-flight tasks."[11]

Interestingly, this argument only suggests an explanation for the need for a broad range of manual control functions on Mercury, while the Soviet preference for complete automation is assumed as a natural solution. Those who used this argument clearly took it for granted that automatic systems were inherently more reliable than human control. Indeed, most Vostok designers viewed the cosmonaut on board as a weak link, a source of potential errors. The leading integration designer Konstantin Feoktistov openly told the cosmonauts, for example, that "in principle, all the work will be done by automatic systems in order to avoid any accidental human errors."[12]

In fact, it is by no means obvious why should one use weight reserves to install redundant sets of equipment instead of building a more flexible and sophisticated manual control system. Soviet space designers admitted that the on-board equipment that they were supplied with was so unreliable that installing extra sets was the only way to ensure an acceptable risk of failure. Boris Chertok acknowledged that the Americans were able to make a much better use of their weight reserves than the Soviets. He wrote: "The weight of Gemini was only 3.8 tons. Vostok weighed almost a ton more, and Voskhod 2 almost 2 tons more than Gemini. Yet Gemini surpassed the Vostoks and the Voskhods in all respects."[13] Gemini had a rendezvous radar, an inertial guidance system with a digital computer, a set of fuel cells with a water regenerator, and many other types of on-board equipment that the first Soviet spacecraft lacked.

The second, "disciplinary" explanation is often put forward by cosmonauts, who tend to blame the "overautomation" of Soviet spacecraft on the professional background of rocket engineers. According to the space historian and former cosmonaut candidate Valentina Ponomareva, "In the United States space technology developed on the basis of aviation, and its traditional attitude toward the pilot was transferred to space technology. In the Soviet Union the base for the space enterprise was artillery and rocketry. Rocketry specialists never dealt with a 'human on board'; they were more familiar with the concept of automatic control."[14] This argument assumes that the Soviet space program was a culturally homogeneous assembly of rocket engineers. In fact, Chief Designer Sergei Korolev, under whose leadership Vostok was

---

11. Golovanov, *Korolev*, p. 604. A similar argument is presented in Ponomareva, "Osobennosti razvitiia," p. 144.

12. Quoted in Vladimir Komarov, Workbook No. 39, 1961, Gagarin Memorial Museum, Town of Gagarin, Russia, *http://hrst.mit.edu/hrs/apollo/soviet/documents/doc-komarov39.pdf* (accessed 21 April 2005).

13. Chertok, *Rakety i liudi*, vol. 3, pp. 256–257.

14. Ponomareva, "Osobennosti razvitiia," p. 161.

constructed, had come into rocketry from aeronautics; in the 1920s and 1930s, he had designed and tested gliders.[15] His deputies, leading spacecraft designers Pavel Tsybin and Sergei Okhapkin, had previously been prominent aircraft designers. Heated debates over the division of function between human and machine often broke out within the space engineering community, and the opponents in those disputes were not necessarily divided along the lines of their disciplinary background. For example, in July 1963, when the leadership of Korolev's design bureau discussed various options for lunar exploration, it was the aviation designer Pavel Tsybin who advocated the use of automatic spacecraft, and it was the rocket designer Mikhail Tikhonravov who insisted on the development of piloted spaceships.[16] Tikhonravov also argued in favor of making Vostok controls completely manual.[17]

Soviet cosmonauts with aircraft piloting background in private tended to blame rocket engineers, nicknamed "artillerymen," for any design flaws. For example, during her training as a cosmonaut, Valentina Ponomareva noticed that yaw and roll in the hand controller on the Vostok spacecraft were rearranged as compared to a typical aircraft hand controller. Fellow cosmonauts told her that it was "because artillerymen had built it."[18] As it turned out, the controller was developed by specialists from the Air Force Flight Research Institute, which specialized in aviation control equipment. Yaw and roll were rearranged because the controller itself was positioned differently (which, in turn, was the result of a different position of the cosmonaut as compared to the aircraft pilot). Moreover, since spacecraft could rotate in all directions, yaw and roll in some cases simply changed places. There was no conspiracy of "artillerymen" here; it was aviation specialists who designed manual control and information display equipment for Soviet spacecraft.[19]

The third, "medical" explanation often cited Soviet doctors' concern that the cosmonaut's mental and physical capacities might be impaired during the flight.[20] In fact, although doctors did study the issue of the cosmonaut's health and working capacity in orbit, they were not pushing for automation. On the contrary, the leading physician, Vladimir Yazdovskii, was in favor of expanding the range of Yuri Gagarin's tasks on the first human flight, while

---

15. See Golovanov, *Korolev*.

16. Vasilii Mishin, diary, 22 July 1963, NASA Historical Reference Collection, Washington, DC.

17. Ponomareva, "Osobennosti razvitiia," p. 147.

18. Valentina Ponomareva interview, Moscow, 17 May 2002, *http://hrst.mit.edu/hrs/apollo/soviet/interview/interview-ponomareva.htm* (accessed 21 April 2005).

19. Yurii Tiapchenko, "Information Display Systems for Russian Spacecraft: An Overview," trans. Slava Gerovitch, *http://hrst.mit.edu/hrs/apollo/soviet/essays/essay-tiapchenko1.htm* (accessed 21 April 2005).

20. Ponomareva, "Osobennosti razvitiia," p. 145.

A pensive Yuri Gagarin is in the bus on the way to the launchpad on the morning of 12 April 1961. Behind him, seated, is his backup, German Titov. Standing are cosmonauts Grigoriy Nelyubov and Andrian Nikolayev. Gagarin began his cosmonaut training in 1960, along with 19 other candidates. On 12 April 1961, Gagarin lifted off in the automated Vostok 1 spacecraft, and after a 108-minute flight, he parachuted safely to the ground in the Saratov region of the USSR. As the first human to fly in space, he successfully completed one orbit around Earth. After his historic flight, Gagarin became an international symbol for the Soviet space program, and in 1963, he was appointed Deputy Director of the Cosmonaut Training Center. In 1966, he served as a backup crew member for Soyuz 1, and on 17 February 1968, he completed a graduate degree in technical sciences. Tragically, during flight training in a UTI-MiG-15 aircraft on 27 March 1968, Gagarin was killed when his plane crashed. *(NASA photo no. Gagarin01)*

Chief Designer Sergei Korolev insisted that Gagarin should limit his actions to visual inspection of on-board equipment and should not touch any controls. Korolev's cautious approach may have been prompted by the responsibility placed on him by the political authorities. It was Nikita Khrushchev himself who on 3 April 1961, just a few days before Gagarin's flight, at a meeting of the Presidium of the Party Central Committee, raised the question about the cosmonaut's working capacity and psychological stability in orbit. Korolev had to give his personal assurances.[21] Not relying entirely on the disciplining force of cosmonaut's written instructions, spacecraft designers took some technological measures to prevent any accidental damage from the cosmonaut's actions in case he did lose his psychological stability. They blocked the manual orientation system for reentry with a digital lock. There was some debate whether to give the combination to the cosmonaut or to transmit it over the radio in case of emergency, and eventually they decided to put the combination in a sealed envelope and to place it on board so that the cosmonaut could open it in an emergency.[22]

In the end, Soviet officials decided to give Gagarin a "broader" set of functions, such as checking equipment before launch, writing down his observations and instrument readings in the on-board journal, and reporting those over the radio. As doctors explained, keeping the cosmonaut busy would help deflect his attention from possible negative emotions during g-loads and weightlessness.[23]

None of the three popular explanations—the reliability of redundant automatics, the disciplinary bias of rocket engineers, and the uncertainty about human performance in orbit—provides an unequivocal argument in favor of automation. All three aspects of the problem of automation—technological, disciplinary, and medical—involved debates and negotiations, whose outcome was not predetermined from the very beginning.

---

21. Nikolai Kamanin, *Skrytyi kosmos*, vol. 1, *1960–1963* (Moscow: Infortekst, 1995), pp. 23 (diary entry of 2 March 1961), 43 (diary entry of 4 April 1961).

22. As it turned out, two people independently told Yuri Gagarin the combination before the launch so that he would not have to waste time on opening the envelope in case of real emergency. See Boris E. Chertok, *Rakety i liudi*, vol. 2, *Fili—Podlipki—Tiuratam*, 3rd ed. (Moscow: Mashinostroenie, 2002), pp. 428–429.

23. Siddiqi, *Challenge to Apollo*, p. 264.

VOSTOK DUAL USE:
MILITARY/CIVILIAN AND AUTOMATIC/MANUAL

Recently published materials suggest another explanation for the Soviet reliance on automation in the design of Vostok, an explanation that emphasizes the social shaping of technology. It suggests that the military context played a decisive role in defining civilian technologies in the Soviet space program.

Vostok was designed at the Experimental Design Bureau No. 1, led by Chief Designer Sergei Korolev, as an add-on to its main specialty, ballistic missiles. In November 1958, the Council of Chief Designers discussed three alternative proposals for a new spacecraft: an automatic reconnaissance satellite, a piloted spacecraft for a ballistic flight, and a piloted spacecraft for an orbital flight. The reconnaissance satellite designers pushed their proposal, stressing its primary importance for defense. This clearly had an appeal to the military, the Design Bureau's main customers. A rival group, led by the integration designer Konstantin Feoktistov, decided to support their proposal for a piloted spacecraft for an orbital flight with what he called a "tactical maneuver": they claimed that their piloted spaceship could be converted into a fully automatic spacecraft and used as a reconnaissance satellite, which would be able to return to Earth not just a small container with film, but a large capsule with the entire camera set. This promised to kill two birds with one stone! Feoktistov drafted a proposal for a piloted spacecraft in the guise of an automatic reconnaissance satellite and submitted it to the Military-Industrial Commission of the Soviet Council of Ministers. Some officials became suspicious when they noticed, for example, that the presumably automatic satellite was equipped with a set of communication devices, and they inquired, "Who is going to talk over this radio? The photo cameras?"[24] But Feoktistov was able to fend off such suspicions, and his proposal was approved.

At this early stage, the competition between automatic satellites and piloted spaceships was resolved by making piloted ships also fully automatic so that they could be flown in both piloted and unpiloted modes. Since the first Soviet piloted spacecraft had to serve a dual purpose—both military and civilian—its controls also had to be dual, both automatic and manual.

Only having a fully automatic spacecraft at hand, spacecraft designers began carving out a role for the cosmonaut to play. By early 1960, Boris Raushenbakh's department at the Experimental Design Bureau No. 1 completed its design of the automatic control system, and after that, they began working on manual control. That is, the issue here was not the automation of certain functions of a human pilot, but the transfer of certain functions from an existing automatic system to a human pilot. What really needs an explanation is not why Vostok

---

24. Konstantin Feoktistov, *Traektoriia zhizni* (Moscow: Vagrius, 2000), p. 62.

was automated, but why it had a manual control system at all. Its purposes were to back up the automatic system in case of malfunction, to expand the window for controlled descent, and, most importantly, to provide psychological support to the cosmonaut. As Raushenbakh put it, "The cosmonaut must be convinced that even if ground control equipment and the on-board automatic system fail, he would be able to ensure his safety himself."[25]

While Gagarin had to limit his in-flight activity to monitoring and reporting, during subsequent Vostok flights, the cosmonauts successfully tested the manual attitude-control system and performed other duties and experiments. In particular, they tested the human ability to carry out military tasks. Korolev had previously suggested that the piloted version of Vostok could be used "to exterminate [enemy] satellites."[26] Tests performed by the cosmonauts Nikolaev and Popovich on Vostok 3 and Vostok 4 demonstrated that the human was "capable of performing in space all the military tasks analogous to aviation tasks (reconnaissance, intercept, strike). Vostok could be used for reconnaissance, but intercept and strike would require the construction of new, more advanced spacecraft." From this information, Kamanin concluded that "man can maintain good working capacity in a prolonged spaceflight. The 'central character' in space is man, not an automaton."[27]

THE VOSKHOD 2 MISSION:
THE COSMONAUT TAKES CONTROL

While the cosmonauts believed that the first spaceflights had demonstrated the human ability to perform in orbit, the engineers largely interpreted the same events as confirming the high reliability of automatic systems. Soviet engineers initially viewed the automatics and the cosmonaut not as a single, integrated system, but as two separate, alternative ways to control a spacecraft. They sought ways to make the automatic control system independently reliable, rather than trying to optimize interaction between human and machine. The probability of a system malfunction that would require resorting to manual control seemed remote, and the manual control system did not seem to have primary importance for spacecraft designers. So when they redesigned Vostok for a three-men crew (the Voskhod mission) and later for a spacewalk (the Voskhod 2 mission), it was the manual control system that got short shrift. To fit in all the new equipment, the designers had to move the main instrument panel and the optical sight from the

25. Aleksei Eliseev, *Zhizn'—kaplia v more* (Moscow: Aviatsiia i kosmonavtika, 1998), p. 15.

26. Sergei Korolev, "Tezisy doklada po kosmosu," June 1960, Russian State Archive of the Economy (RGAE), f. 298, op. 1, d. 1483, l. 246.

27. Kamanin, *Skrytyi kosmos*, vol. 1, pp. 174 (diary entry of 13 September 1962), 149 (diary entry of 16 August 1962).

front to the left side, and the hand controller was also moved.[28] Additional technical measures were taken to ensure the reliability of the automatic control system, and yet when a life-threatening emergency occurred during the Voskhod 2 flight in March 1965, only the cosmonauts' ingenuity and skill saved their lives.

When the Voskhod 2 crew—the commander, Pavel Beliaev, and the first "spacewalker," Alexei Leonov—were preparing for descent, the automatic attitude-correction system failed. Because of an error in the mathematical model, the automatics decided that the orientation engines were malfunctioning and shut them down. Without proper orientation, the firing of the retrorocket was automatically blocked, threatening to leave the crew stranded in the orbit. After some deliberation, the ground control ordered the cosmonauts to perform manual orientation, which was the only option available at that point.

To use the manual system, however, was no easy task. Because of a peculiar cabin layout, the optical sight and the hand controller were located to the left of the commander's seat, rather than in front of it. The cosmonauts could not look through the sight or operate the controller while remaining in their seats. Both cosmonauts had to unbuckle their seatbelts and leave their seats. Beliaev also had to take off his space helmet because he could not bend his neck in it. He had to lie down across both seats, since only while lying down could he use both hands to operate the manual controls. In the meantime, Leonov crawled under his seat and was holding Beliaev by his torso, since in zero gravity, Beliaev tended to float away and block the optical sight. After the orientation, the cosmonauts needed to fire the retrorocket. But before firing it, they had to return to their seats to balance the spacecraft, and they lost 30 or 40 seconds. They spent a few more seconds doublechecking the orientation and then fired the retrorocket. As a result of these delays, the spacecraft overshot its destination. The crew landed in the middle of a thick forest, and before a rescue team was able to reach them, they had to spend two nights on the snow, hiding in their space capsule from hungry wolves.[29]

The Voskhod 2 story also provided an interesting test case for assigning responsibility for various errors to human or machine. The investigating commission noted that the flawed spacecraft design made it impossible for the crew to control the ship manually without leaving their seats, and at the same time, it criticized the crew for violating the rules. In the final report, however, the criticism of spacecraft design was dropped in exchange for removing the criticism of the crew.[30]

---

28. Eliseev, *Zhizn'*, p. 46.

29. Boris E. Chertok, *Rakety i liudi*, vol. 4, *Lunnaia gonka* (Moscow: Mashinostroenie, 2002), p. 418; Eliseev, *Zhizn'*, p. 58; Nikolai Kamanin, *Skrytyi kosmos*, vol. 2, *1964–1966* (Moscow: Infortekst, 1997), p. 190 (diary entry of 22 April 1965); Ponomareva, "Osobennosti razvitiia," pp. 157–158; Siddiqi, *Challenge to Apollo*, p. 458.

30. Kamanin, *Skrytyi kosmos*, vol. 2, pp. 197 (diary entry of 8 May 1965), 199 (diary entry of 13 May 1965).

DESIGNING A COSMONAUT FOR SOYUZ

The second-generation Soviet spacecraft, Soyuz, was designed for a much wider range of missions than Vostok, including Earth-orbit rendezvous and docking. The problem of an efficient division of function between human and machine on Soyuz became the subject of a heated, if closely contained, debate within the Soviet space community. Two groups—the spacecraft designers and the cosmonauts—had very different perspectives on this issue. Briefly put, their positions were as follows.

The spacecraft designers argued that on-board automation had clear advantages. It allowed 1) to test piloted spacecraft in the unpiloted mode, thereby reducing time and expense on ground tests and increasing flight safety; 2) to lower eligibility criteria and reduce training time for cosmonauts; 3) to correct errors in flight.[31] The engineers were willing to assign the cosmonauts a backup function but preferred to keep the automatic mode as nominal.

The cosmonaut corps, on the other hand, tended to view the automation of control functions as excessive and hampering the "progress" of human spaceflight. They argued that a human operator would increase the reliability and effectiveness of a space mission. They especially stressed the human ability to act in unexpected situations, to cope with equipment failures, and to perform in-flight repairs. They argued that full automation alienated the pilot from his craft. They insisted that instead of fitting the human into an existing technological system, one must design human activity first and then determine specifications for the technological components of the system.[32]

The Soviet space program's organizational structure (or lack thereof) gave the spacecraft designers a decided advantage over the cosmonauts in such internal disputes. The Soviet space program was not supervised by a central government agency like NASA, but was scattered over a large number of defense industry, military, and academic institutions. The chief contractor for Soyuz—Korolev's Experimental Design Bureau No. 1—exercised unprecedented control over the course of the space program. Korolev himself, in particular, played a central role in decision-making on a whole range of issues going far beyond engineering, such as spacecraft procurement, cosmonaut training, crew selection, programming of missions, and ground flight control.[33] It was

---

31. Vladimir S. Syromiatnikov, *100 rasskazov o stykovke i o drugikh prikliucheniiakh v kosmose i na Zemle*, vol. 1, *20 let nazad* (Moscow: Logos, 2003), p. 83.

32. See Beregovoi et al., *Eksperimentalno-psikhologicheskie issledovaniia*, pp. 192, 270; Ponomareva, "Nachalo vtorogo etapa"; Ponomareva, "Osobennosti razvitiia."

33. On Korolev, see Golovanov, *Korolev*; Harford, *Korolev*; Boris V. Raushenbakh, ed., *S.P. Korolev i ego delo: svet i teni v istorii kosmonavtiki* (Moscow: Nauka, 1998). In the eyes of Korolev's

continued on page 50

the engineers' vision of the proper division of function between human and machine that was largely implemented in the Soviet space program.

Soyuz designers recognized that manual control would "make it possible to get rid of a number of complex pieces of equipment and to simplify automatic control systems."[34] Compared to Vostok, they significantly broadened the range of manual control functions, but these new functions involved not so much piloting as monitoring numerous on-board systems and dealing with equipment malfunctions. A Soyuz cosmonaut was a different type of cosmonaut, an engineer more than a pilot.

On the Soyuz program, requirements for the skills of the crew, selection criteria for the cosmonaut corps, and the very professional identity of cosmonauts began to change. The first group of Soviet cosmonauts that flew on Vostoks was selected from among young fighter pilots, who had little engineering background and modest flight experience compared to the more educated and experienced test pilots selected for the Mercury astronaut group.[35] Sergei Korolev chose fighter pilots because of their universal skills as pilots, navigators, radio operators, and gunners.[36] On a two- or three-seat Soyuz, these functions could now be divided among the crew members, and narrow specialists, more skilled in one task than another, could be brought on board.

But there was also another, more important factor that precipitated a shift in the cosmonaut professional identity. In the decentralized organizational structure of the Soviet space program, spacecraft design and cosmonaut training were institutionally separated: the design and production of spacecraft was conducted under the Ministry of General Machine-Building, and cosmonaut training was the responsibility of the Air Force. As a result, the cosmonauts had very little input in spacecraft design. They pointed out that in the aviation industry, experienced pilots were regularly consulted during the design phase, while the cosmonaut pilots were entirely left out of spacecraft design.[37] The engineers recognized the problem but came up with a different solution for it. Vasilii Mishin, who replaced Korolev as Chief Designer after his death, argued that "design solutions can only be checked [in flight] by highly qual-

---

continued from page 49
subordinates, he was truly omnipotent. For example, Feoktistov claimed that crucial decisions concerning the Soviet space program were made "not by the Party Central Committee or the Soviet government, but by Korolev and [the defense industry leader Dmitrii] Ustinov (and often by Korolev alone), and later they managed, one way or another, to obtain a retroactive endorsement through an official decree" (Feoktistov, *Traektoriia zhizni*, pp. 36–37).

34. Vasilii Mishin, quoted in Kamanin, *Skrytyi kosmos*, vol. 2, p. 368 (diary entry of 17 August 1966).

35. Siddiqi, *Challenge to Apollo*, p. 246.

36. Gherman S. Titov, "30 let spustia," *Aviatsiia i kosmonavtika*, no. 8 (1991): 26.

37. Chertok, *Rakety i liudi*, vol. 4, p. 149.

ified specialists directly involved in designing and ground testing of the space-craft."[38] Thus, instead of involving cosmonaut pilots in spacecraft design, he proposed to train space engineers as cosmonauts and to let them test new systems in flight.

Soon, Mishin took practical steps toward changing the composition of the cosmonaut corps. In May 1966, the Experimental Design Bureau No. 1 set up a flight-methods department for the training of a civilian group of "cosmonaut testers."[39] This rapidly led to an open confrontation with Air Force officials, who defended their monopoly on cosmonaut selection and training. Wielding his influence with the Soviet leadership, Mishin threatened that only engineers and scientists would fly and that training at the Air Force Cosmonaut Training Center would be simplified or dispensed with altogether.[40] Eventually, a compromise was worked out by which a typical Soyuz crew would include one military pilot as mission commander, one civilian engineer, and one flight researcher, in whose seat military and civilians would alternate.[41]

As spacecraft designers began to enter the cosmonaut corps, they intro-duced elements of engineering design into the planning of cosmonaut activity. The control system engineer and cosmonaut Alexei Eliseev, who took part in a spacewalk during the Soyuz 4–Soyuz 5 mission, applied a genuine engineering skill in designing a step-by-step procedure for the spacewalk, specifying the actions and code words for every crew member. This procedure was recorded on a 4-meter-long scroll of paper.[42] The Experimental Design Bureau No. 1 set up a special department, which designed cosmonaut activity so that it conformed to the logic of on-board automatics. Control system designers worked in close contact with human engineering specialists, who conceptualized the spacecraft control system as a "cybernetic 'human-machine' system."[43] Adapting the cybernetic conceptual framework, they viewed control as a system function that could be performed by both human and machine. Human engineering specialists described the cosmonaut as a "living link"[44] in a human-machine system and analyzed this "link" in terms borrowed from control theory and information theory—the same terms that applied to the other, technical links

---

38. Quoted in Kamanin, *Skrytyi kosmos*, vol. 2, p. 368 (diary entry of 17 August 1966).

39. Siddiqi, *Challenge to Apollo*, p. 566.

40. Chertok, *Rakety i liudi*, vol. 3, p. 242.

41. Eliseev, *Zhizn'*, p. 165.

42. Ibid., p. 91.

43. V. G. Denisov, "Nekotorye aspekty problemy sochetaniia cheloveka i mashiny v slozhnykh sistemakh upravleniia," in *Problemy kosmicheskoi biologii*, ed. N. M. Sisakian and V. I. Iazdovskii, vol. 2 (Moscow: Nauka, 1962), p. 54.

44. V. G. Denisov, A. P. Kuz'minov, and V. I. Iazdovskii, "Osnovnye problemy inzhenernoi psikhologii kosmicheskogo poleta," in *Problemy kosmicheskoi biologii*, ed. N. M. Sisakian and V. I. Iazdovskii, vol. 3 (Moscow: Nauka, 1964), p. 77.

in that system: delay time, perception speed, reaction speed, bandwidth, and so on.[45] They discussed how efficiently a human operator could perform the functions of a logical switchboard, an amplifier, an integrator, a differentiator, and a computer.[46] Spacecraft designers avoided using the word "pilot" and preferred the term "spacecraft guidance operator."[47] The cosmonaut had to fit into an existing technological system, and human performance was effectively evaluated in machine terms.

One of the main criteria for cosmonaut selection was the ability to carry out precisely programmed actions.[48] Subsequent training was geared toward turning the human into a perfect machine. Spacecraft designers took to the heart a piece of advice given by Igor' Poletaev, a leading Soviet cybernetics specialist. He argued that the way to avoid human error was to train the human to operate like a machine. He wrote: "The less his various human abilities are displayed, the more his work resembles the work of an automaton, the less [the human operator] debates and digresses, the better he carries out his task."[49] The cosmonaut training manual explicitly stated that "the main method of training is repetition."[50] Yuri Gagarin recalled how the cosmonauts were "getting used to every button and every tumbler switch, learned all the movements necessary during the flight, making them automatic."[51] The Vostok 5 pilot Valerii Bykovskii was praised in his character evaluation for "the high stability of automation of skill."[52]

The cosmonauts began to resent what they perceived as "excessive algorithmization" of their activity. They argued that the strict regulation of cosmonauts' activity on board forced them "to work like an automaton" and stripped them of the possibility to plan their actions on their own.[53]

---

45. Denisov, "Nekotorye aspekty," p. 55.

46. P. K. Isakov, V. A. Popov, and M. M. Sil'vestrov, "Problemy nadezhnosti cheloveka v sistemakh upravleniia kosmicheskim korablem," in *Problemy kosmicheskoi biologii*, ed. N. M. Sisakian, vol. 7 (Moscow: Nauka, 1967), p. 6.

47. V. N. Kubasov, V. A. Taran, and S. N. Maksimov, *Professional'naia podgotovka kosmonavtov* (Moscow: Mashinostroenie, 1985), p. 278.

48. Siddiqi, *Challenge to Apollo*, p. 244.

49. Igor' A. Poletaev, *Signal: O nekotorykh poniatiiakh kibernetiki* (Moscow: Sovetskoe radio, 1958), p. 281.

50. Kubasov, Taran, and Maksimov, *Professional'naia podgotovka*, p. 138.

51. Yuri Gagarin, *Doroga v kosmos* (Moscow: Pravda, 1961).

52. Quoted in A. N. Babiichuk, *Chelovek, nebo, kosmos* (Moscow: Voenizdat, 1979), p. 209.

53. Beregovoi et al., *Eksperimentalno-psikhologicheskie issledovaniia*, p. 31.

Soyuz Flights: Dividing Glory and Responsibility Between Human and Machine

Several emergency situations that occurred during Soyuz missions underscored the crucial importance of human-machine issues for spacecraft control. As the boundary between human and machine functions was often blurred, so was the responsibility for error. While accident investigators tended to assign the responsibility for error to either human or machine, failures were often systemic. In an emergency, rigid control schemes often had to be reconsidered and human and machine functions had to be redefined. Ground flight controllers frequently stepped in, further complicating the division of responsibility between human and machine. Ultimately, what often decided the success of the mission was not how much or how little the cosmonauts did, but how well they were integrated into the control system, which included both the on-board automatics and mission control.

In April 1967, the Soyuz 1 mission had to be aborted after multiple equipment failures, and the cosmonaut Vladimir Komarov successfully performed manual attitude correction with an ad hoc method invented during the flight. Yuri Gagarin, who served as a CAPCOM on that mission, told the leading control system designer, "What could have we done without a human? Your ion system proved unreliable, a sensor failed, and you still don't trust cosmonauts!"[54] In the end, yet another automatic system—the parachute release— failed, and this time, the cosmonaut had no manual means to override it. The spacecraft hit the ground at full speed, and Komarov died.

In October 1968, the cosmonaut Georgii Beregovoi on Soyuz 3 attempted a manual rendezvous, but he misread the target vehicle indicators and failed to approach the target. Engineers regarded this as a clear human error, yet Nikolai Kamanin, responsible for cosmonaut training, pointed out that the actual manual control system on board in certain respects differed from the version installed on a ground simulator and that the cosmonaut did not have adequate time to adjust to zero gravity. "I did not find my place within a human-machine structure," admitted Beregovoi. He complained that the hand controllers were too sensitive, sending the spacecraft into motion at the slightest touch: "This is good for an automaton, but it creates extra tension for a human."[55] Kamanin interpreted this incident as a systemic failure, rather than simply a human operator error: "If even such an experienced test pilot [as Beregovoi] could not manually perform the docking of two spaceships, this means that the [manual] docking system is too complex to work with in zero gravity."[56]

---

54. Chertok, *Rakety i liudi*, vol. 3, p. 450.
55. Chertok, *Rakety i liudi*, vol. 4, p. 419.
56. Kamanin, *Skrytyi kosmos*, vol. 3, p. 303 (diary entry of 29 October 1968).

Now engineers had to prove that their manual control system was actually operable. Chief Designer Vasilii Mishin insisted on trying manual docking on the Soyuz 4–Soyuz 5 mission in January 1969, even though his boss, the Minister of General Machine-Building, Sergei Afanas'ev, pressured him to resort to the proven automatic docking system.[57] This time the engineers made sure that the cosmonauts received more than sufficient training on the ground. The cosmonaut Vladimir Shatalov had performed 800 simulated dockings in various regimes on a ground simulator before he successfully carried out manual docking of Soyuz 4 and Soyuz 5.[58] Later, for other trainees, the requisite number of simulated dockings was reduced to 150.[59]

In August 1974, the Soyuz 15 crew attempted an automatic rendezvous with the Salyut 3 station, but the automatic system malfunctioned, misjudging the distance to the target and producing an acceleration thrust instead of retrofire. This led to a near collision of the spaceship with the station. Another attempt at automatic approach resulted in another dangerous flyby. The crew suggested to make a third attempt at docking in the manual regime, but ground control did not give permission, due to the low level of remaining propellant. The crew had to return to Earth without completing their mission.[60]

After the flight, heated debates erupted over the question whether the main responsibility for the failed mission should be assigned to human or machine. Engineers argued that the cosmonauts should have recognized the malfunction immediately and should have resorted to manual control. Officials responsible for cosmonaut training replied that this particular type of emergency had not been included in the list and that the cosmonauts had not been trained for it. The investigation was further complicated by the fact that this failure occurred just a year before the scheduled docking of Soyuz with Apollo. The American side, worried about the reliability of the Soviet rendezvous system, requested an explanation of the Soyuz 15 incident.[61] Thus, despite an obvious failure of the automatic docking system, the Soviets preferred to put the blame squarely on the cosmonauts—for not shutting down the malfunctioning system after the first failure.[62] Both cosmonauts were officially reprimanded and never flew into space again.

---

57. Nikolai Kamanin, *Skrytyi kosmos*, vol. 4, *1969–1978* (Moscow: Novosti kosmonavtiki, 2001), p. 11 (diary entry of 10 January 1969), 12 (diary entry of 11 January 1969).

58. Vladimir A. Shatalov, *Trudnye dorogi kosmosa*, 2nd ed. (Moscow: Molodaia gvardiia, 1981), p. 129.

59. Kubasov, Taran, and Maksimov, *Professional'naia podgotovka*, p. 138.

60. Chertok, *Rakety i liudi*, vol. 4, p. 434; Asif A. Siddiqi, "The Almaz Space Station Complex: A History, 1964–1992: Part I," *Journal of the British Interplanetary Society* 54 (2001): 411–414.

61. Dave Shayler, "Soyuz 15 Mission Report," *http://www.astroinfoservice.co.uk/html/soyuz_15_report.html* (accessed 21 April 2005).

62. Rex Hall and David J. Shayler, *Soyuz: A Universal Spacecraft* (Chichester, U.K.: Springer/Praxis, 2003), pp. 186–187; Ponomareva, "Nachalo vtorogo etapa," pp. 169–170.

Rather than being an exclusively human or machine failure, the Soyuz 15 mission illustrated another system failure: a failure to integrate the crew in the control loop in a human-machine system. The crew was kept in "cold reserve," passively monitoring the operations of the automatic docking system. When this system failed, the crew was not ready to take over control operations quickly. Although the engineers switched the blame to the crew, it was the engineers' design of the control system that placed the crew in the role of passive observers. Engineers tacitly admitted that the failure of the Soyuz 15 mission had roots in the overall organization of rendezvous control, including the role of ground control. A special operational group was created as part of Mission Control to develop procedures for automatic and manual rendezvous in various emergency situations and to provide real-time recommendations for the flight director.[63]

After that incident, cosmonaut pilots were assigned responsibility for manual approach from the distance of 200 to 300 meters. In a few years, however, this rule was subjected to a severe test. In October 1977, the Soyuz 25 crew made an attempt at manual docking with the Salyut 6 station, and when the spacecraft almost touched the station, they suddenly realized that they were facing the "bottom" of the station, instead of the docking port. They quickly turned away from Salyut 6 and made several more docking attempts, all of which failed. Having spent much propellant, Soyuz 25, in the end, did not even have enough fuel to back up from the station and remained in close proximity to it for several orbits.[64] As it turned out, what the cosmonauts perceived as the "bottom" of the station was in fact the docking port. Soyuz 25 approached the station from a slightly different angle than was expected, but the cosmonauts were never trained on a ground simulator to recognize the station from that angle. A "conditional reflex" they acquired during incessant training on the simulator prevented them from recognizing the correct position of the station.[65] Although the error was rooted in the inadequate simulator design, the cosmonauts bore their part of the blame. For the first time, the cosmonauts did not receive the honor of the Hero of the Soviet Union, but were awarded "only" the Order of Lenin.[66] Mission planners decided never again to send all-rookie crews into space. Most importantly, it was decided to make the nominal docking regime automatic, and the cosmonauts were allowed to take over manual control only in case of failure of the automatic system.[67] The prolonged struggle for the right to control docking between human and machine began to shift in favor of the latter.

63. Chertok, *Rakety i liudi*, vol. 4, p. 435.
64. Eliseev, *Zhizn'*, pp. 200–204.
65. Chertok, *Rakety i liudi*, vol. 4, p. 439.
66. Iurii M. Baturin, ed., *Mirovaia pilotiruemaia kosmonavtika. Istoriia. Tekhnika. Liudi* (Moscow: RTSoft, 2005), pp. 273–274.
67. Eliseev, *Zhizn'*, p. 209.

THE ROLE OF GROUND CONTROL

The norms of cosmonaut activity included not only following the technical protocol of interaction with on-board equipment, but also following
the social protocol of subordination to their superiors on the ground. Framing
the whole issue as human versus machine is somewhat misleading. The real
issue here was not so much the division of function between human and
machine, but the division of power between the human on the ground and the
human on board.

Boris Chertok acknowledged that the growing complexity of space technology warranted a greater role for the human operator, but his idea of human
participation was to involve "not just an individual, but an entire collective,"[68]
meaning the flight controllers and specialists on the ground. As a result, Soviet
designers adopted the principle that they have followed to this day: all critical
systems had three independent lines of control: automatic, remote (from the
ground), and manual.[69] Control during the three main stages of the flight—
reaching the orbit, orbital flight, and reentry—was automatic; instructions to
switch programs between the stages were given either from the ground or
manually by the cosmonaut. The cosmonaut, however, had to obtain permission from the ground for any critical action. The cosmonaut training manual
clearly stipulated that "all most important decisions are made by Mission
Control."[70] The real control of the mission remained in the hands of engineers:
either through the automatic systems they designed or through their design
and management of cosmonaut activity.

The need to obtain clearance from Mission Control sometimes delayed
critical actions until it was too late. For example, in October 1969, the Soviets
planned a complicated orbital maneuver with three spacecraft: Soyuz 7 and
Soyuz 8 attempted a rendezvous, while Soyuz 6 was to capture the event on
camera. Unfortunately, the automatic approach system on Soyuz 8 failed. At
that moment, the two ships were about 1,000 meters from each other, and the
cosmonauts asked permission to attempt manual approach. While the crew
awaited permission from the ground, the ships drifted apart to the distance of
about 3,000 meters, and manual approach was no longer an option. The next
day, through orbital maneuvers, the ships were brought within 55 feet from each
other, but without any means to determine their relative velocities, all attempts
at manual approach also failed.[71] The crews had to return to Earth without

---

68. B. Evseev (Boris Chertok), "Chelovek ili avtomat?" in *Shagi k zvezdam*, ed. M. Vasil'ev
(Vasilii Mishin) (Moscow: Molodaia gvardiia, 1972), p. 282.

69. Syromiatnikov, *100 rasskazov*, p. 145.

70. Kubasov, Taran, and Maksimov, *Professional'naia podgotovka*, p. 190.

71. Chertok, *Rakety i liudi*, vol. 4, pp. 214–215; Hall and Shayler, *Soyuz*, p. 159.

completing their mission. Nikolai Kamanin subsequently bitterly remarked in his private diary: "Everything [on the Soyuz] is based on the assumption of a flawless operation of automatics, and when it fails, cosmonauts are left without reliable means of control."[72] And yet the responsibility for the failed mission was placed on the cosmonauts.[73] Boris Chertok later admitted, however, that the designers were to blame for overestimating human capabilities and for not providing adequate training on simulators for the situation of failure of the automatic approach system.[74]

On more than one occasion, cosmonauts faced the dilemma: to follow the rules and fail the mission or to take risks and break the rules. Some preferred to break the rules and save the mission. Another emergency that occurred during the Voskhod 2 flight in March 1965 is a case in point. After completing his historic spacewalk, the cosmonaut Alexei Leonov realized that his spacesuit ballooned, his arms and legs did not even touch the inside, and he was unable to reenter the airlock. He was supposed to report all emergencies to the ground and wait for instructions. He later recalled: "At first I thought of reporting what I planned to do to Mission Control, but I decided against it. I did not want to create nervousness on the ground. And anyway, I was the only one who could bring the situation under control."[75] Perhaps, he calculated that instructions from the ground could be delayed because of various bureaucratic procedures and the possible reluctance of some decision-makers to take responsibility, and it would be unwise for him to spend his limited oxygen supply waiting for them. Leonov turned a switch on his spacesuit, drastically reducing the internal air pressure, which allowed him to regain control of his movements. Once he broke one rule, he decided that he would not make things worse by breaking another, and he climbed into the airlock headfirst, in violation of an established procedure.

The Voskhod 2 crew—Alexei Leonov and Pavel Beliaev, both military pilots—were trained to follow the rules and to obey orders from the ground. After more than 150 training sessions on a spacewalk simulator, Leonov was said to have brought his skills "to the point of automatic performance."[76] Yet in a real emergency, Leonov had to perform actions for which he was not trained, to violate explicit rules concerning entry into the airlock, and to make decisions without consulting Mission Control. In other words, his mission was successful precisely because he did not act like a perfect machine.

---

72. Kamanin, *Skrytyi kosmos*, vol. 4, p. 95.

73. Ponomareva, "Nachalo vtorogo etapa," p. 169.

74. Chertok, *Rakety i liudi*, vol. 4, p. 422.

75. David R. Scott and Alexei A. Leonov, *Two Sides of the Moon: Our Story of the Cold War Space Race* (London/New York: Simon & Schuster, 2004), p. 109.

76. N. N. Gurovskii et al., "Trenazhery dlia podgotovki kosmonavtov k professional'noi deiatel'nosti po upravleniiu korablem i ego sistemami," in *Problemy kosmicheskoi biologii*, ed. N. M. Sisakian, vol. 4 (Moscow: Nauka, 1965), p. 6; Siddiqi, *Challenge to Apollo*, p. 451.

THE PARADOX OF DISCIPLINED INITIATIVE

Space engineers believed that flight safety would be best guaranteed by comprehensive automation and by strict following of instructions by the crew, but the cosmonauts pointed out that it was often necessary to break the rules in case of emergency. The engineers often viewed any departure from the standard procedure as a "human error," while it was precisely this ability to deviate from the standard path that made human presence on board so valuable in an emergency situation. Perhaps the main difference between human and machine in a human-machine system is that the machine fails when it does not follow preset rules and the humans fail when they do not recognize that it is time to break the rules.

Valentina Ponomareva, a member of the first women's cosmonaut group, summed up the cosmonauts' vision of the unique human role on board as follows:

> In addition, the cosmonaut must possess such qualities as curiosity and *the ability to break rules* . . . . Regulations work well only when everything goes as planned . . . . The ability to act in extraordinary situations is a special quality. In order to do that, one has to have inner freedom . . . the ability to make non-trivial decisions and to take non-standard actions. In an extreme situation the very life of the cosmonaut depends on these qualities.[77]

Despite her high qualifications as an engineer and a pilot and her excellent test marks, Ponomareva was not selected for the first woman's flight, and she never got a chance to fly. Her independent-mindedness most likely played a role here.

Sonja Schmid, in her study of Soviet nuclear power station operators, observed a similar contradiction in the way the operators were viewed by nuclear reactor designers: both as a "weak link" and as a "reliable cog in the wheel."[78] Both spacecraft designers and nuclear engineers viewed the human operator as part of technology, which must always function according to the rules, and at the same time, they expected the operators to show human qualities such as initiative and inventiveness.

---

77. Valentina Ponomareva, *Zhenskoe litso kosmosa* (Moscow: Gelios, 2002), p. 285.
78. Sonja Schmid, "Reliable Cogs in the Nuclear Wheel: Assigning Risk, Expertise and Responsibility to Nuclear Power Plant Operators in the Soviet Union" (paper presented at Society for the History of Technology [SHOT]-2004, Amsterdam, Netherlands, 7–10 October 2004).

This need for the cosmonauts to be both obedient and creative, to follow the rules and to break them, one might call "a paradox of disciplined initiative." In my view, this paradox reflects one of the fundamental contradictions of the Soviet approach to spacecraft control (and perhaps to social control and government in general).

THE LUNAR PROGRAM:
A TURN TOWARD MANUAL CONTROL

The lunar race further complicated the debates over the human role on board. Lunar mission profiles did not allow ground stations to effectively control the entire flight, and the division of control functions between human, on-board automation, and ground control had to be reevaluated. Initially, it was decided to give the cosmonauts an unusually high degree of control over their spacecraft. Alexei Leonov, who initially trained for a circumlunar mission, recalled that "we had to be able to perform every aspect of the flight manually in case the automatic system failed."[79] Later on, the internal politics of the Soviet lunar program began to erode this principle.

From the very beginning, the Soviet lunar program suffered from the lack of coordination, internal rivalries, duplication of effort, and fracturing of resources. Initially, the heads of two rival design bureaus—Sergei Korolev and Vladimir Chelomey—divided the lunar pie more or less equally: Korolev worked on a lunar landing project, while Chelomey developed a rocket and a spacecraft for a circumlunar flight. After Khrushchev's ouster in October 1964 and the subsequent shakeup in the upper echelons of Soviet power Chelomey lost some of his political support, and Korolev eventually wrestled the circumlunar flight project away from him. In October 1965, a government decree assigned Korolev the responsibility for the development of the 7K-L1, a new spacecraft designed specifically for a circumlunar flight, later publicly named Zond.

One major hurdle in the Soviet lunar program was eliminated: all work on lunar spacecraft was now concentrated in one organization, Korolev's design bureau. Yet the circumlunar flight and the lunar landing remained two separate projects with different goals, independent work schedules, different booster rockets, separate ground infrastructures, and two different types of spacecraft, the L1 and the L3. The addition of the circumlunar project to Korolev's tasks stretched the resources of his design bureau and messed up the lunar landing project schedule. The circumlunar project was given immediate priority in order to complete it by the 50th anniversary of the Great October Revolution in November 1967.

---

79. Scott and Leonov, *Two Sides of the Moon*, p. 189.

Social and political factors influenced the lunar program down to the very technical level. Korolev had to split the responsibility for the development of the control system for the L1 spacecraft with the organization led by his old friend Nikolai Pilyugin. As a result, Pilyugin developed the automatic control system for course corrections and reentry, while Korolev assumed responsibility for manual rendezvous control.[80] The cosmonaut functions on board were thus limited by the division of spheres of responsibility of different design organizations.

The L1 crew consisted of two cosmonauts, whose duties included checking all on-board systems in Earth orbit and then orienting the spacecraft toward the Moon. For the first time in the Soviet piloted space program, the L1 control system included a digital computer, the Argon-11. This computer was part of the automatic control system designed by Pilyugin, and cosmonauts had no access to it.[81] The manual control system included a digital computing device called Salyut 3, which was not reprogrammable; it gave the cosmonauts fixed options for selecting one of the preset programs. According to the control panel designer, Yuri Tiapchenko, the L1 panel was a step backward in comparison with Soyuz: "The functions of cosmonauts were reduced to the simplest operations of entering commands and controlling their execution in accordance with flight instructions and the orders issued by ground control."[82]

In 1967–1968, the Soviets made eight attempts to launch L1 on a circumlunar mission in the unpiloted mode. Only one mission performed a circumlunar flight; all missions were fraught with numerous failures which might have been fatal to a human crew. After the successful Apollo 8 mission in December 1968, the L1 program lost its political rationale, and after another failed L1 mission in January 1969, the plans for a piloted flight were suspended. Eventually the program was canceled without a single attempt for a piloted flight. The cosmonauts unsuccessfully petitioned the Soviet political leadership for continuation of the piloted circumlunar program.[83] The only completely successful L1 mission that would have returned the crew safely to Earth took place on 8 August 1969. The passengers on the spacecraft were four male tortoises. Two cosmonauts, Alexei Leonov and Oleg Makarov, participated in the mission as ground operators.[84]

---

80. Siddiqi, *Challenge to Apollo*, pp. 504–505.

81. V. V. Chesnokov, "Argon-11C computer," *http://www.computer-museum.ru/english/argon11c.htm* (accessed 21 April 2005); Georgii Priss interview, Moscow, 23 May 2002, *http://hrst.mit.edu/hrs/apollo/soviet/interview/interview-priss.htm* (accessed 21 April 2005); Viktor Przhiyalkovsky interview, Moscow, 24 May 2002, *http://hrst.mit.edu/hrs/apollo/soviet/interview/interview-przhiyalkovsky.htm* (accessed 21 April 2005).

82. Iurii A. Tiapchenko, "Sistemy otobrazheniia informatsii pilotiruemykh KA L1 i N1-L3," *http://www.cosmoworld.ru/spaceencyclopedia/publications/index.shtml?tg_moon.html* (accessed 21 April 2005).

83. Scott and Leonov, *Two Sides of the Moon*, p. 252.

84. Siddiqi, *Challenge to Apollo*, pp. 699–700.

That flight took place already after Apollo 11. The Soviet lunar landing project, known as N1-L3, lost its political rationale too, but Chief Designer Vasilii Mishin continued lobbying for it, given the amount of funding and effort already invested in it, and the project was kept afloat for a few more years.

The Soviet lunar landing project was based on a lunar orbit rendezvous scheme similar to Apollo. Because of the limits on the rocket lifting power, however, the weight of the Soviet lunar lander had to be roughly one-third of the weight of the Apollo lander. For this reason, the Soviets planned to send only two cosmonauts on the lunar mission: one cosmonaut landing on the Moon and the other staying on the lunar orbital ship. Severe weight limitations forced Soviet designers to give the cosmonauts a much wider range of functions. In particular, to reduce the bulk of docking equipment and to eliminate extra dockings, the engineers proposed to transfer the cosmonaut from the orbital ship to the lander and back via spacewalk.[85]

Lunar landing was planned to be fully automatic with partial manual backup.[86] Using an on-board computer, a cosmonaut could process information from various sensors, evaluate the condition of the lander according to prepro-grammed algorithms, and choose specific actions. Most importantly, the cosmonaut could manually select a landing site on the lunar surface and give instructions to the computer to produce required landing maneuvers.[87] Lunar landing required extraordinary performance from the cosmonaut: on the Apollo lunar landing module, two astronauts had 2 minutes to make a landing decision, while on the Soviet lander, a single cosmonaut would have only 15 to 20 seconds.[88]

Cosmonauts underwent intensive training, both on simulators and on helicopters, simulating lunar landing. They performed helicopter landings with the engines cut off, a very difficult and dangerous operation.[89] Gradually, however, Chief Designer Vasilii Mishin began to limit the responsibilities of the pilot, placing greater emphasis on automatic systems. This may have had something to do with Mishin's plans to assign a greater role to civilian cosmonauts, engineers from his own design bureau. Cutting on manual control functions made it possible to reduce cosmonaut training time, and civilian cosmonauts, who generally had less training than military pilots, could now compete with the pilots for the lunar landing mission.[90]

---

85. Ibid., pp. 495–497.

86. Chertok, *Rakety i liudi*, vol. 4, pp. 92, 109.

87. Siddiqi, *Challenge to Apollo*, p. 491.

88. Chertok, *Rakety i liudi*, vol. 4, p. 225.

89. Siddiqi, *Challenge to Apollo*, pp. 684–685.

90. Kamanin, *Skrytyi kosmos*, vol. 3, pp. 123–124 (diary entry of 15 October 1967), 312 (diary entry of 13 November 1968), 341 (diary entry of 23 December 1968); Siddiqi, *Challenge to Apollo*, p. 650.

The growing degree of automation on the L3 alarmed the cosmonaut pilots. Alexei Leonov, who trained for lunar landing, commented that "according to the fight plan the automatic system took precedence"; the cosmonauts were allowed to resort to manual control only in case of failure of the automatic system. "I had argued," continued Leonov, "that, as commander of a spacecraft, what I needed once a flight was in progress was as little communication as possible from the ground—since it served mainly to distract me from what I already knew was necessary—and only manual, not automatic, control."[91]

The lunar landing program suffered from a series of setbacks during the failed launches of the giant N1 booster. The last attempt was made in 1972, and soon the program was terminated. The cosmonauts had hoped that they might have a chance to fly the lunar spacecraft during a series of Earth-orbit test flights in 1970–71. The financial difficulties that besieged the Soviet lunar program, however, forced Mishin to eliminate lunar orbiter test flights and to test only the lunar lander, and just in the unpiloted mode. During three tests in Earth orbit, the lunar lander successfully simulated a lunar landing, two liftoff operations with the primary and backup engines, and an entry into lunar orbit. The automatic control system worked perfectly.[92] Whether manual controls would have worked remains unknown. The Soviets kept the existence of their piloted lunar program secret for 25 years. Instead, they cultivated the myth that exploring the Moon with automatic probes was their one and only goal.

DEFINING THE COSMONAUT PROFESSION

The seemingly technical issue of on-board automation raised a larger question of the nature and purpose of human spaceflight. The debates over automation reflected three competing visions of spaceflight: a piloting mission, an engineering task, and a research enterprise.

The first cosmonaut group was composed of military pilots, and they used their growing prestige and political influence to maintain their monopoly on spaceflight. In May 1961, shortly after his historical first flight, Yuri Gagarin sent a letter to the Chief Marshal of Aviation, A. A. Novikov, arguing that "only pilots are capable of carrying out spaceflights. If others want to fly into space, they must learn to fly aircraft first. Aviation is the first step to spaceflight."[93]

---

91. Scott and Leonov, *Two Sides of the Moon*, p. 189.

92. Siddiqi, *Challenge to Apollo*, pp. 734–736.

93. Quoted in Kamanin, *Skrytyi kosmos*, vol. 1, p. 57 (diary entry of 25 May 1961). Later on, Gagarin seemed to have changed his opinion and supported the first civilian engineers who joined the cosmonaut corps; see Georgii Grechko, "Iz-za liubvi k kino ia chut' ne prozeval polet v kosmos!" *Vechernii Omsk*, no. 11 (11 February 2004), *http://epizodsspace.testpilot.ru/bibl/intervy/grechko3.html* (accessed 21 April 2005).

Two N1 Moon rockets appear on the pads at Tyura-Tam in early July 1969. Highly automated, the N1 was designed for the Soviet space program's human lunar missions. In the foreground is booster number 5L with a functional payload for a lunar-orbiting mission. In the background is the IMI ground-test mock-up of the N1 for rehearsing parallel launch operations. After takeoff, the rocket collapsed back onto the pad, destroying the entire pad area in a massive explosion. *(NASA photo no. n1july1969)*

When, in 1962, Korolev for the first time raised the question of including engineers in space crews, Kamanin called this "a wild idea."[94] The military pilots strongly objected to the waiver of "harsh physical tests" for engineers, insisting that the pilots were "the real veterans in the [cosmonaut] corps."[95] A Deputy Minister of Defense said bluntly that "we will select cosmonauts only from among robust young fellows from the military. We don't need those ninnies from civilian science."[96] Kamanin eventually realized the need for a compromise and began lobbying for the inclusion of civilian specialists.

Space engineers, for their part, insisted that they had a legitimate claim for a spacecraft seat. Boris Chertok explained: "We, engineers who designed the control system, believed that controlling a spacecraft is much easier that controlling an aircraft. All processes are extended in time; there is always time to think things over . . . . A good engineer can control a spaceship as well as a pilot, if there are no obvious medical objections."[97] The engineer-cosmonaut Konstantin Feoktistov compiled a chart comparing the professions of the cosmonaut and the pilot and tried to show that piloting skills were unnecessary aboard a spacecraft, but Kamanin interpreted the same chart in the opposite way.[98]

Engineers argued that their presence on board would have dual benefit: a better handling of emergency situations during the flight and a better design of spacecraft resulting from their flight experience. The engineer-cosmonaut Alexei Eliseev reasoned that, as space technology was becoming more and more complex, it would be impossible to write down instructions for all conceivable emergencies. A situation may arise in which only spacecraft designers on board would be able to find the right solution. He also suggested that "one could design on-board equipment for the cosmonauts only with their own participation. Only people who carry out spaceflights can give competent assessments and recommendations with regard to the convenience of use of various types of on-board equipment."[99] Instead of involving cosmonaut pilots in the design process, however, the engineers believed that they themselves should be included in space crews. In April 1967, the engineer-cosmonaut Oleg Makarov met with Chief Designer Vasilii Mishin and proposed a list of measures aimed at changing the role of humans on board. Makarov argued that an engineer must be included in every space crew; that crews must study on-board equipment at the design and production sites, not just on simulators;

94. Kamanin, *Skrytyi kosmos*, vol. 1, p. 105 (diary entry of 19 April 1962).
95. Scott and Leonov, *Two Sides of the Moon*, p. 146.
96. Kamanin, *Skrytyi kosmos*, vol. 1, p. 210 (diary entry of 17 January 1963).
97. Chertok, *Rakety i liudi*, vol. 3, pp. 237, 242.
98. Kamanin, *Skrytyi kosmos*, vol. 3, p. 210 (diary entry of 8 April 1968).
99. Eliseev, *Zhizn'*, pp. 28, 164.

and that cosmonauts must be given the right to take over control in case of malfunction of automatic systems.[100]

Kamanin realized that engineers–turned-cosmonauts might soon replace the military pilots whose training he oversaw. In February 1965, he ordered to organize eight research groups at the Cosmonaut Training Center focused on the following problems: military use of spacecraft; space navigation, life-support and rescue systems; telemetry equipment; scientific orbital stations; circumlunar flight; lunar landing; and weightlessness. Each group would study the assigned problem, formulate the Center's positions on specific issues, and defend those positions before scientists and designers.[101] While spacecraft designers were claiming a seat on board, the cosmonauts began to claim a seat at the designer's workstation.

In the 1970s, with the introduction of orbital stations, mission engineers began playing an ever-growing role in spaceflight. Long-duration missions required such skills as equipment maintenance and repair, observation, and research much more than piloting, which was limited to docking, undocking, and keeping the station in the correct attitude. Although pilots were tradition-ally appointed mission commanders, flight engineers began to demand more authority in decision-making. The engineer-cosmonaut Georgii Grechko summed up the engineers' sentiment as follows: "The time of pilots among cosmonauts is passing. In any case, they are no longer the main agents of the exploration of the Universe. 'Our' era, the era of mission engineers is dawning."[102] Grechko's discussion of these controversial issues with his com-mander, the pilot Yurii Romanenko, during their mission on the Salyut 6 station quickly turned into a heated argument. Eventually, Grechko had to flee into another compartment of the station to avoid violent confrontation.

Maintaining a complex orbital station with its long-term life-support systems devoured most of the cosmonauts' time on board, raising questions about the relative costs and benefits of human flight. The engineer-cosmonaut Valentin Lebedev calculated that during a five-day work week, two cosmonauts spent 111 hours on supporting themselves. Only 9 hours were left for scientific research. "The station is crewed just for the sake of those nine hours."[103] In an interview given after his retirement, Vasilii Mishin similarly estimated that in space, most of a cosmonaut's time on board was spent on preparations for takeoff and landing, on physical exercise, and on sleep: "Only 20 percent of

100. Mishin, diary, 30 April 1967.

101. Kamanin, *Skrytyi kosmos*, vol. 2, p. 134 (diary entry of 2 February 1965).

102. Georgii Grechko, *Start v neizvestnost'* (Moscow: Pravda, 1989), chap. 2.

103. Valentin Lebedev, "U nas velikaia strana. Reshat' ee problemy predstoit novomu pokoleniiu," *Osnova* (Naro-Fominsk), no. 26 (28 May 2004), *http://epizodsspace.testpilot.ru/bibl/intervy/lebedev1.html* (accessed 21 April 2005).

a cosmonaut's time was spent on really productive work." He concluded that the cosmonaut profession as such did not exist and that, at present, piloted flights were "entirely unnecessary."[104]

Konstantin Feoktistov proposed to solve the problem of inefficiency of human spaceflight though automation. "A man assigned to cope only with control functions is an unjustifiable luxury," he argued. "No craft is designed to carry dead weight. It must have a payload that performs a kind of useful work. This can be, for example, research." He proposed to make spacecraft control "simple and executable without high skills and during a minimum time" to allow scientists and engineers to fly space missions. "Every operation that can be automated on board a spaceship should be automated," concluded Feoktistov.[105] Boris Chertok similarly viewed automation as the way to free up the crew from routine functions: "Taken the high degree of automation on Vostok, an even higher degree on Zenit, and totally marvelous automation on future generations of spacecraft, the human on board must engage in research, reconnaissance, and experiments."[106] Feoktistov argued that valuable scientific data could be obtained only if scientists were included in space crews. "Scientists can develop their own experimental agenda, prepare their own instruments and equipment . . . . Cosmonauts [who lack scientific training] do not have this expertise. They are trained for specific mechanical operations: to turn something on, to switch something off, to monitor equipment, etc. If scientists come to space, scientific research would be more productive."[107] Long debates over the question whether scientists should be allowed on board were resolved in favor of a "professional cosmonaut," an engineer or a pilot, who would receive some scientific training and conduct experiments on board in consultation with scientists on the ground. The most the scientists were able to achieve was the privilege of direct communication with the cosmonauts in orbit.[108]

The problem of professional identity of the cosmonaut—a pilot, an engineer, or a scientist—proved inextricably connected with the question of on-board automation. If the first cosmonaut pilots tried to wrestle control functions from the machine, later on, cosmonaut researchers preferred to delegate equipment service functions to automatic systems to free up their own time for experiments and observations. As Valentin Lebedev put it, "Man is not an appendix to a machine. Man is not made for the flight, but the flight is made for man."[109]

---

104. Vasilii Mishin, "I Contend That There Is No Cosmonaut Profession" (English title), *Nezavisimaya gazeta* (13 April 1993), p. 6 (translation, JPRS-USP-93-002, 18 May 1993, p. 28).

105. Quoted in Viktor D. Pekelis, *Cybernetic Medley*, trans. Oleg Sapunov (Moscow: Mir, 1986), p. 287.

106. Chertok, *Rakety i liudi*, vol. 3, p. 242.

107. Konstantin Feoktistov, "'Aliaska' v kosmose," *Voronezhskie vesti*, no. 27 (2 July 2003), *http://epizodsspace.testpilot.ru/bibl/intervy/feoktistov3.html* (accessed 21 April 2005).

108. Eliseev, *Zhizn'*, pp. 172–173.

109. Lebedev, "U nas velikaia strana."

AUTOMATION IN CONTEXT

This brief overview of human-machine issues in the Soviet space program indicates that instead of the binary opposition of manual versus automatic control, we encounter complex human-machine systems, in which both humans and machines depend on one another; manual and automatic functions are not necessarily fixed, but may be redefined during the flight, and human-machine interaction on board becomes part of a vast remote-control network. "Automatic" control operations have some degree of human input, and "manual" control is always mediated by technology. Determining how these lines are negotiated in specific instances provides a glimpse into the internal politics and professional cultures within the space program.

On-board automation appeared as both an instrument and a product of local politics in the Soviet space program. The debates over the proper degree of automation were tied to the definition of cosmonauts' skills as either pilots or engineers. Here, technology, professional identity, and social status were closely intertwined. Soviet cosmonauts were "designed" as part of a larger technological system; their height and weight were strictly regulated, and their actions were thoroughly programmed. Soviet space politics, one might say, was inscribed on the cosmonauts' bodies and minds, as they had to fit, both physically and mentally, into their spaceships.

The existing historiography largely interprets the Soviet approach to human-machine issues as complete reliance on automation. I believe this view misses several important aspects of the story. First, it downplays the intensity of internal debates over the role of the cosmonaut on board. Engineers with their technical notions of reliability, cosmonauts with their piloting aspirations, human engineering specialists with their formulas for optimal division of function between human and machine, industry executives with their aversion to risk-taking, political leaders with their sober calculations of political gains and risks—all these groups had their input in these disputes. The Soviet approach to on-board automation did not appear to have been predetermined; it was developed, refined, and often reshaped in the course of these debates.

The Soviet approach to automation was never fixed; it evolved over time, from the fully automated equipment of Vostok to the semiautomatic analogue control loops of Soyuz to the digital systems of later generations of Soyuz. The role of the cosmonaut also changed, from the equipment monitor and backup on Vostok to the versatile technician on Soyuz to a systems integrator on later missions.

The Soviet approach also changed across various space projects running in parallel. In the late 1960s, while Soyuz was still largely controlled by on-board automatics or by ground operators, the Soviet lunar ships were designed to give the crews a much higher level of autonomy and control over their missions.

The Soviet approach was also flexible in another sense: the division of function between human and machine was not fixed, but was often renegotiated during the flight. Ground flight controllers played a crucial role in deciding whether the crew would be allowed to assume manual control. It is important, therefore, to examine not just the division of technical functions, but also the division of authority between the human on the ground and the human on board.

This analysis suggests that a human-machine system is not a simple dot on a straight line between total automation and complete manual control. This system is not defined by a simple numerical subdivision of function between human and machine. The efficiency of a human-machine system depends on the degree of integration of the human into the technological system, including its social infrastructure. Some space missions failed not because the range of manual functions was too narrow, but because the cosmonauts did not have the authority to use specific functions or because they were not "in the loop" for a timely receipt of crucial information. The efficiency of a human-machine system depends on whether the human in the system can play a truly human role, to have both the authority and the responsibility for decision-making. If a cosmonaut is trained to be a perfect automaton, his nominal role may increase, but this would be achieved at the cost of losing his unique human quality—not to act like a machine.

DIRECTIONS FOR FURTHER RESEARCH

Human-machine issues in the Soviet space program touch upon three large areas of historiography: 1) social history of automation, 2) sociopolitical and cultural history of the Soviet Union, and 3) comparative studies of the American and Soviet space programs.

In the history of technology, automation has traditionally been viewed as a technological implementation of management control resulting in workers' de-skilling and disempowerment.[110] A study of automation in the Soviet space program reveals a more complex story, in which cosmonauts do not simply lose their piloting skills, but adapt to the evolving technological system, making themselves indispensable in emergency situations. A third element—the ground controllers—also enters the equation, reframing the automation issue:

---

110. See David Noble, "Social Choice in Machine Design: The Case of Automatically Controlled Machine Tools," in *The Social Shaping of Technology*, ed. Donald MacKenzie and Judy Wajcman (Buckingham, U.K.; Philadelphia: Open University Press, 1985), pp. 161–176.

instead of a simple binary choice of automatic versus human control, one faces a complex organization in a network of multiple remote-control interactions, mediated by both humans and machines. A study of human–machine issues may provide a new framework for analyzing the social aspects of automation in complex technological systems.

Political historians of the Soviet Union have placed the space program in a larger political context, stressing the growing role of technocracy during the Cold War on both sides of the Iron Curtain.[111] Cultural historians have recently focused on the formation of cultural norms and Bolshevik identity in various periods of Soviet history.[112] The debates over human–machine issues provide a window into the cultural norms and identity of Soviet engineers and cosmonauts during the Cold War. Further studies could identify different political and cultural trends within the broad category of "technical intelligentsia," the backbone of Soviet technocracy; examine the interplay of engineers' and pilots' cultures in the cosmonaut profession; and also explore the tensions between the popular cultural image of the cosmonaut and the cosmonauts' own professional identity.[113]

Comparing the American and Soviet space programs through the prism of automation would help challenge the stereotype of fixed "national styles" in engineering. David Mindell's study of human–machine issues in the U.S. space program provides a thorough analysis of the internal debates between American pilots and space engineers.[114] In both the American and the Soviet cases, different approaches to automation are not predetermined, but emerge out of local negotiations, contingent on the range of available technological alternatives, space policy priorities, and specific configurations of power. What is often perceived as a "natural" technological choice emerges as a historically contingent product of political, socioeconomic, and cultural forces.

After the successful circumlunar mission of Apollo 8, Nikolai Kamanin wrote in his private diary that this flight had confirmed "the primary role of

---

111. See Andrew John Aldrin, "Innovation, the Scientists and the State: Programmatic Innovation and the Creation of the Soviet Space Program" (Ph.D. diss., University of California, Los Angeles, 1996); Barry, "The Missile Design Bureaux"; Walter A. McDougall, . . . *The Heavens and the Earth: A Political History of the Space Age* (New York: Basic Books, 1985).

112. David Hoffmann, *Stalinist Values: The Cultural Norms of Soviet Modernity, 1917–1941* (Ithaca: Cornell University Press, 2003); Oleg Kharkhordin, *The Collective and the Individual in Russia: A Study of Practices*, Studies on the History of Society and Culture, no. 32 (Berkeley and Los Angeles: University of California Press, 1999); Stephen Kotkin, *Magnetic Mountain: Stalinism as a Civilization* (Berkeley: University of California Press, 1995).

113. Two recent studies have adopted a cultural approach: Cathleen Lewis has explored the interplay between the ceremonial openness of Soviet space-related public rituals and the technical secrecy surrounding the investigation of space accidents; Andrew Jenks has examined the connections between the "myth" or "cult" of Yuri Gagarin and the Soviet visions of modernity.

114. See Mindell's article in this volume.

the spacecraft crew in such experiments. Automata can be a hundred times more perfect than man, but they can never replace him"—particularly, stressed Kamanin, in the human space race. "From a larger perspective, our designers are probably right in their intention to create fully automated piloted spaceships," he admitted. "Perhaps in the future, when communism triumphs over the entire planet, people will fly into space on such ships. But in our time one must not forget about the severe struggle between two opposing ideologies."[115] For Kamanin, the human role on board was the central issue of the space race, and the space race a central issue of the Cold War. A challenge for historians is to use analysis of human-machine issues in spaceflight as an entry point into larger questions of modern automation, Cold War, and space history.

---

115. Kamanin, *Skrytyi kosmos*, vol. 3, p. 348 (diary entry of 28 December 1968).

CHAPTER 5

HUMAN AND MACHINE IN
THE HISTORY OF SPACEFLIGHT

David A. Mindell

Astronaut Michael Collins, who orbited the Moon on Apollo 11, remembered
being inspired as a young man by the dashing figure of the barnstormer
pilot Roscoe Turner. "Roscoe had flown with a waxed mustache and a pet
lion named Gilmore," Collins remembered wistfully; "we flew with a rule
book, a slide rule, and a computer." Before being selected for the project that
would change his life and the world, Collins remembered feeling caught
between "the colorful past I knew I had missed and the complex future I did
not know was coming."[1] Collins captures an aspect of the history of spaceflight
little attended to by historians: the relationship between human and machine.
In two sentences, he helps us understand spaceflight and place it within 20th-
century American history and the history of technology.

Roscoe Turner's career peaked just a few decades before Collins's, but the
two seemed worlds apart. Turner, dubbed "Aviation's Master Showman," stunted
and barnstormed his way from rural America into Hollywood in the 1920s and
1930s. He had little training and even less formal education. Yet he self-fashioned
himself as a colorful character, sporting a waxed mustache and a made-up uniform
from a nonexistent military in which he never served. He was married in the
cockpit of his Curtiss Jenny and flew his giant Sikorsky S-29 airplane, dressed up
as a German bomber, in Howard Hughes's film *Hell's Angels*. As Collins noted,
Turner, under the sponsorship of the Gilmore oil company, flew with his pet lion
of the same name. Turner embodied the showy, excited world of aviation in its
"golden age" of transition from dangerous curiosity to commercial service.[2]

This was the world that inspired Collins to enter aviation, but by the time
he had arrived professionally, a great deal had changed. Nearly all astronauts
had college degrees in engineering, some had graduate degrees, and they had
served as test pilots. The technology had changed as well, from simple biplanes

---

1. Michael Collins, *Carrying the Fire: an Astronaut's Journeys* (New York: Farrar, Straus and Giroux,
1974), pp. 16–17.
2. Carroll V. Glines, *Roscoe Turner: Aviation's Master Showman* (Washington, DC: Smithsonian
Institution Press, 1995).

to the complex, high-performance jets Collins had flown. Collins contrasts Turner's pet lion with his "rule book, a slide rule, and a computer." No longer was aviation a world of display and reckless adventure. No longer was the pilot the only master of his craft. Now he shared his authority with flight rules, calculations, and, increasingly in the 1950s, automatic flight controls and computers (not to mention controllers on the ground). At the start of the space program, it seemed to Collins that the world was becoming bureaucratic, technical, and quantitative, with some loss of the pilot's "white scarf" image.

Collins's comments serve as a starting point for examining this critical issue in the history of spaceflight: the relationship between humans and machines.

BETWEEN HUMAN AND MACHINE

Human versus machine—it is not a new story. Indeed, it is one of the great narratives of the industrial world. American history and culture are replete with human-machine conflicts and comparisons. In the Civil War, the crew of the ironclad warship *Monitor* thought themselves well protected by iron armor, but that mechanical contrivances diminished the glory and heroism of their performance in combat.[3] The mythical John Henry won a race with a steam drill at the cost of his life. Factory workers complained that mechanical assembly lines and Frederick Winslow Taylor's "Scientific Management" turned them into unthinking automatons. New combinations of human and machine appeared in the 20th century, from the robots of Fritz Lang's silent film classic, *Metropolis*, to the gas masks and artificial limbs of World War I. Aviation, the technology born with the new century, celebrated the human-machine relationship as never before. Perhaps the most significant of the Wright brothers' innovations was their recognition that an airplane was not a stately ship to be guided by a detached human hand, but an active beast, controlled by an intensely focused, skilled human pilot.[4]

From these diverse histories and technologies, we can distill a few fundamental threads. A good place to begin is the idea of *skill*. Skill is a common enough notion in everyday life, but also a key to understanding the human-machine relationship. On one hand, skill is highly personal—it is practical knowledge; it implies a certain amount of cleverness, perhaps expertise, and we often think about it as residing in our bodies, particularly our hands (e.g.,

---

3. David A. Mindell, *War, Technology, and Experience aboard the USS Monitor* (Baltimore, MD: Johns Hopkins, 2000).

4. David A. Mindell, *Between Human and Machine: Feedback, Control, and Computing Before Cybernetics* (Baltimore: Johns Hopkins, 2002); Anson Rabinbach, *The Human Motor: Energy, Fatigue, and The Origins of Modernity* (New York: Basic Books, 1990); Thomas Crouch, *The Bishop's Boys: A Life of Wilbur and Orville Wright* (New York: W. W. Norton, 1989).

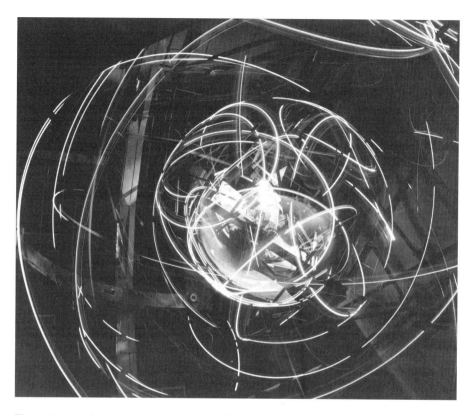

The epitome of human and machine interfaces, this device was formally known as the MASTIF, or Multiple Axis Space Test Inertia Facility, and was located in the Altitude Wind Tunnel in 1959. It was built at Lewis Research Center, now John H. Glenn Research Center, in Cleveland, Ohio, and was designed to train astronauts to regain control of a tumbling spacecraft. *(NASA photo no. C1959-52233)*

"manual skills"). On the other hand, skill is also deeply social—it is not inborn, but acquired, as distinct from an innate quality like talent. Skill implies *training*—the time and effort to learn and master a skill, often with the help of another person. Skill has a social dimension: it garners respect, and the more skill you are perceived to have, the more prestige you seem to earn.

Skilled workers include surgeons, carpenters, and waiters. Obviously, not all skills are equal. Some are more respected than others, and hence there tend to be social and economic differences between their practitioners. Skill also sets people apart. The word itself comes from an Old Norse word meaning *distinction* or *difference*, ideas that remain integral to today's meaning.[5] For any

5. *Oxford English Dictionary*, etymology for *skill*.

skill, some people have it and some people don't. The very notion of skill implies a social group, possibly even an elite. When people with common skills come together, they often form societies, set standards, create and uphold traditions. They also police the boundaries of who is in and who is out, and for high-status skills, this makes them *professions*.[6] Most would agree that surgeons are professionals, but are carpenters, or waiters?

Skills often develop in relation to particular technologies: a blacksmith's skills, for example, are only valuable within a particular mode of production. As technologies change, the skills change as well, sometimes generating social conflicts. For example, as numerically controlled machine tools were developed in the 1950s, some saw them as eliminating the need for skilled machinists. Indeed, the skills required of a machinist did change—and began to require intimacy with numbers and computers as much as with metals and cutting speeds, which favored certain people, or groups of people, over others. The important thing to realize is that technology does not just "change" of its own accord—it is changed by particular people for particular reasons at particular times. In the 20th century, those people were increasingly engineers, who sought to build more "skill" into machines and hence to reduce the requirements on the people who ran the machines, the operators. When those changes derived form computers, they became known as "automation," and they went hand in hand with social changes. Historians of technology, by and large, have focused on ideas of *de-skilling* without attending to the contingent nature of the skills themselves.[7]

In an earlier book, *Between Human and Machine*, I examined human-machine relationships surrounding technologies of control in the first half of the 20th century.[8] During that time, engineers began to understand the idea of the *feedback loop* and began to study the skills of human operators according to new principles of *control theory*. They saw that humans operated machines much like automatic regulators or thermostats—sensing an "error" between the "actual" state of the machine and its "desired" state and directing the machine to close the gap between the two. In the course of that work, it became clear that aviation had always been a rich site of human-machine interaction, and the Apollo landings were in some sense the culmination of the mid-20th-century history of feedback, control, and computing.

Consider the history of instrument flying. When pilots were flying in clouds, they lost the cues from the outside world that allowed them to keep

---

6. Andrew D. Abbott, *The System of Professions: An Essay on the Division of Expert Labor* (Chicago: University of Chicago Press, 1988).

7. David F. Noble, *Forces of Production: A Social History of Industrial Automation* (New York: Alfred A. Knopf, 1984).

8. Mindell, *Between Human and Machine*.

an airplane level, hence their feedback loops broke down (they went unstable). New instruments like directional gyros and artificial horizons replaced the natural cues with technological substitutes, and with some training, the pilots could use their indications as feedback and "fly blind." Of course, a machine could also close this feedback loop, and by no coincidence, the advent of automatic pilots and instrument flying occurred in the same period. Some pilots initially objected to the decline of pilots' "seat of the pants" or "intuitive" flying skills, and instrument flying remains today a compromise between pilot control and ground control. The new technology did change the nature of piloting, but it also allowed pilots new professional prestige and the ability to fly through bad weather on long-range commercial routes. Skill, prestige, training, professionalism, and new technologies are tightly coupled; change one element, and the others evolve as well, though not necessarily in predicable ways.

During World War II, the engineering of feedback control systems led to the emergence of digital computing and its associated sciences. The idea of a "computer" as a general-purpose information system emerged from a number of applications (like radar and gunfire control) which considered human operators and control systems as mathematical calculation. The post–World War II rise of Norbert Wiener's "cybernetics" captured the sense that control and communications were intimately linked with the characteristics of human operators and emphasized the blurring boundaries between human and machine.[9] Wiener's conception, however, elaborated on developments in a variety of engineering fields, particularly aviation.

From its origins, aviation was centrally concerned with the relationship of human and machine. The Wright brothers, by emphasizing the importance of control, created not simply a flying machine, but its human counterpart—the skilled pilot. From the moment Wilbur first flew, this new professional was born.[10] But what kind of person would a pilot be? A variety of models were proposed: soldier, athlete, adventurer, explorer, factory worker, engineer, ship's captain.[11] Which dominated at any given time depended on how the machines were designed, who piloted them, and their social position.

Under a project sponsored by the Sloan Foundation and the Dibner Institute in the late 1990s, a group of students and I began collecting documents, conducting interviews, and defining the boundaries of these issues in manned spaceflight. That project also brought on Slava Gerovitch and supported his early work on the Soviet program that he presents so ably in this volume.

---

9. Norbert Wiener *Cybernetics; or, Control and Communication in the Animal and the Machine*, 2nd ed. (Cambridge, MA: MIT Press, 1961).

10. Crouch, *The Bishop's Boys.*

11. Robert Wohl, *A Passion for Wings: Aviation and the Western Imagination, 1908–1918* (New Haven: Yale University Press, 1994).

Building on the history in *Between Human and Machine*, I began by asking a series of questions about professional identity and its relationship to machinery in human spaceflight:

- Who is in control (human in the cockpit, machine in the cockpit, human on the ground)?

- Who is the pilot/astronaut (i.e., social background and status)?

- Who or what else is in the loop (e.g., copilots, ground controllers, instruments, computers)?

- What is his (or her) training/education (military, university, vocational, etc.)?

- What skills are required (e.g., manual skills, mathematics, design, physical strength)?

- How are they trained (e.g., classrooms, flight training, simulators, experience)?

- How are tradeoffs made between manual and automated tasks?

- Who is responsible for a successful flight, the astronauts or the engineers and controllers on the ground?

- Who is blamed for failure?

- What is the role of computers and automation aboard the spacecraft (automatic pilot, monitoring for failure, primary flight controls)?

- Who is at risk?

- What level of prestige do the astronauts enjoy (e.g., national heroes versus faceless operatives)?

Some of these questions repeatedly arise in discussions and debates about human spaceflight. Others reappear throughout the history but are rarely addressed explicitly. Together, they allow us to make connections in the history of human spaceflight that have not previously been made, to understand historical dynamics, and to open up new research areas and ask new questions. Examining the human-machine relationship in human spaceflight enables us to move beyond the dichotomies of "robotic versus human" to better understand the nature of the human role when it is present, and its interaction with, rather than replacement by, machinery. It also allows us to integrate a variety of historical perspectives into narratives of spaceflight: risk, safety, automation, social relationships, project politics, public perception, gender roles, and cultural iconography.

THE CASE OF APOLLO

A full exploration of human-machine relationships in spaceflight is outside the scope of this paper. Rather, I look at the example of Apollo to support my claim for the larger historical importance of the theme. As defining technological moments of the 20th century, the Moon landings embodied the cooperation of human and machine and the tensions that cooperation embodies. As Michael Collins articulated, the individuals involved had experienced radically different eras in the history of aviation and spaceflight in close proximity (a mere four decades from Lindbergh's flight to Apollo 11). The project spanned the transition from analogue to digital computers, from crude simulators to full virtual environments, from analogue cockpits to digital fly-by-wire. Apollo also provides a unique case, because it combines technical complexity and accomplishment with political and cultural significance—hence we can trace the importance of the human operator from the White House into the machine code, from the public's TV screens to the astronaut's displays. While Apollo exemplifies these issues, human-machine relationships resonate throughout the history of spaceflight, from early science fiction to the new Mars rovers.

Ironically, the human-machine relationship in Apollo has been largely ignored by historians, although much of the existing literature offers tantalizing clues for a larger picture. Existing histories of Apollo are nearly all project-oriented—they begin at Apollo's beginning and end at its end. Other than in memoirs as personal background, little is said about Apollo's connection to larger currents in the history of technology in the 20th century. Such narratives reinforce the project's self-image as something coherent in itself and apart from, outside of, contrary to, other forces in American culture. The histories that do provide context tend to be politically or culturally oriented and don't delve into the machines themselves, the people who built and operated them, or what they meant. Additionally, these histories, certainly the more recent ones, tend to be based on the familiar, public accounts of the Apollo program, or interviews with participants conducted many years afterward. Hence they tend to solidify the canonical narrative of the project around key themes and events: Kennedy's visionary decision, the frenetic engineering efforts, the heroism and skill of the astronauts, the tragic fire, the triumph of Apollo 11, the drama of Apollo 13, etc.[12]

Yet the human-machine relationship, even when synthesized from the existing literature, reveals a different view. From the beginning of Apollo, the

---

12. Two examples are Charles Murray and Catherine Bly Cox, *Apollo: The Race to the Moon* (New York: Simon & Schuster, 1989) and Andrew Chaikin, *Man on the Moon: The Voyages of the Apollo Astronauts* (New York: Viking, 1994).

relative importance of humans and machines was under debate. James Webb argued that the decision to go to the Moon "can and should not be made purely on the basis of technical matters," but rather on "social objectives" of putting people into space. He and Robert McNamara argued that "it is man, not merely machines, in space that captures the imagination of the world."[13] Presidential science adviser Jerome Wiesner famously opposed a manned lunar program because its scientific goals did not justify the cost. In a close reading of the debates leading up to Kennedy's decision, we see an implicit distinction between "exploration," which is manned, and "science," which has a higher prestige value among intellectuals but is best conducted remotely.[14]

Nevertheless, when the decision was made to go to the Moon, there would clearly be a significant human role. Kennedy's 1961 mission statement, "to send a man to the moon and return him safely to earth,"[15] was simple, focused, and included its own schedule. It was also impossible, by definition, to accomplish with a fully automated system. But what role would the astronauts play?

1. The Test Pilots

Apollo came after a decade when the human role in flight had been both celebrated and questioned. The Air Force had struggled with the advent of unmanned missiles to complement its beloved fighters and bombers. As a new elite profession emerged, that of the test pilot, airmen were questioning their own role in flight in general, and in spaceflight in particular. Even in the late 1950s, it was not clear who the new spacefarers would be, what skills they would require, and what social prestige (or derision) they might enjoy.

Tom Wolfe, of course, captured some of this anxiety in *The Right Stuff*. While not scholarly history, the book and subsequent film made sufficient impact in the public imagination that we should consider it here. Focusing on the Mercury program, Wolfe correctly identifies the roots of the astronaut culture in the flight-testing world centered on Edwards Air Force Base. He portrays test pilots as reckless risk-takers, cowboys who could not fit into the traditional professional molds for pilots and who made a living pushing aircraft to their limits, often at the cost of their lives. Perhaps some of them were, and they did place themselves at risk, but Wolfe's image misses the essential feature of the profession: although skilled craftsmen, intimate with the feel of their

---

13. Webb, quoted in John Logsdon, *The Decision to Go to the Moon: Project Apollo and the National Interest* (Cambridge, MA: MIT Press, 1970), pp. 90, 125.

14. Wiesner Committee, "Report to the President-Elect of the Ad Hoc Committee on Space," 10 January 1961, NASA Historical Reference Collection, Washington, DC.

15. John F. Kennedy, "Urgent National Needs," *Congressional Record—House* (25 May 1961), p. 8276 (text of speech can be found in the speech files, NASA Historical Reference Collection, Washington, DC).

aircraft, test pilots worked in a scientific mode. Their goal was to collect data. As the historian Richard Hallion has written, "A research airplane essentially uses the sky itself as a laboratory."[16] Increasingly over the course of the 20th century, what it meant to be a test pilot was not only one trained in flying airplanes, but also one trained in engineering.

Test pilots were always in close touch with controllers on the ground (a feature of flight testing carried to extremes in Apollo). Test pilots understood not only how an airplane flew, but also why it flew. Again to quote Michael Collins,

> A test pilot, more than any other type of aviator, must be objective. It is all right for a squadron pilot to fall in love with his airplane; it is all he has to fly, and he might just as well enjoy it because it has already been designed . . . . The test pilot cannot fall into this trap . . . he must carefully analyze the possible uses to which an airplane might be put and judge it accordingly.[17]

Note that in this passage, Collins emphasizes the judgment of the test pilot—the "pilot opinion," which he must provide as part of the research data. In addition to their cockpit skills, test pilots were also professional storytellers, experts at narrating and recounting their experiences in precise, formal language. Yet the hero of Wolfe's account is Chuck Yeager—an older breed, not college-educated, and without a career-long interest in flight engineering. Nevertheless, despite its limitations, *The Right Stuff* does draw attention to the relationships between machine control and professional identity that were woven throughout the Mercury program.

Looking more seriously at the test pilots' profession reveals even greater historical coherence within Apollo. Much of the time the test pilots flew new aircraft was spent evaluating "stability and control" and "flying qualities," two engineering areas that focused on the match between human and machine. Indeed, this area was pioneered by Robert Gilruth and his group at Langley, which subsequently formed the Space Task Group and the Manned Spacecraft Center (MSC).[18] The Society for Experimental Test Pilots (SETP) formed in 1955, and for the rest of the decade, the group concerned itself with the appropriate role of the pilot—at first in high-performance aircraft with computerized control systems, and then in the space program. One founding member of the SETP would go on to become an astronaut: Neil Armstrong.

---

16. Richard Hallion, *Test Pilots: The Frontiersmen of Flight*, rev. ed. (Washington, DC: Smithsonian Institution Press, 1991), pp. 101, 143.

17. Ibid., p. 238.

18. Renamed Johnson Space Center (JSC) in 1973.

2. Systems Thinking and the Role of the Human

The SETP crystallized the anxiety of pilots in general, especially as they faced the development of unmanned aircraft and ballistic missiles. These technologies not only emerged outside the culture of piloting, they sprang from a new group of engineers: the systems men. Several authors have written of the conflict of cultures that occurred in Apollo between the aeronautics-oriented culture of Langley and Edwards and the systems-oriented culture of the West Coast contractors, embodied in managers like Joe Shea.[19] Looking more deeply at the roots of systems thinking, however, helps connect the project to broader currents and clarifies the alternate view to the tight human-machine coupling advocated by the pilots.

World War II coalesced systems thinking in several arenas. In response to technical problems of radar and automatic gunfire control, engineers began to see that all components of a system needed to be understood together, rather than as glued-together components. Engineers now conceptualized their machines as integrated systems with feedbacks and dynamics, where the behavior of each part helped determine the behavior of the whole.

By 1950, these ideas and techniques began the self-conscious era of systems thinking. The *Oxford English Dictionary* shows that uses of the term *system* exploded after 1950, including *systems engineering, systems analysis, systems dynamics, general systems theory*, and a host of others. Each field had its own innovators, its own emphasis, and its own home institutions and professions, but they shared common concerns with feedback, dynamics, flows, block diagrams, human-machine interaction, signals, simulation, and the exciting new possibilities of computers.[20]

The management aspects of systems engineering formalized in the mid-1950s, when the Air Force stretched its resources to quickly build an intercontinental ballistic missile (ICBM). In the Atlas missile project, management began to move beyond the model that had dominated the aviation industry for decades. Aircraft had always been composed of large numbers of components from a variety of subcontractors, coordinated by the prime contractor, who built the airframe. With a project like Atlas, dynamics, interconnection, and coordination became the dominant aspects of the project, so airframe companies, with their emphasis on structures and manufacturing, lost their central role. Rather, engineers with management experience, comfort with mathematical abstraction,

---

19. Murray and Cox, *Apollo: The Race to the Moon*; Stephen Johnson, *The Secret of Apollo: Systems Management in American and European Space Programs* (Baltimore: Johns Hopkins, 2002); Howard E. McCurdy, *Inside NASA: High Technology and Organizational Change in the U.S. Space Program* (Baltimore: Johns Hopkins, 1993).

20. Louis B. Ridenour, *Radar System Engineering*, vol. 1 of *Radiation Laboratory Series* (New York: McGraw Hill, 1948); Harry Goode and Robert Machol, *Systems Engineering: An Introduction to the Design of Large-scale Systems* (New York: McGraw Hill, 1947).

and insight into dynamics and control coordinated the project. The technical change entailed a social shift; as historian Thomas P. Hughes has written, "the airframe was [now] merely a platform to carry complex, electronic guidance and fire control systems."[21]

Innovators in Cold War systems engineering had their roots at General Electric and AT&T, via the aviation industry. Simon Ramo had cut his teeth at GE and Hughes Aircraft and earned a Ph.D. at Caltech. His friend Dean Wooldridge came out of Bell Labs. In 1953, the two left Hughes Aircraft Corporation to found a systems engineering contractor, Ramo-Wooldridge, that soon became the TRW Corporation and did systems engineering for the Atlas project. Together with the Air Force's Western Development Division, they coordinated contractors and scheduling and oversaw the project's integration. The Navy had a similar project to build a ballistic-missile-firing submarine named Polaris. Here the Navy's "Special Projects Office" performed the systems engineering function.[22]

Ramo became a promoter of systems engineering, which he defined as "the design of the whole from the design of the parts." As Ramo wrote, "Systems engineering is inherently interdisciplinary because its function is to integrate the specialized separate pieces of a complex of apparatus and people—the system—into a harmonious ensemble that optimally achieves the desired end."[23] Atlas included a system of materials, logistics, computers, and ground support, and the missile itself was a system.

In Atlas, Polaris, and other large projects of the 1950s, systems engineering meant coordinating and controlling a variety of technical and organizational elements, from contract specifications to control systems, from computer simulations to deployment logistics. The approaches were diverse, but they shared a common set of assumptions about how the world might be understood in abstract, quantitative terms, and modeled with a series of feedbacks, flows, and dynamics.

Computers, both analogue and digital, figured prominently in the image and the practice of these systems sciences. They could simulate systems and make predictions about the system's behavior in an uncertain environment. Social systems could be modeled with similar techniques as technical systems. Both the computer and the analysts themselves carried the prestige and authority of science: providing dispassionate, expert advice free of political influence. For the

21. Thomas P. Hughes, *Rescuing Prometheus* (New York: Pantheon Books, 1998).

22. Harvey Sapolsky, *The Polaris System Development: Bureaucratic and Programmatic Success in Government* (Cambridge, MA: Harvard University Press, 1972); Benjamin Pinney, "Projects, Management, and Protean Times: Engineering Enterprise in the United States, 1870–1960" (Ph.D. diss., MIT, 2001).

23. For a history of systems thinking in the Atlas project, see Hughes, *Rescuing Prometheus*, chap. 3. Simon Ramo is quoted on p. 67.

strategy to work, the system engineer required a certain amount of authority, a fact that was not lost on the participants. They sold systems engineering as an authoritative, scientific way to transcend "politics" (whether public or military-industrial) with the outside neutrality of the expert. Systems engineering thus elevated the "systems men" to a new level of prestige, creating a new niche for engineers as educated managers of large projects and budgets.

3. X-15 Human and Machine

The successes of Atlas and Polaris gave the systems experts, their companies, and their worldview credibility with the armed services. Furthermore, the expertise they built up in rocketry meant they would be intimately involved in any efforts to send humans into space. For the pilots, however, the systems men could represent a threat—they had engineered a fleet of Air Force weapons that had no pilots at all, and their abstract, analytical approach to engineering could seem to crowd out the "human factor." These issues came to the fore as the test pilots began to contemplate spaceflight.

When the pilots of the SETP reacted to the rise of unmanned missiles, they also reacted to the rise of the social group that built them. In 1960, an author in the SETP *Proceedings* derided

> the great millennium of concentrated effort to design man out of the cockpit to make room for bigger and better "black boxes." There was much gnashing of teeth and waving of arms but alas, the day of the "icy B.M." was upon us. No one wanted the pilot around.[24]

The "icy B.M." is a wonderful triple entendre, referring to an ICBM, the computers of IBM, and a scatological reference to a missile.

One SETP test pilot actually argued that the ICBM was a transitional technology, soon to be replaced when technology allowed humans to pilot the rockets: "The era of the large intercontinental ballistic missile is merely a phase the duration of which is a matter of speculation but the demise of which is nonetheless certain."[25] Indeed, the Air Force had initiated the X-20 "Dyna-Soar" program, a kind of manned orbital space bomber to orbit the Earth. Air Force publicity for the X-20 repeatedly emphasized the man in the loop and that reentry could only be accomplished as a product of human skill. Despite the presence of numerous new technologies, the Air Force declared, "In the end, it takes the cool hand of a skilled pilot to bring his glider in for a

---

24. W. T. Armstrong, "Where do we go from here?" *Cockpit* 4 (May 1965): 7.
25. A. W. Blackburn, "Flight Testing in the Space Age," *SETP Quarterly Review* 7, no. 3 (fall 1957): 17, 10–11.

conventional landing . . . this Dyna Soar project puts an emphasis on the pilot, on the *man*"[26] (emphasis original).

While Dyna–Soar was eventually canceled, another program emerged that sought to demonstrate the importance of human skill for manned spaceflight. The X–15 is of course the best-known of the famous X–planes, but when viewed through the lens of the human-machine relationship, the X–15 takes on great importance for Apollo. In addition to hypersonics, much of the purpose of the X–15 was to evaluate the human role in spaceflight, particularly for reentry, which was considered so dynamic and difficult that it required a human controller. A detailed exploration of these issues is outside the scope of this paper, but roughly half of the publications arising out of the X–15 related to control systems, the role of the pilot, or human-machine interfaces.[27] When an X–15 was donated to the Smithsonian, for example, the press release for the donation read, "One of the major goals of the program which has been most richly achieved was to explore the capabilities and limitations of the human pilot in an aerospace vehicle." And of course, the conclusion was that "the broad positive finding of the program is clear; the capability of the human pilot for sensing, judging, coping with the unexpected, and employing a fantastic variety of acquired skills remains undiminished in all of the key problem areas of aerospace flight."[28] For all of its contributions to hypersonics and related sciences, a major legacy of the X–15 is that of putting human pilots in space and ensuring them a place in the cockpit in future space missions. As it turned out, the skill of reentry was easily mastered, with the help of redundant automated systems. The pilot's primary function evolved to be a monitor, a systems manager, coordinating a variety of controls as much as directly controlling himself.

As a result of his work on the X–15, Neil Armstrong and colleagues conducted a series of simulations which showed that a human pilot could stabilize a multistage vehicle under manual control straight off the launchpad. The pilots saw the tests, and the data they produced, as critical support for the role of the human pilots in orbital operations. Armstrong concluded that the pilots should be allowed to fly the Saturn rocket off the launchpad. He and the simulation

---

26. U.S. Air Force, *This is Dyna-Soar*, film included in CD-ROM published with *Dyna-Soar: Hypersonic Strategic Weapons System*, ed. Robert Godwin (Burlington, Ontario: Apogee Books, 2003).

27. W. H. Stillwell, ed., *X-15 Research Results* (Washington, DC: NASA, 1965). The most complete and prominent example of these is Robert G. Nagel and Richard E. Smith, "An Evaluation of the Role of the Pilot and Redundant Emergency Systems in the X-15 Research Airplane," *SETP Newsletter* 6 (September–October 1962): 12. The SETP publication is a summary of the full study by the same author, "X-15 Pilot-in-the-loop and redundant/emergency systems evaluation," Technical Documentary Report No. 62-20, Air Force Flight Test Center, Edwards Air Force Base, CA, October 1962, NASA Dryden Archives L2-5-1D-3. For a personal account, see Milton O. Thompson, *At the Edge of Space: The X-15 Flight Program* (Washington, DC: Smithsonian Institution Press, 1992).

28. X-15 news release, Edwards Flight Research Center (FRC), 27 April 1969. Reprinted in Goodwin, *X-15 Mission Reports*, pp. 393–394.

engineers argued that pilots could adequately operate the simulation under high g forces—as long as they were provided with adequate information displays to guide their control. "As a passenger, he [the pilot] can be very expensive cargo; but as an integral part of the control loop of the vehicle, he might add materially to the reliability and flexibility of the launch maneuver." Citing the earlier work on flying qualities and aircraft stability, they acknowledged that "the piloting task for these vehicles is certainly more exacting than that of operational aircraft." The simulated rocket was inherently unstable, though just how unstable depended on the amount of fuel it contained and on the external environment. "There is no reason to assume that the pilot cannot control the launch of multistage vehicles . . . it appears to be highly desirable to initiate investigations of the use of the pilot in the control loop of the launch of Saturn boosters."[29]

Armstrong had done other similar tests as well—he flew an aircraft in such a way as to simulate the trajectory of an aborted launch in the Dyna-Soar. Milt Thompson participated in a similar series of trials designed to show that pilots could manually fly the Titan booster into orbit with the Dyna-Soar vehicle on top. "This was a very controversial issue," Thompson recalled; "the booster designers had been using automatic control and guidance systems from day one. In their minds it was the way to go."[30]

The role of the pilot in complex space missions was on the table: the pilots had already lost a battle with the advent of the ballistic missile, in their view little better than a dangerous, unpiloted drone. Would the giant space rockets then under construction be like ballistic missiles, taking a mere "payload" up for a ride, or human-guided machines, directed by keen eyes and hands that could aim it into orbit? Would the X-15 be the way of the future or a forgotten sidelight on a ballistic future?

In the end, they would not fly the rockets off the pad. They would not put the spacecraft into orbit. They would not point toward the Moon and fly there. They would not manually enter lunar orbit, and they would not fly the return to Earth or fly the reentry. These things were all accomplished by computers. What, then, would the astronauts do? They would, in conjunction with a computer, control docking in space, and the lunar landing, and they would monitor and engage various systems throughout the flight. These would be the tasks to showcase human performance and skill and make Apollo a human endeavor.

---

29. E. C. Holleman, N. A. Armstrong, and W. H. Andrews, "Utilization of the Pilot in the Launch and Injection of a Multistage Orbital Vehicle" (presented at the 28th annual meeting of the Institute for Aeronautical Sciences, New York, NY, January 1960); N. A. Armstrong and E. C. Holleman, "A Review of In-Flight Simulation Pertinent to Piloted Space Vehicles," North Atlantic Treaty Organization (NATO), Advisory Group for Aerospace Research and Development (AGARD) Report #403, July 1962.

30. Thompson, *At the Edge of Space*, p. 119.

The Apollo spacecraft would not be built by the people who built the capsules for Mercury and Gemini, but by North American Aviation and the engineering team that built the X-15. The first contract of the Apollo program, however, would not be for a giant rocket, nor for an exotic space vehicle, but for a guidance system and a digital computer. The contract went to the Instrumentation Laboratory at MIT, under the direction of aviation pioneer Charles Stark Draper. Draper's men and women spent the 1950s building guidance systems for nuclear missiles. They had built computers before, but only for automatic systems. They had never built a computer with an interface for a human user.

RETHINKING APOLLO

Using the lens of human-machine relationships, and their prior and subsequent histories, allows us to rethink Apollo and investigate new aspects of the famous project. Now we can consider Apollo through the lens of computing, through training, and through simulation. Each of these topics reveals a project different from the one in the traditional accounts, but one contiguous with larger historical phenomena and with the evolving human-machine relationships of subsequent decades.

In the end, it was not heroic astronauts alone who made the flights to the Moon. They shared their decisions with ground controllers, as well as a small group of software engineers who accompanied them in the form of computer programs that complemented the astronauts' every move. The computer design and the software then emerged to reflect a philosophy of automating the flights and aiding the pilots in critical functions and at critical moments, while not actually replacing them. In the end, the astronauts "flew" a very small part of the mission by hand, but that included the critical lunar landing. Even there, the astronauts flew the lander indirectly—their joystick actually controlled a software program, which then controlled the vehicle, what today we call fly-by-wire.

While the flight technology was being developed, NASA faced a problem: How do you teach astronauts to land on the Moon? How do you train people to do something that has never been done before? Training can be understood as developing the match between human and machine. Again, the human-machine relationship points us toward a much-neglected aspect of the history of spaceflight: simulation. Flight simulators had been built since the 1930s, but to teach pilots how to fly airplanes that already existed, under conditions that were well understood. For the X-15, engineers began building simulators for an airplane before it flew, before it was built, before it was even designed.[31] Apollo took those lessons to heart.

---

31. G. L. Waltman, *Black Magic and Gremlins: Analog Flight Simulations at NASA's Flight Research Center* (Washington, DC: NASA SP-2000-4520, 2000).

All of the human spaceflight missions of the United States require close human support from outside the spacecraft. Here is an overall view of the Mission Control Center (MCC) in Houston, Texas, during the Gemini 5 flight in 1965. Note the screen at the front of the MCC that is used to track the progress of the Gemini spacecraft. *(NASA photo no. S65-28660)*

Apollo simulated everything. There was a simulator for Moon walking, for picking up rocks, for escaping a fire on the launchpad. The critical simulators, however, replicated the spacecraft themselves, simulating not only the physics of their flight, but their internal workings as well. For months before the flight, the astronauts virtually lived inside these strange machines, flying to the Moon under a great variety of conditions, simulating every conceivable kind of failure. Of course, the simulators were built around computers, at first analogue and later digital. But the machines of the time could not replicate the subtle visual cues required for a perfect landing. Instead, NASA engineers built elaborate, finely painted replicas of the Moon and "flew" tiny cameras above the surface to provide accurate images of the Lunar Module's final approach (techniques to be replicated just a few years later in the making of George Lucas's *Star Wars*). Inside the simulated spacecraft, the astronauts used the real guidance computer, programmed with real programs, and became acclimated to their new environment. In the actual lunar landings, the astronauts frequently commented on the simulation, comparing their real experiences to those fabricated in the laboratory. A history of the use of simulation in the space program and its significance for future technology has yet to be written.

Not all simulators were equally virtual. One actually flew, using real gravity and flight dynamics to mimic the lunar lander. Early in the program, a group of NASA engineers who had worked on the X-15 thought up a vehicle that would use a special jet engine to cancel out five-sixths of the Earth's gravity, and would thus fly as though it were on the Moon, which had one-sixth g. The result was the Lunar Landing Research Vehicle, or LLRV, nicknamed "the flying bedstead" because of its extraordinarily strange appearance (later renamed the LLTV, with "training" replacing "research"). In addition to its jet engine, it used a variety of steam jets to control attitude and position, so when it flew, it hissed white jets of steam and whistled like a calliope. The vehicle was complex, unruly, and dangerous. Three of the six built had spectacular crashes; one almost killed Neil Armstrong before his famous flight. NASA wanted to cancel the program, thinking it too risky to the precious astronauts. But when Armstrong returned from the Moon, he insisted that the vehicles remain in use, for they provided the closest approximation of the actual Moon landing. The "flying simulator" further blurred the boundary between real and virtual flight and proved a valuable rehearsal for the human-machine system that would land on the Moon.[32]

Simulation is but one arena where focusing on the human-machine relationship sheds new light on the history. Numerous decisions in Apollo concerned the human-machine relationship in some degree. The famous LOR decision placed great emphasis on human skill in docking and rendezvous. The decision to include three astronauts had to do with how human roles would be allocated. The three were originally dubbed "Pilot," "Co-pilot," and "Systems Engineer" but were later changed to "Commander," "Command Module Pilot," and "Lunar Module Pilot," ensuring that all would be "pilots" even though the "Lunar Module Pilot" would only fly the craft as a backup (and did not train in the LLRV). Decisions about in-flight maintenance and repair traded off human repair skills against mechanical and electronic reliability. Critical functions like navigation could be handled entirely within the capsule but ended up being provided largely by ground stations.

During the actual missions, several key events brought the human-machine issues to the forefront. The "program alarm" in the final minutes of the Apollo landing required human intervention, and the landing ended under manual control, with great success. The incident set off a behind-the-scenes debate about who was to blame. The press reported it as a bug in the

---

32. Christian Gelzer, ed., "LLRV History" (unpublished manuscript, NASA Dryden History Office, 2004). See also "Minutes of Meeting, Flight Review Board, Lunar Landing Training Vehicle," University of Houston, Clear Lake, 12 January 1970, Apollo Chronological File, NASA Historical Reference Collection, for a detailed discussion of why the astronauts found the LLTV valuable.

program (a concept soon to enter popular discourse). MIT engineers pointed out that the astronauts had forgotten to turn off a piece of equipment that was feeding extraneous data to the computer and causing it to overload. Others could point to a problem with procedures that did not correctly direct the astronauts. NASA, by contrast, narrated the landing as the victory of a skilled human operator over fallible automation—a result that highlighted the heroic goals of the program. Who was at fault is less important than the terms of the debate, as the tensions between humans and automated systems refused to go away, even in the triumphant moments of the program.

Other events in the remaining Apollo flights continued to highlight the tensions between the computer, its software, and its human operators. During Apollo 8, astronaut Jim Lovell mistakenly pushed a button that erased the computer's memory—committing an error that NASA swore would never happen. In Apollo 12, the spacecraft was struck by lightning soon after liftoff, causing the system to reboot (imagine if they were running Microsoft!). During Apollo 14, the computer was reprogrammed in flight to help save the astronauts from a sticky abort button. Overall, the computers performed extremely well, and the astronauts spent as much (or more) time on the missions monitoring and managing the computer as they did actually "flying" the spacecraft. Yet on every single landing, for one reason or another, the pilots overrode the automatic systems and landed with their hands on the stick. Manual control of the landings allowed NASA and the public to see the flights as a human accomplishment rather than an automated one.

AN AGENDA FOR RESEARCH

This essay, of course, cannot provide an exhaustive history of the human-machine issues that came to play in Apollo. It merely makes the case that a series of questions about human–machine interaction in the history of spaceflight can open up new research avenues into what some might think is a well-worn historical topic, and indeed these are the kinds of questions I'm currently exploring for a book on Apollo. Research directions include a close reading of the astronaut memoirs, building on Michael Collins's revealing comments, to see how they narrated their own relationships to the computers and how they recalled the human-machine issues in retrospect. I'm also looking carefully at the decisions about how much to automate the landings, how that automation was actually implemented, and at the various parties (engineers, astronauts, managers, etc.) who engaged in the process. Analyzing the actual operations of the flights sheds light on how the human operators performed and what they actually did during the flights.

Of course, these issues extend well beyond Apollo. One can ask about the early planning and decisions on the Space Shuttle and what role pilots played

The Space Shuttle cannot be flown without a human pilot; it is the first piloted spacecraft of the United States that has no capability for automated flight. This fisheye view of the Space Shuttle *Atlantis* is seen from the Russian *Mir* space station during the STS-71 mission. *(NASA photo no. STS071-741-004)*

in developing a spacecraft with a "piloted" reentry. In light of their lost bid to manually fly the Saturn rocket off the pad, the Shuttle decision appears as a victory where pilots again assert their authority and express their love for winged aircraft. Despite the X-15's initial emphasis on the skill required for reentry, only one Shuttle flight has been flown manually from reentry: flight number 2 of *Columbia*, flown by former X-15 pilot Joe Engle from Mach 25 to the ground. Despite the presence of automated landing systems, every single Shuttle flight has ended with a manual landing.

The human–machine relationship, as a meeting point for the social and technical aspects of a system, provides access to a variety of other aspects of space history that are otherwise difficult to integrate. The iconic role of astronauts as American heroes was critically dependent on their roles (real and perceived) in actual piloting of the missions. We can study how such public and political imperatives were incorporated, along with technical considerations, into the actual design of control systems and, conversely, how the technical characteristics of those systems shaped and constrained the public imagery (there was a good technical argument for not allowing the astronauts to fly the Saturn off the pad).

As Slava Gerovitch has explored in his essay in this volume, social and power relationships between different groups involved in the projects—astronauts and ground controllers, engineers versus managers, different groups within a program—manifest themselves in the design of the control systems. Training, as a method of matching of human to machine, is a place where these relationships begin to form, and simulation—as the artificial creation of a human experience or technical system—points to the increasingly blurred line between "real" and "virtual" in our own world. Such a discussion naturally leads into gender history because the issue of the astronaut's control is also an issue of masculinity. Pay attention to how often "manliness" and "sissyness" (especially in jest) arise in conversations about technology and spaceflight, and one realizes that (consciously or unconsciously) gender is never far from operators and designers of control systems. One Apollo guidance engineer still professes his aversion to the use of the term "software" as unmanly.

Beginning with Apollo, and continuing during the 1970s (and certainly into the future), the professional identity of astronauts began to expand—from the exclusive focus on test pilots to scientists and engineers (and even teachers and politicians), with new job titles like "mission specialist" and "payload specialist," coupled with social expansions beyond White men. I recently asked an astronomer-astronaut how much he used his scientific judgment while in orbit—"Not at all," he quickly replied. Most of his time had been spent following well-established procedures to deploy and operate other people's experiments. Under such conditions, what is the necessity for scientific training, or for human presence at all? Still, that same astronaut acknowledged that being able to "speak the same language" as the scientists on the ground proved an important part of his job. Clearly, some level of tacit knowledge, social interaction, and common vocabulary played an important role in space operations (as it did for the CAPCOMs talking to their fellow pilots in Apollo).

It should be possible to do an ethnographic study of space operations examining skill, training, professional identity, automation, divisions of power, and other aspects of human-machine relationships. Where, exactly, are humans in space exercising judgment, tacit knowledge, and creativity? How would the results differ for scientific versus technical operations? Mission transcripts, combined with interviews and a deep analysis of operations, would provide a solid basis for answering these and related questions. Even a cursory look at the Apollo lunar science operations presents rich material, as the astronauts conducted a variety of activities from deploying instruments to collecting samples (where, precisely, did "exploration" occur?). Such an ethnographic analysis, if rigorously done, would have important implications for engineering design, training, mission planning, and safety. It would also likely generate insights into the operation of other complex technical systems whose operations are rarely as well documented or as accessible as those of human spaceflight.

Such research into the human-machine aspects of spaceflight will also help clarify the tensions in human spaceflight between "science" and "exploration." George Bush's January 2004 speech used the word "exploration" more than 25 times, while mentioning "science" only once or twice. In the documents and debates leading up to Kennedy's Apollo decision, the assumption is that "exploration" is manned and "science" is remote or unmanned, and these debates have continued until the present day. What are the critical differences between science and exploration? Exploration, of course, has a long history, although when it has been brought to bear on spaceflight it has tended to take the form of hagiography more than critical analysis. As Steven Pyne's essay in this volume wonderfully demonstrates, however, the large literature in history and the history of science has a great deal to offer current debates. Exploration often includes science, but usually as one component of a broader agenda, and not usually the most important one. For the sake of argument, we might make this oversimplified distinction: science is about collecting data to learn about the natural world, whereas exploration expands the realm of human experience. Sometimes the two overlap, but not always. Exploration has always had significant components of state interest, international competition, technical demonstration, public presentation, national and professional identity, and personal risk. Seen in this light, the prominence of these elements in Apollo seems less an anomaly than sensible in an historical context.

Again, the science versus exploration dichotomy bears on human-machine relationships. McCurdy and Launius provide excellent examples in this volume: Admiral Byrd's use of mechanical aids (i.e., aircraft) in exploring Antarctica raised questions of heroism, manliness, and professional identity. Similar issues arise in ocean exploration today, especially as the role of manned submersibles is questioned in the face of remote—and autonomous—vehicles. Again, the debates over technology often refer to professional identity: are you a *real* oceanographer if you don't descend to the seafloor? Are you a *real* explorer if you never actually set foot in a new world? Must one physically "be there" to be an explorer? How do professional identities adapt to technological change?

My goal here is not to advocate for either side in the debates about whether we should be sending people into space. Rather, I'm arguing that a scholarly, historical understanding of the human-machine relationship will help to clarify the terms of the public debate. And precise, informed public debate is critical if we are to commit significant resources to future projects.

I'll close with a recent anecdote that captures the richness, interest, and relevance of human-machine relationships in spaceflight. In the spring of 2004, the Explorer's Club of New York City held its 100th annual dinner. At this glitzy, black-tie affair, a few thousand people stuffed into the grand ballroom of the Waldorf Astoria. The club has always included scientists, but also a panoply of mountain climbers, Navy captains, pilots, sailors, divers, trekkers, photographers, not a few astronauts, and a host of wannabe adventurers. At this event, on the stage, were some of the "greatest of the great" who rose in turn to give inspiring speeches about their own experiences and the importance of exploration. Bertrand Piccard, heir of the great Swiss exploring family, recounted his balloon circumnavigation of the world. Buzz Aldrin spoke about his journey to the Moon and advocated for a return to the Moon and a venture to Mars. Sir Edmund Hilary recounted the feeling of his first steps on the top of Everest.

The evening's last speaker was Dr. Steven Squyres of Cornell, the chief scientist of the project that had recently landed two robotic rovers on the surface of Mars. I leaned over to my friend and whispered, "This ought to be interesting, because the rest of those guys have actually gone places, where Squyres has done all of his work remotely, from a darkened room." A moment later, Squyres got up there, on the heels of these great explorers, in front of thousands of people, and said (I paraphrase), "I must say I'm a little intimidated, because all of these people have actually gone somewhere, whereas I've done my work from darkened rooms in Ithaca and Pasadena." But he then gave an account of his and his group's remote, robotic exploration of Mars that easily matched the others in excitement and inspiration. He explained how they "live" on Mars, for months at a time, through technologies of remote, virtual presence. He also made a plea for the importance of sending people to Mars, based on the scientific insight a field geologist would generate by actually "being there." Here, as in so many other instances, science, exploration, technology, and professional identity were intertwined, and understanding those relationships is critical not only for the history and future of human spaceflight, but is key to the essence of human-machine relationships, the coupling of the social and technological, at the core of our modern world.

SECTION III

NASA AND EXTERNAL RELATIONS

INTRODUCTION

In achieving its mission over the last 50 years and in pursuit of a variety of goals, NASA has had complex interactions with a large number of external groups. This section discusses three of the most important: the aerospace industry, the Department of Defense, and the international space community. With a few notable exceptions, historians have often submerged these relationships as they concentrated on the internal problems, achievements, and themes of the Agency itself. NASA's relations with any one of these entities would be an enormous topic in its own right; each author in this section has adopted particular case studies that illuminate key issues.

In the first paper, Philip Scranton aims to enhance our understanding of the often contentious interaction between NASA and industry, which has been crucial in designing, testing, and building the hardware necessary to achieve the Agency's mission.[1] This essay gives a vivid accounting of the complexity of the space enterprise at a level that few people outside the space community contemplate. This complexity involves not only the operational relationships between NASA and its prime contractors, but also those among the primes and their thousands of subcontractors, among the subcontractors and the "sub-subs," and so on down the line, all part of the aerospace industry at increasingly diffuse, but real, levels. Scranton points out that while there was (and is) much contention among those in the contracting community, historically all stood together against what they perceived as excessive NASA meddling and oversight. Yet somehow, it all worked (usually) in the end. Drawing on his own work on the fabrication of the Mercury spacecraft; on Bart Hacker and James Grimwood's history of the Gemini program, *On the Shoulders of Titans*;[2] and on Joan Bromberg's *NASA and the Space Industry*, Scranton shows the astonishing array of questions that arise when one considers concrete historical cases.

Beyond his analysis of the problems, Scranton suggests five frameworks for research that might increase our understanding of the relations between NASA and industry, technology and organization, practice and process, and design and production. Two existing frameworks are Stephen Johnson's study of the systems management approach in *The Secret of Apollo* and Howard McCurdy's sociological approach to organizational culture exemplified in *Inside NASA*.[3] Scranton also

---

1. NASA has sponsored one study of the Agency's relationship with the aerospace industry, but there is considerably more work to be done on the subject. See Joan L. Bromberg, *NASA and the Space Industry* (Baltimore, MD: Johns Hopkins, 1999).

2. Barton C. Hacker and James M. Grimwood, *On the Shoulders of Titans: A History of Project Gemini* (1977; reprint, Washington, DC: NASA SP-4203, 2002).

3. See Howard E. McCurdy, *Inside NASA: High Technology and Organizational Change in the U.S. Space Program* (Baltimore, MD: Johns Hopkins, 1993); Stephen B. Johnson, *The Secret of Apollo: Systems Management in American and European Space Programs* (Baltimore, MD: Johns Hopkins, 2002).

proposes that analytical tools be used from the fields of social construction of technology, management theory, and anthropology to attack these problems.

Scranton hopes for a shift in the writing of NASA history in what he sees as a long-overdue direction: the little-understood world of production for NASA. "Retelling NASA stories from the drafting room and shop floor outwards, from the bottom up," he concludes, "has the potential to reorient a universe of NASA-centric histories." He formulates a large number of questions that constitute a research program to this end.

Scranton's essay does not address the Department of Defense, but since the 1980s, DOD has funneled even more money into the space industry than NASA (their respective space budgets were on the order of $19 billion versus $14 billion in 2003). Even before NASA was formed in 1958, DOD, with its growing stock of ballistic missiles, realized the importance of space for military reconnaissance. In the interservice competition to create a scientific satellite for the International Geophysical Year (IGY, 1957–58), the Navy's Vanguard program was given the go-ahead, but it was the Army, with a modified Jupiter C ballistic missile, that launched Explorer 1 on 31 January 1958, the first successful American satellite in the wake of Sputnik. The opening of the Space Age was accompanied by intense discussion as to whether the nation's space program should be military or civilian. NASA's birth signaled the decision for a civilian agency, but the proper role for military and civilian space programs has been debated ever since.

Peter Hays, a policy analyst with 25 years of service in the Air Force, focuses on three key issues and time periods to illuminate NASA-DOD relations. In the first issue, organizing to implement the American space vision in the 1950s, he finds three major activities with bureaucratic interests that endure today: moving the Army Ballistic Missile Agency (ABMA) into NASA, consolidating DOD space activities under the Air Force, and establishing the National Reconnaissance Office (NRO). Once the ABMA was transferred to Marshall Space Flight Center in Huntsville, Alabama, in September 1960, after a protracted struggle, the Army was officially out of the space business; DOD space activities were concentrated in the Air Force. Not trusting reconnaissance satellites to the Air Force, however, President Eisenhower formed what is now known as the NRO in late 1960. DOD and NRO activities became increasingly classified under President Kennedy, a situation that led to widely divergent public and congressional perceptions of the NASA and military space programs and also made the writing of military space history dependent on declassification.

Hays's second issue is the rationale for human spaceflight in the early space program, in particular the competition between NASA and the Air Force for human spaceflight missions. In this competition, NASA was decidedly the winner; the Air Force was rebuffed on its Dyna-Soar effort by the end of 1963 and its Manned Orbiting Laboratory by 1969 (after $1.4 billion in expenditures).

These early interactions among NASA, DOD, and NRO provide deep background for Hays's third issue, the development of the Space Shuttle, which provided "the most focused, longest running, and most intense interplay among these organizations . . . the single most important factor in shaping their interrelationships." As Hays shows and others have suggested before him, in selling the Shuttle project to Congress and the President, and especially once the decision was made that the Shuttle was to be the nation's primary launch vehicle, NASA needed DOD support and DOD needed NASA to launch its large spy satellites.[4] The Air Force component of DOD was essential in determining Shuttle payload and performance criteria and is credited with saving the program during the Carter administration when Vice President Mondale and the Office of Management and Budget tried to cut it. It was the Air Force that successfully argued that four Shuttles were needed. The price exacted from NASA was mission priority for DOD. Yet, because it did not control the Space Shuttle program, the Air Force was never very enthusiastic about it. And in the aftermath of *Challenger*, the Space Transportation Policy underwent a seismic shift, with the Air Force and NRO once again returning largely to expendable launch vehicles. For historians and policy analysts, the Space Shuttle program provides an unparalleled window on the relations among NASA, DOD, and NRO. Hays concludes that it is "an excellent illustration of the general Air Force ambivalence over the military potential of space and military man-in-space as well as evidence of the lack of clear and accepted doctrinal guidance on these issues."

In the third chapter in this section, John Krige asks an intriguing question: why does the most powerful nation on Earth for the last 50 years want or need international space cooperation? As he points out, some have argued that space cooperation was used in the Cold War era and should continue to be used now, under changed circumstances, as an instrument of foreign policy in which to foster and gain allies. But, he notes, blind international cooperation exacts a price: there is a tension among sharing technology, not compromising national security, and remaining industrially competitive. He argues that sharing technology in the interests of international cooperation makes no sense, historically or practically, unless one opens the "black box" of the interaction of technology and foreign policy: "It is crucial to focus on what specific technologies might be available for sharing in the pursuit of specific foreign policy objectives, rather than— as so often happens—to simply lump technology and foreign policy into an undifferentiated whole." Historians must study international collaboration at this fine-grained level, he insists, if the analysis is to be robust.

---

4. See Dwayne A. Day, "Invitation to Struggle: The History of Civil-Military Relations in Space," in John M. Logsdon, gen. ed., *Exploring the Unknown: Selected Documents in the History of the U.S. Civil Space Program*, vol. 2, *External Relationships* (Washington, DC: NASA SP-4407, 1996), esp. pp. 263–270.

In his essay, Krige takes his own advice by analyzing a particular case of attempted technology transfer: the mid-1960s desire by the Johnson administration to collaborate with Western Europe, particularly with the European Launcher Development Organisation (ELDO), on a civilian satellite launcher. This desire was based on the belief that such cooperation would strengthen European unity, close the technology gap between the United States and Europe, and divert ELDO resources from the technology of nuclear weapons delivery by using them in space instead. NASA and the State Department particularly argued the last point: that by sharing launch technology with ELDO, including documentation on the Atlas-Centaur upper stage that would allow European satellites to reach geosynchronous orbit, they would discourage other nations from applying resources to national military programs. In opposition to this desire for cooperation were American national security and business interests. In particular, some felt that American technology transfer might actually benefit the French nuclear weapons program in terms of its delivery system. Others pointed out that the technology transfer might confer commercial advantage to certain countries in terms of competition with INTELSAT, the worldwide communications satellite consortium under U.S. control via COMSAT. Although NASA and the State Department argued for a finer analysis and a case-by-case study rather than the blunt instrument of national security memoranda, in the end, the argument for relaxing constraints on technology transfer lost. Krige explains the reasons, which are deeply rooted in historical events.

Krige suggests that historically, the protection of national security and national industry interests always prevails over foreign policy considerations. His insights into the connections between space and foreign policy open up a new direction in space history and the history of this component of foreign policy.

By no means do the aerospace industry, the Department of Defense, and international relations exhaust even the general categories of NASA's external activities. Other interagency activities, such as interactions with the State Department and the National Oceanic and Atmospheric Administration (NOAA); university relations, as championed by former NASA Administrator James Webb and some of his successors; public and community relations, always important to NASA's image; and congressional relations, so essential to funding, raise their own unique questions as subjects of historical analysis. Nevertheless, taken together, this section highlights how multifaceted NASA history is, as well as how very much remains to be done in a large number of areas and from a variety of new perspectives.

CHAPTER 6

NASA AND THE AEROSPACE INDUSTRY: CRITICAL ISSUES AND RESEARCH PROSPECTS

Philip Scranton

The X-15 was [Harrison] Storms' airplane as much as it was anybody else's airplane. A lot of other people could lay claim to it. The theorists at NACA [National Advisory Committee for Aeronautics] had actually laid out the basic lines and drawn up the specifications. Some of these people thought of [North American's] Storms and his ilk as "tin benders," lowly contractors who simply hammered out the hardware to match the vision of the scientists. But this wasn't hardware. This was jewelry.
— Mike Gray, *Angle of Attack*

As costs rose, schedules slipped. One source of delay was attempted improvements . . . . The Gemini Program Office was less than happy with the course of events . . . . Not only was GPO being bypassed in the process that approved changes Lockheed wanted to make, but the project office was not always even told what those changes were.
  — Bart Hacker and James Grimwood, *On the Shoulders of Titans*

[Reassignment to] Spacecraft Assembly and Test brought me totally down to reality—down and dirty with the thousands of physical details that had to be perfectly crafted, installed, verified, and documented, and face to face with the earnest, hard-working men and women who strove to do their very best to build a spacecraft that would land men on the Moon and bring them back safely . . . . I had seen the effort and concentration by hundreds of skilled craftsmen that was needed to make engineering orders or program decisions take shape in fact, not just on paper.
                — Thomas J. Kelly, *Moon Lander*

In concluding his 1999 essay review of recent works in NASA history, Northeastern University's W. D. Kay noted that however thorough these studies, they "wind up saying very little about the behavior of the private contractors who actually built the rockets, probes, and satellites. With rare exceptions that almost always involve catastrophes . . . the internal workings of the nation's aerospace contractors never receive anywhere near the level of scrutiny routinely accorded to NASA." Tipping his hat to Roger Bilstein's *Stages to Saturn* as a "happy exception" to this pattern, he added his concern that silences on the industrial front obstructed assessment of credit, blame, and "accountability." In this regard, Kay hoped that aerospace companies would disclose the sources that would document their "role(s) in shaping the U.S. space program,"[1] but at least for Mercury, Gemini, and Apollo, mountains of industry documents have been preserved in NASA files and NARA archives, awaiting our attention. Perhaps this essay will encourage scholars to plunge into them bearing questions and agendas that will enrich our appreciation for the business of building space technologies.

During its first years, NASA reluctantly discarded the NACA's "we build it here" philosophy, abandoning its predecessor's approach for an emphasis on design and supervision, project management, and performance review.[2] Rapidly, then durably, the Agency paid out 90 percent of its budget allocations to contractors, chiefly private-sector firms, for engineering, fabrication, testing, redesign, certification, and shipment.[3] These industrial enterprises and their hundreds, perhaps thousands, of subcontractors, constituted the aerospace industry, which commenced in the 1950s chiefly as a series of projects, then divisions, within well-known aircraft companies: North American, Martin, Lockheed, Boeing, Douglas, and McDonnell, supplemented by specialists in electrical or chemical technologies and products (GE, Thiokol).[4] Given the NASA History Office's charge to research Agency plans, programs, and performance, it is understand-

---

1. W. D. Kay, "NASA and Space History," *Technology and Culture* 40 (1999): 120–127. A number of titles partly addressing Kay's concerns appeared later than his January 1999 publication; some of them will be discussed below.

2. George Mueller, NASA's Apollo director, indicated that in the 1950s, NACA depended on the Air Force to do fabrication contracting for them, thus beginning the shift to externalization (NASM Oral History Project, Mueller Interview No. 4, 15 February 1988, p. 13, available at *http://www.nasm. si.edu/research/dsh/TRANSCPT/MUELLER4.HTM*).

3. Howard McCurdy, *Inside NASA: High Technology and Organizational Change in the U.S. Space Program* (Baltimore: Johns Hopkins, 1993), p. 39. Some of this was interagency transfer, I presume, as ABMA built some launch vehicles and assembled others, but the bulk of it was funding to private enterprises.

4. Over time, the number of prime contractors shrank decisively through a series of mergers and acquisitions, notably the creation of McDonnell Douglas (1967) and its amalgamation with North American Rockwell's Aerospace Division in a Boeing-led merger during the 1990s. Martin acquired American Marietta in the 1960s, then merged a generation later with Lockheed, yielding Lockheed Martin in 1994. The rising cost of aerospace projects (and of military aircraft development) and the uncertainty of profitability made failure on a multimillion-dollar bid extremely painful and made

continued on the next page

able that histories to date have fostered far greater appreciation for NASA's managerial, political, and mission-related achievements and conflicts than for its contractors' struggles to fabricate and qualify spaceflight technologies. Hence the epigraphs aim to evoke multiple dimensions of manufacturing for NASA— the tensions between Agency managers/designers and onsite corporate program directors; the extravagant demands spaceware placed on engineering and production capabilities ("jewelry"); the perennial need for improvements and fixes; that work's impact on costs, schedules, and communication; and the substantive gap between management/engineering plans and the grinding detail work on shop floors and in clean rooms across America.[5]

To rephrase this somewhat, an enhanced understanding of industrial practice in relation to NASA projects could benefit from sustained attention to four core but interrelated themes: 1) initial designing and building of technological artifacts; 2) testing, redesigning, and reworking/refabricating such artifacts; 3) alliances among and contests between contractors, as well as contractors' collaboration with or challenges to NASA units; and 4) approaches to conceptualizing complex contracting and managerial relationships in the production of "edge" technologies. Exploring these will help expose their layers and nested problem sets as this discussion moves toward sketching examples which illuminate recurrent situations, some elements of change over time, and key persistent features of the environment for fabricating aerospace innovations. In addition, this essay will briefly review aspects of the literature concerning aerospace production for NASA, will mention preliminary findings from my work with Mercury spacecraft fabrication records, and will close by offering a set of potential research questions in this area.

NASA and Industry: Four Core Issues

1) Initially designing and building aerospace artifacts.

The iconic NASA artifacts were launch vehicles and their payloads (manned capsules, satellites, observatories, etc.), yet a significant class of artifacts never experienced the rigors of the extraterrestrial environment (launch apparatus, testing and simulation devices, ground support and tracking/communications equipment, and much more). While being integral to NASA's ability to reach

---

continued from the previous page

consolidations gradually more attractive. See Joan Bromberg, *NASA and the Space Industry* (Baltimore: Johns Hopkins, 1999), pp. 12–13.

5. The epigraphs reference what Howard McCurdy terms the "first generation" of NASA, the era through 1970. That's the only era about which I can profess anything like detailed knowledge, principally as a result of serving as the Lindbergh Chair at NASM (2003–04) and doing archival research at NARA's Fort Worth branch and at NASA Headquarters on the design and fabrication of the Mercury spacecraft.

space and, not infrequently, reusable,[6] they stood earthbound. Ground equipment, whatever its complexity, arguably faced fewer "unknowns" than that which was launched, suggesting two distinct lines of design and production dynamics. Moreover, as will be indicated below, some aerospace technologies were "merely" complex, whereas others severely "stretched" technological capabilities, another line of differentiation which could profitably be cross-compared with the launched and the grounded artifacts' development.

Nonetheless, virtually all these technological artifacts were custom-designed and purpose-built, although NASA leaders at times urged contractors to use "off-the-shelf" components or items proven in use during earlier projects. The design process was intricate and NASA-led in the early years, at times contentious, and staggeringly demanding in engineering effort and precision. Building was likewise intricate but was contractor-led (with the exception of the Army Ballistic Missile Agency rockets and a few others) and NASA-supervised/-critiqued, while being staggeringly complex in project management, quality control, and shop-floor detail—and yes, often contentious as well.

Moreover, beneath the level of large-object systems (rockets, capsules, launch sites, etc.), complexities in design and building animated the production of components, the parts for components, and the spatial/operational strategies for assembly and integration of components into functional systems (electrical power, fuel delivery, instrumentation) before the further integration of those systems into the large objects. Occasions for error abounded, as all historians of NASA know well, and the challenges of detecting errors' causes varied dramatically—from simply identifying a faulty fuse to reassembling the shattered parts of an exploded Redstone.

The engineering implications of failures were plain: "whenever something broke, we redesigned it."[7] The managerial implications were more ambiguous, for NASA officials, contractors' personnel, subcontractors, veteran Air Force project managers (much involved in NASA efforts), as well as for advocates and critics of the space program, in and out of government. Parts, component, and large-object failures were expected, yet they could (and did) derange budgets, stall schedules, initiate blame games, and hazard careers. Tom Kelly's transfer to Spacecraft Assembly, noted in the third epigraph, was a stark demotion triggered by a dismaying array of leaks in the first Moon lander Grumman had proudly delivered to Cape Kennedy, a shock that led him to a fresh learning curve[8] and leads us to theme two.

---

6. Unlike everything launched before the Shuttle era. On the Shuttle as the first reusable space vehicle, see Diane Vaughan, "The Role of the Organization in the Production of Techno-Scientific Knowledge," *Social Studies of Science* 29 (1999): 919.

7. *Inside NASA*, p. 32.

8. Thomas J. Kelly, *Moon Lander: How We Developed the Apollo Lunar Module* (Washington, DC: Smithsonian, 2001), pp. 165–171. This demoralization is noted by Stephen Johnson in *The Secret of Apollo* (Baltimore: Johns Hopkins, 2002), pp. 145–146.

2) Redesigning, testing, and reworking aerospace artifacts.

In aerospace design and fabrication, three "rules" might be regarded as near universals: a) "the distance between paper and product is greater than you think," b) "nobody gets it right the first time," and c) "learn that failure is your friend." These are applicable in part because space manufacturing has to meet more demanding environmental tests than any other category of production.[9] Zero gravity, temperatures verging on absolute zero, the vacuum of space, launch vibrations and postlaunch rocket oscillations (pogo-ing), combustion instability, the complex interdependencies of functional systems, and the impossibility of most in-mission fixes combined with other hazards to render manufacturing for NASA launches a high-risk, high-stress task. Testing, particularly of components and subsystems, routinely revealed shortcomings in materials, workmanship, capability, or durability, mandating redesign, indeed often multiple redesigns.[10] "Fixes" themselves could create new problems—e.g., a redesigned part impinging more on a nearby component than the prior version, now radiating vibrations that unsettle its neighbors' instrumentation. Recognized insufficiencies in a system could trigger a higher-order redesign (classically, realization that fuel cell reliability was uncertain, yielding a shift to batteries),[11] which then entailed rethinking system integration. Occasionally, interprogram redesigns affected the large objects, which tended to present a stable exterior appearance. For example, the Mercury capsule's system components were largely located in the interior space of the "tin can," crowding one another and the astronaut. They were maddening to adjust or repair (getting at a failed part in one system usually involved removing elements of another, adding possibilities for error and failure). However, in the larger Gemini capsules, designers modularized functional systems (all key parts located together, insofar as was possible) and removed them outside the astronauts' operating space, making them accessible from the exterior of the capsule for maintenance.[12]

---

9. The "rules" are of my devising, derived from (not quoted from) primary sources. Likewise, the "more demanding" claim is arguable, though not pursued here. Comparable, but somewhat less demanding, environments for production, in my view, involve nanotechnologies, biotechnologies, deep underwater artifacts (nuclear submarines), and cryogenic or Arctic/Antarctic processes/places. At the press conference observing the Mercury Project's closure, McDonnell's Walter Burke asserted: "The problem of designing and making work this complex group of systems is one which [required] and did get a degree of attention to detail far surpassing [any] that has ever been evident in any industrial effort up to date." A newsman thoughtfully countered that Admiral Rickover might challenge that claim (transcript, Mercury Project Summary Conference, box 1, "Mercury Final Conference," September–October 1963, entry 196—Subject Files, NASA, Johnson Space Center Files, NARA RG255).

10. As Mission Control's Gene Kranz summarized, "If you were successful, the concept was labeled brilliant, and you could focus your energies on the next step, the next set of unknowns. If you had problems, you found them early and somehow made time to fix them while keeping on schedule. If you failed, a lot of expensive hardware was reduced to junk and the schedule shattered" (Gene Kranz, *Failure Is Not an Option*, New York: Simon & Schuster, 2000, p. 210).

11. Kelly, *Moon Lander*, pp. 83–84.

12. Barton Hacker and James Grimwood, *On the Shoulders of Titans: A History of Project Gemini* (Washington, DC: NASA SP-4203, 1977), pp. 33–34.

In this context, experienced contractors understood that NASA's or their own engineers' blueprint designs represented a preliminary set of parameters for manufacturing, given the multiple uncertainties of testing and use and the unknown unknowns (unk-unks) that could wreak havoc at any point.[13] Thousands of engineering design changes would flow through every large-object project, ripping holes in budgets, but ironically reinforcing the confidence of NASA staff and contractors' engineering and production teams. "As a part of their culture, NASA employees came to believe that risk and failure were normal" and that the anticipation of failure led to its avoidance.[14] Hence the salience of acknowledging the long road from sketch to artifact, the necessity of iterative design and testing, and the value of welcoming failures (though obviously not fatalities).

3) Contests and alliances between/among contractors and NASA units.

One could hardly do better for a starting point in thinking about managerial relationships in high-performance technological production and operation than to revisit W. R. Scott's classic formulation of three central issues:

> We expect *technical complexity* to be associated with structural complexity or performer complexity (professionalization); *technical uncertainty* with lower formalization and decentralization of decision making; and *interdependence* with higher levels of coordination. Complexity, uncertainty, and interdependence are alike in at least one respect: each increases the amount of information that must be processed during the course of a task performance. Thus as complexity, uncertainty, and interdependence increase, structural modifications need to be made that will either 1) reduce the need for information processing, for example by lowering the level of interdependence or lowering performance standards; or 2) increase the capacity of information processing systems, by

---

13. A concise evocation of the "unk-unks" (famously referenced in a 12 February 2002 press conference by Defense Secretary Donald Rumsfeld) can be found in Tom Kelly's analysis of the Apollo Lunar Excursion Module's (LEM) history. Having completed a preliminary design study for Grumman, Kelly's partner Tom Sanial opined: "'I'll bet the real Apollo won't look like any of the vehicles we've studied.' . . . 'Why do you say that? Don't you think we've done a good job,' I challenged. [Sanial replied,] 'Our study was okay as far as it went, but I'm sure we've just probed the obvious. There's still so much we don't know about how to fly to the Moon.' I had to agree with that. 'You're right. We don't even know yet what we don't know'" (Kelly, *Moon Lander*, p. 16).

14. McCurdy, *Inside NASA*, pp. 62–65. For me, at least, it is not clear, in practice, with what reliability anticipation of failure does lead to its avoidance, or indeed how one would know/measure/analyze this. This may be one of those rarely voiced articles of faith that I have elsewhere referred to as "fabrications." See Philip Scranton, "Cold War Technological Complexities: Building American Jet Engines, 1942–60" (unpublished paper presented at SHOT Annual Meeting, Amsterdam, October 2004).

increasing the [flow and carrying] capacity of the hierarchy or by legitimating lateral connection among participants.[15]

Todd La Porte and Paula Consolini appropriated this conceptualizing statement as foundational for their studies of "high-reliability organizations," working a counterpoint to the normalization of complex technology/system failures evident in Charles Perrow's analyses.[16] Having done workplace studies, they argued that with enough attention to detail, procedure, and training, complex organizations can and do manage to handle high-risk situations without catastrophic consequences. Yet the situations their air traffic controllers and aircraft carrier landing technicians mastered were characterized by long-term stable technologies, high-volume repetitions, and thus a restricted, known set of risk-enhancing conditions and emergency-inducing variables (chiefly technical failures and cascading climate problems). Though they partook of Scott's three core features, NASA production and operations did not fit this high-reliability stabilization framework, for these were nearly unique phenomena, lacked technological stability, lacked mastery-inducing repetitions, and thus confronted hazard conditions and variables that could not be fully comprehended, much less defended against by backups and redundancies.[17]

One implication of this difference was that for technological, economic, organizational, and cultural reasons, contracts proved blunt instruments for regulating the production and operational relationships between NASA and its contractors, much less among NASA and primes on one hand and thousands of subcontractors (and sub-subs) on another.[18] Technically, the

---

15. W. R. Scott, *Organizations: Rational, Natural, and Open Systems*, 2nd ed. (Englewood Cliffs, NJ: Prentice Hall, 1987), quoted in Todd La Porte and Paula Consolini, "Working in Practice But Not in Theory: Theoretical Challenges of "High-Reliability Organizations," *Journal of Public Administration Research and Theory* 1 (1991): 30.

16. Charles Perrow, *Normal Accidents*, rev. ed. (Princeton: Princeton University Press, 1999).

17. Vaughan points out that although the Shuttles were reusable, thus superficially identical among existing craft and from mission to mission, in actuality, "no two shuttles were alike; after each mission, the several NASA/contractor work groups made hundreds of changes, so the technical artifact was different for each launch" (Vaughan, "Role," p. 919).

18. In a heroic but doomed effort to "predict changes in NASA satellite contracts," two management analysts secured a NASA grant in the early 1970s and profiled the contract changes for 21 satellite projects. Seeking a predictive formula, they ignored engineering changes below the contract change level (Engineering Change Requests, or ECRs, versus Contract Change Proposals, or CCPs [CCPs were often large-scale shifts in design, whereas ECRs usually were changes in individual components]), identified mean change costs as $100 K–$300 K, and struggled to find something to regress. Yet they did offer an empirical table that suggests the economic foundation for contests and alliances. Focused on 21 contracts between 1959 and 1968, it showed that in the course of the first 10 contracts (1959–62), final costs were 5.1 times initial contract figures on average, though in the final 10 contracts (1964–68), this multiplier fell to 2.1. However, final costs were estimated in half the latter 10, as perhaps cost data

continued on the next page

endless Engineering Change Requests that testing and use generated meant routine contests both over the need for and design of reconfigured components, checkout routines, etc., and over who would bear the costs. Economically, as well, changes (due to incapacities or aimed at improving capabilities) escalated program expenses and generated NASA-corporate alliances between firms when both faced congressional appropriations hurdles. Primes and subs fought over late deliveries and defective products yet stood shoulder to shoulder against persistent NASA "meddling," "intrusive oversight," or "policing."[19]

Varied patterns of clashing cultures stretched back to the space program's earliest days, when, in the course of new and massive contracting for Mercury spacecraft, the inheritance by the Army Ballistic Missile Agency (ABMA) and NASA of "management by detail" from NACA/Peenemünde ran head-on into McDonnell's pride in engineering creativity and independence. Long a principal Air Force aircraft supplier, McDonnell expected a continuation of the arm's-length, consultative style of contract relations crafted over two decades. Instead, NASA designers and managers, who had never held responsibility for a major technologically novel project, locked horns repeatedly with industry specialists who *had* done so.[20] Later, when NASA Administrator James Webb geared up for Apollo in 1963 by reorganizing the Agency's top management, those he brought in had substantial experience in Air Force ballistic missile program management and industrial military contracting (George Mueller, Air Force Generals Samuel Phillips and Edmund O'Connor, and the legendary Joseph Shea).[21] Webb evidently recognized that at NASA, "nobody knew how to do program management or work with industry on large programs."[22]

---

continued from the previous page

remained incomplete at the time of their article's composition. The decline in the overrun due to contract changes does suggest better specifications in the latter period. See William Stephenson and Bruce Berra, "Predicting Changes in NASA Satellite Contracts," *Management Science* 21 (1975): 626–637, table on p. 629. Regarding Apollo, "what began as a $400 million contract would top out at $4.4 billion a decade later. But everybody knew this going in. All of the Apollo bids were smoke and mirrors, because [in 1962] nobody knew what they were talking about" (Gray, *Angle of Attack*, p. 120).

19. Regarding the Shuttle booster, Vaughan observes that NASA saw "Marshall engineering's role" as "policing Thiokol; to find fault, to identify mistakes, to make sure the contractor abided by the contract" (Vaughn, "Role," p. 920). The issue is not that this was not appropriate, but that it was inadequate and ineffectual.

20. Joan Bromberg indicated that NASA core leaders feared loss of design control and shoddy work by companies given too much authority. See Bromberg, *NASA and the Space Industry*, pp. 40, 43. See also McCurdy, *Inside NASA*, pp. 38–42, which includes this gem on p. 41: "In one celebrated instance, contract workers at what became the Kennedy Space Center went out on strike because the von Braun team would not let them alone. The workers were accustomed to Air Force practice, which involved little direct supervision."

21. Shea took personal responsibility for the Apollo fire disaster and resigned from NASA in July 1967 (Kelly, *Moon Lander*, p. 161).

22. McCurdy, *Inside NASA*, p. 92.

McCurdy's judgment on the results of this reorientation is clear: "NASA's success in achieving the goals of the Apollo program was due in large measure to the tension between the Air Force approach to program management and NASA's traditional technical culture."[23]

Organizational structures did create platforms for alliances, however fraught with tension, as well as for clashes. Industry and Agency engineers with similar specialties and backgrounds worked through problem sets in spaces far distant from policy-making and budget authorizations. For example, Space Task Group and McDonnell collaborated in depth to create Project Orbit, the huge vacuum chamber in which an entire Mercury capsule could be tested in as close to space conditions as was then feasible. Later, on the Lunar Lander project, NASA and Grumman co-staffed the Change Control Board to assess modifications and manage configuration (modeled on Air Force practice).[24]

4) Conceptualizing contracting relations and production on technology's edges.

Although these first three items hardly exhaust the potential list of themes linking NASA and industry, technology and organization, practice and process, design and production, it is worth pausing here for a moment to consider the possible conceptual tools and theoretical frameworks with which scholars can map this terrain in ways that increase our understanding. Two existing frameworks stand out, at least in my view: Stephen Johnson's close analysis of systems management's rise to dominion in NASA programs, drawing on Weber, Drucker, and the literature of "knowledge management," and Howard McCurdy's sociological approaches to organizational culture at NASA and its transformations. Johnson's work focuses closely on the struggle to achieve rational control over projects and heighten reliability through devising and enforcing rigorous procedures. McCurdy reaches into the extrarational world of the beliefs and assumptions that underlie (and at

---

23. Ibid. See also Mike Gray, *Angle of Attack: Harrison Storms and the Race to the Moon* (New York: Norton, 1992), pp. 50–52. On p. 50, for example: "Most of [NASA's] key people were creative iconoclasts like Maxime Faget, conceptual thinkers used to a hands-on approach in which they personally supervised every detail . . . . Now they were being asked to create the largest technical organization of all time."

24. Johnson, *Secret*, p. 128; Kelly, *Moon Lander*, p. 102. By contrast, the Apollo program's "powerful Change Control Board," created in 1967 after the astronauts' deaths, seems to have been entirely NASA-staffed, with George Low making final decisions on "changes proposed by NASA or the prime contractors" (Kelly, *Moon Lander*, p. 163). Johnson discusses the collaborative style of early NASA-industry management more fully in *Secret*, pp. 116–120. Superficially, that is, without specific research into the issue, it appears to me that collaborative NASA-industry design and engineering waned and NASA surveillance/policing increased over time, perhaps a shift triggered by the January 1967 deaths of White, Grissom, and Chaffee, as might be inferred from Johnson's review of the postaccident managerial shifts and conflicts (*Secret*, pp. 146–150). If there was such a shift, was it confined to manned space issues, or did it generalize across all NASA projects?

times undermine) practices, offering a dramatically different perspective. Both focus primarily on the Agency, as would be expected, leaving ample room for pursuing questions about the industry and production side of the spacefaring equation.[25]

Three other perspectives, which grapple with practice at the "local" level, strike me as potentially valuable, particularly in thinking about industrial matters:

1) Adapting the social construction of technology (SCOT) framework to encompass ways in which emergent organizations, much like "unruly" technologies, can become "uncertainty multipliers," a notion Diane Vaughan has applied convincingly to "the NASA/contractor organization" for the Shuttle.[26]

2) Exploring management theorists' conceptualization of the interplay between rationality and irrationality within organizations, and its relation to collateral inquiries into organizational disorder and its implications.[27]

3) Developing research questions in relation to work and technology, based on anthropologists' concern for "situated practice" and "communities of practice."[28]

The provocative potential of Vaughan's perspective can be quickly sensed in her opening remarks to a recent discussion paper on organizations and techno-scientific knowledge:

25. Johnson, *Secret*, pp. 1–3; McCurdy, *Inside NASA*, pp. 163–164. Johnson also includes an instructive comparison with the European space agencies (European Space Research Organisation [ESRO]/ELDO, *Secret*, chaps. 6 and 7) but does not appear to have delivered on one significant point. He ends chap. 5 (speaking of the period around 1970) with "The disadvantages of systems management would become apparent later . . ." (pp. 152–153), but so far as I can tell, no discussion of disadvantages appears in the remaining sections of his study. There may be other theoretical frameworks well exemplified in NASA literature, but I'm not yet familiar with them. Both McCurdy and Johnson undertake the explanation of NASA's "decline" and the resurgence of mission failures/disasters two decades after Apollo.

26. Vaughan, "Role," pp. 916–919. Vaughn's inspirations flowed from Clifford Geertz, Charles Perrow, and the "situated action" group (n. 27), as well as from the STS and science studies literatures (see "Role," pp. 935–936, nn. 2–5, 17).

27. Nils Brunsson, *The Irrational Organization: Irrationality as a Basis for Organizational Action and Change* (New York: Wiley, 1985); Massimo Warglien and Michael Masuch, *The Logic of Organizational Disorder* (Berlin: deGruyter, 1996), esp. the editors' introduction and chapters by Bruno Bernardi, Erhard Friedberg, and Nils Brunsson.

28. Lucy Suchman, *Plans and Situated Actions* (Cambridge: Cambridge University Press, 1987); John Seely Brown and Paul Duguid, *The Social Life of Information* (Boston: Harvard Business School Press, 2000); Julian Orr, *Talking About Machines* (Ithaca: Cornell University Press, 1996); Christian Heath and Paul Luff, eds., *Technology in Action* (Cambridge: Cambridge University Press, 2000); Etienne Wenger, *Communities of Practice: Learning, Meaning and Identity* (Cambridge: Cambridge University Press, 1998).

I begin by drawing on organization theory to illustrate the central paradox of organizations: namely, that the characteristics usually associated with the bright side of organizations— the structures and processes designed to assure certainty, order knowledge, and stabilize operations, thereby making coordinated activity possible—also have their dark side— the capacity to generate uncertainty, disordered knowledge, instability and unanticipated outcomes . . . . [T]his paper targets the conjunction of organization and technology that affected the production of knowledge and knowledge claims on a routine basis [at NASA]. The paradox is illustrated by showing the variable effect of the NASA organization on the production of techno-scientific knowledge: 1) the production of disordered and uncertain knowledge on a daily basis; and 2) the fact-hardening mechanisms in place to convert disorder to order when a collective decision was necessary.[29]

Where Johnson sees systems management as generating reliability and certainty, by tracing *Challenger* and other failures to a relaxation of detail discipline,[30] Vaughan sees the ghost as inherent in the great machine and penetrates deeply enough into the everyday life of techno-science to establish that "disordered knowledge is a byproduct of the very organizational mechanisms designed to control it." "Structure creates pockets of meaning systems—distinctive local knowledges . . . —that are by definition contradictory . . . . Structure [also] obscures, so that actions occurring in one part of an organization cannot, for the most part, be observed by people in other parts." Her work echoes in organizational/knowledge terms Perrow's critique of technical complexity, urging that scholars acknowledge that everyday practices and relations have dangerously ambivalent implications for organizational and technical outcomes.[31]

If so, recognizing that nonrational dimensions to organizational and technical practice are routinely yet unevenly present in all action situations can be a valuable step. Nils Brunsson has memorably underscored the presence and significance of nonrational dimensions of organizational practice, especially in regard to innovation. From his perspective, planning creativity is as fruitless as creating a random search for a technical fault, precisely because different modalities of thought and practice inform decision-making versus action-

---

29. Vaughan, "Role," pp. 914–915.

30. Johnson, *Secret*, pp. 228–229, and n. 9, pp. 275–276. McCurdy debits such disasters in fair measure to the attrition of NASA's classic high-performance "technical culture," rising risk aversion, and a politicized intolerance for failure (*Inside NASA*, chaps. 5 and 6).

31. Vaughan, "Role," p. 916, both quotations; Perrow, *Normal Accidents*.

taking. Agents need perennially to be aware that overreliance on rationality can generate stalemates, just as overreliance on intuition and enthusiasm can yield chaos. One central insight Brunsson's exploration of the "irrational organization" offers is that agreement on goals makes conflict difficult to understand in complex environments, whereas failed conflict resolution (organization change) can generate "social deadlock," the outcome when "a group of people have arrived at a situation which satisfies none of them but which they are unable to change."[32] The relevance of these conceptualizations to analyzing patterns of and changes in NASA-contractor relations is hard to miss.[33]

Third, in their anthropology of work and practice, Julian Orr, Lucy Suchman, and their colleagues undertake to reemphasize the importance of informal structures and relations, and of the knowledge and routines they generate, to organizational activity. As Scott noted, even conceptualizations of organization-technology relations that stress contingency, hence situation/place and history "overlook the importance of informal structures as a response to uncertainty and complexity." These are bottom-up processes or, perhaps better, integrative linkages:

> Rather than augmenting hierarchies, they minimize vertical distinctions, and rather than creating new, specialized lateral roles and relations, they encourage more direct, face-to-face communications among any or all participants as required. Decision making and the exercise of control become more decentralized, and organizational roles less formalized.[34]

---

32. Brunsson, *Irrational Organization*, pp. 27, 97, 111. By bringing the irrational into the picture of "normal action," Brunsson generates an array of striking (and testable) insights, namely, "efficiency seldom goes hand in hand with flexibility" (p. 4); it is "important to recognize that decisions can exist without actions and actions without decisions" (p. 21); and that in high-risk situations, those undertaking to reduce uncertainty are "speculators in success" and those trying to lower the stakes at risk are "speculators in failure" (p. 52). The *psychological* dimensions of organizational action are key for Brunsson, and these cannot be reduced to rational propositions.

33. Here's one minor story that shows the power of the nonrational in NASA-business relationships. In early 1963, NASA and North American representatives met 15 hours a day, six days a week in Houston to "hammer out a specific agreement on what North American was going to build and what NASA was going to pay for" in the Apollo program. Yet the NASA team was woefully underexperienced in negotiating contracts. As a NASA designer reflected, "We ought to have known better at the very outset . . . . Not any one of [our] technical guys knew a damn thing about costing. They had no basis to negotiate anything. We locked them up in these rooms [with North American managers and lawyers] and *most of them came out mortal enemies. That set a feeling that lasted a long time*" (Gray, *Angle of Attack*, p. 144, emphasis added).

34. W. R. Scott, *Organizations: Rational, Natural and Open Systems*, 3rd ed. (Englewood Cliffs, NJ: Prentice Hall, 1992), pp. 248–249, both quotations. An excellent ethnography based on this approach is Julian Orr's *Talking About Machines*. For a broader perspective, see Robert J. Thomas, *What Machines Can't Do: Politics and Technology in the Industrial Enterprise* (Berkeley: University of California Press, 1994), and Thomas Davenport, Susan Cantrell, and Robert Thomas, "The Art of Work," *Outlook Journal*, January 2002, *http://www.accenture.com/xd/xd.asp?it=enweb&xd=ideas%5Coutlook%5C1.2002%5Cart.xml*.

In American corporations and state agencies, uncertainty generates managerial hunger for top-down control, but few managers can master the massive knowledge requirements for its exercise, especially in situations where knowledge is emergent and distributed widely, as in complex contracting/subcontracting environments. Moreover, as Vaughan emphasized, the compression/reduction of vast bodies of information and the structural inability of capturing situated practice can readily transform control over uncertainty into a generator of illusion and disorder.[35]

NASA AND INDUSTRY: TWO KEY STUDIES

In identifying the themes and conceptual packages just outlined, both the insights and the silences of previous research bearing on production for NASA proved crucial. Thus far, works by Johnson, Kelly, McCurdy, and Vaughan have been emphasized; here, I'd like to consider the legacy of studies by Bart Hacker and Jim Grimwood (Gemini) and Joan Bromberg (NASA and space industries). First, however, a visit to the shop floor from Mike Gray's and Roger Bilstein's Saturn booster studies will set the stage for underscoring the extravagant technical demands and necessities for innovation that infused production for NASA.

The Apollo program's Saturn artifacts were the largest rockets fabricated in the U.S. in the 1960s (perhaps ever). Yet creating their components was enormously difficult; consider, for example, the propellant tanks for the rocket's lightweight S-2 first stage. Huge (reportedly three railway freight cars could be placed inside them) yet fragile (they couldn't be fabricated horizontally, but had to be built upright), they presented unprecedented challenges in welding. "At a time when a flawless weld of a few feet was considered miraculous, the S-2 called for a half mile of flawless welds." Moreover, the components for the tank's dome—"immense pie-shaped wedges of aluminum eight feet wide at the bottom and twenty feet from there to the apex"—were elaborate spatial forms, "a spherical curve from side to side and a complex double ellipsoid from the base to the apex." Given that no techniques existed for accurately machining such shapes, called gores, North American used explosive forming. Technicians placed the alloy blank on a forming die at the bottom of a 60,000-

---

35. Vaughan, "Role," pp. 926–934. This involves what Vaughan terms "fact-hardening," and the procedures for achieving it here rely substantially on the exclusion of qualitative information. As she notes, "Indeterminacy creates a closure problem." This is resolved by generating quantitatively structured documents and public consensus. "The documents . . . assert consensus through the matter-of-fact tone of the formal mode of discourse, affirming the reality they assert to both the audience and the author. An additional factor that binds people to their actions is 'going public.' When a person participates in and is identified publicly with a decision, that person will resolve inconsistencies to produce attitudes consistent with that choice." Quotations are from pp. 929 and 930.

gallon water tank, then set off a cluster of carefully placed charges on the surface. In an instant the force carried through/by the water pressed the blank into the die-form (trimming followed).[36] These segments in turn were welded by "a new kind of a machine":

> [T]he assemblers . . . were looking at a seam that followed a constantly changing curve over a twenty foot run, and the junction between the [gores] would have to match precisely to within a hundredth of an inch . . . . [T]he ultimate solution looked a little like a Japanese footbridge—a heavily reinforced bow-shaped truss that spanned the width of the dome and carried beneath it a precision track on which the welding machine traveled. The gear-driven welding head, its speed controlled by mathematical formulae, rolled ever so slowly up these rails carrying a tungsten electrode that precisely melted the metal on either side of the joint.[37] [See photo opposite; the footbridge welder is visible at the upper left.]

Thus were intricate demands addressed. Routinely for builders, no obvious means lay available to satisfy the interactive realities of technical complexity, technical uncertainty, and component interdependencies in production for NASA, thus propelling organizational frustration and technological creativity. This pattern is evident in each of the two other studies noted above, to which we now turn.

Industry-NASA relationships are especially prominent in the first 10 chapters of *On the Shoulders of Titans*, the segment authored by Bart Hacker. Like a number of jet engine projects a decade earlier, Gemini was the result of an effort to redesign an existing complex technological artifact, the Mercury capsule. By early 1961, James Chamberlain, Space Task Group's head of Engineering and Contract Administration, determined largely on his own initiative that the Mercury spacecraft needed a redesign "from the bottom up," and thus spent part of February in St. Louis going over possible revisions with McDonnell engineers. Modularizing systems that in Mercury "had been stacked like a layer cake" such that "components of [any] one system had to be

---

36. Gray, *Angle of Attack*, pp. 154–155.

37. Ibid., p. 156. This sequence is also carefully reported by Bilstein in considerably greater detail. See Roger Bilstein, *Stages to Saturn* (Washington, DC: NASA SP-4206, 1980; reprint, Gainesville: University Press of Florida, 2003), pp. 212–222 (page citations are to the reprint edition). For several of the hard-core technological issues, see W. J. Reichenecker and J. Heuschkel, *NASA Contributions to Joining Metal* (Washington, DC: NASA Technology Utilization Division, NASA SP-5064, 1967). This publication includes references to a number of North American reports, as well as reports from Marshall, Pratt & Whitney, Kaiser Aluminum, and others. The figure is drawn from Bilstein, *Stages*, p. 221.

Gores being welded to bulkheads for the S-II stage of the Saturn V. *(Source: Roger Bilstein,* Stages to Saturn *[Washington, DC: NASA SP-4206, 1980; reprint, Gainesville: University Press of Florida, 2003], p. 221)*

scattered about the craft" would "reduce manufacturing and checkout time," Chamberlain argued. Yet as Hacker summarized, "making it better meant making it over." Once Chamberlain and McDonnell's William Blatz collated the redesign elements, they went before the Capsule Review Board, which "seemed staggered by the scope of the changes presented to them" in June.[38] As in jets, what started as a fix, or more accurately, a vector for refining the artifact, morphed into a largely new device, yet here still a one-man capsule.

McDonnell engineers, led by Walter Burke, were the agents who outlined and pushed for the two-man spacecraft, however, as it was the builders who "were pressing for a more radical effort." Indeed, in undertaking the preliminary design work, "McDonnell had not felt obliged to wait until its contract had been amended to provide the extra funds. The company spent its own money," which generated "a good deal of respect in NASA circles." As major spacecraft contract changes arose in order to expand its size, handle

38. Hacker, *Titans*, p. 33–45.

modularization, and create a docking system and (initially) ejections seats, expectations for reusing Mercury technologies in the new developmental trajectory faded as steadily as the project drove forward. This momentum and focus on industry relations were aided by an organizational arrangement which provided the Gemini Project Office and Chamberlain "a degree of autonomy," enabling them "to deal directly with McDonnell and Air Force Space Systems Division" for capsules and boosters respectively. Chamberlain reported only to Marshall Space Flight Center Director Gilruth, chiefly providing him work in process reviews and discussions from coordination meetings, "Gemini's central management device."[39] Thus far, an organizational device giving Chamberlain singular authority (how unusual? with what exact options? how evaluated by Headquarters and by McDonnell?) and decisive redesign innovations from industry engineers and engineering managers facilitated Gemini's emergence.[40]

However, a series of technological disappointments, cost escalations, and budget controversies soon caused massive headaches. In some measure, these derived from the fact that McDonnell "developed and built only the spacecraft structural shell and electrical system"; all else had been subcontracted. Thousands of components made by hundreds of firms flowed into St. Louis; if Gemini mirrored Mercury in this respect, an unknown, sizable subset of those devices would fail on test, fail to meet specifications, or fail to integrate effectively, and thus would need to be redesigned or replaced.[41] In a retrospective overview, Hacker reflected, "Although the precise nature of

---

39. Ibid., pp. 49–82, 95. Even as expectations faded that technical apparatus from Mercury could be duplicated in Gemini, major continuities in personnel between the two programs proved a strength, from Faget, Gilruth, Chamberlain, and McDonnell's Walter Burke down to the shop level, where, for example, NASA plant representative Wilbur Gray shifted gradually from Mercury to Gemini. Gray's memos and reports are a marvelous source for reconstituting, in part, the informal relations and emergent communities of practice mentioned earlier in the essay. Chamberlain's autonomy may have been modeled on the direct relationship NASA's Max Faget and McDonnell's John Yardley had in making "thousands of detailed design decisions" on the Mercury capsules. See Loyd Swenson, "The 'Megamachine' Behind the Mercury Spacecraft," *American Quarterly* 21 (1969): 210–227, quotation from p. 222.

40. This approach in no way intends to overlook issues and pressures *external* to the Gemini project, such as the uncertainties about Apollo's developmental trajectory, funding, and schedule, or the cultural/political pressure to keep performing launches as Mercury was beginning to wind down.

41. Archivists at NARA–Fort Worth indicated that the boxes on technical testing and subcontractor relations I was using in my NASM/Lindbergh-supported research had not previously been pulled. Swenson's *This New Ocean* understandably did not penetrate to this level of source material, some of which, it appears, had not yet been archived or declassified at the time of its writing. NASM's Michael Neufeld suggested to me that the view among space historians is that Gemini was a much less troublesome project than Mercury, due to technological and organizational learning. This is a position that might merit further probing, although Hacker did drive more deeply into industry/production documents than did Swenson (Hacker cites telexes, letters to contractors, and activity, status, and "tiger team" reports, for example).

Gemini's problems could not have been predicted, they did arise *where* they were expected—in those systems that demanded the greatest advances beyond current technology."[42] This is such a basic point that it is worth reinforcing— *innovation generates disorder, and dramatic innovation entails error, failure, and conflict across a broad front.* In some technological environments, a stabilization follows, both of knowledge and technology. When additional requirements are promulgated, extensions of capability are feasible on the basis of retained learning and scalable technique, though the achievements usually are hard-won. In other situations, workable innovations do not provide a foundation for enhancing capabilities, which is to say that stabilization proves illusory and learning less than readily applicable to upgrading. These often involve nonscalable technologies, which are the home for hordes of unk-unks and the sources of persistent frustration and failure in large technological projects.

Two Gemini examples merit recounting: the fuel cell innovation and the recurrent issues surrounding thrusters—both involving subcontractors, here General Electric and Rocketdyne. Fuel cells had the potential to replace batteries as the source of on-board electricity, at a major savings in weight. However, in Gemini, the array of problems cropping up "seemed to suggest that theory had outrun practice." GE researchers knew scientifically that the reaction of hydrogen and oxygen could generate power, and they had devised a clever "solid, ion-exchanging membrane" that dramatically simplified both the device and its operation. Unfortunately, this science-led technology did not operate successfully—the membrane leaked, weakening output, and once this fault was corrected, the cell exhibited "degraded performance" once activated. Technicians traced this to the shortcomings of a fiberglass component and replaced it with a Dacron substitute, which triggered new troubles. Other test failures derived from the cracking of the cell's titanium tubing; these were replaced with a titanium-palladium alloy. Further problems appeared, but they "were never conceptual . . . . The rub came in trying to convert [the] concept into hardware to meet the Gemini specifications." After two years' work, NASA canceled the effort in January 1964, resumed work on battery development, and spent $600,000 to retrofit two capsules outfitted with fuel cells. The same pattern recurred soon after, with the Apollo Moon lander's fuel cell program (this time handled by Pratt & Whitney) canceled early in 1965 following two years of trials and failures, with reversion again to batteries.[43]

Thrusters presented an enduring difficulty. Twice in the Mercury program, their fragility and unreliability caused serious concern. In January 1962, McDonnell was testing Capsule No. 2's Reaction Control System when the

---

42. Hacker, *Titans*, p. 162.
43. Ibid., pp. 103–104, 148–152. For the LEM story, see Kelly, *Moon Lander*, pp. 82–84.

base of the spacecraft caught fire due to leaking thruster propellant, which, when it combusted, caused further leaks, more combustion, and quite a bit of damage to the artifact and to the designers' confidence.[44] Just a month later, during John Glenn's orbital flight, the Automatic Stability Control System, which coordinated the thrusters to maintain proper attitude, went for a walk over Mexico. Glenn explained:

> The capsule started drifting to the right in yaw and it would drift over to about 20 degrees, instead of the normal 30 degree limit, and then the high thruster would kick on and bat it back over to the left. It would overshoot and then it would hunt and settle down again somewhere around zero. The spacecraft would then drift again to the right and do the same thing repeatedly.[45]

Glenn put the system into manual control (then into fly-by-wire), which saved fuel, but the capsule began to yaw to the left, and it was soon apparent that "there was no left low thrust."[46] Glenn discussed how he dealt with the inoperability problem:

> When the fly-by-wire one-pound thruster was not actuating in yaw, I was using a real fast flip of the high thruster in the mode that the one-pound thruster was not operating to control. I couldn't control this as accurately as you can with the one-pound thruster, . . . so what I did several times was, when I would overshoot in rate with the 24-pounder, I would use my one pounder on the other side to bring it back to zero . . . I wouldn't call this desirable.[47]

Unsurprisingly, attention to thruster testing and possible design flaws increased sharply.

With the more ambitious Gemini program's development, thruster problems became more acute and challenging. The smaller of the two propulsion units on Gemini was roughly the size of Mercury's larger unit (25 pounds of thrust), whereas Gemini's big pusher was to yield three times that power (85

---

44. R. H. Lilienkamp, Senior Engineer, McDonnell, "Investigation of the Capsule No. 2 Incident, 9 January 1962," 16 January 1962, MAC Technical Reports, box 27, entry 198C, NASA-JSC, NARA RG 255.

45. R. B. Voas, "Memorandum for Those Concerned, MA-6 Pilots Debriefing," pp. 13–14, Contract Administration Files, box 31, entry 198E, NASA-JSC, RG255.

46. Ibid.

47. Ibid., p. 61.

pounds). The Mercury components had simply managed attitude control; in Gemini, they had to handle spacecraft maneuvering and in-orbit rendezvous. Third, the Gemini fuel was different—monomethylhydrazine and nitrogen tetroxide, which combusted on contact, versus Mercury's simple hydrogen peroxide, which expanded radically on release under pressure. Last, and most troublesome, whereas the Mercury thrusters operated for a few seconds at a time, Gemini's would need to burn steadily for 5 minutes or more, as well as to pulse repeatedly.

The bad news came in waves. Tests early in 1963 showed that the 25-pound Geminis tended to "char through their casings" when run continuously. A redesign at first seemed to remedy this, but pulse testing proved half again more destructive to the casings, and a series of "expedients . . . could only alleviate, not solve, the problem." Most troubling, the nonscalability gremlin soon surfaced, as "new tests revealed that the larger maneuvering thrusters could not be simply enlarged versions" of the 25-pound engines. Therefore, a separate design and testing program for them had to be devised. In October, the hammer dropped—mission simulations showed that astronauts used their thrusters far more than had been anticipated—thus, "thruster life would have to be doubled or tripled."[48]

Rework lasted well into 1964, with the result that Rocketdyne fell far behind schedule and had spent more than double its allotted $30 million. NASA soon demanded a "full scale" audit, which revealed a "badly managed program," for the company had "grossly underestimated the magnitude and complexity" of its engine subcontract. Fewer than half the engines slated for delivery by November 1964 had been received, and McDonnell was far from confident in the thrusters' reliability. Still, by mid-1965, Rocketdyne had reorganized the engine division, recovered its momentum, and begun to meet or exceed schedule expectations.[49] The facts that different-sized and differently purposed engines could not be scaled up or down from existing, workable models and that elaborate fueling and combustion systems were inadequately understood meant that propulsion surprises would continue to arise.[50]

Technological problems solved for a mission having certain requirements did not necessarily spill over to later missions with more demanding require-

---

48. Hacker, *Titans*, pp. 83–84, 154–157. The upgraded demands settled at over 9 minutes for the small thrusters and over 13 minutes for the large.

49. Ibid., pp. 210–211. This happy outcome did not prevent thruster problems from arising on three missions—Gemini V, VII, and VIII. See ibid., pp. 259–260, 292, 314–315.

50. One of the key dilemmas here was combustion instability, which arose when flows of fuels (and oxidizers) failed to generate a steady, focused flame thrust, whether due to cavitation, component performance problems, or other factors. Correcting such instability once it occurred seemed impossible, for the effects were dramatic and instantaneous on missile attitude and trajectory, nor was the science of fluid dynamics sufficiently developed to model these flows mathematically and continuously.

ments. The organizational approaches effective for solving first-generation dilemmas would not assuredly suffice for next-generation challenges. As well, the insufficiencies of science regarding critical, complex phenomena (combustion and fluid dynamics, materials performance under zero gravity, etc.) meant that workable engineering outcomes could not be stripped of their anxiety dimensions, for, as with Mercury, components that worked 10 times could (and did) fail on the 11th, without warning and without obvious (or remediable) cause.[51] In this light, it would perhaps be worthwhile for researchers to explore those domains in which basic science guided NASA technical practice, those where NASA practice extended scientific knowledge and theory, and those where the two remained disconnected in specific situations or for longer periods.

Moving to the industry-NASA relationships depicted in Joan Bromberg's pioneering overview entails a shift in focus, for her work undertakes a long-term analysis. This essay is anchored in thinking through technology and production issues, whereas after its opening sections, *NASA and the Space Industry* (*NSI*) moves toward the second of its two themes—space and the marketplace, for satellites, Shuttle usage, et al.—if you will, the consumption side of NASA. Nonetheless, *NSI*'s first theme, "the innovation process," is clearly germane. Here, Bromberg delineates production for NASA's crucial background conditions, identifies core tensions, and offers two detailed case studies of innovation—satellites at Hughes and Apollo at North American.[52]

Four background items Bromberg highlights are particularly rich with implications:

1) Lockheed's science crisis in the mid-'50s "over whether scientists on a project should have control over advanced development." The firm said no; 15 top scientists left, frustrated that their demand to direct work for which "the skill and technical knowledge [was] beyond the state of the art" had been rejected. Science-engineering and scientist-manager relations are a subplot in NASA-industry relations, though, as a novice, it's not clear to me how much these have been investigated.

---

51. As Hugh Dryden stated in the closing Project Mercury Conference, "We learn how to build things to last longer by trying to build them, by operating them in space, finding out what goes wrong, correcting, learning more about the environment . . . . These are things that we learn by going into space and working there, not from some theory in the laboratory" ("Mercury Final Conference," pp. 1–2, box 1, E196, RG255).

52. At the outset, Bromberg refers to technical professionals' "community of practice" but does not seem to be aware of the communities of practice in literature and research approaches noted here in the section on conceptual frameworks. In a discussion with NASM's Martin Collins (13 January 2005), I came to appreciate that oral history interviewing below the executive level (planned but never completed)—interviewing of design, test, and production engineers, for example—would, in framing novel questions, profit substantively from familiarity with the work of Orr, Suchman, and Lave, and also from thinking closely about Karl Weick's challenging *Sensemaking in Organizations* (Thousand Oaks, CA: Sage, 1995), especially in relation to puzzles, failures, and conflicts over knowledge, interpretation, and practice.

Here, did those resigning create their own firm; move to universities; seek research unit jobs at Mellon, Battelle, or RAND; hire on to other industrial firms; or what? Did such confrontations appear on aerospace's technological edge with some frequency, or was this a rare moment?[53] After all, the role of science and scientists in NASA work is not so obvious at it might seem, given the huge holes in scientific understanding of space environments in this era.

2) The Air Force's creation of Ramo-Wooldridge as a systems engineering and technical management firm (1954). To be sure, this laid the foundation for "weapons system" development and for TRW, but to what extent did valorizing this cluster of sophisticated experts create a template helpful for defining NASA's differences from NACA? Clearly the Air Force was already a contested model in terms of innovation management, so was NASA, in a slightly twisted organizational-lineage sense, Ramo-Wooldridge's unacknowledged or ungrateful offspring?[54]

3) The mid-'50s conflict between the Naval Research Lab and Martin, which prefigured scores of subsequent contretemps. In Project Vanguard, Martin argued that it should be provided "full [technical/managerial?] responsibility," while the NRL demanded the inverse. Martin claimed that the Lab was full of busy fault-finders, "always promoting the 'better' at the expense of the 'good enough,'" whereas the NRL asserted that Martin didn't "grasp how much they were dealing with unknowns, nor the importance of reliability . . . ." This contest, arrayed in just about these exact terms, would be replayed for several decades in NASA-industry relations, so what are we to learn from this early incidence? Was it *that* early, that is, was this just an extension of Navy "control-freakish" patterns, inverse to Air Force (and Army Air Force) delegation of project responsibilities to contractors? Was this "divide" a structural fault in postwar military/space programming, and was it ever resolved? If so, how? If not, with what implications? Or is this whole scenario just an outsider's confused view of the unfolding game?[55]

4) The Army's arsenal system (after its separation from the new Air Force) could not run all its ballistic missile projects inside von Braun's shop, simply because "it did not have the manpower." So was the arsenal system chiefly a managerial/operations framework and, in fair measure,

53. Bromberg, *NSI*, p. 25.
54. These relationships are sketched in Mueller Interview No. 4. See also Bromberg, *NSI*, pp. 26–28.
55. Bromberg, *NSI*, pp. 26–28.

a hollow production system? Did shortcomings in securing adequate manpower (engineering, production, testing?) preview the complexities of producing for NASA? Did contractors learn from ABMA that they needed to resist control moves from their funders in order to protect opportunities for enhancing their own engineers' capabilities? Did "the enmity between the Army and the aircraft industry" bleed through to the space industry–NASA relationship, and if so, to what extent and with what consequences?[56]

Bromberg also details key drawbacks and advantages for companies undertaking production for NASA. On the downside were the small numbers of artifacts ordered, the necessity for expensive experimental development and research (some of which would be self-funded), demands for higher precision than usual in aeronautical engineering and fabrication, and the need to find and hire ever more engineers (and high-skill shop workers). Still, the pluses were substantial, if somewhat more vague: the "chance to learn technologies, develop skills and install production tooling that they could use for other projects," possible spillovers into commercial products, and the excitement of joining the space-race culture.[57]

She also shows that the bases for strain were quite concrete. If industry representatives in the 1950s saw "NACA engineers . . . as researchers, people whose aim was the production of papers and books," the incoming NASA leadership was equally critical. Given the necessity of contracting, Headquarters feared the loss of design control, shoddy work by contractors given too much leeway, and the loss of collective memory (and identity) as project teams formed and disbanded. Specifically in the Mercury capsule case, "Langley engineers mistrusted industry's ability to design something as novel as a spacecraft," whereas "industry and the military were convinced they knew more about space flight than NASA did."[58] This last item, the industry-military connection, reinforced NASA's uncomfortable position as the national novice in major project development and operations. Max Faget may well have had an advantage in being able to conceptualize a blunt-body spacecraft, but McDonnell's Walter Burke and his Air Force Material Command colleagues had learned firsthand how to fabricate complex aerospace technologies, as had von Braun and ABMA. Last, NASA might have considered industry folks immature and arrogant, but, as Bromberg so neatly puts it, "arrogance in proposals is also one of the channels by which creative ideas flow from industry to government."[59]

---

56. Ibid., p. 29.
57. Ibid., pp. 38–39.
58. Ibid., pp. 32, 43.
59. Ibid., p. 43.

When introducing the first of her two case studies (Hughes and satellites), Bromberg poses seven questions which articulate the chief concerns and boundaries of the study, "the relation between U.S. industry and the federal government."[60] Except by inference, none of these questions spotlight the technologies themselves, their design, prototyping, testing, redesign, fabrication, plus the consequent interfirm and contractor-government linkages. One technological-process moment appears when the failure of the first Syncom satellite was traced to a ruptured "gas tank," a problem "corrected" after a "search for a stronger material." The second Syncom "functioned brilliantly," but further questions that might have probed this failure and correction fell outside the study's scope.[61] This set-aside resonates with W. D. Kay's concern about the literature's silences on "the internal workings of the nation's aerospace contractors."[62] It remains for future scholars to address how Hughes designed and built its first three satellites; what the firm learned thereby and through what process; what innovations it embedded in the following four INTELSAT IIs; what machinery, materials, engineers, workers, consultations, conflicts, and compromises were involved.[63]

Similarly with North American, Bromberg's analysis works at the level of policy and program, though the secondary sources drawn on (especially Bilstein) yield a greater frequency of references to technical competencies and fabrication challenges. Thus the confrontation between Air Force General Sam Phillips (working for NASA) and North American leaders over "inadequate engineering, poor fabrication quality, faulty inspections, and cost escalations," all leading to delays and rework, is concisely reviewed, yet the underlying reasons for these multiple failures are not divined. As Bilstein, Kelly, and, to a degree, Mike Gray (*Angle of Attack*) demonstrate, in-depth technical review, appropriately contextualized, generates complex, contingent, and real-time analyses of innovation, critical insights and errors, integration, and technological and organizational learning.[64] This is, however, very difficult without

---

60. The questions are, "How much of the research for the commercial communications satellites would be financed, directed or done by government, and how much by the private sector? Would a private industry arise to launch the satellites or would they be launched by government? Would industry or government own and operate the systems? . . . What private firms would enter into the manufacture and the operation of commercial satellites (comsats)? What strategies would they use to gain market share? How would government policies and actions affect the market positions of private companies? How would these policies and actions affect the technology that was chosen?" (ibid., p. 46).

61. Ibid., p. 53.

62. Kay, "NASA," p. 127.

63. Five years ago, I did an online database search for articles in scholarly and technical journals on the design and fabrication of satellites, which then yielded fewer than a dozen hits. I expect a repeat these days would do much better, although the silences on building aerospace technologies may continue to include these devices.

64. An exceptional source in this regard is Martin Collins's series of interviews with North American Aviation's Lee Atwood, which document the critical role of NASA's detailed oversight in generating

continued on the next page

archival research, which, given its parameters and resources, was not plausible for this study.

Nonetheless, Bromberg skillfully reviews the fabrication and engineering practice changes that followed the Apollo fire deaths: separate managers for each spacecraft, heightened attention to quality control, frequent shop-floor visits (including during night shifts), tightened change controls, along with some of the dilemmas their introduction created. "All changes now had to be funneled first through the program officer at Houston, and then through the manager of that particular spacecraft at NA Rockville. North American engineers were made to adhere rigorously to agreed-on procedures, without any creative flourishes." Moreover, NASA's increased surveillance and micro-management necessitated hiring hundreds of inexperienced technical managers who knew far less about their programs than those they were overseeing, which in turn led to mechanical rule-following and conflicts, very much on the pattern that Vaughn's conceptualizations outline. Pursuing these issues deeply into archival materials, especially those surrounding the astronauts' deaths and their aftermath, could provide valuable understandings of a critical transition in America's space program.[65]

INDUSTRY AND NASA:
MERCURY MOMENTS AND CLOSING QUESTIONS

Scattered about earlier pages are some items derived from my archival work with NASA Mercury sources. I'll mention just two others here focusing on a single matter, engineering changes, and will end by offering questions on other issues which may take on a fresh significance when researched from the contractors' technology and organization viewpoint. These items and issues may have more significance to historians of technology and enterprise (who

---

continued on the next page

masses of change orders and consequent delays and establishes the distinction between projects that were just complex (such as the Apollo Command Module) and those that involved "technological stretching," which ventured into the unknown. (See NASM Oral History Project, Atwood Interviews, no. 4, pp. 3, 10–11; no. 5, pp. 12, 14; no. 6, p. 3; available at *http://www.nasm.si.edu/research/dsh/TRANSCPT/ATWOOD4.HTM, http://www.nasm.si.edu/research/dsh/TRANSCPT/ATWOOD5.HTM,* and *http://www.nasm.si.edu/research/dsh/TRANSCPT/ATWOOD6.HTM.*) It appears that this is the only interview with a contractor official. It would be valuable were someone or some institution to take up Collins's plan for interviews with contractor engineers (and perhaps shop workers) before it is too late to target these sources of work and technology information.

65. Bromberg, *NSI,* pp. 70–73, quotation from 71. NASM's Alan Blinder is currently researching the Apollo 204 fire. For the industry perspective here, Bromberg cites a pamphlet by John L. "Lee" Atwood, NAA president, from NASM's Oral History Working File. Deeply interesting is the extensive oral history interview itself, done by NASM's Martin Collins, noted above. (The first segment is at *http://www.nasm.si.edu/research/dsh/TRANSCPT/ATWOOD1.HTM;* links at each section's end take the reader to the next segment.)

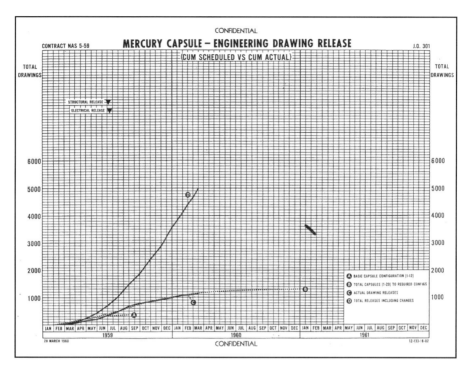

Engineering drawing release for the Mercury capsule, March 1960. *(Source: NASA Contract Administration Files, Procurement Division, box 22, entry 100, RG 255, NARA-Southwest)*

very much need to integrate public-sector innovations and organizations into their private-sector worlds) than for NASA history purposes, unless/until the scope and conceptualization of NASA history shifts in the years ahead.

The figure on this page is a simple graph documenting the engineering drawing releases for the Mercury spacecraft project, from inception through 15 March 1960. Lines A and C indicate that based on component counts, McDonnell had estimated that roughly 1,200 drawings would be needed through early 1961, 500 for the basic configuration and another 700 to include different capsules' mission-specific requirements (e.g., an orbital spacecraft versus one for a ballistic flight). Yet in response to the flow of engineering changes inside the project's first year, the actual number of drawings released reached 5,000 (line D). What significance this volume of redesigns had for project development is evident in Lee Atwood's reflections on Apollo:

> Once your engineering output of drawings and specifications
> gets ragged as far as the schedule is concerned, everything else

gets ragged . . . . An engineering change is really a recall of something that's been released. You stop it, recall your drawing, you get an instruction to change it, bring it back, and the shop is full of that . . . . The things that are most apparent are usually picked up [in] a couple of weeks' surveys, because everybody has some kind of a schedule. Are you on it? Are you not? Well, of course you're not, and the whole place looked like a wreck. It was stop orders, hold orders, missing parts, material procurement had to be modified in many cases.[66]

Change orders were also lightning rods for NASA-industry arm wrestling, as was plainly the case with the Apollo Command Module:

[The CSM] commanded the attention of so many astronauts and so many other people, engineers from Houston and all that. They all had their ideas of how things should be arranged, how controls should be set up, and an awful lot of brouhaha over the actual arrangement [resulted] . . . . One of the astronauts said, in connection with that, "You know, we have a pretty strong union." And they really did. They really did. And Dale [Myers] had to face the problem of arrangement [changes,] plus electrical changes, which came from other parts of the stack and from the ground equipment itself . . . . So there were just infinite refinements and changes, more than the S-II, which was fundamentally structural, a weight problem, . . . whereas the impact on the command module was almost screw by screw, and estimate by estimate and switch by switch.[67]

Researching the dynamics, the politics, the language, and the practices regarding engineering changes, which had pervasive implications for scheduling, cost, and program/artifact reliability and success, demands moving deep within both NASA and contractor organizations, following plant representatives like Wilbur Gray from Mercury to Gemini, chasing the origins and resolutions of

---

66. Atwood Interviews, no. 5, pp. 10–11.

67. Ibid., p. 12. Elsewhere, Atwood added: "Your ideal is to engineer something, put it in the shop, get it built efficiently, and then inspect it carefully and get it out the door and operate. We had an environment that required us to do all those things at once, with much backtracking to make changes. The changes were almost overwhelming. So this was part of the problem of the organization, and it was far from normal. In fact, as Sam Phillips noted, it was to a considerable degree out of control. Parts had to go back for re-engineering, redesign, again and again, re-release, new material, supply and manufacturing and tooling. Yes, it was a struggle" (Atwood Interviews, no. 7, p. 3).

issues that surfaced briefly in configuration control committee minutes, and reconstituting the scale and significance of conflicts over payment for extra work, rework, redesign, supplementary testing and such. Only in this way will historians begin to understand the sadness behind Atwood's crisp aphorism: "If things are done well, NASA succeeded; if things are done poorly, the contractor failed."[68]

A chart issued on the same date as the drawings release graph accounted for the sources of engineering changes through mid-March 1960. I have not yet tallied the total of engineering changes with any precision, as there evidently were several levels of and procedures for requesting and reporting these. However, there were approximately 340 major "contract change orders" in roughly 30 months and at least 6,000 changes to the capsule components and configurations. Key dilemmas included communicating change implementations, authorizing changes, testing implications of changes on other components, identifying failure sources, and updating specifications to reflect changes.

The figure shows that nearly half the ECRs (Engineering Change Requests) emerged from deficiencies detected in testing, here components. A different class of failures, "interferences," was noted under "Manufacturing Coordination," and at that date, my sense is that these were still physical impingements due to the "spaghetti" style of packing in capsule system components. When full capsule testing commenced, a third sort of testing deficiency appeared—system integration and interface problems. These took on yet further ramifications when capsules connected to boosters and to launch-related ground equipment displayed higher-order integration deficiencies. Together, tests and coordination problems represented nearly two-thirds of the ECRs, with improvements, including the famous astronauts' demand for a window, another one-fifth. Engineering studies, the work closest to scientific research, were handled both by NASA Centers and by McDonnell. What significance and impact these studies had on the project is not yet clear, nor do summary documents provide cost figures for the four classes. Still, this simple chart suggests that, from the beginning, waves of engineering changes flowed through manned space projects from multiple directions, generating specialized knowledge, urgent workarounds and overtime labor, unpredictable cost and schedule implications, and fluctuating currents of disorder.[69] In sum, retelling NASA stories from the drafting room and shop floor outwards, from the bottom up, has the potential to reorient a universe of NASA-centric histories.

---

68. Atwood Interviews, no. 4, p. 11.

69. Originals of these two figures may be found in CCP Status Reports, box 20, NAS 5-59, Contract Administration Files, entry 100, NASA-Mercury, RG255.

If such a scheme were to be activated, questions and issues like these, some of which reiterate points sounded earlier, would be tabled, all considering change over time, 1950s–1970s, at least:

1) How were relationships between design revisions and manufacturing practice articulated, in the dual-pressure contexts for extensive changes on one count and design freezes and standardization on another?

2) What implications did NASA contracts have for manufacturers' recruitment, training, and retention of highly skilled workers—engineers, shop-floor workers, and managers—for manufacturers' procurement of machinery and facilities?

3) Considering relationships between primes and subcontractors, what patterns and variations in knowledge exchange, mentoring and monitoring, financial management, etc., emerged in NASA contracts? How were these different from such patterns in military contracts? In commercial contracts? How did they differ when technological stretching was at issue, beyond "routine" complexity?

4) What spatial patterning eventuated in early NASA prime and subcontracts, and did this change? If so, how/when/why? What factors conditioned these outcomes (labor supply, proximities and networks, politics)? How did technological change in communications, creating virtual proximities, affect the spatiality of producing for NASA?

5) How did NASA's fabricators frame practices for identifying/processing/testing new materials, including a) uses in prototyping, b) developing supply lines (titanium being a classic case), and c) adapting existing or creating novel manufacturing procedures? What prior experiences with materials substitution (alloy metals, synthetics) conditioned this process versus what new trajectories of technical knowledge-seeking did the devising of aerospace materials articulate?

6) What historically tested production skills and practices were installed/modified/rejected as shop-floor experience in producing for NASA developed? What occasions for technological learning proved crucial to overcoming obstacles to fabrication, precision, or quality? (Consider candidates like chemical milling, explosive forming, numerically controlled tooling, et al.) What implications for further manufacturing practice did these adaptations/adoptions have, and to what degree were they realized? What conflicts between contractor managers and engineers resulted, between managers/engineers and workers, with what outcomes, including strikes? (N.B.: aircraft/aerospace manufacturing had one of the highest union densities in U.S. manufacturing, 1950–1990.)

7) What would be the breakdown of sources for delays and cost overruns; how would these differ among projects, and why? What links and learning trajectories can be established among projects from the contractors' side—evidence for and significance of knowledge-sharing among aerospace rivals—in terms of materials, electronics, or fabrication shifts? What internal and networked transfers of know-how among projects took place, and how significant were they?[70]

8) What arrays of managerial techniques did contractors deploy in efforts to comprehend and influence fabrication projects that, as Atwood testified, threatened to spin out of control? How did firms assess internally the competence of their production efforts, and to what degree did these evaluations correspond with those authored by NASA overseers? How did such Venn diagrams differ among projects, both over time and across artifact classes?

9) How did primes and subcontractors integrate producing for NASA into their enterprises' overall operations, and how was this integration (or lack of it) evidenced by corporate planning processes, capital funds allocations, career tracks, etc.?

10) What informal practices did contractors' employees devise, at each locus and level of institutional activity, to deal with (make sense of) the persistence of insufficient knowledge, the nonlinearity of testing and performance outcomes, the ubiquity of uncertainty, the stresses of complexity, and the nonrational character of creativity? To what degree were such practices formalized in training procedures or, alternatively, concretized, either spontaneously or in a planned way? Most broadly in this arena, how can we assess the human cost of aerospace innovation to individuals, families, and communities (both of practice and of residence)? How do these practices, trainings, outbursts, quits, and implications compare and contrast with those which materialized in commercial-market enterprises and institutions? Ultimately, how (and to what extent) can producing for NASA be integrated into the experience of American business in the

---

70. Weick makes a provocative comment regarding Westrum's "fallacy of centrality" (the phenomenon of discounting new information because if it were important the individual/organization would already have heard about it): "It is conceivable that heavily networked organizations might find their dense connections an unexpected liability, if this density encourages the fallacy of centrality. 'News' might be discounted if people hear it late and conclude that it is not credible because, if it were, they would have heard it sooner. This dynamic bears watching because it suggests a means by which *perceptions* of information technology might undermine the ability of that technology to facilitate sensemaking. The more advanced the technology is thought to be, the more likely are people to discredit anything that does not come through it. [Thus] the better the information system, the less sensitive it is to novel events" (*Sensemaking*, p. 3, emphasis in original).

Cold War decades, the social life of organizations, the construction of knowledge, and the history of technologies?

These, and surely other, open questions flow from this very partial review of literature and documents concerning NASA-industry relations. Along with the foregoing thoughts on key issues, plausible conceptual frameworks, and implications drawn from that literature, they are offered for reflection and reaction. Perhaps they will encourage what seems a long-overdue vector for research into the distinctive, little-understood world of production for NASA, which exemplifies the intensities, urgencies, joys, and miseries of high-tech, high-pressure, state-sponsored innovation.

CHAPTER 7

NASA AND THE DEPARTMENT OF DEFENSE: ENDURING THEMES IN THREE KEY AREAS

Peter Hays

As with any large government bureaucracies with imprecisely delineated areas of responsibility and potentially overlapping missions, the quality and productivity of the relationship between the National Aeronautics and Space Administration (NASA) and the Department of Defense (DOD) have waxed and waned over the years. The NASA-DOD relationship has been shaped by a series of fundamental issues and questions that accompanied the opening of the Space Age, as well as by subsequent organizational structures, domestic and international politics, technology, and the personalities of key leaders. It is also helpful to consider these relations in terms of the three government space sectors and the bureaucratic roots and culture of the organizations created or empowered to perform these missions: the civil space sector for science and exploration missions performed by NASA, the intelligence space sector for intelligence collection from space by systems procured and operated by the National Reconnaissance Office (NRO), and the defense space sector for military missions enhanced or enabled by space systems procured and operated primarily by the Air Force.[1]

Although relations between these predominant space organizations have usually been quite harmonious and served the United States well, this analysis focuses more attention on periods of uncertainty or tension among these organizations in order to highlight enduring themes that were, and sometimes remain, at stake. Three key issue areas and time periods are examined: organizing to implement America's vision for space in the 1950s, wrestling with the rationale for human spaceflight in the late 1950s and early 1960s, and finding the logical next steps in space transportation and missions in the 1980s. The state of relations between the three predominant space organizations is also an important factor in shaping current issues such as how best to organize and manage national security space activities or implement the President's Vision for Space Exploration.

---

1. The fourth space sector, commercial activities for profit, is regulated by but not performed by government. See the comprehensive discussion of the activities included in each sector in *Report of the Commission to Assess National Security Space Management and Organization* (Washington, DC: Commission to Assess National Security Space Management and Organization, 11 January 2001), pp. 10–14.

DEVELOPING, ORGANIZING, AND IMPLEMENTING AMERICA'S SPACE AGE VISION IN THE 1950S

Following a long and difficult path, the United States Air Force was created as a separate service as a part of the National Security Act of 1947. Its raison d'être was strategic bombing, a mission that had enchanted airmen almost from the inception of flight, provided the foundation for the doctrine that guided America's use of airpower during World War II, and was of even greater concern following the advent of nuclear weapons. The Air Force was organized, trained, and equipped to provide a full range of airpower missions, but strategic bombing, the Strategic Air Command, and bomber pilots formed the institutional core of the new service. The development of long-range ballistic missiles and space systems presented difficult cultural challenges for the Air Force. These new systems held the potential to perform or support the Air Force's core strategic bombing mission, and the service was eager to develop and operate them rather than have them come under the control of the Army or Navy. At the same time, however, the new systems clearly threatened the bombers and bomber pilots at the Air Force's institutional core. The Air Force attempted to walk a difficult organizational tightrope through this situation by pursuing missiles and space strongly enough to keep them out of the grasp of the other services, but not so strongly as to undercut the bomber pilots who ran the service. This Air Force balancing act helps to explain much of its behavior at the opening of the Space Age and continues to be a useful illustration of its ongoing struggles to incorporate space most appropriately in its current and future missions.[2]

Space issues were not primary concerns in the wake of World War II, but America quietly struggled with many questions associated with why it should attempt to go to space and what it might do there. By the mid-1950s, a number of groups and individuals had advanced various reasons for going to space,[3]

---

2. On the evolution of air- and space-power doctrine and their role in Air Force institutional culture see, in particular, Phillip S. Meilinger, ed., *Paths of Heaven: The Evolution of Airpower Theory* (Maxwell Air Force Base [AFB], AL: Air University Press, 1997); Bruce M. DeBlois, ed., *Beyond the Paths of Heaven: The Emergence of Space Power Thought* (Maxwell AFB, AL: Air University Press, 1999); Carl H. Builder, *The Icarus Syndrome: The Role of Air Power Theory in the Evolution and Fate of the U.S. Air Force* (New Brunswick, NJ: Transaction Books, 1994); James M. Smith, *USAF Culture and Cohesion: Building an Air and Space Force for the 21st Century*, Occasional Paper 19 (U.S. Air Force [USAF] Academy: USAF Institute for National Security Studies, June 1998); Mike Worden, *Rise of the Fighter Generals: The Problem of Air Force Leadership, 1945–1982* (Maxwell AFB, AL: Air University Press, 1998).

3. In addition to the space-for-strategic-reconnaissance rationale advocated by RAND, other prominent rationales for space included the scientific imperative that found early expression in the International Geophysical Year (IGY) effort and the exploration imperative perhaps best captured by Wernher von Braun in a series of articles on future space stations published in *Collier's* magazine in the

continued on the next page

but the Eisenhower administration had secretly determined that its primary rationale for going to space was to attempt to open up the closed Soviet state via secret reconnaissance satellites. The RAND Corporation, a think tank sponsored by Army Air Force Commander General Henry H. "Hap" Arnold as a joint project with the Douglas Aircraft Company, was the first to study these issues systematically. RAND's very first report, "Preliminary Design of an Experimental World-Circling Spaceship," was delivered to the Army Air Force in April 1946 and not only detailed the technical design for and the physics involved in launching such a spaceship (the word satellite had not yet come into common usage), but also identified possible military missions for satellites, including communications, attack assessment, navigation, weather reconnaissance, and strategic reconnaissance.[4]

In October 1950, Paul Kecskemeti at RAND produced another comprehensive report on space that Walter A. McDougall believes should "be considered the birth certificate of American space policy."[5] This report highlighted the psychological impact the first satellite would likely have on the public and raised the issue of how the Soviet Union might respond to overflight of their territory and space-based reconnaissance. It even suggested that one way to test the issue of freedom of space would be first to launch an experimental U.S. satellite in an equatorial orbit that would not cross Soviet territory before attempting any satellite reconnaissance overhead the Soviet Union.

The Technological Capabilities Panel and NSC-5520

In March 1954, President Dwight Eisenhower commissioned a secret study and named Dr. James R. Killian, President of the Massachusetts Institute of Technology, as chairman of this Technological Capabilities Panel (TCP). With a thermonuclear standoff looming between the United States and the Soviet Union, Eisenhower wanted the best minds in the country to examine how technology might help to prevent another Pearl Harbor. The TCP report was delivered to the National Security Council (NSC) in February 1955. The report stands out as one of the most important and influential examinations of U.S. national security ever undertaken; it formed the foundation for U.S. national security planning for at least the next two years, made remarkably prescient

continued from the previous page
early 1950s. Several of these articles are reprinted in John M. Logsdon, ed., *Exploring the Unknown*, vol. 1, *Organizing for Exploration* (Washington, DC: NASA SP-4407, 1995), pp. 176–200.

4. Merton E. Davies and William R. Harris, *RAND's Role in the Evolution of Balloon and Satellite Observation Systems and Related Space Technology* (Santa Monica: RAND Corporation, 1988), pp. 6–9. Portions of RAND's first report are reprinted in Logsdon, *Exploring the Unknown*, vol. 1, pp. 236–244.

5. Walter A. McDougall, . . . *The Heavens and the Earth: A Political History of the Space Age* (New York: Basic Books, 1985), p. 108.

predictions about the evolution of the superpowers' strategic nuclear arsenals, and called for crash programs to develop early-warning radars and ballistic missiles, as well as to improve the survivability of Strategic Air Command assets in the face of potential nuclear attack.[6]

The TCP also called for a vigorous program to improve U.S. technological intelligence collection capabilities. Killian and Edwin H. "Din" Land, founder of the Polaroid Corporation and chairman of the intelligence subcommittee of the TCP, were briefed on a wide range of potential collection methods and systems, including satellites, but became most enthused about attempting high-altitude reconnaissance overflights of the Soviet Union via a jet-powered glider that was then on the drawing boards at Clarence "Kelly" Johnson's Lockheed skunk works in Burbank, California. They recommended production of this new aircraft during a series of briefings that culminated in an Oval Office meeting on 24 November 1954, attended by the President, Secretaries of State and Defense, as well as top DOD and Central Intelligence Agency (CIA) officials.[7] The initial programs and structure for a national strategic reconnaissance program were discussed at this meeting; the President verbally authorized the CIA to begin development of the CL-282 (U-2) aircraft program with Air Force support.[8]

---

6. For the text of the TCP report, see John P. Glennon, ed., *Foreign Relations of the United States, 1955–1957*, vol. 19, *National Security Policy* (Washington: Department of State, 1990), pp. 42–55. James R. Killian, Jr., provides details on the workings of the TCP in *Sputnik, Scientists, and Eisenhower: A Memoir of the First Special Assistant to the President for Science and Technology* (Cambridge: MIT Press, 1977), pp. 67–93. On the relationship between the TCP report and subsequent U.S. nuclear strategy, see Lawrence Freedman, *The Evolution of Nuclear Strategy* (New York: St. Martins Press, 1983), pp. 76–90.

7. Stephen M. Rothstein, *Dead on Arrival? The Development of the Aerospace Concept, 1944–58* (Maxwell AFB, AL: Air University Press, November 2000), p. 43; Clarence E. Smith, "CIA's Analysis of Soviet Science and Technology," in *Watching the Bear: Essays on CIA's Analysis of the Soviet Union*, ed. Gerald K. Haines and Robert E. Leggett (Langley, VA: Center for the Study of Intelligence, 2003); Gregory W. Pedlow and Donald E. Welzenbach, *The CIA and the U-2 Program, 1954–1974* (Langley, VA: Center for the Study of Intelligence, 1998). Land wrote a 5 November 1954 letter to CIA Director Allen W. Dulles outlining "A Unique Opportunity for Comprehensive Intelligence" via a specialized high-altitude aircraft; the letter is available electronically from the National Security Archive at *http://www2.gwu.edu/~nsarchiv/NSAEBB/NSAEBB74/U2-03.pdf*.

8. It is not clear from unclassified sources how much RAND reports or the Air Force's nascent WS-117L reconnaissance satellite system was discussed during these meetings. Satellite reconnaissance was strongly advocated by a series of RAND reports during the early 1950s (particularly the 1954 "Project Feed Back Report"; see Logsdon, *Exploring the Unknown*, vol. 1, pp. 269–274). In late 1953, the Air Research and Development Command (ARDC) had published a management "Satellite Component Study" and designated it Weapons System (WS) 117L. On 1 July 1954, the Western Development Division (WDD) of ARDC was established in Inglewood, CA, under the command of Colonel Bernard Schriever (who had participated in Project Feed Back), primarily to speed development of ballistic missiles. WDD formally initiated a program to develop reconnaissance satellites in Weapons System Requirements Number 5 (WS-117L), "System Requirement for an Advanced Reconnaissance System," secretly issued on 27 November 1954. According to Spires, "Focused on Project Aquatone,

continued on the next page

Following the start of these new technical intelligence collection initiatives, in early 1955 the National Academy of Sciences proposal for DOD to support the launch of a scientific satellite for research during the July 1957–December 1958 International Geophysical Year (IGY) landed on the desk of Donald Quarles, Assistant Secretary of Defense for Research and Development. Quarles used this opportunity to tie together various strands of the administration's embryonic policies on satellites, intelligence collection, and ballistic missiles by drafting a space policy for review by the National Security Council. His draft formed the basis for NSC-5520, the most important space policy of the Eisenhower administration. Portions of this document remain classified almost 50 years after it was written, but the basic themes are quite clear: the Space Age would soon open; the TCP "recommended that intelligence applications warrant an immediate program leading to a very small satellite in orbit around the earth" and a reexamination "of the principles or practices of international law with regard to 'Freedom of Space'"; DOD should provide support for launching the IGY satellite so long as such support would not delay or otherwise impede DOD programs; and all U.S. space efforts should be arranged to emphasize peaceful purposes and freedom of space.[9] NRO historian Cargill Hall succinctly summarized how Eisenhower's space policy was put into practice: "The IGY scientific satellite program was clearly identified as a stalking horse to establish the precedent of overflight in space for the eventual operation of military reconnaissance satellites."[10] The final piece of the policy, satellite, and booster puzzle fell into place when Quarles established an advisory committee to decide

---

continued from the previous page

the U-2 project that promised immediate results, the military satellite program received little interest or support from Killian and his experts. At that time, he considered the Air Force's reconnaissance satellite a 'peripheral project.' This attitude from one so influential helps explain the less than enthusiastic administration support of the Air Force's Advanced Reconnaissance Satellite in the two years preceding *Sputnik*. Despite the growing need for strategic intelligence and awareness that the U-2 represented a temporary solution, Killian declined to actively support the military satellite until after the launch of the first *Sputnik*. He believed an American scientific satellite had to precede the launch of a military vehicle to provide the overflight precedent for military satellites to operate with minimum international criticism" (David N. Spires, *Beyond Horizons: A Half Century of Air Force Space Leadership* [Colorado Springs: Air Force Space Command, 1998], p. 39). See Robert L. Perry, *Origins of the USAF Space Program, 1945–1956* (Los Angeles: Space Systems Division, 1961), p. viii, microfiche document 00313 in *U.S. Military Uses of Space 1945–1991: Index and Guide* (Washington, DC: The National Security Archive, and Alexandria, VA: Chadwyck-Healey, Inc., 1991); Spires, *Beyond Horizons*.

9. NSC-5520 was approved at the NSC meeting on 26 May 1955, and Eisenhower signed it the following day. Quotations are from the declassified portions reprinted in Dwayne A. Day, "Invitation to Struggle: The History of Civil-Military Relations in Space," in *Exploring the Unknown*, ed. John M. Logsdon, vol. 2, *External Relationships* (Washington, DC: NASA SP-4407, 1996), p. 241.

10. R. Cargill Hall, "Origins of U.S. Space Policy: Eisenhower, Open Skies, and Freedom of Space," in *Exploring the Unknown*, ed. Logsdon, vol. 1, p. 222.

which military booster should be used, and it recommended the Navy's Viking (Vanguard) booster rather than the Army or Air Force proposals.[11]

This most important but secret process to legitimize overflight spelled out by NSC-5520 was not at all clear at the time, even to many of the senior participants in the development of early U.S. space and missile programs. Indeed, it remained politically expedient to continue obscuring the origins and operation of space-based intelligence collection, America's first and arguably most important space program, for decades into the Space Age.[12] This subtext is, however, critical to understanding the nature of the relationships between NASA, the NRO, and the Air Force.

Responding to the Sputniks and Creating NASA

The Eisenhower administration carefully planned to use the opening of the Space Age to create a new legal regime that would legitimize the operation of reconnaissance satellites, but, despite repeated warnings, it did not prepare well for the psychological implications of this milestone. The worldwide public reaction to the Soviet successes with Sputniks I and II on 4 October and 3 November 1957 precipitated a crisis in confidence in Eisenhower's leadership that was seized upon by opponents of his New Look defense policies and shaped the remainder of his second term. In an attempt to limit the growing crisis, one of Eisenhower's first responses was to appoint Killian to a new position as science adviser to the President. A second major administration response was the establishment of the Advanced Research Projects Agency (ARPA) within DOD on 7 February 1958. ARPA was authorized to direct or perform virtually all United States space research and development efforts but was viewed by many as a stopgap measure and proved insufficient to derail the mounting pressure to create a comprehensive, independent, and civilian space agency.[13]

---

11. The Army Ballistic Missile Agency's Project Orbiter proposal was the most advanced of the proposals presented to the Stewart Committee. On 20 September 1956, a Jupiter-C rose to an altitude of 600 miles while traveling 3,000 miles downrange despite having an inert fourth stage (it was filled with sand) to preclude this vehicle from accidentally launching the first satellite and thereby circumventing the IGY stalking-horse strategy laid out in NSC-5520. See Major General John B. Medaris, U.S. Army (USA) (ret.), *Countdown for Decision* (New York: G. P. Putnam's Sons, 1960), pp. 119–20, 147.

12. The existence of the NRO was first officially acknowledged in September 1992. The importance and uses of United States overhead photoreconnaissance (IMINT), as well as the fact that the United States conducts overhead signals intelligence (SIGINT) and measurement and signature intelligence (MASINT) collection, were first acknowledged in the 19 September 1996 National Space Policy Fact Sheet.

13. Other major responses included authorization for the ABMA to prepare to launch a satellite on the modified V-2 booster known as the Jupiter-C or Juno (this system boosted Explorer I, America's first satellite, into orbit on 31 January 1958), as well as the congressional hearings on satellite and missile programs that were called by Majority Leader Lyndon Johnson and held between 25 November 1957 and 23 January 1958.

Killian was the most important actor in creating NASA as the centerpiece of the organizational structure America developed in response to the Sputniks shock, but he worked very closely with other key actors and organizations such as the President, Senator Lyndon B. Johnson (D-Texas), and the military services. By the end of 1957, the President's Science Advisory Committee (PSAC), under Killian, had decided that a scientifically oriented civil space program, rather than a military program, ought to be the nation's top space priority and that the new civilian space agency ought to be built out of and modeled after the National Advisory Committee for Aeronautics (NACA). This approach was the primary recommendation of the PSAC headed by Edward Purcell; Killian used the Purcell Committee findings to help persuade Eisenhower of the need for a civilian agency and sent proposed legislation to Congress on 2 April 1958.

Both houses held extensive hearings on the civilian space agency proposal during April and May; soon, however, they drifted into positions that differed from one another and from the administration. The most contentious issues revolved around three areas: the relative priority of civil and military space efforts, the appropriate relationship between civilian and military space organizations, and the organizational structure for creating national space policy. Office of the Secretary of Defense (OSD) witnesses included Deputy Secretary Quarles, ARPA Director Roy Johnson, and ARPA Chief Scientist Herbert York. They emphasized that DOD must retain the power to define and control military space programs. Service witnesses generally took the same positions they had over the creation of ARPA. The Navy opposed a strong civilian agency and preferred an organization similar to NACA that would support but not shape military space efforts. The Air Force was confident of its position as the lead service for military space and supported a strong civilian agency as a means to undercut Navy and Army space efforts. By contrast, the Army opposed the creation of a civilian agency or the division of scientific and military space missions; the Army also urged if a civilian space agency were created that it, rather than DOD or the Air Force, should control the national space effort.[14]

Compromises were ironed out following a meeting between Eisenhower and Senator Johnson on 7 July and during Conference Committee meetings later that month. The major compromises included a modified version of the House's Civilian-Military Liaison Committee (CMLC), creation of the National Aeronautics and Space Council (NASC) at the White House, and carefully brokered language in Section 102(b) that was designed to delineate between NASA and DOD space missions. The latter issue was perhaps the most controversial aspect of the entire process. The final language called for NASA to exercise control over all U.S. space activities

---

14. Enid Curtis Bok Schoettle, "The Establishment of NASA," in *Knowledge and Power: Essays on Science and Government*, ed. Sanford A. Lakoff (New York: Free Press, 1966), pp. 162–270.

except that activities peculiar to or primarily associated with the development of weapons systems, military operations, or the defense of the U.S. (including Research and Development necessary to make effective provision for the defense of the U.S.) shall be the responsibility of and shall be directed by the DOD.[15]

Eisenhower signed the National Aeronautics and Space Act into law on 29 July, and NASA was created on 1 October 1958.

Frictions over manned spaceflight, budgets, and organizational structure between NACA and ARPA were evident before NASA was established. Both NACA and ARPA strongly desired to control manned spaceflight, and both organizations fought hard for this mission during a series of meetings with the Bureau of the Budget during the summer of 1958. Once again, Killian was an important player behind the scenes; he helped broker a compromise whereby NASA would design and build the capsules for manned spaceflight and DOD would concentrate on the boosters required for this mission.[16] Killian also pushed to reprogram $117 million from ARPA and the Air Force to NASA, helped ARPA retain $108 million for space programs outside of the WS-117L (see note 8), and steadfastly refused to entertain any suggestions to change the organization or reduce the $186-million budget for the WS-117L.[17] Organizational changes were also looming. The Army's Jet Propulsion Laboratory wished to transition immediately to NASA, and the Army was close to granting this request, but it wanted to use the transfer of JPL as a bargaining chip in its efforts to retain its space crown jewel, the von Braun rocket team at ABMA.

Completing the Organizational Structure

Following the creation of NASA, there were three major tracks of activity that shaped NASA-DOD relations during the remainder of Eisenhower's term and into John Kennedy's administration: moving ABMA into NASA, consolidating DOD space activities under the Air Force, and establishing the NRO. Each of these tracks helped establish the basic organizational structures and bureaucratic interests that endure today.

Army Secretary William Brucker and ABMA Commander Major General John Medaris understood very well how hard the Army had worked to capture and maintain the von Braun group as one of the key spoils of World War II and just how important von Braun's expertise would be to any major U.S.

---

15. Ibid., pp. 260–261.

16. Robert A. Divine, *The Sputnik Challenge* (New York: Oxford University Press, 1993), p. 150. Divine notes that Killian had quickly emerged as Eisenhower's "key post-*Sputnik* advisor."

17. Ibid., pp. 151–152.

space effort—they were not about to give up ABMA without a fight. They had strongly opposed creation of a powerful civilian space agency, and after NASA was established, they redoubled their efforts to retain control of ABMA. NASA had inherited NACA's infrastructure but initially lacked expertise in many space areas such as the development of large boosters. By contrast, ABMA contained arguably the world's best booster development team, but it lacked a specific military rationale for developing large boosters.[18] In October 1958, T. Keith Glennan, NASA's first Administrator, and Deputy Secretary Quarles worked out a deal to resolve this anomalous situation by transferring JPL and ABMA to NASA. Brucker and Medaris successfully blocked transfer of ABMA at this time. But in December, the NASC brokered a second compromise that moved JPL to NASA and left the von Braun team under ABMA while directing that their work on Saturn would be under contract to NASA.

Significant military space organizational restructuring was also under way within DOD. Following creation of NASA and pressure on ABMA, the Navy and the Army, in particular, became increasingly concerned with retaining their military space capabilities, shoring up ARPA, and formulating the proper bureaucratic structure for military space. The Air Force, by contrast, was growing increasingly confident of its inside track for gaining control over military space missions, supported a strong NASA, and continued to oppose ARPA's direction of military space efforts. Another key player that entered the mix at this time was Herbert York, the first Director of Defense Research and Engineering (DDR&E), a position created by the 1958 Defense Reorganization Act.

Debates over DOD's space organizational structure became increasingly heated during 1959 and came to a head in September. In April, Chief of Naval Operations Admiral Arleigh Burke highlighted the indivisibility of space and proposed to the Joint Chiefs of Staff (JCS) creation of a unified (multiservice) command for space. Burke's proposal was supported by the Army but was strongly opposed by the Air Force. Arguing that space systems represented a better way of performing existing missions, the Air Force advocated treating space systems on a functional basis under ARPA or, preferably, under the Air Force. DDR&E York weighed in on this debate and sided strongly with the Air Force, largely because he was eager to consolidate military space efforts under the Air Force as a way to rein in what he considered to be overreaching space proposals on the part of all the services. A memorandum

---

18. ABMA had been tasked by ARPA to study and design a 1.5-million-pound-thrust booster that came to be known as the Saturn B. The Saturn B was, in turn, a primary driver behind the ABMA Project Horizon proposal to use 149 Saturn launches to build a 12-person lunar outpost by 1966. See John M. Logsdon, *The Decision to Go to the Moon: Project Apollo and the National Interest* (Cambridge: MIT Press, 1970), pp. 51–52.

from Secretary Neil McElroy to JCS Chairman General Nathan Twining on 18 September attempted to resolve these disputes and represented a significant bureaucratic victory for the Air Force. McElroy assigned responsibility for most satellite systems, payload integration, and "the development, production, and launching of space boosters" to the Air Force.[19] The memo also found that "establishment of a joint military organization with control over operational space systems does not appear desirable at this time."[20]

For the remainder of the Eisenhower administration and the beginning of the Kennedy administration, the space prospects of the Army continued to decline while those of the Air Force usually continued to rise. Following the transfer of the Redstone program in December 1958 and the Saturn program in November 1959, between March and July 1960 the Army moved the von Braun team and 6,400 other ABMA personnel under NASA control.[21] Eisenhower presided over the 8 September 1960 ceremony in Huntsville, Alabama, that dedicated the Marshall Space Flight Center and officially moved the Army out of the space business. It took decades for the Army to recover from this loss and regain its enthusiasm towards space, but today the Army is the largest user of military space data among the services, and it is eagerly considering a range of significant future enhancements such as Global Positioning System (GPS) III satellites and Blue Force Tracking.

Despite Air Force support for NASA's creation, NASA's role in absorbing the Air Force's most serious competition for developing military space systems, and generally good early relations between America's two largest space organizations, NASA–Air Force relations hit a snag after an internal letter from Air Force Chief of Staff General Thomas White to his staff was leaked to Congressman Overton Brooks (D-Louisiana), Chairman of the House Committee on Science and Astronautics. The bulk of White's 14 April 1960 letter urged the Air Force "to cooperate to the maximum extent with NASA, to include the furnishing of key personnel even at the expense of some Air Force dilution of technical talent."[22] The opening two sentences of White's letter, however, raised questions about the strength and longevity of Air Force support for NASA independence:

---

19. Spires, *Beyond Horizons*, p. 77. ARPA returned responsibility for the WS-117L to the Air Force. By this time, the program consisted of three separate developmental satellite systems: Corona, a recoverable film photoreconnaissance system; Samos, an electro-optical system designed to downlink imagery electronically; and Midas, an infrared satellite sensor system designed to detect ballistic missile launches. The Navy acquired the Transit satellite navigation systems, and the Army gained responsibility for Notus communications satellites. This approach overturned ARPA's monopoly on control over military satellite systems.

20. Ibid.

21. Day, "Invitation to Struggle," p. 253.

22. Ibid., p. 256.

President Dwight D. Eisenhower and Mrs. George C. Marshall unveil the bronze bust of General George C. Marshall during the dedication ceremony of the George C. Marshall Space Flight Center (MSFC) in Huntsville, Alabama, on 8 September 1960. On 21 October 1959, President Eisenhower directed the transfer of personnel from the Redstone Arsenal's Army Ballistic Missile Agency Development Operations Division to NASA. The complex of the new NASA Center was formed within the boundaries of Redstone Arsenal in Huntsville. MSFC began its operations on 1 July 1960 after the transfer ceremony, with Dr. Wernher von Braun as Center Director. *(NASA MSFC photo no. 9131490)*

> I am convinced that one of the major long term elements of the Air Force future lies in space. It is also obvious that NASA will play a large part in the national effort in this direction and, moreover, inevitably will be closely associated, if not eventually combined with the military.[23]

In March 1961, Brooks held hearings to discuss White's letter, the proper balance between military and civil space, and the general direction of U.S. space efforts. Brooks sought and even received clarification from President Kennedy. On 23 March, Kennedy wrote a letter to Brooks that emphasized several key points:

---

23. Ibid.

It is not now, nor has it ever been, my intention to subordinate the activities in space of the National Aeronautics and
Space Administration to those of the Department of Defense. I
believe, as you do, that there are legitimate missions in space for
which the military services should assume responsibility, but
that there are major missions, such as the scientific unmanned
and manned exploration of space and the application of space
technologies to the conduct of peaceful activities, which should
be carried forward by our civilian space agency.[24]

Kennedy's letter helped to delineate space missions between NASA and the
Air Force and indicated Kennedy's growing emphasis on civil missions, an
emphasis that would grow significantly stronger after Yuri Gagarin's orbital
flight some three weeks later.

During the same month as the Brooks hearings, Air Force control over
military space programs was solidified when Secretary Robert McNamara
issued Defense Directive 5160.32, "Development of Space Systems." This directive built on Secretary McElroy's September 1959 memo and the January 1961
recommendations of incoming science adviser Jerome Wiesner. It gave the
Air Force operational control over almost every military space program from
research and development through launch and operations and stopped just short
of naming the Air Force as DOD's executive agent for space. This was, of
course, a welcome development for the Air Force, but McNamara's motivation,
like York's before him, was to consolidate and prune rather than to encourage
Air Force leadership in developing more robust military space activities.

The creation of NRO was the final major organizational response to
the opening of the Space Age and was, like the IGY stalking-horse strategy
in NSC-5520, an official state secret hidden from the public and even many
of the leaders of U.S. civil and military space efforts. Following Sputnik, in
January 1958 the NSC granted highest national priority to development of
an operational reconnaissance satellite, but Eisenhower had doubts about Air
Force management of the WS-117L program and was particularly troubled by
press leaks about the program. Decisions made at meetings on 6–7 February
1958 between the President, Killian, Land, Director of Central Intelligence
Allen Dulles, Secretary of Defense Neil McElroy, and Eisenhower's staff
secretary, Colonel Andrew Goodpaster, created ARPA and publicly gave this
new agency all open military space programs. In secret, these decisions also
gave ARPA direction over the highest priority WS-117L and moved control
of the Corona recoverable film photoreconnaissance system from the Air

---

24. Logsdon, *Exploring the Unknown*, vol. 2, p. 317.

Force to the CIA in an organizational structure that initially mirrored that of the U–2.[25]

U.S. efforts to develop operational spysat systems faced very daunting technological challenges during the late 1950s and early 1960s. Corona was the most mature technology, yet between February 1959 and June 1960, it still suffered a string of 12 consecutive failures of various types that prevented recovery of film imagery from space before achieving its first success in August 1960. These problems with Corona, along with even more serious difficulties with Samos and Midas, prompted Eisenhower, in May 1960, to direct his new science adviser, George Kistiakowsky, to put together a committee to recommend changes to improve these programs. Kistiakowsky and Defense Secretary Thomas Gates decided on the structure and charter of what became known as the Samos Panel and selected members including Under Secretary of the Air Force Joseph Charyk, Deputy DDR&E John Rubel, Killian, Land, York, and Purcell. The Samos Panel reported its recommendations at an NSC meeting on 24 August. Eisenhower and the NSC strongly supported the primary recommendation, immediate creation of an organization to provide a direct chain of command from the Secretary of the Air Force to the officers in charge of each spysat project; this decision was the genesis of the NRO.[26] It represented another vote of no confidence in the Air Force to manage spysat programs through military channels, moved this highest priority space mission and its products out of the military chain of command, and completed America's three-legged organizational structure for space.

In addition to the organizational changes discussed above, beginning in 1961 there was a major change in the way information was released about U.S. military space programs that had a significant effect both on contemporary analyses and the historiography of space. A security clampdown was slowly implemented, first on spy satellite programs and then on all military space efforts. The Samos 2 launch on 31 January 1961 was the first to be affected by the Kennedy administration's new publicity guidelines. Assistant Secretary of Defense for Public Affairs Arthur Sylvester and NRO Director Charyk worked out a very terse statement provided to the press following this launch

25. R. Cargill Hall, "Clandestine Victory: Dwight D. Eisenhower and Overhead Reconnaissance in the Cold War" (paper presented at the "Eisenhower and National Security for the 21st Century Symposium," Industrial College of the Armed Forces, Washington, DC, 26–28 January 2005); Day, "Invitation to Struggle," p. 250; Kenneth E. Greer, "Corona," in *Corona: America's First Satellite Program*, ed. Kevin C. Ruffner (Langley, VA: Center for the Study of Intelligence, 1995).

26. George Kistiakowsky, *A Scientist at the White House: The Private Diary of President Eisenhower's Special Assistant for Science and Technology* (Cambridge: Harvard University Press, 1976); Hall, "Clandestine Victory"; Gerald M. Steinberg, *Satellite Reconnaissance: The Role of Informal Bargaining* (New York: Praeger Publishers, 1983); Jeffrey T. Richelson, *America's Secret Eyes in Space: The U.S. Keyhole Spy Satellite Program* (New York: Harper & Row, 1990).

that contrasted significantly with the large prelaunch publicity packages which had been given out previously.[27] The remainder of 1961 saw a gradual tightening of the security classifications with less and less information provided with each successive launch.[28]

The Air Force chafed at these restrictions, and many officers, including General Schriever, continued publicly to press the case for an increased military space program. This ongoing public discussion of military space programs by the Air Force greatly irritated President Kennedy, and on more than one occasion, he called Sylvester directly, demanding to know why he had "let those bastards talk."[29] Following these calls, Sylvester's office greatly intensified the screening process required for all public releases on space. As a result of this widespread clampdown, planned speeches by Air Force general officers were very carefully screened by civilians in Sylvester's office for any references to the Samos program, and the winter–spring 1960–1961 *Air University Quarterly Review* issue devoted to "Aerospace Force in the Sixties" was heavily censored, including the removal of an article entitled "Strategic Reconnaissance" in its entirety.[30]

The final step in this security-intensification process was the classified DOD directive issued on 23 March 1962 known as the "blackout" directive. According to Stares, this directive

> prohibited advance announcement and press coverage of *all* military space launchings at Cape Canaveral and Vandenberg AFB. It also forbade the use of the names of such space projects as Discoverer, MIDAS and SAMOS. Military payloads on space vehicles would no longer be identified, while the programme names would be replaced by numbers.[31]

While this directive may have made it somewhat more difficult for the Soviets to distinguish between different types of U.S. military space programs and launches, it certainly made it much more difficult for the Air Force to sell its preferred space program to the public or Congress and helped to establish and perpetuate a wide divergence between public knowledge and perceptions of the NASA and DOD space programs.

---

27. Paul B. Stares, *The Militarization of Space: U.S. Policy, 1945–1984* (Ithaca: Cornell University Press, 1985), p. 64. Sylvester and Charyk were mindful of the volume of information provided in the past and deliberately opted for a slow blackout process in the hopes that this would arouse less attention than an abrupt blackout.

28. Richelson, *Secret Eyes*, p. 53. By the time of the Samos 5 launch on 22 December 1961, DOD officials would no longer confirm that the Samos program even existed.

29. Stares, *Militarization of Space*, p. 64.

30. Steinberg, *Satellite Reconnaissance*, p. 43.

31. Stares, *Militarization of Space*, p. 65, emphasis in original.

Wrestling with the Rationale
for Human Spaceflight in the Early Space Program

With the organizational structure for space completed, the majority of issues concerning the relationships and cooperation between NASA, the Air Force, and the NRO revolved around the rationale for human spaceflight, the organizations empowered to perform these missions, and developing and operating space launch vehicles. These issues were, of course, also instrumental in initially shaping and continuing to mold America's space bureaucratic structure.

Jockeying for Human Spaceflight Missions

The period from the opening of the Space Age until completion of NASA's Apollo Moon race was a time of both cooperation and intense competition between NASA and the Air Force. Both organizations were very interested in and believed they would be directed to develop major human spaceflight programs; their intricate dance fundamentally shaped these programs. The Air Force had emerged as the most powerful military space actor, advanced a variety of rationales for manned military spaceflight, and strongly believed—especially at the beginning of the Kennedy administration—that it would be given approval for a major manned spaceflight program. NASA, meanwhile, drew heavily on Army and Air Force expertise to develop its spaceflight programs and struggled to transition from science to prestige as the most important rationale for its manned spaceflight programs. During the 1960s, the Air Force was repeatedly rebuffed in its attempts to gain a foothold in military manned space missions; following the failure of Dyna-Soar, Blue Gemini, and the Manned Orbiting Laboratory (MOL), the Air Force was sufficiently chastened that it remains highly skeptical of manned military missions.

The Air Force had displayed a significant amount of interest in military manned spaceflight well before Sputnik, but, like almost all other space activities, this interest was energized following the Soviet triumph. The Air Force's earliest support was for the dynamic soaring (Dyna-Soar) concept for skipping off the Earth's atmosphere to extend the range of a spaceplane that might be used for a variety of missions including strategic bombing, reconnaissance, and antisatellite attacks.[32] By 1955, the Bell Aircraft Company had received

---

32. The idea of an antipodal bomber that would skip off Earth's atmosphere to achieve intercontinental range was developed by an Austrian, Dr. Eugen Sänger, and was considered in 1943 by von Braun and General Walter Dornberger at Peenemünde. Dornberger worked for Bell Aircraft after the war and was a tireless advocate for Dyna-Soar. The idea of flying to and from space held special appeal to the test pilots who derided the capsule approach to manned spaceflight as "Spam in a can." See Tom Wolfe, *The Right Stuff* (New York: Bantam Press, 1980). The definitive work on Dyna-Soar is Roy F. Houchin, *US Hypersonic Research and Development, 1944–1963: The Rise and Fall of Dyna-Soar* (London: Frank Cass, 2005).

over $1 million in Air Force funding and had raised an additional $2.3 million from six other aerospace firms willing to ante up company funds to support the prospect of a major Air Force manned military space mission.[33]

Following Sputnik, Air Force leaders were among the first to adopt a space-race attitude toward manned spaceflight and supported using either spaceplanes or capsules to achieve rapid results. In a 31 January 1958 letter from Deputy Chief of Staff for Research and Development Lieutenant General Donald Putt to the Air Research and Development Command, Putt advocated rapid development of manned spaceflight and indicated it was "vital to the prestige of the nation that such a feat be accomplished at the earliest technically practicable date—if at all possible before the Russians."[34] Recognizing that congressional deliberations on creating a civilian space agency were under way, the Air Force mounted a full court press to gain approval of its Manned Military Space System Development Plan (MISS) before the civilian agency was established.[35] The MISS plan received support from the highest levels of the Air Force and throughout many DOD offices but was shot down, first by ARPA Director Johnson on 25 July and then by the President a few weeks later, when he formally assigned the role of human spaceflight to NASA.[36]

The Rise and Fall of Dyna-Soar

After its failure to advance its MISS plans and Eisenhower's decision to make NASA primarily responsible for manned spaceflight, the Air Force refocused on the Dyna-Soar program, and it became the service's top space priority. The official start of the program came in November 1957, when Air Research and Development Command issued System Development Directive 464.[37] In May 1958, the Air Force and NACA signed a Memorandum of Understanding (MOU) indicating that Dyna-Soar would be a joint Air Force–NACA project managed and funded along the lines of the X-15 effort.[38] The program took more definite shape during 1959 and 1960, when the Air Force laid out a four-step development program that was designed to achieve full operational capability by 1966. The zenith for the program came early in the Kennedy administration, when the plans were finalized for a small, single-

---

33. McDougall, *Heavens and Earth*, p. 339.

34. "Early AF MIS Activity," microfiche document 00446 in *U.S. Military Uses of Space*.

35. Ibid. The MISS plan had four phases. The first, "Man-In-Space-Soonest," called for the first orbital flight by April 1960, and the last, "Manned Lunar Landing and Return," was to be accomplished by December 1965. The entire program was projected to cost only $1.5 billion.

36. Day, "Invitation to Struggle," p. 252.

37. "Review and Summary of X-20 Military Application Studies," microfiche document 00450 in *U.S. Military Uses of Space*.

38. "Memorandum of Understanding," document II-7 in *Exploring the Unknown*, ed. Logsdon, vol. 2, pp. 284–285.

seat, delta-winged space glider (designated as the X-20 in 1962) that would be launched atop a Titan III and land like an airplane.

Soon, however, the X-20 ran afoul of McNamara's systems analysis approach and his fears of provoking an action-reaction arms race in space. After McNamara refused to accelerate the program, even after receiving an unrequested extra $85.8 million from the House Appropriations Committee for fiscal year (FY) 1962, funding was cut to only $130 million for FY 1963 and 1964, and the first scheduled flight was slipped to 1966.[39] Next, McNamara's systems analysts "showed that a modified Gemini might perform military functions better and more cheaply than the X-20."[40] This finding prompted McNamara to attempt to gain a large role for the Air Force in Project Gemini, a move NASA Administrator James Webb successfully parried by citing the impact of such a restructuring on the nation's highest priority Apollo Program. Instead, on 23 January 1963, Webb and McNamara signed an agreement to allow DOD experiments on Gemini missions. During this time, the Air Force also proposed a plan to procure some of NASA's Gemini spacecraft under a program referred to as Blue Gemini.[41]

The creation of the DOD Gemini Experiments Program and studies on the military usefulness of a space station that would evolve into the Manned Orbiting Laboratory (MOL) program weakened the rationale for the X-20 and placed additional pressures on the troubled program.[42] In October 1963, the

---

39. McDougall, *Heavens and Earth*, p. 340; Stares, *Militarization of Space*, p. 130.

40. McDougall, *Heavens and Earth*, p. 340.

41. Stares, *Militarization of Space*, p. 79. DOD eliminated the Blue Gemini and Military Orbital Development System (MODS) programs from the Air Force budget in January 1963. The NASA-DOD experiment program was officially titled Program 631A, "DOD Gemini Experiments Program," and called for 18 experiments to be run on Gemini flights between October 1964 and April 1967 for a cost of $16 million. The experiments were programmed for areas such as satellite inspection, reconnaissance, satellite defense, and astronaut extravehicular activity. See Colonel Daniel D. McKee, "The Gemini Program," *Air University Review* 16 (May–June 1965): 6–15; Gerald T. Cantwell, "AF in Space, FY 64," pp. 31–36, microfiche document 00330 in *U.S. Military Uses of Space*.

42. NASA and DOD interactions during 1963 over the issue of future manned space stations greatly affected the X-20 and other Air Force man-in-space plans. In November 1962, the Air Force had completed a study on a limited military space station known as the MODS. Based upon the MODS concept, Webb and McNamara discussed the possibility of a joint station project, and on 27 April 1963, they agreed that neither organization would initiate station development without the approval of the other. McNamara pressed Webb for a commitment to a joint program, but Webb did not want to make any pledge that might sidetrack Apollo. Finally, after intervention by Vice President Johnson and the NASC, NASA and DOD agreed in September that, if possible, stations larger and more sophisticated than Gemini and Apollo would be encompassed in a single project. After DDR&E Harold Brown recommended to McNamara on 14 November that the X-20 be canceled and replaced by studies on what would become the MOL program, Brown next attempted, unsuccessfully, to coordinate a joint NASA-DOD station. NASA, wary that the fairly large and sophisticated station Brown favored might threaten its space turf, suggested that DOD pursue a smaller and less sophisticated space *laboratory* rather than a space station. DOD accepted at least the semantic importance of this distinction in initiating MOL studies for an independent military station. See Cantwell, "AF in Space, FY 64," pp. 16–23, microfiche document 00330 in *U.S. Military Uses of Space*.

PSAC compared the relative military utility of the Gemini, X-20, and MOL programs and judged that the X-20 held the least potential.[43] By this time, according to the editor of *Missiles and Rockets*, the X-20 had been "reviewed, revised, reoriented, restudied, and reorganized to a greater extent than any other Air Force program."[44] On 10 December 1963, Secretary McNamara publicly announced cancellation of the X-20 program and, at the same time, assigned primary responsibility for developing MOL to the Air Force.[45]

The MOL Program and the Demise of Military Spaceflight Dreams

Announced at the same time as the cancellation of the X-20, MOL quickly took the place of the X-20 and became the cornerstone of Air Force efforts to build a significant manned military presence in space. The Air Force put a great deal of energy, effort, and funding into MOL, and this project soon emerged as DOD's only manned military space program. Numerous technical and especially political problems beset the program, and MOL was repeatedly cut back and stretched out in the late 1960s. The Nixon administration officially canceled MOL on 10 June 1969. Having been repeatedly thwarted and left without any military man-in-space programs, for many years the Air Force became more resigned to the sanctuary school of thought on space and came to view plans and doctrines calling for the military to help control space or to exploit the high-ground potential of space as increasingly irrelevant.

The roots of the MOL program can be traced back at least to the "Global Surveillance System" proposed by Air Force Systems Command in November 1960.[46] As described above, the more direct inspiration for the MOL came from the MODS space station first proposed by the Air Force in June 1962, the 1963 DOD-NASA deliberations over the possibility of building a joint space station, and the cancellation of the X-20. In his Posture Statement for FY 1965, Secretary McNamara generally remained unconvinced of a specific need for military spaceflight but indicated that the time had come for U.S. military man-in-space efforts to "be more sharply focused on those areas which hold the greatest promise of military utility."[47] Accordingly, he had canceled the X-20, expanded the small-scale testing of the Mach 5-25 flight regime through the unmanned ASSET vehicle, initiated the DOD Gemini Experiments Program, and proposed MOL as a "much more important step" for investigating the possible military utility of man-in-space.[48]

---

43. McDougall, *Heavens and Earth*, p. 340.

44. Ibid., p. 341.

45. Between 1957 and 1963, the X-20 program consumed $400 million, or almost the same amount spent on Project Mercury.

46. Richelson, *Secret Eyes in Space*, p. 83.

47. House Committee on Armed Services, *Fiscal Years 1965–1969 Defense Program and Fiscal Year 1965 Defense Budget*, Hearing before the Committee on Armed Services, 88th Cong., 1st sess., 1964, p. 104.

48. Ibid., pp. 104–106, quotation from page 106.

During 1964 and the first half of 1965, the MOL program was subjected to intense scrutiny by OSD and underwent several design and program application changes. By mid-1965, specific missions and station designs were firmed up. Most importantly, MOL applications added in 1965 were designed to turn MOL into a formidable reconnaissance platform with a large 90-inch telescope and huge signals intelligence (SIGINT) antennas to be assembled on orbit alongside the station.[49] At a press conference on 25 August 1965, President Johnson formally

A 1960 concept image of the United States Air Force's proposed Manned Orbiting Laboratory (MOL), intended to test the military usefulness of having humans in orbit. The station's baseline configuration was that of a two-person Gemini B spacecraft that could be attached to a laboratory vehicle. The structure was planned to launch on a Titan IIIC rocket. The station would be used for a month, and the astronauts could return to the Gemini capsule for transport back to Earth. The first launch of the MOL was scheduled for 15 December 1969, but the program was canceled by Defense Secretary Melvin R. Laird in 1969. *(NASA HQ image no. 2B24070-Fig3)*

49. Stares, *Militarization of Space*, p. 98; Richelson, *Secret Eyes*, p. 85. Richelson indicates that the MOL telescope camera system would have had a resolution of approximately 9 inches and was designated as the KH-10. A depiction of construction of a 100-foot-diameter SIGINT antenna as a proposed MOL experiment is found in J. S. Butz, Jr., "MOL: The Technical Promise and Prospects," *Air Force/Space Digest* (October 1965): 44–45.

approved the development of MOL. The MOL design at this time called for a configuration approximately 54 feet long and 10 feet in diameter consisting of a Gemini B capsule attached to the 41-foot-long laboratory. The station was to be launched into polar orbit from Vandenberg AFB atop a Titan III-C booster.[50] The entire program was originally scheduled to include five manned flights of MOL beginning in 1968 at a cost of $1.5 billion.[51] The overall objectives of the program as approved in August 1965 were to

a) learn more about what man is able to do in space and how that ability can be used for military purposes,

b) develop technology and equipment which will help advance manned and unmanned space flight, and

c) experiment with this technology and equipment.[52]

The Air Force directed the MOL program, and the Navy was a minor partner in the effort.[53] The initial Air Force support for this program was unmistakable. In congressional testimony in early 1965, Deputy Chief of Staff for R&D Lieutenant General James Ferguson indicated that "MOL would provide the space testing and evaluation facility which we have long sought. We consider it to be the keystone of our future space program."[54] Earlier, Ferguson had simply identified the MOL as the Air Force's "most important space program."[55] More generally, Ferguson highlighted the need for MOL due to the Air Force belief "that man is the key to the future in space, and

---

50. Richelson, *Secret Eyes*, p. 85; Executive Office of the President, National Aeronautics and Space Council, *Report to Congress on Aeronautics and Space Activities, 1965* (Washington: GPO, 31 January 1966), pp. 49–50. MOL astronauts would transfer into the shirtsleeve environment of the laboratory via a hatch through the heatshield of the Gemini B capsule. MOL was designed for 30-day missions. At the completion of the mission, the astronauts would transfer back into the capsule and reenter; the station itself would eventually also reenter and burn up. The Titan III-C had originally been developed to launch the canceled X-20.

51. Executive Office of the President, *Aeronautics and Space Activities, 1965*, p. 50.

52. Ibid., p. 49. These three objectives in *Aeronautics and Space Activities, 1965* were considerably less detailed and ambitious than the six MOL objectives that Secretary McNamara and DDR&E Harold Brown had outlined in congressional testimony in early 1965. See, for example, the statement of Brown in U.S. Congress, Senate Committee on Armed Services and Subcommittee on Department of Defense of the Committee on Appropriations, *Military Procurement Authorizations, Fiscal Year 1966*, Hearings before the Committee on Armed Services and the Subcommittee on Department of Defense of the Committee on Appropriations, 89th Cong., 1st sess., 1965, pp. 413–414.

53. Richelson, *Secret Eyes*, pp. 91–92. The original MOL schedule called for Navy MOL astronauts to conduct extensive ocean surveillance and submarine tracking experiments during the fourth mission.

54. U.S. Congress, House Committee on Armed Services, *Hearings on Military Posture, Fiscal Year 1966*, Hearings before the Committee on Armed Services, 89th Cong., 1st sess., 1965, p. 1229.

55. Ibid., p. 1219.

that certain military tasks and systems will become feasible only through the discriminatory intelligence of man."[56]

Soon, however, MOL ran into substantial technical and very difficult political problems. An unmanned Gemini B capsule was successfully tested and recovered from space on 3 November 1966, but design changes and technical difficulties with the laboratory portion of MOL caused delays and weight increases in this portion of the hardware. Due to the greater weight of the laboratory, the booster configuration for MOL was redesigned for more thrust and designated as the Titan III-M.[57] More significantly, the political support for MOL began to erode from all quarters. The Johnson administration was attempting to deal with the effect of the buildup of the war in Vietnam on its Great Society programs and had little time or inclination to focus on MOL. The program also suffered from a lack of strong support within Congress, where space attention was focused on the growing Apollo costs and the upcoming Moon landing. Even within the Air Force, MOL began to face serious questioning as the war in Vietnam heated up and resources were required for this conflict and for more traditional development programs such as the C-5A transport aircraft. With declining political support, funding for MOL began to be cut well below the levels required to keep the program on its original schedule. By early 1969, the first manned MOL mission had been slipped to 1972, while the total projected cost of the program had risen from $1.5 billion to $3 billion.[58] Despite these difficulties, in February 1969 incoming Secretary of Defense Melvin Laird endorsed a comprehensive review of the program that "concluded that the continuance of the program is fully justified by the benefits to our defense posture anticipated from MOL; and that all MOL objectives established by the President in 1965 can now be met with a six- rather than a seven-launch program."[59] Additionally, the Nixon administration initially requested $525 million for MOL in FY 1970.[60]

The Nixon administration quickly and completely reversed its initial support for MOL. President Nixon was eager to limit the budget, and MOL soon emerged as "an ideal target for OMB."[61] The actual decision to terminate MOL was apparently made at a White House meeting of OMB representative Robert Mayo, National Security Advisor Henry Kissinger, and President Nixon.[62] As

---

56. Ibid., p. 1228.

57. Richelson, *Secret Eyes*, p. 90.

58. Ibid., pp. 101–102.

59. Quoted from prepared statement of Air Force Chief of Staff General John McConnell in U.S. Congress, Senate Committee on Armed Services, *Authorization for Military Procurement, Research and Development, Fiscal Year 1970*, Hearings before the Committee on Armed Services, 91st Cong., 1st sess., 1969, p. 956. This cutback meant that MOL would now include only four manned missions rather than the five originally planned.

60. Ibid., p. 957.

61. Quoted from an unnamed "senior Air Force officer" in Stares, *Militarization of Space*, p. 159.

62. Richelson, *Secret Eyes*, p. 102.

they made clear in subsequent congressional testimony, Secretary Laird and the JCS were not consulted prior to this decision.[63] The public announcement of the cancellation of the MOL program came on 10 June 1969. A total of $1.4 billion was spent on the MOL program, making it one of the most expensive military programs ever prematurely terminated as of that date.[64]

The cancellation of MOL must also be viewed within a broader context than just the budgetary concerns of the Nixon administration. Shortly after entering office, Nixon had established a Space Task Group (STG) comprised of Vice President Spiro Agnew, Acting NASA Administrator Thomas Paine, Secretary Laird, and science adviser Lee DuBridge.[65] Nixon tasked the STG to complete a comprehensive review of the future plans of the U.S. space program. The STG national-level review was supported by reports from working groups at the departmental level. The DOD working groups in support of the STG studied future military space plans and budgets and again raised the issue of the military utility of MOL in an era of constrained budgets. More specifically, a report for the STG prepared by Walter Morrow of MIT's Lincoln Laboratory "declared that no significant increase in space spending was necessary to meet DOD requirements and that an annual military space investment of about $2 billion would suffice through the 1970s."[66] In competition for scarce space program funds, MOL did not necessarily do well even in DOD-sponsored analyses.

The most significant factor in the demise of the program, however, was the growing belief that unmanned spy satellites could perform the primary mission of MOL as well as or better than MOL and at a lower cost. According to Richelson, the NRO and CIA had been leery of the idea of a manned reconnaissance system from the outset. They reasoned that a manned system might present more of a provocation to the Soviets, that the contributions of manned operators in space would not be all that significant when balanced against the costs and requirements of life-support systems, and that any accident involving MOL astronauts might set back the whole space-based intelligence-gathering process unacceptably.[67] Moreover, beginning in 1965, NRO had begun development of the United

---

63. Ibid.

64. Ibid.

65. Air Force Secretary Robert Seamans represented Secretary Laird at STG meetings. Seamans had previously been NASA's Associate Administrator.

66. Jacob Neufeld, "The Air Force in Space. 1969–1970," Secret History, Office of Air Force History, July 1972, p. 4, microfiche document 00338 in *Military Uses of Space*. The overall military input to the STG, "DOD Programs, Options, Recommendations," was largely shaped by the Air Force and outlined four primary military space objectives: "(1) information gathering; (2) deterrence; (3) limiting enemy damage to the nation; and (4) support of Allied forces." This report also grouped possible future space efforts into three categories: 1) improvements on existing and planned mainstream space systems, primarily for force enhancement; 2) systems responsive to "significant technological or engineering advances, changes in national policy, or the emergence of new threats" such as a deep space command post, and 3) "undefined" systems such as Earth illumination systems or weather-modification systems (ibid., pp. 2–4).

67. Richelson, *Secret Eyes*, p. 103.

States' fourth-generation photoreconnaissance satellite known as the KH-9 or "Big Bird"—a system originally planned to serve as a backup to MOL.[68] In the late 1960s, with MOL already in jeopardy, the NRO now argued that the projected capabilities of the KH-9 system would make the MOL unnecessary. It is not possible in open sources to trace the exact impact of this argument on the decision to cancel MOL, but it may have been the clincher, given the development paths of both programs and subsequent events. The first KH-9 was launched from Vandenberg AFB atop a Titan III-D on 15 June 1971.[69]

The saga of the demise of the MOL program served as another painful lesson to the Air Force and the military that their preferred military space doctrines and programs would not come to fruition. The loss of MOL hit the Air Force very hard because 1) it was the Air Force's only attempt to establish a major manned military space program during this period, 2) the Air Force had planned to use MOL as the basis to build a larger manned military space presence, and 3) the program had been specifically tailored primarily to support the space-as-sanctuary school but had still been rejected. After the Air Force's plan to use men in space to support the nation's highest priority military space mission was not approved, it was very unlikely that any other military man-in-space program would be approved. For a number of years after the cancellation of the MOL, the Air Force largely lost interest in high-ground and space-control doctrines and basically considered the development of a significant manned military space presence a lost cause. Stares summarizes the organizational impact of the loss of the X-20 and the MOL programs upon the Air Force during this period very well:

> With the cancellation of the Dynasoar and MOL, many believed in the Air Force that they had made their "pitch" and failed. This in turn reduced the incentives to try again and reinforced the bias towards the traditional mission of the Air Force, namely flying. As a result, the Air Force's space activities remained a poor relation to tactical and strategic airpower in its organizational hierarchy and inevitably in its funding priorities. This undoubtedly influenced the Air Force's negative attitude towards the various ASAT modernization proposals put forward by Air Defense Command and others in the early 1970s. The provision of satellite survivability measures also suffered because the Air Force was reluctant to propose initiatives that would require the use of its own budget to defend the space assets of other services and agencies.[70]

68. Ibid., p. 105; Stares, *Militarization of Space*, p. 160; William E. Burrows, *Deep Black: Space Espionage and National Security* (New York: Berkley Books, 1986), pp. 228–229.

69. "Launch Listing," in *U.S. Military Uses of Space*, p. 100. The Titan III-D launch vehicle for the KH-9 was very similar to the Titan III-M designed to launch the MOL.

70. Stares, *Militarization of Space*, p. 242.

DOD AND THE DEVELOPMENT OF THE SPACE SHUTTLE

Interactions between NASA, the NRO, and the Air Force were among the most important inputs in structuring the development and operation of the Space Transportation System (STS) or Shuttle program. STS interactions deserve special attention because they were the most focused, longest running, and most intense interplay among these organizations and became the single most important factor in shaping their interrelationships. NASA's decision to pursue a large shuttle vehicle program to serve as the national launch vehicle was the Agency's primary post-Apollo space program goal. This decision necessitated that the Shuttle design be able to accommodate the most important potential users and satisfy the military in particular. Accordingly, DOD was instrumental in setting Shuttle payload and performance criteria. Even more importantly, when the STS ran into great political and budgetary problems during the Carter administration, DOD stepped in to help save the program—largely due to the Shuttle's projected capability to launch huge spy satellites. Thus, the rationale behind the STS development became increasingly militarized and related to spy satellites. Additionally, STS operations up to the *Challenger* disaster allowed the military to again entertain plans to develop a manned military presence in space.

The question of what the U.S. should focus on in space following its triumph in the Moon race was the overriding issue for U.S. space policy in the late 1960s and early 1970s. President Nixon created the Space Task Group (STG) in February 1969 to examine this issue. On 15 September, the STG presented Nixon with three options for post-Apollo U.S. civil space plans. Option one called for a manned mission to Mars by 1985 supported by a 50-man space station in orbit around Earth, a smaller space station in orbit around the Moon, a lunar base, a space shuttle to service the Earth space station, and a space tug to service the lunar stations. Option two consisted of all of the above except for the lunar projects and delayed the Mars landing until 1986. Option three included only the space station and the space shuttle, deferring the decision on a Mars mission but keeping it as a goal to be realized before the end of the century.[71] The report estimated that option one would cost approximately $10 billion annually, option two would run about $8 billion per year, and option three would be $5 billion annually.[72] Considering that NASA's budget had peaked at the height of the Moon race in 1965 at a little more than $5 billion and that political support for space spectaculars was rapidly eroding, the STG recommendations seemed fiscally irresponsible and politically naive.[73]

---

71. Colonel Cass Schichtle, USAF, *The National Space Program From the Fifties to the Eighties* (Washington: National Defense University Press, 1983), pp. 72–73; McDougall, *Heavens and Earth*, p. 421.

72. McDougall, *Heavens and Earth*, p. 421.

73. Schichtle, *National Space Program*, p. 69.

Meanwhile, the Air Force and NASA had begun coordinating with one another concerning the need for, design criteria, and performance capabilities of a shuttle vehicle. In March 1969, STG Chairman Agnew had directed that a joint DOD-NASA study on a shuttle system be completed to support the overall STG effort.[74] During the spring of 1969, Air Force Chief of Staff General John McConnell was very impressed with the military potential of a shuttle vehicle and even "proposed the Air Force assume responsibility for STS development."[75] Air Force Secretary Robert Seamans was also impressed with the potential of a shuttle but "he vetoed the proposal that the Air Force take charge of STS development, preferring to await additional study results."[76] In June, DOD and NASA submitted to the STG their coordinated report that strongly backed development of a shuttle.[77] By contrast, the Morrow report, which was also prepared for the STG, questioned the technical feasibility of a shuttle and specifically refuted the projected STS launch rates and cost estimates. The Morrow report recommended "the DOD postpone its participation in the system's development pending technical and economic analysis."[78]

DOD and the Air Force acknowledged some of the potential STS difficulties raised by the Morrow report but remained supportive of shuttle development. The military specifications for the shuttle at this time included a 50,000-pound payload capability for launches into a 100-nautical-mile (NM) due-east orbit, a payload compartment measuring 15 by 60 feet, and a cross-range maneuvering capability of 1,500 NM.[79] Some NASA shuttle designs did not meet all of these criteria, but NASA quickly recognized the political necessity for strong Air Force support in attempting to sell the shuttle within the administration and agreed specifically to include the Air Force in future STS design and policy decision-making.

---

74. Neufeld, "Air Force in Space. 1969–1970," p. 5, microfiche document 00338 in *U.S. Military Uses of Space.*

75. Ibid., p. 6.

76. Ibid.

77. Ibid., pp. 6–7. Specifically, "the report concluded that STS development (1) would require no significant 'breakthrough' in technology, (2) could achieve 'a major reduction in the recurring costs of space operations,' and (3) could meet the requirements of both agencies without 'major technical penalty, development risk, limitation on mission flexibility, or cost increase.'" Neufeld's interior quotations are from the report itself. The report recommended a $52-million allocation in FY 1970 for design studies. Moreover, the report also found that the STS could be operational by 1976 for $4–6 billion; projected a launch rate of 30 to 70 flights per year; and estimated that with 100 uses, the STS would lower launch costs per pound into low-Earth orbit to $50–100 and into geostationary transfer orbit to $500.

78. Ibid., p. 7. The Morrow report is also discussed in relation to the MOL in the MOL section above.

79. Ibid., p. 8. The Air Force's weight and volume requirements for the STS seemed to be driven by projected spysat designs, whereas the cross-range maneuverability requirement was apparently a general military requirement relating to safety, survivability, and flexibility considerations. Some critics within NASA and other analysts have charged that these requirements (especially the cross-range criteria) were arbitrarily set too high, caused very significant design changes, and later contributed to STS program delays. See, for example, the positions raised in John M. Logsdon, "The Decision to Develop the Space Shuttle," *Space Policy* 2 (May 1986): 103–119.

To formalize this arrangement, on 17 February 1970 the Air Force signed an agreement with NASA that established the joint USAF/NASA STS Committee.[80]

On the basis of the STG report and the recommendations from other space studies during this period, President Nixon moved to formalize U.S. post-Apollo space policy goals in March 1970.[81] Nixon only endorsed the development of a shuttle and left a space station or a Mars mission contingent upon the successful completion of a shuttle program. Of course, this was far less than NASA had hoped for, and the agency that had conquered the Moon was initially less than enthused about the prospect of building a nonglamorous space truck as its primary post-Apollo mission.[82] Soon, however, NASA came to realize that a space shuttle was the only major program that stood a chance of being approved at this time and the only possible way to preserve at least a part of NASA's integrity in the face of radical cuts in civil space programs and budgets.[83]

Faced with this situation, NASA continued its attempts to design a space shuttle during 1970 and 1971. In late 1970 and early 1971, acting Administrator George M. Low continued Paine's emphasis on the shuttle as a national vehicle by moving NASA from concept towards design of a larger and more capable shuttle. Thus, by 1971, NASA was hard at work on what has been described as a "Cadillac" shuttle system—very large, very capable, and completely reusable, but very expensive to develop.[84] These very capable designs proved to be too expensive, especially after the Office of Management and Budget (OMB) reiterated that NASA could expect no more than $6.5 billion to develop the shuttle.[85] Meanwhile, the Air

---

80. Neufeld, "Air Force in Space. 1969–1970," p. 9. Creation of this committee did not solve all of the Air Force–NASA differences over STS design issues. Powerful elements within NASA, such as Associate Administrator for Manned Spaceflight Dr. George E. Mueller, continued to press for a smaller STS design that would not meet all of the Air Force's criteria.

81. Two of the most important other studies on U.S. post-Apollo space goals that were also completed during this period but not mentioned above were 1) the overall NASA input into the STG, known as the Mueller report after its chairman, George Mueller, and 2) the PSAC report, headed by Lewis Branscomb. The Mueller report stressed a building-block approach for the next major civil space programs and emphasized the general utility of a space shuttle for all other projects. The Branscomb report urged that the U.S. place more emphasis on unmanned versus manned exploration and recommended robotic exploration of Mars. On these two reports and their impact, see Hans Mark, *The Space Station: A Personal Journey* (Durham, NC: Duke University Press, 1987), pp. 31–34.

82. NASA Administrator Thomas Paine resigned in September 1970 over this issue and over his general perceptions of a lack of support for NASA within the Nixon administration. See Joseph J. Trento, *Prescription for Disaster* (New York: Crown Publishers, 1987), pp. 84–99.

83. NASA's budget (in constant dollars) fell to only 36 percent of its 1965 peak by the time of its nadir in 1975. The speed of these reductions meant that NASA's budget often was reduced by more than $500 million, or more than 10 percent, in constant dollars each year. Moreover, the number of jobs in the civil space sector dropped from a peak of 420,000 in 1966 to only 190,000 by 1970 and continued down from that point. See Schichtle, *National Space Program*, p. 73; "NASA Budget History," *Aviation Week & Space Technology* (16 March 1992): 123.

84. Alex Roland, "Priorities in Space for the USA," *Space Policy* 3 (May 1987): 106. Roland is a former NASA historian.

85. Logsdon, "Decision to Develop the Space Shuttle," p. 107.

Force remained adamant on its payload and performance criteria and apparently even raised its maximum payload weight requirement to 65,000 pounds.[86] During the remainder of 1971, NASA came up with a revised shuttle design known as the Thrust-Assisted Orbiter Shuttle (TAOS) that seemed to meet these demanding development cost ceilings and performance criteria better.[87] After very intense scrutiny from the OMB during the fall of 1971, the TAOS design went forward to President Nixon for final approval.[88] Nixon privately decided to approve the full-scale TAOS at the Western White House at San Clemente over the 1971–72 New Year's weekend.[89] James Fletcher, the new NASA Administrator, went to the Western White House to brief the President and to be present when the decision to approve the STS was publicly announced on 5 January.

Other than setting the payload and performance design criteria discussed above, the Air Force was not very involved, financially or otherwise, in the STS program during most of its development period. In 1971, the Air Force agreed that it would not compete against the STS and would forgo the development of any new expendable launch vehicles (ELVs).[90] In April 1972, the Kennedy Space Center (KSC) and Vandenberg AFB were selected as Shuttle launch and landing sites, and the Air Force agreed to reconfigure the planned MOL launch complex at Vandenberg, known as space launch complex (SLC)-6, for STS launches into polar orbit.[91] Interestingly, former NASA Administrator Fletcher claimed in a later interview that the Air Force had verbally committed to him during STS development that they would buy the planned fifth and sixth orbiters.[92]

---

86. Ibid., pp. 108–110. Here, Logsdon discusses the Air Force's payload and performance criteria. He indicates that the most important Air Force weight requirement was for the capability to launch 40,000 pounds into polar orbit and that the 15-foot dimension of the cargo bay was a NASA requirement for possible future station construction rather than an Air Force criterion.

87. The TAOS design moved away from the original designs, which called for a vertically stacked booster-orbiter configuration staging in sequence, as in all previous spacecraft designs, to a horizontally stacked booster-orbiter design where the booster and orbiter engines could be used at the same time. This design also moved the large main fuel tank outside the booster and made this section expendable rather than reusable. The TAOS design lowered the overall size and weight of the vehicle by allowing the Space Shuttle Main Engines (SSMEs) to contribute to takeoff thrust, but it also greatly increased the technological challenges for designing the SSMEs and introduced the problem of asymmetrical thrust on takeoff. This and other design decisions at this time lowered the development costs for the STS but would also contribute significantly to the much higher than desired STS operations costs.

88. Logsdon describes the NASA-OMB exchanges in detail in "Decision to Develop the Space Shuttle," pp. 112–116.

89. Ibid., p. 118.

90. Ibid., p. 110.

91. Major General R. C. Henry and Major Aubrey B. Sloan, "The Space Shuttle and Vandenberg Air Force Base," *Air University Review* 27 (September–October 1976): 19–26. The Aeronautics and Astronautics Coordinating Board formally approved SLC-6 reconfiguration for STS launches in January 1975.

92. Trento, *Prescription for Disaster*, p. 128. I was unable to find any hard evidence of such a commitment. In the wake of the *Challenger* disaster, many varied theories were advanced to determine culpability for the woes of the STS program.

Throughout the remainder of the 1970s, the STS faced difficult technical and political challenges. Three major technical challenges were the most difficult: developing the computer software and interfaces for the orbiter's computer-controlled flight system, designing and especially attaching the ceramic tiles for the orbiter's heat-protection system, and designing and testing the Space Shuttle Main Engines (SSMEs). Politically, the STS faced even more difficult challenges at the outset of the Carter administration. Several powerful individuals and organizations such as Vice President Walter Mondale, the OMB, and the Office of Science and Technology Policy (OSTP) favored drastically cutting back the STS if not canceling the program outright.[93] In the summer of 1977, as the test vehicle *Enterprise* was about to begin STS approach and landing tests at Edwards AFB, President Carter asked newly appointed NASA Administrator Robert Frosch to evaluate comprehensively whether to continue with the STS program.[94] Thus, the stage was set for the most difficult challenge the STS would face during its development process.

At this point, DOD stepped in strongly to defend the STS as a program critical to national security and to play an important role in preserving this program. In July 1977, Dr. Hans Mark, who had been Director of NASA's Ames Research Center, became Under Secretary of the Air Force (and NRO Director). As an avid manned spaceflight enthusiast who believed the STS was an essential step towards a future manned space station and future exploration, Mark was instrumental in lining up DOD support for the STS in its time of peril. During November and December of 1977, OMB called a series of meetings on the future of the STS.[95] The OMB had urged that the STS program be converted into a three-orbiter test project and that only the KSC launch site be built.[96]

According to Mark, Secretary of Defense Harold Brown was persuasive in making the DOD's need for the STS clear at these meetings:

> [Brown] made the case that at least two launch sites (one on the east coast and the other on the west coast) would be required and that at least four Orbiters would be necessary to meet the requirements of national security. This last argument was based on the fact that the first two Orbiters to be built (OV-102, *Columbia*,

93. Mondale had helped to make a name for himself in the Senate both with his attacks on the "bloated" NASA budgets of the late 1960s and as a leader of congressional opposition to building the STS. In 1973, President Nixon had abolished the NASC and moved the science adviser's office out of the Executive Office of the President (EOP). In 1976, President Gerald Ford created OSTP within EOP. Carter's OSTP Director, Dr. Frank Press, saw government funding for all scientific efforts as a zero-sum game and was eager to address the deficiencies he perceived in basic scientific research funding by reducing quasi-scientific efforts such as manned spaceflight.

94. Trento, *Prescription for Disaster*, p. 149.

95. Mark, *Space Station*, pp. 71–73.

96. Ibid., p. 72.

and OV-099, *Challenger*) would be somewhat heavier than the following vehicles and would therefore not be capable of carrying the very heaviest national security related payloads. It was therefore necessary to have at least two Orbiters capable of carrying the very heaviest payloads in order to have a backup in case one of these vehicles was lost. This argument carried the day and the decision was reached to build four Orbiters (OV-103, *Discovery*, and OV-104, *Atlantis*, in addition to the first two) and to continue with construction of the west coast launch site. (The west coast launch site was deemed necessary in order to conduct polar orbiting flights required for national security related missions.)[97]

Although Mark does not highlight another aspect of saving the STS, sometime during this period, perhaps at these OMB meetings, the decision was also taken to make the STS virtually the only launch vehicle for both NASA and DOD.

The outcome of these meetings marked a definite shift in the rationale for the STS program that again illustrates the overriding impact of spysats on all other types of space policy. NASA was publicly selling the STS program as a way to meet U.S. civil space policy goals and on cost-effectiveness grounds, but the rationale that saved it during the Carter administration was its ability to launch huge spy satellites. Moreover, with the pending debate over the ratification of the SALT II Treaty, spy satellites as national technical means of verification took on added significance. On 1 October 1978, President Carter marked the first official break with the blackout policy on spysats promulgated in 1962. In a speech at the KSC, Carter noted that "photoreconnaissance satellites have become an important stabilizing factor in world affairs in the monitoring of arms control agreements. They make an immediate contribution to the security of all nations. We shall continue to develop them."[98] Meanwhile, however, the NRO was ambivalent about the prospects of using the STS as its sole launch vehicle: on

---

97. Ibid.

98. Cited in Stares, *Militarization of Space*, p. 186. According to Richelson, *Secret Eyes*, pp. 140–143, various agencies within the administration debated during early September how far to go in declassifying spysats. The primary motivation behind the desire to loosen the security restrictions on spysats was publicly to provide administration officials with better evidence of U.S. ability to verify SALT II adequately. Those arguing for greater declassification included Secretary of State Cyrus Vance, Arms Control and Disarmament Agency Director Paul Warnke, Director of Central Intelligence Stansfield Turner, National Security Agency Director Bobby Inman, and NRO Director Mark. Secretary Brown, backed by the JCS and the Defense Intelligence Agency, strongly opposed widespread declassification. The most powerful argument raised by DOD (which apparently won the day) was that the release of one spysat photo would lead to a deluge of Freedom of Information Act (FOIA) requests and thereby tie up the manpower of the intelligence agencies in nonproductive activities. On 13 September, the Policy Review Committee (Space) voted for declassification, but only of the fact that the United States conducted photoreconnaissance from space—a "truly minimalist decision," in Richelson's opinion.

the one hand, it was already planning the large spysats that would take advantage of STS capabilities; but on the other hand, it did not want to lose control over its launch vehicles, feared the possible disruption of spysat launchings due to accidents with astronauts, and also chafed at the prospect of the increased media attention that NASA involvement would bring.

General Air Force attitudes towards STS were also ambivalent during this period. While STS was strongly supported by elements within the Space and Missiles Systems Organization and by Mark (who became Secretary of the Air Force in July 1979), other elements such as the Secretary of the Air Force Special Projects Office were less enthusiastic. Mark attempted to push STS and a general space emphasis on the Air Force.[99] These efforts, along with the military potential of the STS, certainly were important in helping to revive Air Force interest in space and in possible military man-in-space applications. At the same time, however, the Air Force was very much a junior partner on STS in terms of funding and effort. Moreover, the Air Force dragged its feet on refurbishing SLC-6 at Vandenberg for STS operations and in developing the Inertial Upper Stage (IUS) to be used for boosting payloads into higher energy orbits than possible with the STS.[100] In sum, then, although the STS program did reignite some Air Force interest in more ambitious space missions, the level of Air Force support for this program by the end of its development did not approach the level of enthusiasm the Air Force had displayed for the X-20 or MOL, and this ambivalent support undoubtedly reflected the fact that the Air Force did not control STS.

The Military, Space Transportation Policy, and STS Operations

The 1980s witnessed both the long-awaited arrival of STS operations and the wrenching reordering of U.S. space transportation policy following the *Challenger* disaster. DOD interactions with the STS program continued to be a very important factor in shaping this program, while DOD's stance on STS provides important insights into the military's space priorities and actual level of commitment to various space programs. Despite the great military potential of the STS and the considerable support for the STS within elements of the Air

---

99. Mark listed "the development of a doctrine and an organization that will permit greatly increased Air Force activities in space in order to take advantage of new technology to enhance communications, reconnaissance, and other vital Air Force functions" as one of the USAF's top priorities. Hans M. Mark, "USAF's Three Top Priorities," *Air Force Magazine* (September 1979), reprinted as appendix 3 in Mark, *Space Station*, pp. 235–236.

100. It is difficult to apportion blame for delays on the STS program; however, STS was originally scheduled to be launched from SLC-6 in December 1982 (after "more than forty launches will have taken place from KSC"!), and SLC-6 would barely have been ready for its rescheduled first launch in March 1986 had the *Challenger* disaster not derailed that plan. In practice, there were only 5 STS flights by December 1982 and a total of only 24 flights prior to the *Challenger* disaster. See Henry and Sloan, "Shuttle and Vandenberg," p. 25; Edgar Ulsamer, "Slick 6," *Air Force Magazine* (November 1985): 47–48.

Force and elsewhere in DOD, several significant points of friction remained between the Air Force, NRO, and NASA concerning STS operations and plans. Even prior to the *Challenger* disaster, the NRO had managed to gain formal approval to build a backup launcher, the Complementary ELV (CELV), for its most important payloads. Following the *Challenger* disaster, U.S. national space transportation policies were completely reordered under the Space Launch Recovery Plan, and the Air Force planned to move almost all DOD payloads onto ELVs. NASA-DOD interactions over STS during the 1980s led to the reversal of several major space transportation policies, abandonment of the original STS program goals, and the demise of yet another potential vehicle for significant military spaceflight.

DOD was instrumental in saving STS from cancellation at the outset of the Carter administration and was again a key player in defending STS late in the Carter administration when the program faced significant political opposition due to successive schedule slips and funding shortfalls requiring supplemental appropriations.[101] DOD support for the STS was critical in maintaining political support for STS within the administration and culminated in a 14 November 1979 White House meeting between the President and all key actors on this issue, where Carter firmly committed his administration to fully funding and rapidly completing STS.[102] DOD support for the national security mission of the STS was also a key factor in pushing the supplemental appropriations through Congress following hearings in March 1980.[103]

DOD exacted a price from NASA for its indispensable support: on 25 February 1980, NASA and DOD signed an extensive MOU on management and operation of the STS which was very favorable to DOD.[104] Specifically, the MOU indicated that "DOD will have priority in mission preparation and operations consistent with established national space policy."[105] Further, the

---

101. In 1979, NASA required supplemental appropriations totaling over $1 billion (1972 dollars) to keep the STS program on track. See Mark, *Space Station*, p. 93.

102. Mark, *Space Station*, pp. 101–103; Trento, *Prescription for Disaster*, p. 169.

103. Representative Edward Boland (D-Massachusetts) was instrumental in gaining approval for these supplemental appropriations as chairman of the NASA appropriations subcommittee. His support for STS stemmed from his position as Chairman of the House Permanent Select Committee on Intelligence, where he learned about the STS-spysat link in detail. See Mark, *Space Station*, p. 105; Trento, *Prescription for Disaster*, pp. 156–157.

104. "NASA/DOD Memorandum of Understanding on Management and Operation of the Space Transportation System," 25 February 1980, microfiche document 00561 in *U.S. Military Uses of Space*. This MOU replaced the 14 January 1977 NASA-DOD MOU on STS and provided the basis for several NASA-DOD subagreements.

105. Ibid., p. 3. The "established national space policy" referenced is presumably Presidential Directive (PD)-37 signed by President Carter on 11 May 1978 (unclassified version available at *http://www.au.af. mil/au/awc/awcgate/nsc-37.htm*). This DOD mission priority on the STS was often referred to as the right of DOD to "bump" other payloads from the STS manifest in favor of top-priority national security

continued on the next page

MOU established two categories of DOD STS missions: 1) national security missions conducted by NASA and 2) "Designated National Security Missions" controlled by the Air Force.[106] Overall, this MOU went a long way towards giving the Air Force the type of operational control over a manned space vehicle it had sought since the late 1950s—an arrangement which was quite remarkable, considering that the Air Force had not paid for the development of the STS.

The initial STS spaceflight took place on 12 April 1981 when *Columbia* was launched from KSC. This marked a bittersweet milestone because it was the world's first reusable spacecraft and signified the return of manned American spaceflight. But the STS was also two years behind schedule and cost $2 billion more to develop than originally projected. Moreover, it rapidly became apparent that due to very intensive and difficult refurbishing requirements following each flight, STS could not come close to meeting its planned flight schedule.[107] However, the military potential of the STS was also apparent from the outset. The second STS mission in November 1981 conducted radar-imaging experiments from orbit that pinpointed an ancient city buried beneath the sands of the Sahara and thereby demonstrated the significant military potential of this type of spaceborne sensor.[108] The first classified military payload was carried into

---

continued from the previous page

payloads. Other significant provisions of this MOU indicated that 1) the Air Force was DOD's "sole point of contact with the NASA for all commitments affecting the STS and its use in matters regarding national security space operations and in international defense activities covered by Government to Government agreements"; 2) the Air Force would "develop, acquire, and operate a dedicated Shuttle mission planning, operations, and control facility for national security missions"; and 3) "an STS mission assignment schedule and plan" would be developed to facilitate the "expendable booster transition and phaseout plans" of NASA and the Air Force.

106. Ibid., pp. 3–4, 6–9. Specifically, for category one DOD STS flights, NASA would exercise flight control from JSC, but "NASA will be responsive to DOD Mission Directors," who would retain "overall responsibility for achieving mission objectives." For these missions, Air Force personnel "will be integrated into NASA line functions for training" in order to "allow the USAF to develop the capability to plan, control, and operate national security missions." For category two DOD STS flights, an Air Force Flight Director "will be responsible for overall mission accomplishment and operational control, including flight vehicle and crew safety, through the Air Force chain of command." Although not specified in this MOU, the implication is that category two DOD STS missions would be controlled from the Shuttle Operations and Planning Complex (SOPC) at the Consolidated Space Operations Center at Falcon (now Schriever) AFB.

107. NASA's STS mission models adopted in the early 1980s were far more realistic than the 60 flights per year originally projected for the Shuttle in the early 1970, but they still called for 24 flights per year from the complete four-orbiter STS fleet. In practice, orbiter turnaround time was approximately 60 days rather than the 7 days originally projected, and the turnaround operation required 6,000 people, nearly four times the expected number. There were only 24 total flights in the nearly five years of STS operations prior to the *Challenger* disaster. See E. C. "Pete" Aldridge, Jr., "Assured Access: 'The Bureaucratic Space War,'" Dr. Robert H. Goddard Historical Essay, n.d., p. 5. Offprint provided to author by the Office of the Secretary of the Air Force.

108. Trento, *Prescription for Disaster*, pp. 200–201; Richelson, *Secret Eyes*, p. 219. These first radar-imaging experiments were conducted with Shuttle Imaging Radar (SIR)-A. SIR-B experiments were conducted with updated hardware on mission 41-G in October 1984. According to Richelson, the SIR-A radar could apparently image objects 16 feet beneath dry sand.

orbit aboard *Columbia* during the STS-4 mission in June–July 1982, which also marked the end of the STS flight-testing phase.[109]

Meanwhile, elements within the Reagan administration and Congress were carefully monitoring early STS developments. On 13 November 1981, President Reagan signed National Security Decision Directive (NSDD)-8 that reaffirmed the space transportation policies of the Ford and Carter administrations by stating, "The STS will be the primary space launch system for both United States military and civil government missions. The transition should occur as soon as practical."[110] According to Mark, NSDD-8 also indicated "that the president had a strong personal interest in the space shuttle program."[111] Reagan's first comprehensive space policy, NSDD-42, was publicly announced by the President himself at a 4 July 1982 ceremony at Edwards AFB marking the beginning of the operational phase of STS operations, with *Columbia* in the background. In terms of space transportation policy, NSDD-42 reaffirmed that the STS was the nation's primary launch system, declared that the United States "is fully committed to maintaining world leadership in space transportation," stated that the "first priority of the STS program is to make the system fully operational and cost-effective in providing routine access to space," and indicated that U.S. "government spacecraft should be designed to take advantage of the unique capabilities of the STS."[112] Additionally, this directive indicated that "for the near-term," the STS would be managed under the terms of the NASA/DOD MOUs but as "STS operations mature, options will be considered for possible transition to a different institutional structure."[113] Finally, NSDD-42 made a concession to the NRO: "Unique national security considerations may dictate developing special-purpose launch capabilities."[114]

Early STS operations presented a variety of challenges and opportunities for the Air Force and NRO. Different elements within the Air Force had particular space priorities and viewpoints on the potential of the Shuttle. The space enthusiasts former Secretary Mark had reenergized within the Air Force were excited about exploring the military potential of STS, especially for military

---

109. Melvyn Smith, *Space Shuttle* (Newbury Park, CA: Haynes Publications, 1989), appendix 7; "Chronology," in *U.S. Military Uses of Space*, p. 52.

110. NSDD-8, "Space Transportation System," 13 November 1981, cited in "Chronology," in *U.S. Military Uses of Space*, p. 51.

111. Mark, *Space Station*, p. 131.

112. NSDD-42, "National Space Policy," 4 July 1982, pp. 2–3, NSC box, National Archives, Washington, DC. Two complete pages and approximately five additional paragraphs are deleted from the sanitized version of this directive. The White House also issued a five-page fact sheet, "National Space Policy," on 4 July 1982, reprinted in NASA, *Aeronautics and Space Report of the President, 1982 Activities* (Washington, DC: GPO, 1983), pp. 98–100.

113. NSDD-42, "National Space Policy," p. 4.

114. Ibid.

man-in-space missions.[115] The NRO was not very happy with being directed to abandon ELVs for STS but was in the process of redesigning and reconfiguring its future payloads to take full advantage of STS's substantial payload capabilities.[116] Other groups within the Air Force were far less excited with space or STS and opposed the substantial Air Force expenditures required to prepare for DOD STS operations. Major Air Force programs designed to support DOD STS operations included the ill-starred Inertial Upper Stage (IUS) program, modifications of SLC-6 at Vandenberg AFB for STS launch, construction of the Space Operations Planning Complex (SOPC) at Falcon AFB, and modifications to the Kennedy, Johnson, and Goddard Space Flight Centers for "controlled mode" DOD STS operations.[117]

---

115. Military uses of STS are not often or fully discussed in open sources. In answering congressional questions in March 1983, DOD drew a distinction between "payload delivery" and "full exploitation" of STS, defining the latter as follows: "In the longer term, when the capabilities of the Shuttle will be routinely available, the DOD envisions use of the enhanced capabilities unique to the Shuttle, such as on-orbit assembly of large structures; checking out payloads prior to deployment; repairing and servicing of satellites on-orbit; retrieving spacecraft for repairs and refurbishment; and performing man in the loop experiments." See House Committee on Appropriations, Subcommittee on the Department of Defense, *Department of Defense Appropriations for 1984*, Hearings before Subcommittee on Department of Defense, 98th Cong., 1st sess., pt. 8, 1983, p. 508. See also Edward H. Kolcum, "Defense Moving to Exploit Space Shuttle," *Aviation Week & Space Technology* (10 May 1982): 40–42. Kolcum notes that DOD's space test program (STP) experiments (e.g., Teal Ruby) would henceforth use STS rather than ELVs.

116. One of the most sensitive points for NASA regarding STS performance is that it never met its original 65,000-pound payload specification as set in conjunction with the Air Force in the early 1970s. The NASA STS performance data in the *President's Space Report* for 1981–87 indicated that the STS was able to boost approximately 65,000 pounds "in full performance configuration." However, the figure in the *Aeronautics and Space Report of the President* for 1988 (after resumption of STS operations) indicated a significant drop in STS full-performance configuration capabilities to approximately 54,895 pounds. Moreover, during congressional testimony in 1981, Air Force Assistant Secretary and NRO Director Robert J. Hermann indicated that "current projections of Shuttle performance show it to be about 8000 lbs lower than the original commitment. DOD missions can profitably use the full capability of the original performance commitment" (Senate Committee on Commerce, Science, and Transportation, Subcommittee on Science, Technology, and Space, *NASA Authorization for Fiscal Year 1982*, Hearing before the Subcommittee on Science, Technology, and Space, 97th Cong., 1st sess., pt. 2, 1981, p. 349). In 1982, Aldridge, Hermann's successor as NRO Director, indicated that the first Vandenberg AFB Shuttle launch scheduled for October 1985 "will require full specification Shuttle performance—as called out in our Performance Reference Mission 4 requirements. Specifically, the Shuttle must be capable of delivering 32,000 pounds to a 98 degree inclined, 150 nautical mile circular orbit and, then, recover another satellite weighing 25,000 pounds and return it to Vandenberg. The Shuttle with its current performance estimate cannot achieve this long standing defense requirement" (prepared statement of Under Secretary Aldridge in Senate Committee on Commerce, Science, and Transportation, Subcommittee on Science, Technology, and Space, *NASA Authorization for Fiscal Year 1983*, Hearing before the Subcommittee on Science, Technology, and Space, 97th Cong., 2nd sess., 1982, p. 166). Later, Aldridge simply indicated that the "final Shuttle capabilities were nearly 20% short" of NASA's originally promised "65,000 pounds of payload to low earth orbit from Kennedy Space Center and 32,000 pounds to a polar orbit from Vandenberg AFB, California." See Aldridge, "Assured Access," p. 3.

117. See Senate Committee on Commerce, Science, and Transportation, *NASA Authorization for Fiscal Year 1982*, pp. 340–341, 346–350, 444, 484. At this time (April 1981), the first STS launch from Vandenberg

continued on the next page

Despite these widespread efforts and considerable expenditures, the Air Force and DOD basic positions on how the STS fit into long-range military space plans or doctrine remained far from clear, at least in the available unclassified material. Undoubtedly, the basic Air Force overall organizational ambivalence towards space missions was a factor in structuring the long-term Air Force relationship with the STS, especially in light of all the rejected military man-in-space programs the Air Force had previously proposed.

In the early 1980s, former astronaut, space enthusiast, and Space Subcommittee Chairman Senator Harrison Schmitt (R–New Mexico) was among those most clearly upset with the apparent lack of Air Force long-range planning for STS use. During exchanges with Air Force and DOD witnesses at congressional hearings in 1981, Schmitt charged that "historic inertia" as well as "the lack of an organizational focus that has [space] as a primary mission" had made the Air Force "relatively slow to grasp the opportunities that the Space Shuttle provides, not only as a launch vehicle, but as a test and operational vehicle in space."[118] Moreover, Schmitt opined that "within a few years, you all are going to come back in and say 'We need a dedicated shuttle fleet.' And it's painted blue that we could use for our purposes."[119] Further, he warned that unless the Air Force pursued space missions more aggressively, "I can almost predict that there is going to be another Department of Something in the Department of Defense. And the Air Force will be flying airplanes, and not Shuttles."[120]

More widespread congressional concern in 1982 focused on Air Force–NASA relations in regard to the question of whether the U.S. should procure a fifth STS orbiter vehicle before the Rockwell orbiter production lines shut

---

continued from the previous page
was scheduled for August 1984. Assistant Secretary Hermann indicated that the term *controlled mode* "signifies that we are protecting the classified information used in the planning and execution of a DOD mission by controlling access to it. The modifications include construction changes to the buildings to isolate certain areas, the procurement of additional equipment, and the shielding of certain equipment to preclude electronic eavesdropping." He also stated, "All defense payloads will have completed their transition to use of the Space Shuttle as the primary launch vehicle by 1987." The SOPC was to "provide the management and control needed for our national security space operations in the post-1985 timeframe." Additionally, the SOPC would provide a backup to the single STS control node at JSC and would "provide a maximum opportunity to fully exploit the Shuttle unique capabilities, in particular the presence of military man in space." At these same hearings, Dr. James Wade, Acting Under Secretary of Defense for Research and Engineering, estimated that all of the DOD STS-related activities would cost approximately $3 billion through FY 1986. In March 1983, DOD provided figures indicating that "DOD's portion ($15.2 billion) of the total STS cost ($51.1 billion) is 30 percent [these figures are projected through FY 1988]." See House Committee on Appropriations, *Defense Appropriations for 1984*, p. 513. On the Air Force's STS-related expenditures and infrastructure, see also William P. Schlitz, "USAF's Investment in the National Space Transportation System," *Air Force Magazine* 65 (November 1982): 106–112.

118. Senate Committee on Commerce, Science, and Transportation, *NASA Authorization for Fiscal Year 1982*, pp. 458–459.

119. Ibid., p. 447.

120. Ibid., p. 460.

down. Many believed that it would be wise to procure a fifth orbiter as a backup and to provide greater STS capability.[121] The Air Force was very interested in producing another of the lighter weight and more capable orbiters but was unwilling to use DOD funds to procure this fifth orbiter.[122] Meanwhile, NASA was less supportive of the need for a fifth orbiter, largely because Administrator James Beggs and Deputy Administrator Mark had privately agreed that NASA should push a permanently manned space station as the nation's new major civil space goal and were therefore unwilling to take on other major new projects at this time.[123] By the end of 1982, despite considerable congressional support for a fifth orbiter, the NASA compromise solution of keeping the Rockwell lines partially open to produce spare parts won out, and the decision to build a fifth orbiter was deferred.[124] This decision was formalized by NSDD-80, issued on 3 February 1983.[125]

During 1983 and 1984, NRO Director Aldridge waged a mostly secret and very difficult, but eventually successful, campaign against NASA to obtain approval to develop a new ELV capable of launching the spy satellites designed to fit into the STS.[126] Building upon the opening in NSDD-42 to consider building "special-purpose launch capabilities" for "unique national security considerations," on 23 December 1983 Aldridge issued a memorandum, "Assured

---

121. Those favoring a decision to build another orbiter at this time also used arguments about the economic impact of keeping the Rockwell production lines open and the lower costs of building a fifth orbiter in sequence. In *Prescription for Disaster*, Trento speculates that a decision to build the fifth orbiter at this time (with the lines open) would have cost approximately $1.2 billion instead of the $2.1 billion that the fifth orbiter (*Endeavour*) actually cost; see p. 205.

122. See, for example, the testimony of Major General James Abrahamson (NASA Associate Administrator for Manned Spaceflight) and Air Force Under Secretary Aldridge in House Committee on Science and Technology, Subcommittee on Space Science and Applications, *The Need For a Fifth Space Shuttle Orbiter*, Hearing before the Subcommittee on Space Science and Applications, 97th Cong., 2nd sess., 15 June 1982.

123. Mark, *Space Station*, pp. 121–122; Trento, *Prescription for Disaster*, pp. 180–181. Following a long NASA campaign within the administration, President Reagan announced in his 1984 State of the Union Address the national goal of building a permanently manned space station (*Freedom*) within 10 years.

124. Trento, *Prescription for Disaster*, p. 205. On congressional support for a fifth orbiter, see, for example, the position of many Representatives in House Committee on Science and Technology, *Need For a Fifth Space Shuttle Orbiter*, as well as the formal recommendation for a fifth orbiter in House Committee on Science and Technology, Subcommittee on Space Science and Applications, *The Need for an Increased Space Shuttle Orbiter Fleet*, 97th Cong., 2nd sess., 1982, Committee Print Serial HH.

125. William Clark, NSDD-80, "Shuttle Orbiter Production Capability," 3 February 1983, NSC box, National Archives, Washington, DC. Specifically, this one-page directive indicated that a warm production line would "be achieved through the production of structural and component spares necessary to insure that the Nation can operate the four Orbiter fleet in a robust manner."

126. The intense NRO-NASA struggles of this period (a "bureaucratic space war") are the primary focus of Aldridge, "Assured Access," pp. 3–15. Naturally, this piece covers the positions of Aldridge and the Air Force far more sympathetically than the positions of Beggs or NASA, but it is by far the most detailed description of developments surrounding the CELV decision uncovered during research for this study.

Access to Space," to Air Force Space Command and Space Division.[127] This memorandum directed these organizations to plan for the procurement of a complementary ELV (CELV) capable of boosting a payload the size of the STS cargo bay and weighing 10,000 pounds into geosynchronous transfer orbit.[128] According to Aldridge, NASA Administrator Beggs "was furious" with these developments and saw them as "only a ploy of the Air Force to abandon the Shuttle."[129] However, in August 1984, Aldridge's position was formally supported by the NSC in NSDD-144 that approved Air Force development of the CELV.[130] Nonetheless, Beggs and NASA continued to oppose the CELV option and enlisted considerable congressional support in opposition to the CELV.[131]

Aldridge notes that the NSC staff hosted "the *critical* meeting" on the CELV issue on 14 February 1985.[132] At this meeting, Aldridge and Beggs finally reached agreement. This agreement was reflected in NSDD-164, issued on 25

---

127. "Chronology," in *U.S. Military Uses of Space*, p. 55. The primary rationale behind developing such a capability was to avoid dependence on a single system for space launch. Additionally, the final Air Force ELV buys were being completed at this time, and the production lines were in danger of being shut down unless new orders were found.

128. Ibid. Secretary Caspar Weinberger outlined a new DOD space launch strategy relying on a mixed fleet of ELVs and the STS in a letter to the President on 7 February 1984; see Aldridge, "Assured Access," p. 6.

129. Aldridge, "Assured Access," p. 6.

130. "Chronology," in *U.S. Military Uses of Space*, p. 56. Presumably, NSDD-144 was the subject of the White House fact sheet "National Space Strategy," issued on 15 August 1984 and reprinted in *Aeronautics and Space Report of the President, Fiscal Year 1984*, pp. 137–139. According to this fact sheet, the directive specified two requirements for "assured launch capability": "the need for a launch system complementary to the STS to hedge against unforeseen technical and operational problems, and the need for a launch system suited for operations and crisis situations." However, there is some confusion about at least the number of this classified directive in open sources. Scott Pace, in "US Space Transportation Policy: History and Issues for a New Administration," *Space Policy* 4 (November 1988): 307, 309, indicates that NSDD-144, "National Security Launch Strategy," was not issued by the EOP until 28 February 1985. Aldridge does not discuss this directive in "Assured Access." NSDD-144 was not available in the NSC box at the National Archives.

131. According to Aldridge, NASA had several concerns with and employed several tactics against the CELV. NASA felt that if DOD moved away from the STS, the costs per launch would increase and NASA would need to charge its commercial customers more for each launch. This, NASA officials thought, would drive more commercial customers towards the Ariane. In an 18 May 1984 letter from Administrator Beggs to Secretary Weinberger, NASA indicated that an STS backup was not necessary but that if DOD was determined to build a new launch vehicle, it should be derived from STS components. Next, NASA supporters in Congress specified that a competition would be run between NASA designs and industry designs for a system to meet Air Force requirements. Aldridge claims in "Assured Access" that NASA put subtle pressure on its suppliers not to compete against its Standardized Launch Vehicle (SLV-X) by indicating that their behavior would have consequences for future NASA purchases. A modified Titan III called a Titan 34D7 was the winner in the industrial competition conducted by the Air Force, while the NASA entry was judged by the Air Force Space Division to be uncontrollable during the boost phase of flight. Finally, as the ELV production lines were beginning to shut down, NASA recommended that several major and lengthy studies be undertaken on the CELV issue as a delaying tactic ("Assured Access," pp. 7–13).

132. Ibid., p. 13, emphasis in original.

February 1985.[133] Specifically, NSDD-164 authorized the Air Force to buy 10 CELVs and to launch approximately 2 CELVs per year in the period 1988–92.[134] Thus, Aldridge won his victory in the bureaucratic space war less than one year prior to the complete reordering of U.S. space transportation policy caused by the *Challenger* disaster.

In hindsight, given large impact of the *Challenger* disaster, it is remarkable that there was such sustained opposition to acquiring a backup capability for the STS. Moreover, while access to space is a prerequisite for any space activity, it is unfortunate that Aldridge and the top levels of Air Force space leadership, as well as much of NASA's leadership, were largely consumed with this issue during the mid-1980s rather than focusing on broader, more important, or more future-oriented space policy issues. Finally, it is also interesting to note that many groups were dissatisfied with STS performance capabilities and especially the mounting STS payload backlog of the mid-1980s but that only the NRO had the clout to develop a new ELV and move its most important payloads off the STS.[135]

The *Challenger* disaster completely reordered U.S. space transportation policy and effectively deferred any Air Force plans to use STS as a vehicle to build a significant manned military presence in space. During 1986 and 1987, NASA, DOD, and the newly formed Office of Commercial Space Transportation (OCST) within the Department of Transportation worked together to produce a new U.S. space launch strategy and the Space Launch Recovery Plan. NSDD-254, "United States Space Launch Strategy," was completed on 27 December 1986.[136] This directive specified that the U.S. would henceforth rely upon a

---

133. NSDD-164, "National Security Launch Strategy," 25 February 1985, NSC box, National Archives, Washington, DC. This unclassified directive was publicly released on 14 November 1985.

134. Ibid., p. 1. NSDD-164 also 1) indicated that a "competitive decision" on a specific CELV would be made by 1 March 1985, 2) directed that "DOD will rely on the STS as its primary launch vehicle and will commit to at least one-third of the STS flights available during the next ten years," 3) directed NASA and DOD to "jointly develop a pricing policy for DOD flights that provides a positive incentive for flying on the Shuttle," and 4) authorized a joint NASA-DOD effort to produce a national security study directive (NSSD) on the development of "a second-generation space transportation system."

135. Some of the strongest opposition to STS "forced busing in space" came from within NASA's own space science community. NASA had directed that all its payloads be launched exclusively by the STS, but by the mid-1980s, the STS backlog and problems with the STS upper stages were causing multiyear delays and significant design changes for key space science projects such as the Galileo Jupiter probe and the Hubble Space Telescope. See, for example, Bruce Murray, "'Born Anew' Versus 'Born Again,'" in "Policy Focus: National Security and the U.S. Space Program After the Challenger Tragedy," *International Security* 11 (spring 1987): 178–182. Even more significantly, because STS was not providing low-cost launch rates (even at its generous pre-*Challenger*-disaster subsidized rates) or reliable service and launch schedules, commercial customers were "voting with their feet" and moving in increasing numbers onto the more commercially viable Ariane ELV.

136. NSDD-254, "United States Space Launch Strategy," 27 December 1986, NSC box, National Archives, Washington, DC. Approximately three sentences of this two-page directive are deleted in the sanitized version. The White House released a fact sheet on this directive on 16 January 1987. NSDD-254 superseded NSDD-164.

"balanced mix of launchers" consisting of the STS and ELVs defined "to best support the mission needs of the national security, civil government and commercial sectors of U.S. space activities."[137] Further, "selected critical payloads will be designed for dual-compatibility, i.e., capable of being launched by either the STS or the ELVs."[138] In order to accomplish these objectives, the directive indicated that DOD "will procure additional ELVs to maintain a balanced launch capability and to provide access to space."[139]

The Space Launch Recovery Plan dealt with the means to implement this new launch strategy in greater detail. The plan focused on the revitalization of the nation's ELV production base and attempted to use government ELV purchases as a means to stimulate the development of a more robust commercial ELV industry. The plan also provided $2.1 billion to NASA for the production of a fifth orbiter, *Endeavour*, to be ready for flight by 1992. In addition, under this plan, the Air Force completely reoriented its future space support infrastructure and plans. The Air Force launched a $12-billion program to initiate or expand four ELV programs.[140] These Air Force ELV programs included expansion of the original 10 booster CELV program to 41 Titan IVs, two medium launch vehicle programs consisting of 20 Delta 2 and 11 Atlas-Centaur 2 ELVs, and refurbishing 14 decommissioned Titan II ICBMs for space launch.[141] Additionally, the Air Force took drastic steps to reconfigure the infrastructure it had developed to operate DOD STS missions, including placing the unused SLC-6 at Vandenberg AFB into "minimum facility caretaker" status in July 1986, eliminating the 32-member-strong Manned Spaceflight Engineer (MSE) program within the Space Division, disbanding the Manned Spaceflight Control Squadron at the JSC as of 30 June 1989, and ending development of the SOPC at CSOC in February 1987.[142] Further, as a result of this plan, the DOD scheduled only seven

---

137. Ibid., p. 1.

138. Ibid.

139. Ibid. Additionally, NSDD-254 specified that NASA would no longer provide commercial or foreign launch services on the STS "unless those spacecraft have unique, specific reasons to be launched aboard the Shuttle." The directive also set a 1995 "commercial contract mandatory termination date." This policy meant that of the 44 commercial and foreign launch commitments NASA had in January 1986, only 20 of these payloads still qualified for STS launch. See *Aeronautics and Space Report of the President, Fiscal Year 1986*, p. 33.

140. Pace, "US Space Transportation Policy," p. 310.

141. Ibid.; William J. Broad, "Military Launches First New Rocket for Orbital Loads," *New York Times* (6 September 1988): 1; Joint Statement of Air Force Secretary Aldridge and Chief of Staff General Larry D. Welch in Senate Committee on Appropriations, Subcommittee on Department of Defense, *Department of Defense Appropriations for Fiscal Year 1988*, Hearings before the Subcommittee on Department of Defense, 100th Cong., 1st sess., pt. 3, 1988, pp. 301–303.

142. William J. Broad, "Pentagon Leaving Shuttle Program," *New York Times* (7 August 1989): A13. Broad estimated the costs for these programs to be "at least $5 billion," the lion's share of which was the $3.3-

continued on the next page

dedicated STS launches for the period 1991–95 and thereafter planned to rely almost exclusively on ELVs.[143]

The relationships between the Air Force, DOD, and NASA over STS operations were clearly marked by great difficulties during the 1980s. The development of military space launch policy during this period provides one of the most powerful instances of organizational behavior inputs shaping U.S. space policy and significantly impacting military space doctrine. Despite building a large and expensive infrastructure for launching and controlling DOD STS missions, the Air Force never fully exercised this capability prior to the *Challenger* disaster, and, following the disaster, the Air Force and NRO were instrumental in leading DOD's rush off the STS in favor of ELVs. The bitter fight with NASA over the CELV and the general desire to fully control its launch vehicles were important factors in motivating this Air Force space launch policy reversal; however, the speed and complete nature of the virtual abandonment of the STS and the significant infrastructure designed to support DOD STS missions is remarkable and not well explained in open sources. The lack of clear and powerful military space doctrine undoubtedly contributed to these false starts, reversals, and lack of clear direction for the DOD STS mission. Cumulatively, this episode seems to be an excellent illustration of the general Air Force ambivalence over the military potential of space and military man-in-space, as well as evidence of its lack of clear and accepted doctrinal guidance on these issues.

---

continued from the previous page

billion SLC-6 at Vandenberg AFB. The SOPC building at CSOC was converted into the National Test Bed (now the Joint National Integration Center) for the Strategic Defense Initiative (SDI) program. As Broad relates, military space critics such as John Pike of the Federation of American Scientists charged that the Air Force went overboard in developing new ELVs and abandoning the STS.

143. Pace, "US Space Transportation Policy," p. 310. The first Titan IV launch took place on 14 June 1989 from Cape Canaveral; see "Chronology," in *U.S. Military Uses of Space*, p. 61.

CHAPTER 8

TECHNOLOGY, FOREIGN POLICY, AND INTERNATIONAL COOPERATION IN SPACE

John Krige

International cooperation has always been part of NASA's mission.[1] But why? Why is it in NASA's and America's interest to collaborate with foreign partners? The question is not as perverse as it sounds. In 1958, the United States was, and probably still remains, the single most important economic and military, but also scientific and technological, as well as industrial and managerial, power on Earth. Those to whom Eisenhower confided the civilian space program drew, though NACA, on a vast and expanding infrastructure of scientists, engineers, and managers, along with the facilities and the budget to match it, especially once President Kennedy committed the country to putting a person on the Moon before 1970. With some important exceptions—like the need for a global network of tracking stations, or sounding-rocket studies of the properties of the upper atmosphere in equatorial regions—there was no overriding scientific or technical (and certainly no financial) reason why NASA and the United States needed to collaborate with any other country in the conquest of space. Unlike small and medium-sized European states, America was rich enough in human and material resources to go it alone, and as such was the envy of all aspirant space powers (except perhaps the Soviet Union, who had to cripple its domestic economy to maintain its military and space capabilities at some sort of parity with those of the U.S.A.).

One classical argument for international collaboration was that it would improve relationships between the United States and the Soviet Union. The decision to establish NASA was, of course, just one of a number of measures taken by the Eisenhower administration to calm the nation in response to the engineered domestic crisis that ensued in the wake of the launch of the Sputniks by the Soviet Union in the fall of 1957. Superpower rivalry was at its height: by the end of the 1950s, each country knew that it could strike a

---

1. For a fine overview of NASA's international program, with supporting key documents, see John M. Logsdon, "The Development of International Space Cooperation," chap. 1 in *Exploring the Unknown: Selected Documents in the History of the U.S. Civil Space Program,* ed. John M. Logsdon, with Dwayne A. Day and Roger D. Launius, vol. 2, *External Relationships* (Washington, DC: NASA SP-4407, 1996).

lethal blow at the other using nuclear-tipped missiles. This balance of terror provided one of the most frequent arguments at the time for international space cooperation. As Lyndon Baines Johnson, then the Majority Leader of the Senate, put it in 1959, "If . . . we proceed along the orderly course of full cooperation, we shall by that very fact of cooperation make the most substantial contribution yet made towards perfecting peace. Men who have worked together to reach the stars are not likely to descend together into the depths of war and desolation."[2] This claim, the conviction that international space cooperation with the Soviets would remove misunderstanding, project a positive image of the U.S. abroad, reduce tension, and advance the cause of world peace was a *leitmotif* of the early arguments for an international component to the space program. It was also used by Richard Nixon, who justified the expansion of U.S.-Soviet space collaboration in the early 1970s as creating "not just a climate for peace," but the "building blocks" for "an actual structure of peace and cooperation."[3]

This rhetoric did not carry much weight with some people, notably Arnold Frutkin. Frutkin, who was responsible for international affairs inside NASA for 20 years, beginning in 1959, was emphatic about this.[4] "Now, I hope it's come through," he said towards the end of a long interview conducted a few years ago, "that I am not soft-headed about dealing with other people—[like] if you knew your neighbor better you'd like him. I never believed that. If you knew your neighbor better," Frutkin went on, "you might conclude that he [was] a worse son of a bitch than you [suspected]."[5] Frutkin spoke from bitter experience: after many years of achieving little more than "arm's-length" cooperation with the Soviets—more may have been possible had Kennedy not been assassinated—he had finally been witness to the famous Apollo-Soyuz "hand shake in space" in July 1975.[6] For him, while international space

---

2. Quoted in Don E. Kash, *The Politics of Space Cooperation* (n.p.: Purdue University Studies, 1967), p. 10.

3. The words are those of Ron Ziegler, the President's press secretary, during a press conference at the White House on the "Agreement Concerning Cooperation in the Exploration and Use of Outer Space for Peaceful Purposes," 24 May 1972, record no. 12594, Presidential Files, NASA Historical Reference Collection, Washington, DC.

4. Arnold W. Frutkin was deputy director of the U.S. National Committee for the International Geophysical Year in the National Academy of Sciences before he joined NASA in 1959 as director of international programs. His official title changed in 1963 to Assistant Administrator for International Affairs. In 1978, Frutkin became Associate Administrator for External Relations. He retired from federal service in 1979.

5. Arnold W. Frutkin interview, Washington, DC, by Rebecca Wright, 11 January 2002, NASA Historical Reference Collection, Washington, DC.

6. In the early years of his presidency, Kennedy made extensive overtures to the Soviets backed by behind-the-scenes negotiations that seemed to be making considerable headway. These were abruptly stopped after his death—see particularly National Security Action Memorandum 271, dated 12 November 1963 and reproduced in Logsdon, "International Space Cooperation," pp. 166–167.

cooperation was a widely endorsed scientific and political objective, it also was also victim of a multitude of "abstractions, moral imperatives, and contrived prescriptions."[7]

Contemporary analyses of the U.S.'s motives for collaborating in space combine a refreshing spirit of *realpolitik* when discussing how the U.S. *has* behaved in the past with a tendency to prescriptive injunctions about how NASA *should* behave in the future, which Frutkin would probably deplore. We shall treat each of these dimensions of this body of literature in turn.

There is something of a consensus that, for the first two or three decades of its existence, NASA, by virtue of America's immense scientific and technological advantage vis-à-vis its partners, could use its power to dictate the terms of any significant international space effort. American hegemony was implicit in the 1958 Space Act which established NASA and which defined the organization's primary objective as being "the preservation of the role of the United States as a leader in aeronautical and space science and technology and in the application thereof."[8] This concept of leadership, we were reminded in 1987 by a task force of the NASA Advisory Council (NAC), chaired by Herman Pollack, meant not simply achieving superior performance in all aspects of space. It also meant "the defining of goals and the establishment of direction that *others w[ould] be willing to make their own or follow*" (emphasis added).[9] To the U.S., according to another group of space activists, for the first two decades after Sputnik, "cooperation was a politically driven means of linking the space programmes of other countries to US goals and activities, rather than having them closely allied with Soviet aspirations in space."[10] Political scientist Joan Johnson-Freese makes a similar point: in the Cold War context of the '60s and '70s, the U.S. actively sought to collaborate with its Western bloc allies and countries that it wanted to attract to the Western alliance. And since it was "dominant in space, it could dictate terms of cooperation to other countries, which they were more than willing to accept in order to gain entrance to the space program."[11]

Scientific research was a privileged site for international collaboration, and Frutkin quickly defined a set of five criteria which guided NASA's policy in this domain and which embodied these precepts.[12] His criteria are well known

---

7. Arnold W. Frutkin, "International Cooperation in Space," *Science* 169 (24 July 1970): 333–339.

8. National Aeronautics and Space Act of 1958 (Unamended), Sec. 102 (c) (5), available online at *http://www.hq.nasa.gov/office/pao/History/spaceact.html* (accessed 27 January 2005).

9. Herman Pollack, "International Relations in Space. A US View," *Space Policy* 4, no. 1 (February 1988): 24–30.

10. Space Policy Institute and Association of Space Explorers, "International Cooperation in Space— New Opportunities, New Approaches," *Space Policy* 8, no. 3 (August 1992): 195–203.

11. Joan Johnson-Freese, *Changing Patterns of International Cooperation in Space* (Malabar, FL: Orbit, 1990), p. 5.

12. Arnold W. Frutkin, *International Collaboration in Space* (Englewood Cliffs, NJ: Prentice Hall, 1965).

and need not be rehearsed here. Suffice it to say that Frutkin's stress on the need for clean interfaces and no exchange of funds between the partners was inspired by the need to limit technological (and managerial) sharing between the U.S. and its partners to a minimum. Even the content of the program had to dovetail with U.S. interests. As Logsdon puts it, being the dominant partner in space science "often meant that NASA and U.S. scientists would define the objectives and content of a scientific mission and only then invite non-U.S. scientists to participate."[13] Even then, NASA sometimes pulled the plug on a well-defined joint international project to meet domestic pressures for budget cuts and the redefinition of priorities.[14]

Scientific collaboration was the most readily available and least controversial instrument of international collaboration, but it was not enough, particularly in dealing with major allies like Western Europe. The U.S. technological lead and the dynamism of American industry allowed the administration to think beyond the limits of scientific collaboration and to use its technological assets, including technological knowledge and skills, as an instrument of foreign policy to consolidate the Atlantic alliance. Put differently, if the U.S. pursued international collaboration, it was because it "sought the political benefits of leadership [while] its partners [sought] the technical and managerial benefits that come from working with the leader."[15] Here lies the soft underbelly of technological collaboration in the space sector. For if the benefit was in foreign policy, as the Pollack Task Force stressed, the cost lay in the risk that technological sharing would subvert U.S. leadership by helping allies to assert themselves, would endanger national security in a sector where almost all satellite and booster technology is "dual-use technology," and would endanger U.S. industry in a crucial high-tech sector.

Once we move beyond scientific collaboration to technological sharing, those who promote international cooperation will be on the defensive. They will have to overcome the opposition of counterforces that stress the threats to the U.S. that such collaboration entails. These critics will point out that if America's allies are willing to be dependent on the U.S. in the short term, it is with the long-term aim of being autonomous. That if those allies accept the hegemonic regime imposed by the U.S., it is in the hope that they will eventually be able to throw off its yoke. And that if they collaborate initially on terms which are not of their own choosing, it is in order later to compete better with the United States as equal partners, or even to become leaders in areas where America was previously supreme. In short, international col-

---

13. Logsdon, "International Space Cooperation," p. 4.

14. For an angry account of this by two ESA insiders, see Roger M. Bonnet and Vittorio Manno, *International Cooperation in Space: The Example of the European Space Agency* (Cambridge, MA: Harvard University Press, 1994).

15. Space Policy Institute and Association of Space Explorers, p. 200.

laboration in space is always a contested policy objective. It will always have to justify itself to critics who will ask, as I did at the start of this paper, "But why collaborate?" and who see little reason for risking national security and industrial competitiveness, which are essential for the long-term strength of the country, in return for the fragile and unpredictable foreign policy benefits that international collaboration putatively enshrines.

This domestic political context informs much of the literature on international cooperation and accounts for the prescriptive dimension alluded to above. It is dominated by activists, administrators, and political scientists who combine their sense of *realpolitik* with a wish to influence the way NASA and the United States behave in current international collaborative projects, notably the negotiations on foreign participation in the International Space Station. All are sensitive to the changed balance of power in the space sector: the collapse of the Soviet Union as a rival superpower (which forced a major reevaluation of one of NASA's original goals) and the technological and managerial maturity achieved by space programs in the U.S.'s traditional allies (notably Western Europe and Japan). All are also convinced that international collaboration is a worthwhile goal and that, to maintain American leadership in at least certain key areas, the U.S. will have to change its attitudes to meet the changed environment of the late 20th century. Thus Joan Johnson-Freese: "Because the United States began as the dominant space power concerning cooperative ventures, it has never had to learn to operate in any manner other than 'the U.S. way'. But things have changed," she goes on. "There are now an increasing number of space 'actors' with varying ranges of capabilities," including the Soviet Union, Japan, and Western Europe, and "the United States is no longer 'the only game in town' in space activities, although in some cases it is still trying to act as though it is."[16] So, too, the Task Force chaired by Pollack in 1987: "The USA will have to adopt [*sic*] its attitude, approach and politics on international cooperation and competition to a new set of realities."[17] And Ken Pederson, who was responsible for NASA's International Affairs Division in the 1980s and who gave some concrete examples of what that meant. "For NASA today," he wrote, "'power' is much more likely to mean the power to persuade than the power to prescribe." This entails 1) that NASA must accept that "leadership does not mean that it must or ought do it all"; 2) that even if it is the provider of major hardware, NASA "may sometimes have to accept the role of junior partner rather than managing partner" and understand that it can still benefit while doing so; and 3) that NASA must "learn to share direct management and operational control in projects where it is the largest hardware and financial contributor, especially when manned flight systems are involved."[18]

---

16. Johnson-Freese, p. 113.

17. Pollack, "International Relations in Space."

18. Ken Pedersen, "The Changing Face of International Space Cooperation. One View of NASA," *Space Policy* 2, no. 2 (May 1986): 120–137.

This stream of modal concepts, this prescriptive discourse is situated at the core of the struggle to define the U.S.'s role in space in the 21st century and intended to reshape its practices in the international domain. These advocates believe that space cooperation is a "good thing" for the United States, and they seek to lay down the ground rules, based on past experience, for what the U.S. "must do" if it wants to retain credibility and leadership as an international partner. And while commendable for their sensitivity to the points of friction which have traditionally irritated America's partners, their proposals also have an air of unreality. *It is indeed striking that, while all of these authors stress that the U.S. international space effort is driven by foreign policy and that technological collaboration is a substantive issue which shapes its physiognomy, none of them deal with foreign policy or technology except in the most generic way.* These are a taken-for-granted backdrop against which their prescriptions are made, a context which, precisely, cannot be taken for granted, for it is the always-contested framework in which stakeholders will decide whether to collaborate internationally at all, let alone on the terms, and respecting the "musts," that the advocates promote so skillfully.

Scientific and technological sharing, and foreign policy concerns, are the material substrates of international collaboration in space. Scientific and particularly technological sharing, both of hardware and of knowledge and skills, are the single most important means that the U.S. has to influence the space programs of other countries, so consolidating and legitimating its leadership and its hegemonic regime. Technological sharing is also the single greatest danger to national security and national industrial competitiveness in a crucial high-technology sector. The onus on those who promote international collaboration in space is to show how the sharing of specific technologies and the knowledge embedded in them will further America's leadership abroad in a particular historical conjuncture and why that objective will not unduly jeopardize national industry or undermine national security. To advance this debate, one cannot "black-box" technology and foreign policy: they are not the context in which international collaboration takes place; they are the stakes that define what is possible.

This paper aims to contribute to our understanding of international collaboration by using an illustrative historical case study to open the black box of technology and of foreign policy.[19] At the risk of oversimplifying an extremely complex debate, I will explain briefly why the Johnson administration decided in the mid-1960s that it was imperative to collaborate with Western Europe in developing a civilian satellite launcher and discuss the kind

---

19. The case study presented here is based on a small subset of a huge number of documents retrieved from the archives preserved in the NASA Historical Reference Collection in Washington, DC, and at the Lyndon Baines Johnson Library in Austin, TX (hereafter LBJ Library). Additional material was acquired from the National Archives and Records Administration in College Park, MD. I would like to thank the archivists for their invaluable help and support.

of technological sharing that some people thought might be used to achieve the President's foreign policy objectives.[20] What I want to emphasize above all is the strong coupling between technology and foreign policy. I also want to insist that, to understand the possibilities of international collaboration in space, it is crucial to focus on what *specific* technologies might be available for sharing in the pursuit of *specific* foreign policy objectives, rather than—as so often happens—to simply lump technology and foreign policy into an undifferentiated whole. Those in the administration who are engaged in working out what can be done with a foreign partner fight over the boundary between what technologies can be shared and what cannot. The advocates of a more open approach are driven by the conviction that the maintenance of American "leadership" and its ability to control the form and content of the space programs of other nations are best achieved by relaxing restrictions in particular areas. Sometimes they win; sometimes, as in the case to be described here, they lose, both because the forces arraigned against them are formidable and because the foreign policy context is never stable and calls forth a different response to changed circumstances. I am convinced that only if historians study international collaboration at this fine-grained level can they help avoid what Frutkin bemoaned over 30 years ago, namely, analyses replete with the "usual quota of abstractions, moral imperatives, and contrived prescriptions."

The Johnson Administration and the ELDO Crisis

On 29 July 1966, Walt W. Rostow, one of LBJ's two national security advisers, signed off on National Security Action Memorandum 354.[21] NSAM 354 was a response to a request from the Department of State that the U.S. "clarify and define" its policy concerning collaboration with the "present and future programs" of ELDO, the European Launcher Development Organisation. The document affirmed that it was "in the U.S. interest to encourage the continued development of ELDO through U.S. cooperation." It referred to the results of an ad hoc working group, established by the State Department and chaired by Herman Pollack, that had prepared a statement "defining the nature and extent of U.S. cooperation with ELDO which the U.S. government is now prepared to extend." This statement was to be "continually reviewed by the responsible agencies," above all, the Department of

---

20. The reactions in the United States to the ELDO crisis in 1966 have received little scholarly attention. For the best analysis, see Lorenza Sebesta, *Alleati Competitivi. Origini e sviluppo della cooperazione spaziale fra Europa e Stati Uniti* (Bologna, Italy: Laterza, 2003), chap. 3. The issue is also described in a project Sebesta worked on with John M. Logsdon. I thank John Logsdon for making a copy of their unpublished manuscript available to me.

21. NSAM 354, "U.S. Cooperation with the European Launcher Development Organization," 29 July 1974, available online at *http://www.lbjlib.utexas.edu* (accessed on 9 March 2005).

Defense and the State Department, along with NASA, "to ensure that it is current and responsive in terms of developing strategies."

The help that the working group proposed was extensive. It was divided into three categories: general, and short-range and long-range assistance.[22] The first contained some standard items—training in technical management, facilitating export licenses, use of NASA test facilities—but also suggested that a technical office be established within NASA "specifically to serve in an expediting and assisting role for ELDO." Short-range help included "technical advice and assistance" in items like vehicle integration, stage separation, and synchronous orbit injection techniques, as well as the provision of unclassified flight hardware, notably a strapped-down "guidance" package used on the Scout launcher which had already been exported to Japan. Long-range assistance was focused on helping with a high-energy cryogenic upper stage of the rocket, currently being considered in ELDO. It was proposed that Europeans be given access to technological documentation and experience available in the Atlas-Centaur systems, that ELDO technical personnel "have intimate touch with the problems of systems design, integration, and program management of a high-energy upper [sic] such as the Centaur," and even that the U.S. consider "joint use of a high-energy upper stage developed in Europe."[23] In short, in mid-1966, the U.S. was considering making a substantial effort to help ELDO develop a powerful launcher with geosynchronous orbit capability by sharing state-of-the-art knowledge and experience and by facilitating the export of hardware which—it should be added—would not normally be available on a bilateral basis to European national launcher programs.

NSAM 354 was catalyzed by a crisis in ELDO in February 1966 and deep concerns in the Johnson administration about the future of the collaborative European effort. ELDO, it must be said, had been a fragile organization from its very inception in 1960–61.[24] It was born of the need by the British government to find a new role for its Blue Streak missile. The liquid-fueled rocket was rendered obsolete by the long time required to prepare it for launch and

---

22. This paragraph is derived from "Policy Concerning US Cooperation with the European Launcher Development Organization (ELDO)," attached to U. Alexis Johnson's "Memorandum," 10 June 1966, folder 15707, International Cooperation and Foreign Countries, NASA Historical Reference Collection, Washington, DC.

23. In summer 1965, ELDO had asked for help from NASA on "designing, testing and launching liquid hydrogen/liquid oxygen upper stages" (Frutkin to Robert N. Margrave, Director, Office of Munitions Control, Department of State, 6 June 1965, record no. 14465, International Cooperation and Foreign Countries, International Cooperation, folder International Policy Manual Material from Code I, NASA Historical Reference Collection, Washington, DC).

24. I describe the launch of ELDO in detail in J. Krige and A. Russo, *A History of the European Space Agency, 1958–1987*, vol. 1, *The Story of ESRO and ELDO, 1958–1973* (Noordwijk, Netherlands: ESA SP-1235, April 2000), chap. 3. See also Michelangelo De Maria and John Krige, "Early European Attempts in Launcher Technology," in *Choosing Big Technologies*, ed. John Krige (Chur, Switzerland: Harwood Academic Publishers, 1993), pp. 109–137.

by the cost, which spiraled to new heights as the expenditures on reinforced concrete silos were factored into the budget. Hence the idea to recycle Blue Streak, stripped of its military characteristics, as the first stage of a multistage civilian satellite launcher, built together with partners in continental Europe. This would save face at home, it would ensure that the money already spent on development was not completely wasted, it would preserve the engineering teams and their skills intact, it would please British industry, and—and this was crucial—it would serve as a gesture of solidarity and good will to the emerging European Common Market, which Britain had previously boycotted, nay, tried to sabotage. Indeed, shortly after the British proposed this joint venture to their continental partners, Prime Minister Harold Macmillan made an official application for his country to join the European Community. Long, drawn-out negotiations ensued before Blue Streak was given a new lease on life. The French would build the second stage atop the British rocket, the Germans would build the third stage, and the Italians would build a test satellite. Clean interfaces were retained to limit technology transfer between firms in different countries to protect competitive advantage and national security (especially in Britain and France, which were both developing independent nuclear deterrents). The ELDO staff had little authority over the separate national authorities and, above all, no power to integrate the three independently built stages of the rocket or to ensure compatibility between the various systems and subsystems built in different countries or in different firms in the same country.[25] By 1966, as many had predicted, ELDO faced the first of many crises that led to its eventual demise in 1972.[26] Development costs had increased from the initial estimate of about $200 million to over $400 million, and no end to the upward spiral was in sight. Blue Streak had been successfully commissioned, while the French and German stages were still under development. What is more, in January 1963, French President de Gaulle had vetoed Britain's application to join the Common Market. For Britain, who was paying almost 39 percent of the ELDO budget, the original technological, industrial, and political rationale for launching the organization had evaporated. In February 1966, her Minister circulated an aide-mémoire to his homologues in the ELDO member states suggesting that it was unlikely that the organization would produce any worthwhile result and that the United Kingdom saw little interest to continue in the program and to contribute financially to it.

This move perturbed the Johnson administration immensely. At the most general level, the U.S. saw ELDO as a technological embodiment of European

---

25. For a fine description of the failure of management in ELDO, see Stephen B. Johnson, *The Secret of Apollo: Systems Management in American and European Space Programs* (Baltimore: Johns Hopkins, 2002), chap. 6.

26. On the crisis, see Krige and Russo, *A History*, vol. 1, chap. 4, sect. 4.3.2.

multilateralism. The withdrawal of the United Kingdom would send a signal that Britain was still not enthusiastic about participating in European integration, which the United States had always regretted. It would also strike a major blow to the gradual movement towards European unity on the continent. This was in a very brittle state at the time. There was a crisis in the European Economic Community (EEC), precipitated by the French, who had begun to boycott the EEC's decision-making machinery so as to liberate the country from its "subordination" to Community institutions and the dilution of sovereignty that that entailed.[27] There was a similar crisis in NATO. The French were not against the Alliance as such but believed that NATO needed reforming. Western European nations were no longer prostrate, as they had been in 1949, and they needed to be prepared to meet a Soviet nuclear threat in Europe with their own independent deterrents (would Washington be prepared to risk New York to defend Paris? it used to be said). "The French have emphasized their dissatisfaction by becoming increasingly an obstructionist force in NATO," one task force wrote, "equating" integration with subordination.[28] In this inauspicious climate, everything possible had to be done to sustain the momentum for European unity. As Under Secretary of State George Ball emphasized, "The United States has a direct interest in the continuation of European integration. It is the most realistic means of achieving European political unity with all that that implies for our relations with Eastern Europe and the Soviet Union . . . and is the precondition for a Europe able to carry its proper share of responsibility for our common defense."[29] While ELDO was not central to European integration, its collapse would provide additional encouragement for those who were increasingly hostile to supranational ventures in Europe.

Saving a European launcher was justified by a second foreign policy concern pressing on the Johnson administration at the time: it would help close the so-called "technological gap" that had opened between the two sides of the Atlantic. Beginning in summer 1965, there were increasingly strident complaints in France, and to some extent Germany, that American business was invading Europe and dominating key sectors of European industry.[30] The U.S. could not

27. Ted Van Dyk to the Vice President, 7 July 1965, folder Germany Erhard Visit [12/65], 12/19–21/65, box 192, National Security Files, Country File Europe and USSR, Germany, LBJ Library.

28. "France and NATO," position paper, 25 September 1965, folder Germany Erhard Visit [12/65] 12/19–21/65, box 192, National Security File, Country File Europe and USSR, Germany, LBJ Library.

29. Department of State to Amembassy Bonn 1209, outgoing telegram, 18 November 1965, signed [George] Ball, folder Germany Erhard Visit [12/65], 12/19–21/65, box 192, National Security File, Country File Europe and USSR, Germany, LBJ Library.

30. SC No. 00666/65B, "US Investments in Europe," CIA Special Report, 16 April 1965, folder Memos [2 of 2], Vol. II, 7/64–7/66, box 163, National Security File, Country File, Europe, LBJ Library. Jean-Jacques Servan-Schreiber's *The American Challenge* (New York: Atheneaum, 1968; translation of *Le Défi américain*) is, of course, the *locus classicus* of this argument.

easily dismiss their concerns. As Frutkin explained, Western Europe's progress in space was "a contribution to the strength of the Free World. An increasing technological gap between us (and them) can only lead to political and economic strains and to weakness."[31] Indeed, the President took this matter so seriously that in November 1966, Johnson personally signed NSAM 357, instructing his science adviser, Donald Hornig, to set up an interdepartmental committee to look into "the increasing concern in Western Europe over possible disparities in advanced technology between the United States and Europe."[32] In its preliminary report, the committee concluded that "the Technological Gap [was] mainly a political and psychological problem" but that it did have "some basis in actual disparities." These included "the demonstrated American superiority in sophisticated electronics, military technology and space systems." Particularly important were "the 'very high technology industries' (particularly computers, space communications, and aircraft) which provide a much greater military capability, are nationally prestigious, and are believed to be far-reaching in their economic, political and social implications."[33] The U.S., Herman Pollack told Sir Solly Zuckerman, Britain's Chief Scientific Adviser, was "seeking new and different ways of expanding cooperation in space because we consider that there is a close connection between [sic] technological gap and the development of space technology."[34]

There was a third, even more fundamental argument for supporting the development of a launcher in the ELDO framework. This was, in fact, the single most important reason why Pollock's ad hoc working group of the NASC was asked to look again at the possibilities of sharing booster technology with foreign nations. It also led directly to the release of NSAM 354, expressing American interest in helping ELDO. The argument, in the words of NASA Administrator James Webb, was that enhanced international collaboration in space would be "a means whereby foreign nations might be increasingly involved in space technology and diverted from the technology of nuclear weapons delivery."[35] More precisely, it was by encouraging multilateral

---

31. Quoted in *Space Business Daily* 25, no. 35 (18 April 1966): 286.

32. NSAM 357, "The Technological Gap," 25 November 1966, available online at *http://www.lbjlib. utexas.edu/johnson/archives.hom/NSAMs/nsam357.gif* (accessed on 9 March 2005). Hornig's official title was the Special Assistant to the President for Science and Technology.

33. "Preliminary Report on the Technological Gap Between U.S. and Europe," attached to David Hornig's letter to the President, 31 January 1967, folder Technological Gap [1 of 2], box 46, Subject File, National Security File, LBJ Library.

34. "Memorandum for the Files. Cooperation with ELDO," 6 May 1966, folder Cooperation in Space—Working Group on Expanded International Cooperation in Space ELDO #1 [2 of 2], box 14, National Security Files, Charles Johnson File, LBJ Library.

35. Webb to Robert McNamara, 28 April 1966, record no. 14459, International Cooperation, International Cooperation and Foreign Countries, folder Miscellaneous Correspondence from CODE I—International Relations 1958–1967, NASA Historical Reference Collection, Washington, DC.

organizations that the nonproliferation of missile technology at the national level could be controlled. A position paper prepared for the very first meeting of Pollack's working group in May 1966 stressed this. Multilateral programs should be encouraged, it asserted, since

> [i]n such a framework rocket programs tend to be more open, serve peaceful uses and are subject to international control and absorb manpower and financial resources that might otherwise be diverted to purely national programs. National rocket programs tend to concentrate on militarily significant solid and storable liquid fueled systems, are less open, and less responsive to international controls. Any break up of ELDO might lead to strengthening national programs tending in the latter direction.[36]

Put differently, since European nations had limited resources to devote to their military and civilian space programs and had to make hard choices about priorities, the U.S. could use the carrot of technological sharing with ELDO to divert human and material resources away from national programs which were more difficult to control and which might see the proliferation of weapons delivery systems.

It was the French national program which particularly bothered the U.S. On 26 November 1965, France had become the third space power by launching its own satellite with its own launcher, Diamant-A, from Hammaguir in Algeria. The feat was repeated in February 1966. This three-stage launcher combined "militarily significant solid and storable liquid fueled systems"—just the kind of technology the U.S. did not want it to develop—in a highly successful vehicle derived from the national missile program.[37] In the light of these achievements and de Gaulle's growing determination to affirm his independence of the EEC and the Atlantic alliance, "The US is concerned that, if ELDO were to be dissolved, France might devote more of its resources to a national, military-related program or that it might establish undesirable bilateral relationships for the construction of satellite launch vehicles"[38]—meaning that unless Britain and America boosted the organization, "the Soviets would

---

36. T. H. E. Nesbitt, "Meeting No. 1, Committee on Expanded International Cooperation in Space Activities. Subject: Cooperation Involving Launchers and Launching Technology," 17 May 1966, folder Cooperation in Space—Working Group on Expanded International Cooperation in Space. ELDO #1 [2 of 2], box 14, National Security Files, Charles Johnson File, LBJ Library.

37. Diamant-A used a mixture of N_2O_4/UDMH (storable liquid fuels) in its first stage and solid fuel in the second and third stages.

38. "US Cooperation with ELDO," position paper, 21 July 1966, folder Cooperation in Space—Working Group on Expanded International Cooperation in Space. ELDO #1 [2 of 2], box 14, National Security Files, Charles Johnson File, LBJ Library.

move into the vacuum if ELDO collapsed."[39] The U.S. had to contain this threat and to ensure that European institutions emerged "from the present crisis with their prestige, power and potential for building a united Europe as little impaired as possible."[40] Developing advanced space technology in Europe and assisting ELDO to develop its launcher, in particular, were some of the many measures considered by the Johnson administration to achieve that objective in 1966.

THE OBSTACLES TO THE SUPPORT FOR ELDO

Two major obstacles stood in the way of these initiatives. Both were enshrined in National Security Action Memoranda. There was NSAM 294 of 20 April 1964, which dealt with "U.S. Nuclear and Strategic Delivery System Assistance to France." The second was NSAM 338 of 15 September 1965, defining "Policy Concerning U.S. Assistance in the Development of Foreign Communications Satellite Capability."[41]

NSAM 294 stated that since the administration opposed the development of a nuclear force outside the framework of NATO and that since France was doing all it could to evade the constraints of the Alliance, nothing should be done to help its nuclear weapons system (France first successfully tested its A-bomb in the Sahara in February 1960), including the "French national strategic nuclear delivery capability." This included "exchanges of information and technology between the governments, sale of equipment, joint research and development activities, and exchanges between industrial and commercial organizations." This obviously made collaboration with ELDO difficult since how could one be sure that technology that was shared with the organization would not leak through to the French military program?[42]

NSAM 338 was less specific, referring instead to the policy guidelines established by General J. D. O'Connell, the President's Special Assistant for Telecommunications, in a memorandum of 25 August 1965. These guidelines effectively extended the military constraints on the transfer of booster technology to cover specific commercial concerns. O'Connell's memo stipulated

---

39. Anonymous, "Memorandum for the Files, Cooperation with ELDO," meeting with Zuckerman, 6 May 1966, folder Cooperation in Space—Working Group on Expanded International Cooperation in Space. ELDO #1 [2 of 2], box 14, National Security Files, Charles Johnson File, LBJ Library.

40. Department of State to Amembassy Bonn 1209, outgoing telegram, 18 November 1965, signed [George] Ball, folder Germany Erhard Visit [12/65], 12/19–21/65, box 192, National Security File, Country File Europe and USSR, Germany, LBJ Library.

41. NSAM 294, "U.S. Nuclear and Strategic Delivery System Assistance to France," 20 April 1964, and NSAM 338, "Policy Concerning U.S. Assistance in the Development of Foreign Communications Satellite Capability," 15 September 1965, both available online at *http://www.lbjlib.utexas.edu* (accessed 9 March 2005).

42. NSAM 294, "U.S. Nuclear and Strategic Delivery System."

that if the U.S. was to help other countries develop a comsat (communications satellite) capability, it had to have guarantees that the foreign program was integrated into the single global system enshrined in the INTELSAT agreements of 1964. INTELSAT was the international consortium that owned and operated the international comsat system. It had 56 member nations in 1967 (though neither China nor the Soviet Union were members). American interests were represented by COMSAT, a private corporation, 50 percent of whose stock was owned by communications carriers (like AT&T). Voting was weighted according to use, which made it "an unusually attractive international vehicle for the U.S."[43] since it had veto power inside INTELSAT at the time (its voice counted for 54 percent). What is more, the 1964 INTELSAT agreements (due to be renegotiated in 1969 to take account of the expected expansion in the use of comsat technology by other nations) stipulated that the U.S. weight could never drop below 50 percent: "in other words, we control."[44] With this power in its pocket, the "core" of NSAM 338, as McGeorge Bundy explained to LBJ, was "to use our technological superiority to <u>discourage</u> commercial competition with COMSAT and/or wasteful investment in several duplicative Free World defense-related systems" (emphasis in the original).[45] To this end, the U.S. should "withhold provision of assistance to any foreign nation in the field of communications satellites which could significantly promote, stimulate or encourage proliferation of communications satellite systems" outside the INTELSAT framework, *including* "the provision of launching services or launch vehicles for communications satellites."[46]

The significance of NSAM 338 for our story is that it extended the provisions of NSAM 294 beyond national security and foreign policy objectives to protect also U.S. business interests.[47] By defining launchers as a component of the "communications satellite *system*," it included delivery systems inside the

---

43. Charles Johnson to Walt Rostow, 13 July 1967, folder NSAM 338, box 7, National Security Files, LBJ Library.

44. Ibid.

45. McGeorge Bundy to the President, "Helping Others to Use Communications Satellites," 13 September 1965, folder NSAM 338, box 7, National Security Files, LBJ Library.

46. "Policy Concerning U.S. Assistance in the Development of Foreign Communications Satellite Capabilities," position paper, unsigned, 23 August 1965, folder NSAM 338, box 7, National Security Files, LBJ Library.

47. It should be stressed that NSAM 338 was not restricted to protecting commercial interests, though it included them. As the memo from McGeorge Bundy that I cited earlier makes clear, there were also national security concerns involved. The United States, he noted, would set up a separate national defense comsat system "where security demands" and would encourage "selected allies" (actually Britain and Canada) to "buy time" on this system for their security needs. Otherwise, he wanted everyone to use the single global system for all purposes. The United States thus wanted to discourage the proliferation of regional comsat systems both to limit international competition for a potentially lucrative market and to limit the spread of parallel regional comsat systems for defense (McGeorge Bundy to the President, "Helping Others to Use Communications Satellites").

policies being defended by COMSAT on behalf of the U.S. in INTELSAT. The sale of launch vehicles and launch services *and* technological assistance with the development of an indigenous launch capability were now conditional on the foreign clients' guaranteeing that such launchers would not be used to subvert a single worldwide commercial satellite communications system then under U.S. control. As one senior administrator put it, "It is difficult to maintain international cooperation on this basis."[48]

FINDING A WAY AROUND THE OBSTACLES

To overcome these obstacles to technology transfer, NASA and the State Department insisted that to promote U.S. foreign policy and business interests, one had to *distinguish between different types of technology* and *the specific foreign policy options* that America wanted to promote. They were convinced that American leadership, and its ability to restrict the proliferation of weapons systems and comsats, was best achieved by treating technology transfer on a case-by-case basis and by "building high walls around small fields," as it is sometimes called today, rather than by blanket restrictions which treated both technology and foreign policy as seamless wholes.

To achieve this, a number of crucial distinctions had to be made. Current U.S. policy was dominated by the "dual-use" aspect of boosters as both ballistic missiles and as stages of satellite launchers. This was too simple, Webb pointed out: "If we could focus our controls on the weapons themselves, we might even hope to free vehicle technology for maximum stimulus of space activity abroad."[49] Consider the constraints on booster technology imposed by NSAM 294. As Webb pointed out to Defense Secretary McNamara, although high-energy, cryogenic, or nonstorable upper stages might conceivably be employed for military purposes, in practice they would probably not be deployed in that way. "Even in the case of France," Webb stressed, "it seems likely that encouragement to proceed with upper stage hydrogen/oxygen systems now under development might divert money and people from a nuclear delivery program rather than contribute to that which is already under way using quite different technology."[50] Guidance and control technology was another gray area. An American company had recently been refused a license to assist France with

---

48. Charles Johnson to Walt Rostow, 13 July 1967.

49. Webb to Johnson, 26 April 1966, record no. 14459, International Cooperation and Foreign Countries, folder Miscellaneous Correspondence from CODE I—International Relations 1958–1967, International Cooperation, NASA Historical Reference Collection, Washington, DC.

50. Webb to McNamara, 28 April 1966, and reply, Bob [McNamara] to Jim [Webb], 14 May 1966, record no. 14459, International Cooperation and Foreign Countries, International Cooperation, folder Miscellaneous Correspondence from CODE I—International Relations 1958–1967, NASA Historical Reference Collection, Washington, DC.

the development of gyro technology. But as Richard Barnes, the Director of Frutkin's Cooperative Projects Division, pointed out to the chair of the NSAM 294 review group, gyros of comparable weight and performance were already available in France. The release of inertial guidance technology to Germany had been officially sanctioned in July 1964 on condition that it was not employed "for ballistic missile use or development."[51] And, as we mentioned earlier, a strapped-down "guidance" package used on the Scout launcher had already been exported to Japan. Here, and in general, wrote Webb to McNamara, rather than a blanket restriction, "we might be better off were we to concentrate on a few very essential restrictions, such as *advanced* guidance and reentry systems" (my emphasis). In a supportive reply to Jim, Bob reassured the NASA Administrator that he strongly supported international cooperation in space and that he had directed his Department of Defense staff "to be as liberal as possible regarding the release of space technology for payloads and other support items."[52]

One important consideration shaping the argument for a revision in policy was that restrictions on the export of some items were now redundant since European booster technology was advancing rapidly without external help. It was also counterproductive to deny a nation a technology if it could easily and quickly be obtained from a source other than the United States: this would not simply be to the detriment of American business, but also to U.S. foreign policy, particularly if that source was the Soviet Union. Thus Barnes suggested (and Webb concurred) that the interpretation of NSAM 294 on the export of booster technology needed to be more specific. The guidelines should deny to a foreign power "only those few critical items which are clearly intended for use in a national program, would significantly and directly benefit that program in terms of time and quality or cost, and are unavailable in comparable substitute form elsewhere than the US" (emphasis in the original). The guidelines should also explicitly recognize that it was in America's interest to promote European space collaboration, so that technology transfer intended for multinational programs like ELDO (and ESRO—the European Space Research Organisation) would "normally be approved" so long as the items were "of only marginal benefit to the national program" or "were available elsewhere than the US without undue difficulty or delay."[53] In short, requests for technology transfer were to be treated on a case-by-case basis and should take into account the kind of technology at issue, its likely uses in practice, the global state of the market for the technology,

51. NSAM 312, "National Policy on Release of Inertial Guidance Technology to Germany," 10 July 1964, available online at *http://www.lbjlib.utexas.edu* (accessed 9 March 2005).

52. Webb to McNamara, 28 April 1966, and reply, Bob [McNamara] to Jim [Webb], 14 May 1966.

53. Richard Barnes to Scott George, Chairman, NSAM 294 Review Group, Department of State, 15 April 1966, record no. 14459, International Cooperation and Foreign Countries, International Cooperation, folder Miscellaneous Correspondence from CODE I—International Relations 1958–1967, NASA Historical Reference Collection, Washington, DC.

and the importance of collaboration from a foreign policy perspective. The last, along with U.S. business interests, was not to be sacrificed on the altar of an overcautious, generalized reluctance to share technology just because it *might* encourage programs which sections of the U.S. administration disapproved of.

Frutkin was also keen to relax the constraints on the sharing of comsat technology that were embodied in NSAM 338. Europeans, he wrote, were persuaded that the United States was "seeking by all means, fair or foul, to maintain political and technical control of Intelsat."[54] He was convinced that, to allay their suspicions, the U.S. had to be prepared to provide launch services on a reimbursable basis for (experimental) foreign communication satellites. This would "extend the market for American vehicles, remove some incentive for independent foreign development of boosters, and assure that we could continue to exercise critical leverage in foreign comsat activities rather than lose such leverage." Frutkin also favored the removal of restrictions on the export of satellite technology as such, including the kick-stage and propulsion technology needed to place a communications satellite in geosynchronous orbit.

An anonymous internal memorandum argued that technological sharing was the best way to enroll foreign firms and their governments in American comsat policy. By allowing "United States firms to enter cooperative arrange-ments with the communications and electronics manufacturing industry in other countries," notably in Western Europe, industries in these countries would develop the technical know-how needed for them "to compete effec-tively for contracts for the space segment of the global communications system." This would "remove a current irritant, primarily expressed by the French but also shared by the British, Italians and Germans, about their inability to supply hardware for the INTELSAT space segment." And even if such technologi-cal sharing did not irreversibly lock these European countries into the single global system favored by the U.S., one could expect them to have a "greater incentive" to collaborate with America in developing that global system. One might also expect them to be more cooperative and sympathetic to the U.S. position during the renegotiation of the INTELSAT agreements scheduled for 1969. Anyway, if the U.S. did nothing to help these nations, they would eventually develop the technology on their own, without American help, and would be quite capable of establishing separate, regional communications satellite systems in due course.[55] As Frutkin explained, "(a) We do need to

---

54. A. W. Frutkin to Mr. Hilburn, "Memorandum for Mr. Hilburn—AAD, Policies Relevant to '69 Revision of Intelsat Agreement," 11 April 1966, record no. 14459, International Cooperation and Foreign Countries, International Cooperation, folder Miscellaneous Correspondence from CODE I—International Relations 1958–1967, NASA Historical Reference Collection, Washington, DC.

55. "Communications Satellite Technology," undated and unsigned memorandum, but obviously written around April 1966, folder Cooperation in Space—Working Group on Expanded International Cooperation in Space, ELDO #1 [2 of 2], box 14, National Security Files, Charles Johnson Files, LBJ Library.

improve our situation in Intelsat with specific reference to the 1969 negotiations. (b) We already have a strong technical lead in the comsat field. (c) We
already have an adequate voting majority in Intelsat. (d) We can rely upon
our technical, moral and financial strength to assure continuing leadership—
without seeking to deny technology to our partners in Intelsat."[56] Rather,
then, use technological sharing as an instrument to divert foreign firms and
governments into working with U.S. industry within the framework of a
single global system where the U.S. was the dominant partner than have them
defiantly develop an independent national or regional comsat capability over
which the U.S. had no control and which could be used to bargain for a major
revision of the INTELSAT agreements against U.S. interests.

I have stressed the pressure which foreign policy concerns played in arguing for technological sharing with ELDO. Implicit in my account is another
dimension of the issue: the need to promote and channel the interests of
American industry. Indeed, NASA officials like Frutkin mediated between
firms who wanted to export technology abroad and the Office of Munitions
Control in the State Department, which authorized them to do so. As Frutkin
explained to Margrave, who directed the Office, American firms were putting NASA, the Department of Defense, and the State Department under
extreme pressure to export nonmilitary vehicle technology to individual
national firms in Europe.[57] By releasing export controls on the transfer of this
technology to ELDO, one could at once satisfy their demands and divert them
from the national to the multilateral level in line with U.S. foreign policy. We
see, then, that arguments for relaxing constraints on booster technology were
intended not simply to advance multinationalism in Europe and to help
ELDO, but also to satisfy pressure for access to the launcher construction market from U.S. business. This stakeholder in international space collaboration is
almost always ignored; it should not be.

DENOUEMENT

Those administrators who were for, and those were against, relaxing constraints on technology transfer to ELDO shared a concern for nonproliferation.
They differed on how best to achieve this. NASA and the State Department
argued that by sharing high-energy nonstorable liquid-fuel launcher technology with ELDO, they could divert resources away from national military
programs for which such fuels were obsolete. Similarly, they argued that by
letting U.S. firms help European industry to build up its comsat capability,

---

56. Frutkin to Hilburn, "Memorandum for Mr. Hilburn."

57. Frutkin to Margrave, 1 April 1965, record no. 14465, International Cooperation and Foreign
Countries, International Cooperation, folder International Policy Manual Material from Code I, NASA
Historical Reference Collection, Washington, DC.

they could more easily engage European governments in the single global system promoted and controlled by Washington at the expense of a proliferation of competing regional communications satellite systems which could serve independent commercial and military needs. The defenders of NSAM 338 were adamant, however, that the U.S. should do nothing to help other countries develop comsats, or the powerful launchers needed to place them in geostationary orbit, without cast-iron guarantees that these would only be used in the INTELSAT framework. For them, technological assistance in either of these domains could only hasten proliferation, not contain it. By summer 1967, it was clear that the latter had won the day.

The reasons for this are complex and will be dealt with very briefly here. Developments in Europe played a role. ELDO (temporarily) survived its crisis and, by September 1966, had reoriented its program unambiguously in favor of developing a launcher called Europa II that achieved geostationary capability by adding a fourth, French-built solid-fuel stage to the previous ELDO-A rocket. In parallel, France and Germany decided to fuse their national comsat projects in a joint experimental telecommunications satellite called Symphonie to be launched by Europa II from the new French base in Guyana.[58] ELDO had moved from an artificial political construct to an organization with a well-defined technical mission and was far less vulnerable to offers of American help.

From the American point of view, to channel this "European fixation on comsats and launch vehicles," as Richard Barnes put it, the U.S. had to make an unambiguous offer for technological assistance in domains which satisfied the interests of both parties.[59] With cryogenic fuels no longer being considered and with France responsible for the kick-stage into geostationary orbit, this was going to be *very* difficult. Divisions within the administration on how best to interpret the requirements of NSAM 338 made it virtually impossible. Frutkin described the state of play in August 1966 to Webb, just before the NASA Administrator was to leave on a crucial European tour to discuss possible collaborative projects. While the "general atmosphere for space cooperation with the United States may have improved slightly," thanks to the initiatives by NASA and the State Department which we have described in this paper, they had done little more than "clear the air somewhat." The Europeans, Frutkin told Webb, "know of no progress in easing US restrictions upon communications satellite technology," and "it may be sometime" before the progress that had been made in Washington could be divulged to them. Webb was therefore to repeat the standard answer to the usual request for comsat launch assistance: "that we could certainly give consideration to such a proposition on the assumption that

---

58. The official agreement between the two governments was signed on 6 June 1967.

59. RJHB to AWF, "The 'Webb Commission,'" 5 May 1967, record no. 14459, International Cooperation and Foreign Countries, International Cooperation, folder Miscellaneous Correspondence from CODE I—International Relations 1958–1967, NASA Historical Reference Collection, Washington, DC.

the European countries take their INTELSAT commitment to a single global system as seriously as we do."[60] By virtue of this approach, there was, to quote Barnes again, a "deterioration of 'climate for cooperation' caused by (1) US policies and actions within the Intelsat, and (2) US export policies in support of the 'single global system.'" This led to "European reaction of suspicion and distrust to US offer to escalate cooperation."[61]

As Barnes remarked, the breakdown in trust between the two sides of the Atlantic was fueled by a very public, high-level offer to "escalate" space collaboration with West Germany and other European allies, which had gained momentum throughout 1966.[62] In an exchange of toasts between President Johnson and Chancellor Ludwig Erhard at a state banquet on 20 December 1965, LBJ suggested that existing scientific cooperation should be extended to embrace "an even more ambitious plan to permit us to do together what we cannot do alone." The President gave two examples of "demanding" and "quite complex" collaborative projects which would "contribute vastly to our mutual knowledge and to our mutual skills": a solar probe and a Jupiter probe. He also announced that NASA Administrator Webb would be traveling to Europe shortly to discuss these ideas in Germany and with other European governments.[63]

The target and timing of Johnson's offer were not coincidental. Erhard was a convinced and reliable American ally and was deeply hostile to de Gaulle's attempts to undermine the existing structures of both NATO and the EEC. As Secretary of State Dean Rusk stressed to James Webb, with the Chancellor boldly resisting this attack on European institutions, "it [was] politically important for the United States to cooperate as closely as possible with Germany." Increasing "the vigor and scope of space cooperation" with the country would be tangible, "positive evidence of constructive American interest in Germany," and it would encourage Erhard to take the lead in advancing U.S. policies in the region.[64]

The fanfare surrounding this offer for expanded scientific cooperation contrasts sharply with the reluctance to disclose publicly the possibility for technological collaboration with ELDO. And it was counterproductive in many respects. The American attempt to isolate de Gaulle was evident for all to see; indeed, Erhard was forced to relinquish his post in November 1966, accused of

---

60. Frutkin to Webb, "Memorandum for Mr. Webb," 11 August 1966, record no. 14618, folder Germany (West), 1956–1990, Foreign Countries, International Cooperation and Foreign Countries, NASA Historical Reference Collection, Washington, DC.

61. RJHB to AWF, "The 'Webb Commission.'"

62. This initiative is worthy of a separate paper; I give only the barest outline here.

63. "Exchange of Toasts Between President Lyndon B. Johnson and Chancellor Ludwig Erhard of the Federal Republic of Germany (In the State Dining Room)," 20 December 1965, folder Germany Erhard Visit [12/65], 12/19–21/65, box 192, National Security Files, Country File Europe and the USSR, Germany, LBJ Library.

64. Dean Rusk to James Webb, 29 August 1966, record no. 14618, folder Germany (West), Foreign Countries, International Cooperation and Foreign Countries, NASA Historical Reference Collection, Washington, DC.

mismanaging the economy and of being too pro-American and anti-French. The cost of the kind of projects discussed (about $100 million) was deemed to be excessive, given the resources available for space science and European priorities (although eventually Germany did embark on a bilateral venture with the U.S., the $100-million Helios project to send two major spacecraft within 45 million miles of the Sun).[65] Finally, with the U.S. publicly insisting on the need to respect the INTELSAT agreements, the American offer was also interpreted by some as a strategy to divert scarce European resources into science and away from applications, notably telecommunications. "All in all," wrote Frutkin to Webb in August 1966, "we must say the President's proposal got off to a poor start due to misunderstandings which are inevitable when a proposition of this sort is made in the headlines without preparation of the ground."[66] Barnes put it pithily: because of European "suspicion and distrust," aggravated by President Johnson's spectacular overtures to Chancellor Erhard, there was "no prospect for escalating cooperation with Europe unless (1) US is willing to modify its present export control policies, and (2) we could offer other possibilities for cooperation in areas of interest to them (i.e., comsats and vehicles)."[67] This was not to be.

CONCLUSION

The defeat of those inside NASA and the State Department who considered sharing communications satellite and booster technology with Europe in mid-1960s was simply the first of a series of setbacks for those in the administration who believed that technological sharing could be used to unite Europeans around projects which were at once useful to them and compatible with the maintenance of U.S. leadership in strategic areas. Indeed, the battle was repeated just a few years later with the same result. European hopes to be integrally engaged at the technological level in the post-Apollo program, sparked by NASA Administrator Tom Paine in the late 1960s, were soon dashed. The compromise that ensued left Germany taking the lead in building a shirtsleeve-environment scientific laboratory that could fit in the Space Shuttle's cargo bay and that, crucially, preserved the basic principles of clean interfaces and no exchange of funds more or less intact. Indeed, Europe's ongoing struggle to be a genuine partner at the level of technological and managerial sharing with NASA and the U.S. might suggest that, when the chips are down, the need by powerful forces in the U.S. to protect national industry and national security will always prevail over foreign policy considerations. For them, American leadership is best preserved by denying sensitive technology, not by finding ways to use technological sharing to orient a partner's program in line with U.S. interests.

---

65. The project is discussed in Frutkin, "International Cooperation in Space."
66. Frutkin, "Memorandum for Mr. Webb."
67. RJHB to AWF, "The 'Webb Commission.'"

The negotiations over the ISS, particularly with Russia, show that this is not always so.[68] Indeed, it is striking that here, NASA *has* departed from past practice in accepting critical-path contributions from Canada and Italy and, more significantly, in accepting that there be a joint U.S.-Russian core and infrastructure as the foundation of the program. Sadeh has enumerated the foreign policy motivations for this move. Some were purely symbolic, e.g., to signal an end to the Cold War and Russia's entry into the club of advanced Western industrial states. Others were fully in line with the use of technology as an instrument of foreign policy as we have described it here. In particular, in these negotiations, as in the debates over the help to ELDO 30 years earlier, technological sharing was an instrument to steer Russia's civilian and military high-tech sectors along paths in line with American interests. Thus, integrating Russia into the core of the Space Station "enhances U.S. efforts to strengthen Russia's commitment to adhere to guidelines of international non-proliferation standards regarding ballistic missiles and nuclear technology, lends support to U.S. efforts to privatize and demilitarize the high-technology sector in Russia . . . and encourages Russian scientists and engineers to work on 'peaceful' projects rather than selling their talents to other, possibly hostile, states."[69] It also, of course, diverts scarce Russian resources away from projects of which the U.S. might not approve. In short, the *kinds of* arguments for technological sharing with ELDO in 1966 were still being used when dealing with Russia in 1996. The difference is that ELDO had nothing to offer at the technical level, while Russia could use its extensive experience in human spaceflight as a bargaining chip to win some key concessions. The lesson is clear: if we want to make sense of international collaboration in space from a U.S. perspective, we need focus carefully not only on what technology the U.S. has to offer, but what its potential partner has to give. In any event, as I have stressed, we simply cannot grasp the dynamics of international cooperation in space if we do not situate the scientific and technological *content* of the collaborative venture at the core of our analysis and relate it to strategies to maintain American "leadership" and some measure of control over the space programs of her partners.

I should like to thank Roger Launius for helpful comments on a previous draft of this paper.

---

68. Two important studies of policy regarding the Space Station are John M. Logsdon's *Together in Orbit: The Origins of International Participation in the Space Station* (Washington, DC: Monographs in Aerospace History, No. 11, November 1998) and Howard E. McCurdy's *The Space Station Decision: Incremental Politics and Technological Choice* (Baltimore: Johns Hopkins, 1990).

69. Eligar Sadeh, "Technical, Organizational, and Political Dynamics of the International Space Station Program," *Space Policy* 20, no. 3 (August 2004): 171–188. Sadeh makes no systematic distinction between the dimensions of the collaboration which were, indeed, symbolic and the far more substantive, material items that I have quoted here. Indeed, quite mistakenly in my view, he reduces *all* these policy considerations to the symbolic level. This evades the question of how the United States uses technology to steer the space and high-tech programs of its partners in particular directions.

Access to Space

~

INTRODUCTION

Nothing has been more significant for the long-term development of the Space Age than the ability to reach Earth orbit. When *Columbia* was lost on Saturday morning, 1 February 2003, one of the issues the accident brought to the fore was the long and complex history of the Space Shuttle's origins, evolution, and operation, as well as the continuing challenge of space access. Even more, the accident opened the issue of space access from the dawn of the Space Age in the 1950s to the present. This is a rich and inviting history, requiring serious inquiry, critical thinking, and hard-edged analysis. The first-generation launchers were all ballistic-missile-derived vehicles that served well; with some upgrades over the years, they are still the backbone of the U.S. space launch fleet. Indeed, Redstone, Atlas, Titan, Delta, and Saturn were all scaled-up variants of the ICBMs, but with notable improvements. The Space Shuttle, the only human-carrying vehicle of the United States since the Apollo program of more than 30 years ago, followed those earlier space launch systems and has served many space-access needs for more than a quarter century.[1]

After more than four decades of effort, access to space remains a difficult challenge. Although space transport services should not be measured by terrestrial standards, if the grand plans of space visionaries and entrepreneurs are to be carried out, there is a real need to move beyond currently available technologies. Unfortunately, the high cost associated with space launch from 1950 to 2005 has demonstrated the slowest rate of improvement of all space technologies. Everyone in space activities shares a responsibility for addressing this critical technical problem. The overwhelming influence that space access has on all aspects of civil, commercial, and military space efforts indicates that it should enjoy a top priority.[2]

Of course, a key element in the spacefaring vision long held in the United States is the belief that inexpensive, reliable, safe, and easy spaceflight is attainable. Indeed, from virtually the beginning of the 20th century, those interested

---

1. For a discussion of the overarching space-access history, see Roger D. Launius and Dennis R. Jenkins, eds., *To Reach the High Frontier: A History of U.S. Launch Vehicles* (Lexington: University Press of Kentucky, 2002).

2. More than 50 space-access studies have reached this conclusion over the last 40 years. See Roger D. Launius and Howard E. McCurdy, *Imagining Space: Achievements, Projections, Possibilities, 1950–2050* (San Francisco: Chronicle Books, 2001), chap. 4; United States Congress, Office of Technology Assessment, *Launch Options for the Future: Special Report* (Washington, DC: Government Printing Office, 1984); Vice President's Space Policy Advisory Board, "The Future of U.S. Space Launch Capability," Task Group Report, November 1992, NASA Historical Reference Collection, Washington, DC; NASA Office of Space Systems Development, *Access to Space Study: Summary Report* (Washington, DC: NASA, 1994).

in the human exploration of space have viewed as central to that endeavor the development of vehicles of flight that travel easily to and from Earth orbit. The more technically minded recognized that once humans had achieved Earth orbit about 200 miles up, the vast majority of the atmosphere and the gravity well had been conquered, and that persons were now about halfway to anywhere they might want to go.[3]

Although a large number of issues could be explored in the history of space access, five central legacies offer tantalizing possibilities for space history and represent critical issues in the field. These include the following:

1. The limitations of chemical rocket technology.
2. The ICBM legacy of space access.
3. The costly nature of space access.
4. Launch vehicle reliability.
5. The value of reusable launch vehicles (RLVs) versus expendable launch vehicles (ELVs).

The two chapters that follow review each of these legacies, sometimes explicitly but more often indirectly, and raise serious policy issues that must inform any debate concerning access to space.[4]

In chapter 9, John M. Logsdon asks the poignant question, why is there no replacement for the Space Shuttle despite the longevity of the issue on the national agenda? From almost the first flight of the Space Shuttle in 1981, NASA realized that planning should begin on an eventual replacement. Most observers in those early years of the program believed that the current fleet could remain operational for about 20 years but that by about the year 2000, replacement would probably be necessary. Understanding that it took most of a decade, sometimes even more, to carry a major spaceflight program to fruition, they thought it important to begin the process of building a successor second-generation reusable space-access vehicle capable of human launch. Yet, as of 2005 and despite a plethora of studies, little has been accomplished.[5]

Logsdon asserts that there was a fundamental "failure of national space policy over the past three plus decades, and that the lack of a replacement for the Space Shuttle is just one of the most obvious manifestations of that policy failure." At sum, he finds that the "lack of a clear 'mandate' for human spaceflight

---

3. G. Harry Stine, *Halfway to Anywhere: Achieving America's Destiny in Space* (New York: M. Evans and Co., 1996).

4. Roger D. Launius, "Between a Rocket and a Hard Place: Legacies and Lessons from 50 Years of Space Launch" (presentation in Lessons Learned Session of the 36th American Institute of Aeronautics and Astronautics [AIAA] Joint Propulsion Conference, sponsored by AIAA Solid Rocket Technical Committee [SRTC], Huntsville, AL, July 17, 2000).

5. See Roger D. Launius, "After Columbia: The Space Shuttle Program and the Crisis in Space Access," *Astropolitics* 2 (July–September 2004): 277–322.

over the past 35 years has meant that the U.S. human spaceflight program, and indeed the NASA program overall, has been sustained by a complex coalition of interests, not by a clearly articulated national goal and a stable political consensus in support of achieving that goal."[6] This is an important observation, for it gets to the heart of the overarching issue of rationales for human space exploration. Those rationales have not proven especially compelling, and NASA and its human spaceflight effort have been forced to deal with a lack of motivating reasons for the Agency's activities since the Apollo program.

Instead of developing a finely honed and convincing rationale for the necessity of humans in space, NASA has cobbled together a loose coalition of government interests, industry contractors, politicians of all stripes who are supportive because of "pork" for districts as well as patriotism, and spaceflight enthusiasts who dream of becoming a multiplanetary species. They came together to support the Shuttle as a means of achieving reliable, assured, and flexible access to space and have continued to support it to the present because of the lack of anything better—however "better" might be defined by the various interest groups—on the horizon.

Logsdon offers the bold assertion that the reason for undertaking human spaceflight was reconsidered by the nation soon after the United States began to fly astronauts in 1961 and that this reflection has led to a less supportive public commitment than NASA or the spaceflight community would like. "The people of the United States and their government have been willing over the past 35 years to continue a human spaceflight program," he writes, "but only at a level of funding that has forced it to constantly operate on the edge of viability." Logsdon concludes, "The lack of a replacement for the Space Shuttle is a symptom of this larger reality."

Logsdon goes on to ask how badly Americans want to fly humans in space and finds that the answer to that is "not very badly." Accordingly, at least by the time of post-Apollo planning, the United States, through the democratic process, had reached the conclusion that spaceflight in general, and human spaceflight particularly, had to stand behind a long list of other national needs. Its funding level would be something less than 1 percent of the federal budget per year, and within that budget, NASA should advance a useful space exploration agenda. Logsdon concludes that spaceflight enthusiasts have failed to align their vision of the future with the democratically arrived-at decisions relative to space policy. In other words, something less than the bold visions of the past are necessary in the realities of the present and the future.

At sum, Logsdon concludes that both the community of spaceflight advocates in the United States and the personnel of NASA have overemphasized

---

6. John M. Logsdon, "'A Failure of National Leadership': Why No Replacement for the Space Shuttle?" chap. 9 in this volume.

human spaceflight's centrality to the modern nation. Instead, he argues for a more realistic perspective that reduces the spaceflight agenda to a realm that might be successful with the funding available. But a question that must be asked is, despite an unwillingness by the public to open the treasury more fully to achieve the human spaceflight vision, would the American public accept a scaled-back program that is far less grandiose? More important for the policy debate concerning a replacement for the Space Shuttle, however, would the American public accept an end to the human spaceflight mission that NASA has conducted since 1961, since failure to replace the vehicle signals that end? Only time will tell if this is how the policy decisions relating to the Shuttle replacement effort will turn out.

In chapter 10, Andrew J. Butrica assesses the historical debate over reusable launch vehicles versus expendable launch vehicles. RLV advocates have been convincing in their argument that the only course leading to "efficient transportation to and from the earth" would be RLVs and have made the case repeatedly since the late 1960s.[7] Their model for a prosperous future in space is the airline industry, with its thousands of flights per year and its exceptionally safe and reliable operations. Several models exist for future RLVs, however, and all compete for the attention—and the development dollars—of the federal government.

Prior to the Mercury, Gemini, and Apollo programs of the 1960s, virtually everyone involved in space advocacy envisioned a future in which humans would venture into space aboard winged, reusable vehicles. That was the vision from Hermann Oberth in the 1920s through Wernher von Braun in the 1950s to the U.S. Air Force's X-20 Dyna-Soar program in the early 1960s.[8] Because of the pressure of the Cold War, NASA chose to abandon that approach to space access in favor of ballistic capsules that could be placed atop launchers originally developed to deliver nuclear warheads to the Soviet Union. NASA developed its human-rated ballistic launch and recovery technology at enormous expense and used it with a 100-percent success rate between 1961 and 1975. As soon as Apollo was completed, NASA chose to retire that ballistic technology, despite its genuine serviceability, in favor of a return to that earlier winged, reusable vehicle. The Space Shuttle was the result.[9]

---

7. This was the argument made to obtain approval for the Space Shuttle. See *The Post-Apollo Space Program: A Report for the Space Task Group* (Washington, DC: National Aeronautics and Space Administration, September 1969), pp. 1, 6.

8. This quest has been well documented in Ray A. Williamson and Roger D. Launius, "Rocketry and the Origins of Space Flight," in *To Reach the High Frontier*, ed. Launius and Jenkins, pp. 33–69.

9. On this issue, see T. A. Heppenheimer, *The Space Shuttle Decision: NASA's Search for a Reusable Space Vehicle* (Washington, DC: NASA SP-4221, 1999); Roger D. Launius, "NASA and the Decision to Build the Space Shuttle, 1969–72," *The Historian* 57 (autumn 1994): 17–34.

Then there is an alternative position that suggests that the most appropriate approach to space access is through the use of throwaway "big, dumb boosters" that are inexpensive to manufacture and operate. Although reusable rockets may seem to be an attractive cost-saving alternative to expendables because they allow repeated use of critical components such as rocket motors and structural elements, ELV advocates claim, they actually offer a false promise of savings. This is because all RLV savings are predicated on maximizing usage of a small number of vehicles over a very long period of time for all types of space launch requirements. Accordingly, cost savings are realized only when an RLV flies many times over many years. That goal is unattainable, they claim, because it assumes that there will be no (or very few) accidents in the reusable fleet throughout its lifespan.[10]

The reality, ELV advocates warn, is that the probability of all RLV components' operating without catastrophic failure throughout the lifetime of the vehicle cannot be assumed to be 100 percent. Indeed, the launch reliability rate of even relatively "simple" ELVs—those without upper stages or spacecraft propulsion modules and with significant operational experience—peaks at 98 percent with the Delta II, and that took 30 years of operations to achieve. To be sure, most ELVs achieve a reliability rate of 90 to 92 percent—again, only after a maturing of the system has taken place. The Space Shuttle, a partially reusable system, has attained a launch reliability rate of slightly more than 98 percent, but only through extensive and costly redundant systems and safety checks. In the case of a new RLV, or a new ELV for that matter, a higher failure rate has to be assumed because of a lack of experience with the system. Moreover, RLV use doubles the time of exposure of the vehicle to failure because the vehicle must also be recovered and be reusable after refurbishment. To counter this challenge, more and better reliability has to be built into the system, and this exponentially increases both R&D and operational costs.[11]

Designing for one use only, those arguing for ELV development suggest, simplifies the system enormously. One use of a rocket motor, guidance system, and the like means that it needs to function correctly only one time. Acceptance of an operational reliability of 90 percent or even less would

10. Barbara A. Luxenberg, "Space Shuttle Issue Brief #IB73091," Library of Congress Congressional Research Service Major Issues System, 7 July 1981, NASA Historical Reference Collection; *Economic Analysis of New Space Transportation Systems: Executive Summary* (Princeton, NJ: Mathematica, Inc., 1971); General Accounting Office, *Analysis of Cost Estimates for the Space Shuttle and Two Alternate Programs* (Washington, DC: General Accounting Office, 1973); William G. Holder and William D. Siuru, Jr., "Some Thoughts on Reusable Launch Vehicles," *Air University Review* 22 (November–December 1970): 51–58; Office of Technology Assessment, *Reducing Launch Operations Costs: New Technologies and Practices* (Washington, DC: U.S. Congress, Office of Technology Assessment, 1988).

11. Stephen A. Book, "Inventory Requirements for Reusable Launch Vehicles" (paper presented at the Space Technology & Applications International Forum [STAIF-99], Albuquerque, NM, copy in possession of the author).

further reduce the costs incurred in designing and developing a new ELV. Indeed, many experts believe that reliability rates cannot be advanced more than another 1.5 percent above the 90-percent mark without enormous effort, effort that would be strikingly cost-inefficient.[12]

The debate is far from decided. As Butrica shows in this essay, human spaceflight advocates seem driven toward RLVs for space access. This has been an enormously costly perspective over time and directly affects the search for a replacement for the Space Shuttle. Butrica recounts the depressing story of failed attempts to build new vehicles and their eventual cancellation.

Collectively, Logsdon and Butrica encapsulate a critical issue for both the history of NASA and the current policy arena as the space agency struggles to deal with an aging Shuttle fleet, a major reorientation of its mission, and prospects for a post-*Columbia*-accident spacefaring future.

---

12. B. Peter Leonard and William A. Kisko, "Predicting Launch Vehicle Failure," *Aerospace America* (September 1989): 36–38, 46; Robert G. Bramscher, "A Survey of Launch Vehicle Failures," *Spaceflight* 22 (November–December 1980): 51–58.

CHAPTER 9

"A FAILURE OF NATIONAL LEADERSHIP": WHY NO REPLACEMENT FOR THE SPACE SHUTTLE?

John M. Logsdon

If the policy for the future of U.S. civilian space activity first laid out by President George W. Bush on 14 January 2004 is pursued, the United States will retire the Space Shuttle from service in 2010. Ending Shuttle flights will leave the United States without its own capability to carry its astronauts into orbit until a replacement crew-carrying vehicle makes its first flight with astronauts aboard. According to the Bush "Vision for Space Exploration," this may not happen until 2014.[1] As leading space historian Roger D. Launius has commented, "The inability to ensure a continued capability for human space access has placed the United States in a situation that is unenviable and unfortunate as the twenty-first century begins."[2]

This essay attempts to set out the reasons why the United States has found itself in this "unenviable and unfortunate" situation, with a focus on why the country had not, by the time of the Space Shuttle *Columbia* accident on 1 February 2003, developed a replacement for the Shuttle as a U.S. means for carrying humans into space. That same question was asked by the Columbia Accident Investigation Board (CAIB) set up in the immediate aftermath of the *Columbia* tragedy. (I was a member of that 13-person group.) In addition to its investigation of the physical and organizational causes of the accident, CAIB, in its 26 August 2003 report, offered brief but pointed observations on the broader policy context within which the accident took place and on "future directions for the U.S. in space."[3] This kind of look ahead was not part of CAIB's original charter; it became part of the CAIB focus after members of Congress asked the Board Chair, retired Admiral Harold Gehman, to have the Board's report "set the stage" for a national debate on the future directions of the U.S. civilian space program. Including a discussion of national space policy in an accident investigation report was unprecedented; neither the internal NASA report following the Apollo 1 fire in January 1967 nor the Rogers Commission

1. White House, "A Renewed Spirit of Discovery," January 2004.
2. Roger D. Launius, "After *Columbia*: The Space Shuttle Program and the Crisis in Space Access," *Astropolitics* 2 (autumn 2004): 279.
3. Columbia Accident Investigation Board, *Report*, vol. 1 (Washington, DC: NASA and GPO, August 2003), p. 209.

investigation of the *Challenger* accident had gone beyond identifying and suggesting remedies for the immediate causes of those tragedies.

The brief section titled "Long-Term: Future Directions for the U.S. in Space" in chapter 9 of the CAIB report has had an impact well beyond the Board's expectations. It is not too grandiose a claim to suggest that it led to a fundamental change in national space policy. Staff members in the Executive Office of the President have confirmed that the Board's observation that there had been a "lack, over the past three decades, of any national mandate providing NASA a compelling mission requiring human presence in space" was the direct catalyst for the White House deliberations in fall 2003 that led to the 14 January 2004 announcement by President George W. Bush of the new space exploration vision. This "Vision for Space Exploration," with its call for a "sustained and affordable human and robotic program to explore the solar system and beyond," is explicitly intended as the "national mandate" that had been missing since Americans landed on the Moon in 1969.

The Board made a second set of general observations. The CAIB report noted that "following from the lack of a clearly-defined long term space mission," there had been no "sustained national commitment over the past decade to improving access to space by developing a second-generation space transportation system." The Board concluded that "*the United States needs improved access for humans to low-Earth orbit as a foundation for whatever directions the nation's space program takes in the future.*" The CAIB report suggested that it was "*in the nation's interest to replace the Shuttle as soon as possible as the primary means for transporting humans to and from Earth orbit.*" Finally, it contained the following indictment: "*previous* [unsuccessful] *attempts to develop a replacement vehicle for the aging Shuttle represent a failure of national leadership*" (all emphasis in original).[4]

In his recent comprehensive and insightful analysis of U.S. policy towards access to space, Launius has used even stronger language than the Columbia Board. He suggests that "the lack of a firm decision to develop a Shuttle replacement represents the single most egregious failure of space policy in history."[5]

This essay will argue that there has been an even more fundamental and "egregious" failure of national space policy over the past three-plus decades and that the lack of a replacement for the Space Shuttle is just one of the most obvious manifestations of that policy failure. The series of decisions regarding a Shuttle replacement must be cast in the broader context of U.S. policy with respect to the reasons for sending people to space in the first place. The lack of a clear "mandate" for human spaceflight over the past 35 years has meant that the U.S. human spaceflight program, and indeed the NASA program overall, has been sustained by a complex coalition of narrow interests, not by a clearly

---

4. CAIB, *Report*, pp. 209–211.
5. Launius, "After *Columbia*," pp. 278–279.

articulated national goal and a stable political consensus in support of achieving that goal. As the CAIB report observed, without such a goal, NASA

> has found it necessary to gain the support of diverse constituencies. NASA has had to participate in the give and take of the normal political process in order to obtain the resources needed to carry out its programs. NASA has usually failed to receive budgetary support consistent with its ambitions. The result . . . is an organization straining to do too much with too little.[6]

It is this situation—"straining to do too much with too little"—that reflects the fundamental failure of U.S. space policy. In the 1969–1970 period, the administration of President Richard M. Nixon made a purposeful decision not to continue in the post-Apollo period the type of space effort that had taken Americans to the Moon. As Nixon stated in March 1970:

> Space expenditures must take their proper place within a rigorous system of national priorities. What we do in space from here on in must become a normal and regular part of our national life and must therefore be planned in conjunction with all of the other undertakings which are important to us.[7]

This declaration was more than rhetorical. The NASA budget was rapidly reduced in the early 1970s to less than 1 percent of the federal budget, approximately one-fifth of its budget share at the peak of Apollo 10 years earlier. Outside of postwar demobilization, few government activities have seen such a rapid decline in the resources devoted to their implementation. More to the point of this essay, this lowered level of budget allocations has persisted to the current time.

What Does "Replacing the Space Shuttle" Mean?

Many people talk of replacing the Shuttle as if the meaning of such an undertaking is quite clear. Such is not the case. There are several meanings that could be attributed to the term "replacing the Space Shuttle." They include the following:

---

6. CAIB, *Report*, p. 209.

7. Richard M. Nixon, "Statement About the Future of the United States Space Program," 7 March 1970, in U.S. President, *Public Papers of the Presidents of the United States: Richard Nixon, 1970* (Washington, DC: GPO, 1971), p. 251.

- Developing an advanced-technology, second-generation vehicle similar in its capabilities to the Shuttle, including the ability to carry both a sizable number of people and large and/or heavy cargo into low-Earth orbit, to provide living and working space for the crew for some period of time, and to be capable of various space operations such as payload deployment and retrieval and in-orbit servicing. Such a vehicle, presumably, would be as reusable as the Shuttle, preferably more so.

- Developing a vehicle that can carry either cargo or passengers to space and deliver its payload to an orbital destination such as the International Space Station; reusability would be a desired, but not necessary, characteristic.

- Developing a vehicle only to carry people to another destination in space and to return them to Earth, with limited or no cargo-carrying capacity. Again, reusability would be a desired, but not necessary, characteristic.

- Developing a vehicle capable of transporting people both to low-Earth orbit and to destinations beyond Earth orbit, such as the Moon, Mars, or a Lagrangian point.

Each of these types of vehicles could be considered a Shuttle replacement, and failure to differentiate among them has caused, and will continue to cause, policy confusion. For the purposes of this essay, the central meaning to be attributed to the term "Space Shuttle replacement" is a vehicle having the capability to transport humans to and from low-Earth orbit. Whether that vehicle would be reusable or not and whether it would be capable of going beyond Earth orbit are secondary considerations. This certainly was what the CAIB had in mind when it judged that "*it is in the nation's interest to replace the Shuttle as soon as possible as the primary means for transporting humans to and from Earth orbit*" (emphasis in original).

What did not happen, either during the CAIB's deliberations or since, was a corresponding adjustment in either the expectations placed on NASA by the nation's leaders or the ambitions of those committed to the vision of an expansive future in space. *The reality that national space policy did not bring ambitions and resources into balance in the 1970s, nor in the subsequent two decades, is the basic policy failure.* Either NASA should have been forced by the White House and Congress to plan and carry out a less ambitious program, or those national leaders should have been willing to provide the resources needed to carry out the ambitious program, with human spaceflight at its core, that NASA has proposed to implement.[8] By allowing NASA to try to "do too much with too little," national leaders failed in their responsibility as stewards of well-conceived national policy. The space sector has suffered as a result, most visibly with two Space Shuttle accidents and the loss of 14 astronaut lives.

An Album of Frustration

How has this "unenviable and unfortunate" situation come to be? The answer to this question can be portrayed by a set of "snapshots" taken at various times during the evolution of the U.S. human spaceflight effort.[9] This "photo album" of the steps towards the current situation will set the stage for a fuller analysis of *why* ensuring reliable, affordable, and safe human access has been a continuing policy problem for the past two decades:

1. From almost the start of serious thinking about human spaceflight, visionaries have expected that people would travel to and from space in a reusable, winged spacecraft; this image has continued to influence thinking about how to send people to space for most of the time since.

2. The pressures of Cold War competition drove the United States and the Soviet Union to abandon a winged approach to spaceflight and to develop instead crew-carrying ballistic capsules launched into space on top of expendable rockets, most of them derived from missiles designed to deliver nuclear warheads over intercontinental distances. Until the Space Shuttle was approved in 1972, only the U.S. Saturn family of boosters was designed from their start in the 1950s as space launch vehicles.

3. Once the United States had won the race to the Moon, the National Aeronautics and Space Administration in 1969 proposed an ambitious post-Apollo space effort beginning with the rapid development of a Saturn V–launched, 12-person space station. As a "logistics vehicle" for such a station, NASA proposed developing a reusable Earth-to-

---

8. In May 1992, then-new NASA Administrator Daniel S. Goldin did recognize this situation and told his senior officials to stop making plans that anticipated future budget increases. This was one of the foundations of Goldin's "faster, better, cheaper" guidance. But Goldin was also impatient and wanted to lay the foundation for human missions to Mars. This made his attempts to limit future ambitions not very effective.

9. In his *Astropolitics* article cited earlier, Roger Launius provides a parallel and well-stated account of this history.

orbit launch vehicle called the Space Shuttle. In NASA's 1970 budget presentation, the space station and Space Shuttle were presented to Congress as a single program. When the Nixon administration refused to approve the space station, NASA, in the fall of 1970, deferred—not canceled—its space station plans and directed its Shuttle contractors to design a vehicle capable of carrying pieces of a space station into orbit. This requirement defined the width of the Shuttle payload bay as no less than 14 feet. Thus the currently unbreakable link between the Space Shuttle and International Space Station programs actually has its roots in decisions taken 35 years ago.

4. In 1971, there was intense debate within the Executive Branch and its advisers of whether to approve Space Shuttle development. This debate led, in January 1972, to approval of Shuttle development as a product of "a series of political compromises that produced unreasonable expectations—even myths—about its performance," with a "technically ambitious design [that] resulted in an inherently vulnerable vehicle."[10] The Space Shuttle program was approved even in the face of a fundamental policy decision, made two years earlier, to reduce the priority of and resultant budget allocations for the civilian space program.[11] Based on that decision, the Office of Management and Budget forced NASA, in May 1971, to accept a $5.15-billion development cost ceiling for the Space Shuttle; this led NASA to abandon hopes for a two-stage, fully reusable vehicle and to quickly examine a wide variety of designs that could be developed within that cost cap.

5. In order to make the case that the investment in developing the Space Shuttle was cost-effective, NASA had to gain the agreement of the military and intelligence communities that when it became operational, the Space Shuttle would be the only launch vehicle for almost all government payloads, both human crews and robotic spacecraft. In order to gain this agreement, NASA had to design a Shuttle with specific performance characteristics that increased its technological risks. CAIB noted that "the increased complexity of a Shuttle designed to be all things to all people created inherently greater risks than if more realistic technical goals had been set from the start."[12] Certainly, if the

---

10. CAIB, *Report*, p. 21.

11. Accounts of the process that led to the decision to develop the Space Shuttle can be found in John M. Logsdon, "The Space Shuttle: A Policy Failure?" *Science* 232 (30 May 1986): 1099–1105; and T. A. Heppenheimer, *The Space Shuttle Decision: NASA's Search for a Reusable Space Vehicle* (Washington, DC: NASA SP-4221, 1999).

12. CAIB, *Report*, p. 23.

Space Shuttle design had been optimized for its crew-carrying role, a less risky vehicle, with more provisions for crew safety, could have been designed.

6. A byproduct of the decisions to develop in the Space Shuttle a vehicle capable of launching all types of payloads was the drying up, beginning in the 1970s, of NASA funding for research and technology development related to any aspect of space transportation not associated with the Shuttle. Thus there was a limited base of technology from which NASA could draw when it did initiate or participate in Shuttle replacement efforts in the 1980s and 1990s.[13]

7. Soon after the first flight of the Space Shuttle in April 1981, the new NASA leadership set as its two top priorities bringing the Shuttle to operational status as soon as possible and getting presidential and congressional approval to develop a (Shuttle-launched) space station. No alternatives to using the Shuttle in this role were considered at the inception of the space station program.[14]

8. Also in 1981, after only two Shuttle flights, President Ronald Reagan approved a formal policy statement saying that the Space Shuttle "will be the primary space launch system for both United States military and civil government missions."[15] This policy was reinforced in a 1982 statement of National Space Policy, which said that "completion of transition to the Shuttle should occur as expeditiously as possible" and that "government spacecraft should be designed to take advantage of the unique capabilities of the STS [Space Transportation System, another designation for the Space Shuttle]."[16]

9. The U.S. Air Force, as the launch agent for both military and intelligence spacecraft, early on recognized the dangers of this "all eggs in one basket" policy. Soon after the Shuttle was declared operational on

---

13. This statement is not quite accurate. There continued to be some low-level efforts within NASA to examine future space transportation vehicles and technologies even as the Shuttle was being developed during the 1970s, but there was very limited financial support of these efforts.

14. For a discussion of the steps leading to President Reagan's approval of a space station program, see Howard E. McCurdy, *The Space Station Decision: Incremental Politics and Technological Choice* (Baltimore, MD: Johns Hopkins, 1990).

15. John M. Logsdon, ed., *Exploring the Unknown: Selected Documents in the History of the U.S. Civil Space Program*, vol. 4, *Accessing Space* (Washington, DC: NASA SP-4407, 1999), pp. 333–334.

16. John M. Logsdon, ed., *Exploring the Unknown: Selected Documents in the History of the U.S. Civil Space Program*, vol. 1, *Organizing for Exploration* (Washington, DC: NASA SP-4407, 1995), pp. 591–592.

4 July 1982, after only four flights, the Air Force began to argue that the risks and costs of the system could be a detriment to its ability to perform its launch responsibilities for critical national security payloads. Most of those payloads had been designed since the late 1970s so that they could only be launched on the Shuttle. Beginning in 1983, the Air Force campaigned for approval of a backup to the Shuttle in order to provide assured access to space for such payloads. NASA fought this move. The dispute between the Air Force and NASA reached the White House in early 1985, where it was decided in favor of the Air Force.[17] This decision led to the development of the Titan IV expendable launch vehicle, which was capable of launching the largest military and intelligence spacecraft. After the 1986 *Challenger* accident, the Titan IV became the primary launcher for large national security missions, and those spacecraft that had been intended for Shuttle launch had to be redesigned at high cost.

10. Discussions within NASA about the need to develop a second-generation replacement for the Space Shuttle began even before the Shuttle was launched.[18] The first public statement of this need came in the report of the National Commission on Space in January 1986 (made public a few days after the *Challenger* accident). The Commission concluded that "the Shuttle fleet will become obsolescent by the turn of the century." It recommended separating cargo and "passenger" (its term) launches and developing, within 15 years, a new system for "passenger transport to and from low Earth orbit."[19] In contrast, an inside-the-government NASA-DOD National Space Transportation and Support Study during 1985–1986, while agreeing that in the future, separate human-carrying and cargo-carrying launch systems were desirable, concluded that "there was not an urgent need for an advanced manned vehicle; incremental improvements to the Space Shuttle would suffice."[20]

11. While NASA during the 1970s and early 1980s allocated only limited funding to advanced space transportation technology, the Department of Defense did support a fair amount of such research and technology

---

17. This dispute can be traced in John M. Logsdon, ed., *Exploring the Unknown: Selected Documents in the History of the U.S. Civil Space Program*, vol. 2, *External Relationships* (Washington, DC: NASA SP-4407, 1996), documents II-40 through II-45.

18. Launius, "After *Columbia*," pp. 287–288.

19. *Pioneering the Space Frontier*, Report of the National Commission on Space, quoted in Launius, "After *Columbia*," p. 288.

20. Ivan Bekey, "Exploring Future Space Transportation Possibilities," in *Exploring the Unknown*, ed. Logsdon, vol. 4, pp. 505–506.

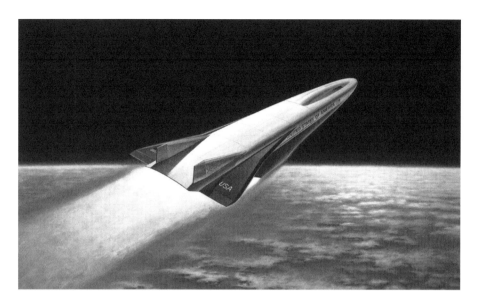

The National Aero-Space Plane (NASP) began as a NASA-DOD joint project in 1982. It called for the development of two vehicles capable of SSTO at Mach 25. It was intended to use a multicycle engine shifting from jet to ramjet to scramjet; it would use liquid-hydrogen fuel with oxygen scooped and frozen from the atmosphere. President Ronald Reagan had high hopes for it, announcing in the State of the Union Address in 1986: "We are going forward with research on a new Orient Express that could, by the end of the decade, take off from Dulles Airport, accelerate up to 25 times the speed of sound attaining low Earth orbit, or fly to Tokyo within two hours." It was canceled in 1992 without ever having flown. *(NASA KSC photo no. EL-2001-00432)*

development related to advanced-technology crew–carrying systems. By the early 1980s, these efforts were focused on a vehicle that used air-breathing engines to accelerate to hypersonic or perhaps even orbital velocity. The Air Force program was focused on a TransAtmospheric Vehicle (TAV), while a separate, highly classified, Advanced Research Projects Agency (ARPA) study was called Copper Canyon.[21] In late 1985, all Department of Defense research and development activity on hypersonic flight was consolidated into a program that became known as the National Aero–Space Plane (NASP); NASA joined the Department of Defense as a minority funder and comanager of the NASP effort. This program was given presidential endorsement in the 1986 State of the Union Address, delivered by President Ronald

---

21. The National Aero-Space Plane Program is discussed in Andrew J. Butrica, *Single Stage to Orbit: Politics, Space Technology, and the Quest for Reusable Rocketry* (Baltimore, MD: Johns Hopkins, 2003), chap. 4.

Reagan on 5 February of that year. In his address, the President spoke of an "Orient Express" that would, "by the end of the decade," be able to "take off from Dulles Airport [near Washington, DC], accelerate up to 25 times the speed of sound attaining low Earth orbit, or fly to Tokyo within two hours."[22]

12. The President's 1986 address came only a few days after the 28 January explosive burning and breakup of the Space Shuttle *Challenger*; seven crew members died in the accident. In the following months, policy toward use of the Space Shuttle came under intense scrutiny. First, the White House, on 15 August, announced that a new Shuttle orbiter would be built to replace *Challenger* but that the Shuttle would no longer be used to launch commercial payloads such as communication satellites. On 27 December, President Reagan signed a directive that established a "mixed fleet" concept for government payloads, with "critical mission needs" supported by both the Shuttle and expendable launch vehicles "to provide assurance that payloads can be launched regardless of specific launch vehicle availabilities." According to this directive, the Space Shuttle would only be used to support programs requiring "manned presence and other unique STS capabilities."[23] These decisions formally reversed the policy that had been one of the foundations of the decision to develop the Space Shuttle—that it could serve as a reliable, affordable launch vehicle for all U.S. payloads. It focused future Shuttle use on missions where the human presence was essential to the mission, not merely crew members delivering cargo to orbit. In 1987, the Air Force announced its support for resuming production of the Delta and Atlas expendable launch vehicles, with the clear implication that the military would in the future use the Space Shuttle only for those few missions that required its specific capabilities. The sum of these post-*Challenger* decisions meant that NASA became not only the operator, but also the main future user, of the Space Shuttle. With fewer missions to fly, with NASA having to pay all the costs of its operation, and with a flat or decreasing NASA budget for most of the 1990s, the Shuttle became a "mortgage" on the NASA budget that had to be paid. Funds for investing in its replacement could be made available only if the NASA budget were increased or the Shuttle program's budget were reduced.

---

22. Quoted in ibid., p. 65.

23. See the essay by Ray A. Williamson, "Developing the Space Shuttle," and documents II-42 and II-43 in *Exploring the Unknown*, ed. Logsdon, vol. 4, for an account of this policy shift.

13. While DOD-NASA work on NASP continued in the late 1980s, with DOD bearing some 80 percent of its costs, NASA gave top priority to returning the Space Shuttle to flight. Leading that effort was Admiral Richard H. Truly, a former Shuttle astronaut who was brought back to NASA in the weeks following the *Challenger* accident as Associate Administrator for Spaceflight.[24] Truly was a firm believer in the value of the Shuttle. When in 1989 the new administration of President George H. W. Bush selected him as NASA Administrator, the Space Shuttle gained a strong proponent at the top of the space agency. Then President Bush proposed an ambitious long-range vision for the nation's space program in July 1989. The NASA plan for implementing that vision did not include a proposal to replace the Shuttle as the means for taking people to orbit, even though the plan extended over several decades.

Administrator Truly's personal embrace of the Shuttle as key to NASA's future was reflected by others in NASA, particularly those working on the Space Shuttle program in NASA Headquarters and at Johnson Space Center and Marshall Space Flight Center. Rather than respond to criticisms of the Shuttle and calls for its replacement, they strove to "impose the party line vision on the environment, not to reconsider it." Central to this behavior was the belief that the Space Shuttle could be made a safe and reliable system and should play a central role in NASA's human spaceflight efforts for many years to come. This behavior, in the judgment of the Columbia Board, led to "flawed decision-making, self deception, introversion and a diminished curiosity" about alternatives to the Shuttle.[25]

14. In 1990, the Advisory Committee on the Future of the U.S. Space Program, usually called the Augustine Committee after its chairman, aerospace executive Norm Augustine, concluded that "we are today overreliant on the Space Shuttle as the backbone of the civil space program." The Committee recommended rapid development of "an evolutionary, unmanned but man-rateable, heavy lift launch vehicle" to replace the Space Shuttle in supporting space station assembly and utilization. Noting that there was no alternative to the Shuttle for human transportation, the Committee recommended "expedited

24. See John M. Logsdon, "Return to Flight: Richard Truly and the Recovery from the Challenger Accident," chap. 15 in *From Engineering Science to Big Science*, ed. Pamela E. Mack (Washington, DC: NASA SP-4219, 1998).

25. Yale University organizational studies scholar Gary Brewer, quoted in CAIB, *Report*, p. 102.

development of a two-way [human] transportation capability" on such a launch vehicle "for use in the event of a Space Shuttle stand-down." The Augustine Committee was critical of the low level of NASA spending on space technology, including that related to advanced propulsion and aerodynamics, and called for a "two-to-three-fold enhancement" of NASA's space technology budget. It recommended an annual increase of 10 percent in the NASA budget if the nation was serious about wanting a successful space program.[26] The Committee concluded its report by recommending that the United States should reduce "dependence on the Space Shuttle . . . for all but missions requiring human presence."[27]

15. After receiving presidential endorsement in 1986, the NASP program over the subsequent several years struggled to achieve its technological and schedule goals. A 1988 Defense Science Board report concluded that the program's advocates had been overly optimistic in their initial promise of an early flight demonstration and suggested that the program should be "realistically presented to its sponsors." A year later, after the Air Force withdrew funding from the program, the White House, in 1989, approved a stretch-out of the program (rather than its cancellation as proposed by Secretary of Defense Richard Cheney), with a flight demonstration of the X-30 test vehicle to come only after relevant technologies had been developed.[28] In the face of competing budget priorities and slow technological progress, the NASP program was canceled in 1992, after $1.7 billion had been spent on it.[29] At that point, the cost of a full X-30 flight-test program was estimated at $17 billion, with another $10–20 billion to develop an operational vehicle.[30] No flight demonstration was attempted, but the program left a technological legacy for future advanced space transportation efforts.

Andrew Butrica observes that "the NASP concept was the wrong road." By pursuing an air-breathing approach to a single stage to orbit

---

26. The administration of George H. W. Bush took to heart the advice that the NASA budget should be substantially increased and proposed significant increases for FY 1992 and FY 1993. However, after coming to the White House in January 1993, the administration of President Bill Clinton reversed this upward trend in the NASA budget, which actually lost more than 10 percent in constant dollars during the eight years that Clinton was President.

27. *Report of the Advisory Committee on the Future of the U.S. Space Program* (Washington, DC: GPO, December 1990), pp. 21, 31, 33–34, 48.

28. Logsdon, *Exploring the Unknown*, vol. 4, documents IV-9, IV-10, quotation from p. 562.

29. Launius suggests that the costs were probably higher since some of the work on the NASP program was classified, and thus not all cost information was readily available ("After *Columbia*," p. 290).

30. Global Security.Org, "X-30 National Aerospace Plane (NASP)," *http://www.globalsecurity.org/military/systems/aircraft/nasp.htm* (accessed 13 January 2005).

vehicle rather than one using rocket power, NASA "propelled the nation into an expensive program that had no chance of success." Its failure "demonstrated unmistakably that an air-breathing, single-stage-to-orbit was not the road to travel." Ivan Bekey adds that "being airplane-like, the NASP concept attracted powerful backing because it was intuitively easy to grasp. The nation fooled itself into believing that because the NASP image was what was desired, the reality itself was therefore attainable."[31] Whatever the reason, the United States had lost several years and almost $2 billion in pursuing a failed path towards a Shuttle replacement.

16. On 1 April 1992, Daniel S. Goldin replaced the fired Richard Truly as NASA Administrator. In contrast to Truly, Goldin would prove to be no fan of the Space Shuttle, viewing its budget demands as a major barrier to initiating new, innovative NASA programs.[32] This was especially the case after 1993, when the new administration of President Bill Clinton retained Goldin as Administrator but declined to increase the NASA budget to both meet the demands of the Space Shuttle and the International Space Station programs and allow significant investments in major new efforts such as a Shuttle replacement. The Space Shuttle budget had peaked at over $5.5 billion per year as NASA recovered from the *Challenger* accident; the Bush administration, in early 1992, had proposed a $4.1-billion allocation. By the time Dan Goldin left office in November 2001, the Shuttle budget had been reduced by another 25 percent, to $3.2 billion per year. Goldin initiated the switch of Shuttle operations to private-sector management both as a cost-savings measure and as a way to encourage NASA engineers to focus on developing new capabilities. Until 1999, when he declared a "space launch crisis," Goldin was unwilling to allocate significant resources to Shuttle upgrades. Even so, Goldin, during his long tenure, came to recognize that successful and safe operation of the Shuttle was critical to political and public support of NASA's programs. His expectation was that by innovative partnerships with the private sector, the technological developments on which to base a Shuttle replacement could be achieved without a multibillion-dollar government investment. This unfortunately proved to be a false hope.

---

31. Butrica, *Single Stage to Orbit*, pp. 66, 81; Bekey, "Exploring Future Space Transportation Possibilities," p. 508.

32. As one indication of his attitude, it is reported that Goldin had removed from the cabin of the NASA Administrator's airplane all the pictures of the Space Shuttle that had been placed there under Richard Truly.

17. In 1992, during the last months of the George H. W. Bush administration, the Vice President's Space Policy Advisory Board, which advised the National Space Council, recommended the development by 2000 of an expendable "Spacelifter" launch vehicle, which would be human-rated, and also the development of a new Personnel Launch System for use with it. This would allow the government to "phase out the Space Shuttle at the earliest opportunity."[33] With the November 1992 election of a new administration, the recommendations contained in the Advisory Board's report were stillborn.

18. In 1993, both NASA Administrator Goldin and the U.S. Congress requested that the NASA staff carry out a comprehensive study of alternate approaches to accessing space through 2030. A principal goal of the study was "to make major reductions in the cost of space transportation (at least 50 per cent), while at the same time increasing safety for flight crews by at least an order of magnitude." The *Access to Space Study* examined three alternatives: 1) an upgraded Shuttle, 2) new expendable vehicles using conventional technologies, and 3) "new reusable vehicles using advanced technologies." The study concluded that "the most beneficial option is to develop and deploy a fully reusable single-stage-to-orbit (SSTO) pure-rocket launch vehicle fleet" that would allow the phasing out of the Space Shuttle, beginning in 2008.[34]

19. This conclusion of the *Access to Space Study* became formalized when President Clinton approved a new statement of National Space Transportation Policy in August 1994. That statement gave NASA the responsibility "to support government and private sector decisions by the end of this decade on the development of an operational next generation reusable launch system." NASA was to focus its research "on technologies to support a decision no later than December 1996 to proceed with a sub-scale demonstration which would prove the concept of single-stage-to-orbit." The policy envisioned that the private sector "could have a significant role in managing the development and operation of a new reusable space transportation system."[35] It was extremely unusual, if not unprecedented, for a specific technological solution such as the SSTO approach to be written into a presidential policy statement on space.

33. Logsdon, *Exploring the Unknown*, vol. 4, document IV-6, p. 550.
34. Ibid., document IV-14, pp. 585–586.
35. Ibid., document IV-16, p. 628.

20. Given the White House policy directive, NASA, over the following two years, organized a competition among potential developers of the subscale flight demonstrator. Three companies submitted proposals. Rockwell International proposed a vehicle that was in many ways a second-generation version of the Space Shuttle; Rockwell had been the prime contractor for the Shuttle. McDonnell Douglas proposed a version of the Delta Clipper vehicle that had been developed under

The McDonnell Douglas Delta Clipper-Experimental (DC-XA) reusable launch vehicle (RLV) was originally developed for DOD. NASA assumed control of the vehicle in the summer of 1995. The DC-XA was to have been an SSTO vertical takeoff/vertical landing launch vehicle concept, whose development would significantly reduce launch cost and provide a test bed for NASA RLV technology. *(NASA MSFC image no. MSFC-9513214)*

the sponsorship of the Strategic Defense Initiative Organization to demonstrate simpler space operations techniques.[36] Lockheed Martin proposed an advanced-technology vehicle based on the use of a linear aerospike engine. On 2 July 1996, Vice President Al Gore announced that NASA had selected the most technologically advanced (and thus the riskiest) of these proposals, that from Lockheed Martin. The reasoning behind this decision has not been adequately explored.

At that time, the plan was to have the first flight of what was christened the X-33 by March 1999 and to complete a 15-flight test program by the end of that year. The goal was to demonstrate the technological foundation for a decision by Lockheed Martin to invest its own funds in a full-scale operational vehicle, which the company named VentureStar.™ The X-33 program would be a cooperative undertaking between NASA and Lockheed Martin, with NASA providing $941 million of the required funding and Lockheed providing $220 million. Once Lockheed Martin developed the VentureStar™ using private capital, the assumption was that NASA would be a major customer for its services, but also that a booming commercial space industry would emerge. The combination of government and commercial demand for access to space, it was claimed, would allow VentureStar™ to be a profitable undertaking.[37]

Although Lockheed Martin, over the following several years, promoted the VentureStar™ project as symbolic of its status on the cutting edge of future technologies, the X-33 program encountered technological difficulties almost from its inception. In November 1999, there was a major test failure of the vehicle's hydrogen fuel tank; by that time, the White House and NASA were losing confidence that the program would be able to overcome its technological hurdles. In March 2001, NASA announced that it would provide no more funding for the X-33, effectively killing it well before a flight demonstration could be attempted. At that point, NASA had spent $912 million on the project, while Lockheed Martin had exceeded its planned investment, having put $356 million into the X-33.[38]

21. Some in the Executive Office of the President and at NASA had, by at least 1998 (if not before), become skeptical that the X-33 pro-

36. See Butrica, *Single Stage to Orbit,* parts III and IV, for a discussion of the origins and fate of the Delta Clipper program.

37. NASA Marshall Space Flight Center, "Lockheed Martin Selected to Build the X-33," news release 96-53, 2 July 1996.

38. Leonard David, "NASA Shuts Down X-33, X-34 Programs," *Space.com,* 1 March 2001, *http://www.space.com/missionlaunches/missions/x33_cancel_010301.html* (accessed 5 February 2005).

gram would be able to overcome its technical challenges and would provide the information needed to decide when and how to replace the Shuttle. In 1998, the Office of Management and Budget asked NASA to fund the aerospace industry to carry out what were called Space Transportation Architecture Studies to determine 1) if the Space Shuttle system should be replaced; 2) if so, when the replacement should take place and how the transition should be implemented; and 3) if not, what is the upgrade strategy to continue safe and affordable flight of the Space Shuttle beyond 2010. Five industry teams examined these questions through 1999 and came up with a variety of approaches to meeting both NASA and commercial-sector launch requirements. Many of the suggested approaches for taking humans to space involved a capsule-type spacecraft launched on top of an expendable launch vehicle. NASA leadership viewed such proposals as not being adequately forward-looking.

22. In 1999, NASA Administrator Daniel Goldin declared a "space launch crisis" and urged the White House to add funds to the NASA budget for necessary safety upgrades to the Shuttle. Substantial funds for this purpose were added to the NASA FY 2001 budget, submitted to Congress in early 2000. However, this upgrade initiative had a short lifespan. Within a year, funding for upgrades was reduced by over one-third in response to rising Shuttle operating costs and the need to stay within a fixed Shuttle budget.[39]

23. Based on the results of the Space Transportation Architecture Studies and the increasingly evident problems with the X-33 program, the NASA FY 2001 budget also contained a new Space Launch Initiative. This effort was to provide some $4.8 billion over five years to conduct studies and technology development to identify the most promising path to replacing the Space Shuttle and meeting other launch requirements. The hope was that this effort could provide the basis for a 2006 decision on what type of Shuttle replacement to develop, with a target date of 2012 for its initial launch. Three contractor teams—Boeing, Lockheed Martin, and a joint team of Orbital Sciences and Northrop Grumman—by early 2002 had identified 15 launcher concepts for detailed study.[40]

---

39. CAIB, *Report*, p. 114.

40. Leonard David, "Plans for Next Generation 'Shuttle' Ends First Phase; 15 Concepts Have Emerged," *Space.com*, 30 April 2002, *http://www.space.com/missionlaunches/sli_firstphase_020430. html* (accessed 5 February 2005).

24. The Space Launch Initiative was also short-lived. By the end of 2002, White House and top-level NASA optimism that it would provide the hoped-for basis for deciding to develop a second-generation, advanced-technology replacement for the Space Shuttle had evaporated. In November 2002, NASA announced that it was terminating the Space Launch Initiative and reallocating its funding to a new Integrated Space Transportation Plan. According to this plan, the Shuttle's life would be extended so that it could fly until 2020, and potentially to 2030. The Shuttle would be used for missions requiring its cargo-carrying and orbital-operations capabilities. However, for missions carrying only crew to and from the International Space Station, a new Orbital Space Plane (OSP) would be developed, but as a complement to, not a replacement for, the Shuttle. The OSP would not be an advanced-technology vehicle; the goal was to have it available for use as an ISS crew-rescue vehicle by 2010, eliminating dependence on the Russian Soyuz spacecraft to perform this function. The OSP would also become a crew-transfer vehicle by 2012, capable of carrying four or more astronauts to the International Space Station. The OSP would be launched either in the Shuttle's cargo bay or atop an expendable launch vehicle. A third element of the plan was funding of technologies and studies for an eventual next-generation vehicle to replace the Shuttle. No date was set for such a replacement vehicle to enter service.

The Integrated Space Transportation Plan was also a reaction to the lack of a long-term plan for U.S. human spaceflight. Without knowing how long the International Space Station would operate, it was not possible to determine how long the Space Shuttle would be needed. Without a post-ISS goal for human spaceflight, particularly given the collapse of the commercial space launch market, it also was not clear what kind of "post-Shuttle" vehicle to develop.

25. On 1 February 2003, Shuttle orbiter *Columbia* broke up over Texas, and all seven crew members aboard died. As noted at the start of this essay, the August 2003 report of the Columbia Accident Investigation Board set off, in the following months, a sweeping review of national space policy. On 14 January 2004, President George W. Bush announced a new "Vision for Space Exploration" centered on "a sustained and affordable program of human and robotic exploration of the solar system."[41] The new Vision had as a key element the decision to retire the Space Shuttle as soon as the assembly of the International Space Station was declared

---

41. White House, "Renewed Spirit of Discovery."

complete, in 2010 or soon thereafter. To replace the Shuttle, the Vision calls for the development of a Crew Exploration Vehicle (CEV) to carry humans into space, first to low-Earth orbit and eventually to the Moon and Mars. This vehicle will house the crew as they travel into space and thus will indeed replace the Shuttle as the means for U.S. human access to space. The CEV is the latest of the many attempts to develop a replacement for the Space Shuttle as a human transport vehicle. One can only hope that it will be become reality, unlike its predecessors.

One cannot escape the conclusion that these 25 "snapshots" add up to a portrait of failure—failure to provide for the United States' "assured access" to space for its citizens. Since 1981, there has been only one way for the United States to send people into space—at least using U.S. hardware. That way, of course, has been the Space Shuttle, and with its two fatal accidents, the United States lost human access to space twice—first for 32 months, and then for more than 30 months. The United States will not have independent access to space for humans between the time the Space Shuttle is retired in 2010 and the CEV begins crewed operations. This interval could be as long as four years, and during that time, the only way for U.S. astronauts to get to and from the International Space Station will be on Russian spacecraft.

It is worth noting that "assured access" for key national security and other robotic payloads *has* been a stated national policy since at least 1988. In its 1988 statement of National Space Policy, the Reagan administration declared that "United States space transportation systems must provide a balanced, robust, and flexible capability with sufficient resiliency to allow continued operations despite failures in any single system." The 1991 National Space Policy of President George H. W. Bush stated that "assured access to space is a key element of U.S. national space policy."[42] This policy continues in force today. President George W. Bush, on 21 December 2004, approved a new National Space Transportation Policy which stated that "'assured access' is a requirement for critical national security, homeland security, and civil missions." To be fair, this most recent statement also suggests that assured access to space for humans is also a desired policy objective. It declares that "access to space through U.S. space transportation capabilities is essential . . . to support government and commercial human spaceflight."[43] If this objective were met, it would signify a strong commitment to human spaceflight on the part of the U.S. government. As the following analysis suggests, such a strong commitment has been missing for many years.

---

42. Thor Hogan and Vic Villhard, "National Space Transportation Policy: Issues for the Future," RAND Science and Technology Working Paper WR-105-OSTP, October 2003, p. 7.

43. Office of Science and Technology Policy, Executive Office of the President, "National Space Transportation Policy," fact sheet, 6 January 2005.

The Root Causes of the Failure To Develop a Shuttle Replacement

There can be no one explanation for why this complex chain of developments has taken place. But certainly it is possible to suggest some of the fundamental reasons for the lack of a Shuttle replacement more than 30 years after the original commitment to the Space Shuttle program.

W. D. Kay, in his book *Can Democracies Fly in Space*, suggests that the "space program's failures, like its earlier successes, have multiple causes, all of them ultimately traceable to the way the American political process operates." Space policy is "a political outcome, a product of the discussion, debates, competition, and compromises that attend all public issues." While there could be alternate frameworks within which to examine the reasons why there has been no replacement for the Space Shuttle, this essay will adopt the political perspective suggested by Kay. He sets out a framework that provides a useful way to analyze this situation. Kay suggests that it is possible to conceptualize the creation of space policy in terms of three levels of analysis:

1. An *organizational output*, produced by the hardware, procedures, and personnel developed and trained by NASA.

2. A *political activity*, an outgrowth of the ongoing debates, compromises, votes, and other decisions involving NASA, its contractors, the Congress, various executive agencies, and a number of other loosely coordinated (and in some cases competing) individuals, institutions, and organizations, both public and private.

3. A *national enterprise*, the product of a society and a people possessing not only a certain level of technical expertise, but also a high degree of consensus and a determination expressed through its political representatives . . . .[44]

These three levels of analysis, and particularly viewing space policy as the foundation of a national enterprise, help to understand was has happened in the space sector over the past three and one-half decades.

---

44. W. D. Kay, *Can Democracies Fly in Space? The Challenge of Revitalizing the U.S. Space Program* (Westport, CT: Praeger Publishers, 1995), pp. 33, 26–27.

Technological Hubris and Organizational Outputs

In the last 20 years, the aerospace community has been given two major opportunities by the national leadership to develop a Shuttle replacement; these opportunities were accompanied by significant (although not adequate) funding commitments. The first of these opportunities, the NASP program, was initially justified on national security grounds; NASA was a junior partner in the undertaking and was not able to continue it as a development effort leading to a flight-test vehicle once Department of Defense funding was withdrawn. The second opportunity was the SSTO effort initiated by NASA in 1996 in response to NASA's internal studies and then the 1994 National Space Transportation Policy.

With the benefit of hindsight, it is possible to see that these two efforts were very likely doomed to failure from their outset. In both cases, the approach selected depended on being able simultaneously to bring to an adequate level of maturity a variety of challenging technologies in areas such as aerodynamics, guidance and control, materials, and propulsion. Those responsible for both efforts within the Department of Defense, NASA, and the aerospace industry assured their leaders that they could overcome these technological challenges and move forward rapidly and with affordable costs. These assurances were at variance with what actually transpired.

As mentioned above, the reality that the NASP program was unlikely soon to result in a flight vehicle became rather quickly evident after President Reagan gave the program national visibility in 1986. By 1988, the Defense Science Board had raised major questions about the technological feasibility of the undertaking. In 1989, the RAND Corporation reported "reservations" with respect to NASP coming "anywhere near its stated/advertised cost, schedule, payload fees to orbit, etc. . . ." and suggested that the "primary NASP X-30 objective—manned single stage to orbit—is exceedingly sensitive to full success in technology maturation."[45] Ivan Bekey, a proponent of a rocket-based approach to space access rather than the NASP air-breathing approach, was less kind; he has characterized the NASP program as "the biggest swindle ever to be foisted on the country," "full of dubious . . . claims" and "hot air."[46]

When Vice President Al Gore announced in July 1996 that NASA had selected Lockheed Martin's proposal to develop an SSTO demonstrator, he made a point of noting that it was the most "technologically advanced" of the

---

45. Bruno Augenstein and Elwyn Harris, "Assessment of NASP: Future Options," RAND Working Draft WD-4437-1-AF, July 1989, p. 2.

46. Quoted in Butrica, *Single Stage to Orbit*, p. 79.

three competing proposals. The story of why this risky choice was made has yet to be told. But once again, an approach to replacing the Shuttle had been chosen that would require simultaneous maturation of challenging technologies.[47] And once gain, achieving that maturation, at least on the original timescale and in the face of an impatient NASA and national leadership, proved impossible.

Why were these two efforts given high-level approval to proceed and widespread publicity when, at the time of approval, their chances of success were known to be low to at least some observers? This is a question deserving of more attention than it has received to date.

In 1989, one veteran aerospace engineer wondered, with respect to NASP, "How could ideas that were so thoroughly explored thirty years ago, and so thoroughly found lacking in sufficient promise twenty years go, have suddenly become once again in vogue?" It was not technological progress that had brought the ideas to the fore, he concluded, but rather "blissful ignorance of the past." Only a few of the instigators of the NASP program had been involved in earlier efforts, and "they were the ones who not only had been infected by the dream of long ago, but who had, in the process, become addicted to it and, therefore, immune to any amount of contrary evidence."[48] One suspects that an informed independent assessment of those who advocated the X-33 program would not be much different in its conclusions.

The costs of a lack of historical perspective and unchecked technological optimism, bordering on hubris, have been high. Roger Launius has suggested that the X-33 program and the NASP program before it "have been enormous detours for those seeking to move forward with a replacement for the Space Shuttle. Expending billions of dollars and dozens of years in pursuit of reusable SSTO technology, the emphasis on this approach ensured the tardiness of development because of the strikingly difficult technological challenges."[49] The Columbia Board agreed, suggesting that one reason for the "failure of national leadership" related to the absence of a replacement for the Space Shuttle was "continuing to expect major technological advances" in a replacement vehicle.[50]

How are nontechnical decision-makers to be protected against the enthusiasm of technological optimists? That is a topic well beyond the scope of this essay, but clearly, in the case of NASP and X-33, the necessary checks and balances were missing or not influential.

---

47. It should be noted that although X-33 and then VentureStar™ were widely perceived as a path to Shuttle replacement, the original designs were for an automated, cargo-carrying vehicle. Presumably, humans could be carried as "cargo," i.e., passengers, as the reliability of VentureStar™ was demonstrated.

48. Carl H. Builder, "The NASP as a Time Machine," RAND Internal Note 25684-AF, August 1989, p. 1.

49. Launius, "After *Columbia*," p. 291.

50. CAIB, *Report*, p. 211.

The Political Process and the Strength
of the Pro-Shuttle Coalition

As noted by the CAIB, the Space Shuttle is "an engineering marvel that enables a wide variety of on-orbit operations."[51] The Shuttle is also a program with a multibillion-dollar annual budget which employs thousands of people in various locations and is the focus of much of the activity at the Johnson Space Center, with a large astronaut corps located there; the Marshall Space Flight Center; and the Kennedy Space Center. Major and smaller aerospace firms across the United States work on the Shuttle program.

It is not surprising, then, that throughout the Shuttle program's history there has grown up a politically active coalition of government, contractor, local, and congressional supporters who argue that the Shuttle is a vehicle that continues to be superior in capabilities to any technologically feasible replacement, and who therefore have suggested that the preferred course of action is to invest scarce funds in upgrading and modernizing the Shuttle rather than seeking an early replacement. From the time when President Jimmy Carter (in 1979) considered terminating the Shuttle program, through the conflicts in the early 1980s with the Air Force on one hand and foreign and domestic competitors on the other, to the aftermath of the *Challenger* and *Columbia* accidents, and perhaps even to the current time, this coalition has argued that it would be a mistake to rush towards a Shuttle replacement. Ten years ago, a report from an advisory group headed by NASA veteran Christopher Kraft argued that the Shuttle was "a mature and reliable system . . . about as safe as today's technology will provide."[52] At the time of the 2003 *Columbia* accident, after the failure of the X-33 program and the Space Launch Initiative, NASA was planning to keep the Shuttle in operation until at least 2020 and potentially beyond.

The existence of an organized coalition of public and private interests with a stake in the Space Shuttle program is an entirely legitimate phenomenon. The whole system design of the American political process is intended to allow organized interests to contend for a favorable policy outcome. In this case, however, there was no organized alternative interest group pushing for an early Shuttle replacement, and thus the default outcome of annual policy debates was likely to favor the pro-Shuttle position, or, at a minimum, not result in outcomes opposing it. While, for example, there was opposition from the scientific community and some members of Congress in the 1980s and 1990s to the space station program, there has been no similar consistent opposition to the Space Shuttle.

---

51. Ibid., p. 25.
52. Quoted in ibid., p. 118.

There were, however, limits to the political strength of the Shuttle support coalition. Although it may have been powerful enough to raise questions about the wisdom of proceeding rapidly towards a Shuttle replacement, it did not have enough power within the political process to influence decision-makers to allocate adequate resources for upgrading the Shuttle and its associated infrastructure. The Shuttle program budget was cut by more than 40 percent in purchasing power between 1991 and 2000. Although some upgrades were introduced into the system, more were not funded or canceled soon after being approved, and the Shuttle's ground infrastructure was "deteriorating."[53] Especially in the decade before the *Columbia* accident, uncertainty about when the Shuttle might be replaced, as the politically weaker and not well organized advocates of such replacement contended with the pro-Shuttle coalition, created an ambivalent policy attitude towards the Shuttle program. This policy outcome was perhaps the worst possible situation—not enough funding for successful operation of the Shuttle, but also inadequate political commitment behind an effort to replace it. It was most fundamentally a reflection of the place that human spaceflight held, and perhaps continues to hold, in the list of national priorities—something that most Americans want to see continue but are unwilling to invest enough resources in to do well.

This is an attitude criticized by those committed to human spaceflight. Launius notes that "if the United States intends to fly humans in space it should be willing to foot the bill for doing so." He suggests that "if Americans are unwilling as a people to make that investment, as longtime NASA engineer and designer of the Mercury capsule spacecraft Max Faget [who died in 2004] recently stated, 'we ought to be ashamed of ourselves.'"[54] These are noble sentiments but do not reflect the long-standing reality of how the space program has been seen in terms of national priorities.

HUMAN SPACEFLIGHT AS A NATIONAL ENTERPRISE

Kay, writing a decade ago, observed that "three decades ago, the United States government made a decision to support space exploration—including human flight—on a rather large scale." He questions whether "our present institutional arrangements and political practices prevent us from carrying out that decision effectively," and thus there may be a need to "rethink our original policy decision."[55]

This essay asserts that at the national leadership level, the decision "to support space exploration—including human flight—on a rather large scale"

---

53. Ibid., p. 114.
54. Launius, "After *Columbia*," p. 295.
55. Kay, *Can Democracies Fly in Space?* p. x.

was rethought soon after it was made and that the outcome of that rethinking was a much more muted commitment to the civilian space program overall, including human spaceflight. The people of the United States and their government have been willing, over the past 35 years, to continue a human spaceflight program, but only at a level of funding that has forced it to constantly operate on the edge of viability. The lack of a replacement for the Space Shuttle is a symptom of this larger reality. In this context, the assertion that the lack of a Shuttle replacement is a "failure of national leadership" is the logical result of the halfhearted U.S. commitment to human spaceflight. If there is a "failure," then, it is the failure to reconcile the reality of limited support with this country's continuing commitment to sending people into space. Human spaceflight may indeed be a "national enterprise"—but it is one that for many years has not been central to important American interests, at least as they are expressed through the political process.

Kay ends his book with the question, "Can democracies fly in space?" His answer to this question is another question: "How badly do they want to?"[56] What will be argued below is that the answer to this second question is "not very badly."

Perhaps the single most convincing piece of evidence in support of this conclusion is the pattern of resources allocated to NASA over its history, as seen in the familiar figure repeated on the following page. Two things are remarkable about this pattern of resource allocation. The one most usually remarked upon is the rapid buildup of resources in the early 1960s in support of Project Apollo. This indeed was a peacetime mobilization of financial (and human) resources on a wartime scale. The Apollo buildup created an image of what a successful space program should be—one developing large-scale, expensive technology to take people into space.

Equally remarkable, however, and more fundamental to the argument of this paper is the rapid builddown of resources allocated to NASA between 1965 and 1974, and even more so the stability of that allocation over the past 30 years. It is impossible to escape the conclusion that, whatever the specific content of the NASA program at a particular time, the American public and their leaders, through the political process, have consistently decided to allocate less that 1 percent of the annual federal budget to the civilian space program as a national enterprise. This decision has been made, and reinforced, as the federal budget for each successive fiscal year has been assembled in the White House and approved or modified by the Congress. Within that allocation, national leaders have expected NASA to carry on a successful program of human spaceflight as well as its other activities. The result, as the CAIB

---

56. Ibid., p. 193.

NASA Budget as a Percentage of Federal Budget

TQ= Transition Quarter

observed with respect to the *Columbia* accident, has been an agency striving to "do too much with too little."

The basic decision that the United States, after succeeding in being first to land humans on the Moon, would not continue an ambitious program of human spaceflight in Earth orbit and beyond was made in 1969–1970 as the administration of President Richard Nixon formulated its post-Apollo policy for the civilian space program. It is a decision that has been reinforced by Presidents Ford, Carter, Reagan, and Clinton.

Up to 2004, only President George H. W. Bush (in 1989) suggested a reinvigoration of the human spaceflight program. Between President Bush's 1989 proposal for a "Space Exploration Initiative" and the time he was defeated in the 1992 election, it became clear, through the operation of the political process, that the country was not interested in a higher priority, more expensive human spaceflight effort.[57]

The first step in the process of formulating a policy to guide the space program after the end of the Apollo program was the creation in February 1969 of the Space Task Group, chaired by Vice President Spiro T. Agnew. This group was charged with preparing "definitive recommendations on the direction

---

57. See Thor Hogan, "Mars Wars: A Case History of Policymaking in the American Space Program" (Ph.D. diss., George Washington University, 2004), for a careful account of the origins and fate of the 1989 Space Exploration Initiative.

which the U.S. space program should take in the post-Apollo period."[58] In its 15 September 1969 report, the Space Task Group set out several options for the future and, "as a focus for the development of new capability," recommended that "the United States accept the long-term option or goal of manned planetary exploration with a manned Mars mission before the end of the century as the first target." This recommendation was actually a watered-down version of what the Group intended to recommend. President Nixon's advisers had intervened at the last minute, as the report was going into print, to make sure that the report did not contain the Group's planned recommendation that the initial mission to Mars be carried out in the 1980s, a recommendation that was politically unacceptable. The report proposed that whatever option was chosen by the President, the NASA budget by 1980 should be anywhere from the same as to twice that at the peak of the Apollo program.[59]

Accepting the Space Task Group's recommendations would have meant accepting a long-term national commitment to a robust program of human spaceflight, with repeated trips to the Moon and, eventually, forays to Mars. This was not at all what Richard Nixon and his advisers had in mind for the post-Apollo space effort. Rather than reward NASA for the success of the Apollo 11 landing, between October 1969 and January 1970, the NASA budget for fiscal year 1971 was severely reduced. In October, NASA requested White House approval of a $4.5-billion budget which would allow it to begin to implement the recommendations of the Space Task Group; by the time the President's budget was sent to Congress the following January, that amount had been reduced to $3.3 billion, a cut of over 25 percent from NASA's request and even $400 million less than the previous year's budget.

This outcome was not just the result of the Nixon administration's desire to submit a balanced budget; it reflected a major space policy choice. As Nixon's top adviser on space policy Peter Flanigan told the President in a 6 December 1969 memorandum:

> The October 6 issue of *Newsweek* took a poll of 1,321 Americans with household incomes ranging from $5,000 to $15,000 a year. This represents 61% of the white population of the United States and is obviously the heart of your constituency. Of this group, 56% think the government should be spending less money on space exploration, and only 10% think that the government should be spending more money.[60]

---

58. Logsdon, *Exploring the Unknown*, vol. 1, document III-22, p. 513.

59. Ibid., document III-25, p. 524.

60. Ibid., document III-27, p. 546.

NASA Administrator Thomas Paine, who had been touring both the United States and foreign countries to promote a post-Apollo space program as set out in the Space Task Group report, met with President Nixon on 22 January 1970 to make one last attempt to keep NASA on a path towards the approach laid out in the report. He had no success; Nixon told Paine that although he regretted the severe cuts to the NASA budget, "they were necessary in view of the overall budget situation—the reduced revenues and inflation." Nixon discussed "the mood of the country," which in the President's judgment "was for cuts in space and defense." Paine, ever an optimist, felt that the President "honestly would like to support a more vigorous space program if he felt the national mood favored it." But that was not the case, and Nixon wanted to make sure that he was not put in a position where "the opposition could invidiously compare his positive statements on space to problems in poverty and social programs here on Earth." Nixon did not want to appear to be "taking money away from social programs and the needs of the people here to fund spectacular crash programs out in space." Paine also noted that in their meeting, "the President didn't mention the Space Task Group Report."[61]

On 7 March 1970, the White House released a presidential statement on the future of the U.S. space program; Richard Nixon never addressed the subject in a public address. The statement was cast both as a response to the Space Task Group report and as an evaluation of where space fit into the country's future. Its message was clear:

> Space expenditures must take their proper place within a rigorous system of national priorities. What we do in space from here on in must become a normal and regular part of our national life and must therefore be planned in conjunction with all of the other undertakings which are important to us.[62]

The 1969–1970 interactions between NASA and the Nixon White House have been given detailed attention because they reflect a fundamental policy decision that has not been given adequate historical attention. In the months following the apex of U.S. success in human spaceflight with the Apollo 11 mission, the American President decided that it was neither in his political interest nor, more important, consistent with the desires of the American public to continue with a well-funded program of human spaceflight. This was not, as has been suggested, a case in which "the budget begat space policy

---

61. Thomas Paine, "Meeting with the President, January 22, 1970," memo for record, 22 January 1970, Apollo Files, University of Houston–Clear Lake Library, Clear Lake, TX.

62. Richard M. Nixon, "Statement About the Future of the United States Space Program," 7 March 1970, in U.S. President, *Public Papers of the Presidents of the United States: Richard Nixon, 1970* (Washington, DC: GPO, 1971), p. 251.

instead of space policy begetting the budget."[63] Rather, it reflected a deliberate, purposeful reversal of the space policy adopted by the Kennedy administration that had led to Project Apollo. That policy held that success in highly visible space projects was "part of the battle along the fluid front of the cold war"; that "dramatic achievements in space . . . symbolize the technological power and organizing capacity of a nation"; that it was "man, not machines, that captures the imagination of the world"; and that *the nation needs to make a positive decision to pursue space projects aimed at national prestige*"[64] (emphasis in original). To Richard Nixon and his advisers, this was not an acceptable rationale for a post-Apollo space program. They did not want to put an end to human spaceflight, but they were unwilling to set an ambitious goal to guide that effort. Instead, they approved development of a means—the Space Shuttle—without stating clearly the objectives it was to serve.

The decision on the future of the space program, and particularly on the future of its most visible element, human spaceflight, taken by the Nixon administration 35 years ago has remained the core national space policy until recently. That decision viewed the space program as a national enterprise, to use Kay's term, but one of secondary priority compared to other areas of national activity such as a strong defense, adequate social welfare, and, since 2001, homeland security. Based on the priority assigned to space efforts in this policy, for more than 30 years there has been a remarkably consistent share of the federal budget allocated to NASA.

That budget share has also been consistently inadequate to support the aspirations of NASA and the space community. Neither the space agency nor its supporters have adjusted their aspirations to that reality. Instead, they have continued to hold on to the hope that either a technological breakthrough on the order of NASP or VentureStar™ or a shift in the national priority assigned to space will allow them to make their dreams reality.

It is understandable that those most directly involved in the space sector harbor expansive ambitions for the future. What is not acceptable as a basis for government policy is to allow those ambitions to remain unchecked when the resources for achieving them are not, and are not likely to be, available. It is up to the leaders of NASA and to those to whom they report in the White House and Congress to steer the organization in a direction consistent with its place in the public's priorities. As suggested earlier, those leaders have failed to do so.

---

63. This is the argument put forth by Joan Hoff in her essay "The Presidency, Congress, and the Deceleration of the U.S. Space Program in the 1970s," in *Spaceflight and the Myth of Presidential Leadership*, ed. Roger D. Launius and Howard E. McCurdy (Urbana: University of Illinois Press, 1997), p. 106.

64. This quotation comes from the 8 May 1961 memorandum, signed by NASA Administrator James E. Webb and Secretary of Defense Robert S. McNamara, recommending that President Kennedy set a human lunar landing as a national goal. The memorandum can be found in Logsdon, *Exploring the Unknown*, vol. 1, p. 444.

This analysis seems to have wandered rather far from the focus of this essay on explaining why no replacement for the Space Shuttle has yet been developed. On the contrary—the answer to that question depends on understanding the context within which the human spaceflight program has operated for at least the last 35 years. Beginning with the Nixon administration (or perhaps even earlier),[65] the political process by which the United States sets priorities among various government activities has assigned a consistently secondary priority to the NASA space program. Operating within that priority, NASA was able to develop the Space Shuttle during the 1970s only by retiring all of the systems that had been developed for Project Apollo, with the exception of using surplus equipment for the 1973 Skylab and the 1975 Apollo-Soyuz missions. With these two exceptions, NASA accepted a lengthy hiatus in human spaceflight as an acceptable price to pay for being permitted to develop the Space Shuttle.

Once the Space Shuttle started flying in 1981 and a space station was approved in 1984, NASA has had no similar opportunity to stop what it was doing and invest the funds thereby made available in developing a Shuttle replacement. Instead, it has had to try both to continue its ongoing, Shuttle-based human spaceflight program and to develop new spaceflight capabilities within an unvarying share of the federal budget. This has, to date, proven an impossible challenge to surmount. Therein lies the fundamental reason why there is, today, no replacement for the Space Shuttle; it is a product of a space policy decision made many years ago and not reversed since.

SO HAS THERE REALLY BEEN A FAILURE?

Calling the lack of a replacement for the Space Shuttle "a failure of national leadership" is based on the assumption, as stated in the CAIB report, that "America's future space efforts must include human presence in Earth orbit, and eventually beyond."[66] If the United States is to continue human spaceflight, so this line of argument goes, it is essential to develop a Shuttle replacement rather than continue to rely on the aging and expensive Shuttle. To have come so far in space and not to have such a replacement ready or on the horizon must indeed be the result of a failure on the part of those responsible for allocating national resources to provide the support needed.

There is an alternative perspective: that a program of continuing human spaceflight, eventually leading to travel beyond Earth orbit, does serve the national interest. The rationales in support of human spaceflight are diffi-

---

65. The NASA budget actually began its rapid decline from the 1965 peak of spending on Apollo while Lyndon B. Johnson was President. Although Johnson was committed to completing Apollo, he apparently gave post-Apollo spaceflight lower priority in the context of the other issues facing him in the 1965–1968 period.

66. CAIB, *Report*, p. 210.

cult to articulate to the unconvinced in convincing fashion; Launius calls the rationale for human spaceflight "highly problematic."[67] For example, one member of the space community recently commented that taking "as axiomatic that space's highest and true calling is achieving societal goals of research and exploration into the unknown" is the "burdensome baggage of an aristocratic calling, now bankrupt both ideologically and financially."[68]

What appears to be needed is some form of a national debate on the future of human spaceflight that will allow these and other conflicting perspectives to be fully articulated and the long-standing policy of assigning space efforts a secondary priority as a national enterprise to be reassessed. As suggested above, the current policy that assigns space such a priority has resulted in a human spaceflight effort that has struggled now for many years to be a viable undertaking. As one recent analysis suggests, the fact that the vision of human spaceflight, including the resumption of human voyages of exploration, has not resonated "with the American public to the point where it inspires action is a reflection of a larger problem: the U.S. currently has no larger shared vision" into which a space exploration vision can fit.[69]

The policy of assigning secondary priority to space is thus not a "failure" in a basic sense; the policy is the consistent result of a democratic political process and thus can be said to represent the will of the American public. It is also difficult to say that national leaders have failed when they have acted in accordance with the public will as expressed through established institutions and processes.

Who then—or what—has failed? As suggested above, there has been a leadership failure in the sense that space ambitions and the resources to accomplish them have not been brought into balance. But perhaps the failure also lies with those who continue to advocate the original space dream, which was based on "adventure, mystery, and exploration." To date, they have failed to convince enough others that this dream is worth realizing to make it a focus of a higher priority national (or international) enterprise. Most Americans appear not to care very much about a future that includes a vigorous space effort. Advocates have not adjusted their hopes to reflect the resources society is willing to provide them. Rather, "the dreams continue, while the gap between expectations and reality remains unresolved."[70]

---

67. For a discussion of the difficulty in stating a compelling rationale for human spaceflight, see John M. Logsdon, "A Sustainable Rationale for Human Spaceflight," *Issues in Science and Technology* (winter 2004); Launius, "Beyond *Columbia*," quotation from p. 308.

68. Rick Fleeter, "Contemplating Which Direction in Space," *Space News* (18 October 2004): 7.

69. Center for Cultural Studies & Analysis, "American Perception of Space Exploration: A Cultural Analysis for Harmonic International and the National Aeronautics and Space Administration," report to NASA, 1 May 2004, p. 3.

70. Howard E. McCurdy, *Space and the American Imagination* (Washington, DC: Smithsonian Institution Press, 1997), p. 243.

EPILOGUE: AN ACHIEVABLE VISION?

On 14 January 2004, President George Bush laid out what has become known as the Vision for Space Exploration. In his speech announcing this new vision, the President called for a "journey, not a race." In the formal language of the policy directive underlying the Vision, the objective is a "sustained and affordable program of human and robotic exploration of the solar system and beyond."[71]

Those planning this new approach to the U.S. space program appear to have recognized the reality described in this essay: any major new space initiative, if it is to be achievable, must be planned so that it can be carried out within a level of funding consistent with the pattern of more than three decades. The Vision gives highest priority within the NASA program to those activities related to exploration; other activities will receive lower priority and thus less funding in the future. A firm deadline has been set for retiring the Space Shuttle from service, and NASA's activities aboard the International Space Station will be gradually phased out. A replacement for the Space Shuttle in its role of carrying Americans into space, the Crew Exploration Vehicle, is a key part of the new Vision. In order to stay within a politically feasible budget, the first crew-carrying flight of the CEV is not scheduled until the 2012–2014 timeframe, and the first human mission to the Moon is planned for 2018–2020. A several-year period during which the United States will have to depend on Russia for human access to space is accepted. Cost of achieving the Vision will be minimized by substantial international and private-sector involvement. According to the Vision's financial projections, the NASA budget between 2004 and 2020 will increase only by 1.5 percent in the first five years of the new effort and not at all in constant dollars in the subsequent decade.

Is this a vision that the country will support on a stable basis? Can its objectives be achieved within the resources projected?[72] These are questions that cannot be answered now. What can be said is that the Vision for Space Exploration in its conception reflects the realities described in this essay. Whether its aspirations can become reality remains to be seen.

---

71. White House, "Renewed Spirit of Discovery."

72. See U.S. Congress, Congressional Budget Office, "A Budgetary Analysis of NASA's New Vision for Space Exploration," September 2004, for a skeptical response to this question.

CHAPTER 10

REUSABLE LAUNCH VEHICLES OR EXPENDABLE LAUNCH VEHICLES? A PERENNIAL DEBATE

Andrew J. Butrica

The decades-long debate over reusable launch vehicles (RLVs) versus expendable launch vehicles (ELVs) has been less a reasoned debate than a sustained argument for the building of reusable launchers instead of the standard throwaway rocket. The single greatest touted advantage of reusable launch vehicles is that they reduce launch costs.[1] Comparing reusable and expendable rockets is not simple; it is a rather complicated task not unlike the proverbial comparing of apples and oranges. To compare the costs of the two types of rockets, we must consider two types of costs, recurring and nonrecurring. Nonrecurring costs entail those funds spent on designing, developing, researching, and engineering a launcher (called DDR&E costs). Recurring costs fall into two categories: expenses for building the launcher and the costs of its operation and maintenance.

Outlays for designing, developing, researching, and engineering reusable launchers are necessarily higher than those for expendable launchers because reusable rockets are technologically more challenging. For example, a reusable launch vehicle must have advanced heat shielding to allow it to reenter the atmosphere not once, but many times. Throwaway rockets have no need for such heat shielding. In addition, we possess a profound knowledge of expendable rocket technologies thanks to our long experience (over a half of a century) with ICBMs and other single-use rockets, while many of the technologies needed to build a fully reusable launcher remain in the elusive future. Construction costs, however, favor reusable launchers. For each launch, the cost of building a new expendable rocket is a recurring expense. For reusable launchers, construction costs are part of the upfront costs amortized over each launch.

Because reusable launch vehicles must fly many times in order to amortize startup costs, they have to be a lot more reliable than throwaway rockets,

---

1. Another cost-comparison method, but one that applies to specific launchers rather than launcher types and is considered to be more like comparing apples to apples (rather than oranges), is to determine the cost of delivering a pound of payload into orbit using a given launch system.

as well as more robust, so that on any given flight the craft does not suffer significant deterioration. The reliability of throwaway launchers is about 95 percent—that is, on average, 1 launch in 20 fails. A reusable launcher with equal reliability would not be able to recoup the higher investment needed to develop and build it. Achieving the necessary increased robustness and reliability also increases the cost and decreases the useful payload weight for reusable launchers.

The result of these intrinsic differences between the two launcher types leads to a tradeoff between the lower development costs of expendable rockets and the lower recurring costs of reusable launchers. In making that trade-off, one must take into account a number of other realistic factors that favor expendable launchers. For example, although one can amortize reusable vehicle construction costs over many flights, they are far more expensive to build than expendable rockets. Building a full-scale version of the VentureStar™, Lockheed Martin's failed attempt at a reusable, single stage to orbit (SSTO) launch vehicle, would have cost (conservatively) more than the $1 billion NASA spent on the X-33 program, the intent of which was to build a prototype of the VentureStar™ craft.[2] That same amount of money might have bought 10 expendable rockets at $100 million each. Also, the knowledge gained in manufacturing a large number of a given type of disposable launcher actually can help to lower construction costs. Thus, in order to compete with the low development and construction costs of the established expendable industry, a reusable launcher would have to fly more than 50 times.

The gamble of the reusable launcher is that a small fleet of three to five vehicles could put payloads into orbit for less than the cost of the number of expendable rockets required to lift similar payloads. A commercial builder and operator of reusable launchers, however, would be burdened by the need to amortize development and construction costs over each mission. An obvious solution would be to have the government pay for most or all of the development costs and for government (NASA and the Air Force) to buy one or two reusable launchers for its exclusive use.

The preceding discussion applies to a comparison of expendable rockets with fully reusable launchers. The economics of launching a reusable vehicle atop an expendable booster are rather different. Such hybrid systems are technologically more achievable than fully reusable single-stage or two-stage rockets. A variety of launchers that combine reusable and expendable stages have been under development by companies and government, and they appear to promise reductions in the cost of placing payloads in orbit. Throughout

---

2. NASA canceled plans to have a history of the X-33 written. To date, the best brief description of the project's evolution is General Accounting Office, *Status of the X-33 Reusable Launch Vehicle Program*, GAO/NSIAD-99-176 (Washington, DC: GPO, August 1999), pp. 2–8.

the decades-long quest for reusability, the configuration of a reusable reentry vehicle atop a throwaway booster (a so-called boost-glide system) has dominated launcher thinking. In these boost-glide systems, the upper stage vehicle, once released from its booster rocket, climbs into orbit on its own power, then glides to a landing. Some reusable suborbital vehicles launch from a large jet, such as a B-52 or an L-1011.

Cost has not been the only factor favoring one launch technological system over another. Emotional and political considerations are certainly key, as is the pull on the imagination exercised by the promise of reusable launchers. RLV enthusiasts believe that a fully reusable rocket would provide the low-cost, reliable transport to space necessary to realize the seemingly endless possibilities of exploiting space—the "final frontier"—for colonization, mining, tourism, manufacturing, or just exploration.

The history of the debate over reusable versus expendable launchers is complex, and one can explore it from a variety of perspectives. The most obvious is a narrative of the enduring endeavor to conceive and develop a reusable launch vehicle. This chapter begins with such an account, then discusses the evolution of space transportation policy regarding reusable and expendable launchers. A third section raises historiographical questions about launch vehicle history as well as space history in general.

THE SPACEPLANE CONCEPT

One of the earliest reusable vehicle concepts was that of the spaceplane.[3] They are like airplanes in a rather simplistic and literal way. They have wings and take off and land horizontally like an airplane; a pilot and copilot sit in a cockpit. They usually (but not always) feature a kind of air-breathing engine known as a scramjet.[4] Their appeal is rather similar to that of jet aircraft, namely, the urge to go faster and higher than before that permeates the history of flying. Indeed, spaceplanes are little more than aircraft that fly into space.

One of the first spaceplane concepts was that of the American rocketeer Robert Goddard. In a *Popular Science* article published in December 1931, he described a spaceplane ("stratosphere plane") with elliptically shaped wings and propelled by a combination air-breathing jet and rocket engine. The rocket engine drove the vehicle while it was outside the atmosphere, and two turbines moved into the rocket's thrust stream to drive two large propel-

---

3. I am excluding all of those reusable launch vehicles described in science fiction literature.

4. *Scramjet* is a truncation of "supersonic combustion ramjet." Ramjets are jet engines that propel aircraft at supersonic speeds by igniting fuel mixed with air that the engine has compressed. Scramjets achieve hypersonic velocities.

lers on either wing, thereby powering the vehicle while in the atmosphere.[5] German researcher Eugen Sänger, in his 1933 book on rocket flight, described a rocket-powered suborbital spaceplane known as the *Silbervogel* (Silver Bird), fueled by liquid oxygen and kerosene and capable of reaching a maximum altitude of 160 kilometers (100 miles) and a speed of Mach 10. Later, working with his future wife, the mathematician Irene Bredt, and a number of research assistants, Sänger designed the Rocket Spaceplane, launched from a sled at a speed of Mach 1.5. A rocket engine capable of developing 100 tons of thrust would boost the craft into orbit, where it could deploy payloads weighing up to 1 ton.[6]

The appearance of ideas for craft capable of flying into space is not surprising. They reflected the interwar enthusiasm for the airplane, as well as excitement over rocketry, and projected those technological enthusiasms into space. New technologies often look like older technologies. For example, James Prescott Joule's electric motor resembled a steam engine, and Samuel F. B. Morse built his first telegraph from a canvas stretcher, a technology he knew as an artist.[7] Inventors necessarily proceed from the known to the technologically unknown. The passion for spaceplanes continued for decades more, feeding off the exciting advances in technology that propelled aircraft faster and faster to supersonic, then to hypersonic, speeds.

Spaceplanes remained largely fictional concepts until 1957, when the Air Force initiated what became the Aerospaceplane program to develop a single stage to orbit vehicle powered by an air-breathing engine. By 1959, the project had evolved into the Recoverable Orbital Launch System (ROLS), an SSTO design that would take off horizontally and fly into a 300-mile-high (483-meter-high) orbit. The ROLS propulsion system collected air from the atmosphere, then compressed, liquefied, and distilled it in order to make liquid oxygen, which mixed with liquid hydrogen before entering the engines.

---

5. Russell J. Hannigan, *Spaceflight in the Era of Aero-Space Planes* (Malabar, FL: Krieger Publishing Company, 1994), p. 71. Materials in file 824 of the NASA Historical Reference Collection at NASA Headquarters, Washington, DC, indicate that the article appeared in the December 1931 issue, pp. 148–149, and was titled "A New Turbine Rocket Plane for the Upper Atmosphere."

6. Irene Sänger-Bredt, "The Silver Bird Story: A Memoir," file 7910, NASA Historical Reference Collection, Washington, DC; Hannigan, *Spaceflight in the Era of Aero-Space Planes*, pp. 71–73; Michael J. Neufeld, *The Rocket and the Reich: Peenemünde and the Coming of the Ballistic Missile Era* (New York: The Free Press, 1995), pp. 7–10; Richard P. Hallion, "In the Beginning Was the Dream . . . ," in *The Hypersonic Revolution: Eight Case Studies in the History of Hypersonic Technology*, ed. Richard P. Hallion, vol. 1, *From Max Valier to Project Prime, 1924–1967* (Dayton, OH: Special Staff Office, Aeronautical Systems Division, Wright-Patterson AFB, 1987), pp. xi–xv.

7. Brooke Hindle, *Emulation and Invention* (New York: New York University Press, 1981), pp. 85–108, 120–121; Lewis Coe, *Telegraph: A History of Morse's Invention and Its Predecessors in the United States* (Jefferson, NC: McFarland, 1993); J. M. Anderson, "The Invention of the Telegraph: Samuel Morse's Role Reassessed," *IEEE Power Engineering Review* 18 (July 1998): 28–29.

This complicated propulsion system, dubbed LACES (Liquid Air Collection Engine System), later renamed ACES (Air Collection and Enrichment System), as well as various scramjet engine concepts, underwent Air Force evaluation over time. Faced with the uncertainties of the single-stage design, the Air Force shifted the focus of the Aerospaceplane to two stage to orbit concepts in 1962, and following the program's condemnation by the Scientific Advisory Board, the Aerospaceplane died in 1963. Congress cut fiscal 1964 funding, and the Pentagon declined to press for its restoration.[8]

Dyna-Soar

A rather different reusable vehicle concept was the boost-glide system. The Peenemünde rocket group under Wernher von Braun originally planned to develop a much larger missile, the A-10/A-9, capable of delivering a 1-ton bomb over 5,000 kilometers (3,125 miles) away. The A-10 first stage was a conventional booster rocket, while the A-9 upper stage was a winged vehicle that could glide at supersonic speeds before hitting its target. Other Peenemünde work, kept secret from the Nazis, included a piloted version of the A-9 that would launch vertically and land horizontally, like the Space Shuttle. An even larger vehicle, the A-12, was a fanciful three-staged launcher whose top stage was a reusable winged reentry vehicle.[9] None of these concepts, however, were orbital vehicles.

At the end of World War II, as is widely known, Wernher von Braun and much of the German rocket program became a vital part of the United States' own missile program and contributed to the development of boost-glide systems.[10] Walter Dornberger, a key Nazi rocketeer and later a consultant for Bell Aircraft, persuaded that firm to undertake a study of boost-glide technology. In 1952, that study led to the joint development by Bell and the Wright Air Development Center, Dayton, Ohio, of a piloted bomber missile and reconnaissance vehicle called BoMi. A two-stage rocket would lift BoMi, which would operate at speeds over Mach 4. By 1956, the BoMi study work had evolved into a contract for Bell to develop Reconnaissance System 459L, commonly known as Brass Bell, a piloted two-stage boost-

---

8. Hannigan, *Spaceflight in the Era of Aero-Space Planes*, pp. 77–78; T. A. Heppenheimer, *The Space Shuttle Decision: NASA's Search for a Reusable Space Vehicle* (Washington, DC: NASA SP-4221, 1999), pp. 75–78; Hallion and James O. Young, "Space Shuttle: Fulfillment of a Dream," in *The Hypersonic Revolution: Eight Case Studies in the History of Hypersonic Technology*, ed. Hallion, vol. 2, *From Scramjet to the National Aero-Space Plane* (Dayton, OH: Special Staff Office, Aeronautical Systems Division, Wright-Patterson AFB, 1987) pp. 949–951.

9. Neufeld, *Rocket and the Reich*, pp. 92–93, 121, 138–139, 156–157, 283; Hallion, "In the Beginning Was the Dream . . . ," p. xviii; Hannigan, *Spaceflight in the Era of Aero-Space Planes,* p. 73.

10. Linda Hunt, *Secret Agenda: The United States Government, Nazi Scientists, and Project Paperclip, 1945 to 1990* (New York: St. Martin's Press, 1991).

glide reconnaissance system, while the bomber part of the BoMi work became RoBo, a piloted hypersonic, rocket-powered craft for bombing and reconnaissance missions.[11]

A major step in orbital boost-glide systems was the Dyna-Soar (for Dynamic Soaring) program. It was the final stage of a three-stage study of rocket-powered hypersonic flight initiated by the National Advisory Committee for Aeronautics (NACA) with Air Force participation. The study used a series of experimental aircraft ("X" vehicles) lifted into the sky by reusable aircraft. "Round One," to use the NACA nomenclature, consisted of the Bell X-1 series, the Bell X-2 series, and the Douglas D-588-2 Skyrocket. "Round Two" was the series of flights eventually undertaken by the X-15. "Round Three" called for testing winged orbital reentry vehicles.[12]

The Air Force's Dyna-Soar program emerged from a 1957 consolidation of the NACA's "Round Three" and several military hypersonic flight programs. Eventually, NASA participated in the project as well. Launched on an expendable booster, the Dyna-Soar X-20 would fly orbital or suborbital trajectories, perform reconnaissance at hypersonic speeds, and land horizontally like an aircraft at many U.S. air bases. Although the Dyna-Soar vehicle was never built, a prototype was near completion when Secretary of Defense Robert McNamara terminated the program on 10 December 1963, only eight months before drop tests from a B-52. The first piloted flight had been scheduled for 1964.[13]

Dyna-Soar had a lot to offer the Air Force and the nation and might have changed history. The military might have benefited economically by possessing the world's first reusable orbital vehicle, and the Pentagon would not have

---

11. Clarence J. Geiger, "Strangled Infant: The Boeing X-20A Dyna-Soar," in *The Hypersonic Revolution*, ed. Hallion, vol. 1, pp. 189, 191–198, a manuscript copy of which is in file 11326, NASA Historical Reference Collection, Washington, DC, as Geiger, "History of the X-20A Dyna-Soar," October 1963; additional items from files 495 and 11923, NASA Historical Reference Collection, Washington, DC; Hallion, "Editor's Introduction," in *The Hypersonic Revolution*, ed. Hallion, vol. 1, p. II-xi.

12. Hallion, "In the Beginning Was the Dream . . . ," p. xxi; Hallion, "Editor's Introduction," in *The Hypersonic Revolution*, ed. Hallion, vol. 1, pp. I-iv–I-v, II-xi.

13. R&D Project Card Continuation Sheet, 23 August 1957, file 11325, NASA Historical Reference Collection, Washington, DC; additional items in file 11340, NASA Historical Reference Collection, Washington, DC; Geiger, "Strangled Infant," pp. 198–199, 201–204, 261, 263, 266, 276–278, 296–297, 299–301, 305, 308–309. A number of studies are available on the Dyna-Soar program. See, for instance, Terry Smith, "The Dyna-Soar X-20: A Historical Overview," *Quest: The History of Spaceflight Magazine* 3, no. 4 (1994): 13–18, 23–28; Matt Bacon, "The Dynasoar Extinction," *Space* 9 (May 1993): 18–21; Roy Franklin Houchin II, "The Rise and Fall of Dyna-Soar: A History of Air Force Hypersonic R & D, 1944–1963" (Ph.D. diss., Auburn University, 1995); Houchin, "The Diplomatic Demise of Dyna-Soar: The Impact of International and Domestic Political Affairs on the Dyna-Soar X-20 Project, 1957–1963," *Aerospace Historian* 35 (December 1988): 274–280.

Artist's concept of a Dyna-Soar manned space glider being launched into space by a modified Titan ICBM. The glider, riding on the nose of the Titan, would be separated from its booster, leaving the spacecraft in piloted, near-orbital flight. The pilot could glide to a conventional landing at an Air Force base. The Boeing Company was the prime contractor for the glider, which was a U.S. Air Force program. Only a prototype of the glider was built before the program was terminated on 10 December 1963. *(Boeing drawing S-5938, dated 22 September 1960)*

been forced to become NASA's political ally in the space agency's political struggle to win funding for its Space Shuttle program. Also, Dyna-Soar could have provided NASA a less expensive, but two-stage, orbital shuttle. The knowledge gained from the research program, which included over 14,000 hours of wind tunnel tests, could have been applied to a number of applications from glide bombers to future spacecraft. Moreover, after termination of the program, Boeing carried out a small "X-20 continuation program" for several more years that involved testing various X-20 components and design features both in ground facilities and on flight research vehicles. The René 41 high-temperature nickel alloy developed for the X-20 reappeared in the 1970s as part of the airframe structure and heat shielding for Boeing's Reusable Aerodynamic Space Vehicle (RASV).[14]

Lifting Bodies

Also of note among these early boost-glide systems was a group of reusable suborbital vehicles known as lifting bodies. A lifting body is a wingless aerodynamic shape that develops lift—the force that makes winged craft fly—because of its peculiar body shape. Research on lifting bodies began in early 1957 at the NACA's Ames Aeronautical Laboratory (now NASA's Ames Research Center). Following NASA's success with its wooden M2-F1, the Air Force joined NASA at Edwards AFB in the test-flight program of the rocket-powered M2-F2, launched from a B-52 from 1966 until its crash in 1967.[15]

The most prominent of these lifting-body craft was the Air Force's X-24B, built by Martin Marietta in 1972. A modified X-24B powered by aerospike engines became Lockheed's Space Shuttle design concept in the latter 1960s, the StarClipper, while the X-24B's shape also inspired the design of what eventually became Lockheed skunk works' X-33 launch vehicle. Despite the apparent name similarity, the X-24B had rather different shapes and distinct origins from the X-24A lifting body built for NASA, though both had a role in the Air Force's lifting-body program.[16]

The RASV

Even as NASA and industry were building the Space Shuttle, the search for a reusable Shuttle replacement was under way. As with lifting-body research,

14. Geiger, "Strangled Infant," pp. 319–320, 369; Andrew K. Hepler interview, tape recording and transcript, Seattle, WA, by Butrica, 11 July 2000, NASA Historical Reference Collection; Hepler and E. L. Bangsund, Boeing Aerospace Company, Seattle, WA, *Technology Requirements for Advanced Earth Orbital Transportation Systems*, vol. 1, *Executive Summary* (Washington, DC: NASA Contractor Report CR-2878, 1978).

15. R. Dale Reed, *Wingless Flight: The Lifting Body Story* (Washington, DC: NASA SP-4220, 1997), pp. 9, 67, 69–72, 75, 87, 91, 96–98, 102, 106–109, 116; John L. Vitelli and Hallion, "Project PRIME: Hypersonic Reentry from Space," in *The Hypersonic Revolution*, ed. Hallion, vol. 1, p. 529.

16. Vitelli and Hallion, "Project PRIME," pp. 558, 566, 571, 577–596, 694–695, 699, 702–704, 711.

NASA led the way. In 1972, the Langley Research Center, with the approval of NASA Headquarters, set up a small group to study the possibility of growing an aircraft known as the Continental/SemiGlobal Transport (C/SGT) into a single stage to orbit vehicle. The C/SGT would take off, almost attain orbit, then land, delivering people or cargo to any place on Earth in less than 2 hours. Langley researchers' analyses of the vehicle suggested that with just a little bit more speed, the C/SGT could achieve orbit.[17]

Using Shuttle technology as the starting point for their study of the structures, materials, and engines needed for a Shuttle replacement, the Langley analysis team evaluated the impact of improving structures and materials (such as composites) beyond the Space Shuttle on various configurations. The improved materials promised to reduce overall vehicle weight significantly, thereby seeming to bring SSTO transport within the realm of the possible.[18] Then, in 1975, Langley funded two industry studies of SSTO rocket concepts carried out by teams from Martin Marietta Denver and Boeing Seattle. The stated purpose of the study was to determine the future technology development needed to build an operational rocket-powered, single stage to orbit Space Shuttle replacement by the year 1995. Each team concluded that such a vehicle was feasible using technology available in the near term.[19]

Next, Boeing tried to sell their vehicle design from the Langley 1975 study to the Air Force. The company's interest in the reusable SSTO vehicle was "based on the belief that the reusable airplane type operation of earth orbit transportation vehicles will allow considerable improvement in cost per flight and flexibility."[20] The vehicle would have incorporated both proven and unproven technologies. The cylindrically shaped, delta-winged, reusable single stage to orbit craft, powered by Space Shuttle Main Engines, would have take off with the help of a sled and land horizontally on a conventional runway. It would have used a combination of aluminum-brazed titanium and René 41, a high-temperature nickel alloy developed for the Dyna-Soar X-20, for both its structure and heat shielding. The vehicle would have stored liquid-hydrogen fuel in its body and liquid oxygen in its wings. The integration of the liquid-hydrogen and liquid-oxygen tanks into the load-carrying

17. Alan Wilhite interview, tape recording and transcript, NASA Langley Research Center, Hampton, VA, by Butrica, 22 May 1997, NASA Historical Reference Collection, Washington, DC.

18. Charles H. Eldred interview, tape recording and transcript, NASA Langley Research Center, Hampton, VA, by Butrica, 20 May 1997, NASA Historical Reference Collection, Washington, DC.

19. The two studies were Rudolph C. Haefeli, Earnest G. Littler, John B. Hurley, and Martin G. Winter, Denver Division, Martin Marietta Corporation, *Technology Requirements for Advanced Earth-Orbital Transportation Systems: Final Report* (Washington, DC: NASA Contractor Report CR-2866, October 1977); and Andrew K. Hepler and E. L. Bangsund, Boeing Aerospace Company, Seattle, WA, *Technology Requirements for Advanced Earth Orbital Transportation Systems*, vol. 1, *Executive Summary*, and vol. 2, *Summary Report* (Washington, DC: NASA Contractor Report CR-2878, 1978).

20. Hepler and Bangsund, *Technology Requirements for Advanced Earth Orbital Transportation Systems*, 1:13–14.

structure (that is, the wings and the main body of the craft), combined with the metallic shell made of honeycomb panels, went far in reducing overall vehicle weight.[21]

Boeing soon interested the Air Force Space and Missiles System Organization (Los Angeles Air Force Station) in this vehicle concept. The Air Force dubbed it the Reusable Aerodynamic Space Vehicle (RASV) and, in 1976, provided funding for a seven-month preliminary feasibility study of the RASV concept. It concluded (not surprisingly) that the RASV was feasible and that it would fulfill Air Force requirements. Among those requirements were flying 500 to 1,000 times "with low cost refurbishment and mainte-nance as a design goal" from a launch site in Grand Forks, North Dakota, into a polar orbit or once around the planet in a different orbit. The vehicle would have to reach "standby status within 24 hours from warning. Standby to launch shall be three minutes."[22]

In all, the Air Force invested $3 million in the project for technology development. The service had become convinced that the RASV potentially could provide a manned platform that could be placed above any point on the planet in less than an hour and could perform a variety of missions, includ-ing reconnaissance, rapid satellite replacement, and general space defense. In December 1982, Boeing Chairman T. A. Wilson gave the RASV effort the go-ahead to propose a $1.4-billion prototype vehicle to the Air Force.[23] Boeing, however, would not build the RASV.

The problem was not the steep technological hurdles that the firm would have to leap, such as development of the sled to accelerate the RASV to a speed of 600 feet per second or achievement of fast turnaround time (24 hours or perhaps as short as 12 hours) for the Strategic Air Command (SAC).[24] The Air Force ordered two classified studies of single stage to orbit technologies, "Science Dawn" (1983–1985) and "Have Region" (1986–1989), conducted by industry partners Boeing, Lockheed, and McDonnell Douglas. They inter-

---

21. Ibid., 1:14–16, 2:191; Hepler interview.

22. Boeing Aerospace Company, *Final Report on Feasibility Study of Reusable Aerodynamic Space Vehicle*, vol. 1, *Executive Summary* (Kent, WA: Boeing Aerospace Company, November 1976), pp. 5, 35.

23. Hallion, "Yesterday, Today, and Tomorrow: From Shuttle to the National Aero-Space Plane," in *The Hypersonic Revolution*, ed. Hallion, vol. 2, p. 1334; P. Kenneth Pierpont, "Preliminary Study of Adaptation of SST Technology to a Reusable Aero-space Launch Vehicle System," NASA Langley Working Paper NASA-LWP-157, 3 November 1965; Boeing RASV proposal, December 1982, file 256, X-33 Archive, record group 255, accession number 255-01-0645, Washington National Records Center, Suitland, MD (hereafter, X-33 Archive); Jess Sponable interview, tape recording and transcript, NASA Headquarters, Washington, DC, by Butrica, 19 January 1998, NASA Historical Reference Collection, Washington, DC; Gary Payton and Jess Sponable, "Designing the SSTO Rocket," *Aerospace America* (April 1991): 40.

24. Hepler interview.

preted the study results as demonstrating the technological feasibility of the RASV for SAC.[25] But instead of proceeding with further RASV studies, the Air Force chose to develop a space vehicle that not only operated like an aircraft, as the RASV did, but had air-breathing jet engines, too. That space vehicle would be known as the National Aero-Space Plane (NASP).

The National Aero-Space Plane

With NASP, the spaceplane quest returned.[26] The milestone moment was President Ronald Reagan's State of the Union Address, delivered on 4 February 1986, just days after the *Challenger* disaster. Reagan declared: "We are going forward with research on a new Orient Express that could, by the end of the decade, take off from Dulles Airport, accelerate up to 25 times the speed of sound attaining low Earth orbit, or fly to Tokyo within two hours."[27] As portrayed by the President, the Orient Express would be both a high-speed aircraft and a single stage to orbit vehicle, powered by air-breathing engines. The program merged two existing efforts.

One was the TransAtmospheric Vehicle (TAV) program, set up in 1982 as an Air Force study of Space Shuttle replacement concepts. Air-breathing engines were a serious, though not exclusive, consideration. The program considered a variety of both single- and two-stage vehicle configurations, powered by either rocket or jet engines.[28] Interest in the TransAtmospheric Vehicle grew as a direct result of the increased need for launchers driven

---

25. Raymond L. Chase, "Science Dawn Overview," March 1990, file 235, X-33 Archive; Major Stephen Clift, "Have Region Program: Final Brief," September 1989, file 235, X-33 Archive; Sponable interview.

26. For background information on NASP, see the materials in file 106, box 4, X-33 Archive; Larry Schweikart, "Command Innovation: Lessons from the National Spaceplane Program," in *Innovation and the Development of Flight*, ed. Roger D. Launius (College Station: Texas A&M University Press, 1999), pp. 299–323; Hannigan, *Spaceflight in the Era of Aero-Space Planes*, passim; Schweikart, "The National Spaceplane: Evolving Management Approaches to a Revolutionary Technology Program," *Essays in Economic and Business History* 12 (1994): 118–33; Alan W. Wilhite, Richard W. Powell, Stephen J. Scotti, Charles R. McClinton, S. Zane Pinckney, Christopher I. Cruz, L. Robert Jackson, James L. Hunt, Jeffrey A. Cerro, and Paul L. Moses, "Concepts Leading to the National Aero-Space Plane Program" (paper read at the 28th Aerospace Sciences Meeting, Reno, NV, 8–11 January 1990), file 703, box 23, X-33 Archive.

27. Quoted in Scott Pace, "National Aero-space Plane Program: Principal Assumptions, Findings, and Policy Options," RAND publication P-7288-RGS, December 1986, p. 1. Reagan's speechwriters confused the NASP reusable single stage to orbit vehicle with the Orient Express, a McDonnell Douglas hypersonic aircraft design in which Federal Express had shown interest. The confusion probably screened the flight vehicle's military mission, though the McDonnell Douglas prototype claimed to be capable of performing either a NASP single stage to orbit or an Orient Express mission, depending on the vehicle's propulsion system. See Paul Czysz interview, tape and transcript, NASA Langley Research Center, Hampton, VA, by Erik M. Conway, 17 July 2001, pp. 1–5, 8–9, 11.

28. Hallion, "Yesterday, Today, and Tomorrow," pp. 1337, 1340–1341, 1345.

by the Strategic Defense Initiative (SDI) and Space Station *Freedom*.[29] The second program was the classified three-phase Copper Canyon program of the Advanced Research Projects Agency (ARPA), which funded research on scramjet hypersonic vehicles.[30] The Copper Canyon and TransAtmospheric Vehicle efforts merged to form a larger program that comprised the gamut of government agencies involved in hypersonic air-breathing engine studies at one time or another: NASA, ARPA, the Air Force, the Navy, and the Strategic Defense Initiative Organization (SDIO). On 1 December 1985, the title National Aero-Space Plane (NASP) replaced all earlier designations.[31]

The NASP program initially proposed to design and build two research craft, the X-30, at least one of which was to achieve orbit by flying in a single stage through the atmosphere at speeds up to Mach 25. The X-30 would use a multicycle engine that shifted from jet to ramjet and scramjet speeds as the vehicle ascended, burning liquid-hydrogen fuel with oxygen scooped and frozen from the atmosphere. The engine and vehicle designs had come from Tony DuPont, an aerospace designer who had developed a multicycle jet and rocket engine under contract with NASA, then ARPA.[32] DuPont's vehicle design rested on a number of highly questionable assumptions, optimistic interpretations of results, and convenient omissions (such as landing gear).[33]

NASP, like the Aerospaceplane program, fell victim to budget cuts, but this time as a result of the end of the Cold War. Congress canceled NASP in 1992, during fiscal 1993 budget deliberations. Although the program never came near to building or flying hardware, NASP contributed significantly to the advance of materials capable of repeatedly withstanding high temperatures (on the vehicle's nose and body) or capable of tolerating repeated exposure to extremely low temperatures (the cryogenic fuel tanks).[34]

---

29. Ibid., pp. 1336–1337, 1340–1341.

30. John V. Becker, "Confronting Scramjet: The NASA Hypersonic Ramjet Experiment," in *The Hypersonic Revolution*, ed. Hallion, vol. 2, pp. VI.xii, VI.xiv, 765, 786–789, 824, 841; Heppenheimer, *The National Spaceplane* (Arlington, VA: Pasha Market Intelligence, 1987), p. 14; Hallion, "Yesterday, Today, and Tomorrow," p. 1361; Larry Schweikart, "The Quest for the Orbital Jet: The National Aerospace Plane Program, 1983–1995," manuscript, pp. I.30–I.31, NASA Historical Reference Collection, Washington, DC. For background on these and other hypersonic research projects, see Erik Conway, *High-Speed Dreams: NASA and the Technopolitics of Supersonic Transportation, 1945–1999* (Baltimore, MD: Johns Hopkins, 2005).

31. Heppenheimer, *The National Spaceplane*, p. 14; Hallion, "Yesterday, Today, and Tomorrow," pp. 1334, 1362–1364; Schweikart, "The Quest for the Orbital Jet," pp. I.30–I.31; Becker, "Confronting Scramjet," in *The Hypersonic Revolution*, ed. Hallion, vol. 2, p. VI.xv.

32. Robert Jones interview, tape and transcript, NASA Langley Research Center, Hampton, VA, by Erik M. Conway, 25 June 2001, pp. 8–9; Conway to Butrica, e-mail message, 5 April 2002; Schweikart, "The Quest for the Orbital Jet," pp. I.19–I.20, I.23, I.28, III.31, III.43–III.44; Hallion, "Yesterday, Today, and Tomorrow," pp. 1346, 1351, 1379.

33. Schweikart, "The Quest for the Orbital Jet," pp. I.11–I.12, I.19–I.20, I.23, I.28, III.43.

34. Ibid., pp. III.37–III.38, III.41–III.42.

The Delta Clipper

The end of NASP was not the end of efforts to realize a fully reusable launch vehicle. In parallel with, but never in competition with, NASP was the SSTO Program of the SDIO. This program differed radically from its predecessors that had attempted to develop flight technology; instead, it tested the flight operations of a single stage to orbit vehicle, the Delta Clipper Experimental (DC-X). Its intent was not to develop technology, but to demonstrate "aircraft-like" operations, which included autonomous operations, minimal launch and operational crews, ease of maintenance, abort capability, and short turnaround time. The novelty of the SSTO Program also was to combine the goal of "aircraft-like" operations with the use of an "X" vehicle and a "lean" management approach by both government and industry in the hope of expediting the project and keeping costs low.

In early 1990, the Strategic Defense Initiative Organization started the SSTO Program. The 10-month-long Phase I consisted of design studies and the identification of critical technologies by Boeing, General Dynamics, McDonnell Douglas, and Rockwell International.[35] In June 1991, following a review of Phase I concepts by NASA's Langley Research Center, the SDIO solicited proposals for Phase II. The Statement of Work described the capabilities of the full-scale operational single stage to orbit vehicle—which would loft SDI Brilliant Pebbles payloads into orbit—and the Phase II small suborbital "X" vehicle, its support infrastructures (such as the launchpad), and operational concepts.[36] Of the three contractors competing—General Dynamics, McDonnell Douglas, and Rockwell International—the SDIO selected McDonnell Douglas in August 1991 to build its Delta Clipper Experimental (DC-X) in 24 months. The firm clearly understood the need to demonstrate operations rather than develop technology.[37]

McDonnell Douglas rolled out the 111-foot (34-meter) DC-X in record time, four months ahead of schedule, in April 1993. The company built the Delta Clipper out of modified existing hardware, some of which, such as welding rods and hinges, they purchased literally from local hardware stores. Pressure regulators and cryogenic valves came from Thor missiles formerly positioned in Europe, and the manufacturer of the alu-

---

35. McDonnell Douglas Space Systems Company, "Single Stage to Orbit Program Phase I Concept Definition," 13 December 1990, file 267, X-33 Archive; General Dynamics Space Systems Division, "Concept Review Technical Briefing," 13 December 1990, file 265, X-33 Archive; Space Transportation Systems, Boeing Defense and Space Group, "Single Stage to Orbit Technology Demonstration Concept Review Technical Briefing," 12 December 1990, file 264, X-33 Archive; Rockwell International, "SDIO Single Stage to Orbit Concept Review," 12 December 1990, file 259, X-33 Archive.

36. "NASA Evaluation of SDIO Phase I SSTO Concepts," n.d., file 294, X-33 Archive.

37. Sponable interview.

minum liquid-oxygen and -hydrogen tanks was not an aerospace firm, but Chicago Bridge and Iron (CBI) of Birmingham, Alabama.[38] More importantly, McDonnell Douglas sought to achieve SSTO Program operational goals. The Flight Operations Control Center at the White Sands Missile Range, New Mexico, consisted of a compact, low-cost, 40-foot (12-meter) mobile trailer. Three people operated the ground support equipment and launched the DC-X, not the hundreds typically used for NASA or military rocket launches. Former astronaut Pete Conrad was the "flight manager." McDonnell Douglas designed the DC-X so that they could fly it again after only three days. Eventually, on 8 June 1996, the Clipper team demonstrated a one-day (26-hour) turnaround.[39]

By the time the DC-X undertook its first flight on 18 August 1993, the world had changed dramatically. The Cold War was over, and defense cuts were the order of the day. As DC-X flight trials took place, the future of funding for those flights, as well as for completion of the program, grew less certain. Money for Phase III disappeared, and various bureaucratic maneuvers stymied White House and congressional approval of financing. The predicament grounded the Clipper after only three flights, until the NASA Administrator intervened financially in January 1994.[40]

NASA's "X" Vehicles

By January 1994, NASA Administrator Daniel S. Goldin had become interested in single stage to orbit and other kinds of reusable launchers. His interest did not arise from any internal NASA studies, such as those conducted by the Langley Research Center as early as the 1970s, nor from the influence of high-level individuals at NASA Headquarters, such as Ivan Bekey, Director

---

38. Paul L. Klevatt interview, tape recording and transcript, Tustin, CA, by Butrica, 14 July 2000, NASA Historical Reference Collection, Washington, DC; William Gaubatz interview, tape recording and transcript, Huntington Beach, CA, by Butrica, 25 October 1997, NASA Historical Reference Collection, Washington, DC; Klevatt, "Design Engineering and Rapid Prototyping for the DC-X Single Stage Rocket Technology Vehicle," AIAA-95-1425 (paper read at AIAA-ASME-ASCE-AHS-ASC Structures, Structural Dynamics, and Materials Conference, New Orleans, LA, 10–12 April 1995).

39. Klevatt interview; McDonnell Douglas Space Systems Company, "Single Stage to Orbit Program Phase I Concept Definition," 13 December 1990, file 267, X-33 Archive; Charles "Pete" Conrad interview, tape recording and transcript, Rocket Development Company, Los Alamitos, CA, by Butrica, 22 October 1997, NASA Historical Reference Collection, Washington, DC; Luis Zea, "The Quicker Clipper," *Final Frontier* (October 1992): 4, file 267, X-33 Archive; Mark A. Gottschalk, "Delta Clipper: Taxi to the Heavens," *Design News* (September 1992), file 292, X-33 Archive; Leonard David, "Unorthodox New DC-X Rocket Ready for First Tests," *Space News* (11–17 January 1993): 10.

40. George E. Brown, Jr., to Les Aspin, 31 January 1994, file 293, X-33 Archive; Ben Iannotta, "DC-X Hangs by Thin Thread Despite Short-term Reprieve," *Space News* (7–13 February 1994): 4; Iannotta, "Pentagon Frees Funds for More DC-X Flights," *Space News* (9–15 May 1994): 4; Warren E. Leary, "Rocket: Program Faces Budget Ax," *New York Times* (31 January 1994): 13A.

of Advanced Programs in the Office of Space Flight, although Bekey was to play a role.[41] Rather, the Administrator was reacting to a September 1992 mandate from Congress to assess national space launch requirements, particularly in light of declining federal budgets.[42]

The NASA *Access to Space Study* considered NASA, military, and commercial launch needs for the period between 1995 and 2030. It examined three different launcher alternatives ("options")[43] and strongly concluded in favor of pursuing the development of a single stage to orbit replacement for the Space Shuttle, especially because it appeared to be the best approach to reducing overall launch costs.[44] Indeed, the single stage to orbit zeal of the *Access to Space* team was so strong that they proposed a NASA technology development program using an "X" vehicle—the X-2000 (for the program's final year of operation)—to be built entirely by NASA with joint funding from the Pentagon. The X-2000, not by chance, closely resembled the Phase III vehicle of the Delta Clipper program.[45]

NASA, however, was not going to build the X-2000. In April 1994, the White House released a draft National Space Transportation Strategy that made NASA "the lead agency for technology development and demonstration for advanced next generation reusable launch systems."[46] It also decreed, in section III, paragraph 2(b): "Research shall be focused on technologies to support a decision, no later than December 1996, to proceed with a subscale flight demonstration which would prove the concept of single-stage to orbit."[47] In this way, the new space transportation policy committed NASA to the development of reusable and single stage to orbit space launch vehicles.

Because that policy designated NASA as the lead agency for reusable launchers and the Department of Defense as the lead agency for expendable

41. Ivan Bekey interview, tape recording and transcript, NASA Headquarters, Washington, DC, by Butrica, 2 March 1999, NASA Historical Reference Collection, Washington, DC.

42. U.S. House of Representatives, *Conference Report*, 102nd Cong., 2nd sess., Report 102-902 (Washington, DC: GPO, 1992), pp. 69–70.

43. Arnold D. Aldrich and Michael D. Griffin to Daniel S. Goldin, "Implementation Plan for 'Access to Space' Review," 11 January 1993, file 197, X-33 Archive; Office of Space Systems Development, NASA, "Access to Space Study: Summary Report," January 1994, pp. 2–5, 8–58, file 100, X-33 Archive; Access to Space Study Advanced Technology Team, "Final Report," vol. 1, "Executive Summary," July 1993, pp. iii, 38, file 85, X-33 Archive. According to Bekey in the aforementioned interview, the study initially was to compare Space Shuttle upgrades and a new expendable, or partially reusable, launcher. These alternatives ultimately became Option 1 and Option 2.

44. Bekey interview.

45. Ben Iannotta, "Winged X-2000 Project Considered," *Space News* (15–28 November 1993): 14; "Single Stage to Orbit Advanced Technology Demonstrator (X-2000)," briefing, August 1993, file 122, X-33 Archive; "Single Stage to Orbit: Advanced Technology Demonstrator: SSTO Concept Proposal, X-2000," August 1993, file 162, X-33 Archive.

46. Draft, National Space Transportation Strategy, April 26, 1994, file 153, X-33 Archive.

47. Cited in NASA news release 95-1, 12 January 1995.

systems,[48] the DC-X was transferred to NASA, where it formed the initial component of the Agency's Reusable Launch Vehicle (RLV) Program. While NASA's DC-XA (where "A" stood for Advanced) tested certain key operational concepts, such as a critical rotational maneuver and a 72-hour turnaround time, the vehicle also was a technology demonstrator.[49]

In addition to the DC-XA, NASA's new RLV Program consisted of two additional "X" vehicles. One, the X-34, also known as the Reusable Small Booster Program, would demonstrate certain technologies and operations useful to smaller reusable vehicles launched from aircraft. Among those were autonomous ascent, reentry, and landing; composite structures; reusable liquid-oxygen tanks; rapid vehicle turnaround; and thermal-protection materials.[50] The other was the X-33, known also as the Advanced Technology Demonstrator Program, which proved far more challenging technologically. Among the operations and technologies it would demonstrate were reusable composite cryogenic tanks, graphite composite primary structures, metallic thermal-protection materials, reusable propulsion systems, autonomous flight control, and certain operating systems, such as electronics for monitoring vehicle hardware.[51]

The X-33 program experienced insurmountable difficulties. After seeming to overcome weight and control problems, the X-33 project encountered one delay after another because of complications and obstacles encountered in the design and construction of the linear aerospike engines and the construction and testing of the composite liquid-hydrogen tanks. The vehicle's launch was postponed from the original March 1999 date to sometime in 2003. However, with program expenditures totaling over $1.4 billion, construction of the vehicle halted and the components were divided up among NASA and the contractors.[52]

---

48. Department of Defense, "Space Launch Modernization Plan: Executive Summary," April 1994, p. 29; Iannotta, "Congress, NASA Dueling Over Reusable Rocket Management," *Space News* (23–29 May 1994): 25.

49. After the death of General Graham, the DC-XA took on the name Clipper Graham. The DC-XA differed from the DC-X in six main areas: 1) a switch from an aluminum oxygen tank to a Russian-built aluminum–lithium alloy cryogenic oxygen tank with external insulation, 2) an exchange of the aluminum cryogenic hydrogen tank for a graphite-epoxy composite liquid-hydrogen tank with a low-density reinforced internal insulation, 3) a graphite-epoxy composite intertank structure, 4) a graphite-epoxy composite feedline and valve assembly, 5) a gaseous-hydrogen and -oxygen auxiliary power unit to drive the hydraulic systems, and 6) an auxiliary propulsion system for converting liquid hydrogen into gaseous hydrogen for use by the vehicle's reaction control system. See Delma C. Freeman, Jr., Theodore A. Talay, and R. Eugene Austin, "Reusable Launch Vehicle Technology Program," IAF 96-V.4.01 (paper read at the 47th International Astronautical Congress, Beijing, China, 7–11 October 1996), p. 3, file 92, X-33 Archive.

50. John W. Cole, "X-34 Program," in "X-33/X-34 Industry Briefing, October 19, 1994," file 12, X-33 Archive, especially slide 1A-1216.

51. X-33 announcement in *Commerce Business Daily* (29 September 1994), file 276, X-33 Archive.

52. Several other serious troubles emerged along the way, but I have mentioned only the best known of the numerous X-33 problems. See NASA news release 00-157, 29 September 2000;

continued on the next page

This artist's concept shows the X-33 Advanced Technology Demonstrator, a subscale prototype reusable launch vehicle (RLV), in its 1997 configuration. Named the VentureStar™, this vehicle was to have been manufactured by Lockheed Martin's "skunk works." The VentureStar™ was one of the earliest versions of the RLVs developed in an attempt to replace the aging Shuttle fleet. The X-33 program was discontinued in 2001 without flight. *(NASA MSFC image no. MSFC-9711197)*

Shortly after the start of the RLV Program, NASA also initiated the Pathfinder and Trailblazer programs to develop low-cost reusable space transport. Pathfinder involved technology experiments conducted on existing flight vehicles, such as the Space Shuttle. Trailblazer, on the other hand, entailed the construction of entirely new "X" vehicles to demonstrate advanced space transport technologies and operations. In August 1998, NASA solicited proposals for Future-X, the first of the Trailblazer vehicles,[53] and, in December, announced that it had entered into negotiations with

continued from the previous page

"Development Troubles Push First X-33 Flight Back to July '99," article 34208 in *Aerospace Daily* (24 June 1997, electronic edition), hard copy in file 225, X-33 Archive; Brian Berger, "Activists Say Lockheed Should Not Compete for X-33 Funds," *Space News* 11 (16 October 2000): 21.

53. NASA news release 98-141, 3 August 1998.

Boeing to design and build the Advanced Technology Vehicle (ATV), the first "X" vehicle to fly in orbit and to reenter the atmosphere.[54]

The Advanced Technology Vehicle soon became the X-37. The Shuttle would carry the craft into space, then release it. The X-37 would orbit the planet, then return to Earth through the atmosphere, testing heat shielding and other advanced space materials and technologies. The vehicle's shape derived from that of the X-40A, an unpowered Air Force craft designed and built by Boeing's Phantom Works. In August 1998, the Air Force drop-tested the X-40A from an Army Black Hawk helicopter above Holloman Air Base, New Mexico, and the vehicle landed under remote control on a runway. The Air Force provided partial funding for the X-37 in the hope of realizing some of the objectives of its Space Maneuver Vehicle (SMV), a reusable winged craft capable of deploying satellites, weapons, and antisatellite devices; inspecting enemy satellites; and other military missions. The Space Maneuver Vehicle could have remained in orbit for up to a year and would have been capable of a 72-hour turnaround.[55]

No discussion of NASA's reusable "X" vehicles would be complete without at least a mention of the defunct Crew Recovery Vehicle (CRV), which would have served as a lifeboat for the International Space Station (ISS). Drop tests of the X-38, an experimental 80-percent scale version of the vehicle, at increasing altitudes from a B-52 began in 1999. The basic design for the X-38 and CRV originated at NASA's Langley Research Center as the HL-10 (Horizontal Lander) lifting body. The initial HL-10 design derived from photographs of the BOR-4 (Unpiloted Orbital Rocketplane in Russian), a Russian reusable rocket, that had landed in the Indian Ocean. Renamed the HL-20 by NASA Headquarters, the vehicle concept subsequently became popular in NASA launcher studies.[56]

---

54. NASA news release c98-w, 8 December 1998.

55. NASA news release 99-139, 14 July 1999; Frank Sietzen, Jr., "Air Force's Needs Shape Newest NASA X Rocket," *Space.com*, 25 August 1999, *http://www.space.com/businesstechnology/ business/x37_briefing.html*, hard copy in file 386, X-33 Archive; "Space Maneuver Vehicle Drop Test Planned for Early August," article 110718 in *Aerospace Daily* (21 July 1998, electronic edition), hard copy in file 226, X-33 Archive; "USAF Sets Aug. 4 Test of Space Maneuver Vehicle," article 111407 in *Aerospace Daily* (30 July 1998, electronic edition), hard copy in file 226, X-33 Archive; "Competition Likely for Space Maneuver Vehicle Demonstrator," article 111904 in *Aerospace Daily* (6 August 1998, electronic edition), hard copy in file 226, X-33 Archive.

56. Theodore A. Talay interview, tape recording and transcript, NASA Langley Research Center, by Butrica, 21 May 1997, NASA Historical Reference Collection, Washington, DC; Doug Stanley interview, tape recording and transcript, Orbital Sciences Corporation, Dulles, VA, by Butrica, 25 February 1999, NASA Historical Reference Collection, Washington, DC; "NASA's X-38 Station Lifeboat Testbed Completes a Drop Test," article 124222 in *Aerospace Daily* (9 February 1999, electronic edition), hard copy in file 386, X-33 Archive; Andrew Bridges, "Space Station Lifeboat Sails to Success in Desert Test," *SpaceViews* (2 November 2000), hard copy available in file 854, X-33 Archive.

Commercial Launchers

NASA and the Air Force were not the only developers of reusable launchers during the 1990s. As the global market for satellite launches grew throughout the decade, small startup companies entered the field with plans for a variety of two-stage reusable vehicles. Among those was Kelly Space & Technology, initially headed by Michael S. Kelly. Starting in 1993, with funding from NASA and the Air Force, the firm began developing the Astroliner, a reusable glider towed to launch altitude by a Boeing 747 aircraft using patented Eclipse towing technology. An expendable stage launched from the Astroliner would place payloads in orbit. Subsequently, Kelly received NASA funding to develop its reusable launcher.[57]

A comparable two-stage system that combined a reusable first stage with a throwaway second stage was Pioneer Rocketplane's Pathfinder. The two-seat Pathfinder aircraft powered by air-breathing and (RD-120) rocket engines would have taken off from Vandenberg AFB, taken on additional liquid oxygen in midair from a Boeing 747 freighter, then climbed outside the atmosphere, where it would release an upper stage and its payload, then reenter the atmosphere and land like an aircraft.[58] Pursuing development of a different two-stage launch system known as the K-1 is the Kistler Aerospace Corporation. The K-1 was an unpiloted vehicle powered by surplus Russian NK-33 and NK-43 engines. It would launch vertically and be capable of a turnaround of nine days. A system of parachutes and air bags (field-tested in 1998) would allow the company to recover and reuse both the booster and orbital stages.[59]

The only single stage to orbit vehicle under commercial development—Rotary Rocket Company's Roton—also was the only one that did

---

57. Kelly news releases for 7 October 1996, 22 May 1997, and 2 February 1998, file 373, X-33 Archive.

58. "RLV Startups Have Enough Capital, But Worry About Regulation," article 37503 in *Aerospace Daily* (13 February 1998, electronic edition), hard copy in file 226, X-33 Archive; "Rocketplane System," *Pioneer Rocketplane* Web site, *http://www.rocketplane.com*.

59. "RLV Startups Have Enough Capital"; "Kistler May Shift Flight Tests to Australia," article 37615 in *Aerospace Daily* (23 February 1998, electronic edition), hard copy in file 226, X-33 Archive; "Developments in the Field of Space Business are Briefly Noted," article 109711 in *Aerospace Daily* (7 July 1998, electronic edition) and article 111101 (27 July 1998), hard copies in file 226, X-33 Archive; Frank Morring, Jr., "Tight Money Forces Slowdown at Kistler Aerospace," article 122111 in *Aerospace Daily* (8 January 1999, electronic edition), hard copy in file 386, X-33 Archive; "Northrop Grumman Increases Stake in Kistler's K-1 Vehicle," article 127002 in *Aerospace Daily* (22 March 1999, electronic edition), hard copy in file 386, X-33 Archive; "Kistler Has a Line on Remaining Financing, But Much Rests on Contingent Funds," article 132104 in *Aerospace Daily* (2 June 1999, electronic edition), hard copy in file 386, X-33 Archive; "NASA Taps Kistler to Evaluate ISS Access Options," article 163106 in *Aerospace Daily* (28 August 2000, electronic edition), hard copy in file 854, X-33 Archive; additional materials in file 179, X-33 Archive.

not receive NASA funding. The firm's founder, Gary Hudson, with funding from the private sector, has pursued single stage to orbit concepts since the 1980s. A staunch believer in private enterprise, Hudson received substantial backing for the Roton from author Tom Clancy, along with other investors. Like the Delta Clipper, the Roton would take off and land vertically but would use rocket-powered rotors for the final descent and touchdown, much like a helicopter.[60]

Analysis of a Perennial Debate

The quest for reusability certainly has had its losses, mistakes (NASP), overly ambitious projects (X-33), and seemingly fruitful routes taken but abandoned (Dyna-Soar, RASV). Success has been partial for three major reasons: 1) the major technological challenges of achieving full reusability and "aircraft-like" operations; 2) the lack of an ongoing technology development program; and 3) the toll on the search for a new launch system taken by past space policy and political decisions. Current policy does not redress these issues, but rather appears to exacerbate, not assuage, them.

POLICY

The Era of Space Transportation

Space transportation policy obviously did not begin to include reusable launch vehicles until reusable launchers were about to become a reality. The evolution of launchers as a means for transporting people was gradual, beginning with the recoverable, but not reusable, craft used for the Mercury and Gemini missions.[61] Similarly, the means for transporting astronauts to the Moon were the recoverable, single-use Apollo spacecraft. These vehicles differed from ordinary transportation in that they could not be used more than once. Aircraft, for instance, can fly over and over again, and that reusability is an essential characteristic of any form of transportation. We therefore can think of the advent of the Space Shuttle as ushering in a new era or phase of space history, as well as a new period of space policy that would address issues related to space transportation.

In this new era, everything—whether reusable or expendable—that carried a payload conceptually was transportation. The Shuttle held a privileged

---

60. Materials relating to Gary Hudson and the Roton rocket are in file 348, X-33 Archive.

61. Starting in 1959, the Air Force's ASSET (Aerothermodynamic/elastic Structural Systems Environmental Tests) boost-glide system involved lofting small, reusable hypersonic gliders from Cape Canaveral on top of expendable rockets. The gliders were recovered, and though they potentially were reusable, none ever flew more than once. See Hallion, "ASSET: Pioneer of Lifting Reentry," in *The Hypersonic Revolution*, ed. Hallion, vol. 1, pp. 449–450, 510, 512–513, 515–516, 518, 523–524.

place in the constellation of space transporters. It was not only the only reusable launch vehicle, but also *the* Space Transportation System (STS). Despite the de facto mix of expendable and reusable launchers, government policy leaned toward domination by the reusable Space Shuttle. Driving this policy were claims and assurances—made as early as the 1960s[62]—that the Shuttle would be a low-cost, reliable launcher (a space "bus" or space "truck"). In addition, NASA aggressively marketed the Space Shuttle as a vehicle that could place any satellite into orbit.[63] Ironically, the Shuttle would not only inspire and empower space policy, it would impede it as well.

President Ronald Reagan made this "one-size-fits-all" strategy national policy through National Security Decision Directive 8, "Space Transportation System," dated 13 November 1981. It stated, succinctly, that "the STS will be the primary space launch system for both United States military and civil government missions." Moreover, its language, that the Shuttle would "service all authorized space users," left the door open for a subsequent enlargement of this basic space policy.

The issuance of National Security Decision Directive 42, "National Space Policy," on 4 July 1982, reiterated the "one-size-fits-all" policy and, more importantly, defined the "authorized space users" of the Space Shuttle as "domestic and foreign, commercial, and governmental."[64] In effect, the new space policy called for making the Shuttle available to all commercial users, provided no conflicts with national security resulted. The directive marked a dramatic policy shift, indeed, a redefinition of space policy, not seen since the launch of Sputnik in 1957, because for the first time in the history of the U.S. space program, a high-level official document made a direct reference to the American business community.[65] Between November 1982 and January 1986, the Space Shuttle carried 24 communication satellites into orbit on 11 flights. Five were for private corporations: Westar 6, two Telstars, and two SATCOMs. Others were for foreign clients, including Canada (four Aniks),

---

62. *The Post-Apollo Space Program: A Report for the Space Task Group* (Washington, DC: NASA, September 1969), pp. 1, 6.

63. Hans Mark, *The Space Station: A Personal Journey* (Durham: Duke University Press, 1987), pp. 61–65; Heppenheimer, *Space Shuttle Decision*, pp. 275–280; David M. Harland, *The Space Shuttle: Roles, Missions and Accomplishments* (Chichester, U.K.: Praxis Publishing, Ltd., 1998), pp. 411–412.

64. Christopher Simpson, *National Security Directives of the Reagan and Bush Administrations: The Declassified History of U.S. Political and Military Policy, 1981–1991* (Boulder, CO: Westview Press, 1995), pp. 136–143 (classified version) and pp. 144–150 (unclassified version); "National Space Policy," 4 July 1982, file 386, X-33 Archive. An NSDD 42 innovation of at least equal significance was the establishment of the National Security Council Senior Interagency Group (Space), usually referred to as simply SIG (Space), as the primary forum for the formulation of space policy. Chaired by the Assistant to the President for National Security Affairs, SIG (Space) was the locus of policy-making throughout the two terms of Ronald Reagan's presidency.

65. W. D. Kay, "Space Policy Redefined (Again)," chap. 7 in *Defining NASA: The Historical Debate over the Agency's Mission* (Albany: State University of New York Press, 2005).

Australia (two AUSSATs), Indonesia (two Palapas), India (INSAT), and Saudi Arabia (ARABSAT).[66]

The 1972 decision by President Richard Nixon to build the Space Shuttle short-circuited debate on the desirability of investing in new expendable launch vehicles and facilities and froze them in 1970s technologies. NASA no longer ordered Delta or Atlas launches, and the Air Force began shutting down production lines for the Titan.[67] Expendable launch systems began to age and became increasingly expensive to build and operate (which added to the cost of military and NASA space programs) because needed improvements in launch technology had been set back some two decades. The Shuttle already was expensive to operate and soon would show its grounding in yesterday's technology. Space transportation came to be perceived as consuming too large a share of the federal budget, thereby shutting out opportunities for new science and technology initiatives. Eventually, the government would have to spend over $12 billion to restore abandoned ELV operations and to transfer satellites designed for the Shuttle back to these aging launchers.[68]

A Mixed Fleet

National space transportation policy, however, soon crashed on the rocks of reality—and on the launchpad. Following a launch failure of a Titan 34D on 28 August 1985, the Air Force temporarily suspended Titan launches until after an investigation.[69] Five months later, the *Challenger* accident, on 28 January 1986, grounded the STS for two years, a watershed moment for the U.S. space program, for NASA, for the Department of Defense, and for space commerce. What made the accident so damaging, aside from the loss of human life, was the policy that placed NASA, military, and commercial payloads aboard the Shuttle. The dependence on the Space Shuttle as the nation's "primary" launch system impaired the ability of the nation's defense and intelligence agencies to place payloads into orbit, and it stymied the development of a commercial launch industry which had been struggling against both the Shuttle and its European ELV competitor, Ariane.

---

66. Dennis R. Jenkins, *Space Shuttle: The History of Developing the National Space Transportation System* (Marceline, MO: Walsworth Publishing Co., 1992), pp. 286–287.

67. For a discussion of the process leading up to Nixon's decision, see Heppenheimer, *Space Shuttle Decision*; Dorsey Oles Boyle, "The Nixon Space Policy, 1969–1974" (M.A. thesis, University of Maryland at Baltimore County, 1993).

68. National Space Council, "Final Report to the President on the U.S. Space Program," January 1993, pp. 5, 33, file 017, box 1, X-33 Archive; John M. Logsdon and Craig Reed, "Commercializing Space Transportation," in *Exploring the Unknown: Selected Documents in the History of the U.S. Civil Space Program*, ed. John M. Logsdon, vol. 4, *Accessing Space* (Washington, DC: NASA SP-4407, 1999), pp. 408–409.

69. William Boyer, "Titan IV Explosion Halts Launch Program," *Air Force Times* 54 (16 August 1993): 33.

Shortly after the *Challenger* tragedy, additional expendable launcher failures took place. A more disastrous Titan 34D launch accident on 18 April 1986 effectively grounded military space operations on both coasts until the military and industry could ensure the Titan's reliability. The rocket exploded only 8 seconds after lifting off. Upper sections of its solid rockets and fuel showered the launchpad, causing severe damage to nearby launch facilities. In some instances, large steel fragments were blown 3,000 feet from the explosion, which also created a toxic cloud that rose to an altitude of 8,000 feet before being blown over the Pacific Ocean. The following month, on 3 May 1986, a Delta carrying the $57-million GOES-G weather satellite broke up about 90 seconds after liftoff from Cape Canaveral, Florida. The root cause of the failure (a lightning strike) needed to be determined before more Deltas could fly.

The lessons learned (or that ought to have been learned) from these various launch accidents were that NASA needed to reduce its dependence on the Space Shuttle and that the nation needed a variety of launchers, both reusable and expendable, as well as a variety of disposable rockets. Collectively, these incidents brought home the dangers of relying on one or two launch systems. Subsequently, National Security Decision Directive 254, "United States Space Launch Strategy," 27 December 1986, took NASA and the Space Shuttle out of competition with potential commercial launch providers. Specifically, the directive stipulated that "NASA shall no longer provide launch services for commercial and foreign payloads subject to exceptions for payloads that: (1) are Shuttle-unique; or (2) have national security or foreign policy implications." By "Shuttle-unique," the directive meant payloads requiring either human intervention or facilities available only on the Space Shuttle.[70]

President Reagan approved a revised national space policy on 5 January 1988. It too overthrew the long-standing notion of the Shuttle as the nation's "primary" launch system and established the de facto mixed fleet of launchers as policy.[71] Essentially, NASA henceforth would use the (partially) reusable Space Shuttle, and the Department of Defense would rely on expendable

---

70. James A. Baker III to Economic Policy Council, "Presidential Policy Directive—Space Commercialization," 6 October 1986, file 387, X-33 Archive; "Presidential Directive on National Space Policy," fact sheet, 11 February 1988, p. 9, file 386, X-33 Archive.

71. The classified space policy was not released until 11 February 1988, following completion of a parallel review of commercial space policy being conducted by the Economic Policy Council. See NSDD 293, "Presidential Directive on National Space Policy," 5 January 1988; "Presidential Directive on National Space Policy," fact sheet, 11 February 1988, pp. 7, 9–10, file 386, box 15, X-33 Archive. The essential parts of the 1988 national space policy that attempted to foster a domestic launch industry made their way into legislation as the Commercial Space Launch Act Amendments of 1988 (Public Law No. 100-657). See Kim G. Yelton, "Evolution, Organization, and Implementation of the Commercial Space Launch Act and Amendments of 1988," *Harvard Journal of Law & Technology* 4 (1989): 34.

launchers.[72] This institutional division between expendable and reusable launchers based on whether or not the launcher carried humans remained in effect over the following years, buttressed by intervening space policy declarations, despite partisan and ideological changes in White House leadership. The policy was based not on any study of expendable versus reusable launch vehicles, but on the exigencies of national security and the promotion of (space) business, not to mention the underlying assumption (and fact) that the only "human-rated" launcher was the partially reusable Space Shuttle.

A New World (Dis)Order?

The period of George H. W. Bush's presidency, 1989–1993, was marked more by change than by continuity with the past. The biggest change—the winding down of the decades-long Cold War—had many consequences for space transportation, especially for the use of reusable and expendable launchers, as well as for the federal budget, the economy, and strategic planning. For starters, the budget reality that emerged at the end of the Cold War meant that fewer government dollars were available for space transportation. The government would have to find cheaper ways to launch payloads. The pressure to reduce launch costs was reflected in the December 1992 study "A Post Cold War Assessment of U.S. Space Policy." It called for the scaling back of all NASA, Defense Department, and Department of Energy space facilities, whether operated by the government or a contractor; the elimination of all duplication within governmental agencies with space programs; and the formation of a nonpartisan commission modeled after the Base Closure Commission to suggest consolidation measures.[73]

The end of the Cold War also raised new questions about the usefulness of President Reagan's quixotic Strategic Defense Initiative, which had its own launcher needs. Additionally, with the Soviet Union no longer a military foe, to what extent was it now feasible (or legal) for the United States government and launch industry to acquire Russian technology, such as rocket engines, or even Russian launchers? By the end of George H. W. Bush's presidency, space policy also began to accommodate new space launch trade agreements with Russia as well as China.[74]

---

72. There have been exceptions. Typically, Defense Support Program (DSP) satellites are launched into geosynchronous orbit by a combination of a Titan IV booster and an Inertial Upper Stage. However, one DSP satellite was launched using the Space Shuttle on mission STS-44 (24 November 1991). Also, policy excluded NASA specifically from maintaining its own expendable launchers. If the Agency wanted to launch on an ELV, it would have to turn to the Pentagon or industry.

73. Vice President's Space Policy Advisory Board, "A Post Cold War Assessment of U.S. Space Policy," December 1992, pp. 39–43, file 016, box 1, X-33 Archive.

74. See National Space Policy Directive 2, "Commercial Space Launch Policy," 5 September 1990, in National Space Council, "Final Report to the President on the U.S. Space Program."

Similarly, a surfeit of now-useless missiles and hardened silos became available for nonmilitary uses. Could those Minuteman II ICBMs be used to conduct scientific research, as the United States had done with V-2 rockets brought back from Germany after World War II?[75] That is exactly what the Universities Space Research Association wanted to do with the surplus missiles. Specifically, the association proposed conducting a pilot program to demonstrate low-cost, short-duration, small scientific satellite missions in support of university research and technology development. The initial problem was getting the missiles transferred from the military to NASA.[76]

Into this mix of questions and problems President Bush threw a new space program that would require the development of its own launch system. The Space Exploration Initiative (SEI) was a grandiose plan to return to the Moon, set up a lunar base, and send astronauts to Mars by 2019. Like space station *Freedom*, it would require development of a heavy-lift expendable rocket.[77] As a result, both NASA and the Defense Department were in the market for an expendable launcher, but the Senate Commerce Committee essentially zeroed out its funding before the program even began.[78]

In addition to supporting the development of medium- and heavy-lift ELVs by and for both NASA and the Defense Department, the Bush administration funded two programs to create innovative reusable launch vehicles: the National Aero-Space Plane and the SDIO's Single Stage to Orbit Program (DC-X). Both were the most technologically challenging kind of reusable transport to build: single stage to orbit launchers. Technological change generally occurs incrementally, not in giant leaps, and an operational single stage to orbit vehicle is too much of a leap. To date, no single stage to orbit craft has taken off or landed on this planet. These launchers likely will remain in the domain of science fiction and fantasy for a long time into the future, like the *Star Trek* transporter or the Stargate.

---

75. See David H. DeVorkin, *Science with a Vengeance: How the Military Created the U.S. Space Sciences after World War II* (New York: Springer, 1992); William R. Corliss, *NASA Sounding Rockets, 1958–1968: A Historical Summary* (Washington, DC: NASA SP-4401, 1971).

76. Materials in file 130, box 5, X-33 Archive, relate to the use of excess DOD ballistic missiles by the Universities Space Research Association.

77. Synthesis Group on America's Space Exploration Initiative, "America at the Threshold: America's Space Exploration Initiative," n.d., p. 31, file 104, box 4, X-33 Archive; William Piland, "Space Transportation in the Future: Practical Considerations," presentation to Access to Space Red Team, 4 June 1992, file 430, box 16, X-33 Archive. Piland pointed out that SEI could be accomplished with existing ELVs and the Russian Energia launcher. National Space Policy Directive 6, Space Exploration Initiative Strategy, dated 13 March 1992, dealt with Bush's SEI. A copy is in National Space Council, "Final Report to the President on the U.S. Space Program," appendix III, "National Space Policy Directives."

78. Lyn Ragsdale, "Politics Not Science: The U.S. Space Program in the Reagan and Bush Years," in *Spaceflight and the Myth of Presidential Leadership*, ed. Launius and Howard E. McCurdy (Urbana: University of Illinois Press, 1997), pp. 161, 163–164.

The Vision Thing

The undertaking of these single stage to orbit, as well as expendable launch vehicle, programs required for the Strategic Defense Initiative, the Space Exploration Initiative, and Space Station *Freedom* shaped space transportation policy during George Bush's presidency. In addition, the search for a Space Shuttle replacement continued, and the nation's aging launchers and launch facilities—the heritage of the "one-size-fits-all" Shuttle policy—demanded attention.[79] The basis for the institutional division that made NASA responsible for reusable launchers and the Defense Department responsible for single-use rockets continued to be the implicit assignment of the role of human spaceflight to NASA and its Space Shuttle.[80] In the future, however, those roles might change, as reusable launchers began to supply the nation's launch needs.

Bush's National Space Launch Strategy, released 24 July 1991, laid the groundwork for that change to take place. The strategy charged the Defense Department and NASA with joint development, funding, and management of a new suite of expendable rockets capable of lifting medium and heavy pay-loads for both civil and military use and set the first flight of the new system for 1999. Reflecting the stringent budgetary environment and the new direction of space commercialization, the space launch strategy called for the two agencies to explore potential participation by the private sector.[81] The 10-year space launch technology plan mandated by the space launch strategy, issued in October 1991 by NASA and the Departments of Defense and Energy, painted a picture of what the nation's fleet of launchers would look like a decade later, as well as the technologies needed to get there.

By then, the United States would have a new family of expendable launch-ers, known as the National Launch System (NLS), including a heavy-lift rocket for the Space Exploration Initiative. Reusable launchers continued to be the technological system of choice for human spaceflight, although the expend-able launchers under development would have the capability and high reliabil-ity required to boost a crew into orbit as part of a Space Shuttle–replacement launch system. Starting in 2005, Reusable Aerospace Vehicles, in the language

---

79. See, for example, National Research Council, *From Earth to Orbit: An Assessment of Transportation Options* (Washington: National Academy Press, 1992), p. 3, copy available in file 102, box 4, X-33 Archive.

80. "National Space Policy," 2 November 1989, file 374, box 15, X-33 Archive; White House, Office of the Press Secretary, "U.S. National Space Policy," fact sheet, 16 November 1989, file 374, box 15, X-33 Archive; National Space Policy Directive 3, "U.S. Commercial Space Policy Guidelines," issued 12 February 1991 in National Space Council, "Final Report to the President on the U.S. Space Program," appendix III, "National Space Policy Directives."

81. Interagency Working Group on Space Transportation, "Current National Space Policy on Space Transportation," p. 2; National Space Policy Directive 4, "National Space Launch Strategy," in National Space Council, "Final Report to the President on the U.S. Space Program," appendix III, "National Space Policy Directives," pp. III-27–III-28.

of the plan, would complement and later replace the Shuttle. The plan included a reusable military launcher known as the Military Aerospace Vehicle, which also would be operable around 2005, just in time to replace the Space Shuttle. Initially, a robotic version of the craft could be launched to address commercial launch needs, and a later version could be equipped to carry a crew. By merging NASA, military, and commercial launch needs, the 10-year plan envisioned the possibility of a low-cost-per-flight reusable vehicle that would satisfy all of the nation's launcher needs.[82] In effect, the plan for implementing Bush's launcher strategy would have committed the same mistake as his predecessor's space policy, which put all of its launch eggs in a single, reusable basket.

The NASA *Access to Space Study*

The election of William Jefferson Clinton as President in November 1992 opened the door to a significant change in launcher policy. The new Democratic administration would want to shape space policy to suit its own agendas, which were certain to be different from those of its Republican predecessors. Three studies formed the basis for the new space transportation policy, and they came to different conclusions about the future of reusable launchers, especially single stage to orbit rockets. The most important of those was NASA's *Access to Space Study*. Mandated by the House Subcommittee on Space of the Committee on Science, Space, and Technology in 1992, *Access to Space* focused on future launch systems, analyzed the launcher needs of NASA, Defense, and industry, and developed various alternatives for addressing those needs for the period 1995 to 2030.[83]

Option 1 involved retaining the Space Shuttle until 2030. The Option 1 team endorsed fresh studies of flyback, fully reusable liquid-fueled Shuttle boosters in order to increase safety and to reduce costs. Option 2 replaced the Shuttle in 2005 with a new expendable launcher using state-of-the-art technology. Option 3 was more daring. It would replace the Space Shuttle in 2030 with "an unspecified . . . next-generation, advanced technology system . . . a 'leapfrog' approach, designed to capitalize on advances made in the NASP and SDI [the DC-X] programs to achieve order-of-magnitude improvements in the cost effectiveness of space transportation."[84]

---

82. National Space Council, "Ten-Year Space Launch Technology Plan," October 1992, pp. ES-1, ES-2, 1-1, 2-8, 2-10, 2-11, file 103, box 4, X-33 Archive.

83. Ivan Bekey, "Access to Space," IAF-94-V.1.515 (paper read at the 45th Congress of the International Astronautical Federal, Jerusalem, Israel, 9–14 October 1994), p. 3, copy available in file 098, box 3, X-33 Archive, summarizes the *Access to Space Study* more succinctly than the study's own executive summary.

84. Office of Space Systems Development, NASA Headquarters, "Access to Space Study: Summary Report," January 1994, p. 71, file 100, box 3, X-33 Archive; Arnold D. Aldrich, "NASA's

continued on the next page

The Option 3 team considered three launcher architectures. The first was a rocket-powered SSTO ship. The second was a single stage to orbit craft powered by a combined rocket and air-breathing propulsion system. A combination of rocket and air-breathing engines propelled the third architecture, which was a two stage to orbit launcher. As part of the Option 3 study, the team specifically compared a generic rocket-powered single stage to orbit launcher with the NASP, looking at such factors as cost, risk, and development schedule. They concluded against NASP and all other air-breathing vehicles because their technological difficulty would drive up costs and require a longer period of development. The Option 3 team report concluded that reusable launchers could replace medium-load throwaway rockets, leaving expendable launchers to lift heavy payloads in the short term, and that in time, reusable vehicles would replace even those.[85]

Once each team selected the best vehicle design from the range of alternatives considered, the *Access to Space Study* then compared all of the winning designs. This comparison necessarily included weighing expendable rockets against reusable launchers. The study concluded that the most beneficial option was to develop and deploy a fleet of fully reusable, rocket-powered single stage to orbit vehicles and recommended phasing out current throwaway rockets—as well as the Shuttle—beginning around 2008. The new reusable launch vehicles would be able to accommodate all conceivable NASA, military, and commercial payloads, and—despite their need for a large upfront investment, especially in technological development—they would cut government launch costs by up to 80 percent while increasing vehicle reliability and safety by about an order of magnitude.[86]

After the *Access to Space Study*, several of the NASA officials involved in it began to proselytize their belief in the near-term feasibility of SSTO rockets in various venues, including such popular journals as *Aerospace America*.[87] Furthermore, the Space Frontier Foundation—dedicated to human colonization

---

continued from the previous page
Access to Space Study," 21 November 1993, file 101, box 4, X-33 Archive; Arnold D. Aldrich and Michael D. Griffin to Daniel S. Goldin, "Implementation Plan for 'Access to Space' Review," 11 January 1993, file 197, box 7, X-33 Archive. According to Ivan Bekey in his interview cited earlier, the study initially was to compare Shuttle upgrades and a new expendable, or partially reusable, launcher. These alternatives ultimately became Option 1 and Option 2.

85. Access to Space Advanced Technology Team, "Final Report," vol. 1, "Executive Summary," July 1993, file 85, box 3, X-33 Archive; Office of Space Systems Development, NASA Headquarters, "Access to Space Study: Summary Report," p. 71; "Integration of Existing ELVs into the Option 3 Architecture," in Access to Space Advanced Technology Team, "Final Report," vol. 1, "Executive Summary."

86. An order of magnitude is a tenfold increase. See Aldrich, "NASA's Access to Space Study," pp. 8–12.

87. See, for example, Bekey, "SSTO Rockets: A Practical Possibility," *Aerospace America* 32 (July 1994): 32–37; Robert E. Austin and Stephen A. Cook, "SSTO Rockets: Streamlining Access to Space," *Aerospace America* 32 (November 1994): 34; and Austin, "Studies Show SSTO Plan is Feasible," *Space News* (31 July 31–6 August 1995), among others in file 352, box 14, X-33 Archive.

of space—organized a congressional briefing in the spring of 1996 that they called Cheap Access to Space. The message to Congress was to support single stage to orbit vehicle programs as the only way to get low-cost space launchers, and in particular to fund the DC-X (then a NASA program) and NASA's X-33. With generous funding from NASA Headquarters, the foundation organized the Cheap Access to Space symposium in July of 1997 with the same message.[88]

Defense Department Studies

The NASA *Access to Space* enthusiasm for reusable and single stage to orbit rockets was missing from the two Defense Department studies that contributed to the formulation of Clinton administration space transportation policy. Instead, they proposed to keep launching the existing disposable rockets. Such, for instance, was the conclusion of the so-called "Bottom-Up Review." Completed in 1993, the "Bottom-Up Review" of military launchers, like NASA's *Access to Space*, considered three alternatives. Alternative 1 was to extend the life of current military expendable rockets, while Alternative 2 was to develop a new launch system. Alternative 3 funded the development of advanced reusable launch vehicle technologies and maintained current expendable launchers until the Pentagon could switch to reusable vehicles. Alternative 3 evaluated four reusable launcher concepts chosen for their level of increasing technological complexity, ranging from a flyback first stage to a fully reusable two stage to orbit craft, plus two different single stage to orbit designs, one powered by rockets and the other by a combination of rockets and air-breathing engines. Ultimately, the study team eliminated Alternative 3 but shifted the SSTO rocket to Alternative 2 for consideration. Unlike NASA's *Access to Space*, the "Bottom-Up Review" did not embrace single stage to orbit rockets or reusable launchers in general. Rather, it concluded that the current fleet of expendable boosters was fulfilling the Defense Department's launcher needs and selected Alternative 1.[89]

The other key Defense Department launcher study stemmed from a congressional mandate, like NASA's *Access to Space*. Section 213 of the National Defense Authorization Act for 1994 directed the Defense Secretary to develop a plan for modernizing its launchers and launch facilities, lowering the costs of manufacturing current single-use rockets, and developing a new launch sys-

---

88. The presentations from the Cheap Access to Space congressional briefing are in file 360, box 14, X-33 Archive. Notes and other materials from the Cheap Access to Space Symposium held in Washington on 21–22 July 1997, including the $100,000 in underwriting from NASA, are in file 705, box 24, X-33 Archive, and more detailed information on the NASA underwriting is in file 842, box 28, X-33 Archive.

89. Director, Strategic & Space Systems, "Space Launch Systems Bottom-Up Review," 4 May 1993, file 233, box 8, X-33 Archive; "Executive Summary," in *Space Launch Modernization Study*, April 1994, pp. 5–6, file 142, box 5, X-33 Archive.

tem. Issued in April 1994, the *Space Launch Modernization Study*, better known as the Moorman Report after its chairman, Air Force Lieutenant General Thomas S. Moorman, Jr., considered four launcher options.[90]

Option 1 would have maintained the current fleet of ELVs—Delta, Atlas, Titan—and the Space Shuttle while NASA funded a technology program that eventually would lead to the development of a reusable launcher to replace the Shuttle. In Option 2, NASA also funded development of an RLV and continued using the Shuttle, but the current throwaway rockets were upgraded. Option 3 involved developing a new expendable launcher. One version would launch only cargo and eventually would replace current systems, while the other would carry either cargo or passengers, one day replacing both the current expendable rockets and the Space Shuttle. Option 4 involved developing a reusable vehicle in cooperation with NASA, plus setting up a government-mandated launch corporation. The arrangement would bring together public and private financing; government and contractors would share the costs.[91]

Although directed to select the "most attractive" option, the Moorman Report simply presented the four options without stating a preference for any of them.[92] Despite its apparent ambiguity, the report contained a number of suggestions that soon became part of national space policy. For instance, it recommended that NASA—because of its need to continue human space-flight and to replace the Shuttle—be assigned the lead for developing RLVs, with the Defense Department maintaining a cooperative reusable launcher program that would include experimental flight demonstrations. The X-33 program embodied that suggestion. Meanwhile, the Defense Department would take the lead in developing single-use rockets, and each agency would manage and fund efforts within their area of responsibility. That recommendation became policy. The Moorman Report, however, was not immune to the raging enthusiasm for reusable launch vehicles, especially for the growing commercial launch industry. It proclaimed that once reusable vehicles reduced launch costs by a factor of 10, they would "ignite a commercial space boom."[93] They were not alone in that belief.

---

90. "Executive Summary," in *Space Launch Modernization Study*, April 1994, pp. 1–2, 15–23, file 142, box 5, X-33 Archive; Lieutenant General Thomas S. Moorman, Jr., "DoD Space Launch Modernization Plan," briefing to the Commercial Space Transportation Advisory Committee (COMSTAC), 10 May 1994, file 588, box 29, X-33 Archive; Bekey, "Access to Space," p. 14.

91. "Executive Summary," in *Space Launch Modernization Study*, pp. 15–19; Moorman, "DoD Space Launch Modernization Plan"; Bekey, "Access to Space," p. 14.

92. Nonetheless, on the question of developing a new launcher, it recommended that the Defense Department develop a heavy-lift launcher. See "Executive Summary," in *Space Launch Modernization Study*, p. 25.

93. "Executive Summary," in *Space Launch Modernization Study*, p. 29; Moorman, "DoD Space Launch Modernization Plan."

The 1994 Space Transportation Policy

The Moorman Report, the "Bottom-Up Review," and the *Access to Space* studies quickly became the foundation for the preparation of a new space launch policy by the Clinton White House Office of Science and Technology Policy (OSTP), which had absorbed the duties of the National Space Council.[94] Its goal was to piece together a single, coherent space transportation policy[95] that addressed the various launch vehicle needs of NASA, the Pentagon, and industry, while taking into account the changing character of the era following the Cold War. Signed by President Clinton in August 1994, the new space transportation policy addressed the range of ills afflicting the country's launchers and facilities.

It ruled, for instance, on the use of excess Minuteman missiles[96] and gave Russian launch vehicles a larger role by involving that country in the space station program.[97] The policy also proposed the modernization of existing launch systems (both expendable rockets and the Shuttle) and facilities and the development of a new reusable launch vehicle that would reduce "greatly" the cost of putting payloads in orbit. In addition, it extended and expanded the standing policy of fostering the commercialization of space, as well as the international competitiveness of the U.S. commercial launch industry.[98]

---

94. The actual work of preparing the policy was carried out by the Interagency Working Group on Space Transportation. Established by the Office of Science and Technology Policy, it consisted of representatives of the various agencies with an interest in space policy: NASA; the Defense Department; the Joint Chiefs of Staff; the National Security Council; the Director of Central Intelligence; the Departments of State, Commerce, Treasury, and Transportation; the Council of Economics Advisors; the Nuclear Energy Commission; the Office of Management and Budget; the Office of the Vice President; and the United States Trade Representative. See Office of Science and Technology Policy, Executive Office of the President, "Interagency Working Group on Space Transportation Representatives," May 1994, file 147, box 5, X-33 Archive.

95. The National Space Transportation Policy replaced National Space Policy Directive (NSPD) 2, NSPD 4, and National Security Directive (NSD) 46, "Cape York," as well as the portions that pertain to space transportation of NSPD 1/NSD 30, "National Space Policy"; NSPD 3, "U.S. Commercial Space Policy Guidelines"; and NSPD 6, "Space Exploration Initiative Strategy." See Interagency Working Group on Space Transportation, "Current National Space Policy on Space Transportation," p. 1; National Space Transportation Policy, draft, 10 May 1994, file 147, box 5, X-33 Archive.

96. Office of Science and Technology Policy, White House, "Statement on National Space Transportation Policy," 5 August 1994, file 147, box 5, X-33 Archive; Presidential Decision Directive National Science and Technology Council (NSTC) 4, 5 August 1994, file 147, box 5, X-33 Archive.

97. The Joint Statement on Cooperation in Space, signed by Vice President Albert Gore, Jr., and the Russian Prime Minister in September 1993, laid the foundation for the two countries to cooperate on the Station project. The 1 November 1993 addendum approved by President Clinton declared that the Russian launchers (as well as the Shuttle) would carry the various Station segments and that Russia was a full partner in the project. See "Use of foreign launch vehicles for the Space Station has already been approved by the President," file 149, box 6, X-33 Archive.

98. Richard DalBello, Office of Science and Technology Policy, White House, to multiple addressees, "May 17, 1994 Meeting of the Interagency Working Group on Space Transportation," 11 May 1994, file 147, box 5, X-33 Archive.

The 1994 National Space Transportation Policy continued the standing decision to utilize a mixture of expendable and reusable launchers but added the notion of a lead agency for each type of launch technological system, as the Moorman Report had recommended. The new language shifted the basis for distinguishing institutional responsibilities from the nature of the payload (human spaceflight) to the type of technological system utilized (expendable versus reusable launch vehicle). Thus, NASA would be the lead agency in developing the "next generation" of reusable launchers—including single stage to orbit rockets—while the military would implement improvements in expendable rockets on behalf of the entire national security sector.[99]

Even though the Space Transportation Policy made NASA the lead agency for the development of reusable launchers, individuals within the Air Force, such as Simon P. Worden, and Congress, especially Representative Dana Rohrabacher (R–California), wanted to continue work on such reusable military craft as the TransAtmospheric Vehicle and the Space Maneuver Vehicle.[100] The position of the Defense Department, however, was that the 1994 Space Transportation Policy clearly gave NASA the responsibility for reusable launchers, not the Department, and the Pentagon preferred to split the funding the same way. The Air Force recently had started the Evolved Expendable Launch Vehicle (EELV) program to develop a low-cost heavy-lift expendable rocket in collaboration with Boeing and Lockheed Martin. As a result, Paul G. Kaminski, Under Secretary of Defense for Acquisition and Technology, explained, the Department had "no requirement to initiate an additional program." NASA Administrator Dan Goldin agreed with Kaminski on splitting launch vehicle funding in the same way that the space transportation policy divided up launch vehicle responsibilities.[101]

---

99. Office of Science and Technology Policy, White House, "Statement on National Space Transportation Policy," 5 August 1994; Presidential Decision Directive NSTC 4, 5 August 1994. The DOD, in cooperation with NASA, could use the Shuttle to meet national security needs. Launch priority would be provided for national security missions as governed by appropriate NASA/DOD agreements. Launches necessary to preserve and protect human life in space would have the highest priority except in times of national emergency. NASA would maintain the Shuttle until a replacement became available.

100. Rohrabacher to members of the House Appropriations National Security Subcommittee, "A request for assistance on this week's markup," 11 July 1995, file 506, box 19, X-33 Archive; "Department of Defense Appeal: FY 1996 Defense Authorization Bill," 15 June 1995, file 506, box 19, X-33 Archive; Jeffrey M. Lenorovitz, "Reusable Launcher Backers Push X-Plane Test Program," *Aviation Week & Space Technology* (25 July 1994): 24–25, copy available in file 180, box 7, X-33 Archive; Warren Ferster, "U.S. Air Force Awards 2 Study Contracts for Space Plane," *Space News* 8 (8–14 September 1997): 19, copy available in file 192, box 7, X-33 Archive; James Cast to Gary Payton, e-mail message, 4 September 1997, copy available in file 192, box 7, X-33 Archive.

101. Paul G. Kaminski, Under Secretary of Defense for Acquisition and Technology, to Goldin, 4 May 1995, file 506, box 19, X-33 Archive; Goldin to Kaminski, 12 June 1995, file 506, box 19, X-33
continued on the next page

What is striking about the 1994 Space Transportation Policy is that it was the first space policy statement to contain language regarding a specific program, NASA's X-33 project. That peculiarity was the direct result of strong NASA lobbying. One set of proposed language made NASA focus on developing technologies "to support a decision no later than December 1996 to proceed with a subscale flight demonstration which would prove the concept of SSTO."[102] Later, the Agency suggested wording that supported its single stage to orbit project by authorizing technology development leading up to a June 1997 decision to proceed with a subscale flight demonstration to "prove the concept of Single Stage To Orbit (SSTO)."[103] Another iteration of draft policy added: "The technology development program will lead to the full-scale development of a next generation reusable space transportation system by the end of the decade."[104] NASA subsequently made a point of holding back the release of the Cooperative Agreement Notice for the X-33 program until after the White House reviewed NASA's plans for implementing the 1994 space policy and responded to NASA in writing.[105] Thus, the Space Transportation Policy represented a clear victory for NASA's pursuit of single stage to orbit launchers and reusable launch vehicles in general.

The RLV Bubble Bursts

The same enthusiasm for reusable launchers translated to the commercial launch industry, too. Government policy—the 1994 Space Transportation Policy—and government investment in such projects as the NASP and the DC-X, followed now by the X-33, favored the development of reusable launch vehicles. In part, too, this enthusiasm resulted from one of the touted advantages of reusable launch vehicles, namely, their lower operating costs. This advantage took on new importance because of the considerable, in fact unprecedented, number of launches projected to take place in the near future. Setting up the Milstar, Teledesic, Orbcomm, Intermediate Circular Orbit (ICO), Globalstar, and Iridium networks would involve launching literally hundreds of satellites.

continued from the previous page

Archive; "Memorandum of Agreement between Air Force Space Command, the Air Force Research Laboratory, and the National Aeronautics and Space Administration for Cooperative Technology Development Support of NASA Reusable Launch Vehicles and Air Force Military Spaceplanes," 12 October 1997, file 506, box 19, X-33 Archive.

102. Office of Science and Technology Policy, White House, "Statement on National Space Transportation Policy," 5 August 1994; Presidential Decision Directive NSTC 4, 5 August 1994.

103. "NASA Comments on the Draft National Space Transportation Strategy Directive and on May 17 Interagency Comments," 19 May 1994, file 147, box 5, X-33 Archive.

104. Gary Krier to Jeff Hofgard, "NASA Comments to the OSTP National Space Transportation Strategy Draft of 8 April 1994," 20 April 1994, file 151, box 6, X-33 Archive.

105. Richard DalBello, Technology Division, OSTP, to Jack Mansfield, NASA, 8 November 1994, file 153, box 6, X-33 Archive.

Commercial launch firms' enthusiasm for reusable launch vehicles was reflected in the technological shift that took place between 1989 and 1999 within the industry. In 1989, when the Department of Transportation issued the first commercial launch licenses,[106] expendable rockets based on 1950s technology and established companies with deep roots in the military-industrial complex dominated the industry. These included Martin Marietta, manufacturer of the Titan; McDonnell Douglas, maker of the Delta rocket; and General Dynamics, which built the Atlas-Centaur. The nation's smaller startup launch providers also were utilizing expendable launchers: the Conestoga rocket of Space Services, Inc. (SSI); the Industrial Launch Vehicle (ILV) of the American Rocket Company (AmRoc); and Conatec, Inc., and E'Prime Aerospace Corporation used various sounding rockets.[107]

The picture in 1999 was quite different. Reusable vehicles were now the space launcher *du jour*, thanks mainly to the enthusiasm of a half dozen relatively small startup launcher companies that were developing RLVs for commercial and government payloads. Among these were the Astroliner of Kelly Space and Technology, the K-1 of Kistler Aerospace Corporation,[108] the Pathfinder of Pioneer Rocketplane, Rotary Rocket Company's Roton C-9, Space Access's SA-1, and Vela Technology Development's Space Cruiser System.[109] Meanwhile, with NASA funding, Lockheed Martin was developing its single stage to orbit VentureStar™, as well as the X-33 prototype; Orbital Sciences Corporation was building and testing the X-34; Boeing was working on the Future X Trailblazer; and Scaled Composites was involved in the X-38 Crew Return Vehicle program.[110] The Space Maneuver Vehicle, moreover, was under development by the Air Force Space Command in conjunction with McDonnell Douglas, Lockheed Martin, and the Boeing Phantom

---

106. Stephanie Lee-Miller, "Message from the Director," October 1989, in Department of Transportation Office of Commercial Space Transportation, *The U.S. Office of Commercial Space Transportation Fifth Annual Report* (Washington, DC: GPO, 1990), copy available in file 393, box 15, X-33 Archive.

107. U.S. Department of Transportation, Office of Commercial Space Transportation, "Annual Report to Congress: Activities Conducted under the Commercial Space Launch Act," 1987, pp. 5–6, file 391, box 15, X-33 Archive.

108. Walter Kistler, Bob Citron, and Thomas C. Taylor, "A Small, Reusable Single Stage to Orbit Rocketship," IAF-94-V.3.536 (paper read at the 45th Congress of the International Astronautics Federation, Jerusalem, Israel, 9–14 October 1994), file 179, box 7, X-33 Archive; Kistler Aerospace Corporation, "K-1 Aerospace Vehicle Overview," December 1997, file 179, box 7, X-33 Archive.

109. Unless indicated otherwise, the following discussion of RLV projects is from Associate Administrator for Commercial Space Transportation, "1999 Reusable Launch Vehicle Programs & Concepts," January 1999, pp. 7, 22–29, file 564, box 20, X-33 Archive; and Bill Sweetman, "Rocket Planes," *Popular Science* 232 (February 1998): 40–45, file 180, box 7, X-33 Archive.

110. The vehicle would be attached to the International Space Station as a means of returning to Earth if an emergency required an immediate evacuation of the Station, if an astronaut had a medical emergency, or if the Shuttle were grounded and the astronauts had to return to Earth. Strictly speaking, the X-38 was not an RLV; that is, it was not intended to be a launch vehicle but was capable of multiple flights nonetheless.

Works.[111] Nor was RLV fever confined to the United States. Similar efforts were under way in the United Kingdom, India, and Japan.[112]

This RLV bubble burst in 2000, just as various high-technology industries were beginning to soften. Space commerce, because of its high capital requirements, was one of the first to falter, starting with the failure of Motorola's Iridium communication satellite constellation. The possibility of winning the Ansari X Prize encouraged some firms to keep trying, however.[113] Meanwhile, NASA terminated its RLV programs: the X-33 and the X-34 on 1 March 2001, followed by the Future X Trailblazer,[114] and the X-38 prototype Crew Return Vehicle on 29 April 2002. The space agency was out of the business of developing reusable launchers.

The New Bush

George W. Bush brought about major changes in Clinton space policy largely through his appointee to head NASA, Sean O'Keefe. Within a month of taking charge, O'Keefe embarked on a series of measures that brought NASA and the Defense Department into closer collaboration on technology development, including a possible jointly developed reusable launch vehicle.[115] While O'Keefe was drafting NASA once again into military service, Defense Secretary Donald Rumsfeld announced the revival of President Reagan's space-based missile defense system and elevated the agency's status from the Ballistic Missile Defense Organization to the Missile Defense Agency (MDA) on 4 January 2002 in recognition of the high national priority that the President gave to missile defense.[116] Bush, however, did not give space commercialization the same status, perhaps because his policy advisers believed that the major downturn in the market for commercial launch services had undermined the

---

111. The Air Force gave study contracts to both Lockheed Martin and McDonnell Douglas Space Division to develop concept designs for the suborbital vehicle. McDonnell Douglas based its design on the DC-X. The Boeing Phantom Works was developing the SMV.

112. Associate Administrator for Commercial Space Transportation, "1999 Reusable Launch Vehicle Programs & Concepts," January 1999, pp. 7, 22–29, file 564, box 20, X-33 Archive; Sweetman, "Rocket Planes," pp. 40–45, file 180, box 7, X-33 Archive.

113. The X Prize was a $10-million prize offered to the first entrant able to launch a vehicle capable of carrying three people to a 100-kilometer suborbital altitude and repeating the flight within two weeks. See Associate Administrator for Commercial Space Transportation, "1999 Reusable Launch Vehicle Programs & Concepts," pp. 30–32; Rebecca Anderson and Michael Peacock, "Ansari X-Prize: A Brief History and Background," NASA History Division Web site, *http://history.nasa.gov/x-prize.htm* (accessed 24 March 2005).

114. The goal of Future X was to develop vehicles more technologically advanced than the X-33. It consisted of a series of experimental flight demonstrators called the Pathfinder and Trailblazer series. Material on the Future X program can be found in file 184, box 7, X-33 Archive.

115. Marc Selinger, "Air Force, NASA Studying Joint Development of New Reusable Launch Vehicles," article 197714 in *Aerospace Daily* (25 January 2002, electronic edition), hard copy in file 854, X-33 Archive.

116. "BMDO's Name Changed to Missile Defense Agency," article 196406 in *Aerospace Daily* (7 January 2002, electronic edition), hard copy in file 854, X-33 Archive.

ability of industry to recoup the considerable investments needed to develop launch systems.[117] Instead, on 14 January 2004, he revived his father's failed Space Exploration Initiative as the Vision for Space Exploration.[118]

Later that year, on 21 December 2004, the White House released a new space transportation policy. It raised more questions than it answered. The policy made no basic changes in existing space commerce policy, but it did throw up barriers to the commercial launch industry by allowing the government to use excess ballistic missiles when their use was cheaper than flying on a commercial launcher. It also made it harder for companies to put payloads on foreign launchers (despite the reliance on Russian launchers following the *Columbia* disaster). Furthermore, the new space transportation policy did not make reusable and expendable launcher responsibility the basis for distinguishing the institutional responsibilities of NASA and the Defense Department. Instead, it made the Defense Secretary responsible for national security launchers and facilities, and the NASA Administrator responsible for "the civil sector," without any mention of reusable or expendable launchers or even which agency had responsibility for human spaceflight.

The central issue addressed by the policy was the need for launchers to achieve the Vision for Space Exploration. It declared that the Space Shuttle would return to flight, complete assembly of the Space Station by the end of the decade, then retire. Concurrently, NASA would develop a new "crew exploration vehicle" for human spaceflight.[119] Furthermore, it declared that the Evolved Expendable Launch Vehicle (EELV) program was now "the foundation for access to space" for intermediate and heavy payloads serving both military and civilian missions. The policy also directed NASA and DOD to develop jointly a version of the EELV suitable for "space exploration."

In January 2004, NASA announced that it would begin developing the Crew Exploration Vehicle, a piloted vehicle to carry humans into orbit "and beyond," as well as to ferry astronauts to and from the Space Station following the retirement of the Shuttle. Different versions of the vehicle could operate in Earth orbit or near the Moon or even on the surface of Mars. The Crew Exploration Vehicle effort was part of what the space agency was calling its Constellation Systems Theme, a set of projects to develop, test, and deploy the various systems needed to prosecute the Vision for Space Exploration. In addition, NASA planned to use an established military acquisition process known as spiral or evolutionary acquisition to develop space exploration hardware.

---

117. "U.S. Space Transportation Policy," fact sheet, 6 January 2005, p. 2.

118. Office of the Press Secretary, White House, "Executive Order President's Commission on Implementation of United States Space Exploration Policy," 30 January 2004.

119. The following section is from "U.S. Space Transportation Policy," fact sheet, 6 January 2005, except where noted.

The first spiral or stage would deliver humans to orbit in a Crew Exploration Vehicle by 2014. The second would land humans on the Moon's surface by 2020, followed by extended lunar visits in the third stage.[120] All of these proposed systems would be launched on top of an EELV.

In the end, the 2004 Space Transportation Policy and its implementation seemed to assign reusable vehicles the same role played by Mercury, Gemini, and Apollo capsules: sitting atop expendable boosters. This time, though, the rocket of choice was the Evolved Expendable Launch Vehicle and its future variants. Implicit in the decisions underlying the latest space transportation policy was the assumption of a reduced launch rate. Reusable launch vehicles only make economic sense if they have numerous payloads to launch, and their absence in the 2004 Space Transportation Policy can be interpreted as an admission (or at least an assumption) that launch rates for the foreseeable future will be low. One must wonder, then, what the thinking is that lies behind the current Russian effort to build the Kliper reusable launch vehicle for transporting crew and cargo to the Space Station. Do they see launch rates rising? Is the purpose of the Kliper just to bring down launch costs below those for the Soyuz for the cash-starved Russian space effort?[121]

In the relatively brief period between 1980 and 2005, the status of reusable launch vehicles in national space transportation policy waxed and waned more than once. The perception that there was something called space transportation began as people started to fly into space on a reusable, rather than a recoverable, craft; that is, the notion of transportation involved both reusability and human spaceflight. Thus, the advent of the Space Shuttle engendered and dominated (monopolized) space transportation policy. Beginning in 1986, however, reusable craft took their place alongside expendable launchers in a mixed fleet. The dividing line between NASA and Defense Department institutional responsibilities was human spaceflight, but that did not give NASA responsibility for all reusable and the Pentagon responsibility for all expendable launchers. Nonetheless, the 1994 Space Transportation Policy explicitly did enunciate that technological separation of institutional responsibilities, and it created the framework within which a tremendous commercial and governmental enthusiasm for reusable launch vehicles thrived. That policy also broke new ground by mentioning—for the first time—specific space programs. Following the collapse of enthusiasm that began in 2000, reusable launch vehicles disappeared from space transportation policy.

---

120. NASA, fiscal year 2006 budget request, *http://www.nasa.gov/pdf/107488main_FY06_low. pdf*, pp. SAE 5-2, SAE 5-3, SAE 6-1, SAE 6-4 to 6-6. On spiral acquisition, see, for example, Alexander R. Slate, "Evolutionary Acquisition: Breaking the Mold—New Possibilities from a Changed Perspective," *Program Manager* 31 (May–June 2002): 6–13. It is from the lingo of spiral acquisition that NASA has picked up the phrase "system of systems."

121. Anatoly Zak, "Russians Propose a New Space Shuttle," *IEEE Spectrum* 42 (February 2005): 13–14.

HISTORIOGRAPHY

A Question

The history of air travel in the United States can be traced back to a time over two centuries ago. A symposium held at the National Air and Space Museum attempted to deal with the subject, a sort of "bicentennial survey" held in the year of the U.S. bicentennial, specifically on 4 November 1976.[122] The history of motorized winged flight is much shorter, of course, and the first Sputnik launches took place scarcely two decades before the symposium. Several of the speakers lamented the chore of condensing 15 or 70 years of history into 20 minutes. In placing their talks in a broader context, historian Thomas Parke Hughes noted that 70 years was not a large amount of time. Nor did he find aeronautics and astronautics to be "an overwhelmingly significant" subject. "We are dealing here with a very short period of time and one episode in a long history of man and technology."[123]

Little has changed in the intervening two decades since Hughes made that observation. The year 2007 will mark only the 50th anniversary of the Sputnik launches, followed by NASA's 50th anniversary. Fifty years is a short historical span; it is certainly not *histoire à longue durée*. Furthermore, during the past two decades, the amount of printed literature and unpublished talks on space history has multiplied swiftly, confirming once again the de Solla Price curve.[124] Despite this growth, we lack a "big picture" understanding of space history. A different, but associated, question is how space history fits into general histories, such as those of the United States, or into specialized histories, such as the history of transportation. Is space history such a peculiar topic of study that it does not lend itself to integration into other histories, into larger historical questions?

A recent joint publication of the American Historical Association and the Society for the History of Technology[125] that surveyed U.S. transportation history ended with a chapter on "airways," but not a mention of space travel. Is going into space such a peculiar human endeavor that its history must be segregated from the other categories into which we parse history? Is it because many space and space history enthusiasts act as if the space program were a nontheistic religion? Or should we be asking whether space transportation is

122. Eugene M. Emme, ed., *Two Hundred Years of Flight in America: A Bicentennial Survey*, AAS History Series, vol. 1 (San Diego: Univelt, Inc., for the American Astronautical Society, 1977). The symposium sponsors were NASM, the AIAA, SHOT, and AAS, which published the proceedings.

123. Hughes, "Perspectives of a Historian of Technology: A Commentary," in *Two Hundred Years of Flight in America*, ed. Emme, p. 257.

124. Derek de Solla Price, *Science Since Babylon*, 1st edition (New Haven, CT: Yale University Press, 1961).

125. Robert C. Post, *Technology, Transport, and Travel in American History* (Washington, DC: American Historical Association, 2003).

really a form of transportation? Was there anything of substance to the transportation references common to space travel—such as the Space *Transportation* System and National Space *Transportation* Policy—or were they just figures of speech, similar to the analogies with aircraft and ships reflected by the terms *spacecraft, spaceplane, rocket ship,* and *spaceship*[126] or, say, the maritime analogies used by presidential speechwriters[127] and space advocates?[128]

One of the peculiar aspects of space launch vehicles is their origins in rocketry, which for centuries served largely military purposes. The aerospace engineer Maxwell W. Hunter II captured the difference between the two uses of rocket technology with his use of the terms "ammunition" and "transportation." Expendable rockets, he wrote, were ammunition, while reusable launch vehicles were transportation.[129] The shift from "ammunition" to "transportation" was not just one of application, but also a change of perception that occurred once people replaced the bombs, electronic instrumentation, and other inanimate objects that had served for decades as the sole payloads carried into space or the uppermost reaches of the atmosphere. The transformation of a military technology into a mode of transport is rather unique in world history, perhaps as unique as turning swords into plowshares.

The reverse, turning transportation into a weapon, is certainly not unique, but rather a common occurrence in history. In recent times, we have witnessed aircraft turned into weaponry on 7 December 1941 and 11 September 2001, for example. Automobiles and trucks also have become bomb delivery systems in the hands of Timothy McVeigh and colleagues on 19 April 1995, against the Murrah Federal Building in Oklahoma City, and Ramzi Yousef and his fellow coconspirators on 26 February 1993, against the World Trade Center in New York City. Any form of transpor-

---

126. Another term that evokes the maritime analogy is *spacefaring*. Much can be written on the analogy between moving through outer space and sailing, as I suggested in *Single Stage to Orbit: Politics, Space Technology, and the Quest for Reusable Rocketry* (Baltimore, MD: Johns Hopkins 2003), pp. 21–22, 217.

127. For example, President Kennedy told a crowd at the Rice University stadium, "We intend to be first . . . to become the world's leading space-faring nation" (John F. Kennedy, address at Rice University, 24 September 1962, *Public Papers of the Presidents* [Washington, DC: National Archives and Records Service, 1963], p. 329).

128. Lieutenant Colonel Daniel O. Graham, the well-known proponent of what became the Strategic Defense Initiative, believed that a U.S. space-based global defense system would bring about a Pax Americana similar to the Pax Britannica induced by Britain's domination of the world's oceans. See Erik K. Pratt, *Selling Strategic Defense: Interests, Ideologies, and the Arms Race* (Boulder, CO: Lynne Rienner Publishers, 1990), p. 96.

129. See, for example, Hunter to E. P. Wheaton, vice president for research and development, Lockheed, "Orbital Transportation," 28 October 1965, pp. 1–2, file 338, box 13, X-33 Archive. The distinction between ammunition and transportation appears throughout Hunter's oeuvre. See, for instance, Hunter, "The SSX: A True Spaceship" (manuscript, 2000), pp. 17, 18, 22; and Hunter, "The SSX: A True Spaceship" (manuscript, 4 October 1989), pp. 2, 6, both in file 338, box 13, X-33 Archive.

tation can undergo this transformation, yet we cannot imagine any bomb delivery system or other form of weapon system being turned into a form of transportation. Although certain military-use vehicles have found civilian applications—such as the Jeep of World War II and the High Mobility Multipurpose Wheeled Vehicle (more commonly known as the Hum Vee or Hummer)—they always served as military transport vehicles, never as weapon systems. One could stretch the point and argue that the Bradley M2 Fighting Vehicle or the Abrams M1 tank could be turned into transport, but their high maintenance and operational costs, frequent need for maintenance and repairs, lack of reliability, and poor performance only highlight the absurdity of the proposition.

If we define space transportation as human flight into space via reusable launch vehicles (the key being the combination of *reusability* and humans in space), then the real question historians need to answer is not whether space transportation is really transportation and therefore part of transportation history. Space travel clearly has many characteristics in common with the various forms of terrestrial transportation. One can point to numerous aspects of space transportation shared by other forms of transportation, from the model-building of amateurs to the carrying of cargo and passengers (both astronauts and tourists) to desired destinations. Even the inherent danger of space travel has had its precedents in the boiler explosions that pervaded early steam-powered transporters. Like other forms of transportation, travel to places off the planet requires a complex infrastructure.

For instance, one can compare the launch infrastructure required by rocketry with the infrastructures that support automobile or truck travel. In addition to the nation's vast network of roads, signage (and the systems needed to maintain and operate them), and facilities for refueling and repairing vehicles (gas and repair stations), these include such legal and regulatory elements as driving rules and laws, driver license and registration facilities, driver education, vehicle inspections and inspection stations, and various regulatory agencies from the local motor vehicle agency to the Interstate Commerce Commission and the Department of Transportation. Infrastructure issues also are relevant to the choice between using a solid-fueled or a liquid-fueled rocket. Similarly, in the early history of the automobile, different engine types (electric, steam, gas) required a dedicated infrastructure. Reusable and expendable launch vehicles similarly have different infrastructure needs.

Historians often claim that one properly cannot write the history of a subject until the passage of a certain amount of time. The subject, like a bottle of wine, must age and somehow achieve a certain degree of ripeness before it is suitable for historical inquiry. Space history, as measured from Goddard's first liquid-fueled rocket near Auburn, Massachusetts, on 16 March 1926 to the present, does not span a very long period, just eight decades—even less if one counts from Sputnik forward. In comparison, Georg Agricola, nearly a half millennium ago, recounted the use of railways in mining operations,[130] and the Appian Way is centuries older still. And yet, histories that cover periods as short as five years or less have been—and are continually being—written. The challenge is not the relatively short length of the space travel era nor its topical nature. Historians routinely research and write about events that have taken place only a few years earlier—or investigate history as it happens.[131] The real question is a challenge, the challenge for space historians to integrate their work into the larger historical context, with its rich fabric of political, economic, social, and cultural threads.

---

130. Georg Agricola, *De Re Metallica* (Basil, Switzerland: H. Frobenium and N. Episcopium, 1556), trans. and annotated by Herbert Clark Hoover and Lou Henry Hoover (London: *The Mining Magazine*, 1912).

131. This was the subject of a recent panel, "Doing the History of the Recent Past: Historiography, Sources, Disciplinary Boundaries . . . ," held by the Society for the History of Technology in 1997. The panel's contributors consisted of Joseph N. Tatarewicz, "In from the Cold or Out in the Cold? Warriors and Nuclear Weaponeers Search for their Place in History"; Pascal Griset, "Oral History and Recent Evolutions in the History of French Industry"; and Butrica, "From the X-Files: Some Source and Historiographical Problems of the X-33 History Project, or 'History Made While You Wait.'"

NASA Cultures

Introduction

One of the main conclusions of the Columbia Accident Investigation Board (CAIB) was that "the organizational causes of this accident are rooted in the Space Shuttle Program's history and culture" and that over many years at NASA, "cultural traits and organizational practices detrimental to safety and reliability were allowed to develop . . . ."[1] The idea of organizational culture is therefore a critical issue, though, as La Porte points out in this section, it is a "slippery concept" with a "high degree of operational ambiguity, its use subject to stiff criticism." Although organizational culture may in fact mean many things, all three authors in this section find the concept useful, for lack of a better term, to refer to what La Porte characterizes in the NASA context as "the norms, shared perceptions, work ways, and informal traditions that arise within the operating and overseeing groups closely involved with the systems of hazard." Slippery as it may be as a concept, organizational culture is important to understanding real-world questions, such as those that Vaughan (a sociologist by profession and a staff member of the CAIB) enumerates in her article: How do organizations gradually slide into negative patterns? Why do negative patterns persist? Why do organizations fail to learn from mistakes and accidents? Although human and technical failures are important, she finds their root causes in organizational systems. In order to reduce accidents, therefore, organizational systems and their cultures must be studied and understood.

The first two papers in this section concentrate on organizational culture as it relates to accidents in human spaceflight, here restricted to those in NASA's space program. Vaughan focuses on the Space Shuttle *Challenger* and *Columbia* accidents in 1986 and 2003, respectively, while Brown adds the ground-based Apollo 204 (also known as Apollo 1) fire in 1967. Altogether, 17 astronauts were killed in these accidents, triggering massive criticism, investigations, official reports, and personal and organizational soul-searching. Vaughan finds that, due to overly ambitious goals in an organization strapped for resources, NASA's Apollo-era technical culture was turned into a "culture of production" by the time of the *Challenger* accident, a culture that persisted through *Columbia* and was characterized by "cultural mandates for business-like efficiency, production pressures, allegiance to hierarchy, and rule-following." The result was what she calls "the normalization of deviance"—in other words, over time, that which was deviant or anomalous incrementally became redefined as normal, most notably Solid Rocket Booster (SRB) O-ring behavior in cold weather for *Challenger* and foam hits from the External

---

1. Columbia Accident Investigation Board, *Report* (Washington, DC: NASA and GPO, August 2003), chap. 8. Chapter 8 was largely written by Diane Vaughan.

Tank (ET) to the wing of the Shuttle in the case of *Columbia*. Lack of communication, which she terms "structural secrecy," within layers of NASA administration compounded the problem.

Vaughan believes that the thesis of "history as cause" in the CAIB report demonstrates how the history of decisions made by politicians and by NASA engineers and managers combined twice to produce disaster. She warns that economic strain and schedule pressure still exist at NASA and that in such circumstances, system effects, including accidents, tend to reproduce. It is important to note that it is not possible to prevent all accidents, but, she concludes, the *Challenger* and *Columbia* accidents, with their long incubation periods, were preventable. In her view, reducing the probability of accidents means changing NASA's culture as well as externally imposed expectations and limitations, a difficult and ongoing process, one in which social scientists must play a role in a systematic way.

Brown, a historian of technology in the Science, Technology and Society program at MIT, takes another approach by analyzing the "disjunctures" in the three fatal NASA accidents. In the case of Apollo 204, the disjuncture is between the engineers designing and managing the spacecraft and the technicians manufacturing it. For the two Shuttle accidents, the disjuncture is between managers controlling the Shuttle program and engineers maintaining and analyzing the spacecraft. By way of explaining these disjunctures, he analyzes the three accident reports and relates their styles and conclusions to the engineering practices of NASA and its contractors. Whereas the Apollo 204 report concluded that poor engineering practice was the sole cause of the fire, the Challenger Commission, by contrast, emphasized secondary causes in addition to the technical O-ring failure, including the decision to launch, schedule pressure, and a weak safety system. As emphasized in Vaughan's paper, the *Columbia* report went even further, pointing (partly at her urging) to equal importance for technical and social causes.

Reading the three accident reports to gain historical insights, Brown finds that they suggest a growing separation between management and engineering over the period under review. They reveal an asymmetry assumed by the accident investigators, in the sense that the technical/engineering causes are to be understood as "context-free and ahistorical activity," while management causes are to be understood in a complex historical and cultural framework. Brown therefore asks two questions: what historical processes caused this separation between management and engineering? And what changes in engineering over the quarter century covered by the accident reports might be important for placing engineering in its own historical and cultural context? In answer to the latter, he enumerates three changes: widespread use of computers, changes in engineering education, and the move away from systems engineering as an organizing philosophy. During the period 1967 to 2003, modeling, testing, and simulation had changed from hand calibration to computer-based calculations, resulting in loss of transparency. For example, Boeing engineers who used a computer model known as "Crater" to predict the effects of foam impacts on the Shuttle were unaware of its limitations precisely because the process had been computerized; this ignorance greatly affected their ability to make engineering judgments. Over the same period, engineering education, which was moving toward science and away from design, rendered engineering more abstract and less connected to reality. The *Challenger* and *Columbia* reports criticized the lack of engineering design expertise in some of the contractors involved. Finally, whereas systems engineering was the guiding philosophy of the space program at the time of the Apollo 204 fire, Total Quality Management and the "faster, better, cheaper" approach replaced system engineering during the 1990s for senior management, while engineers still used the tools of system management.

La Porte takes a broader view, tackling the issues of high-reliability systems that must operate across decades or generations, as NASA must do in planning and implementing its vision to take humans to the Moon and Mars. Drawing on a variety of empirical studies in the social and management sciences, including nuclear power plant operation and waste disposal, he undertakes this analysis of highly reliable operations that take place over decades, and he assumes high levels of public trust over that time. Such long-term operations also involve issues of institutional constancy. He finds, among other things, that high-reliability organizations (HROs) must have technical competence, stringent quality-assurance measures, flexibility and redundancy in operations, decentralized decision-making, and an unusual willingness to reward the discovery and reporting of error without assigning blame. Maintaining an organizational culture of reliability exhibiting these characteristics is difficult, but important. Nor can HROs become overly obsessed with safety; they must strive equally for high levels of production and safety. If the Shuttle never launches, NASA fails its mission in equal measure as it does when it has accidents. La Porte also emphasizes the importance of external "watchers," including congressional committees and investigating boards, to sustaining high-reliability organizations, a factor also evident in Vaughan's and Brown's analyses of the accident reports.

La Porte notes that, for obvious reasons, maintaining these characteristics over long-term, even trans-generational, efforts is the least-understood process in terms of empirical studies. In an attempt to shed light on this problem, he examines the idea of "institutional constancy" and concludes that in order for such long-term efforts to be successful, an agency such as NASA must demonstrate to the public and to Congress that it can be trusted to keep its word long into the future, and it must "show the capacity to enact programs that are faithful to the original spirit of its commitments." La Porte discusses the characteristics associated with institutional constancy, summarized in his table 13.2. He, too, calls for further empirical and analytical study, especially to delineate requirements for long-term institutional constancy and trustworthiness.

Implicitly or explicitly, these papers also deal with the question of risk. The Challenger Commission found that its managers and engineers understood risk in very different ways, with the engineers seeing it as quantifiable and the managers as flexible and manageable. The Columbia Accident Investigation Board noted similar differences in the perception of risk. La Porte broaches the question of risk averseness and the public's risk-averse demand for very reliable operations of intrinsically hazardous systems. He suggests research on the conditions under which the public would be willing to accept more risk, given that such operations can never be risk-free. NASA's "Risk and Exploration" symposium, held in late 2004 in the midst of the Hubble Space Telescope controversy and with the Shuttle still grounded, came to a similar conclusion: the public needs to be made aware that accidents are not completely preventable.[2]

Nevertheless, the three views in this section, by a sociologist, a historian, and a political scientist, shed important light on NASA cultures and, if one accepts their arguments, on ways to reduce accidents in what inevitably remains a high-risk endeavor. How to balance risk and exploration is the key question.

---

2. Steven J. Dick and Keith Cowing, *Risk and Exploration: Earth, Sea and Sky* (Washington, DC: NASA SP-2005-4701, 2005).

CHAPTER 11

CHANGING NASA: THE CHALLENGES OF ORGANIZATIONAL SYSTEM FAILURES

Diane Vaughan

In both the *Columbia* and *Challenger* accidents, NASA made a gradual slide into disaster. The history of decisions about the risk of Solid Rocket Booster O-ring erosion that led to *Challenger* and the foam debris that resulted in *Columbia* is littered with early warning signs that were misinterpreted. For years preceding both accidents, technical experts defined risk away by repeatedly normalizing technical anomalies that deviated from expected performance. The significance of a long incubation period leading up to an accident is that it provides greater opportunity to intervene and to turn things around, avoiding the harmful outcome. But that did not happen. The Columbia Accident Investigation Board's report concluded that NASA's second Shuttle accident resulted from an organizational system failure, pointing out that the systemic causes of *Challenger* had not been fixed.[1] In fact, both disasters were triggered by NASA's organizational system: a complex constellation of factors including NASA's political/economic environment, organization structure, and layered cultures that affected how people making technical decisions assessed risk. These three aspects of NASA's organizational system interacted, explaining the origins of both accidents.

The amazing similarity and persistence of these systemic flaws over the 17 years separating the two accidents raise several questions: How do organizations gradually slide into negative patterns? Why do negative patterns persist? Why do organizations fail to learn from mistakes and accidents? In this chapter, I examine NASA's experience to consider the challenges of changing NASA's organizational system and to gain some new insight into these questions. My data for this analysis are my *Challenger* research, experience as a researcher and writer on the staff of the Columbia Accident Investigation Board, conversations and meetings with NASA personnel at Headquarters and a NASA "Forty Top Leaders Conference" soon after the CAIB report release, and, finally, a content analysis of the two official accident investigation

---

1. Columbia Accident Investigation Board, *Report* (Washington, DC: NASA and GPO, August 2003).

reports.[2] Summarizing from my testimony before the CAIB, I begin with a brief comparison of the social causes of *Challenger* and *Columbia* to show the systemic causes of both, how the two accidents were similar and different, and how and why NASA twice made an incremental descent into disaster.[3] I then review the conclusions of the Presidential Commission investigating the *Challenger* accident and their recommendations for change, the changes NASA made, and why those changes failed to prevent the identical mistake from recurring in *Columbia*.[4] Next, I contrast the Commission's findings with those of the CAIB report and discuss the CAIB's recommendations for changing NASA, the direction NASA is taking in making changes, and the challenges the space agency faces in preventing yet a third Shuttle accident.

Robert Jervis, in *System Effects*, considers how social systems work and why so often they produce unintended consequences.[5] He stresses the importance of dense interconnections and how units and relations with others are strongly influenced by interactions at other places and at earlier periods of time. Thus, disturbing a system produces chains of consequences that extend over time and have multiple effects that cannot be anticipated. I will argue in this chapter for the importance of analyzing and understanding the dynamics of organizational system failures and of connecting strategies for change with the systemic causes of problems. The "usual remedy" in the aftermath of a technological accident is to correct the causes of a technical failure and alter human factors that were responsible so that they, too, can be fixed. However, the root causes of both human and technical failure can be found in organizational systems. Thus, remedies targeting only the technology and individual error are insufficient. Neither complacency, negligence, ignorance, poor training, fatigue, nor carelessness of individuals explains why, in the face of increasing in-flight damage, NASA made flawed decisions, continuing to fly. The lessons to be learned from NASA's experience are, first, in order to reduce the potential for gradual slides and repeating negative patterns, NASA and other organizations dealing with risky technologies must go beyond the search for technical causes and individual error and search the full range of social causes located in the organizational system. Second, designing and implementing solutions that are matched to those causes is a crucial but challenging step in preventing a recurrence.

---

2. Diane Vaughan, *The Challenger Launch Decision: Risky Technology, Culture, and Deviance at NASA* (Chicago: University of Chicago Press, 1996); Diane Vaughan, "History as Cause: *Columbia* and *Challenger*," chap. 8 in Columbia Accident Investigation Board, *Report*; Presidential Commission on the Space Shuttle *Challenger* Accident, *Report to the President by the Presidential Commission on the Space Shuttle Challenger Accident*, 5 vols. (Washington, DC: GPO, 1986).

3. Vaughan, "History as Cause," pp. 185–204.

4. Presidential Commission on the Space Shuttle *Challenger* Accident, *Report to the President.*

5. Robert Jervis, *System Effects: Complexity in Political and Social Life* (Princeton: Princeton University Press, 1997).

NASA's Slippery Slope: O-Rings, Foam Debris, and Normalizing Deviance

In a press conference a few days after the *Columbia* tragedy, NASA's Space Shuttle Program Manager, Ron Dittemore, held up a large piece of foam approximately the size of the one that fatally struck *Columbia* and discounted it as a probable cause of the accident, saying, "We were comfortable with it." Prior to the *Challenger* accident in 1986, that phrase might have been said about O-ring erosion by the person then occupying Dittemore's position. The O-ring erosion that caused the loss of *Challenger* and the foam debris problem that took *Columbia* out of the sky both had a long history. Neither anomaly was permitted by design specifications, yet NASA managers and engineers accepted the first occurrence, then accepted repeated occurrences, concluding after examining each incident that these deviations from predicted performance were normal and acceptable. In the years preceding NASA's two

This photograph of the Space Shuttle *Challenger* accident on 28 January 1986 was taken by a 70-millimeter tracking camera at site 15, south of Pad 39B, at 11:39:16.061 EST. One of the Solid Rocket Boosters can be seen at the top of the view. *(Image no. STS-51L 10181; Kennedy Space Center alternative photo no. is 108-KSC-86PC-147)*

accidents, managers and engineers had normalized recurring technical anomalies—anomalies that, according to design specifications, were not allowed. How—and why—was the normalization of technical deviations possible?

We must avoid the luxuries of retrospection, when all the flawed decisions of the past are clear and can be directly linked to the harmful outcomes, and instead see the events preceding each accident as did the personnel making risk assessments, as the problems unfolded. As managers and engineers were making decisions, continuing to launch under the circumstances they had made sense to them. The immediate social context of decision-making was an important factor. Although NASA treated the Shuttle as if it were an operational vehicle, it was experimental: alterations of design and unpredictable flight conditions led to anomalies on many parts on every mission. Because having anomalies was normal, neither O-ring erosion nor foam debris was the signal of danger it seemed in retrospect. In both cases, engineering decisions were made incrementally, anomaly by anomaly. Accepting the first deviation set a precedent on which future decisions were based. After inspection and analysis, engineers calculated a safety margin that placed initial damage within a safety margin showing that the design could tolerate even more.

In addition, the pattern of information had an impact on how managers and engineers were defining and redefining risk. As the anomalies began to occur, engineers saw signals of danger that were mixed—an anomalous incident would be followed by a mission with none or a reduced level of damage, so they believed they had fixed the problem and understood the parameters of cause and effect. Or signals were weak—incidents that were outside what had become defined as the acceptable parameters were not alarming because their circumstances were so unprecedented that they were viewed as unlikely to repeat. And finally, signals became routine, occurring so frequently that the repeating pattern became a sign that the machine was operating as predicted. The result was the production of a cultural belief that the problems were not a threat to flight safety, a belief repeatedly reinforced by mission success. Both erosion and foam debris were downgraded in official systems categorizing risk over time, institutionalizing the definition of these problems as low-level problems.

Although these patterns are identical in the two accidents, two differences are noteworthy. First, for O-ring erosion, the first incident of erosion occurred on the second Shuttle flight, which was the beginning of problem normalization; for foam debris, the normalization of the technical deviation began *even before the first Shuttle was launched*. Damage to the thermal-protection system—the thousands of tiles on the orbiter to guard against the heat of reentry—was expected due to the forces at launch and during flight, such that replacement of damaged tiles was defined from the design stage as a maintenance problem that had to be budgeted. Thus, when foam debris damage was observed on the orbiter tiles after the first Shuttle flight in 1981, it was defined as a maintenance problem, not a flight hazard. This early definition of the foam problem as routine and normal perhaps explains a second difference. Before the *Challenger* disaster, engineering concerns about proceeding with more frequent and serious erosion were marked by a paper trail of memos. The foam debris problem history also had escalations in occurrence but showed no such paper trail, no worried engineers.

These decisions did not occur in a vacuum. To understand how these two technical anomalies continued to be normalized, we need to grasp the important role that NASA's political and budgetary environment played and how the history of the Space Shuttle program affected the local situation. Decisions made by leaders in the White House and Congress left the space agency constantly strapped for resources to meet its own sometimes overly ambitious goals. The Agency's institutional history was one of competition and scarcity, which created a "trickle-down effect."[6] Thus, the original, pure technical culture of NASA's Apollo era was reshaped into a *culture of production* that existed at the time of *Challenger* and persisted over 50 launches later, for *Columbia*. NASA's original technical culture was reshaped by new cultural mandates for business-like efficiency, production pressures, allegiance to hierarchy, and rule-following.

This culture of production reinforced the decisions to proceed. Meeting deadlines and schedule was important to NASA's scientific launch imperatives and also for securing annual congressional funding. Flight always was halted to permanently correct other problems that were a clear threat to take the Shuttle out of the sky (a cracked fuel duct to the Space Shuttle main engine, for example), but the schedule and resources could not give way for a thorough hazard analysis of ambiguous, low-lying problems that the vehicle seemed to be tolerating. Indeed, the successes of the program led to a belief that NASA's Shuttle was an operational, not an experimental, system, thus affirming that it was safe to fly. Finally, the fact that managers and engineers obeyed the cultural mandates of hierarchy and protocol reinforced the belief that the anomalies were not a threat to flight safety because NASA personnel were convinced, having followed all the rules, that they had done everything possible to assure mission safety.

Both problems had gone on for years. Why had no one recognized what was happening and intervened, halting NASA's two transitions into disaster? The final piece of the organizational system contributing to both accidents was *structural secrecy*. By this I refer to how organization structure concealed the seriousness of the problems from people with responsibility for technical oversight who might have turned the situation around prior to both

---

6. Diane Vaughan, "The Trickle-Down Effect: Policy Decisions, Risky Work, and the *Challenger* Accident," *California Management Review* 39 (winter 1997): 1–23.

accidents. Organization structure affected not only the flow of information, a chronic problem in all organizations, but also how that information was interpreted. Neither NASA's several safety organizations nor the four-tiered Flight Readiness Review (FRR), a formal, adversarial, open-to-all structure designed to vet all engineering risk assessments prior to launch, called a halt to flying with these anomalies. Top administrators and regulators alike were dependent upon project groups for engineering information and analysis. As managers and engineers reinterpreted warning signs as weak, mixed, and routine signals, normalizing deviance, that diagnosis was what got passed up the hierarchy. Instead of reversing the pattern of flying with erosion and foam debris, Flight Readiness Review ratified it.

The structure of safety regulation also affected understandings about risk. NASA's internal safety system—both times—a) had suffered safety personnel cuts and de-skilling as more oversight responsibility was shifted to contractors in an economy move and b) was dependent upon the parent organization for authority and funding, so it had no ability to independently run tests that might challenge existing assessments. NASA's external safety panel had the advantage of independence but was handicapped by inspection at infrequent intervals. Unless NASA engineers defined something as a serious problem, it was not brought to the attention of safety personnel. As a result of structural secrecy, the cultural belief that it was safe to fly with these two anomalies prevailed throughout the Agency in the years prior to each of NASA's tragedies.

TWO ACCIDENTS: THE REPRODUCTION OF SYSTEM EFFECTS

I have shown how the organizational system worked in the years preceding both accidents to normalize the technical anomalies: the immediate context of decision-making—patterns of information; the context of multiple problems; mixed, weak, and routine signals—the culture of production, and structural secrecy all interacted in complex ways to neutralize and normalize risk and keep NASA proceeding with missions. To show how NASA's organizational system affected the crucial decisions made immediately before both accidents, I now revisit the unprecedented circumstances that created yet new signals of potential danger: an emergency teleconference held on the eve of the 1986 *Challenger* launch, when worried engineers recommended not launching in unprecedented cold temperatures predicted for the next day, and the events at NASA after the 2003 *Columbia* foam debris strike, when engineers again expressed concerns for flight safety. I selectively use examples of these incidents to show similarities and differences, recognizing that doing so greatly simplifies enormously complicated interactions.[7] An initial difference

---

7. For details, see Vaughan, *Challenger Launch Decision*, chap. 8; and CAIB, *Report*, chap. 6.

that mattered was the window of opportunity for decision and number of people involved. The *Challenger* teleconference was held prelaunch, involved 34 people in three locations, consuming several hours of one day, the proceedings unknown to others at NASA. *Columbia*'s discussion was postlaunch, with a window of 16 days before reentry, and videos of the foam debris strike were widely circulated, involving people throughout the Agency. They can be called crisis situations only in retrospect because at the time these events were unfolding, many participants did not define it as a crisis situation, which was, in fact, one of the problems.

In both scenarios, people facing unprecedented situations came to the table with a cultural belief in the risk acceptability of O-ring erosion and foam debris based on years of engineering analysis and flight experience. Thus, both the history of decision-making and the history of political and budgetary decisions by elites had system effects. As these selected examples show, the mandates of the culture of production for efficiency, schedule, hierarchy, and protocol infiltrated the proceedings. Also, structural secrecy acted as before, feeding into the tragic outcomes.

- Schedule pressure showed when *Challenger*'s Solid Rocket Booster Project Manager and *Columbia*'s Mission Management Team (MMT) Head, responsible for both schedule and safety, were confronted with engineering concerns. Both managers repeated that preexisting definition of risk, sending to others a message about the desired result. Schedule pressure on managers' thinking also showed when engineers proposed a temperature criterion for *Challenger* that would jeopardize the launch schedule for all launches, and for *Columbia* when obtaining satellite imagery would require the orbiter to change its flight orientation, thus prolonging the mission and likely jeopardizing the timing of an important future launch. Believing the safety of the mission was not a factor, both managers focused on future flights, making decisions that minimized the risk of delay.

- In both cases, hierarchy and protocol dominated; deference to engineering expertise was missing. In the *Challenger* teleconference, unprecedented and therefore open to innovation, participants automatically conformed to formal, prelaunch, hierarchical Flight Readiness Review procedures, placing engineers in a secondary role. The postlaunch *Columbia* Mission Management Team operation, intentionally decentralized to amass information quickly, also operated in a hierarchical, centralized manner that reduced engineering input. Further, engineering attempts to get satellite imagery were blocked for not having followed appropriate protocol. In both cases, norms requiring quantitative data were pushed, rendering engineering concerns insufficient; they were asked to prove that it was unsafe to fly, a reverse of the normal situation, which was to prove it was safe to fly. Engineers animated by concern took the issue to a certain level, then, discouraged and intimidated by management response, fell silent. A difference for *Columbia*: the rule on rule-following was inoperative for management, whose definition of risk was influenced by an "informal chain of command"—one influential person's opinion, not hard data.

• Organization structure created structural secrecy, as people structurally peripheral to the technical issue, either by location or expertise or rank, had information but did not feel empowered to speak up. Thus, critical input was lost to the decision-making. The weakened safety system was silent. No safety representative was told of the *Challenger* teleconference. Present at the *Columbia* MMT meeting but weak in authority, safety personnel interjected no cautions or adversarial challenges; information dependence and organizational dependence gave them no recourse but to follow the management lead.

This overview shows these accidents as the unanticipated consequences of system effects, the causes located in the dynamic connection between three layers of NASA's organizational system:

1) Interaction and the Normalization of Deviance: A history of decision-making in which, incrementally, meanings developed in which the unacceptable became acceptable. The first decisions became a basis for subsequent ones in which technical anomalies—signals of danger—were normalized, creating a cultural belief in the safety of foam and O-ring anomalies.

2) The Culture of Production: History was important in a second way. Historic external political and budgetary decisions had system effects, trickling down through the organization, converting NASA's original, pure technical culture into a culture of production that merged bureaucratic, technical, and cost/schedule/efficiency mandates that, in turn, reinforced decisions to continue flying with flaws.

3) Structural Secrecy: These same external forces affected NASA's organization structure and the structure of the safety system, which in turn affected the interpretation of the problem, so that the seriousness of these two anomalies was, in effect, unknown to those in a position to intervene. Instead, before the crisis events immediately preceding

the accidents, a consensus about these anomalies existed, including among agents of social control—top administrators and safety personnel—who failed to intervene to reverse the trend.

With these systemic social causes in mind, I now turn to the problem of repeating negative patterns and learning from mistake by considering the "Findings" and "Recommendations" of the report of the Presidential Commission on the Space Shuttle *Challenger* Accident, NASA's changes in response, and why the changes NASA implemented failed to prevent a second tragedy.[8]

THE PRESIDENTIAL COMMISSION: CONNECTING CAUSES AND STRATEGIES FOR CONTROL

Published in June 1986, the Presidential Commission's report followed the traditional accident investigation format of prioritizing the technical causes of the accident and identifying human factors as "contributing causes," meaning that they were of lesser, not equal, importance. NASA's organizational system was not attributed causal significance. However, the report was pathbreaking in the amount of its coverage of human factors, going well beyond the usual focus on individual incompetence, poor training, negligence, mistake, and physical or mental impairment.

Chapters 5 and 6 examine decisions about the O-ring problems, adhering to the traditional human factors/individual failure model. Chapter 5, "The Contributing Cause of the Accident," examines the controversial eve-of-the-launch teleconference. A "flawed decision making process" is cited as the primary causal agent. Managerial failures dominate the empirical "Findings": the teleconference was not managed so that the outcome reflected the opposition of many contractor engineers and some of NASA's engineers; managers in charge had a tendency to solve problems internally, not forwarding them to all hierarchical levels; the contractor reversed its first recommendation for delay "at the urging of Marshall [Space Flight Center] . . . to accommodate a major customer."[9]

Chapter 6, "An Accident Rooted in History," chronicled the history of O-ring decision-making in the years preceding the teleconference. Again, the empirical Findings located cause in individual failures.[10] Inadequate testing was done; neither the contractor nor NASA understood why the O-ring anomalies were happening; escalated risk-taking was endemic, apparently "because they got away with it the last time"; in a thorough review at Headquarters in 1985, information "was sufficiently detailed to require corrective action prior

---

8. Presidential Commission on the Space Shuttle *Challenger* Accident, *Report to the President*.

9. Ibid., p. 104.

10. Ibid., p. 148.

to the next flight"; managers and engineers failed to carefully analyze flight history, so data were not available on the eve of *Challenger*'s launch to properly evaluate the risks.[11] The system failure cited was in the anomaly tracking system, which permitted flight to continue despite erosion, with no record of waivers or launch constraints, and paid attention only to anomalies "outside the data base."

Both chapters described decision-making, focusing on interaction, but did not explain why decisions were made as they were. Chapter 7, "The Silent Safety Program," turned to organizational matters, initially addressing them in the traditional accident investigation frame. The Commission noted the failures: "lack of problem reporting requirements, inadequate trend analysis, misrepresentation of criticality and lack of involvement in critical discussions."[12] For example, they found so many problems listed on NASA's Critical Items List that the number reduced the seriousness of each. Acknowledging that top administrators were unaware of the seriousness of the O-ring problems, the Commission labeled the problem a "communication failure," thus deflecting attention from organization structure as a cause of the problems. In evaluating NASA's several safety offices and panels, however, the Commission made a break with the human factors approach by addressing the structure of regulatory relations. The Commission found that in-house safety programs were dependent upon the parent organization for funding, personnel, and authority. This dependence showed when NASA reduced the safety workforce even as the flight rate increased. In another economy move, NASA had increased reliance upon contractors, relegating many NASA technical experts to safety oversight of contractor activities, becoming dependent on contractors rather than retaining safety control in-house.

In chapter 8, "Pressures on the System," the Commission took an unprecedented step by examining schedule pressure and its effects on the NASA organization. However, this pressure, according to the report, was NASA-initiated, with no reference to external demands or restrictions on the Agency that might have contributed to it. The fault rested with NASA's own leaders. "NASA began a planned acceleration of the Space Shuttle launch schedule . . . . In establishing the schedule, NASA had not provided adequate resources for its attainment. As a result, the capabilities of the system were strained . . . ."[13] The system being analyzed is the flight production system: all the processes that must be engaged and completed in order to launch a mission. The report states that NASA declared the Shuttle "operational" after the fourth experimental flight even though the Agency was not prepared to meet the demands of an

---

11. Ibid., p. 148.
12. Ibid., p. 152.
13. Ibid., p. 164.

operational schedule. This belief in operational capability, according to the Commission, was reinforced by NASA's history of 24 launches without a failure prior to *Challenger* and to NASA's legendary "can-do" attitude, in which the space agency always rose to the challenge, draining resources away from safety-essential functions to do it.[14]

Next consider the fit between the Commission's "Findings," above, and their "Recommendations" for change, summarized as follows.[15] Many of the changes, if properly implemented, would reduce structural secrecy. The Commission mandated a review of Shuttle Management Structure because Project Managers felt more accountable to their Center administration than the Shuttle Program Director, thus vital information bypassed Headquarters. The Commission targeted "poor communications" by mandating that NASA eliminate the tendency of managers not to report upward, "whether by changes of personnel, organization, indoctrination or all three"; develop rules regarding launch constraints; and record Flight Readiness Reviews and Mission Management Team Meetings. Astronauts were to be brought into management to instill a keen awareness of risk and safety.[16]

Centralizing safety oversight, a new Shuttle Safety Panel would report to the Shuttle Program Manager. It would attend to Shuttle operations, rules and requirements associated with launch decisions, flight readiness, and risk management. Also, an independent Office of Safety, Reliability and Quality Assurance would be established, headed by an Associate NASA Administrator, with direct authority over all safety bodies throughout the Agency, and reporting to the NASA Administrator. With designated funding to give it independence, SR&QA would direct reporting and documentation of problems and trends affecting flight safety. Last, but by no means least, to deal with schedule pressures, the Commission recommended that NASA establish a flight rate that was consistent with its resources.

These were the official lessons to be learned from *Challenger*. The Commission's "Findings" and "Recommendations," in contrast to those later forthcoming from the CAIB, were few and very general, leaving NASA considerable leeway in how to implement them. How did the space agency respond? At the interaction level, NASA addressed the flawed decision-making by following traditional paths of changing policies, procedures, and processes that would increase the probability that signals of danger would be recognized. NASA used the opportunity to make changes to "scrub the system totally." The Agency rebaselined the Failure Modes Effects Analysis. All problems tracked by the Critical Items List were reviewed, engineering

14. Ibid., pp. 171–177.
15. Ibid., pp. 198–201.
16. Ibid., p. 200.

fixes implemented when possible, and the list reduced. NASA established Data Systems and Trend Analysis, recording all anomalies so that problems could be tracked over time. Rules were changed for Flight Readiness Review so that engineers, formerly included only in the lower-level reviews, could participate in the entire process. Astronauts were extensively incorporated into management, including participation in the final prelaunch Flight Readiness Review and signing the authorization for the final mission "go."

At the organizational level, NASA made several structural changes, centralizing control of operations and safety.[17] NASA shifted control for the Space Shuttle program from Johnson Space Center in Houston to NASA Headquarters in an attempt to replicate the management structure at the time of Apollo, thus striving to restore communication to a former level of excellence. NASA also initiated the recommended Headquarters Office of Safety, Reliability and Quality Assurance (renamed as Safety and Mission Assurance), but instead of the direct authority over all safety operations, as the Commission recommended, each of the Centers had its own safety organization, reporting to the Center Director.[18] Finally, NASA repeatedly acknowledged in press conferences that the Space Shuttle was and always would be treated as an experimental, not operational, vehicle and vowed that henceforth, safety would take priority over schedule in launch decisions. One step taken to achieve this outcome was to have an astronaut attending Flight Readiness Reviews and participating in decisions about Shuttle readiness for flight; another was an effort to bring resources and goals into alignment.

Each of these changes addressed causes identified in the report, so why did the negative pattern repeat, producing the *Columbia* accident? First, the Commission did not identify all the social causes of the accident. From our post-*Columbia* position of hindsight, we can see that the Commission did not target NASA's institutional environment as a cause. The powerful actors whose actions precipitated "Pressures on the System" by their policy and budgetary decisions do not become part of the contributing-cause scenario. NASA is obliged to bring resources and goals into alignment, although resources are determined externally. NASA took the blame for safety cuts, which were attributed to NASA's own "perception that less safety, reliability and quality assurance activity would be required during 'routine' Shuttle operations."[19] The external budgetary actions that forced NASA leaders to impose such efficiencies were not mentioned. Most of the Commission's recommended changes aimed at the organization itself, in particular, changing interactions

---

17. CAIB, *Report*, p. 101.

18. Ibid.

19. Presidential Commission on the Space Shuttle *Challenger* Accident, *Report to the President*, p. 160.

structure. The Commission did not name culture as a culprit, although production pressure is the subject of an entire chapter. Also, NASA's historic "can-do" attitude (a cultural attribute) is not made part of the "Findings" and "Recommendations." Thus, NASA was not sensitized to possible flaws in the culture or that action needed to be taken. The Commission did deal with the problem of structural secrecy; however, in keeping with the human factors approach, the report ultimately places responsibility for "communication failures" not with organization structure, but with the individual middle managers responsible for key decisions and inadequate rules and procedures. The obstacles to communication caused by hierarchy and consequent power that managers wielded over engineers, stifling their input in crucial decisions, are not mentioned. These obstacles originate in organization structure but become part of the culture.

Second, consider NASA's response to these "Recommendations" and the challenges they faced. Although NASA's own leaders played a role in determining goals and how to achieve them, the institutional environment was not in their control. NASA remained essentially powerless as a government agency dependent upon political winds and budgetary decisions made elsewhere. Thus, NASA had little recourse but to try to achieve its ambitious goals—necessary politically to keep the Agency a national budgetary priority—with limited resources. The intra-organizational changes that NASA did control were reviewed in the CAIB report.[20] It found that many of NASA's initial changes were good. However, a critical one—the structural changes to centralize safety—was not enacted as the Commission had outlined. NASA's new Headquarters Office of Safety, Reliability and Quality Assurance did not have direct authority, as the Commission mandated; further, the various Center safety offices in its domain remained dependent because their funds came from the activities that they were overseeing.[21]

The CAIB also found that other changes—positive changes—were undone by subsequent events stemming from political and budgetary decisions made by the White House and Congress. The new, externally imposed goal of the International Space Station (ISS) forced the Agency to mind the schedule and perpetuated an operational mode. As a consequence, the culture of production was unchanged; the organization structure became more complex. This structural complexity created poor systems integration; communication paths were not clear. Also, the initial surge in post-*Challenger* funding was followed by cuts, such that the new NASA Administrator, Daniel Golden, introduced new efficiencies and smaller programs with the slogan "faster, better, cheaper." As a result of the squeeze, the initial increase in NASA safety

---

20. CAIB, *Report*.
21. Ibid., pp. 101, 178–179.

personnel was followed by a repeat of pre-accident economy moves that again cut safety staff and placed even more responsibility for safety with contractors. The accumulation of successful missions (defined as flights returned without accident) also reproduced the belief in an operational system, thus legitimating these cuts: Fewer resources needed to be dedicated to safety. The loss of people and subsequent transfer of safety responsibilities to contractors resulted in a deterioration of post-*Challenger* trend analyses and other NASA safety oversight capabilities.

NASA took the report's mandate to make changes as an opportunity to make others it deemed necessary, so the number of changes actually made is impossible to know and assess, much less report in a chapter of this length. The extent to which additional changes might have become part of the problem rather than contributing to the solution is also unknown. Be aware, however, that we are assessing these changes from the position of post-*Columbia* hindsight, tending to identify all the negatives associated with the harmful outcome.[22] The positive effects, the mistakes avoided by post-*Challenger* changes,

The 13-member Columbia Accident Investigation Board poses for a group photo taken in the CAIB boardroom. The official STS-107 insignia hangs on the wall in the center of the boardroom. From left to right, seated, are Board members G. Scott Hubbard, Dr. James N. Hallock, Dr. Sally Ride, Chairman Admiral Hal Gehman (ret.), Steven Wallace, Dr. John Logsdon, and Dr. Sheila Widnall. Standing, from left to right, are Dr. Douglas D. Osheroff, Major General John Barry, Rear Admiral Stephen Turcotte, Brigadier General Duane Deal, Major General Kenneth W. Hess, and Roger E. Tetrault. *(CAIB photo by Rick Stiles, 2003)*

---

22. William H. Starbuck, "Executives' Perceptual Filters: What They Notice and How They Make Sense," in *The Executive Effect*, ed. Donald C. Hambrick (Greenwich, CT: JAI, 1988).

tend to be lost in the wake of *Columbia*. However, we do know that increasing system complexity increases the probability of mistake, and some changes did produced unanticipated consequences. One example was NASA's inability to monitor reductions in personnel during a relocation of Boeing, a major contractor, which turned out to negatively affect the technical analysis Boeing prepared for NASA decision-making about the foam problem.[23] Finally, NASA believed that the very fact that many changes had been made had so changed the Agency that it was completely different from the NASA that produced the *Challenger* accident. Prior to the CAIB report release, despite the harsh revelations about organizational flaws echoing *Challenger* that the CAIB investigation frequently released to the press, many at NASA believed no parallels existed between *Columbia* and *Challenger*.[24]

THE CAIB: CONNECTING CAUSES WITH STRATEGIES FOR CONTROL

Published in August 2003, the Columbia Accident Investigation Board report presented an "expanded causal model" that was a complete break with accident investigation tradition. Turning from the usual accident investigation focus on technical causes and human factors, the CAIB fully embraced an organizational systems approach and was replete with social science concepts. Further, it made the social causes equal in importance to the technical causes, in contrast to the Commission's relegation of nontechnical causes to "contributing causes." Part 1 of the CAIB report, "The Accident," addressed the technical causes; part 2, "Why the Accident Occurred," examined the social causes; part 3 discussed the future of spaceflight and recommendations for change.

In the executive summary, the CAIB report articulated both a "technical cause statement" and an "organizational cause statement." On the latter, the Board stated that it "places as much weight on these causal factors as on the more easily understood and corrected physical cause of the accident."[25] With the exception of the "informal chain of command" operating "outside the organization's rules," this organizational cause statement applied equally to *Challenger*:

> The organizational causes of this accident are rooted in the Space Shuttle Program's history and culture, including the

23. CAIB, *Report*.

24. Michael Cabbage and William Harwood, *CommCheck . . . The Final Flight of Shuttle Columbia* (New York: Free Press, 2004), p. 203.

25. CAIB, *Report*, p. 9.

original compromises that were required to gain approval for the Shuttle, subsequent years of resource constraints, fluctuating priorities, schedule pressures, mischaracterization of the Shuttle as operational rather than developmental, and lack of an agreed national vision for human space flight. Cultural traits and organizational practices detrimental to safety were allowed to develop, including reliance on past success as a substitute for sound engineering practices (such as testing to understand why systems were not performing in accordance with requirements); organizational barriers that prevented effective communication of critical safety information and stifled professional differences of opinion; lack of integrated management across program elements; and the evolution of an informal chain of command and decision-making processes that operated outside the organization's rules.[26]

The part 2 chapters described system effects. In contrast to the Commission's report, the CAIB explained NASA actions as caused by social factors. Chapter 5, "From *Columbia* to *Challenger*," began part 2 with an analysis of NASA's institutional environment. Tracking historic decisions by leaders in NASA's political and budgetary environment and the effect of policy decisions on the Agency after the first accident, it showed how NASA's external environment caused internal problems by shaping organization culture: the persistence of NASA's legendary can-do attitude, excessive allegiance to bureaucratic proceduralism and hierarchy due to increased contracting out, and the squeeze produced by "an agency trying to do too much with too little" as funding dropped so that downsizing and sticking to the schedule became the means to all ends.[27] The political environment continued to produce pressures for the Shuttle to operate like an operational system, and NASA accommodated. Chapter 6, "Decision Making at NASA," chronicled the history of decision-making on the foam problem, showing how the weak, mixed, and routine signals behind the normalization of deviance prior to *Challenger* also precipitated NASA's second gradual slide into disaster. Chapter 6 presented evidence that schedule pressure directly impacted management decision-making about the *Columbia* foam debris hit. Also, it showed how NASA's bureaucratic culture, hierarchical structure, and power differences created missing signals, so that the depth of engineer concerns and logic of their request for imagery were not admitted to poststrike deliberations.

---

26. Ibid.
27. Ibid., pp. 101–120.

Chapter 7, "The Accident's Organizational Causes," stepped back from the reconstruction of the decision history to examine how the organizational context affected the decisions traced in chapter 6. The chapter set forth an analysis of NASA's organizational culture and structure. The focal point was the "broken safety culture" that resulted from a weakened safety structure that, in turn, caused decision-makers to "miss the signals the foam was sending."[28] Organization structure, not communication failure, was responsible for problems with conveying and interpreting information. Systems integration and strong independent NASA safety systems were absent. Incorporating the social science literature from organization theory, theories of risk, and accidents, this chapter surveyed alternative models of organizations that did risky work, posing some safety structures that NASA might consider as models for revamping the Agency. Then, in the conclusion, it connected these organizational factors with the trajectory of decision-making after the *Columbia* foam strike. Chapter 8, "History as Cause: *Columbia* and *Challenger*," compared the two accidents. By showing the repeating patterns, it established the second accident as an organizational system failure, making obvious the causal links within and between the three preceding chapters. It demonstrated that the causes of *Challenger* had not been fixed. By bringing forward the thesis of "history" as cause, it showed how both the history of decision-making by political elites and the history of decision-making by NASA engineers and managers had twice combined to produce a gradual slide into disaster.

Now consider the fit between the Board's expanded causal model and its "Findings" and its "Recommendations." Empirically, the CAIB found the same problems as did the Presidential Commission and in fact recognized that in the report: schedule pressure; dependent and understaffed safety agents; communication problems stemming from hierarchy, power differences, and structural arrangements; poor systems integration and a weakened safety system; overburdened problem-reporting mechanisms that muted signals of potential danger; a can-do attitude that translated into an unfounded belief in the safety system; a success-based belief in an operational system; and bureaucratic rule-following that took precedence over deference to the expertise of engineers.[29] The data interpretation and causal analysis differed, however, because the CAIB report integrated social science analysis and concepts throughout part 2: culture, institutional failure, organizational system, history as cause, structure, the normalization of deviance, and the causal linkages between the three empirical chapters. Thus, the CAIB targeted for change each of the three layers of NASA's organizational system. A second difference

---

28. Ibid., p. 164.
29. Ibid., p. 100.

was that the number of findings and recommendations was greater and each was more detailed and specific than those of the Commission. A few of those illustrative of the organization system approach to change follow.

Chapter 5, "From *Challenger* to *Columbia*," tracing historic decisions by leaders, included neither findings nor recommendations about NASA's external environment. However, in contrast to the Commission's report, the CAIB specifically implicated decision leaders by the data in chapter 5, and in the introduction to part 2, the CAIB report stated that the Agency

> accepted the bargain to operate and maintain the vehicle in the safest possible way. The Board is not convinced that NASA has completely lived up to the bargain, or that Congress and the Administration have provided the funding and support necessary for NASA to do so. This situation needs to be addressed— if the nation intends to keep conducting human space flight, it needs to live up to its part of the bargain.[30]

Policy and budgetary decisions by leaders again show up in the "Findings" and "Recommendations" in chapters 6 and 7. Chapter 6, "Decision Making at NASA," makes three Recommendations, primary among them the adoption of "a Shuttle flight schedule that is consistent with available resources."[31] Also, it advocated training the Mission Management Team, which did not operate in a decentralized mode or innovate, instead adhering to an ill-advised protocol in dealing with the foam strike. As Weick found with forest-fire fighters in a crisis, the failure "to drop their tools," which they were trained to always carry, resulted in death for most.[32] The CAIB recommendation was to train NASA managers to "drop their tools," responding innovatively rather than bureaucratically to uncertain flight conditions and to decentralize by interacting across levels of hierarchy and organizational boundaries.[33]

Chapter 7, "The Accident's Organizational Causes," asserts the important causal role of a broken safety culture and NASA's cultural "blind spot" that kept them from getting the signals the foam was sending. The "Recommendations" advocated changes in the structure of NASA's safety system: the broken safety culture was to be fixed by changing the safety structure. The Commission charged NASA to create an "independent Technical Engineering Authority" with complete authority over technical issues, its independence guaranteed by funding directly from NASA Headquarters, with no responsibility for sched-

---

30. CAIB, *Report*, p. 97.
31. Ibid., p. 139.
32. Karl E. Weick, "The Collapse of Sensemaking in Organizations: The Mann Gulch Disaster," *Administrative Science Quarterly* 38 (1993): 628–652.
33. CAIB, *Report*, p. 172.

ule or program cost.[34] After *Challenger*, cost, schedule, and safety were all the domain of a single office. Second, NASA Headquarters' Office of Safety and Mission Assurance would have direct authority and be independently resourced. Finally, to assure that problems on one part of the Shuttle (e.g., the foam debris from the External Tank) took into account ramifications for other parts (e.g., foam hitting the orbiter wing), the Space Shuttle Integration Office would be reorganized to include the orbiter, previously not included.

Chapter 8, "History as Cause," presented general principles for making changes, rather than concrete recommendations. These principles incorporate the three layers of NASA's organizational system and the relationship between them. First, decision-making patterns that normalize deviance should be altered by "strategies that increase the clarity, strength, and presence of signals that challenge assumptions about risk," which include empowering engineers, changing managerial practices, and strengthening the safety system.[35] Second, this chapter reiterates the accountability at higher levels, stating, "The White House and Congress must recognize the role of their decisions in this accident and take responsibility for safety in the future."[36] Later and more specifically, "Leaders create culture. It is their responsibility to change it .... The past decisions of national leaders—the White House, Congress, and NASA Headquarters—set the *Columbia* accident in motion by creating resource and schedule strains that compromised the principles of a high-risk technology organization."[37] Third, at the organizational level, culture and structure are both targets for change. Understanding culture should be an ongoing research-based project. Necessary changes to organization structure must be carefully considered because of the law of unintended consequences: change and increased complexity produce mistake; changing structure can change culture in unpredictable ways.

The report made it imperative that NASA respond to many of these recommendations prior to the Return to Flight Evaluation in 2005.[38] Although change is still under way at NASA, it is appropriate to examine the direction NASA is taking and the obstacles the Agency is encountering as it goes about implementing change.

Signals of Danger and the Normalization of Deviance

Because the Space Shuttle is and always will be an experimental vehicle, technical problems will proliferate. In such a setting, categorizing risk will

---

34. Ibid., p. 193.
35. Ibid., p. 203.
36. Ibid., p. 196.
37. Ibid., p. 203.
38. Prior to the resumption of Shuttle launches, progress on these changes was monitored and approved by a NASA-appointed board, the Covey-Stafford Board, and also by the U.S. Congress House Committee on Science, which has official oversight responsibilities for the space agency.

always be difficult, especially with low-lying, ambiguous problems, like foam debris and O-ring erosion, where the threat to flight safety is not readily apparent and mission success constitutes definitive evidence: calculations and lab experiments are approximations, but flight outcome is considered the final test of engineering predictions. The decision problem is not only how to categorize the many elements and variations in risk, but how to make salient early warning signs about low-lying problems that, by definition, will be seen against a backdrop of more serious problems.

The new NASA Engineering and Safety Center (NESC), created after the *Columbia* accident, is to be a safety resource for engineering decisions throughout the Agency. NESC will review recurring anomalies that engineering had determined do not affect flight safety to see if those decisions were correct.[39] Going back to the start of the Shuttle program, NESC will create a common database, looking for missed signals, reviewing problem dispositions, and taking further investigative and corrective action when deemed necessary. However, as we have seen from *Columbia* and *Challenger,* what happens at the level of everyday interaction, interpretation, and decision-making does not occur in a vacuum, but in an organizational system in which other factors affect problem definition, corrective actions, and problem dispositions.

The Culture of Production: NASA's Political/Economic Environment

NASA remains a politically vulnerable agency, dependent on the White House and Congress for its share of the budget and approval of its goals. After *Columbia*, the Bush administration supported the continuation of the Space Shuttle program and supplied the vision for NASA's future that the CAIB report concluded was missing: the space program would return to exploration of Mars. However, the funds to make the changes required for the Shuttle to return to flight and simultaneously accomplish this new goal were insufficient. Thus, NASA, following the CAIB prescription, attempted to align goals and resources by phasing out the Hubble telescope program and, eventually, planning to phase out the Shuttle itself. Further, during the standdown from launch while changes are implemented, the International Space Station is still operating and remains dependent upon the Shuttle to ferry astronaut crews, materials, and experiments back and forth in space. Thus, both economic strain and schedule pressure still persist at NASA. How the conflict between NASA's goals and the constraints upon achieving them will unfold is still unknown, but one lesson from *Challenger* is that system effects tend to reproduce. The Board mandated independence and resources for the safety system, but when

---

39. Frank Morring, Jr., "Anomaly Analysis: NASA's Engineering and Safety Center Checks Recurring Shuttle Glitches," *Aviation Week & Space Technology* (2 August 2004): 53.

goals, schedule, efficiency, and safety conflicted post-*Challenger*, NASA goals were reined in, but the safety system also was compromised.

The Organization: NASA Structure and Culture

In the months preceding the report release, the Board kept the public and NASA informed of some of the recommended changes so that NASA could get a head start on changes required for Return to Flight. With the press announcement that the CAIB would recommend a new safety center, and pressed to get the Shuttle flying again, NASA rushed ahead to begin designing a center despite having no details about what it should entail. When the report was published, NASA discovered that the planned NASA Engineering and Safety Center (NESC) it had designed and begun to implement was not the Independent Technical Authority that the Board recommended. Converting to the CAIB-recommended structure was resisted internally at NASA, in large part because the proposed structure a) did not fit with insiders' ideas about how things should work and where accountability should lie and b) was difficult to integrate into existing operations and structures. NESC is in operation, as described above, but NASA is now working on a separate organization, the Independent Technical Authority, as outlined by the CAIB.

Whereas CAIB recommendations for changing structure were specific, CAIB directions for changing culture were vague. The CAIB was clear about implicating NASA leaders, making them responsible for changing culture. What was the role of NASA leaders in cultural change, and how should that change be achieved? The report's one clear instruction for making internal change was for correcting the broken safety culture by changing the structure of the safety system. From my participation in meetings at NASA, it was clear that NASA leaders did not understand how to go about changing culture. To these leaders, who were trained in engineering and accustomed to human factors analysis, changing culture seemed "fuzzy." Many NASA personnel believed that the report's conclusion about Agencywide cultural failures wrongly indicted parts of NASA that were working well. More fundamentally, they had a problem translating the contents of the report to identify what changes were necessary and what actions they implied. Each of the three causal chapters contained explicit information about where necessary cultural changes were needed:

1) Chapter 5 shows actions by leaders in OMB, Congress, the White House, and NASA made cost and schedule a part of the organization culture, competing with safety and technical and scientific innovation as goals.

2) Chapter 6 shows how the technical anomaly became normalized, experience with the foam debris problem leading to a cultural belief that foam was not a threat to flight safety.

3) Chapter 7 points out a gap; administrators' belief in NASA's strong "safety culture" was contradicted by the way the organization actually operated in this accident. Layers of structure, hierarchy, protocol, power differences, and an informal chain of command in combination stifled engineering opinion and actions, impeding information gathering and exchange, showing a culture where deference to engineering technical expertise was missing. The belief that operations were safe led NASA to buy as much safety as they felt they needed; cutbacks were made in safety personnel accordingly.

So changes that targeted the cause of NASA's cultural problems had to be three-pronged. But how to do it? NASA's approach was this: On 16 December 2003, NASA Headquarters posted a Request for Proposals on its Web site for a cultural analysis to be followed by the implementation of activities that would eliminate cultural problems identified as detrimental to safety. Verifying the CAIB's conclusions about NASA's deadline-oriented culture, proposals first were due 6 January; then the deadline was extended by a meager 10 days. Ironically, the CAIB mandate to achieve cultural change itself produced the very production pressure about which the report had complained. Although the study was to last three years, NASA required data on cultural change in six months (just in time for the originally scheduled date of the Return to Flight Evaluation, later deferred several times), then annually.

The bidders were corporate contractors with whom NASA frequently worked. Details are not available at this writing, but the awardee conducted a "cultural analysis" survey to gather data on the extent and location of cultural problems in the Agency. The ability of a survey to tap into cultural problems is questionable because it asks insiders, who can be blinded to certain aspects of their culture. A better assessment results when insider information is complemented by outside observers who become temporary members, spending sufficient time there to be able to identify cultural patterns, examine records, and interview asking open-ended questions. A further problem is implied in the initial response rate of 40 percent, indicating that insider viewpoints tapped will not capture Agencywide cultural patterns. Further, this survey was to be followed by plans to train and retrain managers to listen and decentralize and to encourage engineers to speak up. Thus, the Agency response would be at the interactional level only, leaving other aspects of culture identified in the CAIB report—such as goals; schedule pressures; power distribution across the hierarchy and between administrators, managers, and engineers—unaddressed. The agency that had always been expected to do too much with too little was still struggling with that all-too-familiar situation.

CONCLUSION: LESSONS LEARNED

The dilemmas of slippery slopes, repeating negative patterns, and learning from mistake are not uniquely NASA's. We have evidence that slippery slopes are frequent patterns in manmade disasters.[40] We also know that slippery slopes with harmful outcomes occur in other kinds of organizations where producing and using risky technology is not the goal: think of the incursion of drug use into professional athletics, U.S. military abuse of prisoners in Iraq, and Enron—to name some sensational cases in which incrementalism, commitment, feedback, cultural persistence, and structural secrecy seem to have created an organizational "blind spot" that allowed actors to see their actions as acceptable and conforming, perpetuating a collective incremental descent into poor judgment. Knowing the conditions that cause organizations to make a gradual downward slide, whether the manmade disasters that result are technical, political, financial, public relations, moral, or other, does give us some insight into how it happens that may be helpful to other managers hoping to avoid these problems.

In contradiction to the apparent suddenness of their surprising and sometimes devastating public outcomes, mistakes can have a long incubation period. How do early warning signs of a wrong direction become normalized? A first decision, once taken and met by either success or no obvious failure (which also can be a success!), sets a precedent upon which future decisions are based. The first decision may be defined as entirely within the logic of daily operations because it conforms with ongoing activities, cultural norms, and goals. Or, if initially viewed as deviant, the positive outcome may neutralize perceptions of risk and harm; thus, what was originally defined as deviant becomes normal and acceptable as decisions that build upon the precedent accumulate. Patterns of information bury early warning signs amidst subsequent indicators that all is well. As decisions and their positive result become public to others in the organization, those making decisions become committed to their chosen line of action, so reversing direction—even in the face of contradictory information—becomes more difficult.

The accumulating actions assume a taken-for-granted quality, becoming cultural understandings, such that newcomers may take over from others without questioning the status quo; or, if objecting because they have fresh eyes that view the course of actions as deviant, they may acquiesce and participate upon learning the decision logic and that "this is the way we do it here." Cultural beliefs persist because people tend to make the problematic nonprob-

---

40. Barry M. Turner, *Man-made Disasters* (London: Wykeham, 1978); Scott A. Snook, *Friendly Fire: The Accidental Shootdown of U.S. Black Hawks over Northern Iraq* (Princeton: Princeton University Press, 2000).

lematic by defining a situation in a way that makes sense of it in cultural terms. NASA's gradual slides continued because 1) the decisions made conformed to the mandates of the dominating culture of production and 2) because organization structure impeded the ability of those with regulatory responsibilities—top administrators, safety representatives—to critically question and intervene.

Why do negative patterns repeat? Was it true, as the press concluded after *Columbia*, that the lessons of *Challenger* weren't learned? When we examined the lessons of *Challenger* identified in the "Findings" and "Recommendations" of the Commission's 1986 report, they located cause primarily in individual mistakes, misjudgments, flawed analysis, flawed decision-making, and communication failures. The findings about schedule pressures and safety structure were attributed also to flawed decision-making, not by middle managers but by NASA leaders. In response, the Commission recommended adjusting decision-making processes, creating structural change in safety systems, and bringing goals and resources into alignment. NASA acted on each of those recommendations; thus, we could say that the lessons were learned. The *Columbia* accident and the CAIB report that followed taught different lessons, however. They showed that an organizational system failure, not individual failure, was behind both accidents, causing the negative pattern to repeat. So, in retrospect, we must conclude that from *Challenger* NASA learned incomplete lessons. Thus, they did not connect their strategies for control with the full social causes of the first accident.

Events since *Columbia* teach an additional lesson: we see just how hard it is to learn and implement the lessons of an organization system failure, even when the CAIB Report pointed them out. Further, there are practical problems. NASA leaders had difficulty integrating new structures with existing parts of the operation; cultural change and how to go about it eluded them. Some of the CAIB recommendations for change were puzzling to NASA personnel because they had seen their system working well under most circumstances. Further, understanding how social circumstances affect individual actions is not easy to grasp, especially in an American ethos in which both success and failure are seen as the result of individual action.[41] Finally, negative patterns can repeat because making changes has system effects that can produce unintended consequences. Changing structure can increase complexity and, therefore, the probability of mistake; it can change culture in unpredictable ways.[42]

---

41. After a presentation in which I translated the cultural change implications of the CAIB report to a group of administrators at NASA Headquarters, giving examples of how to go about it, two administrators approached me. Drawing parallels between the personalities of a *Columbia* engineer and a *Challenger* engineer who both acted aggressively to avert an accident but, faced with management opposition, backed off, the administrators wanted to know why replacing these individuals was not the solution.

42. Charles B. Perrow, *Normal Accidents* (New York: Basic Books, 1994); Diane Vaughan, "The Dark Side of Organizations: Mistake, Misconduct, and Disaster," *Annual Review of Sociology* 25

continued on the next page

Even when the lessons are learned, negative patterns can still repeat. The process and mechanisms behind the normalization of deviance make incremental change hard to detect until it's too late. Change occurs gradually, the signs of a new and possibly harmful direction occurring one at a time, injected into daily routines that obfuscate the developing pattern. Moreover, external forces are often beyond a single organization's ability to control. Cultures of production, whether production of police statistics, war, profits, or timely Shuttle launches, are a product of larger historical, cultural, political, ideological, and economic institutions that produce them. Making organizational change that contradicts them is difficult to implement but, in the face of continuing and consistent institutional forces, even more difficult to sustain as time passes. The extent to which an organization can resist these conditions is likely to vary as its status and power vary. Although compared to some, NASA seems a powerful government agency, its share of the federal budget is small compared to other agencies. In the aftermath of both accidents, NASA changes were undermined by subsequent events, many of which they could not control. Political and budgetary decisions of elites created new goals, resulting in new structures, making the system more complex; by not giving sufficient support, they reproduced a culture dominated by schedule pressures, deadlines, resource scarcity, bureaucratic protocols, and power differences that made it difficult to create and sustain a different kind of NASA where negative patterns do not repeat. It may be argued that under the circumstances, NASA's Space Shuttle program has had a remarkable safety record.

But even when everything possible is done, we cannot have mistake-free organizations because system effects will produce unanticipated consequences. Because the Shuttle is unprecedented and flight conditions unpredictable, NASA will always have many postflight anomalies to deal with, and low-lying problems with hard-to-decipher, uncertain outcomes like O-ring erosion and foam debris will always be a challenge. Part of the remedy is to increase the power and effectiveness of the safety system, but the critical piece to this puzzle is changing the culture of production. For *Columbia*, as for *Challenger*, resources—both time and money—were not available for thorough hazard analysis to fully explore why these two technical problems were occurring and the implications of continuing to fly with flaws. The reason they were not thoroughly analyzed and fixed was that the level of risk assigned to these problems was low. The definition of risk precluded the dedication of time and money to problems that had no clear potential for high costs. Further, all contingencies can never be predicted; most people don't understand how social

continued from the previous page

(1999): 271–305; Diane Vaughan, "Organisational Rituals of Risk and Error," in *Organisational Encounters with Risk*, ed. Bridget M. Hutter and Michael K. Power (Cambridge: Cambridge University Press, 2005).

context affects individual action and so cannot create strategies of control that connect with the social causes of a problem; organizational changes that correct one problem may, in fact, have a dark side, creating unpredictable others; and external environments are difficult to control.

Jervis describes the unintended consequences and harmful outcomes that result from complex interactions in social systems.[43] When complex, interactive technical systems, like the Space Shuttle, are run by complex organizations, like NASA, the probability of accidents is increased. Thus, system effects force us to recognize that it is not possible to prevent all accidents. However, it is important to remember that both of NASA's accidents had a long incubation period, and thus *were preventable*. By addressing the social causes of gradual slides and repeating negative patterns, organizations *can reduce the probability* that mistakes and accidents will occur. To do so, connecting strategies for correcting organizational problems with their social causes is crucial. Social scientists can play a significant role. First, we have research showing the problem of the slippery slope is perhaps more frequent than we now imagine, but less is known about cases where this pattern, once begun, is reversed.[44] Building a research base about organizations that make effective cultural change and reverse downward slides is an important step. Further, by their writing, analysis, and consulting, social scientists can 1) teach organizations about the social sources of their problems, 2) advise on strategies that will address those social causes, and 3) explore the system effects of planned changes, helping to forestall unintended consequences.[45]

---

43. Robert Jervis, *System Effects: Complexity in Political and Social Life* (Princeton: Princeton University Press, 1997).

44. Turner, *Man-made Disasters*; David Miller, *The Icarus Paradox: How Exceptional Companies Bring About Their Own Downfall* (New York: Harper, 1990), but see Rosabeth Moss Kanter, *Confidence: How Winning Streaks and Losing Streaks Begin and End* (New York: Simon & Schuster, 2004).

45. See, e.g., Rosabeth Moss Kanter, *The Changemasters* (New York: Simon & Schuster, 1983), and *Confidence: How Winning Streaks and Losing Streaks Begin and End* (New York: Simon & Schuster, 2004); Karlene H. Roberts, "Managing High Reliability Organizations," *California Management Review* 32, no. 4 (1990): 101–114; Karl E. Weick, Kathleen Sutcliffe, and David Obstfeld, "Organizing for High Reliability," *Research in Organizational Behavior* 21 (1990): 81–123; Todd R. La Porte and Richard Consolini, "Working in Practice but not in Theory: Theoretical Challenges of High-Reliability Organizations," *Journal of Public Administration Research and Theory* 1 (1991): 19–47; Diane Vaughan, "The Trickle-Down Effect: Policy Decisions, Risky Work, and the *Challenger* Accident," *California Management Review* 39 (winter 1997): 1–23; Lee Clarke, *Mission Improbable: Using Fantasy Documents to Tame Disaster* (Chicago: University of Chicago Press: 1999); Anita L. Tucker and Amy C. Edmondson, "Why Hospitals Don't Learn from Failures: Organizational and Psychological Dynamics that Inhibit System Change," *California Management Review* 45 (winter 2003): 55–72; Karen Marais, Nicolas Dulac, and Nancy Leveson, "Beyond Normal Accidents and High Reliability Organizations: The Need for an Alternative Approach to Safety in Complex Systems," (unpublished manuscript, Massachusetts Institute of Technology, 2004); Amy C. Edmondson, Michael Roberto, and Richard Bohmer, *The Columbia's Last Flight* (multimedia business case, Harvard Business School, 2005).

Second, NASA's problem of the cultural blind spot shows that insiders are unable to identify the characteristics of their own workplace structure and culture that might be causing problems. This suggests that rather than waiting until after a gradual slide into disaster or repeat of a negative pattern to expose the dark side of culture and structure, organizations would benefit from ongoing cultural analysis by ethnographically trained sociologists and anthropologists giving regular feedback, annually replaced by others to avoid seduction by the cultural ethos and assure fresh insights. Bear in mind this additional obstacle: the other facet of NASA's cultural blind spot was that the Agency's success-based belief in its own goodness was so great that it developed a pattern of disregarding the advice of outside experts.[46] To the extent that the CAIB report's embrace of an organizational system approach becomes a model for other accident investigation reports, other organizations may become increasingly aware of the social origins of mistakes and of the need to stay in touch with how their own organizational system is working.

---

46. CAIB, *Report*, chap. 5.

CHAPTER 12

ACCIDENTS, ENGINEERING, AND HISTORY AT NASA, 1967–2003

Alexander Brown

Section 203(a)(3) of the National Aeronautics and Space Act directs NASA to "provide for the widest practicable and appropriate dissemination of information concerning its activities and the results thereof."[1] To fulfill that mandate, NASA Administrator T. Keith Glennan instituted the NASA History Office in 1959.[2] The office has stayed open ever since, collecting archival materials for NASA staff and outside researchers, writing history, and commissioning a wide range of works on NASA's history. Over the last decade, the budget of NASA's history office has remained constant at around $335,000 per annum, although funds allocated to the history office from project offices vary from year to year. Even assuming such a level over the lifetime of the office, and not adjusting for inflation, NASA's commitment to telling its own history has cost the organization at least $15 million. But this figure is dwarfed by three official histories of NASA not commissioned by the history office. In 1967, 1986, and 2003, NASA spent $31 million, $75 million, and $152.4 million to produce multivolume accounts of fatal accidents in the manned space program.[3] These three accident reports examined the fatal fire in Apollo 204 (Apollo 1) in 1967, the explosion of the Solid Rocket Booster in STS-51L (*Challenger*) in 1986, and the destruction of the orbiter in STS-107 (*Columbia*).

Fatal accidents in publicly funded systems catch particular media and public attention.[4] Governments become compelled to conduct wide-ranging

---

1. John M. Logsdon et al., eds., *Exploring the Unknown: Selected Documents in the History of the U.S. Civil Space Program*, vol. 1, *Organizing for Exploration* (Washington, DC: NASA SP-4407, 1995), p. 337.

2. Roger D. Launius, "NASA History and the Challenge of Keeping the Contemporary Past," *Public Historian* 21, no. 3 (summer 1993): p. 63.

3. For Apollo 1, see Ivan D. Ertel and Roland Newkirk, with Courtney G. Brooks, *The Apollo Spacecraft: A Chronology*, vol. 4 (Washington, DC: NASA SP-4009, 1978); for *Challenger*, see Frank Oliveri, "NASA gets $50 million for Shuttle Investigation," *Florida Today* (21 February 2004); for *Columbia*, see Paul Recer, "NASA: Columbia Cleanup costs near $400M," *Newsday* (11 September 2003).

4. Thomas White, Jr., "Establishment of Blame as a Framework for Sensemaking in the Space Policy Subsystem: A Study of the Apollo 1 and Challenger Accidents" (Ph.D. diss., Virginia Polytechnic

continued on the next page

investigations to reassure the public of the safety of the system and the integrity of the funding process. Accidents at NASA are particularly public and so demand an investigation process that is accountable not only to the Congress but also to the American people. NASA accident investigation boards are forced to draw connections between national politics and engineering design and operations. The process of writing a final report also forces an accident investigation body to tell one coherent story about the accident—how the accident happened, what and who was at fault, and how steps can be taken to ensure the accident cannot happen again.

But as Peter Galison has observed in his study of aircraft accidents in the 1980s, accident reports are inherently unstable. They are multicausal in their historical explanations, and yet embedded in the very process of investigation is a drive for a single point of culpability upon which to base moral responsibility and recommendations for corrective action. Accident reports, then, are always ambiguous about the appropriate explanatory scale, so that it is never clear which is the right scale for analysis—whether the small scale or the large scale, the inflexible O-ring or the schedule pressure imposed on NASA by the White House and Congress.[5]

Galison is certainly correct to assert that reports show an explanatory tension, but this instability between frames of analysis is not just a function of the particular genre of accident reports. Engineering has changed such that there is now a social and epistemological gap between the management of engineering and engineering practice. The analytical tension in the investigation reports mirrors the real gap between engineers and managers at NASA. Furthermore, the reports are analytically asymmetrical, treating engineering as a context-free activity while explaining management in a sophisticated historical and cultural framework.

These gaps are not just a phenomenon inherent to accident reports, but the outcome of a set of historical and historiographical changes. The Apollo 204 accident shows the disjuncture between the engineers designing and managing the project and the technicians manufacturing the spacecraft. The *Challenger* and *Columbia* accidents show that disjuncture has shifted to the gap between managers controlling the project and engineers maintaining and analyzing the spacecraft. Similarly, since the 1980s, the organizational theory and organizational communications communities have joined the aeronautical engineering community in paying significant scholarly attention to

---

continued from the previous page

Institute and State University, 2000). White's thesis analyzes the ways in which blame was allocated in these two accidents but also makes it very clear that public and political concern and outrage were extremely high in both cases.

5. Peter Galison, "An Accident of History," in *Atmospheric Flight in the Twentieth Century*, ed. Peter Galison and Alex Roland (Dordrecht, Netherlands: Kluwer Academic Publishers, 2000), pp. 3–43.

accidents at NASA. Their engagement has shifted attention to the historical and organizational context of management decision-making surrounding the accidents. No historians of engineering and technology have matched this contextualization of management with a history of the engineering involved in the accidents or an attempt to integrate the two.

This paper will briefly lay out the accidents and discuss the findings of their investigative bodies. The changing historiographical styles, frameworks, and conclusions of the reports will be analyzed. These changes will be linked to changes in the practice of engineering by NASA and its contractors. Finally, some suggestions will be made for future research into accidents and changes in engineering.

Apollo 204

On 27 January 1967, Spacecraft 012, assigned to the Apollo 204 mission, was undergoing a Plugs-Out Integrated Test on Pad 34 at Kennedy Space Center in Florida. The internal power systems of the newly delivered Command and Service Module were being tested, and so the crew cabin was pressurized to 16 pounds per square inch (psi) of pure oxygen. There were three astronauts on board: Gus Grissom, Ed White, and Roger Chaffee. At around 6:31 p.m. EST, the crew reported a fire in the spacecraft. Less than 20 seconds later, the spacecraft heatshield had ruptured and flame had burst into the service tower. The crew in the Command and Service Module (CSM) level of the support tower immediately evacuated the area but quickly returned with what firefighting and protective gear they could find. However, they were unable to extinguish the fire immediately or remove the crew from the cabin. Meanwhile, the crew had attempted to remove the middle hatch of the spacecraft but had been overcome before doing so. Firefighting crews and medical support arrived approximately 20 minutes later.

NASA Deputy Administrator Robert Seamans had already considered the possibility of an accident in the manned spaceflight program, after Neil Armstrong and Dave Scott in Gemini VIII had lost control of their capsule after docking with an Agena booster.[6] In the aftermath of Gemini VIII, he developed a set of procedures to be followed should an accident ever occur. On the evening of 28 January, he followed those procedures and immediately convened an accident review board.[7] The board convened at Kennedy Space Center in Florida and was

---

6. Robert C. Seamans, *Aiming at Targets* (Washington, DC: NASA SP-4106, 1996), pp. 135–136; Barton C. Hacker and James M. Grimwood, *On the Shoulders of Titans: A History of Project Gemini* (Washington, DC: NASA SP-4203, 1977), pp. 308–319.

7. Apollo 204 Review Board, appendix a-G, "Board Minutes," in *Report of Apollo 204 Review Board to the Administrator, National Aeronautics and Space Administration* (Washington, DC: GPO, 1967), pp. 1-5–1-6.

The mission officially designated Apollo/Saturn 204 is more commonly known as Apollo 1. This close-up view of the interior of the Command Module shows the effects of the intense heat of the flash fire that killed the prime crew during a routine training exercise. While they were strapped into their seats inside the Command Module atop the giant Saturn V Moon rocket, a faulty electrical switch created a spark that ignited the pure-oxygen environment. The speed and intensity of the fire quickly exhausted the oxygen supply inside the crew cabin. Unable to deploy the hatch due to its cumbersome design and the lack of breathable oxygen, the crew lost consciousness and perished. They were astronauts Virgil I. "Gus" Grissom (the second American to fly into space), Edward H. White II (the first American to "walk" in space), and Roger B. Chaffee (a "rookie" on his first space mission). *(JSC image no. S-67-21294, 28 January 1968)*

chaired by Floyd "Tommy" Thompson, Director of NASA's Langley Research Center.[8] The board was made up of three senior NASA engineers, a chemist from the Bureau of Mines, an Air Force officer from the Inspector General's office, NASA Langley's general counsel, and an astronaut.[9]

On 5 April 1967, the Apollo 204 Review Board presented its report to NASA Administrator James Webb. They concluded that the fire was caused

---

8. James R. Hansen, *Engineer in Charge: A History of the Langley Aeronautical Laboratory, 1917–1958* (Washington, DC: NASA SP-4305, 1987), pp. 387–391.

9. Apollo 204 Review Board, appendix a-G, "Board Minutes," pp. 2-1–2-17.

by an unknown source of electrical arc, probably malfunctioning wire insulation around the environmental control unit on the floor of the spacecraft, although the cause would never be definitively known. The spark then ignited nylon netting, Velcro strips, and other combustible materials inside the spacecraft. These materials would have been removed before spaceflight, but under test conditions were not seen as hazardous. The coolant inside the spacecraft, water-glycol, was flammable and left a flammable residue in the cabin after evaporation. As the pipes melted, coolant leaked and ignited, further fueling the fire. The fire was rendered particularly dangerous by the high-pressure, pure-oxygen environment inside the spacecraft during the test. The crew was unable to use the inward-opening inner hatch under the pressurized conditions. The Board determined that the crew had died from asphyxiation caused by fumes from the fire.[10]

The Board told a story of engineering failure, identifying six conditions that led to the fire, and provided recommendations to fix the engineering problems they identified. These conditions were a sealed cabin with a pressurized atmosphere, extensive distribution of flammable materials in the cabin, vulnerable wiring carrying spacecraft power, vulnerable plumbing containing combustible and corrosive coolant, inadequate escape provisions, and an inadequate provision for rescue or medical assistance.[11]

After the Board made their engineering recommendations, they spoke briefly about the larger circumstance surrounding the accident:

> Having identified the condition that led to the disaster, the Board addressed itself to the question of how these conditions came to exist. Careful consideration of this question leads the Board to the conclusion that in its devotion to the many difficult problems of space travel, the Apollo team failed to give adequate attention to certain mundane but equally vital questions of crew safety. The Board's investigation revealed many deficiencies in design and engineering, manufacture and quality control. When these deficiencies are corrected the overall reliability of the Apollo Program will be increased greatly.[12]

On 27 February 1967, the Senate Committee on Aeronautical and Space Sciences started to hold hearings on the Apollo 204 fire, and on 11 April, the House Committee on Science and Astronautics started to hold hearings into the Apollo 204 fire.

---

10. Ibid., pp. 5-1–5-12.
11. Ibid., p. 5-12.
12. Ibid., p. 5-12.

On the first day of the hearings before the Senate, NASA Administrator James Webb, Deputy Administrator Seamans, and Associate Administrator George Mueller were sandbagged by the Democratic Senator from Minnesota, Walter Mondale. Mondale asked them about a report that Apollo Program Director Major General Sam Phillips had prepared in 1965 after visiting North American Aviation (NAA), manufacturers of the spacecraft.[13] Mueller first denied any knowledge of the report, arguing that Phillips had prepared many reports on many NASA contractors. Webb then argued that he was not going to release the report for reasons of commercial confidentiality, as it contained details of contract negotiations between NASA and NAA.[14] Senators Brooke, Percy, and, in particular, Mondale became highly critical of NASA's unwillingness, as they saw it, to be accountable to elected officials.[15]

The Phillips report was damning. Phillips had written:

> I am definitely not satisfied with the progress and outlook of either program and am convinced that the right actions now can result in substantial improvement of position in both programs in the relatively near future.
>
> Even with due consideration of hopeful signs, I could not find a substantive basis for confidence in future performance. I believe that a task group drawn from NAA at large could rather quickly verify the substance of our conclusions, and might be useful to you in setting the course for improvements.[16]

Phillips recommended that NAA thoroughly revise (and in many cases implement) systems management and engineering procedures. He called for them to implement a program management system and to significantly improve their manufacturing and quality control.[17]

The House hearing subcommittee was chaired by Representative Olin Teague of Texas, a long-term supporter of the space program. The hearings were contentious—with a Republican from Illinois, Donald Rumsfeld, taking particular aim at NASA senior officials Webb, Seamans, and Faget. Rumsfeld took objection to the constitution of the Board, arguing that it

---

13. Senate Committee on Aeronautical and Space Sciences, *Apollo Accident*, Hearing, 90th Cong., 1st sess., 7 February 1967, pp. 125–127.

14. Ibid., pp. 131–132.

15. Ibid., pp. 217, 331–332; Senate Committee on Aeronautical and Space Sciences, *Apollo 204 Accident. Report of the Committee on Aeronautical and Space Sciences, United States Senate, with Additional Views* (Washington, DC: GPO, 1968), pp. 13–16.

16. Samuel Phillips, cover letter to Lee Atwood, in "NASA Review Team Report," 1965.

17. Ibid., pp. 1–20.

was made of people responsible for the areas of work whose failure they were investigating. NASA was, in effect, investigating itself. Rumsfeld was also concerned about the narrow focus of the Board's report, suggesting that they had defined their terms very specifically to avoid investigating larger problems within NASA management. Finally, he wanted to know why NASA did not have a separate and independent safety organization.[18] Webb and Seamans gave fairly weak responses to Rumsfeld's questions and were only saved by Teague's interruptions.

But the worst was still to come for NASA. It was revealed that in the initial awarding of the CSM contract to NAA, NAA had scored lower in the technical assessment than Martin. The Congressmen used this revelation to imply some sort of improper relationship between NASA and NAA.[19] In the final days of the House hearing, Thomas Baron, a quality-assurance inspector from NAA, presented to the Committee a detailed report of deficiencies, official malfeasances, and general complaints about the standard of workmanship and care at NAA.[20] While the Baron report was eventually proved to be largely personal grievances and unproven accounts of interactions between workers at NAA, it all contributed to a larger picture of poor management and workmanship at NAA and poor supervision at NASA.

Although the Apollo 204 Board did not blame any individuals for the fire, there were consequences. Joseph Shea, manager of the Apollo Spacecraft Program Office, and Harrison Storms, NAA's vice president in charge of the Space and Information Division, were both moved out of their positions.[21] Deputy Administrator Seamans also resigned soon after the investigation had concluded, his personal relationship with James Webb having deteriorated dramatically over the fire.[22]

CHALLENGER

On 28 January 1986, the Space Shuttle *Challenger* launched from Kennedy Space Center on mission 51-L. There were seven astronauts on board: Dick Scobee, Michael Smith, Ellison Onizuka, Judith Resnik, Ronald McNair, Christa McAuliffe, and Gregory Jarvis. Their mission was to deploy and

---

18. House Committee on Science and Astronautics, *Investigation into Apollo 204 Accident*, Hearings before the Subcommittee on NASA Oversight, 90th Cong., 1st sess., 10 April 1967, pp. 10–14.

19. Ertel and Newkirk, *The Apollo Spacecraft: A Chronology*, vol. 4, entry for 11 May 1967.

20. Thomas Ronald Baron, "An Apollo Report," in House Committee on Science and Astronautics, *Investigation into Apollo 204 Accident*, pp. 483–500.

21. Ertel and Newkirk, *The Apollo Spacecraft: A Chronology*, vol. 4, entry for 7 April 1967; Mike Gray, *Angle of Attack: Harrison Storms and the Race to the Moon*, 1st ed. (New York: W. W. Norton, 1992), pp. 254–255.

22. Seamans, *Aiming at Targets*, pp. 145–147.

recover a satellite in orbit and to conduct flight-dynamics experiments.[23] Christa McAuliffe, a teacher from New Hampshire, was to conduct a science lesson in orbit.[24] The 28th of January was a very cold morning. The temperature at Kennedy Space Center in Florida had dropped below freezing overnight, and ice teams had been sent out three times to examine potential damage. Parts of the Space Shuttle, including the Solid Rocket Boosters, were still below freezing point at launch. The ambient air temperature was 36°F, 15 degrees lower than any previous flight.[25]

Less than a second after launch, at 11:38 a.m. EST, a puff of gray smoke emerged from the right Solid Rocket Booster (SRB). Over the next 2 seconds, eight more puffs of smoke, blacker and more dense, emerged from the same place on the SRB. Thirty-seven seconds after launch, the Shuttle experienced a 27-second period of severe wind shear, stronger than any other Shuttle launch had experienced. Fifty-eight seconds after launch, a small flame appeared on the aft field joint of the right SRB. Over the next 14 seconds, the flame grew rapidly, burning through the lower strut holding the SRB to the External Tank. Seventy-two seconds after launch, the strut burned through and the right SRB rotated around the upper strut, crashing into the External Tank. The tank collapsed, venting the hydrogen fuel into the atmosphere. The fuel immediately ignited, and the entire Shuttle flew into the fireball. The orbiter entered the fireball, broke up under severe aerodynamic load, and fell back into the Atlantic Ocean. There were no survivors.[26]

On 3 February 1986, President Ronald Reagan appointed the Presidential Commission on the Space Shuttle *Challenger* Accident.[27] The Commission was chaired by William Rogers, Secretary of State under Richard Nixon and an attorney by training and experience. The Commission included two astronauts, a test pilot, two physicists, another attorney, three engineers, a senior Air Force officer, an aerospace journalist, and an astronomer. Another engineer was executive director. The Commission conducted public and private hearings over the early part of 1986 and presented its report to President Reagan on 6 June 1986.

Like the Apollo 204 Review Board, the Commission understood its objectives to be investigating the accident and providing a series of rec-

23. Presidential Commission on the Space Shuttle *Challenger* Accident, *Report to the President*, 5 vols. (Washington, DC: GPO, 1986), p. 16; "John F. Kennedy Space Center—51-L Shuttle Mission," *http://www-pao.ksc.nasa.gov/kscpao/shuttle/missions/51-l/mission-51-l.html*.

24. For Christa McAuliffe's official NASA biography, see *http://www.jsc.nasa.gov/Bios/htmlbios/mcauliffe.html*.

25. Presidential Commission on the Space Shuttle *Challenger* Accident, *Report to the President*, vol. 1, pp. 16–21.

26. Ibid., pp. 20–21.

27. Ibid., pp. 212–213.

The STS-51L crew members. In the back row, from left to right: mission specialist Ellison S. Onizuka, Teacher in Space participant Sharon Christa McAuliffe, payload specialist Greg Jarvis, and mission specialist Judy Resnik. In the front row, from left to right: pilot Mike Smith, commander Dick Scobee, and mission specialist Ron McNair. *(JSC image no. S85-44253,15 November 1985)*

ommendations for a return to safe flight.[28] And like the Apollo Board, the Commission examined the physical causes of the accident but was also critical of NASA and its contractors as organizations:

> The genesis of the Challenger accident—the failure of the joint of the right Solid Rocket Motor—began with decisions made in the design of the joint and in the failure by both Thiokol (manufacturer of the Solid Rocket Motors) and NASA's Solid Rocket Booster project office to understand and respond to facts obtained during testing.[29]

28. Ibid., p. 1.
29. Ibid., p. 166.

The Commission determined that a combustion gas leak through the aft field joint on the right Solid Rocket Motor caused the flame plume. The field joint was designed to be sealed by O-rings. On STS-51L, the O-rings failed to work because ambient temperature was too cold and the O-rings lost resilience and hence their ability to seal quickly.[30] The Commission's report took aim at poor management decisions, arguing that schedule- and cost-conscious managers misunderstood and overruled the safety judgments of engineers. They concluded that flaws existed in NASA's decision-making process and that these flaws had caused NASA to decide to launch STS-51L when there was reason to believe that launching would be risky and potentially catastrophic. NASA's safety system was indicted as silent and ineffective in the face of increasing pressure on the launch schedule. Finally, the Commission suggested that these flaws were rooted in the history of the Space Shuttle program and the history of NASA.[31]

Commissioner Richard Feynman went further in appendix F to the report. This appendix contained Feynman's personal observations from his service on the Commission and particularly addressed the difference he had observed between NASA and Thiokol engineers and managers. Feynman observed that managers and engineers tended to calculate risk in very different ways—managers determining risk from a number of qualitative factors, whereas engineers calculated risk quantitatively, using standard statistical methods. He also observed that these two methods tended to produce widely divergent results. Managers generally understood risks to be orders of magnitude less than engineers.[32] Feynman was highly critical of this gap, arguing that there were only two ways to understand it. The first was dishonesty on the part of managers, designed to ensure a continuous flow of funding for the Shuttle. The second was an incredible lack of communication between engineers and managers.[33] He argued that to ensure safe operation of the Shuttle, NASA managers needed to understand the realities of risk involved in flying high-performance vehicles like the Shuttle. After all, he concluded, "for a successful technology, reality must take precedence over public relations, for Nature cannot be fooled."[34]

---

30. Ibid., chaps. 3, 4.

31. Ibid. Chapter 5 discusses management decisions; chapter 6, the historical background of the accident; and chapter 7, NASA's safety program.

32. Richard Phillips Feynman and Ralph Leighton, *What Do You Care What Other People Think? Further Adventures of a Curious Character*, 1st ed. (New York: Norton, 1988), pp. 179–184. This volume also contains a version of appendix F edited for clarity, pp. 220–237. For the original version, see Presidential Commission on the Space Shuttle *Challenger* Accident, *Report to the President*, pp. F-1–F-5.

33. Feynman and Leighton, *What Do You Care What Other People Think?* pp. 236–237.

34. Presidential Commission on the Space Shuttle *Challenger* Accident, *Report to the President*, p. F-5.

The Commission's report echoed Feynman's findings, even though he felt upset that his opinions had not been adequately incorporated into the final document.[35] The report suggested that NASA management and NASA engineers saw the material world in very different ways—the engineers understanding risk as quantifiable and determined by the material world, whilst managers understood risk as flexible and manageable in commercial and political contexts. The cause of the accident, the report concluded, was the failure of communication between these two perspectives. The ultimate expression of this philosophy was the statement by Jerald Mason of Morton Thiokol telling Robert Lund, vice-president of engineering, "You've got to put on your management hat, not your engineering hat" in order to determine whether the *Challenger* would launch the next day despite engineers' concerns over the safety of the Solid Rocket Motor.[36] In its final recommendations, the Commission wanted design changes to the Solid Rocket Motor, reform of the Shuttle program management structure, and the establishment of a Shuttle Safety Panel and an independent Office of Safety, Reliability and Quality Assurance.

The House Committee on Science and Technology started holding hearings on the *Challenger* accident on 10 June 1986. As in Apollo 204, from which the Committee drew its precedent, hearings were delayed until the Commission report was published. The Committee conducted 10 days of hearings, questioning senior NASA and Morton Thiokol officials, as well as members of the Commission, astronauts, and Morton Thiokol engineers.[37] While the Committee endorsed the findings of the Commission, their report went further:

> The Committee feels that the underlying problem which led to the Challenger accident was not poor communication or inadequate procedures as implied by the Rogers Commission conclusion. Rather the fundamental problem was poor technical decision-making over a period of several years by top NASA and contractor personnel, who failed to act decisively to solve the increasingly serious anomalies in the Solid Rocket Booster joints.[38]

Neither the Commission nor the Committee explicitly laid blame at the feet of any individuals. However, their criticisms of management at NASA's

---

35. Feynman and Leighton, *What Do You Care What Other People Think?* pp. 199–205.
36. Presidential Commission on the Space Shuttle *Challenger* Accident, *Report to the President*, p. 94.
37. House Committee on Science and Technology, *Investigation of the Challenger Accident: Report of the Committee on Science and Technology, House of Representatives*, 99th Cong., 2nd sess., 1986, pp. 37–38.
38. Ibid., p. 5.

Marshall Space Flight Center and at Morton Thiokol were duly noted by those organizations. Most of Morton Thiokol management involved in the launch decision were reassigned, retired, or resigned, including Jerald Mason and Robert Lund. At NASA, Associate Administrator for Space Flight Jesse Moore resigned, while MSC Director William Lucas and booster project manager Lawrence Mulloy both retired early.[39]

COLUMBIA

On 16 January 2003, the Space Shuttle *Columbia* launched from Kennedy Space Center on mission 107. There were seven astronauts on board: Rick Husband, William McCool, Michael Anderson, David Brown, Kalpana Chawla, Laurel Clark, and Ilan Ramon. Fifty-seven seconds after launch, at around 10:40 a.m. EST, the *Columbia* entered a period of unusually strong wind shear, which created a low-frequency oscillation in the liquid oxygen in the External Tank.[40] At 81.7 seconds after launch, at least three pieces of Thermal Protection System foam detached from the left bipod ramp of the External Tank and fell backwards at between 416 and 573 miles per hour, smashing through the leading edge of the left wing of the orbiter. The largest piece of foam was around 2 feet long and 1 foot wide. The launch was otherwise without incident, and *Columbia* arrived in orbit by 11:39 a.m. EST.

On 23 January, Mission Control e-mailed commander Husband and pilot McCool to inform them of the foam strike, informing them that some foam had hit the orbiter but reassuring them that "we have seen this phenomenon on several other flights and there is absolutely no concern for entry."[41]

On 1 February 2003, after a successful 17-day mission, the orbiter reentered the Earth's atmosphere for a landing at Kennedy Space Center. As the orbiter reentered, superheated air penetrated the left wing through the foam strike in the leading edge and started to melt away the wing from the inside. At around 9:00 a.m. EST, the orbiter broke up under severe aerodynamic load and disintegrated over the Southwest of the United States. There were no survivors.

Around 10:00 a.m. on 1 February 2003, NASA Administrator Sean O'Keefe declared a Shuttle Contingency and, acting under procedures set in place after the *Challenger* accident, established the International Space Station

---

39. Claus Jensen, *No Downlink: A Dramatic Narrative About the Challenger Accident and Our Time*, 1st ed. (New York: Farrar, Straus and Giroux, 1996), pp. 354–355; Richard S. Lewis, *Challenger: The Final Voyage* (New York: Columbia University Press, 1988), pp. 222–223.

40. Columbia Accident Investigation Board, *Report*, vol. 1 (Washington, DC: NASA and GPO, 2003), pp. 33–34.

41. Ibid., p. 159.

and Space Shuttle Mishap Interagency Board.[42] O'Keefe named Admiral Harold Gehman as Chair of the Board. Gehman was retired from the Navy and had recently headed the investigation into the terrorist attack on the USS *Cole*.[43] Ex officio, there were immediately seven Board members: four military officers with responsibilities for safety in their home services, a Federal Aviation Administration representative, a Department of Transportation representative, and a NASA Center Director. O'Keefe soon thereafter named both NASA's Chief Engineer and the counsel to Glenn Research Center to the Board. Over the next six weeks, five more members were appointed to the renamed Columbia Accident Investigation Board. They included an aeronautical engineer and former Air Force Secretary, a physicist, a former astronaut and *Challenger* Commission member, a space policy expert, and the retired CEO of a major defense contractor.[44] Over the first six months of 2003, the Board held hearings and conducted investigations into the *Columbia* accident and, on 26 August 2003, released its report.

The CAIB report identified the physical cause of the accident as the foam strike on the left wing leading edge. But unlike the Apollo 204 Board, which briefly mentioned organizational and other factors, or the *Challenger* Commission, which described these factors as contributory, the CAIB emphasized that factors other than the proximate physical cause were as, if not more, important in understanding the *Columbia* accident:

> Many accident investigations make the same mistake in defining causes. They identify the widget that broke or malfunctioned, then locate the person most closely connected with the technical failure: the engineer who miscalculated an analysis, the operator who missed signals or pulled the wrong switches, the supervisor who failed to listen, or the manager who made bad decisions. When causal chains are limited to technical flaws and individual failures, the ensuing responses aimed at preventing a similar event in the future are equally limited: they aim to fix the technical problem and replace or retrain the individual responsible. Such corrections lead to a misguided and potentially disastrous belief that the underlying problem has been solved. The Board did not want to make these errors. A central piece of our expanded cause model involves NASA as an organizational whole.

---

42. Ibid., pp. 231–232.
43. William Langewiesche, "Columbia's Last Flight," *Atlantic Monthly* (November 2003): 65–66.
44. CAIB, *Report*, p. 232.

The organizational causes of this accident are rooted in the Space Shuttle Program's history and culture, including the original compromises that were required to gain approval for the Shuttle Program, subsequent years of resource constraints, fluctuating priorities, schedule pressures, mischaracterizations of the Shuttle as operational rather than developmental, and lack of an agreed national vision. Cultural traits and organizational practices detrimental to safety and reliability were allowed to develop, including: reliance on past success as a substitute for sound engineering practices (such as testing to understand why systems were not performing in accordance with requirements/specifications); organizational barriers which prevented effective communication of critical safety information and stifled professional differences of opinion; lack of integrated management across program elements; and the evolution of an informal chain of command and decision-making processes that operated outside the organization's rules.

In the Board's view, NASA's organizational culture and structure had as much to do with this accident as the External Tank foam.[45]

Seventeen years after *Challenger,* the Board concluded that many of the findings of the *Challenger* Commission were still applicable to the Space Shuttle program in the early 21st century. They were critical of the similarities between the *Challenger* and *Columbia* accidents, noting in the *Columbia* accident flawed decision-making processes, a silent safety program, and schedule pressure. The Board also observed that the causes of these failures were rooted in NASA's history and culture; the history of the Space Shuttle program had been a history of the normalization of deviance. Increasingly large engineering problems that had not caused catastrophic failures had been incorporated into NASA's experience base instead of raising safety concerns. NASA had come to rely on past success (or lack of past catastrophe) rather than rigorous testing and analysis. NASA's safety system was still silent. Decision-making was still flawed, with managers and engineers still unable to communicate effectively about risk.

The Commission recommended design changes to the Thermal Protection System on the External Tank, reform of the Space Shuttle Integration Office, training for the Mission Management Team, the establishment of an indepen-

---

45. Ibid., p. 177.

dent Technical Engineering Authority with safety responsibilities, and rendering the NASA Office of Safety and Mission Assurance independent and with total oversight of the Space Shuttle program safety organization.[46]

READING ACCIDENT REPORTS AS HISTORY

The Apollo 204 report is almost exclusively devoted to an analysis of the engineering problems that the Board argued caused the fire. It divides its analysis into two parts, parts IV and V of the report.[47] Part IV, "History of the Accident," provides a chronology of the accident from August 1964 until 28 January 1967. The sections discussing the fabrication, delivery, and inspection of the CSM spacecraft, which cover the period from August 1964 until December 1966, take up less than 10 percent of the report. The remainder of the history of the accident is a detailed chronology of the Plugs-Out Integrated Test of CSM 012, starting around 5 hours and 30 minutes before the accident. Part V, "Investigation and Analysis," has four sections: "Inspection and Disassembly," "Chronology," "Data Analyses," and "Cause of the Fire." Both the "Inspection and Disassembly" and "Chronology" sections are strictly narrative. "Data Analyses" discusses analyses of spacecraft telemetry data and crew voice transmissions from less than a minute before the accident, while the "Causes of the Fire" section notes deficiencies in electrical equipment and wiring insulation, the effects of electrical arcs on wiring and coolant on other equipment, and the effects of a cabin environment of pure oxygen under pressure. The sole mention of other, larger contributory factors is the final paragraph, noting that these engineering problems came about through deficiencies in design and manufacturing.[48]

But none of the political circumstances surrounding the Apollo program—its iconic status as the martyred President Kennedy's legacy, as a visible symbol of American technical prowess, as a marker of position in the Cold War—were identified as contributory. Nor was NASA's organizational structure or its culture. No individuals were identified as bearing particular responsibility for the accident. The report makes clear that poor engineering practice, whether design, management, or operation, was to blame.

The report of the Presidential Commission on the Space Shuttle *Challenger* Accident is a striking contrast to the Apollo 204 report. Even superficially, the reports are dissimilar. The Apollo 204 report looks like a report—it is

---

46. Ibid., chap. 11.

47. Parts I, II, and III describe the Board's legal authority, the biographies of its members, and the proceedings of the Board.

48. Apollo 204 Review Board, *Report to the Administrator, National Aeronautics and Space Administration* (Washington, DC: GPO, 1967), pp. 5–12.

monochromatic, printed in standard Government Printing Office format, and appears very similar to a multitude of other NASA reports. The report on the *Challenger* accident looks more like a magazine or coffee table book. It has large sections of color photographs used as visual evidence by the Commission, was printed on glossy paper, and was written in a narrative form familiar to readers of nonfiction. It opens with a preface and an introduction, outlining the task of the Commission and contextualizing the development of the Space Shuttle. The report goes on to outline the events of 28 January 1986 and from there leads into its analysis of the physical cause of the accident in a chapter simply titled "The Cause of the Accident."[49] The remainder of the report analyzes the series of events that contributed to the accident: the chain of decisions that led to the decision to launch, the history of design problems with the O-ring system, the political and organizational pressures to launch, and the failure of the safety system.[50] In seeking to understand the contributory causes of the accidents, the Commission's report does not explicitly draw on any theoretical work. The report's footnotes are to transcripts of Commission hearings or to original NASA and Morton Thiokol documents, rather than any other writings on accidents or safety.

The Presidential Commission was clear that there were physical causes for the accident—in this case, the failure of the O-rings to seal correctly. But unlike the Apollo 204 Review Board, the Commission saw secondary contributing causes. These secondary causes were the flawed launch decision, political pressures on the launch schedule, and a silent safety system. The 1967-model report, setting out an understanding of engineering failures to be fixed with engineering solutions, was changed into a critique of both engineering and management with separate solutions for each area of endeavor.

The report of the Columbia Accident Investigation Board (CAIB) was even more like a magazine. Unlike the Apollo 204 and *Challenger* reports, the CAIB report has its own logo and its own page headers and footers. The report contains sidebars to provide contextual or background material and is illustrated with images of the *Columbia* in preparation and in flight and images of the *Columbia* crew both before and during the 107 mission.

Like the *Challenger* report, the CAIB report devotes only one chapter, chapter 3, to the proximate physical cause of the accident—the separation of Thermal Protection System (TPS) foam from the External Tank and its subsequent impact on the leading edge of the orbiter. But the report has four chapters, chapters 5 to 8, discussing the context of the decision-making that led to the breakup of the orbiter on reentry. Chapter 3 discusses the engineering

---

49. Presidential Commission on the Space Shuttle *Challenger* Accident, *Report to the President*, vol. 1, chap. 4.

50. Ibid., chaps. 5 through 8, respectively.

analyses the Board performed, the history of External Tank design decisions, and the conclusions to be drawn from these, but it does so without using any theory, simply presenting this engineering section as needing no context or justification. It is only where the Board starts to examine the decision-making of NASA engineers and managers that led to the *Columbia* disaster that more sophisticated explanatory frameworks are needed. The Board drew on a variety of theoretical perspectives, considering Charles Perrow's theory of normal accidents and the work of both Scott Sagan and Todd La Porte on high-reliability theory.[51]

Perhaps most interestingly, the CAIB report drew heavily on the work of Diane Vaughan. Vaughan's 1996 book, *The Challenger Launch Decision*, set out a sociological explanation for the flawed decision, arguing that, far from the managerial misconduct identified by the *Challenger* report, the accident can best be understood in terms of the normalization of deviance, the culture of production at NASA and Morton Thiokol, and structural secrecy.[52] Vaughan argued:

> This book explicates the sociology of mistake. It shows how mistake, mishap and disaster are socially organized and systematically produced by social structures. No extraordinary actions by individuals explain what happened: no intentional managerial wrongdoing, no rule violations, no conspiracy. The cause of disaster was a mistake embedded in the banality of organizational life.[53]

This perspective informed chapter 8 of the CAIB report, where the Board drew explicit links between the *Challenger* and *Columbia* accidents, applying the components of Vaughan's analysis to *Columbia*. The Board concluded:

> First, the history of engineering decisions on foam and O-ring incidents had identical trajectories that "normalized" these anomalies, so that flying with these flaws became routine and acceptable. Second, NASA history had an effect. In response to White House and Congressional mandates, NASA leaders took actions that created systemic organizational flaws at the time of Challenger that were also present for Columbia.[54]

---

51. CAIB, *Report*, p. 180.
52. Diane Vaughan, *The Challenger Launch Decision: Risky Technology, Culture, and Deviance at NASA* (Chicago: University of Chicago Press, 1996).
53. Ibid., p. xiv.
54. CAIB, *Report*, p. 195.

Unlike the *Challenger* report, the CAIB report gives equal weight to the organizational causes of the accident, arguing that while mistakes were made, the organizational structure of NASA was more to blame that any individual failings.

The three reports suggest a story of growing separation of management and engineering. As Peter Galison has suggested, this may simply be a result of the instability between frames of analysis: the desire both to localize and to diffuse the locus of causation, to find a single physical cause, and to explain the accident in terms of larger organizational and cultural problems.[55] But it is interesting to note that these two activities are not only juxtaposed as possible sources of accidents, but also understood and analyzed in different ways. There has been a growing sophistication in the ways that decision-making and its contexts have been understood. There is a transition from Apollo 204's one-paragraph analysis of larger causes, to *Challenger*'s inclusion of organizational and political factors as contributory, to *Columbia*'s equal pairing of technical and social causes. There is a corresponding increase in the contextualization of these social elements of the analysis, from rudimentary mentions in Apollo 204 to a full examination and consideration of sociological and organizational theory literature in *Columbia*.

But there is an interesting asymmetry in these reports as well. As analyses of decision-making and its historical and cultural contexts have grown ever more sophisticated in these accident reports, the discussions of physical causes have remained remarkably similar. In each accident report, a number of possible causes are considered and eliminated before attention is turned to the actual cause. In each of the sections of the reports dealing with physical cause, there is little or no contextualization of engineering and design decision-making and no attempt to locate the discussion in a body of literature. This separates the physical and technical causes of accidents from their contexts and sets up the two activities—engineering and decision-making about engineering—as two quite different activities, to be understood and analyzed in different terms. In this formulation, engineering seems to be understood on its own terms, as a context-free and ahistorical activity, whereas management decision-making is understood as contingent and located within a complex historical and cultural framework.

This asymmetry immediately opens two questions. First, what historical processes caused the separation of engineering and management in the manned space program from 1967 to the present day? Second, what changes in engineering over the same period can be seen in the three accident reports and might provide the basis for understanding engineering in its own historical and cultural context? The disciplines of the history of technology and the history of science provide some directions to go look for answers to these questions.

---

55. Galison, "An Accident of History," p. 4.

Engineering accidents can be understood in a similar way to scientific controversies. A scientific controversy is resolved when the winners declare that their account is true and opponents are no longer taken seriously by the relevant scientific community.[56] Just as scientific controversies open up the inner workings of a laboratory or research group, so accidents open up the internal practices and politics of engineering. But accidents also provide a way to examine how engineers go about activities other than design and innovation. Most studies of engineers and engineering focus on design because it is the most creative and innovative element of the engineer's craft.[57] However, the vast majority of time spent by engineers is taken up with the development and operation of technologies rather than their design. Accident investigations take a comprehensive look at the design, manufacture, and operation of the broken artifact or system and so provide a way to look at engineering work at the routine, everyday level, as well as at the creative design level. The process of investigating an accident results in the extensive description of these everyday routines, routines that are often seen as so mundane as to leave little trace in the documentary record of the project. Thus, if these NASA accident reports are examined as a historian might examine them, they can trace changes in both design and routine engineering.

By treating accidents and their investigations as windows into engineering at NASA, there are at least three aspects of engineering at NASA that have changed since the 1960s—the widespread use of computers in engineering, the emergence of astronautical engineering as a new discipline, and a move away from systems engineering as an organizing philosophy for large projects.

Computing

Since the 1960s, computers have become ubiquitous, and there is a growing literature that points to the ways in which interaction with computers reshapes

---

56. This particular interpretation of scientific controversy is taken from the works of Bruno Latour and Wiebe Bijker in particular. See Bruno Latour, *Science in Action: How to Follow Scientists and Engineers through Society* (Cambridge, MA: Harvard University Press, 1987); Bruno Latour and Steve Woolgar, *Laboratory Life: The Construction of Scientific Facts* (Princeton, NJ: Princeton University Press, 1986); Wiebe E. Bijker, Thomas Parke Hughes, and T. J. Pinch, *The Social Construction of Technological Systems: New Directions in the Sociology and History of Technology* (Cambridge, MA: MIT Press, 1987).

57. This point was well made by John Staudenmaier in his surveys of the field of history of technology. See John M. Staudenmaier, *Technology's Storytellers: Reweaving the Human Fabric* (Cambridge, MA: Society for the History of Technology and the MIT Press, 1985); John M. Staudenmaier, "Recent Trends in the History of Technology," *American Historical Review* 95, no. 3 (1990). For examples of this focus on design to the exclusion of other aspects of engineering, see, for example, Walter G. Vincenti, *What Engineers Know and How They Know It: Analytical Studies from Aeronautical History*, Johns Hopkins Studies in the History of Technology (Baltimore: Johns Hopkins, 1990); Henry Petroski, *To Engineer Is Human: The Role of Failure in Successful Design* (New York: St. Martin's Press, 1985).

the ways people live and work.[58] Just like scientists, the engineering profession has adopted computing extensively, with almost all elements of engineering activity now mediated through computers—design, simulation modeling, communications, logistics, financial management, and administration.[59] Over the period 1967–2003, modeling, testing, and simulation moved from being largely hand-calibrated to being almost exclusively computer-mediated.[60] The Columbia Accident Investigation Board report shows, however, that this process involved the loss of much of the transparency of older techniques.

A brief history of the modeling tool Crater illustrates this process well. Crater was originally built in 1966 by Allen Richardson at Rockwell. It was designed in conjunction with NASA engineers to predict the effects of hyper-velocity impacts on multilayer surfaces like those of the Apollo CSM. Crater was a curve fit from a data set generated in part from Gemini experience and in part from testing performed by General Motors and NASA on aluminum honeycomb materials. Crater could predict threshold velocities and penetration damage but was complex to use; the number and complexity of calculations needed to derive a result made it time-consuming and prone to error. Crater was validated using small pieces of foam and ice on single tiles. During the process of turning empirical data into a predictive equation, the limitations and contingencies of these initial data sets were lost.[61] Furthermore, the process of computerization of Crater rendered the uncertainties inherent in the tool even more invisible, and the specific mode of computerization, a plug-in-the-numbers spreadsheet, gave a false sense of clarity and certainty to the results. Thus, an engineer unaware of the history of the tool and its limita-

---

58. See Sherry Turkle, *The Second Self: Computers and the Human Spirit* (New York: Simon & Schuster, 1984); and Sherry Turkle, *Life on the Screen: Identity in the Age of the Internet* (New York: Simon & Schuster, 1995), for an examination of how interaction through the mediation of computers changes identity. More specifically, Dominique Vinck and Eric Blanco, *Everyday Engineering: An Ethnography of Design and Innovation*, Inside Technology (Cambridge, MA: MIT Press, 2003), and Susan Leigh Star, *The Cultures of Computing* (Oxford, U.K., and Cambridge, MA: Blackwell Publishers, 1995), start to address how engineering and scientific work has changed.

59. For a general overview of computing since World War II, see Paul E. Ceruzzi, *A History of Modern Computing*, 2nd ed. (Cambridge, MA: MIT Press, 2003). Paul E. Ceruzzi, *Beyond the Limits: Flight Enters the Computer Age* (Cambridge, MA: MIT Press, 1989), provides a good outline of the introduction of computers into aerospace, although the focus of the work is on-board computers rather than ground equipment or design tools. Gary Lee Downey, *The Machine in Me: An Anthropologist Sits among Computer Engineers* (New York: Routledge, 1998), and Louis L. Bucciarelli, *Designing Engineers*, Inside Technology (Cambridge, MA: MIT Press, 1994), both provide ethnographies of engineering that discuss the effects of the ubiquity of computers in the workplace.

60. For a good overview of this topic, see Sergio Sismondo and Snait Gissis, "Practices of Modelling and Simulation," *Special Issue of Science in Context* 12 (1999). George E. Smith, "The Dangers of Cad," *Mechanical Engineering* (February 1986), gives an early warning of the dangers of increasingly closed simulation tools.

61. Allen Richardson interview, by A. Brown, 15 February 2005.

tions, as was the Boeing engineer who did the *Columbia* analysis, could not know that the predictive powers of Crater were unknown outside a limited range of values. The piece of foam that fell from *Columbia*'s external tank was 640 times larger than Crater's valid range. The Crater model predicted that the foam strike would have broken entirely through the Thermal Protection System of the Shuttle and exposed the aluminum wing structure.[62] But because the engineers were aware that there were limitations to the tool, but not aware of how to correct or modify the model, they dismissed their results as too conservative and not predictive of a problem.

This example shows that the Boeing engineers were working in a mode of engineering where their relationships to the materials and objects that they build and study were profoundly mediated through a computer and profoundly dependent on the uncritical acceptance of the findings and assumptions of previous generations of engineers. In January 2003, Boeing engineers and NASA's Debris Assessment Team had no choice but to accept the results of their Crater analysis. Their reliance on a computer model, with the inherent lack of access to the mechanics of the model, let alone the assumptions and uncertainties underlying it, had profoundly affected their ability to make engineering judgments. A similar story can be told about the External Tank bolt catchers—their safety margin, flagged by the Board as dangerously low and a possible source of disaster, was computed using ancient data sets whose origins and limitations had been obscured by computerization.[63]

Engineering Education

There is a growing trend in the history of science to look towards pedagogy as a lens through which to understand how science and scientists come to be.[64] David Kaiser writes, "Scientists are not born, they are made. The ways in which this happens bears the marks of time and place."[65] This observation holds equally true for engineers. Engineering education has changed since Apollo 1. In the late 1960s, engineering schools started to move towards

---

62. CAIB, *Report*, pp. 144–145.

63. Ibid., pp. 86–88.

64. David Kaiser, *Pedagogy and the Practice of Science: Historical and Contemporary Perspectives* (Cambridge, MA: MIT Press, 2005), is a collection of essays examining science pedagogy over a variety of disciplines, places, and times. Andrew Warwick, *Masters of Theory: Cambridge and the Rise of Mathematical Physics* (Chicago: The University of Chicago Press, 2003), a study of mathematical training in 19th-century Cambridge and its relationship to 19th-century physics in Britain, is perhaps the most sustained development of the argument for the value of the study of pedagogy. Sharon Traweek, *Beamtimes and Lifetimes: The World of High Energy Physicists* (Cambridge, MA: Harvard University Press, 1988), examines contemporary Japanese physicists and identifies education as critical in the formation of a distinctively Japanese way of doing physics.

65. Kaiser, *Pedagogy and the Practice of Science,* p. 1.

engineering science and away from engineering design as a model for the discipline.[66] Engineering students were required to take classes in physics, math, and chemistry to give them a thorough grounding in the physical sciences before going on to engineering classes. The ongoing effects of the National Defense Education Act of 1958 meant changes towards more easily teachable and assessable modes of learning as educators struggled to manage massive expansions in class sizes.[67] The combination of these two trends meant that for many freshmen and sophomores in the 1970s, engineering meant doing physics and math problem sets rather than sketching, building, and working with their hands.[68] This mode of learning fit well with the growing presence of computers in education, providing students with the mathematical tools needed to build and use their own software. As computers became ubiquitous, so engineering schools brought computing into engineering education.

These changes served to both render engineering more abstract and arcane, less connected to its objects of study, and to make it more automated. Both the *Challenger* and *Columbia* reports are critical of the relationships between NASA and its contractors, and particularly critical of the lack of engineering design and development capacity amongst some of the contractors.[69] Embodying engineering judgment in computer programs can devalue that judgment when embodied in engineers, leading to downgrading of the institutional value placed on engineers as employees. This leaves engineers and their skills more vulnerable to privatization and commodification and hence leads to the downgrading of the engineering design capacity of commercial organizations.

The new discipline of astronautics or astronautical engineering was also emerging over this period, intertwined with the development of NASA as an

---

66. Rosalind H. Williams, *Retooling: A Historian Confronts Technological Change* (Cambridge, MA: MIT Press, 2002), pp. 40–42.

67. Barbara Barksdale Clowse, *Brainpower for the Cold War: The Sputnik Crisis and National Defense Education Act of 1958* (Westport, CT: Greenwood Press, 1981), examines the initial responses to the Sputnik crisis. David Kaiser, "Scientific Manpower, Cold War Requisitions, and the Production of American Physicists after World War II," *Historical Studies in the Physical and Biological Sciences* 33 (fall 2002), looks specifically at the relationship between Cold War geopolitics and changing styles of science and education during this period.

68. Both Kathryn Henderson, *On Line and on Paper: Visual Representations, Visual Culture, and Computer Graphics in Design Engineering, Inside Technology* (Cambridge, MA: MIT Press, 1999), and Eugene S. Ferguson, *Engineering and the Mind's Eye* (Cambridge, MA: MIT Press, 1992), examine the changes in engineering brought about by changes in the ways in which students learn to interact with the material world. Ferguson discusses the loss of a visual intuitiveness amongst young engineers brought about by a move to a more analytic style of engineering in the 1960s and 1970s. Henderson looks at the ways in which engineering knowledge and practices are transformed when computer visualization tools are introduced into the workshop and drafting room.

69. Presidential Commission on the Space Shuttle *Challenger* Accident, *Report to the President*, pp. 194–195; CAIB, *Report*, pp. 110–118.

organization.[70] The new discipline drew heavily on the principles of aeronautical engineering but taught students how to apply these principles in higher stress environments—at higher temperatures and pressures, with higher aerodynamic loads, in high-radiation environments, using finer tolerance manufacturing, and with larger and more complex vehicle systems. The new discipline of astronautical engineering had to learn how to manage problems with testing the massive vehicles it built. In many cases, it was physically impossible to adequately test astronautical hardware, and so new methods of producing knowledge about complex systems like computer modeling and simulation were developed. The Apollo 204 report illustrates the engineering challenges that accompanied the transition from designing and developing craft to operate within the atmosphere to craft designed to operate in the space environment. As the report makes clear, the levels of both precision and complexity needed to build a spacecraft grew dramatically, perhaps beyond the capacity of North American Aviation engineers to keep up. As astronautics developed, engineering scale, engineering knowledge, and engineering management changed.

The Systems Approach

Systems engineering as a philosophy emerged from the complex military defense projects of the 1950s. It can be best described as a "set of organizational structures and processes to rapidly produce a novel but dependable technological artifact within a predictable budget."[71] Systems engineering was one element in a long history of the application of scientific and engineering principles to complex commercial or organizational problems, a history that started with Taylorism and scientific management in the late 19th century.[72] Systems engineering involved the use of engineering ideas to organize large engineering projects—most profoundly, systems engineering defines project

70. W. Henry Lambright, *Powering Apollo: James E. Webb of NASA* (Baltimore: John Hopkins, 1995); W. Henry Lambright, Edwin A. Bock, and Inter-university Case Program, *Launching NASA's Sustaining University Program* (Syracuse, NY: Inter-university Case Program, 1969); Howard E. McCurdy, *Inside NASA: High Technology and Organizational Change in the U.S. Space Program* (Baltimore: Johns Hopkins, 1993).

71. Stephen B. Johnson, *The Secret of Apollo: Systems Management in American and European Space Programs* (Baltimore: Johns Hopkins, 2002), p. 17.

72. See James R. Beniger, *The Control Revolution: Technological and Economic Origins of the Information Society* (Cambridge, MA: Harvard University Press, 1986), and JoAnne Yates, *Control through Communication: The Rise of System in American Management* (Baltimore: Johns Hopkins, 1989), for a brief introduction to the literature on system and scientific management. Robert Kanigel, *The One Best Way: Frederick Winslow Taylor and the Enigma of Efficiency* (New York: Viking, 1997), and Hugh G. J. Aitken, *Scientific Management in Action: Taylorism at Watertown Arsenal, 1908–1915* (Princeton, NJ: Princeton University Press, 1985), both provide excellent introductions to Taylor and scientific management.

management as an engineering problem best solved by engineers and engineering practice. In this philosophy, management becomes a subset of engineering practice. The large SAGE (Semi-Automatic Ground Environment) air defense and Atlas missile projects trained a generation of engineers how to apply systems engineering ideas to complex research, development, and manufacturing projects.[73] Systems management experts from the Air Force and the aerospace industry were brought into NASA to manage the Apollo program as it grew in the 1960s.[74] The Apollo 204 accident marks the moment of transition into a full acceptance of systems engineering as the guiding philosophy of the space program, whereas throughout the early part of the 1960s, there was tension between the aircraft manufacturers and the missile-program-trained NASA engineering managers. Indeed, the most common historiographical interpretation of the larger significance of Apollo 204 is simply that—the fire forced NASA and its contractors to find new ways of managing the complexity of the Apollo program, and systems management was the new way.[75]

The manned spaceflight community within NASA made the transition from research and development to being primarily an operational organization in the 1980s and 1990s, as the focus of the U.S. manned spaceflight program moved from exploration to ready access to low-Earth orbit. Systems engineering as an overarching philosophy for the management of complexity was replaced with new approaches drawn from both the business and government worlds. This does not mean that the tools that collectively made up systems engineering—configuration control boards, integrated management

73. Agatha C. Hughes and Thomas Parke Hughes, *Systems, Experts, and Computers: The Systems Approach in Management and Engineering, World War II and After, Dibner Institute Studies in the History of Science and Technology* (Cambridge, MA: MIT Press, 2000); Thomas Parke Hughes, *Rescuing Prometheus*, 1st ed. (New York: Pantheon Books, 1998); and Kent C. Redmond and Thomas M. Smith, *From Whirlwind to Mitre: The R&D Story of the SAGE Air Defense Computer* (Cambridge, MA: MIT Press, 2000), all discuss the origins of systems management in the ballistic missile and air defense programs of the 1950s.

74. Johnson's *The Secret of Apollo: Systems Management in American and European Space Programs* is by far the most comprehensive examination of the rise of systems management in the U.S. space program. See also Arnold S. Levine, *Managing NASA in the Apollo Era* (Washington, DC: NASA SP-4102, 1982); John M. Logsdon, *Managing the Moon Program: Lessons Learned from Project Apollo: Proceedings of an Oral History Workshop* (Washington, DC: NASA SP-4514, 1999).

75. For examples of this type of interpretation, see Andrew Chaikin and Tom Hanks, *A Man on the Moon: The Voyages of the Apollo Astronauts* (New York: Penguin Books, 1998), chap. 1; Charles A. Murray and Catherine Bly Cox, *Apollo, the Race to the Moon* (New York: Simon & Schuster, 1989), chaps. 15–16; William E. Burrows, *This New Ocean: The Story of the Space Age*, 1st ed. (New York: Random House, 1998), pp. 406–415. Astronaut and flight controller biographies make a similar point. See, for example, Frank Borman and Robert J. Serling, *Countdown: An Autobiography*, 1st ed. (New York: W. Morrow, 1988), chap. 9; Michael Collins, *Carrying the Fire: An Astronaut's Journeys* (New York: Farrar, 1974), pp. 269–275; Christopher C. Kraft, *Flight: My Life in Mission Control* (New York: Dutton, 2001), pp. 269–278; Gene Kranz, *Failure Is Not an Option: Mission Control from Mercury to Apollo 13 and Beyond* (New York: Simon & Schuster, 2000), pp. 208–214.

systems, resident program offices at contractors—ceased to be used, but rather that the philosophy that a collection of these tools was the best way to manage a program was replaced by other ways of thinking.[76]

Total Quality Management, reengineering, and "faster, better, cheaper" took the place of systems engineering in the 1990s, part of a larger cultural trend in the United States that valorized the business approach to organization and emphasized the merits of private free-enterprise solutions to problems previously thought the realm of government.[77] The idea of using scientific and engineering principles to solve business and organizational challenges was replaced by the application of business and commercially derived management philosophy to an engineering organization.

Changes in engineering practice over the 1970s, 1980s, and 1990s meant that engineers in the manned space program were working in the increasingly mediated environment of computer-based engineering whilst working on technological systems that were becoming increasingly complex, difficult to test, and designed to operate at an increasingly high performance envelope. Margins for error grew ever smaller, whilst the computer-based tools being used to manage that margin grew increasingly less transparent. At the same time, the shared organizational philosophy of systems engineering was being abandoned by senior management in favor of more commercially oriented ideas, while engineers still used the tools of systems management.

FURTHER RESEARCH

There are several areas that call for further research in order to put together a picture of changes in engineering in the U.S. manned space program. The first area is studies of engineering in practice in the late 20th century. Although the genre of engineering ethnographies is growing, it is still small. Some of these studies examine the impact of computers in the engineering workplace, but none do so in the context of aeronautics or astronautics. Howard McCurdy's work on NASA culture provides an excellent base to work from but focuses on organizational change rather than engineering change from the 1970s onwards.[78] Furthermore, the field needs not just in-depth studies of engineering practice, but broad-scope surveys comparable to Sylvia Fries's *NASA Engineers in the Age of Apollo*.[79] We do not yet know

---

76. See CAIB, *Report*, pp. 105–110.

77. Howard E. McCurdy, *Faster, Better, Cheaper: Low-Cost Innovation in the U.S. Space Program* (Baltimore: Johns Hopkins, 2001).

78. McCurdy, *Inside NASA: High Technology and Organizational Change in the U.S. Space Program*; McCurdy, *Faster, Better, Cheaper: Low-Cost Innovation in the U.S. Space Program*.

79. Sylvia Doughty Fries, *NASA Engineers and the Age of Apollo* (Washington, DC: NASA SP-4104, 1992).

enough about the educational and demographic characteristics of NASA engineers from the 1970s onwards.

There is a need for a body of literature on the recent institutional and cultural history of engineering comparable to the literature on the rise of the engineering profession in the later half of the 19th century. We know much about the ways in which engineers developed a clearly articulated professional identity, created a standardized curriculum and accreditation process, and made themselves middle-class in the late 19th century.[80] We know much about the engineering triumphs of the early 20th century and the involvement of engineers in the winning of World War II and the Cold War, both as producers of military technology but also as the creators of the consumer society.[81] But we know little about how engineers have responded to changing economic and cultural circumstances since the 1960s.

We need more nuanced histories of the NASA of the 1970s, 1980s, and 1990s. Reflecting the ongoing cultural legacy of the Apollo program, much of the literature on the U.S. manned spaceflight program focuses on the triumphs of the 1960s. Those histories that do attempt to cover the entire history of the program tend to fall into a declensionist mode of writing, discussing NASA's decline and fall from Apollo. A more nuanced understanding of the legacy of the Apollo program, including a more realistic assessment of the relative safety of Apollo and Shuttle missions, might serve to provide a new framework in which to understand the history of NASA over this period.

---

80. For example, see Edwin T. Layton, *The Revolt of the Engineers: Social Responsibility and the American Engineering Profession* (Cleveland: Press of Case Western Reserve University, 1971); George S. Emmerson, *Engineering Education: A Social History* (New York: David & Charles; Crane, 1973); David F. Noble, *America by Design: Science, Technology, and the Rise of Corporate Capitalism*, 1st ed. (New York: Knopf, 1977); Brendan Patrick Foley, "Fighting Engineers: The U.S. Navy and Mechanical Engineering, 1840–1905" (Ph.D. thesis, MIT, June 2003).

81. See Thomas Parke Hughes, *American Genesis: A Century of Invention and Technological Enthusiasm, 1870–1970* (New York: Penguin Books, 1990); David A. Hounshell, *From the American System to Mass Production, 1800–1932: The Development of Manufacturing Technology in the United States* (Baltimore: Johns Hopkins, 1984); Terry S. Reynolds, *The Engineer in America: A Historical Anthology from Technology and Culture* (Chicago: University of Chicago Press, 1991).

CHAPTER 13

INSTITUTIONAL ISSUES FOR CONTINUED SPACE EXPLORATION: HIGH-RELIABILITY SYSTEMS ACROSS MANY OPERATIONAL GENERATIONS— REQUISITES FOR PUBLIC CREDIBILITY[1]

Todd R. La Porte

Highlighting critical issues arising from the evolution of a large government enterprise is both important and occasionally painful and sometimes provides a basis for exciting next steps. Calling out critical technical issues from past developments inspires engineers and makes visible to policymakers likely requests for program funding to address them. A "critical issues" focus also holds the promise of exploring other sorts of issues: those that arise in deploying technologies.[2] These are particularly interesting when they entail large-scale organizations that are judged to be highly hazardous.

This paper highlights the challenges and issues involved when we wish large, technically rooted organizations to operate *far more* effectively, with *much less* error than they should be expected to exhibit—given what we know about organizations more generally. Recall that "Murphy's Law" and trial-and-error learning are reasonably accurate descriptors of how all organizations generally behave. Routinely expecting otherwise is quite remarkable.

First, let us set a context. In your mind's eye, imagine space-related activities two or three decades into the future. President George W. Bush's current vision for NASA focused the Agency's efforts in the early 21st century, and

---

1. This paper draws on presentations to the Workshop on Space Policy held by the National Academies of Science in Irvine, CA, 12–13 November 2003; the National Academies' Board on Radioactive Waste Management Panels on "Principles and Operational Strategies for Staged Repository Systems," 27 June 2001, and "Long-Term Institutional Management of Hazards Sites," 7 August 2001, both held in Washington, DC; and the American Association for the Advancement of Science (AAAS) symposium, "Nuclear Waste: File and Forget? Institutional Challenges for High-Reliability Systems Across Many Operational Generations—Can Watchfulness Be Sustained?" held in Denver, CO, 18 February 2003. Since these presentations were given to quite different, nearly mutually exclusive audiences, the various conference sponsors have agreed to this repetition.

2. This conference on "Critical Issues" casts a wider net and includes issues relevant to the understanding of policy development, technical operations as well as systems safety, and the conduct of historical studies of large systems per se.

our reach has extended to periodic flights to the Moon and to an international space platform.[3] With international cooperation, three to four major launches and recoveries a year have become more or less routine. Another six or seven unmanned launches resupply the Station and various probes for scientific programs. Assume that national intelligence and communications demands require another half dozen annually. And imagine that commercial spaceflight enthusiasts have found enough "venture capitalists" and adventurers to sustain several highly visible, elite space experiences. This is edging toward 20 launches a year and evokes images of science fiction and early *Star Trek* tableaux.

This sort of future moves us well beyond the sharply defined, novel images of machinery and spectacularly framed astronauts spacewalking against the black of the heavens. It conjures the extraordinary organizations that these activities imply. There would be the early vestiges of, say, a U.S.–European Union space traffic control—analogous to the existing global air traffic control system—alert to tracking both space vehicles and the detritus of former flights, closely concentrating on bringing each flight to rest without encountering objects aloft or mishaps of human or mechanical origin. Operational scope would be widespread and expected to continue indefinitely. This organizational reach is extraordinary. It immediately raises the question of the "operational sustainability" of NASA's space missions, especially those that propel humans into space.

The missions and the technologies that typify NASA and its industrial contractors prompt demands that NASA programs exhibit highly reliable, humanly safe operations, often projected to continue for a number of management generations (say some 10 to 15 years each). NASA has, in the past, taken up these challenges emphasizing both engineering controls and administrative controls that embrace safety and effective performance.

This paper highlights a third emphasis: the organizational relationships and safety culture of the Agency and its contractors that would manage an astonishing array of complicated technical systems and far-flung facilities making up a global space complex. It draws on work examining the operations of several mature, large-scale technical systems. Then it considers in this light the qualities likely to be necessary in the evolution of NASA's humans-in-space activities if they are routinely to achieve a high degree of public acceptance and sustained credibility.

Putting the question directly: What organizational conditions have arisen when the operating technologies are so demanding or hazardous that trial-

---

3. President George W. Bush, "A Renewed Spirit of Discovery: The President's Vision for U.S. Space Exploration," 14 January 2004, folder 12886, NASA Historical Reference Collection, Washington, DC. For NASA's most recent expression of this declaration, see NASA, "The New Age of Exploration: NASA's Direction for 2005 and Beyond," February 2005, same folder. The operative portion from the mission: "To understand and protect our home planet, To explore the universe and search for life, To inspire the next generation of explorers."

and-error learning, while likely, no longer seems to be a confident mode of learning and when the next error may be your last trial?

What can be said about managing large-scale technical systems, responsible for often highly hazardous operations on missions that imply operational stability for many, many years? The institutional design challenges are to provide the work structures, institutional processes, and incentives in such ways that they assure highly reliable operations[4] over the very long term—perhaps up to 50 years[5]—in the context of continuously high levels of public trust and confidence.[6] My purpose here is less to provide a usable explication of these concepts (see the supporting references) and more to demonstrate, by a blizzard of lists, the complexity and range of the institutional conditions implied by NASA's program reach. I foreground properties that are especially demanding, keeping these questions in mind: How often and at what effort does one observe these characteristics in the organizational arenas you know best? Could one imagine such an ensemble within NASA in the foreseeable future?

PURSUING HIGHLY RELIABLE OPERATIONS

Meeting the challenges of highly reliable operations has been demonstrated in enough cases to gain a rough sense of the conditions that seem associated with extraordinary performance. These include both internal processes and external relations. What can be said with some confidence about the qualities NASA managers and their overseers could seek?[7] (See table 13.1.)

---

4. Initial empirical work included close study of the operations of U.S. Air Traffic Control, aircraft carriers at sea, and nuclear power plants. For summaries, see G. I. Rochlin, "Reliable Organizations: Present Research and Future Directions," and T. R. La Porte, "High Reliability Organizations: Unlikely, Demanding and at Risk," both in *Journal of Crisis and Contingency Management* 4, no. 2 (June 1996): 55–59 and 60–71, respectively; T. R. La Porte and P. M. Consolini, "Working in Practice but not in Theory: Theoretical Challenges of High Reliability Organizations," *Journal of Public Administration Research and Theory* 1, no. 1 (January 1991): 19–47; K. H. Roberts, "New Challenges to Organizational Research: High Reliability Organizations," *Industrial Crisis Quarterly* 3 (1989): 111–125.

5. Prompting the concept of "institutional constancy." See discussion later in this chapter, along with T. R. La Porte and A. Keller, "Assuring Institutional Constancy: Requisites for Managing Long-Lived Hazards," *Public Administration Review* 56, no. 6 (November/December 1996): 535–544.

6. In the context of this paper, sustaining public trust and confidence, while a very important consideration, takes second seat to the issues of reliable operations across multiple generations. Public trust is a condition that evokes high institutional demands and calls for a discussion that extends beyond the limitations of this paper. See, for example, U.S. Department of Energy (DOE), "Earning Public Trust and Confidence: Requisite for Managing Radioactive Waste. Report of the Task Force on Radioactive Waste Management, Secretary of Energy Advisory Board," November 1993, available online at *http://www.seab.energy.gov/publications/trust.pdf*; T. R. La Porte and D. Metlay, "Facing a Deficit of Trust: Hazards and Institutional Trustworthiness," *Public Administration Review* 56, no. 4 (July–August 1996): 341–347.

7. Draw generalized inferences from this discussion with care. These findings are based mainly on three types of organizations, each with a limited number of cases, and bits from others (e.g., K. H.

continued on the next page

Table 13.1. Characteristics of Highly Reliable Organizations (HROs)

Internal Processes

1. Strong sense of mission and operational goals, commitment to highly reliable operations, both in production and safety.

2. Reliability-enhancing operations.

    A. Extraordinary technical competence.

    B. Sustained, high technical performance.

    C. Structural flexibility and redundancy.

    D. Collegial, decentralized authority patterns in the face of intense, high-tempo operational demands.

    E. Flexible decision-making processes involving *operating teams*.

    F. Processes enabling continual search for improvement.

    G. Processes that reward the discovery and reporting of error, *even one's own.*

3. Organizational culture of reliability, including norms, incentives, and management attitudes that stress the equal value of reliable production and operational safety.

External Relationships

1. External "watching" elements.

    A. Strong superordinate institutional visibility in parent organization.

    B. Strong presence of stakeholding groups.

2. Mechanisms for "boundary spanning" between the units and these watchers.

3. Venues for credible operational information on a timely basis.

---

continued from the previous page

Roberts, "Some aspects of organizational cultures and strategies to manage them in reliability enhancing organizations," *Journal of Managerial Issues* 5 [1993]: 165–181). Though these organizations operate in quite different institutional milieus, we cannot say they represent a systematic sample. No one now knows what the population of HROs might be. And highly reliable operations are keenly sought for situations that are not so dramatically hazardous in the physical sense, e.g., HRO operations in financial transactions or in the performance of sophisticated computer chips or large software programs. See K. H. Roberts and C. Libuser, "From Bhopal to banking: Organizational design can mitigate risk," *Organizational Dynamics* 21 (1993): 15–26. In these situations, motivation stems from fear of serious financial losses that are seen as amounting to institutional, not physical, death.

Internal Processes[8]

Organizationally defined intention. High-reliability organizations (HROs) exhibit a strong sense of mission and operational goals that stress assuring ready capacity for production and service with an *equal* commitment to reliability in operations and a readiness to invest in reliability-enhancing technology, processes, and personnel resources. In cases such as our space operations, these goals would be strongly reinforced by a clear understanding that the technologies upon which the organizations depend are intrinsically hazardous and potentially dangerous to human and other organisms. It is notable that for U.S. space operations, there is also high agreement within the operating organizations and in the society at large about the seriousness of failures and their potential costliness, as well as the value of what is being achieved (in terms of a combination of symbolic, economic, and political factors). This consensus is a crucial element underlying the achievement of high operational reliability and has, until recently, increased the assurance of relatively sufficient resources needed to carry out failure-preventing/quality-enhancing activities. Strong commitment also serves to stiffen corporate or agency resolve to provide the organizational status and financial and personnel resources such activities require. But resolve is not enough. Evidence of cogent operations is equally crucial.

Reliability-enhancing operations. These include the institutional and operational dynamics that arise when extraordinary performance must be the rule of the day—features that would be reinforced by an organizational culture of reliability, i.e., the norms and work ways of operations.[9] A dominant quality of organizations seeking to attain highly reliable operations is their intensive technical and social interdependence. Characterized by numerous specialized functions and coordination hierarchies, this prompts patterns of complexly related, tightly coupled technical and work processes which shape HROs' social, structural, and decision-making character.[10]

---

8. This section draws strongly from La Porte and Consolini, "Working in Practice but not in Theory"; Rochlin, La Porte, and Roberts, "The self-designing high-reliability organization: Aircraft carrier flight operations at sea," *Naval War College Review* 40, no. 4 (1987): 76–90; La Porte, "High Reliability Organizations"; Rochlin, "Reliable Organizations: Present Research and Future Directions," pp. 55–59; T. R. La Porte, "High Reliability Organizations: Unlikely, Demanding and at Risk," pp. 60–71; K. H. Roberts, "Some characteristics of high reliability organizations," *Organization Science* 1, no. 2 (1990): 160–177; P. R. Schulman, "Negotiated Order of Organizational Reliability," *Administration & Society* 25, no. 3 (November 1993): 356–372.

9. K. E. Weick, "Organizational culture as a source of high reliability," *California Management Review* 29 (1987): 112–127; K. H. Roberts, "Some aspects of organizational cultures and strategies to manage them in reliability enhancing organizations," *Journal of Managerial Issues* 5 (1993): 165–181.

10. La Porte and Consolini, "Working in Practice but not in Theory"; Rochlin, "Reliable Organizations: Present Research and Future Directions"; C. Perrow, *Normal Accidents: Living With High-Risk Technologies* (New York: Basic Books, 1984); K. H. Roberts, K. H. and G. Gargano, "Managing a High Reliability Organization: A Case for Interdependence," in *Managing Complexity in High Technology Industries: Systems and People*, ed. M. A. Von Glinow and S. Mohrmon (New York: Oxford University Press, 1989), pp. 147–159.

The social character of the HRO is typified by high technical/professional competence and performance, as well as thorough technical knowledge of the system and awareness of its operating state.

1. Extraordinary technical competence almost goes without saying. But this bears repeating because continuously attaining very high quality requires close attention to recruiting, training, staff incentives, and ultimately the authority relations and decision processes among operating personnel who are, or should be, consummately skilled at what they do. This means there would be a premium put on recruiting members with extraordinary skills and an organizational capacity to allow them to burnish these skills in situ via continuous training and an emphasis on deep knowledge of the operating systems involved. Maintaining high levels of competence and professional commitment also means a combination of elevated organizational status and visibility for the activities that enhance reliability. This would be embodied by "high reliability professionals"[11] in positions with ready access to senior management. In aircraft carrier operations, this is illustrated where high-ranking officers are assigned the position of Safety Officer reporting directly to the ship's captain.

2. HROs also continuously achieve high levels of operational performance accompanied by stringent quality assurance (QA) measures applied to maintenance functions buttressed by procedural acuity.[12] Extensive performance databases track and calibrate technical operations and provide an unambiguous description of the systems' operating state. NASA's extraordinary investment in collecting system performance data is a prime example of this characteristic. These data inform reliability statistics, quality-control processes, accident modeling, and interpretations of system readiness from a variety of perspectives. In some organizational settings, the effectiveness of these analyses is enhanced by vigorous competition between groups formally responsible for safety.[13]

---

11. P. Schulman, E. Roe, M. van Eeten, and M. de Bruijne, "High Reliability and the Management of Critical Infrastructures," *Journal of Crisis and Contingency Management* 12, no. 1 (March 2004): 14–28. Also see David Mindell's chapter in this book and his attention to the self "identity" of technical operators.

12. Schulman, "Negotiated Order of Organizational Reliability"; M. Bourrier, "Organizing Maintenance Work at Two American Nuclear Power Plants," *Journal of Crisis and Contingency Management* 4, no. 2 (June 1996): 104–112.

13. T. R. La Porte and C. Thomas, "Regulatory Compliance and the Ethos of Quality Enhancement: Surprises in Nuclear Power Plant Operations," *Journal of Public Administration Research and Theory* 5, no. 4 (December 1994): 250–295.

HROs' operations are enabled by structural features that exhibit operational flexibility and redundancy in pursuit of safety and performance, and overlapping or nested layers of authority relationships.

3. Working with complex technologies is often hazardous, and operations are also carried out within quite contingent environments. Effective performance calls for flexibility and "organizational slack" (or reserve capacity) to ensure safety and protect performance resilience. Such structural flexibility and redundancy are evident in three ways: key work processes are designed so that there are parallel or overlapping activities that can provide backup in the case of overload or unit breakdown and operational recombination in the face of surprise; operators and first-line supervisors are trained for multiple jobs via systematic rotation; and jobs and work groups are related in ways that limit the interdependence of incompatible functions.[14] NASA has devoted a good deal of attention to aspects of these features.

The three characteristics noted so far are, in a sense, to be expected and command the attention of systems engineering and operational managers in NASA and other large-scale technical programs. There is less explicit attention to understanding the organizational relationships that enhance their effectiveness. I give these a bit more emphasis below.

4. Patterns of formal authority in large organizations are likely to be predominately hierarchical (though this may have as much to do with adjudicative functions as directive ones). And, of course, these patterns are present in HROs as well. Top-down, commandlike authority behaviors are most clearly seen during times of routine operations. But importantly, two other authority patterns are also "nested or overlaid" within these formal relations. Exhibited by the same participants who, during routine times, act out the roles of rank relations and bureaucrats, in extraordinary times, when the tempo of operations increases, another pattern of collegial and functionally based authority relationships takes form. When demands increase, those members

---

14. For work on functional redundancy, see especially M. Landau, "Redundancy, Rationality, and the Problem of Duplication and Overlap," *Public Administration Review* 27 (July/August 1969): 346–358; A. W. Lerner, "There is More Than One Way to be Redundant: A Comparison of Alternatives for the Design and Use of Redundancy in Organizations," *Administration & Society* 18 (November 1986): 334–359; D. Chisholm, *Coordination Without Hierarchy: Informal Structures in Multi-organizational Systems* (Berkeley: University of California Press, 1989); C. F. L. Heimann, "Understanding the *Challenger* Disaster: Organizational Structure and the Design of Reliable Systems," *American Political Science Review* 87 (June 1993): 421–435.

who are the most skilled in meeting them step forward without bidding to take charge of the response, while others who may "outrank" them slip informally into subordinate, helping positions.

And nested within or overlaid upon these two patterns is yet another well-practiced, almost scripted set of relationships that is activated during times of acute emergency. Thus, as routine operations become high-tempo, then perhaps emergencies arise, observers see communication patterns and role relationships changing to integrate the skills and experience apparently called for by each particular situation. NASA has had dramatic experience with such patterns.

Within the context of HROs' structural patterns, decision-making dynamics are flexible, dispersed among operational teams, and include rewards for the discovery of incipient error.

5. Decision-making within the shifting authority patterns, especially operating decisions, tends to be decentralized to the level where actions must be taken. Tactical decisions often develop on the basis of intense bargaining and/or collegial interaction among those whose contributions are needed to operate effectively or problem-solve. Once determined, decisions are executed, often very quickly, with little chance for review or alteration.[15]

6. Due in part to the irreversibility of decisions once enacted, HROs put an unusual premium on assuring that decisions will be based on the best information available. They also try to insure that their internal technical and procedural processes, once put in motion, will not become the sources of failure. This leads, as it has within NASA, to quite formalized efforts, continually in search of improvement via systematically gleaned feedback, and periodic program and operational reviews. These are frequently conducted by internal groups formally charged with searching out sources of potential failure, as well as improvements or changes in procedures to minimize the likelihood of failure. On occasion, there may be several groups structured and rewarded in ways that puts them in direct competition with each other to discover potential error, and, due to their formal attachment to different reporting levels of the management hierarchy, this encourages the quick forwarding of information about potential flaws to higher authority.[16]

---

15. Roberts, "Some characteristics of high reliability organizations"; Schulman, "Negotiated Order of Organizational Reliability."

16. La Porte and Thomas, "Regulatory Compliance and the Ethos of Quality Enhancement"; Diane Vaughan, *The Challenger Launch Decision: Risky Technology, Culture, and Deviance at NASA* (Chicago: University of Chicago Press, 1990).

Notably, these activities, due to their intrinsic blame-placing potential, while they may be sought by upper management in a wide variety of other types of organizations, are rarely conducted with much enthusiasm at lower levels. In response, HROs exhibit a most unusual willingness to reward the discovery and reporting of error without peremptorily assigning blame for its commission at the same time. This obtains even for the reporting of one's own error in operations and procedural adherence. The premise of such reward is that it is better and more commendable for one to report an error immediately than to ignore or to cover it up, thus avoiding untoward outcomes as a consequence. These dynamics rarely exist within organizations that operate primarily on punishment-centered incentives, that is, most public and many private organizations.

Organizational culture of reliability. Sustaining the structural supports for reliability and the processes that increase it puts additional demands on the already intense lives of those who operate and manage large-scale, advanced technical systems. Operating effectiveness calls for a level of personal engagement and attentive behavior that is unlikely to be manifest merely on the basis of formal rules and economic employee contracts. It requires a fully engaged person responding heedfully to norms of individual and group relations that grow out of the particular demands and rewards of the hazardous systems involved.[17] For lack of a better concept to capture these phenomena, let us accept the slippery concept of "organizational culture" as a rough ordering notion.[18] A culture of organizational reliability refers to the norms, shared perceptions, work ways, and informal traditions that arise within the operating and overseeing groups closely involved with the systems of hazard.[19]

Recall that HROs strive equally for high levels of production and safety.[20] HROs face the challenge of being reliable both as producers (many under all manner of demanding conditions) *and* as safety providers (under conditions of high production demands). While most organizations combine varying

17. Weick, "Organizational culture as a source of high reliability"; Roberts, "Some aspects of organizational cultures."

18. The concept of organizational culture captures the sense that there are norms, values, and "taken for granted" modes of behavior and perceptions that shape interpersonal and group relations. At the same time, the concept retains a high degree operational ambiguity, its use subject to stiff criticism. See J. S. Ott, *The Organizational Culture Perspective* (Chicago: Dorsey Press, 1989); Roberts, "Some aspects of organizational cultures"; G. I. Rochlin, "Les organizations 'a' haute fabilite': bilan et perspective de recherche" (Highly Reliable Organizations: Exploration and Research Perspectives), chap. 2 in *Organiser la fiabilite*, ed. M. Bourrier (Paris: L'Harmattan, 2001).

19. Roberts, "Some characteristics of high reliability organizations"; "Nuclear Power Operations: A Cross-Cultural Perspective," *Annual Review of Energy and the Environment* 19 (1994): 153–187.

20. Cf. Rochlin, "Reliable Organizations: Present Research and Future Directions"; Schulman, "Negotiated Order of Organizational Reliability."

degrees of production plus service/safety emphasis, HROs have continuously to strike a balance. In times of routine, safety wins out formally (though watchfulness is harder to sustain); in times of high tempo/surge, this becomes reordered (though watchfulness is much more acute). This suggests an organizational culture integrating the familiar norms of mission accomplishment and production with those of the so-called safety culture.[21]

Elements of the results are operator/member élan, operator autonomy, and intrinsic tension between skilled operators and technical experts.

- Operating personnel evince an intense élan and strongly held expectations for themselves about the value of skilled performance. In the face of hazard, it takes on a kind of prideful wariness. There are often intense peer-group pressures to excel as a highly competitive team and to cooperate with and assist each other in the face of high operating demands. This includes expectations of fulfilling responsibilities that often go well beyond formal role specifications. For example, there is a view that "whoever spots a problem owns it" until it is mitigated or solved in the interest of full, safe functioning. This sometimes results in operators realizing that, in the face of unexpected contingencies, they may have to "go illegal," i.e., to go against established, formal procedures if the safety operating procedures appear to increase the difficulty of safely meeting the service demands placed on the organization. Operator élan is reinforced by clearly recognized peer-group incentives that signal high status and respect, pride in one's team, emphasis on peer "retention" and social discipline, and reward for contributing to quality-enhancing, failure-preventing activities.

- Hazardous operations are often time-critical, where effectiveness depends on keen situational awareness. When it becomes clear that speedy, decisive action must be taken, there is little opportunity for assistance or approval from others.[22] Partly as a result, HRO operators come to develop, indeed insist upon, a high degree of discretion, autonomy, and responsibility for activities "on their watch."[23] Often typified as being "king of my turf," this is seen as highly appropriate by both other operators and supervisors.

---

21. See G. I. Rochlin, "Safe operations as a social construct," *Ergonomics* 42, no. 11 (1999): 1549–1560; cf. Weick, "Organizational culture as a source of high reliability."

22. See K. E. Weick, K. M. Sutcliffe, and D. Obstfeld, "Organizing for high reliability: Processes of collective mindfulness," *Research in Organizational Behavior* 21 (1999): 81–123, for a related perspective.

23. K. H. Roberts, D. M. Rousseau, and T. R. La Porte, "The culture of high reliability: Quantitative and qualitative assessment aboard nuclear powered aircraft carriers," *Journal of High Technology Management Research* 5, vol. 1 (spring 1994): 141–161.

- But operator autonomy is often bought at a moderate price. The HROs we studied all operated complex technical systems that put a premium on technical engineering knowledge as well as highly skilled operating knowledge and experience. These two types of skills are usually formally distinguished in the occupational roles designations within HROs. Each has a measure of status; each depends on the other for critical information in the face of potential system breakdown and recovery if problems cannot be contained. But in the operators' eyes, *they* have the ultimate responsibility for safe, effective operation. They also have an almost tactile sense of how the technical systems actually function in the organization's operating environments, environments that are likely to be more situationally refined and intuitively more credibly understood than can be derived from the more abstract, cognitively based knowledge possessed by engineers. The result is an intrinsic tension between operators and technical experts, especially when operators judge technical experts to be distant from actual operations, where there is considerable confidence placed on tacit knowledge of system operations based on long operating experience.[24]

These dominant work ways and attitudes about behavior at the operating levels of HROs are prompted by carrying out activities that are closest to the hazards and suggest the important affective nature of HRO dynamics. These patterns provide the basis for the expressive authority and "identitive compliance"[25] norms that sustain the close cooperation necessary when facing the challenges of unexpected high-tempo/high-surge situations with minimum internal harm to people and capital equipment. But HROs operate in the context of many interested outsiders: sponsors, clients, regulators, and surrounding neighborhoods. Relations with outside groups and institutions also play a crucial role.

External Relationships

HRO performance is clearly dependent on extraordinarily dense patterns of cooperative behavior within the organization. These are extensive, often quite intense, and unusual both in terms of achieving continuous reliability and in higher costs. As such, they are difficult to sustain in the absence of external reinforcement. Continuous attention both to achieving organizational missions and to avoiding serious failures requires repeated interactions with—one might

---

24. G. I. Rochlin and A. von Meier, "Nuclear Power Operations: A Cross-Cultural Perspective," pp. 153–187; Rochlin, "Safe operations."

25. See A. Etzioni, "Organizational Control Structure," chap. 15 in *Handbook of Organizations*, ed. J. G. March (Chicago: Rand McNally, 1965), pp. 650–677.

say pressures from—elements in the external environment, not only to insure resources, but, as importantly, to buttress management resolve to maintain the internal relations outlined above and to nurture HROs' culture of reliability. These cultural characteristics are the most important of all the properties of HROs, for if they are absent, the rest are difficult to achieve and sustain.

NASA has certainly learned how external interests—we will call them "the watchers"—can enter into the Agency's everyday life, especially when major failures are seized upon as a chance to ventilate concerns about operational reliability.[26] "Watchers" include externally situated, independent public bodies and stakeholding interest groups and the institutional processes that assure their presence, efficacy, and use of tools for external monitoring in the interest of hazard evaluations.

Aggressive, knowledgeable "watchers" increase the likelihood that a) reliability-enhancing operations and investments will be seen as legitimate by corporate and regulatory actors, b) such costs *should* be absorbed, and c) regulations and internal social demands should be allowed in the interest of safety. This may mean investing, on one hand, in developing and training external review groups and in some instruments of behavioral surveillance, e.g., random drug tests, and, on the other, assuring these "watchers" that HRO leaders will quickly be held accountable for changes that could reduce reliability in service or safety. These watching groups may be either formal or informal and are found both within the HRO's immediate institutional environment, e.g., congressional committees, and outside it.

It is crucial that there be clear institutional interests in highly reliable performance. This should be evident in strong, superordinate institutional elements of the parent organization, such as agency and corporate headquarters or command-level officers (e.g., utility corporate headquarters, higher military command, and Washington agency headquarters), and sometimes industrial association watchdogs (e.g., the nuclear industry's Institute for Nuclear Power Operators, or INPO).[27]

At the same time, the persistent presence of external stakeholding groups assures attentiveness (and occasional resentment). These groups range from quite formal public watchers, such as regulatory overseers (e.g., state Public Utility Commissions, Nuclear Regulatory Commissions, the Environmental Protection Agency, the Federal Emergency Management Agency, and the Occupational Safety and Health Administration), user and client groups (e.g., instrument-rated pilots using air traffic control services and Congresspersons), to a wide sweep of "public interveners" (e.g., state, local governments, land-

---

26. Diane Vaughan's work (cited above) and conference paper contrasting the *Challenger* and *Columbia* accident reports gives eloquent testament to the dynamics of intense external scrutiny.

27. T. Rees, *Hostages to Each Other* (Chicago: University of Chicago Press, 1994).

use advocates, and citizen interest groups). Finally, this important function is also played by professional peer bodies and by HRO alumni who are seen as operationally knowledgeable observers. They are likely to be accorded respect both by other outsiders and by the HRO operators themselves.

An abundance of external watchers seems crucial in attaining continuous, highly reliable operations and a culture of reliability. So are boundary-spanning processes through which encouragement and constraints are exercised in the interest of product/safety reliability. Two types are evident. First, there are formally designated positions and/or groups who have external oversight responsibilities. Two examples of formalized channels are Nuclear Regulatory Commission On-site Residents, two or three of whom are assigned to each nuclear power plant, with nearly complete access to power plant information, review meetings, etc., and, second, military liaison officers who are permanently assigned to air traffic control centers. Sometimes these boundary-spanning activities are expressed in aircraft carriers' operations via dual reporting requirements for nuclear engineering officers to report problems immediately, not only to the ship's captain, but to a central nuclear affairs office at naval headquarters in Washington, DC, as well.

Boundary spanning, and with it increased transparency, also occurs intermittently in the form of periodic formal visits from "check" or review groups, who often exercise powerful sanctions if their reviews do not measure up. These activities come in a number of forms, for example, phased inspections and training checks in aircraft carrier combat preparations, as well as the more familiar Inspector General reviews, and nuclear power utilities requirements to satisfy rigorous performance in responding to the NRC-mandated, biannual activation of power plant emergency scenarios in which all the relevant local and state decision-makers engage in a daylong simulation leading to possible regional evacuation under the watchful eye of NRC and FEMA inspectors.[28]

Finally, external watchers, however well provided with avenues of access, must have available full, credible, and current information about system performance. This almost goes without saying, for these data, often in the form of annual evaluations, hazard indices, statistical summaries noted above, and indicators of incipient harm and the early onset of danger, become a crucial basis for insightful reviews and public credibility.

This is a formidable array of conditions for any organization to seek or to sustain, even for the short term. To what degree would they suffice over the long term? This will become a major challenge for NASA as missions take on multiyear scope and programs are premised on a long-term human presence in space.

---

28. La Porte and Thomas, "Regulatory Compliance and the Ethos of Quality Enhancement."

ASSURING INSTITUTIONAL CONSTANCY AND
FAITHFULNESS IN THE FUTURE

Many highly reliable organizations operate systems whose full range of positive and negative outcomes can be perceived more or less immediately.[29] When this happens, organizational leaders can be rewarded or held accountable. But when operating systems are also capable of large-scale and/or widely distributed harm which may not occur or be detected for several operational generations, our familiar processes of accountability falter and overseers and the public are likely to be concerned that such HROs be worthy of the trust placed in them across several generations. In NASA's case, these challenges stem from the extraordinary reach of the administration's vision for the Agency's future.

NASA is contemplating missions that will send humans in space for several years to facilities that are likely to be designed to last 10 to 20 years (two management generations). Add to this any of half a dozen hoped-for lunar and exploratory missions. In a much more extreme case, the management of nuclear materials, obligations can be expected to continue for at least 50 to 100 years, perhaps centuries.[30] These cases suggest that shouldering an obligation to demonstrate the faithful adherence to a mission and its operational imperatives for a remarkably long time is inherent in accepting the program—even in the face of a variety of social and institutional environmental changes. As the longer term effects of such technologies become more clear, trying to take into account their transgenerational nature presents particularly troublesome challenges for managers and for students of organization.[31] And it is this aspect of highly reliable operations about which the social and management sciences have the least to say.

---

29. This section draws from portions of T. R. La Porte and A. Keller, "Assuring Institutional Constancy: Requisite for Managing Long-Lived Hazards," *Public Administration Review* 56, no. 6 (November/December 1996): 535–544. It is also informed by my work at Los Alamos National Laboratory (LANL) exploring the organizational challenges posed for the laboratory by its missions of science-based stockpile stewardship (of nuclear weapons), nuclear materials stewardship, and sometimes environmental stewardship. While the operations of the first two, contrasted to the latter, are very different, the challenges provoked by the longevity of the materials involved prompt very similar organizational puzzles. For a similar rendering, see T. R. La Porte, "Fiabilite et legitimaite soutenable" (Reliability and Sustainable Legitimacy), chap. 3 in *Organiser la fiabilite*, ed. M. Bourrier (Paris: L'Harmattan, 2001).

30. Readers can add other technically oriented programs or activities that have a similar extraordinary property, say in the environmental or public works domain.

31. Two conditions, noted here, increase the public demands for constancy because they undermine our typical means of ensuring accountability and are sometimes characteristic of hazardous technical systems. These two are 1) when the information needed to provide unequivocal evidence of effects is so extensive and costly that the public comes to expect that it will not be forthcoming and 2) if harmful effects occur, they are unlikely to be unequivocally detected for some time into the future due to the intrinsic properties of the production processes and their operating

continued on the next page

A partial remedy is to consider what we might call "institutional constancy." More formally, institutional constancy refers to "faithful, unchanging commitment to, and repeated attainment of performance, effects, or outcomes in accord with agreements by agents of an institution made at one time as expressed or experienced in a future time."[32] An organization exhibits constancy when, year after year, it achieves outcomes it agreed in the past to pursue in the spirit of the original public policy bargain.[33]

Conditions Encouraging Institutional Constancy[34]

What little systematic examination of this remarkable intention there is suggests that institutional constancy requires demonstrating to the public or its major opinion leaders that the agency, public contractors, or firms in question (for example, NASA operating very reliably) can both be trusted to keep its word—to be steadfast—for long into the future and to show the capacity to enact programs that are faithful to the original spirit of its commitments.[35] What conditions signal continued political and institutional will, steadfastness in "keeping the faith"? What conditions assure the capacity to follow through for many years, i.e., the organizational infrastructure of institutional constancy?

Institutional purpose. Constancy is about future behavior, and the organization must signal its collective resolve to persist in its agreements, especially

---

continued from the previous page

environments. While the mind's eye turns quickly to public organizations for examples, the argument applies with nearly equal force to the private sector in the United States, especially to those firms responding to the strong economic incentives for short-term gain with the systematic deferral of costs for some time.

32. T. R. La Porte and A. Keller, "Assuring Institutional Constancy."

33. Think, for example, of the FAA's air traffic control operations, together with air carriers. They have consistently achieved high levels of flight safety and traffic coordination in commercial aviation and flight operations at sea. And the Navy has a long-term record of exceptional safety aboard nuclear submarines. Electrical utilities have made remarkably high levels of electrical power available. Great universities exhibit constancy in commitments to intellectual excellence, generation after generation, through producing very skilled undergraduates and professionals as well as pathbreaking research.

34. Note: There are strong analytical and practical limitations to attaining institutional constancy over many generations, especially a) weak analytical bases for confidently predicting the outcomes of institutional activities over long periods of time, b) limited means to reinforce or reward generations of consistent behavior, and c) scanty knowledge about designing institutional relationships that improve rather than degrade the quality of action-taking in the future that is faithful to the spirit of present commitments and agreements. Incentives to improve conditions that would assure constancy of institutional capacities are scant. And so is interest in analysis that would improve our understanding of institutional and administrative design. Indeed, there is almost nothing insightful in the literature about *increasing* institutional inertia or constancy. It is still an analytical puzzle.

35. While these two qualities are closely related, one can imagine succeeding at one without achieving the other. An HRO might be able to persuade the public that it was firmly committed to certain objectives but actually turn out to be in no position to realize them. Conversely, an HRO could very well be situated, motivated, and structured to carry out its commitments for years to come but be unable to convince the public of its steadfastness.

with strong commitments to trusteeship in the interests of future generations. Measures that reinforce this perception are as follows:

- The necessary formal, usually written goal of unswerving adherence to the spirit of the initial agreement or commitment; documents that can be used in the future to hold each generation's organizational leaders accountable for their actions.

- Strong, public articulation of commitments to constancy by high-status figures within an agency or firm, calling especially on professional staff and perhaps key labor representatives to emphasize the importance of constancy. Coupled with formal declarations, consistent emphasis upon steadfastness within an organization reinforces the otherwise difficult commitments of energy and public witness that are needed by key members of the technical staff and workforce.

- Strong evidence of institutional norms and processes that nurture the resolve to persist across many work generations, including, in the public sector, elements in labor contracts that extend over several political generations.[36] When these exist, they bind workers and their leaders to the goals of the agency, often transcending episodes of leadership succession. The content of these norms and the processes that reinforce them are now not well calibrated, though examples are likely to be found in public activities that draw the deep loyalty of technical staff and former members. This seems to be the case for elite military units, e.g., the U.S. Marine Corps and Navy Seals; groups within the Centers for Disease Control (CDC) and some other public health activities; and some elements within U.S. air traffic control circles. A close examination of the internal processes of socialization the produce such loyalty is warranted.[37]

- Commitments to courses of action, particularly those where benefits may be delayed until a succeeding management or political genera-

---

36. This point is akin to the arguments made classically by P. Selznick, *Leadership in Administration* (New York: Harper & Row, 1957), and J. Q. Wilson, *Bureaucracy: What Government Agencies Do and Why They Do It* (New York: Basic Books, 1989), pp. 99–102, about the importance of institutional leadership and the character of the organization's sense of mission.

37. For an early exploration of this aspect, see Selznick, *Leadership in Administration*, and his discussion of the transformation of an instrumental organization into one that has been "infused with value," i.e., that becomes an "institution." For a recent project attempting to address these questions, see A. Boin, "The Early Years of Public Institutions: A Research Agenda" (paper issued by the Department of Public Administration, Leiden University, Netherlands, 2004).

tion, are difficult to sustain in the face of U.S. political metabolism. Therefore, vigorous external reinforcement from both regulatory agencies and "public watching" groups must be present to assure that the relevant agencies and their contractors will not flag in attending to the performance promised by one generation to the next. This would include reinforcing the vigor of outside groups by regularly assuring their formal involvement and providing sufficient resources to sustain their expectations and prompt their demands for consultation if the next generation of leaders wavers in its resolve. The optimum would be when these measures lead to laws, formal agreements, and foundation/nongovernmental funding and infrastructure for continual encouragement and sanctions for "keeping the faith."

The infrastructure of constancy. While strong motivations and earnestness are necessary, they alone do not carry the day. Other conditions should also be present to assure interested outsiders that actions will, in fact, be carried out in realizing important commitments across multiple generations. As I outline

Table 13.2. Characteristics Associated with Institutional Constancy
(i.e., Organizational Perseverance, Faithful Adherence to the Mission and Its Operational Imperatives)

1. Assurance of steadfast political will.

    A. Formal goal of unswerving adherence to the spirit of the initial agreement.

    B. Strong articulation of commitments by high-status agency leaders calling on staff in achieving constancy.

    C. Clear evidence of institutional norms that nurture the persistence of commitments across many generations.

    D. Vigorous external reinforcement from regulatory agencies and public watching groups.

2. Organizational infrastructure of constancy.

    A. Administrative and technical capacity to carry out constancy-assurance activities reinforced by agency rewards.

    B. Adequate resources to assure the "transfer" of requisite technical and institutional knowledge across worker and management generations.

    C. Analytical and resource support for "future impact analyses."

    D. Capacity to detect and remedy the early onset of likely failure that threatens the future, with the assurance of remediation if failures occur.

these, return in your mind's eye to the U.S. space community and the many organizations revolving satellite-like around the central sun/star of NASA. How many of the conditions I will suggest below already exist within NASA? How difficult would their introduction and persistence likely be? If these seem sparse, or absent, this points to a "critical institutional issue."

These conditions of constancy include the following:

- The technical capabilities and administrative infrastructure which are needed to assure performance, along with agency or contractor rewards and incentives for articulating and pursuing measures that enhance constancy and intergenerational fairness. These would include executive socialization and training processes to reinforce commitments and long-term perspectives to nurture a culture of constancy. Such processes and resources are rarely provided in today's institutional environments. Rather, perspectives and rewards are intensely generation-centric, characterized by quite short-term evaluations, and strongly reinforced by contemporary business and legislative cycles.

- In addition to assuring consistency in organizational culture, the resources and activities needed to "transfer" or "pass on" the organization's critical operating, technical, and institutional knowledge from one work and management generation to the next are crucial. This includes systematic capture of critical skills and operating histories, as well as continuous training and evaluation of each generation's capabilities. Some portion of each future generation should be present in the current one.

The remaining conditions point to keen powers of analysis in service to the future.

- Analytical supports should be evident for analysis and decision-making which take into account the interests of the future and enable work, such as "future impact analyses," that seeks to identify the effects of present institutional actions on future capabilities. Something like this goes on during budgetary planning efforts, but, in the U.S. system, the timeframes are invariably merely short-term, tied to legislative or corporate profit reporting cycles. Scanning further into an institution's future—at least beyond the present generation—is also called for. Analytical capabilities to do this are likely to require at least a small cadre of highly skilled professionals, systems for rewarding their efforts, and organizational and agency venues where their reflections will have a respected voice.

- And, perhaps most important, publicly obvious, effective capacity would be in place to detect the early onset of likely failures related to the activities that could threaten the future. This analytical capacity should then be joined with institutional capabilities to initiate remedies, along with the assurance of remediation resources in the event failures should occur.[38] Without quite visible, publicly evident, and well-exercised capacity for early warning and *preemptive* remediation, the public is likely to remain skeptical, potentially suspicious, and ripe for mobilization into recalcitrant opposition.[39]

This suite of conditions intended to assure institutional constancy is very demanding and costly. Whether leaders would consider developing them is likely to be contingent upon external demand. Pressure to try is increased when programs exhibit three characteristics. There will be particularly aggressive insistence on faithfulness when agency programs a) are perceived to be large-scale efforts whose activities may occur across broad spatial and temporal spans and seem to pose potentially irreversible effects; b) are seen as intensely hazardous, even if the likelihood of failure is small and accompanied by substantial gains for the program's prime beneficiaries; and c) pose significant risks whose costs are likely to be borne by future generations who receive little benefit.

This third characteristic—temporal asymmetry of benefits and costs—raises a particularly difficult dilemma. Put in question form: should current populations endure costs today so that future populations will not have to?[40] In NASA's case, this would include investing to avoid future risks against the accrual of present benefits, say, in symbolic returns, or perhaps knowledge that is potentially useful in providing novel artifacts. These long-term benefits

---

38. See, for example, T. R. La Porte and C. Thomas, "Regulatory Compliance and the Ethos of Quality Enhancement: Surprises in Nuclear Power Plant Operations," *Journal of Public Administration Research and Theory* 5, no. 4 (December 1994): 250–295. Cf. K. Shrader-Frechette, "Risk Methodology and Institution Bias," *Research in Social Problems and Public Policy* 5 (1993): 207–223; and L. Clarke, "The Disqualification Heuristic: When Do Organizations Misperceive Risk?" *Research in Social Problems and Public Policy* 5 (1993): 289–312, for discussions of the conditions that result in operator misperception of risk, conditions that would require strong antidotes if constancy is to be assured.

39. This seems clearly to be the case for the many years of political and legal travail experienced by the Department of Energy. See DOE, "Earning Public Trust and Confidence."

40. See, for example, R. M. Green, "Inter-generational Distributive Justice and Environmental Responsibility," in *Responsibilities to Future Generations: Environmental Ethics*, ed. E. D. Partridge (Buffalo: Prometheus Books, 1980); R. Howarth, "Inter-generational Competitive Equilibria Under Technological Uncertainty and an Exhaustible Resource Constraint," *Journal of Environmental Economics and Management* 21 (1991): 225–243; B. Norton, "Environmental Ethics and the Rights of Future Generations," *Environmental Ethics* (winter 1982): 319–338; P. Wenz, "Ethics, Energy Policy, and Future Generations," *Environmental Ethics* 5 (1983): 195–209.

would have to be balanced against present costs and, as importantly, future industrial environmental damage from large-scale facilities, or having to abandon teams of astronauts due to the inability to retrieve them, and, more remotely, infecting terrestrial populations with extraterrestrial organisms.

Uncertainty about the knowledge and technological capacity of future generations exacerbates the problem. An optimistic view assumes that difficult problems of today will be more easily solved by future generations.[41] No problem today is too big for the future. Skepticism about this, however, makes it an equivocal basis for proceeding with multigenerational programs. An inherent part of assuring constancy would be an agreed-upon basis, an "ethic," of how costs and benefits should be distributed across generations. This is especially true when operational effects extend well into the future, for it demands that generation after generation respond to new information and changing value structures in coping with long-term effects.

This array of constancy-enhancing characteristics raises serious, unresolved operational, political, and ethical questions. If an organization's program provokes demands for nearly error-free operations, then assurances of institutional constancy in meeting the conditions for reliability are likely to be demanded as a substitute for accountability.[42] Apprehensive publics seek assurances that these institutions, such as NASA, will be uncompromising in their pursuit of highest quality operations through the relevant lifetimes of the systems in question.

When harmful effects may be visited upon future generations, assurances of continuity or institutional constancy take on increasing importance.[43] Why would this be the case? Those who implement such programs could quite probably escape accountability for failures. They would have retired,

---

41. For comment on how responsibility should be divided between generations that accounts for changes in knowledge, see W. Halfele, "Energy from Nuclear Power," *Scientific American* 263, no. 3 (September 1990): 136–144; C. Perrings, "Reserved Rationality and the Precautionary Principle: Technological Change, Time and Uncertainty in Environmental Decision Making," in *Ecological Economics: The Science and Management of Sustainability*, ed. R. Costanza (New York: Columbia University Press, 1991).

42. For those HROs whose technical operations and consequences of failure can be seen as having constancy-evoking characteristics, ignoring "constancy magnets" is an institutionally risky business. This is especially the case for the combination of uneven distribution of benefits and costs among generations and the potential for a long lag in discovering information about possibly grievous damages. Setting these matters aside allows festering seeds of suspicion to multiply, and, if coupled with conditions that also evoke "reliability and regulatory magnets," they are likely grounds for political opposition and demands for increasing rigorous regulation as a condition for even initial approval for new projects. But if organizational remedies are called for, how much additional effort and evolution of institutional capabilities could be entailed?

43. While the mind's eye turns quickly to public organizations, the argument applies equally to the private sector in the United States, especially those firms responding to the strong economic incentives for short-term gain and deferral of costs.

died, or moved on. Leaders of such institutions, therefore, are quite likely to be pressed to assure the public (especially able opinion leaders) that, as a condition of winning approval and resources to initiate or continue programs, agencies and corporate contractors involved should credibly be expected to keep agreements and commitments with potentially affected communities far into the future.

CONCLUDING REFLECTIONS

The reach of NASA's space programs continues to levy remarkable operational demands, for the programs imply very long-term management of both the unmanned and manned aspects of space exploration and possibly commercial and security exploitation. This rather cryptic application to NASA's space exploration programs of work done in other technical domains hints at the challenges involved when we insist on extraordinary levels of reliability that should go on for a number of management generations. It suggests an array of conditions that would become increasingly salient as NASA seeks to regularize and sustain its space traffic regime.

These are very demanding conditions for organizational leaders to consider, much less actively insist upon, encourage, and nurture, even if we knew how to establish organizational patterns I have summarized.[44] It is notable that my discussion is based on work dealing with operations that, *unlike NASA spaceflights*, were quite mature, pretty routine, and had managed to continue for some time. Although the HRO field work involved nearly 10 years of observing and intensive subjective onsite experience with each of three large technical systems in the study, it was not so intensive as discovering the process through which these organizations had gone to result in the variegated patterns that were described. We do not know exactly how they got there.

If the constructs I have outlined here are taken seriously, it is likely to pose unwelcome challenges to agency and program leaders. Our workshop discussions called out a range of critical institutional (as well as historiographical) issues and point toward matters of serious design examination. But the analytical bases for designing and assuring institutional forms at substantial

---

44. They are also conditions that are not likely to flourish without a high degree of public trust and confidence in operating and overseeing institutions—something that is in increasingly short supply in contemporary American culture. NASA has skated across the increasingly thin ice of waning public confidence in programs involving humans in space. The several high-profile congressional investigations and the Agency's agony over the past decade have eroded a general sense of public confidence in future operations. This in itself should be seen as a major critical institutional issue. For an earlier consideration of this, see T. La Porte, "Institutional Challenges for Continued Space Exploration: High-reliability systems across many operational generations. Are these aspirations publicly credible?" (presented at the Workshop on Space Policy, National Academies of Science, Irvine, CA, 12–13 November 2003).

scale are limited at best.[45] For example, there is scant work on effecting institutional constancy per se, and only limited study of the evolution of highly reliable organizations. A remedy to these important gaps in understanding requires both analytical and experimental efforts to calibrate the dynamics of highly reliable operations, and especially probing the requisites for long-term institutional constancy and trustworthiness.

At least three additional aspects of this challenge are apparent; each prompts a demanding set of research imperatives (see table 13.3).

First, we need to improve our knowledge about the wider institutional currents within U.S. patterns of public and corporate governance that provoke repeated, stubborn resistance to the organizational changes needed to sustain very reliable operations, and reassure citizens that the responsible institutions will be able to keep their word through the relevant program timeframes—and do so in ways that enhance their trustworthiness. Even if there is a reasonably benign political and social environment, these are qualities that are very difficult to establish and maintain. In answering "Why can't we do

---

Table 13.3. Research Directions: When Highly Reliable Operations, Long-Term Institutional Constancy, and Trustworthiness Are Indicated

Q: Why can't we do it?

A: Institutional impediments to conditions sustaining very reliable operations, institutional constancy, and trustworthiness.

Q: Why do we have to?

A: Technical imperatives requiring very reliable operations over multiple political generations. (Seek technical design alternatives having equivalent physical and organic effects without HRO or institutional constancy imperatives.)

Q: Why do we need to?

A: Alternatively, there are institutional activities that reduce the public's

    1. risk-averse demand for very reliable operations of intrinsically hazardous systems,

    2. worry about the longer term consequences of operational errors, and

    3. sense of vulnerability that fosters a demand for trustworthy public institutions.

---

45. Some of these are highlighted in the chapters by Diane Vaughan and Philip Scranton.

it?" historical insight surely can be brought to bear. NASA is a particularly visible case, certainly not the only instance in which a public agency seems unable to alter its internal dynamics so that it avoids repeating what outsiders perceive (invariably after a serious mishap) to be dysfunctional organizational patterns. Observers of the Department of Energy's Radioactive Waste Programs are also likely to regard these efforts as deeply flawed. In these and other cases, such evaluations arise during nearly each generation of new management. For NASA, it is observed that dysfunctions have afflicted each of the last seven Administrators with repeated problems in the evolution of NASA's institutional culture. The conference papers contributed by Scranton and Vaughan give witness to many of these debilitating dynamics. Some of this is internally self-inflicted, to be sure. But for my part, I suspect more important sources lurk in NASA's relations with Congress and the Agency's extensive contractual community. In the early pages of the *Columbia* report, these sources of dysfunction were noted. They then escaped detailed examination thereafter. In the future, these should be the objects of as much analysis as NASA's internal dynamics. The historical community seems particularly positioned to furnish keen insight into what—in repeated instances—seems likely to be the result of a much deeper structural relationship than merely a series of very able people somehow succumbing to individual weakness and local bureaucratic perversity.

Second, we need to deepen our understanding of the technical sources that drive systems operators toward "having to" attain very high reliability. Technologies vary in the degree they require closely harmonized operator behavior. They also vary in their intrinsic hazardousness. Both of these characteristics can be shaped by the engineering design teams who provide the technical heart of operating systems. What is it about technical communities that prompts their members to propose technologies that require extraordinary behavior as a *condition* of delivering the hoped-for benefits? Is this intrinsic to some technical domains and not others? This suggests studies that calibrate the degree to which present technical and operational directions in the development of, at least, environmentally sensitive operations, materials management, and transportation and biological technologies a) require highly reliable operating organizations, b) imply long-term operating trajectories and potentially negative effects, and hence c) produce a requirement for high levels of public trust and confidence. In-depth sociological and historical studies could, one imagines, shed light on these matters.

A better understanding of these relationships can be crucial in democratic societies. It can be argued that the more the requirements for HRO, institutional constancy, and public trust and confidence are present, the more demanding the institutional challenges will be in sustaining public legitimacy. A closely related emphasis follows: what changed within technical design communities would be necessary for them aggressively to seek technical design alternatives that provide equivalent physical and organic effects varying the degree to which they produce demands for high-reliability operations over many work generations.

But wait, wait! Is there an alternative to the two research and development vectors just noted? They are very demanding R&D domains. Actually realizing the organizational imperatives that lurk within such designs is even more difficult to assure within private or public enterprises in the U.S. and abroad. Indeed, even entertaining the desirability of such changes is disputed by institutional leaders and provokes strong managerial reluctance to consider them seriously. So why are we trying? "Why do we need to?"

The need to try (or act as if we were trying) stems, importantly, from the public's expressed worry about their own exposure to what they perceive to be "risky systems." They worry and appear to have a very low tolerance for risk-taking. It could be argued, we need to try because "they" demand it. However, an alternative program of research and activities could be launched.

What activities could be carried out which would *reduce* the public's risk-averse demand for very reliable operations of intrinsically hazardous systems, *reduce* the public's worry about the longer term consequences of operational errors, and *lessen* the public's sense of vulnerability that nourishes a deep longing for trustworthy public institutions? As far as I know, there is very little systematic work exploring the grounds upon which alert publics would come to understand the rationality of accepting the likelihood of increased exposure to malfunctions of hazardous technical systems in the interest of smoothing production flows or stabilizing revenue streams for major investors. Nor do I know of any efforts to understand the basis for convincing the public explicitly that it would be acceptable to engage in developments that promise attractive short-term benefits which would export severe costs across several future generations to their grandchildren's children. Worries about the potential for immediate exposure to personal injury or environmentally derived insult, and a more diffuse concern that important dangers may await our children some years from now, continue to spawn irritating (probably irrational) objections to developing and deploying exciting new technical possibilities. Well, perhaps they could produce untoward surprises, but they are (probably) manageable. We can count on clever technical solutions.

"Why can't they trust us?" Indeed, this deserves analytical attention as well. Why do alert publics feel so vulnerable that they increasingly wish for trustworthy institutions? What developments could be devised that publics would relax their demand for trustworthiness and accept technical leaders and provide support for the technical future we designers see? In effect, "Why," as Henry Higgins and one technical designer put it, "why can't they be more like us?"

Space History:
State of the Art

~

INTRODUCTION

What is the current state of space history as the 21st century commences and the Space Age reaches its 50th anniversary? Is it a vibrant marketplace of ideas and stimulating perspectives? Is it a moribund backwater of historical inquiry with little of interest to anyone and nothing to offer the wider historical discipline? As the four essays in this section demonstrate, space history is at neither extreme of this dichotomy. It has been energized in the last quarter century by a constant stream of new practitioners and a plethora of new ideas and points of view. A fundamental professionalization of the discipline has brought to fruition a dazzling array of sophisticated studies on all manner of topics in the history of spaceflight. Yet, as the collective authors of the section argue, there is much more to be done, and each offers suggestions for how historians might approach the field in new and different ways, each enriching what already exists.

This section opens with an essay by Asif A. Siddiqi assessing the state of U.S. space history. He asserts that scholars have concentrated their work in one of four subfields that collectively may be viewed as making up the whole. As Siddiqi writes, "Some saw the space program as indicative of Americans' 'natural' urge to explore the frontier; some believed that the space program was a surrogate for a larger struggle between good and evil; others wrote of a space program whose main force was modern American technology; and others described a space program whose central actors were hero astronauts, representing all that was noble in American culture." He notes that space history started as a nonprofessional activity undertaken by practitioners and enthusiasts, always viewing the field from the top down and producing an exceptionally "Whiggish" perspective on the past. In the 1980s, the field began to broaden, deepen, and expand through the entry of a number of professionally trained historians who brought new skills and new interests to the subject. A number of pathbreaking works have emerged, especially in the realm of the history of space policy, and the current state of the subdiscipline is vibrant.

Lest space historians become complacent, however, Siddiqi concludes that there is still much to be done. He points to several specific areas that offer tantalizing possibilities for future research. These include studies on political, social, technological, and cultural history using themes and methodologies borrowed from the larger historical community. Good work has already been done, and Siddiqi analyzes some of this work, but many opportunities for additional study are present. Siddiqi expends considerable effort documenting a future research agenda and goes far toward identifying potentially fruitful avenues of research for new scholars seeking entrance into the field.

Siddiqi also comments on the interesting and unusual circumstance of government sponsorship of space history and the possibility that this might taint the published product in some way. Many of the historians working in the field have been sponsored in some measure by NASA, the United States Air Force, or the Smithsonian Institution, either as employees of these entities or as contractors or fellows. What does this connection mean for the work done, Siddiqi asks? Clearly it plays a role, but what role? This issue itself might be a useful avenue of study, relating as it does to the concerns raised in the work of Peter Novick and others about the pursuit of objectivity in historical studies.[1]

For many years, a stigma has existed among some academic historians against sponsored history; such a view is usually misplaced and not a little naive. Those who criticize such work invariably invoke the characterization "court historian" to damn the effort. There are, of course, some instances of influence that all can point to. But the reality is that historical truth is elusive in any setting. Historians usually have a clientele, whether writing for other academic specialists in whatever field is under investigation; or for groups bound together by religion, ethnicity, labor, etc.; or for any number of identifiable groups that have an interest in the subject.[2] Consciously or unconsciously, historians—even if they have not been formally hired to prepare histories for the group—shape their discourses to provide understanding about the past in relationship to ideas already present among those with an interest in the subject. If one strays too far afield from the major streams of understanding about the subject, the historian may be unable to find an outlet for publication, may be censured in reviews, may have his or her livelihood destroyed by not receiving tenure, or may lose whatever reputation he or she had. All of that takes place, even without serving some formal client that may have a vested interest in ensuring that a historian tells a story in a certain way.[3] Still, a study of the influence of government sponsorship on the field of space history would prove a fascinating subject of study.

In chapter 15, Stephen B. Johnson presents a lengthy discussion of the historical study of military space history from the 1950s to the present. In this exhaustive review, Johnson divides his analysis into major sections, conforming to the various missions that the Department of Defense undertakes relative to space operations. After a review of overview sources of military history, the author undertakes an analysis of the development and fielding of intercontinental ballistic missiles (ICBM) and space launchers, which defined the strategic defense capabilities of the United States during the Cold War.

---

1. Peter Novick, *That Noble Dream: The "Objectivity Question" and the American Historical Profession* (New York: Cambridge University Press, 1988).

2. See Roger D. Launius, "NASA History and the Challenge of Keeping the Contemporary Past," *Public Historian* 21 (summer 1999): 63–81.

3. See Roger D. Launius, "Mormon Memory, Mormon Myth, and Mormon History," *Journal of Mormon History* 21 (spring 1995): 1–24.

From there, Johnson moves on to discussions of early-warning and space sur-veillance; command and control; communications; ballistic-missile defense; robotic intelligence and reconnaissance; military human spaceflight; weather and science; navigation; antisatellite and space warfare; organization, man-agement, and acquisition of space systems; and space power theory. Johnson concludes with a gap analysis of "holes in the literature" and offers suggestions for future historical study.

Margaret A. Weitekamp follows with a discussion of how historians working in space history might consider the topic with new "tools" drawn from social and cultural studies. Indeed, one of the most exciting areas of historical inquiry in the last 20 years has been the postmodern analysis of history. Weitekamp acknowledges that richness, which ensures within space history, as it has elsewhere, "the proliferation of subject areas created when historians wrestling with questions of race, class, ethnicity, and gender chal-lenged the artificial nature of the consensus school's master narrative." She then surveys the field, noting important developments in the application of themes in social and cultural studies to the subject of space history, but more importantly, Weitekamp then explores the relationship between space history and this larger discourse. She finds that "space history exists both in 'relation to' other history subdisciplines (a terminology which implies separation from the other subfields and an internal cohesion within space history, two points that deserve questioning in their own right), and in a continually evolving 'relationship with' the rest of the discipline."

Weitekamp also finds that the application of "critical theory" to the his-tory of spaceflight may offer uniquely useful perspectives on the subdisci-pline. She defines "critical theory" as "an umbrella term that encompasses the diverse and often divergent theoretical schools of structuralist, poststructural-ist, feminist, Marxist, postmodern, and psychoanalytic theory that emerged since the 1970s in literary and anthropological analysis." Already intriguing possibilities for this area have been opened through the work of Jodi Dean, Constance Penley, M. G. Lord, De Witt Douglas Kilgore, and others.[4] Greater use of these methods of historical inquiry has the potential to transform the field of study.

This section closes with an intriguing and stimulating essay by David H. DeVorkin on the importance of the artifact in the study of the history of tech-nology. Most historians, he asserts, do not pay much attention to the objects

---

4. Jodi Dean, *Aliens in America: Conspiracy Cultures from Outerspace to Cyberspace* (Ithaca, NY: Cornell University Press, 1998); M. G. Lord, *Astro Turf: The Private Life of Rocket Science* (New York: Walker & Co., 2005); De Witt Douglas Kilgore, *Astrofuturism: Science, Race, and Visions of Utopia in Space* (Philadelphia: University of Pennsylvania Press, 2003); Constance Penley, *NASA/TREK: Popular Science and Sex in America* (New York: Verso, 1997).

that they write about. They use quite traditional sources—manuscript materials and other written work—but fail to observe carefully the actual spacecraft, rocket, or other physical object that performed the work under study. He asks the important question, "Are artifacts historical evidence?" Of course they are, he notes, but few historians exploit them effectively in their own work. Perhaps that is because they fail to grasp their significance, but more importantly, it is probably because they do not understand how they work and why they were constructed in the way they were. DeVorkin argues for a greater appreciation of the artifact in the enterprise of historical study and the centrality of it in the narratives fashioned by historians of spaceflight.

Collectively, these four essays point up the richness of the study of the history of the American effort to fly in space since the 1950s. As such, they represent a report from the field of its status and possibilities for the future. Most important, each essay points the direction for future efforts.

CHAPTER 14

AMERICAN SPACE HISTORY:
LEGACIES, QUESTIONS, AND OPPORTUNITIES
FOR FUTURE RESEARCH[1]

Asif A. Siddiqi

In the 35 years since astronauts Neil A. Armstrong and Buzz Aldrin set foot on the Moon, no space achievement has quite captured people's imaginations as Apollo. Thirty-five years after that singular event, the specter of Apollo still looms large as a benchmark for all that came later. In the context of the current inertia of the American space program—the Space Shuttle temporarily grounded while astronauts take to orbit in Russian rockets for unimaginative tours of the International Space Station—Apollo retains an even stronger pull to those seeking adventure and exploration.[2] Given Apollo's centrality in popular conceptions of the history of the space program, it is not surprising that historical writing—both popular and academic—has been shaped profoundly by the experience of the Moon landings. Even those areas of space history that have no apparent connection to Apollo, such as military space history, for example, assume their historical places in our memory in relation to Apollo. Because of the project's status as being emblematic of a lost, young, and adventurous America, space historians negotiating the delicate boundaries between memory and nostalgia have typically veered from the former to the latter with an ease that underscores more about the state of the current space program than the one that actually happened. In addition, Apollo's huge shadow has helped to marginalize many important but unexplored areas of space history.

In the past 40 years of space history, historians have worked within several interpretive approaches to space history, all of them defined and demarcated by the shadow of Apollo and its political backdrop, the Cold War. This essay is an attempt to revisit that historiography in search of some common unify-

1. I would like to thank Dwayne A. Day, Steven J. Dick, Roger D. Launius, and Michael J. Neufeld for their helpful comments.

2. For the current crisis, see Roger D. Launius, "After Columbia: The Space Shuttle Program and the Crisis in Access to Space," *Astropolitics* 2 (July–September 2004): 277–322.

ing themes.[3] The goal is to identify certain interpretive and narrative patterns and then elaborate on areas where scholarship is lacking or where important questions remain unexplored.[4] A close reading of the literature shows that historians have located their work within four different narratives based around exploration, competition, technology, and the astronauts. These interpretive paradigms continue to dominate and define our understanding of the origins, evolution, and nature of the American space program. The categories were not mutually exclusive, and the approaches have overlapped over time, but these four guiding themes have remained as important explanatory devices. Some saw the space program as indicative of Americans' "natural" urge to explore the frontier; some believed that the space program was a surrogate for a larger struggle between good and evil; others wrote of a space program whose main force was modern American technology; and others described a space program whose central actors were hero astronauts, representing all that was noble in American culture.[5]

In all of the four schools, which continue to flourish today, historians have typically examined the history from the top looking down, describing only the tallest trees of a vast forest of society and culture. The first generation of scholarship was distinguished by a focus on linear, narrow, and progress-oriented narratives unencumbered by context, critique, or culture. Historians also shared a nostalgic yearning for the 1960s, the halcyon period of American space exploration. Like the space program itself, historians repeatedly romanticized the claimed victories of Apollo without questioning many of the incontrovertible motivations and repercussions of the space program.

Starting in the 1980s but really coming to fruition in the 1990s, a "new aerospace history" began to emerge. Building on a few notable works published during the late Cold War, a new generation of historians tackled the history of American space exploration from different perspectives involving politics, society, and culture. These new works distinguished themselves from the older canon because they revisited, cajoled, and questioned some of the basic foundational notions of the received space history. Some did so explic-

---

3. For earlier works on the historiography of American space exploration, see Richard P. Hallion, "A Source Guide to the History of Aeronautics and Astronautics," *American Studies International* 20, no. 3 (1982): 3–50; Hunter A. Dupree, "The History of the Exploration of Space: From Official History to Contributions to Historical Literature," *Public Historian* 8 (1986): 121–128; Pamela E. Mack, "Space History," *Technology and Culture* 30 (1989): 657–665; Roger D. Launius, "The Historical Dimension of Space Exploration: Reflections and Possibilities," *Space Policy* 16 (2000): 23–38.

4. In the paper, I do not distinguish between the often false dichotomy of academic versus popular works. Important contributions to space history have come from both ends of the spectrum, and both have had their strengths and weaknesses. I also do not explore the study of international cooperation in space history, a vast topic covered by others in this volume. Finally, due to limitations of length, I omit discussion of those histories dedicated to the events of the pre-Sputnik era.

5. I list and describe representative examples from each group in the main body of the essay.

itly, others more implicitly. The new history also moved beyond the lenses of competition, exploration, technology, or astronauts. In some cases, the literature built upon the older models, while in others, it made a clean break from the older canon.

Historians also moved into new areas of political, technological, social, and cultural history benefiting from a shared interest in new sources and new methodological approaches. Simultaneously, the old Cold War paradigm of historiography continues to flourish, propagated especially in several syntheses, creating an interpretive tension between the old and new writing that may promote a middle ground in the future. Whether this mix will generate new, interesting, and challenging ideas remains to be seen, but it has been healthy for the field to expand beyond the previously narrow borders, if for nothing else to link and relocate space history, not as something peculiar and unique, but as part of a broader inquiry into American history.

EXPLORATION

The most common motif in space historiography has been that of locating space exploration as part of an eons-long human urge to push the geographical frontiers of existence. Prescriptive works on space exploration published in the pre-Sputnik era—some of which assumed iconic status in later years—firmly established such an approach to history. A harbinger of this paradigm was Willy Ley, a veteran of early amateur German rocketry groups from the 1930s. Updating a book he had first authored in 1944 through 21 printings, Ley's *Rockets, Missiles, and Man in Space* (1968) was a landmark publication that former NASA Chief Historian Roger D. Launius has called "one of the most significant textbooks available in the mid-twentieth century on the possibilities of space travel."[6] A popular historical narrative tracing the evolution of rocket technology from the ancient Babylonians to the mid-1960s, Ley's work weaved together human imperatives and technical evolution in a seamless whole. From the beginning, he described his book as "the story of the idea that we possibly could, and if so should, break away from our planet and go exploring to others, just as thousands of years ago men broke away from their islands and went exploring to other coasts."[7] By focusing on a few scattered, talented individuals with a vision of space travel, Ley delineated the history of space

---

6. Roger D. Launius, *Frontiers of Space Exploration* (Westport, CT: Greenwood Press, 1998), p. 190; Willy Ley, *Rockets, Missiles, and Men in Space* (New York: Viking Press, 1968). Ley also published an abridged and slightly updated version of his book the following year as *Events in Space* (New York: D. McKay, 1969).

7. Willy Ley, *Rockets: The Future of Travel Beyond the Stratosphere* (New York: The Viking Press, 1945), p. 3. In popular history, others have connected space history to the exploration paradigm. See, for example, Daniel J. Boorstin, *The Discoverers* (New York: Random House, 1983).

exploration as essentially one with an individualistic character. In Ley's world, technology, i.e., the means to fulfill these singular visions, was subordinated to the needs and whims of resourceful scientists or engineers whom he called "Prophets of Some Honor." Thus, the principal actors behind space exploration were neither nations nor states, but noble visionaries. Ley also established a pantheon of icons for the future history of space; by giving currency to such names as Konstantin Tsiolkovskiy, Hermann Oberth, and Robert Goddard, he gave a face to the technology.[8] German rocketry pioneer Wernher von Braun's *History of Rocketry and Space Travel* (1966) (cowritten with Frederick I. Ordway III) built upon Ley's work and cemented a number of unquestioned narratives about the origins of the "Space Age," including the centrality of von Braun's V-2 "rocket team" in the postwar American rocket and space program, thus marginalizing a number of other equally important indigenous innovators in the American context such as the Guggenheim Aeronautical Laboratory at Caltech (GALCIT) and the American Rocket Society.[9] So powerful was this synthesis that to this day, almost all history books on space exploration begin by invoking Tsiolkovskiy, Oberth, and Goddard—and then move to von Braun's rocket team.

What these pioneers had in common was a sustained belief that the human spirit was possessed of an indomitable urge to explore and, as a corollary, to seek knowledge. In one of his most oft-repeated quotes, the Russian theoretician Konstantin Tsiolkovskiy (1857–1935) had written that "the earth is the cradle of reason, but one cannot live in a cradle forever."[10] For the historian of the American space program, reason was combined with a modern version of manifest destiny, a marriage of the near-spiritual urge to explore new frontiers and the cold, hard rationale of technology. One of the earliest scholarly works to equate the idea of the American West with the space fron-

---

8. For biographies, see Helen B. Walters, *Hermann Oberth: Father of Space Travel* (New York: Macmillan, 1962); Hans Barth, *Hermann Oberth: Vater der Raumfahrt: autorisierte Biographie* (Esslingen: Bechtle, 1991); David A. Clary, *Rocket Man: Robert H. Goddard and the Birth of the Space Age* (New York: Hyperion, 2003); Milton Lehman, *This High Man: The Life of Robert H. Goddard* (New York: Farrar, Straus, 1963); A. Kosmodemiansky, *Konstantin Tsiolkovsky, 1857–1935* (Moscow: Nauka, 1985).

9. Wernher von Braun and Frederick I. Ordway III, *History of Rocketry and Space Travel* (New York: Thomas Y. Cromwell Company, 1966). The book was published in revised editions in 1969, 1975, and 1985. The final edition was published as *Space Travel: A History* (New York: Harper & Row, 1985).

10. K. Tsiolkovskii, "Issledovanie mirovykh prostranstv reaktivnymi priborami (1911–1912 gg.)," in *Izbrannye trudy*, ed. B. N. Vorob'ev and V. N. Sokol'skii (Moscow: Nauka, 1962), p. 196. The original phrase was "Планета есть колыбель разума, но нельзя вечно жить в колыбели," or *"Planeta est' kolybel' razuma, no nel'zia vechno zhit' v kolybeli."* For typical references to the quote, see A. A. Kosmodemyansky, *K. E. Tsiolkovsky—His Life and Work* (Moscow: Nauka, 1960), p. 153; William Shelton, *Soviet Space Exploration: The First Decade* (New York: Washington Square Press, 1968), pp. 12–13; Roger D. Launius, *Space Stations: Base Camps to the Stars* (Washington, DC: Smithsonian Books, 2003), p. 9.

tier was *The Railroad and the Space Program: An Exploration in Historical Analogy* (1965), a collection of essays which used the American railroad as a metaphor for the slow human migration into space.[11] These early works foreshadowed and exemplified an important thread in the future of space history, equating the American frontier in the West with the space frontier beyond the Earth.

Through the past 50 years, those looking ahead, such as policy-makers and spaceflight advocates from John F. Kennedy to Wernher von Braun to Mars Society President Robert Zubrin, have used Frederick Jackson Turner's frontier motif to inspire, justify, and advocate space exploration on a grand scale.[12] Those looking back, especially space historians, have also invoked the frontier thesis to explain the majesty of the early years of American space exploration; they have explained not only how engagement with the frontier has shaped American society and culture, but also how the foundations of American society and culture—particularly democracy and individualism— have shaped space exploration. The frontier ideal resonated partly because, like space explorers, many of the original explorers of the West shared uto- pian ideals.[13] The space program represented a potent union of two powerful strands of American culture, the search for utopia and the belief in the power of technology, a manifestation of 20th-century technological utopianism.[14] In the 1960s, at a time when the emerging reevaluation of the frontier thesis and its attendant costs to both the environment and the native peoples of the continent had yet to enter the mainstream discourse in American history, the use of the West as a guiding analogy for space exploration implied expansion, development, freedom, and ultimately liberation from the chains of previous existence. If there were pitfalls in exploration, they were minimal at best.[15] These markers of frontier exploration resonated deeply with many histori-

---

11. Bruce Mazlish, ed., *The Railroad and the Space Program: An Exploration in Historical Analogy* (Cambridge, MA: MIT, 1965).

12. For Frederick Jackson Turner's original works on the frontier thesis, see John Mack Faragher, ed., *Rereading Frederick Jackson Turner: The Significance of the Frontier in American History and Other Essays* (New Haven, CT: Yale University Press, 1994); George Rogers Taylor, *The Turner Thesis: Concerning the Role of the Frontier in American History*, 3rd ed. (Lexington, MA: Heath, 1972). For the frontier's resonance in modern times, see Richard Slotkin, *Gunfighter Nation: The Myth of the Frontier in Twentieth Century America* (New York: Atheneum, 1992). Roger D. Launius gives some notable examples of prominent advocates invoking the frontier thesis in the 1960s in his "Historical Dimension of Space Exploration."

13. Roger D. Launius, "Perfect Worlds, Perfect Societies: The Persistent Goal of Utopia in Human Spaceflight," *Journal of the British Interplanetary Society* 56 (2003): 338–349.

14. For an excellent look at the origins of technological utopianism in American culture, see Howard P. Segal, *Technological Utopianism in American Culture* (Chicago: University of Chicago Press, 1985).

15. For critiques of the frontier thesis, see Patricia Nelson Limerick, Clyde A. Milner II, and Charles E. Rankin, eds., *Trails: Toward a New Western History* (Lawrence: University Press of Kansas, 1991); Richard White, *It's Your Misfortune and None of My Own: A New History of the American West* (Norman: Oklahoma University Press, 1991).

ans, enough that many still invoke them in the 21st century. Describing the parallel paths of the Russian and American space programs, author Robert Zimmerman, in *Leaving Earth: Space Stations, Rival Superpowers, and the Quest for Interplanetary Travel* (2003), compared them to colonization of Earthly landscapes: "The ancestors of both peoples were pioneers . . . . The land both groups settled was harsh, brutal, and unyielding. Death was omnipresent. Out of these two pioneer struggles have risen nations able to forge in the sky the first rockets, the first spacecraft, and the first tentative and grand attempts to colonize the stars."[16] Similar notions run through Bruce C. Murray's *Journey into Space: The First Three Decades of Space Exploration* (1989) and William E. Burrows's *Exploring Space: Voyages in the Solar System and Beyond* (1990), both of which explicitly deal with deep space exploration by robotic probes.[17] That Earthly exploration remains a powerful motif for making sense of space exploration is exemplified best by *Where Next, Columbus? The Future of Space Exploration* (1994), a collection of meditations by prominent historians that link Columbus's seabound trip to the early years of space exploration.[18]

Once the landing of Apollo astronauts on the Moon in July 1969 effectively ended the "space race" for the United States, historians took up the challenge of chronicling this extraordinary technological achievement in a multitude of works, many of which framed the project as part of the human exploration imperative. Unlike many other programs of the 1960s, or indeed since, the Apollo program represented a perfect distillation of interpretive approaches that focused on exploration since the Apollo missions had geographical delimiters that paralleled exploration of the West: beginning from the known, the Earth, voyagers set out in a very physical way for the unknown, the Moon. In contrast, the hundreds of Earth-orbital missions since 1972, while risky and adventurous, have not represented physical movement in the same way Apollo did.[19] NASA managers early on recognized Apollo's exceptionalist nature within the space program. In the introduction to one of the first volumes to reflect on Apollo, then–NASA Administrator James C. Fletcher explicitly located the Apollo expeditions as part of a tradition stretch-

---

16. Robert Zimmerman, *Leaving Earth: Space Stations, Rival Superpowers, and the Quest for Interplanetary Travel* (Washington, DC: Joseph Henry Press, 2003), p. 460.

17. Bruce C. Murray, *Journey into Space: The First Three Decades of Space Exploration* (New York: W. W. Norton, 1989); William E. Burrows, *Exploring Space: Voyages in the Solar System and Beyond* (New York: Random House, 1990).

18. Valerie Neal, ed., *Where Next, Columbus? The Future of Space Exploration* (New York: Oxford University Press, 1994). See also Peter Bond, *Reaching for the Stars: An Illustrated History of Manned Spaceflight*, 2nd ed. (London: Cassell, 1996).

19. Deam argues that "this shift has essentially emptied the [space] program of its public character, moving spaceflight from an open embrace of political action to closed concerns with economics and technological determinism" (Dirk Deam, "Public Space: Exploring the Political Dimensions of the American Space Program" [Ph.D. diss., University of Iowa, 1999]).

Since the time of the Apollo 11 Moon landing in 1969, space history has matured into a much more rigorous and complex area of study, one with which the theme of exploration has long been associated. No photograph better illustrates this connection than the image of Buzz Aldrin on the Moon. It has assumed iconic proportions in modern society. *(NASA image no. AS11-40-5903)*

ing back to the Pilgrims at Plymouth and Darwin's voyages on the HMS *Beagle*; both were "ventures into uncharted waters."[20] Similarly, Harry Hurt III, in his *For All Mankind* (1988), compared the Apollo missions to Earthly explorations, specifically invoking "Christopher Columbus's daring voyage to the New World."[21]

---

20. James C. Fletcher, "Foreword," in *Apollo Expeditions to the Moon*, ed. Edgar M. Cortright (Washington, DC: NASA, 1975).

21. Harry Hurt III, *For All Mankind* (New York: The Atlantic Monthly Press Book, 1988), p. xiii.

Beyond linking the great Earthly explorations and migrations with the Apollo expeditions, early works on Apollo, such as the Apollo 11 astronauts' (ghostwritten) *First on the Moon* (1970) and Richard Lewis's *The Voyages of Apollo: The Exploration of the Moon* (1974), focused predominantly on the people at the tip of the iceberg, i.e., the astronauts who performed the missions.[22] Two decades later, Andrew Chaikin's landmark *A Man on the Moon* (1994) continued in this vein, merging the exploration motif with the astronauts' perspectives on the project while omitting any interpretive look at the broader political, social, or cultural factors behind Apollo.[23] By focusing exclusively on the thoughts of the astronauts, the details of the missions, and the nuances of the technology, Chaikin masterfully conveyed the experience of Apollo as if it were one in which only a few dozen people were involved. Context was provided only to the extent that the news media reported it at the time of the Apollo missions.[24] Thus, in one sense, in the historiography of the space program, Apollo became a national, even global experience that was conceived, executed, and directly experienced by a few chosen ambassadors. This contradiction may not be as irreconcilable as it appears, for Apollo was a unique artifact of its time. Millions of people witnessed the first landing of humans on another celestial body through their black-and-white TVs in the comfort of their homes. Such vicarious exploration had no precedent. If the import of Apollo was ultimately global, signaling human migration off the planet, then its immediate communicative power was ultimately largely private, in homes and offices.

Historically, many of those who advocated space exploration emphasized science as an important rationale for exploration. The literature on the history of space-based science has, however, not been significant. Several factors explain the weakness of a unified tradition of writing on space science history. These include the fragmentary nature of the field, where much of the work is generated from other history-of-science subdisciplines such as the history of physics, astronomy, life sciences, meteorology, and oceanography. The contributions in two volumes of essays separated by 10 years, *Space Science Comes of Age: Perspectives in the History of the Space Sciences* (1981) and *A Spacefaring Nation: Perspectives on American Space History and Policy* (1991), underline the difficult struggles of nascent space-based science constituencies (within solar science and planetary science) to escape the shadow of their parent communi-

---

22. Neil Armstrong, Michael Collins, and Edwin E. Aldrin, Jr., with Gene Farmer and Dora Jane Hamblin, *First on the Moon* (Boston: Little, Brown, and Company, 1970); Richard S. Lewis, *The Voyages of Apollo: The Exploration of the Moon* (New York: Quadrangle, 1974).

23. Andrew Chaikin, *A Man on the Moon: The Voyages of the Apollo Astronauts* (New York: Viking, 1994).

24. For media treatments of the space program, see Andrew A. Klyukovski, "The Space Race as the American Dream: Fantasy Theme Analysis of 'The New York Times' Coverage" (Ph.D. diss., University of Missouri-Columbia, 2002).

ties (physics and astronomy).[25] Additionally, science has traditionally played a secondary (if not tertiary) role in the American space program, behind political and military imperatives. For space historians who have chronicled the American space program as political, nationalistic, or technological enterprises, space science has been a corollary theme rather than a central one.[26] Two volumes of NASA's *Exploring the Unknown* series chronicling the history of American civilian space exploration are the most important contributions to space science history, but the editors' consignment of space sciences to volumes 5 and 6 in the series underscores the subfield's priority in the schematic of space history overall.[27] Finally, historians have frequently seen space science as deeply connected to rationales of militarization or exploration. As such, space science history remains embedded with these other narratives. For example, in his *Science with a Vengeance: How the Military Created the US Space Sciences after World War II* (1992), David DeVorkin argued that space science was created largely due to the existence of the German V-2 missile, a weapon of war whose development had nothing to do with either the search for scientific knowledge or exploration.[28]

25. Paul A. Hanle and Del Chamberlain, eds., *Space Science Comes of Age: Perspectives in the History of Space Sciences* (Washington, DC: Smithsonian Institution Press, 1981). See also Karl Hufbauer, "Solar Observational Capabilities and the Solar Physics Community Since Sputnik, 1957–1988"; Joseph N. Tatarewicz, "Space Technology and Planetary Science, 1950–1985," in *A Spacefaring Nation: Perspectives on American Space History and Policy*, eds. Martin J. Collins and Sylvia D. Fries (Washington, DC: Smithsonian Institution Press, 1991), pp. 77–114, 115–132.

26. Two important works on science performed during Apollo are framed as part of programmatic "mission-oriented" histories. See William David Compton, *Where No Man Has Gone Before: A History of Apollo Lunar Exploration Missions* (Washington, DC: NASA SP-4214, 1989); David M. Harland, *Exploring the Moon: The Apollo Expeditions* (London: Springer, 1999). A third, lesser-known but more accomplished work focuses exclusively on the science rather than the missions: Donald A. Beattie, *Taking Science to the Moon: Lunar Experiments and the Apollo Program* (Baltimore: Johns Hopkins, 2001).

27. See particularly the excellent introductory essays in John M. Logsdon, ed., *Exploring the Unknown: Selected Documents in the History of the U.S. Civil Space Program*, vol. 5, *Exploring the Cosmos* (Washington, DC: NASA SP-2001-4407, 2001); John M. Logsdon et al., eds., *Exploring the Unknown: Selected Documents in the History of the U.S. Civil Space Program*, vol. 6, *Space and Earth Science* (Washington, DC: NASA SP-2004-4407, 2004). For the few other notable works on the history of space science, see Charles A. Lundquist, *Skylab's Astronomy and Space Sciences* (Washington, DC: NASA, 1979); John A. Pitts, *The Human Factor: Biomedicine in the Manned Space Program to 1980* (Washington, DC: NASA SP-4213, 1985; John E. Naugle, *First Among Equals: The Selection of NASA Space Science Experiments* (Washington, DC: NASA SP-4215, 1991); David Leverington, *New Cosmic Horizons: Space Astronomy from the V-2 to the Hubble Space Telescope* (New York: Cambridge University Press, 2001).

28. David H. DeVorkin, *Science with a Vengeance: How the Military Created the US Space Sciences after World War II* (New York: Springer-Verlag, 1992); David H. DeVorkin, "Military Origins of the Space Sciences in the American V-2 Era," in *National Military Establishments and the Advancement of Science and Technology*, eds. Paul Forman and José M. Sánchez-Ron, Studies in Twentieth Century History (Boston: Kluwer Academic Publishers, 1996). See also DeVorkin's "Solar Physics," in *Exploring the Unknown*, vol. 6, pp. 1–37.

COMPETITION AND NATIONAL SECURITY

The exploration motif overlaps with a second theme running through the historiography of space exploration, that of competition. Richard Lewis, in his *From Vinland to Mars: A Thousand Years of Exploration* (1976), eloquently illustrated the ways in which competition over resources and land spurred exploration. He found a common imperative existing from the Greenland and Vinland voyages of the Viking Eric the Red all the way to the Viking spacecraft landings on Mars in the bicentennial year of 1976. Framing his narrative around this coincidence of names, Lewis focused on competition as a guiding metaphor for space exploration:

> The common denominator [in all exploration] is intraspecific competition . . . : deadly competition among men and families for land, among nations for power and wealth. This is the force that drove the have-nots in medieval Scandinavia across uncharted seas, impelled Renaissance Europe to seek the wealth of the Indies and circumnavigate the planet, urged Amundsen and Scott on the tragic race to the geographic south pole, and launched Americans to the Moon.[29]

Like Lewis, many space historians have used competition—specifically, the Cold War—as a second defining lens to understand space history. Most popular accounts of the space race, and many from an academic perspective, have framed the American adventure in space as competition with an adversary who did not share the same moral commitment to freedom and equality. In the canon, both Sputnik and Apollo emerge, at least implicitly, as material representations embedded with notions of two ideologically opposed systems of governance. To a large degree, such evaluations of Apollo reflected rhetoric from the 1960s—from American politicians, the American media, and from participants in the Apollo project itself. But because accounts of the space race have been typically undergirded by implicit claims about morality of national cultures, historians rarely engaged in critiques of Apollo or the space program in general, since such methodological approaches would be tantamount to challenging the moral authority of the United States. In his recent *Apollo: The Epic Journey to the Moon*, an engaging and awe-inspiring account of the Apollo project, David West Reynolds distills this rationale succinctly and emotionally:

---

29. Richard S. Lewis, *From Vinland to Mars: A Thousand Years of Exploration* (New York: Quandrangle, 1976), p. xii.

[The Moon race] was a Cold War battle to demonstrate the
superior ability of the superior system, capitalism versus communism . . . . And the battle did prove out the more capable
system . . . . The reasons are many, but among them the power
of free enterprise ranks high . . . . Free competition motivated
American workers whose livelihoods were related to the quality and brilliance of their work, and we saw extraordinary,
impossible things accomplished by ordinary Americans. The
American flag on the Moon is such a powerful symbol because
it is not a vain one. America, like no other nation, *was* capable
of the Moon.[30]

Beyond linking Cold War competition to celebratory nationalistic
impulses, others used competition to revisit seminal events in space history.
John M. Logsdon's *The Decision to Go to the Moon: Project Apollo and the
National Interest* (1970), the classic study of the original imperatives that gave
rise to Apollo, was one of the earliest.[31] Kennedy's actual decision to go to
the Moon stemmed from a series of politically inopportune precipitates,
including the aborted Bay of Pigs invasion and Yuri Gagarin's historic first
flight into space in April 1961. Keen to respond to the unending humilia-
tions in the new space frontier, Kennedy enlisted the aid of Vice President
Lyndon B. Johnson to formulate an ambitious but realistic response to the
Soviets. By the end of May, after extensive consultations with their advisers,
Kennedy and Johnson had their goal: send Americans to the Moon before
the end of the decade, an announcement the President made to a joint session of Congress on 25 May 1961. By synthesizing the disparate threads of
the events of 1961 using primary documentation, Logsdon laid the
groundwork for understanding a seminal event in U.S. space policy and thus
built the foundation for a new interpretive school of space history, space
policy history.[32]

Cold War competition has loomed large in the vast subgenre of space
policy history, and a number of works have sought to explain the twists and
turns of American space policy through its interdependence with Cold War

---

30. David West Reynolds, *Apollo: The Epic Journey to the Moon* (New York: Tehabi, 2002), p. 257.

31. John M. Logsdon, *The Decision to Go to the Moon: Project Apollo and the National Interest*
(Cambridge, MA: MIT Press, 1970).

32. For collections that include essays on the history of space policy, see Radford Byerly, Jr.,
ed., *Space Policy Reconsidered* (Boulder, CO: Westview Press, 1989); Radford Byerly, Jr., ed., *Space
Policy Alternatives* (Boulder, CO: Westview Press, 1992); Roger D. Launius, ed., *Organizing for the
Use of Space: Historical Perspectives on a Persistent Issue* (San Diego: Univelt, 1995); Eligar Sadeh,
ed., *Space Politics and Policy: An Evolutionary Perspective* (Dordrecht, Netherlands: Kluwer Academic
Publishers, 2003).

politics on an international scale.[33] The results of several history conferences in the 1980s—hosted by NASA and the National Air and Space Museum— broke new ground in the field of space policy history by going beyond the original Cold War competition dynamic.[34] A number of these papers departed from much of the early historiography by focusing on post-Apollo efforts including the space station *Freedom* and the Hubble Space Telescope. In exploring, for example, how NASA's Space Station Task Force convinced a lukewarm White House to support the original *Freedom* proposal in the early 1980s, Howard McCurdy highlighted the influence of government agencies over governmental policy.[35] Others explored the dynamics of space policy through specific presidential administrations, thus analyzing the causes why some space projects survive and others don't, depending on politics at the highest level.[36]

A number of space policy histories took an overtly critical stance to NASA and its mission, focusing often on the lack of foresight exhibited by policymakers and managers at NASA, the Congress, and the Executive Branch.[37] Amitai Etzioni's *The Moon Doggle: Domestic and International Implications of the Space Race* (1964), although not a history book, was one such early critique which called the entire enterprise of Apollo into doubt since he believed that

---

33. See, for example, William H. Schauer, *The Politics of Space: A Comparison of the Soviet and American Space Programs* (New York: Holmes & Meier Publishers, 1976); Xavier Pasco, *La Politique Spatiale des Etats-Unis: 1958–1995: Technologie, inérêt national et débat public* (Paris: L'Harmattan, 1997); Matthew J. Von Bencke, *The Politics of Space: A History of U.S.-Soviet/Russian Competition and Cooperation in Space* (Boulder, CO: Westview Press, 1997); Dale L. Hayden, *The International Development of Space and Its Impact on U.S. National Space Policy* (Maxwell AFB, AL: Airpower Research Institute, College of Aerospace Doctrine, Research and Education, Air University, 2004).

34. For the proceedings of the 1981 and 1987 conferences, see Hanle and Chamberlain, *Space Science Comes of Age*; Collins and Fries, *Spacefaring Nation*. The proceedings of a similar conference hosted by Yale University in 1981 were published as Alex Roland, ed., *A Spacefaring People: Perspectives on Early Spaceflight* (Washington, DC: NASA SP-4405, 1985).

35. Howard E. McCurdy, "The Space Station Decision: Politics, Bureaucracy, and the Making of Public Policy," in *Spacefaring Nation*, ed. Collins and Fries, pp. 9–28.

36. Linda T. Krug, *Presidential Perspectives on Space Exploration: Guiding Metaphors From Eisenhower to Bush* (New York: Praeger, 1991); Derek W. Eliott, "Finding an Appropriate Commitment: Space Policy Development Under Eisenhower and Kennedy, 1954–1963" (Ph.D. diss., George Washington University, 1992); Howard E. McCurdy, *The Space Station Decision: Incremental Politics and Technological Choice* (Baltimore: Johns Hopkins, 1990); Mark Damohn, *Back Down to Earth: The Development of Space Policy for NASA During the Jimmy Carter Administration* (San Jose, CA: Authors Choice Press, 2001).

37. Erik Bergaust, *Murder on Pad 34* (New York: Putnam, 1968); Erlend A. Kennan and Edmund H. Harvey, Jr., *Mission to the Moon: A Critical Reexamination of NASA and the Space Program* (New York: Morrow, 1969); Hugo Young, Brian Silcock, and Peter Dunn, *Journey to Tranquillity: The History of Man's Assault on the Moon* (London: Cape, 1969); Roger Handberg, *Reinventing NASA: Human Spaceflight, Bureaucracy and Politics* (Westport, CT: Praeger, 2003); Greg Klerkx, *Lost in Space: The Fall of NASA and the Dream of a New Space Age* (New York: Pantheon Books, 2004).

it represented a cynical public relations exercise diverting attention away from more pressing domestic issues such as the War on Poverty.[38]

Since the mid-1980s, a number of important works used the Cold War competition paradigm but focused specifically on national security programs, which constituted about half of all national expenditures on spaceflight yet received relatively little scrutiny from historians. The earliest academic work in this subfield was Paul B. Stares's *The Militarization of Space: U.S. Policy, 1945–1984* (1985), which examined the rise of the American space weapons program and its largely unrecorded but substantial influence over American military policy.[39] Writing during a time of extreme tension between the Soviet Union and the United States, Stares argued that the arms race was migrating to the arena of space by the mid-1980s. Equally groundbreaking was journalist William E. Burrows's *Deep Black: Space Espionage and National Security* (1986), in which he focused on the development of highly classified photoreconnaissance satellites which spy on other nations. Using anonymous sources and declassified materials, he wove a story of a secret world that in fact consumed a substantial share of the American space budget but whose very existence was never explicitly acknowledged by the U.S. government.[40]

The early work of Stares and Burrows was overshadowed by CIA-sponsored post–Cold War declassification initiatives. In 1995, the U.S. government revealed details of one of the biggest secrets of the Cold War, the United States' first operational spy satellite system, CORONA, whose satellites flew dozens of missions in the 1960s over secret targets in the Soviet Union, China, Vietnam, and elsewhere. If earlier writing on the genesis of the U.S. space effort emphasized civilian programs such as Vanguard and Explorer, the CORONA revelations helped to reframe the early years of the American space program as parallel and sometimes interconnected civilian and military

---

38. Amitai Etzioni, *The Moon Doggle: Domestic and International Implications of the Space Race* (Garden City, NY: Doubleday, 1964). For other contemporary works, see Edwin Diamond, *The Rise and Fall of the Space Age* (Garden City, NY: Doubleday, 1964); Vernon van Dyke, *Pride and Power: The Rationale of the Space Program* (Urbana: University of Illinois Press, 1964).

39. Paul B. Stares, *The Militarization of Space: U.S. Policy, 1945–1984* (Ithaca: Cornell University Press, 1985). I differentiate here between military space programs and intelligence space programs, both of which fall under national security programs. The former include weapons development, while the latter include reconnaissance satellites. The earliest open work to explore the American military and intelligence space programs was Phillip Klass's *Secret Sentries in Space* (New York: Random House, 1971). Anthony Kenden was another pioneering scholar in the field. See his "U.S. Reconnaissance Satellite Program," *Journal of the British Interplanetary Society* (July 1978), and "A New U.S. Military Space Mission," *Journal of the British Interplanetary Society* (October 1982).

40. William E. Burrows, *Deep Black: Space Espionage and National Security* (New York: Berkley Books, 1986). For a Cold War–era look at space weaponization, see Curtis Peebles, *Battle for Space* (Dorset, U.K.: Blandford, 1983). Another important contribution in the pre-CORONA-revelation era was Jeffrey T. Richelson's *America's Secret Eyes in Space: The U.S. Keyhole Spy Satellite Program* (New York: Harper & Row, 1990).

projects. Where civilian efforts, especially the human spaceflight program, had assumed center stage in the historiography, CORONA highlighted how much of the old history had told only half the story. The CIA's first deputy director for science and technology, Albert D. "Bud" Wheelon, who managed the CORONA program in the mid-1960s, wrote in *Eye in the Sky: The Story of the CORONA Spy Satellites* (1998):

> When the American government eventually reveals the full range of reconnaissance systems developed by this nation, the public will learn of space achievements every bit as impressive as the Apollo moon landings. One program proceeded in utmost secrecy, the other on national television. One steadied the resolve of the American public; the other steadied the resolve of American presidents.[41]

Photoreconnaissance satellite programs such as CORONA and its successors, such as the KH-9 HEXAGON and KH-11 KENNAN, consumed a lion's share of the U.S. "black" space program and, in fact, drove much of early U.S. space policy. Historical details of other important programmatic elements of American national security projects, such as early-warning systems, signals intelligence, military communications, meteorology, navigation, antisatellite, and (abandoned) human military spaceflight projects, have come to light owing to the research of several historians including R. Cargill Hall, Jeffrey T. Richelson, and Dwayne A. Day, whose works represented a major shift in the scholarship on military space programs, moving from speculative works based on rumor, leaks, and analysis of orbital parameters to using primary documentation.[42] Day's work has been particularly groundbreaking, opening up previ-

---

41. Albert D. Wheelon, "CORONA: A Triumph of American Technology," in *Eye in the Sky: The Story of the CORONA Spy Satellites*, ed. Dwayne A. Day et al. (Washington, DC: Smithsonian Institution Press, 1998), p. 38.

42. For a discussion of early warning, see Jeffrey T. Richelson, *America's Space Sentinels: DSP Satellites and National Security* (Lawrence: University Press of Kansas, 1999); R. Cargill Hall, "Missile Defense Alarm: The Genesis of Space-Based Infrared Early Warning," *Quest: The History of Spaceflight Quarterly* 7, no. 1 (1999): 5–17. For naval strategy and military space programs, see Norman Friedman, *Seapower and Space: From the Dawn of the Missile Age to Net-Centric Warfare* (Annapolis, MD: Naval Institute Press, 2000). For manned military programs, see Roy F. Houchin II's "Why the Air Force Proposed the Dyna-Soar X-20 Program" and "Why the Dyna-Soar X-20 Program Was Cancelled," both in *Quest: The History of Spaceflight Magazine* 3, no. 4 (1994): 5–12 and 35–37, respectively; Steven R. Strom, "The Best Laid Plans: A History of the Manned Orbiting Laboratory," *Crosslink* 5, no. 2 (2004): 11–15. For weather satellite programs, see Dwayne A. Day, "Dark Clouds: The Classified Origins of the Defense Meteorological Satellite Program," *Spaceflight* 43 (2001): 382–385; R. Cargill Hall, "A History of the Military Polar Orbiting Meteorological Satellite Program," *Quest: The History of Spaceflight Quarterly* 9, no. 2 (2002): 4–25. For navigation satellites, see Bradford W. Parkinson et al., "A History of Satellite Navigation," *Navigation: Journal*

continued on the next page

ously hidden aspects of geodetic, signals intelligence, and photoreconnaissance satellite projects.[43] His recent work on the Air Force's interest in developing a dual human space capsule and reconnaissance satellite in the late 1950s adds to our understanding of the motivations and strategies institutions used to achieve specific goals in the early days of space exploration.[44]

This substantive (and generational) shift in scholarship, made possible by post–Cold War declassifications, has allowed the study of American military space history to focus on questions common to the study of American military history and intelligence collection, such as civil-military relations, interservice and interorganizational rivalry, and the relationship between technological development and mission requirements. Day, for example, produced important scholarship on the uses of satellite intelligence in monitoring the supersecret Soviet human lunar landing project in the 1960s, thus illuminating the hitherto unknown ways in which the civilian NASA interacted with the intelligence community.[45] Richelson's groundbreaking *The Wizards of Langley* (2001), a history of the CIA's Directorate of Science and Technology which developed and deployed both photoreconnaissance and signals intelligence systems during the Cold War, also exemplifies this new generation. Weaving

---

continued from the previous page
of the Institute of Navigation 42, no. 1, special issue (1995): 109–164; Chris Banther, "A Look into the History of American Satellite Navigation," *Quest: The History of Spaceflight Quarterly* 11, no. 3 (2004): 40–48. For antisatellite projects, see Wayne R. Austerman, *Program 437: The Air Force's First Antisatellite System* (Peterson AFB, CO: Office of History, 1991); Dwayne A. Day, "Arming the High Frontier," *Spaceflight* 46 (2004): 467–471. For organizational histories, see Harold M. Sapolsky, *Science and the Navy: The History of the Office of Naval Research* (Princeton: Princeton University Press, 1990); David N. Spires, *Beyond Horizons: A Half Century of Air Force Space Leadership* (Peterson AFB: Air Force Space Command, 1997); James Bamford, *Body of Secrets: Anatomy of the Ultra-Secret National Security Agency* (New York: Anchor Books, 2001); Jeffrey T. Richelson, *The Wizards of Langley: Inside the CIA's Directorate of Science and Technology* (Boulder, CO: Westview Press, 2001). For command and control, see David C. Arnold, *Spying from Space: Constructing America's Satellite Command and Control Systems* (College Station: Texas A&M University Press, 2005).

43. Day has published a series of articles on these topics. For geodetic projects, see "Mapping the Dark Side of the World: Part 1: The KH-5 ARGON Geodetic Satellite" and "Mapping the Dark Side of the World: Part 2: Secret Geodetic Programmes after ARGON," both in *Spaceflight* 40 (1998): 264–269 and 303–310, respectively. For signals intelligence satellites, see "Tinker, Tailor, Radar, Spy: Early American Ferret and Radar Satellites," *Spaceflight* 43 (2001): 288–293; "Ferrets Above: American Signals Intelligence Satellites During the 1960s," *International Journal of Intelligence and Counterintelligence* 17, no. 3 (2004): 449–467. For photoreconnaissance, see "A Sheep in Wolf's Clothing: The Samos E-5 Recoverable Satellite, Part 1," *Spaceflight* 44 (2002): 424–431; "A Square Peg in a Cone-Shaped Hole: The Samos E-5 Recoverable Satellite, Part 2," *Spaceflight* 45 (2003): 71–79; "From Cameras to Monkeys to Men: The Samos E-5 Recoverable Satellite, Part 3," *Spaceflight* 45 (2003): 380–389.

44. Day, "From Cameras to Monkeys to Men."

45. Dwayne A. Day and Asif A. Siddiqi, "The Moon in the Crosshairs: CIA Intelligence on the Soviet Manned Lunar Programme," *Spaceflight* 45 (2003): 466–475 and 46 (2004): 112–125; Dwayne A. Day, "From the Shadows to the Stars: James Webb's Use of Intelligence Data in the Race to the Moon," *Air Power History* 51, no. 4 (winter 2004): 30–39. See also Roger D. Launius, "NASA Looks to the East: American Intelligence Estimates of Soviet Capabilities and Project Apollo," *Air Power History* (fall 2001): 5–15.

an intricate story of various projects that "represented a quantum leap in U.S. intelligence capabilities," he locates the development of these systems in a broader context involving relationships with influential scientists outside the agency, the necessity to fill gaps in intelligence collection, and the connections between satellite development and intelligence production.[46]

The two most important works on CORONA, Day et al.'s *Eye in the Sky* and McDonald's *CORONA*, included contributions from individuals who participated in CORONA development in the late 1950s and early 1960s; as such, they can be characterized as semiofficial histories.[47] Both unequivocally extolled the technological, managerial, and operational successes of the project. Its history was framed as part of a singularly powerful story about the efficacy of good management and high technology to benefit the national interest of the United States, which was synonymous with engendering peace and freedom abroad. Writing about CORONA's use in monitoring compliance with arms control agreements, historian Ernest R. May concluded his essay by suggesting that "probably . . . the best one-line epitaph for CORONA would read: 'It helped keep peace in the nuclear age'."[48]

The end of the Cold War—specifically the collapse of the Soviet empire—validated, to some degree, the moral ground for historians writing of American military space programs. The writing on CORONA echoed a powerful strand of post-1991 historiography of the Cold War in general, which celebrated American motives over ideologically and morally suspect Soviet intentions. The post–Cold War self-congratulatory climate insulated the history of CORONA or other U.S. military space programs from critiques of their relationship to the Cold War military-industrial complex or

---

46. Richelson, *Wizards of Langley*, p. 287. For a poor example of the "new" history—on the understudied topic of intelligence analysis—see David T. Lindgren, *Trust But Verify: Imagery Analysis in the Cold War* (Annapolis, MD: Naval Institute Press, 2000). For civil-military interactions, see John Cloud, "Imaging the World in a Barrel: CORONA and the Clandestine Convergence of the Earth Sciences," *Social Studies of Science* 31, no. 2 (2001): 231–251; John Cloud, "Re-Viewing the Earth: Remote Sensing and Cold War Clandestine Knowledge Production," *Quest: The History of Spaceflight Quarterly* 8, no. 2 (2001): 4–16; Ronald E. Doel, "Constituting the Postwar Earth Sciences: The Military's Influence on the Environmental Sciences in USA After 1945," *Social Studies of Science* 33, no. 5 (2003): 635–666.

47. Day et al., *Eye in the Sky*; Robert McDonald, ed., *CORONA: Between the Sun & the Earth: The First NRO Reconnaissance Eye in Space* (Bethesda, MD: American Society for Photogrammetry and Remote Sensing, 1997). For derivative works based on the above, see Curtis Peebles, *The Corona Project: America's First Spy Satellites* (Annapolis, MD: Naval Institute Press, 1997); Philip Taubman, *Secret Empire: Eisenhower, the CIA, and the Hidden Story of America's Space Espionage* (New York: Simon & Schuster, 2003). For an overview of the literature on CORONA, see Dwayne A. Day, "Rashomon in Space: A Short Review of Official Spy Satellite Histories," *Quest: The History of Spaceflight Quarterly* 8, no. 2 (2000): 45–53.

48. Ernest R. May, "Strategic Intelligence and U.S. Security: The Contributions of CORONA," in *Eye in the Sky*, p. 28.

as part of American interventionist aims in global conflicts played out in the developing world (in, for example, Southeast Asia and Central America).[49]

While the contextual touchstone of U.S. military space history is the Cold War, the literature has remained woefully disconnected from many of the broader intellectual debates that have characterized the historiography of the Cold War through the past 40 years and now in the post–Cold War era. Beginning with the historians who defended the policy of containment against expansionist Soviet intentions, to the generation of revisionists who argued the left-liberal position that American economic interests on a global level contributed to the Cold War, to the postrevisionists who emphasized misperception and misunderstanding to explain much of the Cold War, the canon has passed through many transformations.[50] From the 1980s, and especially in the post–Cold War period, several new threads emerged as diplomatic, social, and cultural historians contributed richly to understanding not only international relations, but also domestic American cultural currents that formed part of the mosaic of the country's trajectory through the Cold War. For example, a new generation of historians is now looking at how domestic culture affected foreign policy.[51]

In terms of international competition—the principal context for the origins of the American space program—the biggest public splash was made by John Lewis Gaddis's *We Know Now: Rethinking Cold War History* (1997), which harked back to the original view that Stalin's personality, Soviet authoritarianism, and communist ideology were principal reasons for the Cold War.[52]

---

49. For a rare example on the strategic dimension of space support during wartime, see Henry W. Brandli, "The Use of Meteorological Satellites in Southeast Asia Operations," *Aerospace Historian* 29, no. 3 (1982): 172–175.

50. For useful summaries of the enormous transformations in Cold War historiography, see Melvyn P. Leffler, "The Cold War: What Do 'We Know Now'?" *American Historical Review* 104, no. 2 (1999): 501–524; Timothy J. White, "Cold War Historiography: New Evidence Behind Traditional Typographies," *International Social Science Review* 1, no. 1 (fall–winter 2000).

51. David Campbell, *Writing Security: United States Foreign Policy and the Politics of Identity* (Minneapolis: University of Minnesota Press, 1992); Brenda Gayle Plummer, *Rising Wind: Black Americans and U.S. Foreign Affairs, 1935–1960* (Chapel Hill: University of North Carolina Press, 1996); Akira Iriye, *Cultural Internationalism and World Order* (Baltimore: Johns Hopkins, 1997); Frank Costigliola, "'Unceasing Pressure for Penetration': Gender, Pathology, and Emotion in George Kennan's Formation of the Cold War," *Journal of American History* 84 (1997): 1309–1339; Robert D. Dean, "Masculinity as Ideology," *Diplomatic History* 22 (1998): 29–62.

52. John Lewis Gaddis, *We Know Now: Rethinking Cold War History* (Oxford: Clarendon Press, 1997). See also Gaddis, "Rethinking Cold War History: A Roundtable Discussion," in *At the End of the American Century: America's Role in the Post–Cold War World*, ed. Robert L. Hutchins (Baltimore: Johns Hopkins, 1998), pp. 52–66; Douglas J. Macdonald, "Communist Bloc Expansion in the Early Cold War: Challenging Realism, Refuting Revisionism," *International Security* 20 (1995–1996): 152–188. For similar perspectives on the Soviet side, see Vladislav Zubok and Constantine Pleshakov, *Inside the Kremlin's Cold War: From Stalin to Khrushchev* (Cambridge, MA: Harvard University Press, 1996); Vojtech Mastny, *The Cold War and Soviet Insecurity: The Stalin Years* (New York: Oxford University Press, 1996).

Gaddis's arguments were countered by many who emphasized and explored ideology on both sides, the organization of overseas propaganda by both governments, transnational global relations, the relationship between military capabilities and diplomatic policies, the end of colonialism, and conflicts played out between "strong" and "weak" powers.[53] Military space historians whose objects of study are firmly embedded in the Cold War have yet to evolve through these larger debates. The recent works on CORONA, for example, implicitly and closely follow the "Gaddis school," remaining disconnected from equally compelling but entirely different narratives of the history of the Cold War.[54] In *The Devil We Knew: Americans and the Cold War* (1993), respected diplomatic historian H. W. Brands argued that the battle with the Soviet Union served a spectrum of psychological, economic, strategic, and political imperatives. He claimed that the United States subverted some of the nation's best principles to win the Cold War. Thus any proclaimed victory was, at best, ambiguous.[55] How does the success of CORONA fit into such thinking? We may have much to learn from an exploration of this question.

ARTIFACTUAL AND PROGRAMMATIC HISTORIES

Beyond exploration and competition, a third large body of space history represents history centered on artifacts and/or programs. Willy Ley's early works—as well as those of David Lasser, Chas G. Philp, and P. E. Cleator—pioneered the artifact-centered history by merging the canon of popular science with popular history.[56] This school focused mainly on explaining how particu-

---

53. See for example, Thomas Borstelmann, *Apartheid's Reluctant Uncle: The United States and Southern Africa in the Early Cold War* (New York: Oxford University Press, 1993); Robert J. McMahon, *The Cold War on the Periphery: The United States, India, and Pakistan* (New York: Columbia University Press, 1994); David Holloway, *Stalin and the Bomb: The Soviet Union and Atomic Energy, 1939–1954* (New Haven, CT: Yale University Press, 1994); Thomas Risse-Kappen, *Cooperation Among Democracies: The European Influence on U.S. Foreign Policy* (Princeton, NJ: Princeton University Press, 1995); Ilya Gaiduk, *The Soviet Union and the Vietnam War* (Chicago: I. R. Dee, 1996); Steven J. Zaloga, *The Kremlin's Nuclear Sword: The Rise and Fall of Russia's Strategic Nuclear Forces, 1945–2000* (Washington, DC: Smithsonian Institution Press, 2002).

54. For a balanced view of American military space policy within the broader international context, see Michael E. O'Hanlon, *Neither Star Wars nor Sanctuary: Constraining the Military Uses of Space* (Washington, DC: The Brookings Institution, 2004). The few explicit critiques of the U.S. military space program, unfortunately, have been shrill and largely without value. See, for example, Jack Manno, *Arming the Heavens: The Hidden Military Agenda for Space, 1945–1995* (New York: Dodd, Mead, & Co., 1984); Loring Wirbel, *Star Wars: US Tools of Space Supremacy* (London: Pluto Press, 2004).

55. H. W. Brands, *The Devil We Knew: Americans and the Cold War* (New York: Oxford University Press, 1993).

56. David Lasser, *The Conquest of Space* (New York: The Penguin Press, 1931); Chas G. Philp, *Stratosphere and Rocket Flight (Astronautics)* (London: Sir Isaac Pitman & Sons, Ltd., 1935); P. E. Cleator, *Rockets Through Space: The Dawn of Interplanetary Travel* (New York: Simon & Schuster, 1936).

lar technologies worked, how they were developed, how they were tested, and finally, how they behaved during operational flights. De Witt Douglas Kilgore, in his recent *Astrofuturism: Science, Race, and Visions of Utopia in Space* (2003), calls the authors of this subgenre "scientists, engineers, and writers [who were] public apologists for the value of science."[57] Their works, grounded in scientific laws and mathematics, were not only accounts of past technological developments, but also contained narratives about the immense potential of engineers and managers to solve engineering problems; on a fundamental level, they are narratives about the "myth of [technological] progress."[58]

The programmatic histories typically encompass an arc from the conception of the project (the first chapter) to the final successful mission (the last) while maintaining a perspective that renders extraprogrammatic perspectives invisible. By rejecting contingency and context and embracing narratives of chronology and progress, they represent the distillation of teleology and Whiggish notions in space history.[59] The central actors in programmatic histories have typically been the artifact—the rocket engine, the launch vehicle, the spacecraft, and the ground complex. Such a focus reflects the organizational approach of the early American space program, where any new space technologies—such as liquid-hydrogen propulsion technology, for example—were developed under discrete NASA programs (in this case, Centaur).[60] As a result, programmatic histories have been frequently indistinguishable from artifactual histories.

Building on the tradition of Ley, Lasser, and others, beginning in the 1960s and continuing to the present, the NASA History Office has produced a series of works that have focused on particular programs. Although these studies were largely divorced from broader political, social, or cultural concerns, they served as important foundations for future historians to study how and why particular technologies emerged and how states and institutions arbitrate over questions of technology and management. An exemplary and excellent first step in the field was *The History of Rocket Technology: Essays on Research, Development, and Utility* (1964), a collection of essays on the development of ballistic missiles and spacecraft by a number of important architects

---

57. De Witt Douglas Kilgore, *Astrofuturism: Science, Race, and Visions of Utopia in Space* (Philadelphia: University of Pennsylvania Press, 2003).

58. For a critique of the "myth of progress" in the history of technology, see John Staudenmaier, *Technology's Storytellers: Reweaving the Human Fabric* (Cambridge, MA: MIT Press, 1985).

59. The term "Whig history" originally comes from Herbert Butterfield's *The Whig Interpretation of History* (London: G. Bell and Sons, 1931), where, in his examination of British constitutional history, he found a historical canon that framed history from a presentist stance without taking into account the viewpoints prevailing during the times of the figures under study. His was also an early critique of narratives centered on the "march of progress."

60. For Centaur, see Virginia P. Dawson and Mark D. Bowles, *Taming Liquid Hydrogen: The Centaur Upper Stage Rocket, 1958–2002* (Washington, DC: NASA SP-2004-4230, 2004).

of the U.S. rocketry and space program, including Walter R. Dornberger, Frank J. Malina, and Wernher von Braun. In his preface, then–NASA Chief Historian Eugene M. Emme argued that rocket technology was of fundamental importance to Western society, in effect restating the Cold War paradigm but linking it to the development of modern science and technology: "The eminence of Western science and technology—and all that this means, including but also beyond the connotations of national power—is not a little dependent upon the short and long-term success of technological progress in rocketry and astronautics."[61] All of these essays reflected prevailing interpretive trends in the relatively new field of history of technology, whose practitioners were fascinated with inventors, their inventions, and the effect of these inventions on society. In other words, these histories approached technology through deterministic and unidirectional perspectives where technology had profoundly impacted societies; the possibility of a reverse relationship was left unexplored. In his introduction to the 1964 volume, Emme encapsulated this view, suggesting that "rocketry has influenced the entire structure and conduct of national and international politics and economics."[62]

Since the Emme volume, NASA has sponsored numerous works in the canon, many of which have contributed to recording and chronicling important aspects of the country's efforts to explore space. The biggest subgroup—on human spaceflight—includes Swenson, Grimwood, and Alexander's *This New Ocean: A History of Project Mercury* (1966); Hacker and Grimwood's *On the Shoulders of Titans: A History of Project Gemini* (1977); Benson and Faherty's *Moonport: A History of Apollo Launch Facilities and Operations* (1978); Brooks, Grimwood, and Swenson's *Chariots for Apollo: A History of Manned Lunar Spacecraft* (1979); Compton and Benson's *Living and Working in Space: A History of Skylab* (1983); and Compton's *Where No Man Has Gone Before: A History of Apollo Lunar Exploration Missions* (1989).[63] Other NASA or NASA-sponsored books have focused on robotic missions, including NASA's extraordinarily successful and impressive deep space and interplanetary programs.[64] A recent work,

---

61. Eugene M. Emme, ed., *The History of Rocket Technology: Essays on Research, Development, and Utility* (Detroit: Wayne State University Press, 1964), p. 1.

62. Emme, *History of Rocket Technology*, p. 1.

63. Loyd S. Swenson, Jr., et al., *This New Ocean: A History of Project Mercury* (Washington, DC: NASA SP-4201, 1966); Barton C. Hacker and James C. Grimwood, *On the Shoulders of Titans: A History of Project Gemini* (Washington, DC: NASA SP-4203, 1977); Charles D. Benson and William Barnaby Faherty, *Moonport: A History of Apollo Launch Facilities and Operations* (Washington, DC: NASA SP-4204, 1978); Courtney G. Brooks et al., *Chariots for Apollo: A History of Manned Lunar Spacecraft* (Washington, DC: NASA SP-4205, 1979); W. David Compton and Charles D. Benson, *Living and Working in Space: A History of Skylab* (Washington, DC: NASA SP-4208, 1983); Compton, *Where No Man Has Gone Before.*

64. Richard Fimmel et al., *Pioneer Odyssey* (Washington, DC: NASA SP-394/396, 1977); Henry C. Dethloff and Ronald A. Schorn, *Voyager's Grand Tour: To The Outer Planets and Beyond* (Washington, DC: Smithsonian Institution Press, 2003).

No aspect of space travel is more exciting or has received greater historical attention than the human component. Too many observers, however, are too enthralled with the spectacle of flight to probe the history of the activity deeply. Here is the Return to Flight launch of Space Shuttle *Discovery* and its five-man crew from Pad 39B at 11:37 a.m., 29 September 1988, as *Discovery* embarked on a mission of 4 days and 1 hour. *(NASA image no. 88PC-1001)*

To Reach the High Frontier: The History of U.S. Launch Vehicles (2002), updated Emme's earlier seminal work by adding a number of essays on the technological development of the major American satellite launchers derived from Cold War–era warhorses such as the Atlas and Titan ICBMs.[65] The book was a timely update on the history of efforts to develop efficient access to space.

Beyond NASA, unofficial historians have devoted an enormous amount of ink and paper to the early American human spaceflight program. These works, which exploded in number in the late 1990s and the first decade of the 21st century, represent the perfectly idealized form of the programmatic and artifactual history. Many of the artifactual histories, such as Dennis Jenkins's

---

65. Roger D. Launius and Dennis R. Jenkins, eds., *To Reach the High Frontier: A History of U.S. Launch Vehicles* (Lexington: University Press of Kentucky, 2002). See also the essays on launch vehicles and access to space in John M. Logsdon et al., eds., *Exploring the Unknown: Selected Documents in the History of the U.S. Civil Space Program*, vol. 4, *Accessing Space* (Washington, DC: NASA SP-4407, 1999).

Space Shuttle: The History of the National Space Transportation System: The First 100 Missions (2001), comprise extremely thorough and informative narratives, providing an engineer's perspective on the many technical decisions during design, testing, and operations of particular projects.[66] Because of their distance from the original events, the prevailing context of a directionless American space program, and perceptions of American greatness compromised by liberals and social programs, these works communicate not only nostalgia, but also regret.[67] In *Leaving Earth* (2003), Robert Zimmerman notes, "Can we no longer imagine a future where humanity goes out and settles the far-flung stars? Have we become so small-minded that we cannot envision a tomorrow as idealistic and hopeful as that imagined by men like Ley, Korolev, and von Braun?"[68]

Histories of robotic exploration have been less mired in the betrayal of the post-Apollo times. Like their human spaceflight counterparts, they are coherent and useful accounts of humanity's first efforts to probe beyond circumterrestrial space. There exist comprehensive and technically detailed histories of Voyager, Galileo, Ulysses, and Mars Pathfinder, as well as broader histories of lunar and planetary exploration.[69] As part of its *Exploring the Unknown* series, NASA has also sponsored studies on scientific research by robotic probes.[70] The study of applications satellites (communications, weather, remote sensing, etc.) remains relatively neglected within the space history community, because it lacks the cachet of both human and deep space exploration, in part

66. Dennis R. Jenkins, *Space Shuttle: The History of the National Space Transportation System: The First 100 Missions* (Cape Canaveral, FL: D. R. Jenkins, 2001). See also Richard S. Lewis, *The Voyages of Columbia: The First True Spaceship* (New York: Columbia University Press, 1984).

67. See, for example, Robert Zimmerman, *Genesis: The Story of Apollo 8: The First Manned Flight to Another World* (New York: Four Walls Eight Windows, 1998); Harland, *Exploring the Moon*; John Catchpole, *Project Mercury: NASA's First Manned Space Programme* (London: Springer, 2001); David Shayler, *Gemini: Steps to the Moon* (London: Springer, 2001); David Shayler, *Skylab: America's Space Station* (London: Springer, 2001); David Shayler, *Apollo: The Lost and Forgotten Missions* (London: Springer, 2002); Reynolds, *Apollo*; Reginald Turnill, *The Moonlandings: An Eyewitness Account* (Cambridge, U.K.: Cambridge University Press, 2003).

68. Zimmerman, *Leaving Earth*, p. 463.

69. Henry S. F. Cooper, Jr., *Imaging Saturn: The Voyager Flights to Saturn* (New York: Holt, Rinehart, and Winston, 1982); Murray, *Journey into Space*; Burrows, *Exploring Space*; Robert Reeves, *The Superpower Space Race: An Explosive Rivalry Through the Solar System* (New York: Plenum Press, 1994); Donna Shirley and Danelle Morton, *Managing Martians* (New York: Broadway Books, 1998); Robert S. Kraemer, *Beyond the Moon: A Golden Age of Planetary Exploration, 1971–1978* (Washington, DC: Smithsonian Institution Press, 2000); David M. Harland, *Jupiter Odyssey: The Story of NASA's Galileo Mission* (London: Springer, 2000); Judith Reeves-Stevens et al., *Going to Mars: The Untold Story of Mars Pathfinder and NASA's Bold New Missions for the 21st Century* (New York: Pocket Books, 2000); David M. Harland, *Mission to Saturn: Cassini and the Huygens Probe* (London: Springer, 2002); Andrew Mishkin, *Sojourner: An Insider's View of the Pathfinder Mission* (New York: Berkeley Books, 2003); Paolo Ulivi, *Lunar Exploration: Human Pioneers and Robotic Surveyors* (London: Springer-Verlag, 2003); Ben Evans with David M. Harland, *NASA's Voyager Missions: Exploring the Outer Solar System and Beyond* (London: Springer, 2004).

70. Logsdon et al., *Exploring the Unknown*, vol. 5.

because these satellites carry no people and go nowhere. In contrast to human and deep space robotic spaceflight, the services offered by applications satellite systems deeply shape social, political, and cultural dimensions of societies. The objectives, capabilities, and design of such systems are in turn profoundly shaped by social, political, and cultural needs. Although many such "civilian" technological systems developed from firm connections with military projects, few historians have produced scholarship on their origins, performance, and ramifications.[71]

A number of historians and journalists have explored aspects of the many large-scale technological systems that were part of the American space program. These include management-focused histories such as Arnold S. Levine's *Managing NASA in the Apollo Era* (1982) and Stephen B. Johnson's *The Secret of Apollo: Systems Management in American and European Space Programs* (2002).[72] Two biographical works have enriched our understanding of the success of Apollo: Henry W. Lambright's *Powering Apollo: James E. Webb of NASA* (1995) and Robert C. Seamans's *Aiming at Targets: The Autobiography of Robert C. Seamans* (1996).[73] Both Webb and Seamans played critical roles in facilitating one of the most impressive and largest technological systems in 20th-century America. Their own words will be crucial for future historians interested in relocating Apollo in the same kind of social, political, and cultural context that Thomas P. Hughes did for electrical systems in his landmark *Networks of Power* (1983).[74]

T. A. Heppenheimer's multivolume history of the Space Shuttle is an important contribution to the programmatic space history genre. Although it

71. For the few works on applications projects, see Pamela E. Mack, *Viewing the Earth: The Social Construction of the Landsat System* (Cambridge, MA: MIT Press, 1990); David J. Whalen, *The Origins of Satellite Communications, 1945–1965* (Washington, DC: Smithsonian Institution Press, 2002); Donald H. Martin, *Communications Satellites*, 4th ed. (El Segundo, CA: Aerospace Press, 2000); Donna A. Demac, ed., *Tracing New Orbits: Cooperation and Competition in Global Satellite Development* (New York: Columbia University Press, 1986); P. Krishna Rao, *Evolution of the Weather Satellite Program in the U.S. Department of Commerce: A Brief Outline* (Washington, DC: NOAA, 2001); James M. Allen and Shanaka de Silva, "Landsat: An Integrated History," *Quest: The History of Spaceflight Quarterly* 12, no. 1 (2005): 6–22. See also the essays on satellite communications and remote sensing in John M. Logsdon et al., *Exploring the Unknown: Selected Documents in the History of the U.S. Civilian Space Program*, vol. 3, *Using Space* (Washington, DC: NASA SP-4407, 1998).

72. Arnold S. Levine, *Managing NASA in the Apollo Era* (Washington, DC: NASA SP-4102, 1982); Stephen B. Johnson, *The Secret of Apollo: Systems Management in American and European Space Programs* (Baltimore, MD: Johns Hopkins, 2002).

73. Henry W. Lambright, *Powering Apollo: James E. Webb of NASA* (Baltimore, MD: Johns Hopkins, 1995); Robert C. Seamans, *Aiming at Targets: The Autobiography of Robert C. Seamans* (Washington, DC: NASA SP-4106, 1996). See also the essays by Seamans, Webb, and other Apollo-era NASA managers, including Robert R. Gilruth, Wernher von Braun, George M. Low, Rocco A. Petrone, Samuel C. Phillips, and George E. Mueller, in *Apollo Expeditions to the Moon*, ed. Cortright.

74. Thomas P. Hughes, *Networks of Power: Electrification in Western Society, 1880–1930* (Baltimore: Johns Hopkins, 1983).

skirts social issues and references no literature from the academic historiography of American technology, it represents a fleshed-out narrative that expertly describes the interplay between politics and technology that affected key milestones in the Shuttle program, including the requirements for such a system and how those requirements evolved over time depending on claims made by constituencies within NASA and the Department of Defense.[75] Similarly, Roger D. Launius's *Space Stations: Base Camps to the Stars* (2003) looks thematically at the historical development of space stations and their central role in the evolution of both prescriptive and practical plans for the exploration of space, entrenched partly by what Dwayne A. Day has called the dominant "von Braun" paradigm of space exploration.[76]

NOSE CONE HISTORY

The astronaut memoir (or, more broadly, the astronaut-centered history) constitutes one of the largest historical subgenres in the field of space history. I call these works "nose cone histories" since they describe a narrowly circumscribed circle of events visible only to the astronauts and in which only the astronauts were visible. For the millions who followed the space program in the 1960s, astronauts—not engineers nor servicepersons nor managers—were the most visible human representations of the technological accomplishments of the early Space Age. Our natural urge to distill all the meaning of the space program—in particular its avatar Apollo—was embodied potently by the astronauts. As Tom Wolfe described in *The Right Stuff* (1979), these young, able, athletic, and short-haired men each seemed an idealized version of an American everyman, with a wife, a picket fence, a shiny car—and yet simultaneously wrapped in myth and mystery.[77]

Some of the nose cone histories have added important dimensions of the story of the American human spaceflight program. For example, Apollo 11 astronaut Michael Collins, in his fascinating memoir *Carrying the Fire: As Astronaut's Journeys* (1974), shows a deep empathy and understanding of the role of astronauts in the halcyon days leading up to the epic Moon landing in 1969. Collins's narrative provided the first glimpse behind the iconogra-

75. T. A. Heppenheimer, *The Space Shuttle Decision: NASA's Search for a Reusable Space Vehicle* (Washington, DC: NASA SP-4221, 1999); Heppenheimer, *Development of the Space Shuttle, 1972–1981* (Washington, DC: Smithsonian Institution Press, 2002).

76. Launius, *Space Stations*. See also the equally fine Giovanni Caprara, *Living in Space: From Science Fiction to the International Space Station* (Willowdale, Ontario: Firefly Books, 2000). Less successful is Zimmerman's *Leaving Earth*, which is a sprawling and flawed attempt to locate the development of space stations in domestic and international politics. For the "von Braun paradigm," see Dwayne A. Day, "The Von Braun Paradigm," *Space Times* 33 (November–December 1994): 12–15.

77. Tom Wolfe, *The Right Stuff* (New York: Farrar, Straus and Giroux, 1979).

phy of the astronaut-as-unidimensional-hero of popular American culture, a self-sustaining myth given birth after the "original seven" Mercury astronauts were presented to the American media in 1959.[78] Collins described his colleagues as a complex group with diverse personality traits spanning the whole gamut: overachieving, academic, adventurous, risk-averse, emotionally distant, publicity-seeking, insecure, and brilliant. All were fully ready to do the job they were given. Further astronaut memoirs, particularly Walt Cunningham's *All-American Boys* (1977) and Gene Cernan's *The Last Man on the Moon* (1999), were, like Collins's pioneering work, candid about the singularly unique experiences of the NASA astronauts of the 1960s, especially their relationship to top management, their competitiveness among themselves, and their often complicated private lives.[79] Astronaut Donald "Deke" Slayton, the man responsible for selecting every American space crew between 1965 and 1975, added important historical details to how astronaut crews were picked—including Armstrong and Aldrin for the first lunar landing—in his posthumously published memoir, *Deke! An Autobiography* (1995).[80]

Fully fleshed, well-researched, and contextual biographies can say something profound not just about an individual, but also the period under study; yet most nose cone space histories have been narrow, hagiographic, or self-serving. They reinforce rather than explore the mythmaking associated with the astronaut as icon. They also continue to marginalize the many thousands who also worked on the space program; in other words, fetishization of the astronaut has been a potent barrier against a social history of the space program since, in the popular consciousness, the history of the American space program remains inseparable from the biographies and heroism of astronauts.[81]

---

78. Michael Collins, *Carrying the Fire: An Astronaut's Journeys* (New York: Farrar, Straus and Giroux, 1974). Soon after their selection in 1959, the original seven astronauts signed deals with *Life* magazine for exclusive rights to bring their lives to the public. Apart from the many *Life* stories, one major output of this agreement was the very clinical book by W. Scott Carpenter et al., *We Seven, By the Astronauts Themselves* (New York: Simon & Schuster, 1962).

79. Walt Cunningham with Mickey Herskowitz, *The All-American Boys* (New York: Macmillan, 1977); Eugene A. Cernan and Donald A. Davis, *The Last Man on the Moon: Astronaut Eugene Cernan and America's Race in Space* (New York: St. Martin's Press, 1999).

80. Donald K. "Deke" Slayton and Michael Cassutt, *Deke! An Autobiography* (New York: St. Martin's Press, 1995). See also Joseph D. Atkinson, Jr., and Jay M. Shafritz's *The Real Stuff: A History of NASA's Astronaut Recruitment Program* (New York: Praeger, 1985) for a more academic perspective on astronaut selection.

81. Important exceptions to the bland astronaut-centered histories include two works by Henry S. F. Cooper: *Before Liftoff: The Making of a Space Shuttle Crew* (Baltimore: Johns Hopkins, 1987) and *A House in Space* (New York: Holt, Rinehart, and Winston, 1976). The former is an excellent study on the dynamics of forming and training crews for human spaceflight, while the latter explores the interactions of crew members on the long-duration *Skylab* missions. Jim Hansen's biography of Neil Armstrong, *First Man: The Life of Neil A. Armstrong* (New York: Simon & Schuster, forthcoming) also promises to be an important contribution to the field.

A new generation of space enthusiasts (affectionately called "space cadets" by some) has taken up the job of producing a slew of astronaut biographies. The first few published in the 1980s and 1990s provided unique viewpoints to the history of the American human space program, but by the early 2000s, their utility as history texts has diminished.[82] Many astronauts continue to write their own memoirs, usually ghost-written with others. The memoirs of some would suggest that travel through space engendered profound spiritual transformations—or often crises of the spirit—that led them to unexpected pathways.[83] The ones who achieved important management or advisory positions in the space program—such as Gemini and Apollo astronaut Thomas P. Stafford—have more to say than others. But all ponder, explore, and frequently advocate specific policies to give direction to a space program evidently lacking one since the golden age of Apollo.[84]

NEW HISTORY

In an article in 2000, then–NASA Chief Historian Roger D. Launius identified a "New Aerospace History" that emerged in the 1980s that was

---

82. See, for example, Colin Foale, *Waystation to the Stars: The Story of Mir, Michael, and Me* (London: Headline, 1999); Evelyn Husband with Donna Van Liere, *High Calling: The Courageous Life and Faith of Space Shuttle Commander Rick Husband* (Nashville, TN: Thomas Nelson, 2003); Colin Burgess et al., *Fallen Astronauts: Heroes Who Died Reaching the Moon* (Lincoln: University of Nebraska Press, 2003); Ray E. Boomhower, *Gus Grissom: The Lost Astronaut* (Indianapolis: Indiana Historical Society Press, 2004); Neal Thompson, *Light This Candle: The Life & Times of Alan Shepard—America's First Spaceman* (New York: Crown Publishers, 2004); Leon Wagener, *One Giant Leap: Neil Armstrong's Stellar American Journey* (New York: Forge, 2004); Nancy Conrad and Howie Klausner, *Rocketman: Astronaut Pete Conrad's Incredible Ride to the Moon and Beyond* (New York: New American Library, 2005).

83. Edwin E. Aldrin, Jr., with Wayne Warga, *Return to Earth* (New York: Random House, 1973); James Irwin and Williams Emerson Irwin, *To Rule the Night* (Philadelphia: A. J. Holman, 1973); Kathleen Maughn Lind, *Don Lind: Mormon Astronaut* (Salt Lake City: Deseret Book, 1985); Charlie Duke and Dotty Duke, *Moonwalker* (Nashville: Oliver-Nelson Books, 1990); Edgar D. Mitchell, *The Way of the Explorer: An Apollo Astronaut's Journey Through the Material and Mystical Worlds* (New York: G. P. Putnam's Sons, 1996); Gordon Cooper and Bruce Henderson, *Leap of Faith: An Astronaut's Journey into the Unknown* (New York: Harper Collins, 2000).

84. Armstrong et al., *First on the Moon*; Frank Borman with Robert J. Serling, *Countdown: An Autobiography* (New York: W. Morrow, 1988); Wally Schirra and Richard N. Billings, *Schirra's Space* (Boston: Quinlan Press, 1988); Jim Lovell and Kluger Jeffrey, *Lost Moon: The Perilous Voyage of Apollo 13* (Boston: Houghton Mifflin, 1994); Mike R. Mullane, *Liftoff! An Astronaut's Dream* (Parsippany, NJ: Silver Burdett Press, 1995); Bill Nelson with Jamie Buckingham, *Mission: An American Congressman's Voyage to Space* (San Diego: Harcourt Brace Jovanovich, 1988); Alan Bean with Andrew Chaikin, *Apollo: An Eyewitness Account by an Astronaut* (Shelton, CT: Greenwich Workshop Press, 1998); John Glenn and Nick Taylor, *John Glenn: A Memoir* (New York: Bantam Books, 1999); Jerry Linenger, *Off the Planet: Surviving Five Perilous Months Aboard the Space Station Mir* (New York: McGraw-Hill, 2000); Scott Carpenter, *For Spacious Skies: The Uncommon Journey of a Mercury Astronaut* (Orlando, FL: Harcourt, 2002); Thomas P. Stafford and Michael Cassutt, *We Have Capture: Tom Stafford and the Space Race* (Washington, DC: Smithsonian Institution Press, 2002).

"intrinsically committed to relating the subject to larger issues of society, politics, and culture and taking a more sophisticated view," a history that "move[d] beyond a fetish for the artifact."[85] More generally, Launius characterized these works as being in the middle ground between critique and celebration of the space program. I would modify Launius's typology by expanding the parameters to include a wider range of intellectual inquiry that often includes *both* critiques and celebration of the space program. They are, however, distinguished from the more traditional canon in two important ways: first, they do not rely on singular approaches to interpreting the history of space exploration, such as exploration, competition, technology, and astronauts. Instead, these works combine different elements of each and firmly locate their narratives in broader political, social, technological, and/or cultural contexts; i.e., they function as political, social, technological, and/or cultural histories. Second, they attempt to link to other historical subdisciplines such as the history of the Cold War, diplomatic history, and the history of science and technology.

In analyzing the new history, I describe important examples from each of four categories of new history—political, social, technological, and cultural history—and summarize opportunities for future research in each subgenre.

Political History

In the new history, political history has led the way in important reevaluations of the American space program. Walter A. McDougall's Pulitzer Prize–winning . . . *The Heavens and the Earth: A Political History of the Space Age* (1985) remains the most important and influential work in the genre. The book contributed to relocating the early years of the American space program in the broader context of postwar American politics. McDougall's main argument was that after World War II, and especially after Sputnik, the U.S. government marshaled resources on an unprecedented scale to promote advancements in science and technology, in effect, transforming the country into a new kind of 20th-century state, the technocracy. He noted:

> In those years [of the Sputnik challenge] the fundamental relationship between the government and new technology changed as never before in history. No longer did state and society react to new tools and methods, adjusting, regulating, or encouraging their spontaneous development. Rather, states took upon themselves the primary responsibility for generating new technology. This has meant that to the extent revolution-

---

85. Launius, "The Historical Dimension of Space Exploration," p. 23.

ary technologies have profound second-order consequences in the domestic life of societies, by forcing new technologies, *all* governments have become revolutionary, whatever their reasons or ideological pretensions.[86]

In McDougall's formulation, the rise of a postwar technocracy was inseparable from the rise of the national security state, since federal policies on science and technology—especially after Sputnik—were closely related to countering the perceived intellectual and military power of the Soviet Union. McDougall's overarching thesis substantively redefined the way in which historians viewed the space program. If they had previously resorted to invoking the "natural" human urge to explore, technological fetishization, or international competition, his work redirected attention to a magnitude of changes on the domestic political and institutional stage associated with the origins of the space program.

McDougall also argued that the Eisenhower administration's concerns over establishing a "freedom of space" rationale guided its initial formulations of American space policy. According to McDougall, neither the White House nor the Department of Defense emphasized a policy of being first to launch an artificial satellite of the Earth; instead, national security considerations—such as establishing the "freedom of space" precedent, developing a military space program under the cover of a civilian one, and not diverting resources from the concurrent ICBM program—trumped any drive to beat the Soviets. McDougall's work challenged readers to reevaluate the ingrained notion of the Eisenhower administration's space policy as confused and ineffectual.[87] Besides facilitating a shift in the tone of historical scholarship on American space exploration, . . . *The Heavens and the Earth*'s Pulitzer Prize validated historical scholarship on the space program as worthy of serious academic study.

---

86. Walter A. McDougall, . . . *The Heavens and the Earth: A Political History of the Space Age* (New York: Basic Books, 1985), pp. 6–7, emphasis in original.

87. Stephen E. Ambrose, in his multivolume biography of President Eisenhower, was the first to reframe the Eisenhower administration's role in the origins of the U.S. space program, but McDougall fully developed the idea. See Stephen E. Ambrose, *Eisenhower*, vol. 2, *The President* (New York: Simon & Schuster, 1983). The reevaluation of the Eisenhower administration's role in early U.S. space policy was fleshed out further in Rip Bulkeley, *The Sputniks Crisis and Early United States Policy: A Critique of the Historiography of Space* (Bloomington: Indiana University Press, 1991). For pathbreaking research on the "freedom of space" issue, see also Dwayne A. Day, "New Revelations about the American Satellite Programme Before Sputnik," *Spaceflight* 36 (1994): 372–373; R. Cargill Hall, "Origins of U.S. Space Policy: Eisenhower, Open Skies, and Freedom of Space," in *Exploring The Unknown: Selected Documents in the History of the U.S. Civil Space Program,* ed. John M. Logsdon et al., vol. 1, *Organizing for Exploration* (Washington, DC: NASA SP-4407, 1995), pp. 213–229; Dwayne A. Day, "Cover Stories and Hidden Agendas: Early American Space and National Security Policy," in *Reconsidering Sputnik: Forty Years Since the Soviet Satellite*, ed. Roger D. Launius et al. (Amsterdam: Harwood Academic Publishers, 2000), pp. 161–195.

Following in the footsteps of . . . *The Heavens and the Earth*, innovative scholarship by space policy scholar Howard E. McCurdy and historian Roger D. Launius advanced a reinterpretation of the "golden age" of Apollo at a 1993 symposium on presidential leadership and its influence on U.S. space policy. Instead of seeing Apollo as a "normal" stage in the evolution of American space policy, several historians argued that "the Apollo decision was . . . an anomaly in the history of the U.S. space program."[88] The implication was that policy-makers of the future could not use Apollo as a model of how to explore space since Apollo was intrinsically a unique product of its time that existed only because of exceptional circumstances, primarily national prestige and Cold War competition. Although this was not a new viewpoint, for the first time, space historians placed this notion as the key to understanding the early direction of American space exploration. In the conference proceedings, published as *Spaceflight and the Myth of Presidential Leadership* (1997), historians also argued that the role of presidential leadership in general may have been overestimated by advocates of space exploration after the Kennedy era. Recent reexaminations of Kennedy's historical 1961 decision to go to the Moon bolstered such a contrasting perspective.[89]

A 1997 conference on the 40th anniversary of Sputnik provided an opportunity for new and exciting scholarship on the origins and repercussions of the early American and Soviet space programs. Using recently declassified documents, historians amplified a number of important topics, including the "freedom of space" rationale for the beginning of the American space program, the selection of the Vanguard satellite project as the first civilian program, the formulation of the National Aeronautics and Space Act that led to the formation of NASA, and the effects of the National Defense Education Act that fundamentally altered the role of science and engineering in higher education in the United States. The collected papers from this conference, published as *Reconsidering Sputnik: Forty Years Since the Soviet Satellite* (2000), remain the most important set of intellectual inquiries into the origins of the American space program, complementing Robert Divine's systematic study of the Eisenhower administration's response to Sputnik, *The Sputnik*

---

88. Roger D. Launius and Howard E. McCurdy, eds., *Spaceflight and the Myth of Presidential Leadership* (Urbana: University of Illinois Press, 1997), p. 9. See also W. D. Kay, *Can Democracies Fly in Space? The Challenge of Revitalizing the U.S. Space Program* (Westport, CT: Praeger, 1995).

89. See also James L. Kauffman, *Selling Outer Space: Kennedy, the Media, and Funding for Project Apollo, 1961–1963* (Tuscaloosa: University of Alabama Press, 1994); Michael R. Beschloss, "Kennedy and the Decision to Go to the Moon" in *Spaceflight and the Myth of Presidential Leadership*, pp. 51–67; Stephen J. Garber, "Multiple Means to an End: A Reexamination of President Kennedy's Decision to Go to the Moon," *Quest: The History of Spaceflight Quarterly* 7, no. 2 (1999): 5–17; Andrew Chaikin, "White House Tapes Shed Light on JFK Space Race Legend," *Space.com*, 22 August 2001, *http://www.space.com/news/kennedy_tapes_010822.html*; Roger D. Launius, "Kennedy's Space Policy Reconsidered: A Post-Cold War Perspective," *Air Power History* 50, no. 4 (2003): 16–29.

Challenge (1993).[90] Similar reevaluations have been focused on other presidential administrations and their positions on initiatives within the civilian space program.[91]

The new political history suggests six broad areas ripe for future scholarship. These include the following:

1) Revisiting the early American space program in light of the complex debates within the canon of Cold War history, including studies of the space program as an adjunct for the less savory dimensions of American foreign policy; additionally, historians could explore not only how the Cold War shaped the contours of the civilian and military space programs, but also how the latter shaped aspects of the former; Giles Alston's dissertation on the influence of Apollo on international relations points to further avenues of research.[92]

2) Further study of the ways in which different administrations have used specific initiatives and programs as part of political agendas unrelated to the stated goals of the initiatives or programs;[93] surprisingly, there exist no systematic studies of the Nixon or Reagan administration's stance towards civilian and military space policy.

90. Launius et al., *Reconsidering Sputnik*; Robert A. Divine, *The Sputnik Challenge: Eisenhower's Response to the Soviet Satellite* (New York: Oxford University Press, 1993); Lafayette P. Temple III, "Organizing Space: The Political-Bureaucratic Dynamics Through 1961" (Ph.D. diss., George Washington University, 1999). See also Matt Bille and Erika Lishock, *The First Space Race: Launching the World's First Satellites* (College Station: Texas A&M University, 2004), which assembled all the new research into a single volume; Roger D. Launius, "Eisenhower, Sputnik, and the Creation of NASA: Technological Elites and Public Policy Agenda," *Prologue* 28 (summer 1996): 127–143; Peter J. Roman, *Eisenhower and the Missile Gap* (Ithaca, NY: Cornell University Press, 1995).

91. McCurdy, *The Space Station Decision*; Mark Damohn, *Back Down to Earth*; Krug, *Presidential Perspectives on Space Exploration*; Thor Nels Hogan, "Mars Wars: A Case History of Agenda Setting and Alternative Generation in the American Space Program" (Ph.D. diss., Public Policy and Public Administration Department, George Washington University, 2004). In addition, Launius and McCurdy's *Spaceflight and the Myth of Presidential Leadership* includes a number of important essays on Eisenhower, Kennedy, Johnson, Reagan, and George H. W. Bush.

92. Giles Alston, "International Prestige and the American Space Programme" (Ph.D. diss., Queen's University of Belfast, 1989).

93. For some examples, see Dwayne A. Day, "Space Policy-Making in the White House: The Early Years of the National Aeronautics and Space Council," in *Organizing for the Use of Space*, ed. Launius, pp. 117–154; Joan Hoff, "The Presidency, Congress, and the Deceleration of the U.S. Space Program in the 1970s," and Robert H. Ferrell, "Presidential Leadership and International Aspects of the Space Program," both in *Spaceflight and the Myth of Presidential Leadership*, ed. Launius and McCurdy, pp. 92–132 and 172–204, respectively. For a comparative study of NASA under two different administrations, see John D. Kelley, "An Organizational History of the National Aeronautics and Space Administration: A Critical Comparison of Administrative Decision Making in Two Pivotal Eras" (Ph.D. diss., University of Southern California, 2002).

3) The relationship, exchanges, and competition between the civilian and military/intelligence space programs, in terms of intelligence, hardware, and managerial and engineering expertise;[94] for example, how does the movement of high administrators (such as Dan Goldin and Michael Griffin) from one sector affect NASA policies?

4) The connections between foreign policy and domestic space policy, a vast topic which has been studied piecemeal, but not in any systematic and long *durée* approach.

5) The relationship between domestic political transactions (congressional politics, redistricting, lobbying, policy papers, advisory boards, etc.) and the making of space policy.

6) The role of institutions in the making of civilian and military space policy; the scholarship would encompass the study of why certain institutions are created, others are dissolved, what kind of inertia they carry through their history, and the ways in which particular institutions relate to others.

History of Technology

The second broad field of new history has emerged from within the bounds of the history of technology. Most artifactual histories of space programs tend to accept implicitly notions of technological determinism, especially that the space program exists as autonomous technology, affecting society around it but not being affected by it. There have been many works on the societal impacts of space exploration;[95] the field of space exploration has, however, largely been insulated from the paradigmatic revolution in the history of

---

94. For general perspectives, see Dwayne A. Day, "Invitation to Struggle: The History of Civilian-Military Relations in Space," in *Exploring the Unknown: Selected Documents in the History of the U.S. Civil Space Program*, ed. John M. Logsdon, vol. 2, *External Relations* (Washington, DC: NASA SP-4407, 1996), pp. 233–270; Mark A. Erickson, "The Evolution of the NASA-DoD Relationship from Sputnik to the Lunar Landing" (Ph.D. diss., George Washington University, 1997). For exchanges of hardware between "black" and civilian space projects, see Dwayne A. Day's "Not So Black and White: the Military and the Hubble Space Telescope," *Space Times* 34 (March–April 1995): 20–21, and "From Above the Iron Curtain to Around the Moon," *Spaceflight* 47 (2005): 66–71. For an excellent work on the relationships between private industry, government-funded intelligence satellite programs, and technological innovation, see Jonathan E. Lewis, *Spy Capitalism: Itek and the CIA* (New Haven, CT: Yale University Press, 2002).

95. See, for example, Lillian A. Levy, ed., *Space, Its Impact on Man and Society* (New York: Norton, 1965); Raymond A. Bauer et al., *Second-Order Consequences: A Methodological Essay on the Impact of Technology* (Cambridge, MA: MIT Press, 1969); Charles P. Boyle, *Space Among Us: Some Effects of Space Research on Society* (Washington, DC: AIAA, 1974); Tim Greve et al., eds., *The Impact of Space Science on Mankind* (New York: Plenum Press, 1976).

technology in the 1980s that redirected focus from technological determinism to the social construction of technology (and technological systems).[96] A few notable exceptions include Pamela E. Mack's *Viewing the Earth: The Social Construction of the Landsat System* (1990) and Donald A. Mackenzie's *Inventing Accuracy: A Historical Sociology of Nuclear Missile Guidance* (1990).[97] In the latter, Mackenzie argued that missile accuracy was not an inevitable consequence of technical change, but rather part of a process involving negotiation between a wide range of actors. His use of missile guidance as a window into exploring how accuracy was socially constructed suggests important future avenues of further research on the space program, including studies of the ways in which crew safety, mission success, or risk assessments in the human space program have been negotiated and socially constructed.

The social constructivist approach is to some degree related to the influential shift in the literature on technological systems. In moving the study of the history of technology from artifacts to systems, historian Thomas P. Hughes's work fundamentally altered the ways in which historians conceived of the relationship between technology and society.[98] Tentative steps towards a view of space projects as large-scale technological systems were taken in important works such as R. Cargill Hall's *Lunar Impact: A History of Project Ranger* (1977) and Roger E. Bilstein's *Stages to Saturn: A Technological History of the Apollo/Saturn Launch Vehicles* (1980).[99] Similarly, Charles Murray and Catherine Bly Cox's excellent *Apollo: The Race to the Moon* (1989) describes the Apollo project as a system whose primary actors were managers, engineers, politicians, and organizations rather than astronauts. Based on documentation and interviews with the remaining living actors of the endeavor, their reconstruction of the Apollo project as a milestone in the history of management makes it probably the single best historical overview of Apollo.[100]

Beyond social constructivism, others have begun the work of looking at the space program as a case study in technological culture. In *Goals in Space:*

96. For seminal early works on the social construction of technology, see Wiebe J. Bijker et al., eds., *The Social Construction of Technological Systems: New Directions in the Sociology and History of Technology* (Cambridge, MA: MIT Press, 1987); Wiebe J. Bijker, *Of Bicycles, Bakelites, and Bulbs: Toward a Theory of Sociotechnical Change* (Cambridge, MA: MIT Press, 1995); Merritt Roe Smith and Leo Marx, eds., *Does Technology Drive History? The Dilemma of Technological Determinism* (Cambridge, MA: MIT Press, 1994).

97. Mack, *Viewing the Earth*; Donald A. Mackenzie, *Inventing Accuracy: A Historical Sociology of Nuclear Missile Guidance* (Cambridge, MA: MIT Press, 1990).

98. Hughes, *Networks of Power*; Thomas P. Hughes, *Rescuing Prometheus* (New York: Pantheon Books, 1998).

99. R. Cargill Hall, *Lunar Impact: A History of Project Ranger* (Washington, DC: NASA SP-4210, 1977); Roger E. Bilstein, *Stages to Saturn: A Technological History of the Apollo/Saturn Launch Vehicles* (Washington, DC: NASA SP-4206, 1980).

100. Charles Murray and Catherine Bly Cox, *Apollo: Race to the Moon* (New York: Simon & Schuster, 1989).

American Values and the Future of Technology (1991), William Sims Bainbridge used sociological methods to investigate how actors in American culture have used language in popular discussions on space exploration. On the institutional and organizational side, Diane Vaughan, in *The Challenger Launch Decision: Risky Technology, Culture, and Deviance at NASA* (1996), used interdisciplinary approaches derived from sociology and communications theory to analyze the culture of NASA in the 1980s.[101] Her research illustrates the ways in which organizations develop their own culture that, depending on the scarcity of resources, fosters an environment that finds high risk acceptable without breaking any major rules. Her conception of the "normalization of deviance" suggests important avenues of further research, especially for studying space projects that did not achieve any significant successes.[102]

Others have explored more esoteric approaches to the technological history of the space program. In *The Religion of Technology* (1997), David F. Noble investigates the role of scripture and definable Christian symbolism in the "dreaming" for space exploration in the pre-Sputnik days and the invocation of God as a transcendental element in the rhetoric of modern-day managers, activists, and astronauts.[103] If not all of his ruminations are convincing, his findings on the prehistory of space travel suggest as-yet-unexplored opportunities for scholarship on the relationship between religion and spaceflight in the early 20th century, furthered recently by Roger D. Launius in a meditation on utopianism and space advocacy.[104] David E. Nye, in his essay "Don't Fly Me to the Moon: The Public and the Apollo Space Program," also contributes to the move away from technological determinism. He challenges the near-sacred notions among the "space cadet" community that the history of space exploration was of any significance in the history of humanity; he also questions the notion that "experiencing outer space transformed inner consciousness," a claim which hinged on the images of a fragile Earth as seen from deep space by the Apollo astronauts. He concludes that retrospect has made Apollo a unifying memory when in reality, during its execution, the polity and populace remained fractured over its symbolic and material benefits. He concludes, "Just as all Americans revere their Revolution, even though less than half the population actively supported it in 1776, the Apollo Program appears to be gaining sanctity in retrospect."[105]

---

101. Diane Vaughan, *Challenger Launch Decision: Risky Technology, Culture, and Deviance at NASA* (Chicago: University of Chicago Press, 1996).

102. Vaughan's analysis, of course, also influenced the work of the Columbia Accident Investigation Board.

103. David F. Noble, *The Religion of Technology: The Divinity of Man and the Spirit of Invention* (New York: Alfred A. Knopf, 1997), pp. 115–142.

104. Launius, "Perfect Worlds, Perfect Societies."

105. David E. Nye, *Narratives and Spaces: Technology and the Construction of American Culture* (New York: Columbia University Press, 1997), p. 160.

Robotic spaceflight has yielded significant new understandings about the solar system. This is the first contiguous, uniform, 360-degree color panorama taken by the Imager for Mars Pathfinder (IMP) over the course of sols 8, 9, and 10 (Martian days) in 1997. Different regions were imaged at different times over the three Martian days

These new works underscore that, collectively, historians need to move beyond methodological approaches that embrace technological determinism, Whiggish history, and program-centered histories. They suggest six areas for further research:

1) Despite nearly 40 years of writing space history, we still do not have a substantive history of space technology, work focused not on programs but on the technologies that constitute a complete system capable of spaceflight, including rocket engines, solar cells, fuel cells, communications equipment, thermal protection, guidance systems, materials, etc.[106] We need histories that are neither programmatic nor artifact-centered; for example, a history of satellite-based optical systems (cameras, lenses, mirrors, data recovery, etc.) could shed light on the relationship between a particular technology, commercial industry, or the military and the way in which consumers can shape technologies.

2) An important but unexplored aspect of the space industry is the economic history of space manufacturing—in particular of rockets, engines, and satellites, which would illuminate issues of government-industry relations, quality control, and labor practices; it is also

---

106. For works on discrete technologies, see Lillian D. Kozloski, *U.S. Space Gear: Outfitting the Astronaut* (Washington, DC: Smithsonian Institution Press, 1993); Eldon C. Hall, *Journey to the Moon: The History of the Apollo Guidance Computer* (Reston, VA: AIAA, 1996); Gary L. Harris, *The Origins and Technology of the Advanced Extravehicular Space Suit* (San Diego, CA: Univelt, 2001); James A. Dewar, *To the End of the Solar System: The Story of the Nuclear Rocket* (Lexington: University of Kentucky Press, 2004).

to acquire consistent lighting and shadow conditions for all areas of the panorama. At left is a lander petal and a metallic mast that is a portion of the low-gain antenna. Deflated air bags are visible at the perimeters of all three lander petals. *(NASA image no. PIA00752)*

necessary to locate this history within the broader history of mass production in America.[107]

3) Journalists have devoted much attention to the various disasters of the Space Age, but besides one significant exception—David Shayler's *Disasters in Manned Spaceflight* (2000)—they have been focused narrowly on particular incidents.[108] Because the literature on space history has had a triumphalist arc (introduction, plot thickens, crisis, triumph over adversity), it has ignored accounts of long-range technological failures, which can also shed light on abandoned lineages of technologies and the contingencies that shaped our adoption of certain systems over others.[109]

107. For mass production in general, see David A. Hounshell's seminal *From the American System to Mass Production* (Baltimore: Johns Hopkins, 1984). For a brief essay on the economics of the space program, see Henry R. Hertzfeld, "Space as an Investment in Economic Growth," in *Exploring the Unknown*, ed. Logsdon, vol. 3, pp. 385–400.

108. David Shayler, *Disasters and Accidents in Manned Spaceflight* (New York: Springer, 2000). For various disaster-focused works, see Henry S. F. Cooper, *Thirteen, the Flight That Failed* (New York: Dial Press, 1972); Malcolm McConnell, *Challenger: A Major Malfunction* (Garden City, NY: Doubleday, 1987); Joseph Trento, *Prescription for Disaster* (New York: Crown, 1987); Richard S. Lewis, *Challenger: The Final Voyage* (New York: Columbia University Press, 1988); Claus Jensen and Barbara Haveland, *No Downlink: A Dramatic Narrative about the Challenger Accident and Our Time* (New York: Farrar, Straus and Giroux, 1996); Michael Cabbage and William Harwood, *Comm Check: The Final Flight of Shuttle Columbia* (New York: Free Press, 2004).

109. For technological failure, see Neil Schlager, ed., *When Technology Fails: Significant Technological Disasters, Accidents and Failures of the Twentieth Century* (Detroit: Gale Research, 1994); Azriel Lorber, *Misguided Weapons: Technological Failure and Surprise on the Battlefield* (Washington, DC: Brasseys, 2002).

4) The social constructivist approach remains a powerful methodological tool for in-depth studies of any number of rocket and spaceflight systems, including, for example, the Space Shuttle, which is an excellent case for studying how different actors can shape the form and function of a technological system; such an approach would help to avoid the deterministic historical narratives that assume, for example, that the liquid-propellant rocket was the obvious method to reach space without questioning the social and cultural forces that led Tsiolkovskiy, Goddard, Oberth, and others to arrive at the rocket as the propulsive force for access to space.

5) A relatively unexplored area is the social construction of risk in space technological systems; for example, we know little in a systematic way about the manner in which risk has been constructed, defined, and invoked in human versus robotic systems, in different human spaceflight programs, among engineers and flight directors, etc. An important unexplored question remains the historical evolution of what it means to "man-rate" a vehicle.

6) We still do not have well-researched histories on the continuing tension between robotic and human spaceflight; specific areas of inquiry could include the interplay between technology, policy, and organizational culture in determining choices for robotic versus human spaceflight; what role economics plays in these choices; and the ways in which we measure "output" for given space projects (whether human or robotic) and how these evaluations may or may not be contingent upon premiums placed upon human or robotic spaceflight. Finally, a useful avenue of research may be to explore why and how, during the early space era (especially in the pre-Sputnik years), policy-makers overwhelmingly emphasized human spaceflight in their public advocacy.

Social History

Beyond political history, several historians and sociologists have taken up the job of moving beyond nose cone history into broader social themes. An early progenitor of this subgenre was William S. Bainbridge's *The Spaceflight Revolution: A Sociological Study* (1976). Although his focus was primarily on spaceflight visionaries from the late 19th and early 20th centuries, Bainbridge argued that the advancement of technology was not necessarily deterministic. In fact, in cases of revolutionary technology such as the rocket, the principal actors (such as von Braun) maneuvered the government and military into facilitating resources to implement their goals of spaceflight. Thus, instead of being co-opted by the state, scientists and engineers opportunistically took advantage of the state.[110]

Historians have also investigated a number of methodological issues related to the study of the early space program, including the problem of doing contemporary or near-contemporary history. Because of the recent nature of the history of space exploration, participants can play a large role in the way space history is chronicled. Participants provide evidence for historians, write history books, and sometimes dismiss nonparticipant history with a "you-weren't-there" rationale; historians respond by condescending to the participants by invoking "that noble dream" of objectivity and distance.[111] Space historians must explicitly address these methodological concerns if their goal is to produce history without baggage.

Beyond methodological concerns, an important aspect of the social dimension of spaceflight has been the relationship between public opinion and the space program. Mark E. Byrnes, in his *Politics and Space: Image Making by NASA* (1994), traced the effects of NASA's image-building policy on popular perceptions of the organization as well as broader support for the cause of space travel. He argued that NASA primarily used three images—nationalism, romanticism, and pragmatism—to create and consolidate political support across the nation for its major endeavors in space.[112] Similar work by others has helped to challenge many accepted notions about public advocacy for the space program. Using quantitative data, for example, Herbert E. Krugman found that "given the extensive media coverage of the space events throughout [the Apollo program], favorable publicity did not seem to have generated equally favorable public support for the Apollo program."[113] Roger D. Launius found that popular support for the space program remained at the same relative level both during and after the Apollo program, undercutting the received notion

---

110. William S. Bainbridge, *The Spaceflight Revolution: A Sociological Study* (New York: Wiley, 1976).

111. For a history of the search for objectivity in the discipline of history in American academia, see Peter Novick's *That Noble Dream: The 'Objectivity Question' and the American Historical Profession* (Cambridge, U.K.: Cambridge University Press, 1988). For some of the methodological considerations in writing space history, see Joseph N. Tatarewicz, "Writing the History of Space Science and Technology: Multiple Audiences with Divergent Goals and Standards," in *The Historiography of Contemporary Science and Technology*, ed. Thomas Söderqvist (Amsterdam: Harwood Academic Publishers, 1997), pp. 71–89.

112. Mark E. Byrnes, *Politics and Space: Image Making by NASA* (Westport, CT: Praeger, 1994). See also James L. Kauffman, *Selling Outer Space: Kennedy, the Media, and Funding for Project Apollo, 1961–1963* (Tuscaloosa: University of Alabama Press, 1994); Lynn Marie Disbrow, "A Metaphorical Analysis of the Evolution of NASA's Public Image, 1962–1986" (Ph.D. diss., Wayne State University, 1989).

113. Herbert E. Krugman, "Public Attitudes Toward the Apollo Space Program, 1965–1975," *Journal of Communication* 27, no. 4 (1977): 87–93.

of a "golden age" of mass support for the space program.[114] Expanding the frontier on social histories of the Space Age, recent studies have also focused on hitherto unexplored but crucial elements of the history of spaceflight such as the pro-space movement, the impact of the space program on geographical locales, and engineers as a mass demographic.[115]

Beyond these important exceptions, social history, which revolutionized mainstream American history beginning in the 1960s, has not made many inroads into space history. I identify five areas for further study concerning the relationship between society and space:

1) The history of the space program remains incomplete unless we explore the lived experiences and backgrounds of large demographic groups such as engineers, servicemen and -women, military and intelligence personnel involved in programs, launch personnel, staff workers, spouses and families of engineers in both the civilian and military space programs, etc.

2) Further exploration is necessary on the relationship between public advocacy and political commitment in the context of the space program, extending the work already done; such approaches would require explorations of the efficacy of formal and informal lobby groups.

3) In the past few years, a number of historians have taken steps into exploring the place of gender in the history of the space program; all of the work so far has focused on early women contenders for the astronaut corps, the so-called FLATs (First Lady Astronaut Trainees); most of these are narrow "surgical" histories that say little beyond recounting their life histories. The one exception, Margaret Weitekamp's superb *Right Stuff, Wrong Sex: America's First Women in Space Program* (2004), uses the FLATs story to revisit the social and cultural codes that guided broader American views on women, technology, and exploration in late-20th-century America.[116] Yet these

---

114. Roger D. Launius, "Public opinion polls and perceptions of US human spaceflight," *Space Policy* 19 (2003): 163–175.

115. Michael A. G. Michaud, *Reaching for the High Frontier: The American Pro-Space Movement, 1972–84* (New York: Praeger, 1986); William Barnaby Faherty, *Florida's Space Coast: The Impact of NASA on the Sunshine State* (Gainesville: University Press of Florida, 2002); Sylvia D. Fries, *NASA Engineers and the Age of Apollo* (Washington, DC: NASA SP-4104, 1992).

116. Margaret A. Weitekamp, *Right Stuff, Wrong Sex: America's First Women in Space Program* (Baltimore: Johns Hopkins, 2004). See also Bernice Trimble Steadman with Jody M. Clark, *Tethered Mercury: A Pilot's Memoir: The Right Stuff—but the Wrong Sex* (Traverse City, MI: Aviation

continued on the next page

works still leave much to be done since we still do not have any systematic studies of the role of women in much larger demographics who participated in the space program—in engineering, medicine, administration, and staff positions, as well as the thousands who were spouses in a predominantly male-dominated project;[117] we also need histories of women astronauts who actually flew in space, as opposed to those who never did.

4) We need more studies of how the growth of the space industry has affected particular geographical locales, particularly Texas, Alabama, California, and Florida; space historians need to rise up to the challenge to link subdisciplines such as urban history to space history by chronicling, for example, the transformation of urban sites through development and abandonment cycles or the motivations of many young scientists and engineers to pursue a career in the space program.[118]

5) The American space program was most identified with a White male demographic which reflects the natural distribution of those who managed and participated in the endeavor, yet it is important that we have a good understanding of the role and place of the space program demographic through broader—and, in some ways, cataclysmic—changes in the social fabric of American society from the 1960s to the 1990s in terms of racial relations and immigration.[119]

---

continued from the previous page

Press, 2001); Pamela Freni, *Space for Women: A History of Women with the Right Stuff* (Santa Ana, CA: Seven Locks Press, 2002); Stephanie Nolen, *Promised the Moon: The Untold Story of the First Women in Space Race* (New York: Four Walls Eight Windows, 2003); Bettyann Kevles, *Almost Heaven: The Story of Women in Space* (New York: Basic Books, 2003); Martha Ackmann, *The Mercury 13: The Untold Story of 13 American Women and the Dream of Spaceflight* (New York: Random House, 2003).

117. For recent autobiographical works that touch on broader issues of the role of women engineers in the American space program, see Shirley and Morton, *Managing Martians*; M. G. Lord, *Astro Turf: The Private Life of Rocket Science* (New York: Walker & Co., 2004). See also the piece on women who worked at Australia's Woomera Rocket Range: Kerrie Dougherty, "Calculating Women: A Brief History of the LRWE/WRE Computing Team," *Quest: The History of Spaceflight Quarterly* 9, no. 4 (2002): 31–39.

118. A recent pathbreaking article on the influence of postwar suburbanization on physicists' selection of professional topics is exemplary of the kinds of new work in other fields. See David Kaiser, "The Postwar Suburbanization of American Physics," *American Studies* 56, no. 4 (2004): 851–888.

119. Like the gender issue, the role of race in the American space program has been explored only through the focus of astronauts. See for example, J. Alfred Phelps, *They Had A Dream: The Story of African-American Astronauts* (Novato, CA: Presidio, 1994); Stanley P. Jones, *African-American Astronauts* (Mankato, MN: Capstone High/Low Books, 1998); Mae Jemison, *Find Where the Wind Goes: Moments From My Life* (New York: Scholastic, 2001); Betty Kaplan Gubert et al., *Distinguished African Americans in Aviation and Space Science* (Westport, CT: Oryx Press, 2002). There is also a large canon of juvenile literature on African American astronauts.

Cultural History

The cultural history of spaceflight is the most recent subgenre in the field and also the most heterogeneous. A survey of the key works shows deep and broad work encompassing everything from relatively orthodox studies of the place of spaceflight in American culture to more postmodern meditations on modernity, masculinity, and machines. Perhaps the earliest work in the field was Norman Mailer's *Of a Fire on the Moon* (1969), which, coming as it did in the year of Apollo 11, contrasted sharply with other contemporary accounts of Apollo.[120] Using field research, Mailer constructed a narrative that illustrated the clash—and sometimes rapprochement—between the young counterculture of the late 1960s and the pseudomilitary culture of NASA. Mailer implicitly critiqued what he believed was the militarized and regimented culture of NASA, with its middle-class values that cherished patriotism and encouraged unquestioned adherence to the dominant political culture.

A few authors have explored how the space program has resonated in modern literature. In the insightful *Seeing Earth: Literary Responses to Space Exploration* (1985), Ronald Weber deconstructed many of the attendant metaphors that cultural commentators—writers, poets, scholars, philosophers, theologians, astronauts, and others—have used to invoke, explain, extol, and critique the American space program, locating their meditations between the broad themes of "liberating leap into a mysterious future" and a new appreciation of the Earth itself.[121] William D. Atwill, in *Fire and Power: The American Space Program as Postmodern Narrative* (1994), adopts a similar methodological approach but takes a more critical stance towards the American space program, specifically Apollo. His thought-provoking explorations, which touch on domestic shocks of the Vietnam War, try to unpack "the difficulty so many writers had telling [the] story of a technocratic enterprise simultaneously central and antithetical to the time and place that produced it."[122]

Dale A. Carter also referenced American literature—in his case, Thomas Pynchon's classic novel *Gravity's Rainbow* (1973)—but had a more ambi-

---

120. Norman Mailer, *Of a Fire on the Moon* (New York: New American Library, 1969). See also W. David Lewis, "Buzz Aldrin's Return to Earth: The Astronaut and Social Values in Apollo Era America," *Quest: The History of Spaceflight Quarterly* 6, no. 1 (1998): 40–43.

121. Ronald Weber, *Seeing Earth: Literary Responses to Space Exploration* (Athens: Ohio University Press, 1985). For other, similar explorations, see Laurence Goldstein, *The Flying Machine and Modern Literature* (Bloomington: Indiana University Press, 1986); George Held, "Men on the Moon: American Novelists Explore Lunar Space," *Michigan Quarterly Review* 18 (spring 1979): 318–342; Laurence Goldstein, "'The End of All Our Exploring': The Moon Landing and Modern Poetry," *Michigan Quarterly Review* 18 (spring 1979): 192–217. For a look at space and the visual medium, see Laura M. Andre, "Lunar Nation: The Moon and American Visual Culture, 1957–1972" (Ph.D. diss., University of North Carolina, 2002).

122. William D. Atwill, *Fire and Power: The American Space Program as Postmodern Narrative* (Athens, GA: University of Georgia Press, 1994), p. 11.

tious goal: to rewrite the postwar history of the American space program as a critique of American expansionist military and economic aims. In Carter's worldview, the American space program represented a "Rocket State," a confluence of civilian and military interests with little or no moral code. The book remains one of the most important synthetic cultural histories of the American space program.[123] Other, similar critiques of the American space program have emerged from the new cultural history and include David Lavery's *Late for the Sky: The Mentality of the Space Age* (1992), which rejects one of the most fundamental assumptions of space mythology, taken as gospel by other cultural commentators such as Wyn Wachhorst, that humans are propelled by unknown and innate forces to explore space.[124]

New work has also focused on popular culture. While not strictly a cultural history, Howard E. McCurdy's *Space and the American Imagination* (1997) remains one of the most powerful studies on how popular conceptions of space exploration in American culture helped to shape national space policy.[125] The iconography of space exploration in the 1950s, McCurdy argued, tapped deeply into some of America's most entrenched cultural ideals such as the "limitless frontier," the "heroic explorer," the romance of aviation through Lindbergh and Earhart, and ultimately the utopian ideal of progress through technology.[126] Space enthusiasts and advocates such as Wernher von Braun used many of the same cultural representations in their lobbying but added the fear of the Soviet threat during the Cold War. By invoking the specter of world domination in the late 1950s and early 1960s, they were able to influence major policy decisions, including Kennedy's historic decision to go to the Moon in 1961.[127] Marina Benjamin's eloquent *Rocket Dreams: How the Space Age Shaped Our View and the Future of Technology* (2003) is the view from the other side, i.e., how the space program has affected popular culture. Her exploration of how popular culture has relegated the "space age" to a cultural

---

123. Dale Carter, *The Final Frontier: The Rise and Fall of the American Rocket State* (London: Verso, 1988).

124. David Lavery, *Late for the Sky: The Mentality of the Space Age* (Carbondale: Southern Illinois University Press, 1992); Wyn Wachhorst, *The Dream of Spaceflight: Essays on the Near Edge of Infinity* (New York: Basic Books, 2000).

125. Howard E. McCurdy, *Space and the American Imagination* (Washington, DC: Smithsonian Institution Press, 1997).

126. See also James A. Spiller, "Constructing America at the Peripheries: The Cultural Politics of United States Science and Exploration in Outer Space and Antarctica, 1950s–1990s" (Ph.D. diss., University of Wisconsin, 1999); Susan L. Mangus, "Conestoga Wagons to the Moon: The Frontier, the American Space Program, and National Identity" (Ph.D. diss., Ohio State University, 1999).

127. See also Mike Wright, "The Disney–Von Braun Collaboration and Its Influence on Space Exploration," in *Inner Space/Outer Space: Humanities, Technology and the Post-Modern World*, ed. Daniel Schenker, Craig Hanks, and Susan Kray (Huntsville, AL: Southern Humanities Press, 1993), pp. 151–160.

hinterland in the post-Apollo era is a powerful investigation into why the "space age" resonated in the first place to so many.[128]

Along with the works of McCurdy and Benjamin, De Witt Douglas Kilgore's *Astrofuturism* (2003) represents one of the three most important books on the cultural history of spaceflight to appear thus far.[129] Marshaling an impressive array of source material, Kilgore investigates the conflicting ideals embedded in America's vision of the future as represented in intellectual, scientific, artistic, and political discourse of the late 20th century. The power of Kilgore's work lies not only in his explication of how and why a whole progress-oriented and futuristic space discourse resonated with so many in American culture, but also why Americans have found certain values in knowledge, politics, and art so desirable. The work depicts the history of futures propagated, struggled over, and, in some cases, lost.[130]

These recent works point to six different areas within the cultural history of spaceflight fertile for future scholarship:

1) The role of memory, myth, and nostalgia in shaping current understanding of the history of spaceflight remains unexplored; deconstructing the Apollo myth in popular discourse—particularly its resale as cultural cachet via what Michael L. Smith has called "commodity scientism"—may deepen our understanding of why Apollo retains such a grip on the collective memory.[131]

2) Going beyond hagiographical treatments of astronauts, cultural historians should devote attention to the complex role astronauts play as part of the iconography of heroism in American culture; further exploring the groundwork laid by Tom Wolfe in his seminal *The Right Stuff* (1979) as well as focusing on astronauts in the post-Apollo era

128. Marina Benjamin, *Rocket Dreams: How the Space Age Shaped Our Vision of a World Beyond* (New York: Free Press, 2003). See also Paul Levinson, *Realspace: The Fate of Physical Presence in the Digital Age, On and Off the Planet* (London: Routledge, 2003), a similar meditation on the ways in which the digital age may have dampened humanity's urge to explore space.

129. Kilgore, *Astrofuturism*. See also his "Engineers' Dreams: Wernher von Braun, Willy Ley, and Astrofuturism in the 1950s," *Canadian Review of American Studies* 27, no. 2 (1997): 103–131.

130. See also Roger D. Launius, "Perceptions of Apollo: Myth, Nostalgia, Memory, or All of the Above?" *Space Policy* 21 (May 2005): 129–139; Roger D. Launius and Howard E. McCurdy, *Imagining Space: Achievements, Predictions, Possibilities, 1950–2050* (San Francisco: Chronicle Books, 2001); Bruce Horrigan, "Popular Culture and Visions of the Future in Space, 1901–2001," in *New Perspectives on Technology and American Culture*, ed. Bruce Sinclair (Philadelphia: American Philosophical Society, 1986), pp. 49–67.

131. Michael L. Smith, "Selling the Moon: The U.S. Manned Space Program and the Triumph of Commodity Scientism," in *The Culture of Consumption: Critical Essays in American History, 1880– 1980*, ed. Richard Wrightman Fox and T. J. Jackson Lears (New York: Pantheon Books, 1983), pp. 175–209.

would add significantly to understanding the shaping and evolution of the astronaut icon from hero and explorer in the 1960s to mechanic and experimenter in the 21st century.[132] Susan Faludi's *Stiffed: The Betrayal of the American Man* (1999), where she argues that the emasculation of the astronaut in the post-Apollo era in part contributed the "betrayal" of the "American Man," suggests that the fall of the astronaut icon was as salient as its rise, but the extant scholarship remains woefully incomplete.[133]

3) A cultural history of the Space Age would be incomplete without fully researched scholarship on the rituals that have shaped the lives of not only participants in the space program but also those who witnessed it as viewers;[134] similarly, we need to revisit the history of space travel through the lens of popular scientific culture.[135] An area ripe for investigation is the ways in which popular space culture shaped the lives of adolescents in the 1960s through science fiction, popular magazines, toys, models, and clubs.[136]

4) The recent graphic anthology *2001: Building For Space Travel* (2001) was an important step in connecting space culture with the history of the built environment on Earth, particularly architecture;[137] there still remains much to be done in terms of connecting the history of space exploration with the history of material culture—automobiles, toys, home appliances—to name only a few examples.

5) Essential for studying the history of space exploration is the role of particular ideologies—whether utopian, spiritual, millenarian, excep-

---

132. Wolfe, *The Right Stuff*.

133. Susan Faludi, *Stiffed: The Betrayal of the American Man* (New York: William Morrow & Co., 1999). See also Debra Benita Shaw, "Bodies Out of this World: The Space Suit as Cultural Icon," *Science as Culture* 13 (March 2004): 123–144.

134. For early explorations on this field, see several articles by Colin Fries in *Quest: The History of Spaceflight Quarterly*: "Space Age Legends: Urban Folk Tales Collected by the NASA Headquarters History Office" (vol. 8, no. 1 [2000]: 18–23), "Flying for Us: Space Age Milestones Celebrated in Music" (vol. 9, no. 3 [2002]: 30–36), "Sports Milestones in Space" (vol. 10, no. 2 [2003]: 37–40), and "Traditions of the Space Age" (vol. 11, no. 1 [2004]: 31–39).

135. For a notable exception, which primarily covers the media of TV and film, see Robert A. Jones, "They Came in Peace for all Mankind: Popular Culture as a Reflection of Public Attitudes to Space," *Space Policy* 20 (2004): 45–48.

136. For a brief look at the relationship between the proliferation of science fiction and the cause of spaceflight in the U.S., see the essay "Rockets to the Moon, 1919–1944: A Debate Between Reality and Fiction," in Paul A. Carter, *Politics, Religion, and Rockets: Essays in Twentieth Century American History* (Tucson: The University of Arizona Press, 1991), pp. 181–195.

137. John Zukowsky, ed., *2001: Building for Space Travel* (New York: Harry N. Abrams, 2001).

tionalist, modernist, humanist, atheistic, technological, environmental, or other—that motivated advocates, critics, and participants
(direct and vicarious) of spaceflight in the 20th century.[138]

6) A few have begun to revisit the history of space exploration through
the theoretical framework of feminist studies, some through a reading
of such sources as female-written "slasher" novels. Constance Penley's
NASA/TREK: Popular Science and Sex in America (1997) critically
tackles, among many topics, the role of sexuality in spaceflight culture and also discusses NASA's "inability to manage the meanings of
women in space";[139] additionally, Yaakov Jerome Garb's ecofeminist
approach to reevaluating the famous photograph of the whole Earth
from lunar distance focused not on the epiphany of (re)discovering
"one world" for all of humanity, but rather on how that iconic image
of the Earth helped to entrench a more negative view, one of the
dispassionate gaze of omniscient science as a masculine epistemology controlling all of nature, knowledge, and humanity.[140] Finally,
in *Cosmodolphins: Feminist Cultural Studies of Technology, Animals and
the Sacred* (2000), authors Mette Bryld and Nina Lykke used a critical
feminist approach to unpack the relationships between the Space Age,
the "New Age," and the ecological symbolism of nature (represented
through the icon of the dolphin). In taking a feminist approach to
rewriting the master narratives of spaceflight, they identified what
I believe is an important topic for future historians, the relationship
between national identity and the making of history. They write:

> The early space race was, amongst other things, a discur-
> sive battle over entitlement to represent Universal Man in
> the biggest story told in modern times. Who was going
> to be the script writer and the protagonist of the master
> narrative of mankind's cosmic exodus? This was and is a

---

138. For an unusual look at space culture through "posthuman theory," see Melanie A. R.
Brown, "Posthumanity's Manifest Destiny: NASA, Its Contradictory Image and Promises, and
Popular Culture" (Ph.D. diss., University of Central Florida, 2004).

139. Constance Penley, *NASA/TREK: Popular Science and Sex in America* (New York: Verso,
1997), p. 3.

140. Yaakov Jerome Garb, "The Use and Misuse of the Whole Earth Image," *Whole Earth Review*
no. 45 (March 1985): 18–25, and "Perspective or Escape? Ecofeminist Musings on Contemporary
Earth Imagery," in *Reweaving the World: The Emergence of Ecofeminism*, ed. Irene Diamond and
Gloria Feman Orenstein (San Francisco: Sierra Club Books, 1990), pp. 264–278. See also Jonathan
Bordo, "Ecological Peril, Modern Technology and the (Post)Modern Sublime," in *Shadow of Spirit:
Postmodernism and Religion*, eds. Phillipa Berry and Andrew Wernick (New York: Routledge, 1992),
pp. 165–178.

question that matters a great deal when the official story of spaceflight is retold [separately in the U.S. and Russia].[141]

Their conclusions hint at further opportunities for research on national claims for the history of space travel: which was more "important" in the history of space exploration, the first time a human left the planet Earth (Yuri Gagarin) or the first time a human set foot on another celestial body (Neil Armstrong)? Ask a Russian and then an American, and one would get different responses. In both cases, historians use extraordinary metaphors to imbue them with gravity, comparisons that typically center on the movement of Earthly life from the oceans to land. The parallel narratives are contradictory but exist simultaneously in multiple national discourses, buttressed by masculine notions of rationalism, exploration, and evolution. In some sense, space historians need to question how "thematic consensus" in space historiography was shaped by national identity.

CONCLUSIONS

The flavor of American space history has also been profoundly shaped by the location and sponsorship of its primary practitioners. In other words, American *space* history largely remains "court history." For the past 40 years, it has been predominantly sponsored, written, and issued as a result of funding from sources who direct and operate the space program, i.e., the U.S. government (through NASA, the Smithsonian's National Air and Space Museum, and the Department of Defense) or major corporations. Because there has been no vibrant nongovernmental or noncorporate space history community (in academia, public history positions, or elsewhere), American space history has been much more conservative than other historical subdisciplines. The field has typically had a romance with the power and progress inherent in technology; it eulogizes and deifies a few important men; and it eschews any position that would criticize celebratory, jingoistic, or militaristic elements of the space program. The works of those who have broken this mold despite their connections to official organizations—Launius, Logsdon, McCurdy, and Neufeld, for example—collectively represent an important and positive, albeit minority, trend in the field of space history.[142]

---

141. Mette Marle Bryld and Nina Lykke, *Cosmodolphins: Feminist Cultural Studies of Technology, Animals and the Sacred* (London: Zed Books, 2000).

142. Michael J. Neufeld's work, particularly his seminal *The Rocket and the Reich: Peenemünde and the Coming of the Ballistic Missile Era* (Cambridge, MA: Harvard University Press, 1995), revolutionized the history of the wartime German ballistic missile program by providing a balanced treatment of the development of the V-2 that did not gloss over the terrible human costs of its development.

The mainstream academic community has devoted very little attention to the space program, partly because academics tend to be narrowly focused on topics such as race, ethnicity, and gender. Typically, academics have had a condescending attitude towards fields such as the history of technology or space history, partly because they see in these fields little of interest to such contemporary conceptual lenses such as poststructuralism; postcolonial studies; feminist studies; and issues revolving around gender, ethnicity, power, transnationalism, and sexuality. Academics have often refused to see the complexities of the space program, relying instead on unidimensional, weak, and often lazy interpretations of the space program as a bankrupt and militaristic element of American society.

The publication of syntheses can say much about a particular discipline. On the one hand, in a field that is very young, one might expect most works to be somewhat of a synthesis given the paucity of subject matter. On the other hand, maturity and longevity of a discipline and its attendant accumulation of source material might also engender the writing of syntheses. Since the beginnings of the field of space history, journalists and historians have tackled the problem of the synthesis with various degrees of success. Von Braun and Ordway's *History of Rocketry and Space Travel* (1966) was an early attempt that emphasized some of the key motifs of Cold War historiography such as exploration, competition, and the social welfare of all humankind. The work focused on great figures, civilian space exploration, and the potential benefits of the project.[143] More comprehensive works appeared in the 1980s and 1990s that benefited from post–Cold War revelations. T. A. Heppenheimer's *Countdown: A History of Spaceflight* (1997) traced the evolution of rocketry from pioneering theoreticians in the late 19th century to the mid-1990s. Heppenheimer's marshaling of information is masterful, and his use of inspiring language complements his view that Apollo was "a drive toward a new human future."[144] Tom Crouch's *Aiming for the Stars: The Dreamers and Doers of the Space Age* (1999) is an eloquent exegesis on innovators in the 20th century who tried to translate their visions of space exploration—both successfully and unsuccessfully—into reality.[145] Although focused on great men and great technology, Heppenheimer's and Crouch's works remain the most successful syntheses in the traditional style of space history.[146]

---

143. Von Braun and Ordway, *History of Rocketry and Space Travel*. The monograph was published in several updated versions up to 1985.

144. T. A. Heppenheimer, *Countdown: A History of Spaceflight* (New York: John Wiley & Sons, 1997), p. 2.

145. Tom D. Crouch, *Aiming for the Stars: The Dreamers and Doers of the Space Age* (Washington, DC: Smithsonian Institution Press, 1999).

146. For other syntheses, see Andrew Wilson, *The Eagle Has Wings: The Story of American Space Exploration, 1945–1975* (London: British Interplanetary Society, 1982); David Baker, *The History*

continued on the next page

Other recent syntheses remain flawed by their dated interpretations. William E. Burrows, in his *This New Ocean: The Story of the First Space Age* (1998), used an array of recently declassified material from both the United States and former Soviet Union to produce an otherwise eloquent narrative of the entire Space Age.[147] Burrows's work, however, derives solidly from the Cold War framework of space exploration as a battle of noble proportions against a morally untrustable adversary. In demonizing communism as "more insidious" than Nazism, he describes the former as a "cancer, a disease that surreptitiously rode the bloodstream of the world, attacking and devouring every healthy organism in its path and growing bigger and more dangerous as it did so."[148] By dismissing all of Soviet society as cancerous yet eulogizing such men as Sergei Korolev, such works inevitably end up in contradictions since we are left with no insight into how the former managed to produce the likes of the latter. Similarly, Mike Gruntman, in *Blazing the Trail: The Early History of Spacecraft and Rocketry* (2004), provides a well-researched and comprehensive tale of the history of rocketry and spaceflight, with lucid explanations of technologies, but does Burrows one better by repeatedly denigrating not only the Russians but also American and Western liberals who questioned the American space program.[149]

With the rise of the new history, two threads of historiography now exist. One remains celebratory and internalist and the other questioning and externalist. Although there has been spillover from the former to the latter, the reverse, as evident in the works of Burrows and Gruntman, has been less common. It is clear, though, that both traditions have very important contributions to make. The old internalist history, focused on important men and singular artifacts, provided the backbone of our conception of the history of the space program. The new externalist history contributes the rationale, explication—and the critiques—that make the old history meaningful. Despite the large canon of space history, those who have written syntheses have not man-

---

continued from the previous page

of *Manned Spaceflight*, 2nd ed. (New York: Crown Publishers, 1985); Michael Collins, *Liftoff: The Story of America's Adventure in Space* (New York: Grove Press, 1988); H. P. Arnold, ed., *Man in Space: An Illustrated History of Spaceflight* (New York: Smithmark, 1993); Roger D. Launius, *NASA: A History of the U.S. Civil Space Program* (Malabar, FL: Krieger Publishing Company, 1994); Helen Gavaghan, *Something New Under the Sun: Satellites and the Beginning of the Space Age* (New York: Copernicus, 1998).

147. William E. Burrows, *This New Ocean: The Story of the First Space Age* (New York: Random House, 1998).

148. Ibid., p. 148.

149. In describing the development of the Woomera missile test range in Australia in the 1960s, for example, Gruntman notes that "pacifists and communists tried to interfere with the construction, as their counterparts invariably did with defense initiatives in other countries of the free world, thus serving willingly or unwittingly as a Soviet fifth column" (Mike Gruntman, *Blazing the Trail: The Early History of Spacecraft and Rocketry* [Reston, VA: AIAA, 2004], p. 425).

aged to combine the two in any coherent fashion. One way to engender such a union would be for historians of spaceflight to engage much more actively with the mainstream American history community.[150] Unlike the literature on American history, the writing on American space history is very young, but by engaging with a bigger audience—not only the broader public but also the academic history community—we might benefit from a rich vista of viewpoints that would move us forward from a fledgling subdiscipline to one that is vibrant, mature, and complex. And with maturity, we might yet see a powerful work that brings together the dictates of policy, the forces of society, and the nuances of culture into a grand narrative that chronicles the romance and the reality of this country's efforts to explore space.

---

150. It is of some importance that in the "list of upcoming meetings" section of the past four issues of *News & Notes*—the regular newsletter issued to the aerospace history community by the NASA History Office—one would find announcements for the many meetings of professional aerospace organizations but none for the annual meetings of the American Historical Association (AHA) or the Organization of American Historians (OAH). See the last four newsletters: NASA History Office, *News & Notes* 21, nos. 1–4 (2004).

CHAPTER 15

THE HISTORY AND HISTORIOGRAPHY OF NATIONAL SECURITY SPACE[1]

Stephen B. Johnson

The intent of this essay is to provide space historians with an overview of the issues and sources of national security space so as to identify those areas that have been underserved. Frequently, ballistic missiles are left out of space history, as they only pass through space instead of remaining in space like satellites. I include ballistic missiles for several reasons, not the least of which is that they pass through space en route to their targets.

Space programs originated in the national security (NS) arena, and except for a roughly 15-year period from the early 1960s through the mid-1970s, NS space expenditures in the United States (U.S.), let alone the Union of Soviet Socialist Republics (USSR), have equaled or exceeded those of civilian programs. Despite this reality, the public nature of government-dominated civilian programs and issues of security classifications have kept NS space out of the limelight. The recent declassification of the early history of the National Reconnaissance Office (NRO) and the demise of the Soviet Union have led to a recent spate of publications that have uncovered much of the "secret history" of the early Cold War. Nonetheless, much of NS space history has received little attention from historians.

One feature of military organizations that is of great value for historians is their penchant to document their histories, and space organizations are no exception. Most military organizations have historians assigned to them, with professional historians at many of the positions documenting events as they occur.

Unfortunately, this very positive feature is countered by the requirements of secrecy and classification (and, in the case of the Naval Research Laboratory, the loss of its archives by fire). It is unfortunately true that much of this treasure trove of documentation created by historians within space organizations will remain classified for years to come. Some of the earlier

---

1. Many thanks to David Arnold, Donald Baucom, Matt Bille, Dwayne Day, Steve Dick, R. Cargill Hall, and Rick Sturdevant, all of whom provided many useful comments and provided me with many more sources than I would ever have been able to find on my own.

material is being declassified now or could be declassified if someone would request it and if sufficient priority were assigned to the task. This is a field where outsiders can be of great service.

To exploit the mass of documents that exist requires that historians have a basic grasp of the subject, what has been published to date, and what is yet to be done. This article aims to perform these functions by surveying the various military space programs and issues, giving a very brief sketch of their histories, and identifying the main sources that historians have created and used.

OVERVIEW SOURCES

While there is no single comprehensive overview history of NS space, several works cover a variety of areas. Walter McDougall's Pulitzer Prize–winning . . . *The Heavens and the Earth*, written in 1985, thoroughly discussed the NS aspects of the space race; it is getting dated but remains useful for an introduction to the politics of the 1950s and 1960s.[2] William Burrows's *This New Ocean* integrates NS space issues nicely into his acclaimed overview space history.[3] Mike Gruntman's *Blazing the Trail* is an overview history of space technology, accounting for military contributions.[4] So, too, does Asif Siddiqi's authoritative *Challenge to Apollo* for the Soviet program up to the mid-1970s, which also has a fine essay on Soviet space history sources.[5] Peter Hays[6] and Dwayne Day[7] provide overviews of military and intelligence space, respectively, in Eligar Sadeh's *Space Politics and Policy*.

An earlier, short review of the state of national security space research is provided by Day's 1997 article, which focuses on issues as opposed to a bibliographic treatment.[8] Day provided an overview of U.S. military space

---

2. Walter McDougall, . . . *The Heavens and the Earth: A Political History of the Space Age* (New York: Basic Books, 1985).

3. William E. Burrows, *This New Ocean: The Story of the First Space Age* (New York: The Modern Library, 1998).

4. Mike Gruntman, *Blazing the Trail: The Early History of Spacecraft and Rocketry* (Reston, VA: American Institute of Aeronautics and Astronautics, 2004).

5. Asif A. Siddiqi, *Challenge to Apollo: The Soviet Union and the Space Race, 1945–1974* (Washington, DC: NASA SP-2000-4408, 2000).

6. Peter L. Hays, "Space and the Military," in *Space Politics and Policy, an Evolutionary Perspective*, ed. Eligar Sadeh (Dordrecht, Netherlands: Kluwer Academic Publishers, 2002), pp. 335–370.

7. Dwayne A. Day, "Intelligence Space Program," in *Space Politics and Policy, an Evolutionary Perspective*, ed. Sadeh, pp. 371–388.

8. Dwayne A. Day, "The State of Historical Research on Military Space," *Journal of the British Interplanetary Society* 50 (1997): 203–206. See also Roger D. Launius, "The Military in Space: Policy-Making and Operations in a New Environment," in *A Guide to the Sources of United States Military History: Supplement IV*, ed. Robin Higham and Donald J. Mrozek (North Haven, CT: Archon Books, 1998), pp. 488–522.

operations from 1987 to 1995 in *Journal of the British Interplanetary Society* in December 1993, as well as an updated and extended version of the article in *Countdown*.[9] Cargill Hall and Jacob Neufeld wrote an early work that gives a flavor of USAF activities.[10] David Spires's overview history of the USAF in space is the best single place to start for the USAF portion of NS space history.[11] Curtis Peebles's *High Frontier* is a much shorter introduction to USAF space history.[12] USAF Space Command recently published a two-volume set of basic documents that are of great value to military space historians.[13]

Steven Zaloga's *The Kremlin's Nuclear Sword* is the best overview of Soviet control of and defense against nuclear forces.[14] Nicholas Daniloff's 1972 *The Kremlin and the Cosmos* is an early but important source on the Soviet program,[15] as is Christian Lardier's *L'Astronautique Soviétique*,[16] which is excellent for the technical aspects of Soviet space systems. Gerald Borrowman wrote a short overview of Soviet military space activities in 1982.[17] Nicholas Johnson created yearly assessments of the Soviet space program, some of which are summarized in *Soviet Space Programs, 1980–1985*.[18] His 1987 *Soviet Military Strategy in Space* was also a major work at the time.[19] Finally, Johnson's books *Europe and Asia in Space: 1993–1994* and *Europe and Asia in Space: 1991–1992* are outstanding sources for those two regions.[20]

9. Dwayne A. Day, "A Review of Recent American Space Operations," *Journal of the British Interplanetary Society* 46, no. 12 (1993): 459–470; Dwayne A. Day, "Capturing the High Ground: The U.S. Military in Space, 1987–1995, Part 1," *Countdown* 13, no. 1 (1995): 30–45; Dwayne A. Day, "Capturing the High Ground: The U.S. Military in Space, 1987–1995, Part 2," *Countdown* 13, no. 3 (1995): 17–31.

10. R. Cargill Hall and Jacob Neufeld, *The U.S. Air Force in Space: 1945 to the 21st Century* (Washington, DC: USAF History and Museums Program, 1998).

11. David N. Spires, *Beyond Horizons: A Half Century of Air Force Space Leadership* (Peterson AFB, CO: Air Force Space Command, 1997).

12. Curtis Peebles, *High Frontier: The United States Air Force and the Military Space Program* (Washington, DC: Air Force History and Museums Program, 1997).

13. David N. Spires, *Orbital Futures: Selected Documents in Air Force Space History*, vol. 1 (Peterson AFB, CO: Air Force Space Command, 2004); David N. Spires, *Orbital Futures: Selected Documents in Air Force Space History*, vol. 2 (Peterson AFB, CO: Air Force Space Command, 2004).

14. Steven J. Zaloga, *The Kremlin's Nuclear Sword: The Rise and Fall of Russia's Strategic Nuclear Forces, 1945–2000* (Washington, DC: Smithsonian Institution Press, 2002).

15. Nicholas Daniloff, *The Kremlin and the Cosmos* (New York: Alfred A. Knopf, 1972).

16. Christian Lardier, *L'Astronautique Soviétique* (Paris: Armand Colin, 1992).

17. Gerald L. Borrowman, "Soviet Military Activities in Space," *Journal of the British Interplanetary Society* 35, no. 2 (1982): 86–92.

18. Nicholas L. Johnson, *Soviet Space Programs, 1980–1985* (San Diego: Univelt Press, 1987).

19. Nicholas L. Johnson, *Soviet Military Strategy in Space* (Coulsdon, U.K.: Jane's Information Group, 1987).

20. Nicholas L. Johnson, *Europe and Asia in Space: 1993–1994* (Kirtland AFB, NM: USAF Phillips Laboratory, 1995; Colorado Springs, CO: Kaman Sciences Corporation, 1995); Nicholas L. Johnson, *Europe and Asia in Space: 1991–1992* (Kirtland AFB, NM: USAF Phillips Laboratory, 1993; Colorado Springs, CO: Kaman Sciences Corporation, 1993).

Some encyclopedic sources are useful. The latest *Cambridge Encyclopedia of Space* has significant information about military space, particularly in providing summaries of all programs and launches up to 2000.[21] Shirley Thomas's eight-volume *Men of Space* from the 1960s remains a useful source.[22] The forthcoming space history encyclopedia *Space Exploration and Humanity* will have a major section on NS space history.[23]

Samuel Miller's *An Aerospace Bibliography* is a good starting point to search for space history articles prior to 1978,[24] as is John Looney's 1979 bibliography for NASA.[25] So, too, is the Smithsonian bibliography edited by Dominic Pisano and Cathleen Lewis, *Air and Space History: An Annotated Bibliography*, which takes researchers up to 1988.[26] Jeffrey Richelson edited *Military Uses of Space, 1946–1991*, a useful bibliographic source.[27]

With the explosion of the World Wide Web in the 1990s, no discussion of sources can avoid online sources. An excellent online source for aerospace history, including defense space matters, is the government site for the U.S. Centennial of Flight Commission. This contains a plethora of short essays on a variety of aerospace history topics.[28] The NASA History Division also has an excellent site with many online publications, including many that involve NASA-DOD relations. The Air War College Gateway is another excellent resource of past and current military space activities.[29] Other credible sites include those for USAF Space Command, the National Security Archives of George Washington University, and the Federation of American Scientists. Several declassified USAF works are now online.[30] Mark Wade's online

21. Fernand Verger, Isabelle Sourbès-Verger, and Raymond Ghirardi, with contributions by Xavier Pasco, *The Cambridge Encyclopedia of Space* (Cambridge: Cambridge University Press, 2003).

22. Shirley Thomas, *Men of Space: Profiles of the Leaders in Space Research, Development, and Exploration*, 8 vols. (Philadelphia: Chilton Company, 1960–68).

23. Stephen B. Johnson et al., eds., *Space Exploration and Humanity: A Historical Encyclopedia* (Santa Barbara, CA: ABC-CLIO, forthcoming, expected publication 2006–07).

24. Samuel Duncan Miller, *An Aerospace Bibliography* (Washington, DC: Office of Air Force History, USAF, 1986).

25. John J. Looney, *Bibliography of Space Books and Articles from Non-Aerospace Journals, 1957–1977* (Washington, DC: NASA History Office, 1979).

26. Dominick A. Pisano and Cathleen S. Lewis, eds., *Air and Space History: An Annotated Bibliography* (New York: Garland Publishing, 1988).

27. Jeffrey Richelson, ed., *U.S. Military Uses of Space, 1945–1991: Index and Guide* (Washington, DC: The National Security Archive; Alexandria, VA: Chadwyck-Healey, Inc., 1991).

28. United States government, Centennial of Flight Web site, *http://www.centennialofflight.gov*.

29. Air War College Gateway to Space Operations and Resources, *http://www.au.af.mil/au/awc/awcgate/awc-spc.htm*.

30. Mark C. Cleary, *The 6555th: Missile and Space Launches through 1970* (Patrick AFB, FL: 45th Space Wing, 1991); Mark C. Cleary, *The Cape: Military Space Operations, 1971–1992* (Patrick AFB, FL: 45th Space Wing, 1994); Harry Waldron, *Historical Overview of the Space and Missile Systems Center, 1954–2003* (Los Angeles AFB, CA: Space and Missile Systems Center, 2003).

Encyclopedia Astronautica has become a popular Internet source for space history. Unfortunately, while it contains a great deal of information, not all of it is correct. Space historians have noticed a variety of factual problems, and unfortunately these problems have not been consistently repaired. Since this is not a peer-reviewed source and historical errors have not always been fixed, this cannot be considered a reliable source, despite its impressive appearance. Many other online sources have the same problems.[31]

Since reactions to the launch of Sputnik encompassed a variety of areas and actions, it is appropriate to mention a few key sources about that event and its ramifications here. The best recent overview is Roger Launius, John Logsdon, and Robert Smith's *Reconsidering Sputnik*.[32] Important earlier works on the topic include those by Robert Divine[33] and Rip Bulkeley.[34]

BALLISTIC MISSILES AND MILITARY SPACE LAUNCHERS

Ballistic missiles originated from the rocketry experiments of amateurs in the 1920s and 1930s, which then gained the interest of military organizations, particularly in Germany, the Soviet Union, and the United States. These stories have been described in a variety of books and articles through the years, as they account for the origins of space programs around the world.

The story of the German V-2 project is perhaps the best known, both because it led to the world's first operational ballistic missile and because of its leader, Wernher von Braun, who became famous in the United States after World War II. American forces captured most of von Braun's team at the end of World War II, along with parts and plans to rebuild the Nazi program on American soil. Most of the team came to the United States, where they assisted American contractors and the U.S. military to develop their own ballistic missile capabilities. The United States already had its own rocketry programs, with the Navy working with physicist Robert Goddard and members of the American Rocket Society, and the Army funding the Jet Propulsion Laboratory. Missile efforts proliferated after the war but did not gain priority until the early 1950s. Only then did the Air Force's Atlas ICBM project, soon followed by the Thor, Titan, and other ballistic missile programs, push forward at a rapid pace. These liquid-propellant rockets were soon displaced as weapons by solid-propellant

---

31. *Encyclopedia Astronautica* is available online at *http://www.astronautix.com/*.

32. Roger D. Launius, John M. Logsdon, and Robert W. Smith, eds., *Reconsidering Sputnik: Forty Years Since the Soviet Satellite* (Amsterdam: Harwood Academic Publishers, 2000).

33. Robert A. Divine, *The Sputnik Challenge: Eisenhower's Response to the Soviet Satellite* (Oxford: Oxford University Press, 1993).

34. Rip Bulkeley, *The Sputniks Crisis and Early United States Space Policy* (Bloomington: Indiana University Press, 1991).

ballistic missiles such as Minuteman and Polaris, which were much more useful militarily because they did not require a time-consuming and dangerous
liquid fueling process. Once the Cold War ended, ballistic missile forces in the
United States shrank rapidly along with the Soviet threat. Other nations each
developed their own nuclear and ballistic missile programs.

Ballistic missiles were the technical progenitors of the first-generation
space launchers. The Atlas, Titan, and Thor missiles led to the Atlas, Titan,
and Delta families of launchers, while the R7 became the Soyuz launcher.
Similarly, early Chinese ballistic missile programs derived from the Nazi V-2
through the Soviet R1 and R2 programs evolved into the Long March series
used for military and civilian launches.

Finally, the military also developed hypersonic technologies from the
1950s to the present, some of which evolved into craft capable of going into
space. The X-series aircraft went faster and higher, culminating in the X-
15 and X-20 Dyna-Soar programs of the early 1960s. Later efforts included
the X-24, involvement with the Space Shuttle program, and the National
Aerospace Plane, and they continue today with a variety of studies and tests.

The early history of ballistic missile programs in Germany, the United
States, and the Soviet Union is well documented. Nazi efforts on the V-2 program are the subject of many books with a variety of perspectives. The single
best volume on the V-2 development program is Michael Neufeld's *The Rocket
and the Reich*,[35] thoroughly researched from the German-language original
documents. Overview space histories, such as Burrows's *This New Ocean* and
Heppenheimer's *Countdown*, also provide good descriptions of the V-2 project, as well as both Soviet and American ballistic missile programs through
the 1950s.[36] Older histories stemmed mainly from von Braun supporters, such
as Frederick Ordway's *The Rocket Team* and Walter Dornberger's *V-2: The
Nazi Rocket Weapon*.[37] Less well known is the actual V-2 rocket campaign
against Britain and British countermeasures, well documented in King and
Kutta's *Impact: The History of Germany's V-Weapons in World War II*.[38] R. V.
Jones's *The Wizard War* gives an earlier description of British espionage efforts
in World War II, including against the V-2 offensive.[39] Revisionist histories

---

35. Michael J. Neufeld, *The Rocket and the Reich: Peenemünde and the Coming of the Ballistic Missile
Era* (New York: The Free Press, 1995).

36. Burrows, *This New Ocean*; T. A. Heppenheimer, *Countdown: A History of Space Flight* (New
York: John Wiley & Sons, 1997).

37. Walter Dornberger, *V-2: The Nazi Rocket Weapon,* trans. James Cleugh and Geoffrey Halliday
(New York: Viking, 1954); Frederick I. Ordway III and M. Sharpe, *The Rocket Team: From the V-2
to the Saturn Moon Rocket* (New York: Thomas Y. Crowell, 1979).

38. Benjamin King and Timothy Kutta, *Impact: The History of Germany's V-Weapons in World War
II* (Rockville Centre, NY: Sarpedon, 1998).

39. R. V. Jones, *The Wizard War: British Scientific Intelligence, 1939–1945* (New York: Coward,
McCann, and Geoghegan, 1978).

looking skeptically at von Braun and at the use of slave labor in World War II began to appear in the late 1990s. The two best of these sources are Andre Sellier's *A History of the Dora Camp* and Jean Michel's *Dora*.[40] Others include Yves Beon's *Planet Dora* and Dennis Piszkiewicz's *Wernher von Braun: The Man Who Sold the Moon*.[41] The journey of von Braun's team to the United States and other nations is the subject of a variety of literature, including works by Huzel, Lasby, Bower, Freeman, and Vilain.[42]

Early overviews of rocketry, which unavoidably discuss military involvement, include Zim's *Rockets and Jets*; Vaeth's *200 Miles Up*; Caidin's *Rockets and Missiles*; Emme's edited *History of Rocket Technology*; Baker's *The Rocket*; von Braun, Ordway, and Dooling's *History of Rocketry and Space Travel*; Winter's *Rockets into Space*; and Alway's *Rockets of the World*.[43]

The origins of American rocket and ballistic missile programs are well documented. The best overview of the early USAF missile programs remains Jacob Neufeld's internal Air Force history, *Ballistic Missiles in the United States Air Force, 1945–1960*. Older works also discuss the early ballistic missile programs, such as Schwiebert's *A History of the U.S. Air Force Ballistic Missiles*, Bergaust's *Rockets of the Armed Forces*, Neal's popular work on Minuteman, Chapman's early history of Atlas, Rosen's narrative of the Navy's Viking, Green and Lomask's history of Vanguard, and Hartt's story of the Thor missile. Thor and Atlas are described by Wambolt. Martin's series on Atlas is informative. A more recent work is Stine's 1991 *ICBM*. Greene's early internal history of Titan is still valuable. The most detailed recent historical study of a single program is Stumpf's *Titan II*. Titan's evolution is also described by

40. Andre Sellier, *A History of the Dora Camp* (Chicago: Ivan R. Dee, 2003); Jean Michel, *Dora* (New York: Holt, Rinehart, and Winston, 1980).

41. Yves Béon, *Planet Dora: A Memoir of the Holocaust and the Birth of the Space Age* (Boulder, CO: Westview Press, 1998); Dennis Piszkiewicz, *Wernher von Braun: The Man Who Sold the Moon* (Westport, CT: Praeger Publishers, 1998).

42. D. K. Huzel, *Peenemünde to Canaveral* (Englewood Cliffs, NJ: Prentice-Hall, 1962); Clarence G. Lasby, *Project Paperclip: German Scientists and the Cold War* (New York: Atheneum, 1971); Tom Bower, *The Paper Clip Conspiracy* (Boston: Little, Brown and Company, 1987); Marsha Freeman, *How We Got to the Moon: The Story of the German Space Pioneers* (Washington, DC: 21st Century Associates, 1993); J. Vilain, "France and the Peenemunde Legacy," in *History of Rocketry and Astronautics*, ed. P. Jung, American Astronautical Society History Series, vol. 21 (San Diego: Univelt Press, 1997), pp. 119–161.

43. Herbert H. Zim, *Rockets and Jets* (New York: Harcourt Brace & Company, 1945); J. Gordon Vaeth, *200 Miles Up: The Conquest of the Upper Air* (New York: The Ronald Press Company, 1951); Martin Caidin, *Rockets and Missiles: Past and Future* (New York: The McBride Company, 1954); Eugene Emme, ed., *The History of Rocket Technology* (Detroit: Wayne State University Press, 1964); David Baker, *The Rocket: The History and Development of Rocket and Missile Technology* (New York: Crown Books, 1978); Wernher von Braun, Frederick I. Ordway III, and Dave Dooling, *History of Rocketry and Space Travel* (New York: Thomas Y. Crowell, 1986); Frank H. Winter, *Rockets into Space* (Cambridge, MA: Harvard University Press, 1990); Peter Alway, *Rockets of the World* (Ann Arbor, MI: Saturn Press, 1992).

Falconer, as well as Richards and Powell. Reed's dissertation is an outstanding study of Minuteman. The Navaho, although a cruise missile, was crucial for rocket engine technology and is analyzed by Gibson. Two early works focused on ballistic missile operations are by Hunter, and Baar and Howard. Powell describes Blue Scout, a military research vehicle, Project Farside, an early USAF balloon rocket program, and the obscure Draco launcher. The Association of Air Force Missileers publishes a newsletter and has a Web site that frequently contains missile stories and historical information.[44]

Older political studies started analytical assessments of ballistic missiles and remain useful, such as the works of Armacost, Beard, and Sapolsky[45] on the 1950s American intermediate-range ballistic missile (IRBM), ICBM, and submarine-launched ballistic missile programs. Reed's dissertation on the politics of Minuteman is valuable.[46] Lonnquest and Winkler coauthored *To Defend*

---

44. Jacob Neufeld, *Ballistic Missiles in the United States Air Force, 1945–1960* (Washington, DC: Office of Air Force History, USAF, 1990); Ernest G. Schwiebert, *A History of the U.S. Air Force Ballistic Missiles* (New York: Frederick A. Praeger, 1964); Erik Bergaust, *Rockets of the Armed Forces* (New York: Putnam, 1966); Roy Neal, *Ace in the Hole: The Story of the Minuteman Missile* (Garden City, NY: Doubleday, 1962); John L. Chapman, *Atlas: The Story of a Missile* (New York: Harper & Brothers, 1960); Milton Rosen, *The Viking Rocket Story* (New York: Harper & Brothers, 1955); Constance McLaughlin Green and Milton Lomask, *Vanguard: A History* (Washington, DC: NASA SP-4202, 1970); Julian Hartt, *The Mighty Thor* (New York: Duell, Sloan, and Pearce, 1961); Joseph F. Wambolt, "Medium Launch Vehicles for Satellite Delivery," *Crosslink* 4, no. 1 (winter 2002/2003): 26–31; Richard E. Martin, "A Brief History of the Atlas Rocket Vehicle, Part 1," *Quest: The History of Spaceflight Quarterly* 8, no. 2 (2000): 54–61; Richard E. Martin, "A Brief History of the Atlas Rocket Vehicle, Part 2," *Quest: The History of Spaceflight Quarterly* 8, no. 3 (2000): 40–45; Richard E. Martin, "A Brief History of the Atlas Rocket Vehicle, Part 3," *Quest: The History of Spaceflight Quarterly* 8, no. 4 (2000): 46–51; G. Harry Stine, *ICBM: The Making of the Weapon that Changed the World* (New York: Orion Books, 1991); W. E. Greene, *The Development of the SM-68 Titan*, AFSC Historical Publications Series 62-23-1 (Wright-Patterson AFB, OH: Air Force Systems Command, 1962); David K. Stumpf, *Titan II: A History of a Cold War Missile Program* (Fayetteville: The University of Arkansas Press, 2000); Art Falconer, "Epic Proportions: The Titan Launch Vehicle," *Crosslink* 4, no. 1 (winter 2002/2003): 32–37; G. R. Richards and J. W. Powell, "Titan 3 and Titan 4 Space Launch Vehicles," *Journal of the British Interplanetary Society* 46, no. 4 (1993): 123–144; George A. Reed, "U.S. Defense Policy, U.S. Air Force Doctrine and Strategic Nuclear Weapon Systems, 1958–1964: The Case of the Minuteman ICBM" (Ph.D. diss., Duke University, 1986); James N. Gibson, *The Navaho Missile Project: The Story of the "Know-How" Missile of American Rocketry* (Atglen, PA: Schiffer Military/Aviation History, 1996); Mel Hunter, *The Missilemen* (Garden City, NY: Doubleday, 1960); James J. Baar and William E. Howard, *Combat Missilemen* (New York: Harcourt, 1961); Joel Powell, "Blue Scout—Military Research Rocket," *Journal of the British Interplanetary Society* 35, no. 1 (1982): 22–30; Joel W. Powell, "Project Farside, America's First Space Venture," *Journal of the British Interplanetary Society* 35, no. 10 (1982): 462–466; Joel W. Powell, "The Curious Case of Draco and the 'Secret' Cape Canaveral Launches of 1959," *Quest: The History of Spaceflight Quarterly* 6, no. 1 (1998): 44–46.

45. Michael H. Armacost, *The Politics of Weapons Innovation: The Thor-Jupiter Controversy* (New York: Columbia University Press, 1969); Edmund Beard, *Developing the ICBM: A Study in Bureaucratic Politics* (New York: Columbia University Press, 1976); Harvey M. Sapolsky, *The Polaris System Development: Bureaucratic and Programmatic Success in Government* (Cambridge, MA: Harvard University Press, 1972).

46. Reed, "U.S. Defense Policy."

and Deter,[47] which provides technical details and overviews of all major U.S. programs. Lonnquest's dissertation was a focused study on General Bernard Schriever's role in Atlas.[48] Koppes's history of the Jet Propulsion Laboratory (JPL) remains a good introduction to its Army-funded rocketry and ballistic missile programs.[49] Spinardi provides an overview of the U.S. Navy's submarine-based ballistic missile programs,[50] as does Fuhrman.[51] Friedman's *The Evolution of Nuclear Strategy* remains a valuable work about nuclear warfare in general,[52] as is Kaplan's *The Wizards of Armageddon.*[53] There are no major publications on recent U.S. ballistic missile history beyond 1970, although there are many political science and politically motivated studies of arms control and disarmament.

Soviet ballistic missile history has gotten a major boost since the end of the Cold War. The foremost work is currently Zaloga's outstanding study, *The Kremlin's Nuclear Sword,*[54] which provides an overview of Soviet nuclear forces from 1945 to 2000. Zaloga's earlier study *Target America* also remains useful.[55] Also useful is Podvig's *Russian Strategic Nuclear Forces.*[56] Siddiqi's *Challenge to Apollo*, originally published by NASA and now published commercially, covers in depth the early ballistic missile development of Korolev's design bureau from the R1 to the R7.[57] Siddiqi also covers the development and deployment of a Soviet Fractional Orbiting Bombardment System (FOBS).[58] The Yangel design bureau was selected to build the R–36–O FOBS over competing proposals by the Korolev and Chelomey design bureaus. This system, which deployed 18 missiles from 1971 to 1983, placed a nuclear warhead in

---

47. John C. Lonnquest and David F. Winkler, *To Defend and Deter: The Legacy of the United States Cold War Missile Program*, Special Report 97/01 (Champaign, IL: U.S. Army Construction Engineering Research Laboratories, 1996).

48. John Lonnquest, "The Face of Atlas: General Bernard Schriever and the Development of the Atlas Intercontinental Ballistic Missile, 1953–1960" (Ph.D. diss., Duke University, 1996).

49. Clayton R. Koppes, *JPL and the American Space Program: A History of the Jet Propulsion Laboratory* (New Haven, CT: Yale University Press, 1982).

50. Graham Spinardi, *From Polaris to Trident: The Development of U.S. Fleet Ballistic Missile Technology* (New York: Cambridge University Press, 1994).

51. R. A. Fuhrman, "The Fleet Ballistic Missile System: Polaris to Trident," *Journal of Spacecraft* 15, no. 5 (1978): 265–286.

52. Lawrence Friedman, *The Evolution of Nuclear Strategy* (New York: St. Martin's Press, 1983).

53. Fred Kaplan, *The Wizards of Armageddon* (Stanford: Stanford University Press, 1983).

54. Steven Zaloga, *The Kremlin's Nuclear Sword: The Rise and Fall of Russia's Strategic Nuclear Forces, 1945–2000* (Washington, DC: Smithsonian, 2002).

55. Steven J. Zaloga, *Target America: The Soviet Union and the Strategic Arms Race, 1945–1994* (Novato, CA: Presidio, 1993).

56. P. Podvig, *Russian Strategic Nuclear Forces* (Cambridge, MA: MIT Press, 2001).

57. Asif A. Siddiqi, *Sputnik and the Soviet Space Challenge* (Gainesville: University of Florida Press, 2003); Asif A. Siddiqi, *The Soviet Space Race with Apollo* (Gainesville: University of Florida Press, 2003).

58. Asif A. Siddiqi, "The Soviet Fractional Orbiting Bombardment System (FOBS): A Short Technical History," *Quest: The History of Spaceflight Quarterly* 7, no. 4 (1999): 22–33.

temporary orbit, going over the South Pole to evade American early-warning radars and then deorbiting quickly to hit the United States. Harford's *Korolev* also has a significant amount of information about the early ballistic missile programs.[59] Barry's Ph.D. dissertation, "The Missile Design Bureaux and Soviet Piloted Space Policy," describes some political aspects of early design bureaus.[60] Zak wrote a short piece on the origins of the Cosmos launcher.[61]

China's early ballistic missile program is tied to the story of Tsien Hsue-Shen, which is chronicled in Chang's *Thread of the Silkworm*.[62] Harvey's *The Chinese Space Programme* provides an overview of ballistic missile and launcher developments.[63] Lewis also describes the Chinese ballistic missile programs.[64]

Histories of other nations' ballistic missile programs and their transformation to launchers remain far less documented. The British program is the one major exception, with Morton's *Fire across the Desert*, Twigge's *The Early Development of Guided Weapons in the United Kingdom, 1940–1960*, Hill's *A Vertical Empire*, and Martin's *De Havilland Blue Streak*.[65] A recent article on early French missile and launcher efforts is by Huwart.[66]

The single best source for the history of U.S. space launchers is Launius and Jenkins's edited work, *To Reach the High Frontier*, which has articles on all major American launch programs.[67] This work also has an overview of the evolution of the Minuteman ICBM program by Hunley. Isakowitz is now up to the fourth edition of his *International Reference Guide to Space Launch Systems*; tracing the evolution of these editions provides historians with a thorough grounding in the technical aspects of the subject.[68] Hall provides an overview of the military ori-

---

59. James Harford, *Korolev: How One Man Masterminded the Soviet Drive to Beat America to the Moon* (New York: John Wiley & Sons, 1997).

60. William P. Barry, "The Missile Design Bureaux and Soviet Piloted Space Policy, 1953–1974" (Ph.D. diss., University of Oxford, 1995).

61. Anatoly Zak, "Cosmos Launcher: The Story of the Soviets' Space Workhorse," *Spaceflight* 38, no. 12 (1996): 416–418.

62. Iris Chang, *Thread of the Silkworm* (New York: Basic Books, 1995).

63. Brian Harvey, *The Chinese Space Programme: From Conception to Future Capabilities* (New York: John Wiley & Sons, 1998).

64. J. D. Lewis and H. Di, "China's Ballistic Missile Programs," *International Security* 17, no. 2 (1992): 5–40.

65. Peter Morton, *Fire across the Desert: Woomera and the Anglo-Australian Joint Project, 1946–1980* (Canberra: Australian Government Publishing Services, 1989); Stephen Robert Twigge, *The Early Development of Guided Weapons in the United Kingdom, 1940–1960* (Chur, Switzerland: Harwood Academic Publishers, 1993); C. N. Hill, *A Vertical Empire: The History of the UK Rocket and Space Programme, 1950–1971* (London: Imperial College, World Scientific, 2001); Charles H. Martin, *De Havilland Blue Streak: An Illustrated Story* (London: British Interplanetary Society, 2002).

66. Olivier Huwart, "Du V-2 à Veronique: Les Premières Recherches Spatiales Militaires Françaises," *Review Historiques des Armées* 3 (1997): 113–126.

67. Roger D. Launius and Dennis R. Jenkins, eds., *To Reach the High Frontier: A History of U.S. Launch Vehicles* (Lexington: The University Press of Kentucky, 2002).

68. Steven J. Isakowitz, Joshua B. Hopkins, and Joseph P. Hopkins, Jr., *International Reference Guide to Space Launch Systems*, 4th ed. (Reston, VA: AIAA, 2004).

gins of Agena in the CORONA program.[69] Siddiqi chronicles some of the conversions of Soviet ballistic missiles to launchers in *Challenge to Apollo*.[70] Harvey's *Russia in Space* gives a good overview of Russian launch systems.[71] Bille and Lishock describe early military launchers, including the obscure NOTSNIK, a designation combining the acronym for Naval Ordnance Test Station and Sputnik.[72] NOTSNIK received attention earlier from Pesavento and Powell.[73]

Military involvement with space transportation also includes the development of hypersonic and reusable systems. Overviews of hypersonics include Caidin's early *Wings into Space*, the two volumes of *The Hypersonic Revolution*, and Miller's *The X-Planes*.[74] The X-15 story dominates the early history of military reusable systems, and has garnered significant attention in the last two years. These include works by Jenkins, by Jenkins and Landis, Thompson, the reprint of Tregaskis, and Godwin.[75] *Quest* issue 3, number 1, has a number of articles on the X-15.

The Air Force's abortive Dyna-Soar program, later renamed the X-20, is discussed in Spires's *Beyond Horizons* and received historical attention in *Quest* issue 3, number 4, with articles by Houchin and Smith.[76] Houchin's work is

---

69. R. Cargill Hall, "The Air Force Agena: A Case Study in Early Spacecraft Technology," in *Technology and the Air Force: A Retrospective Assessment*, ed. Jacob Neufeld, George M. Watson, Jr., and David Chenoweth (Washington, DC: Air Force History and Museums Program, 1997), pp. 231–244.

70. Asif A. Siddiqi, *Challenge to Apollo*.

71. Brian Harvey, *Russia in Space: The Failed Frontier?* (Chichester, U.K.: Praxis Publishing, 2001).

72. Matt Bille and Erika Lishock, *The First Space Race: Launching the World's First Satellites* (College Station: Texas A&M Press, 2004).

73. Peter Pesavento, "US Navy's Untold Story of Space-Related Firsts: Space Projects of the Naval Ordnance Test Station (NOTS)," *Spaceflight* 38, no. 7 (1996): 239–243; Peter Pesavento, "Secrets Revealed About the Early US Navy Space Programme," *Spaceflight* 38, no. 7 (1996): 243–245; J. Powell, "The NOTS Air-Launched Satellite Programme," *Journal of the British Interplanetary Society* 50, no. 11 (1997): 433–440.

74. Martin Caidin, *Wings into Space: The History and Future of Winged Space Flight* (New York: Holt, Rinehart, and Winston, 1964); *The Hypersonic Revolution: Case Studies in the History of Hypersonic Technology*, vol. 1, *From Max Valier to Project PRIME (1924–1967)* (Bolling AFB, Washington, DC: USAF History and Museums Program, 1998); *The Hypersonic Revolution: Eight Case Studies in the History of Hypersonic Technology*, vol. 2, *From Scramjet to the National Aero-Space Plane* (Dayton, OH: Special Staff Office, Aeronautical Systems Division, Wright-Patterson AFB, 1987); Jay Miller, *The X-Planes: X-1 to X-45* (Stillwater, MN: Voyageur Press, 2001).

75. Dennis R. Jenkins, *Hypersonics Before the Shuttle: A Concise History of the X-15 Research Airplane* (Washington, DC: NASA SP-2000-4518, 2000); Dennis R. Jenkins and Tony Landis, *Hypersonic: The Story of the North American X-15* (North Branch, MN: Specialty Press, 2003); Milton O. Thompson, *At the Edge of Space: The X-15 Flight Program* (Washington, DC: Smithsonian Institution Press, 2003); Richard Tregaskis, *The X-15 Diary: The Story of America's First Space Ship* (New York: Dutton, 1961; reprint, Lincoln: University of Nebraska Press, 2004); Robert Godwin, *X-15—The NASA Mission Reports Incorporating Files from the USAF* (Burlington, Ontario: Apogee Books, 2000).

76. Roy F. Houchin II, "Why the Air Force Proposed the Dyna-Soar X-20 Program," *Quest: The History of Spaceflight Magazine* 3, no. 4 (winter 1994): 5–12; Roy F. Houchin II, "Why the Dyna-

continued on the next page

based on his dissertation, and he also has a more recent article on Dyna-Soar in the *Journal of the British Interplanetary Society*.[77] Strom has a short introduction to Dyna-Soar in *Crosslink*.[78] Apogee's series of historic space document publications includes Godwin's collection for Dyna-Soar.[79]

Russell Hannigan's *Spaceflight in the Era of Aero-Space Planes* was the first general work on the topic.[80] Reed and Thompson both describe USAF involvement with lifting-body research.[81] Schweikart describes the USAF's efforts for an orbital reusable system in his *Quest for the Orbital Jet*.[82] Butrica documents later military efforts to build reusable systems in his *Single Stage to Orbit*.[83] It is also important to note the military's involvement with the Space Shuttle program, both in its design and in its operations. These are currently best documented in T. A. Heppenheimer's two recent volumes and are also noted in David Spires's overview of the U.S. Air Force in space, *Beyond Horizons*.[84] Tomei discusses the USAF Space Shuttle program.[85] The Inertial Upper Stage, developed to support the Space Shuttle program, is described by Dunn.[86]

The military was also crucial in the development of the various technologies of rocketry. Military funding of liquid-propellant and solid-propellant engines was the starting point for rocketry. The various stories of rocket pio-

---

continued from the previous page

Soar X-20 Program was Cancelled," *Quest: The History of Spaceflight Magazine* 3, no. 4 (winter 1994): 35–37; Terry Smith, "The Dyna-Soar X-20: A Historical Overview," *Quest: The History of Spaceflight Magazine* 3, no. 4 (winter 1994): 13–18; Terry Smith, "Dyna-Soar X-20: A Look at Hardware and Technology," *Quest: The History of Spaceflight Magazine* 3, no. 4 (winter 1994): 23–28.

77. Roy Franklin Houchin II, "The Rise and Fall of Dyna-Soar: A History of Air Force Hypersonic R&D, 1944–1963" (Ph.D. diss., Auburn University, 1995); Roy F. Houchin II, "Air Force-Office of the Secretary of Defense Rivalry: The Pressure of the Political Affairs in the Dyna-Soar (X-20) Program, 1957–1963," *Journal of the British Interplanetary Society* 50 (May 1997): 162–168.

78. Steven R. Strom, "Jurassic Technology: The History of the Dyna-Soar," *Crosslink* 5, no. 1 (winter 2003/2004): 6–9.

79. Robert Godwin, *Dyna-Soar: Hypersonic Strategic Weapon System* (Burlington, Ontario: Apogee Books, 2001).

80. Russell J. Hannigan, *Spaceflight in the Era of Aero-Space Planes* (Malabar, FL: Krieger Publishing, 1994).

81. R. Dale Reed with Darlene Lister, *Wingless Flight: The Lifting Body Story* (Washington, DC: NASA SP-4220, 1997); Milton O. Thompson and Curtis Peebles, *Flying Without Wings: NASA Lifting Bodies and the Birth of the Space Shuttle* (Washington, DC: Smithsonian Institution Press, 1999).

82. Larry Schweikart, *The Quest for the Orbital Jet* (Washington, DC: USAF History and Museums Program, 1998).

83. Andrew J. Butrica, *Single Stage to Orbit: Politics, Space Technology, and the Quest for Reusable Rocketry* (Baltimore, MD: Johns Hopkins, 2003).

84. T. A. Heppenheimer, *The Space Shuttle Decision: NASA's Search for a Reusable Space Vehicle* (Washington, DC: NASA SP-4221, 1999); T. A. Heppenheimer, *Development of the Space Shuttle, 1972–1981: History of the Space Shuttle*, vol. 2 (Washington, DC: Smithsonian Institution Press, 2002); Spires, *Beyond Horizons*.

85. E. J. Tomei, "The Air Force Space Shuttle Program: A Brief History," *Crosslink* 4, no. 1 (winter 2002/2003): 22–25.

86. W. Paul Dunn, "The Evolution of the Inertial Upper Stage," *Crosslink* 4, no. 1 (winter 2002/2003): 38–42.

neers (not repeated here), who were mostly funded by the military, invariably describe the early travails in the development of liquid and solid propellants. Volume 13 in the AAS History Series, edited by Doyle, provides a number of papers on the history of liquid-propellant rocketry.[87] Heppenheimer describes the key role of the Navaho program in American liquid-propellant rocketry.[88] The best work on solid-propellant rocketry in the United States has been done by Hunley.[89] McKenzie's sociological study of nuclear missile guidance, *Inventing Accuracy*, remains the best study of this aspect of ballistic missiles.[90] Martin describes the development of the balloon tank structure of Atlas.[91] The evolution of reentry systems is described by Hartunian.[92]

Cleary provides two volumes on military operations at Cape Canaveral.[93] Guillemette describes the history of Space Launch Complex 6 at Vandenberg AFB.[94] Day provides an unusual look at the archaeology of Vandenberg Air Force Base in a two-part series in *Spaceflight*.[95] Powell and Scala tell story of White Sands Missile Range, and Powell describes its Green River Annex.[96] With the end of the Cold War, there have been a number of Historic American Engineering Record surveys of U.S. missile and space sites, such as Lauber and Hess's survey of the Denver Titan site.[97] Boxx describes the development of Woomera.[98]

87. S. E. Doyle, ed., *History of Liquid Rocket Engine Development in the United States: 1955–1980*, American Astronautical Society History Series, vol. 13 (San Diego: Univelt Press, 1992).

88. Thomas A. Heppenheimer, "The Navaho Program and the Main Line of American Liquid Rocketry," *Air Power History* 44, no. 2 (1997): 4–17.

89. J. D. Hunley, "The Evolution of Large Solid Propellant Rocketry in the United States," *Quest: The History of Spaceflight Quarterly* 6, no. 1 (1998): 22–39. See also Hunley's article on Minuteman in Launius and Jenkins, *To Reach the High Frontier*.

90. Donald McKenzie, *Inventing Accuracy: A Historical Sociology of Nuclear Missile Guidance* (Cambridge, MA: MIT Press, 1990).

91. R. E. Martin, "The Atlas and Centaur 'Steel Balloon' Tanks. A Legacy of Karel Bossart," IAA Paper 82-738 (40th Congress of International Astronautical Federation, Malaga, Spain, 7–12 October 1989).

92. Richard A. Hartunian, "Ballistic Missiles and Reentry Systems: The Critical Years," *Crosslink* 4, no. 1 (winter 2002/2003): 5–9.

93. Cleary, *The 6555th*; Cleary, *The Cape*.

94. Roger Guillemette, "Vandenberg: Space Shuttle Launch and Landing Site, Part 1, Construction of Shuttle Launch Facilities," *Spaceflight* 36, no. 10 (1994): 354–357; Roger Guillemette, "Vandenberg: Space Shuttle Launch and Landing Site, Part 2, Abandoned in Place," *Spaceflight* 36, no. 11 (1994): 378–381.

95. Dwayne A. Day, "Relics of the Space Race: Space Archeology at Vandenberg Air Force Base, Part 1," *Spaceflight* 42, no. 2 (2000): 59–62; Dwayne A. Day, "Relics of the Space Race: Space Archeology at Vandenberg Air Force Base, Part II," *Spaceflight* 42, no. 3 (March 2000): 120–122.

96. J. W. Powell and K. J. Scala, "Historic White Sands Missile Range," *Journal of the British Interplanetary Society* 47, no. 3 (1994): 83–98; Joel W. Powell, "Green River, Utah: a Forgotten Annex of White Sands Missile Range," *Spaceflight* 43, no. 3 (2001): 123–125.

97. John F. Lauber and Jeffrey A. Hess, *Glenn L. Martin Company Titan Missile Test Facilities, Denver, Colorado*, Historic American Engineering Record (HAER) #CO-75 (Minneapolis, MN: Hess, Roice and Company, December 1993).

98. Isaac G. Boxx, "Woomera, Part 1," *Spaceflight* 37, no. 6 (1995): 200–202; Isaac G. Boxx, "Woomera, Part 2," *Spaceflight* 37, no. 7 (1995): 243–247.

EARLY WARNING AND SPACE SURVEILLANCE

Response to an attack by ballistic missiles first requires warning that an attack is under way and the ability to discriminate between these and other natural or humanmade objects that reenter the atmosphere. Given that the flight time of intercontinental ballistic missiles from the U.S. to the USSR and vice versa is about 30 minutes and that defenses against missiles have remained extremely difficult, the main purpose of these systems was to send warning to the political and military leaders to command a retaliatory strike. In practice, this meant launching ballistic missiles, getting bombers into the air, and sending signals to submarine forces. Both the United States and Soviet Union developed ground-based and space-based systems for these purposes at the same time as ballistic missiles became viable as operational weapons.

During World War II, radar systems in the United States were developed mainly at Massachusetts Institute of Technology's (MIT) Radiation Laboratory, which developed a variety of ground-, ship-, and aircraft-based radar systems to detect enemy aircraft and submarines and also to aid strategic bombing. After the war, the threat of Soviet nuclear-armed bombers spurred the creation of progressively more powerful radar systems, along with the need to connect the many radar systems together across increasingly larger regions, eventually to protect the entire North American continent. The problem of rapidly correlating these data as aircraft speeds increased led researchers at the MIT Radiation Laboratory and at the University of Michigan to develop computer-based technologies to integrate the variety of data for each air defense sector. The USAF ultimately selected MIT's system, which became known as the Semi-Automatic Ground Environment (SAGE) system. SAGE became the most expensive computer and largest software programming effort of the 1950s. Unfortunately, the Soviet Union quickly made it obsolete by creating ballistic missiles.

To detect ballistic missiles, the SAGE system was inadequate. What the United States needed was a large, over-the-horizon radar that could pick up ballistic missile launches as early as possible in their flight trajectories. The new system, called the Ballistic Missile Early Warning System (BMEWS), whose first radar system in Thule, Greenland, began operation on 31 December 1960, could detect ballistic missiles launched from the Soviet Union 15 minutes prior to impact. This provided a bare minimum of time for the United States to retaliate by getting its bombers and ballistic missiles into the air before impact. Phased-array radar systems, including the PAVE PAWS and COBRA DANE systems of the 1970s and 1980s, were later implemented to improve capabilities to track multiple objects and to detect submarine-launched ballistic missiles.

Such a short response time was problematic, and the USAF sought any means to extend it. By the late 1950s, satellites beckoned as a possibility. Building off of infrared sensor technologies developed in Nazi Germany, Lockheed

An Agena A spacecraft for an early MIDAS launch undergoes a weight test in 1960 at Lockheed's plant in Sunnyvale, California, before shipment to Cape Canaveral for launch. *(Official USAF photo. Air Force Space Command, Office of History)*

Corporation proposed a variant of its military satellite project, Weapon System 117L (WS-117L), that could detect the infrared signature of a ballistic missile's rocket exhaust plume in the first few minutes of flight. This experimental project, called the Missile Defense Alarm System (MIDAS), placed infrared detectors on polar-orbiting satellites. Despite many failures, to the surprise of its many skeptics, MIDAS proved that the technology was viable. Improvements in the detector technologies allowed the USAF to put out requests for an operational geosynchronous system of three satellites that could monitor the entire globe. Eventually called the Defense Support Program (DSP), this program has gone through several upgrades since the early 1970s and remains functional today. DSP gained notoriety during the Gulf War of 1991 when it detected Iraqi short-range ballistic missile launches. Based on this experience, DSP has been tied more closely to tactical users, as shown in the Iraq War of 2003, when it relayed missile launch data to U.S. Central Command. It is currently to be replaced in the late 2000s by the Space-Based Infrared System (SBIRS).

The Soviet Union went through a similar evolution from local to continental radars for air defense, and then ballistic missile detection, and finally to space-based systems. In the 1960s, the Soviets developed the Dnestr and Dnepr systems. The late 1970s and 1980s saw the deployment of the more powerful Daryal radars into operation, one of which was the Krasnoyarsk system that became a focus of controversy when the United States accused the Soviet Union of violating the Anti-Ballistic Missile Treaty by aiming this radar east across Siberia instead of across national borders as the treaty required. The Soviets also deployed three powerful over-the-horizon Duga-2 systems in the 1970s. Finally, the Lavotchkin design bureau developed early-warning satellites, first a constellation of Molniya orbit satellites called Oko, in the 1970s, and a geosynchronous system called Prognoz, first deployed in the 1980s. Oko deployed a nine-satellite constellation with its apogee above North America and Europe to ensure satellites were deployed over these regions at all times. The fall of the Soviet Union has caused major problems with the early-warning system, as some of the ground-based radar sites were located in newly independent Baltic States that refused to operate them. In addition, the financial crises associated with the fall of the communist empire meant that the Oko and Prognoz constellations have not been fully maintained. The combination of these problems means that the now-Russian system has significant gaps in coverage.

The American and Soviet navies both came to rely on space-based surveillance of the oceans to identify the location of each other's fleets for both strategic and tactical purposes. Significantly outgunned by the U.S. Navy, the Soviet Union relied far more on submarines and ground-based aircraft for its naval goals and developed naval surveillance satellites to augment these capabilities. Its US-A (active radar—RORSAT, Radar Ocean Reconnaissance Satellite) and US-P (passive radar—EORSAT, Electronic Intelligence Ocean Reconnaissance Satellite) systems, designed by Vladimir Chelomey's OKB-52, were deployed in the 1970s. The United States also saw the utility in a naval satellite system, also developing and deploying its White Cloud satellites in the 1970s. White Cloud, US-P, and their descendants remain active in the early 21st century, but US-A's last launch was in 1988, and the program is now defunct.

Both the United States and the Soviet Union also had to distinguish between ballistic missiles and natural or artificial debris reentering the atmosphere. Neither side desired to launch a nuclear strike to retaliate against a meteor or old spacecraft burning up in the atmosphere. Starting in the late 1950s, both sides began to develop space surveillance networks that used combinations of active radar and passive optical and electronic sensors to monitor the trajectories of Earth-orbiting satellites and associated debris.

Early-warning systems are most frequently encountered in books with larger goals. The best starting point to understand radar's development from prior to World War II into the early Cold War is Buderi's *The Invention that Changed the World*.[99] The best source for an overview of the U.S. systems is Spires's *Beyond Horizons*,[100] which contains descriptions of the USAF ground- and space-based early-warning systems. Schaffel's *The Emerging Shield* gives the prehistory of the air defense systems from the end of World War II to 1960, including the various radar systems.[101] Winkler gives an overview of both air defense and missile warning radar systems.[102] Needell's biography of Lloyd Berkner contains a chapter on his role in the development of the Distant Early Warning (DEW) line in the Arctic.[103] Klass was among the first to discuss MIDAS in his *Secret Sentries in Space* in 1971.[104] Sprague's 1985 study of MIDAS at Air University is another early work.[105] The National Reconnaissance Office recently declassified Hall's history of MIDAS, originally written in 1989, but which was publicly published in 1999 both by the NRO and in *Quest*.[106] N. W. Watkins published a short history of MIDAS after Hall's work was written but before it was publicly released.[107] Day published a three-part series on DSP in 1996.[108] Richelson's *America's Space Sentinels* is one

99. Robert Buderi, *The Invention that Changed the World* (New York: Simon & Schuster, 1996).

100. Spires, *Beyond Horizons*.

101. Kenneth Schaffel, *The Emerging Shield: The Air Force and the Evolution of Continental Air Defense, 1945–1960* (Washington, DC: Office of Air Force History, 1991).

102. David F. Winkler, *Searching the Skies: The Legacy of the United States Cold War Defense Radar Program* (Langley AFB, VA: HQ Air Combat Command, 1997).

103. Allan A. Needell, *Science, Cold War and the American State: Lloyd V. Berkner and the Balance of Professional Ideals* (Amsterdam: Harwood Academic Publishers, 2000), chap. 9.

104. Philip J. Klass, *Secret Sentries in Space* (New York: Random House, 1971).

105. Major Barkley G. Sprague, *Evolution of the Missile Defense Alarm System, 1955–1982* (Maxwell AFB, AL: Air Command and Staff College, Air University, 1985).

106. R. Cargill Hall, *Missile Defense Alarm: The Genesis of Space-Based Infrared Early Warning* (Washington, DC: NRO History Office, July 1988); R. Cargill Hall, "Missile Defense Alarm: The Genesis of Space-based Infrared Early Warning," *Quest: The History of Spaceflight Quarterly* 7, no. 1 (spring 1999): 5–17.

107. N. W. Watkins, "The MIDAS Project Part 1: Strategic and Technical Origins and Political Evolution, 1955–1963," *Journal of the British Interplanetary Society* 50, no. 6 (1997): 215–224.

108. Dwayne A. Day, "Top Cover: The Origins and Evolution of the Defense Support Program, Part 1," *Spaceflight* (January 1996): 22–26; Dwayne A. Day, "Top Cover: The Origins and Evolution of the Defense Support Program, Part 2," *Spaceflight* (February 1996): 59–63; Dwayne A. Day, "Top Cover: The Origins and Evolution of the Defense Support Program, Part 3," *Spaceflight* (March 1996): 95–99.

of the few books devoted to the topic, in this case to the genesis and evolution of American early-warning systems, starting with MIDAS, but focusing on the DSP system.[109] Since DSP had a ground control center in Australia, Ball's *Base for Debate* was an early monograph that described DSP, among other systems.[110] An obscure but useful source produced when the Woomera DSP facility was closed is Erickson's *The History of the JDFN (Joint Defence Facility Nurrungar)*.[111] Rosolanka created a short pictorial history of DSP.[112]

The best source for the Soviet and Russian program is Zaloga's *The Kremlin's Nuclear Sword*, which contains descriptions and development history of all Soviet and Russian ground- and space-based early-warning systems.[113] Another good overview is part 2 of Whitmore's "Red Bear on the Prowl."[114] Harvey's *Russia in Space* provides a brief description of Oko and Prognoz.[115] Kagan also describes Soviet early-warning satellites, as does Forden.[116] A description of the various post–Cold War gaps in the Russian system is given in Forden, Podvig, and Postol's "False Alarm, Nuclear Danger"[117] and in Clark's "Decline of the Russian Early Warning System."[118]

United States and Soviet/Russian naval surveillance satellites are discussed, along with their implications for naval strategy and tactics, in Friedman's dense and informative *Seapower and Space*.[119] Siddiqi discusses the Soviet programs in a 1999 article in the *Journal of the British Interplanetary Society*.[120] Muse provides another recent treatment of RORSAT.[121] Teal Ruby, the failed Defense

---

109. Jeffrey T. Richelson, *America's Space Sentinels: DSP Satellites and National Security* (Lawrence: University Press of Kansas, 1999).

110. Desmond Ball, *A Base for Debate: The U.S. Satellite Station at Nurrungar* (North Sydney: Unwin Hymna, 1988).

111. Mark Erickson, ed., *The History of the JDFN (Joint Defence Facility Nurrungar), 1970–1999* (Woomera, Australia: 5th Space Warning Squadron and No. 1 Joint Communications Unit, 1999).

112. James J. Rosolanka, *The Defense Support Program (DSP): A Pictorial Chronology, 1970–1998* (Los Angeles AFB, CA: SBIRS System Program Office, 1998).

113. Zaloga, *The Kremlin's Nuclear Sword*.

114. Paul H. Whitmore, "Red Bear on the Prowl: Strategic Warning in the Soviet Union, Part 2," *Quest: The History of Spaceflight Quarterly* 10, no. 1 (2003): 54–62.

115. Harvey, *Russia in Space: The Failed Frontier*.

116. Boris Kagan, *Soviet ABM Early Warning System: Satellite-Based Project M* (Falls Church, VA: Delphic, 1991); Geoffrey Forden, "Russia's Early Warning System: Which Came First, Technology or Doctrine?" *Breakthroughs* 10, no. 1 (spring 2001): 9–16.

117. Geoffrey Forden, Pavel Podvig, and Theodore A. Postol, "False Alarm, Nuclear Danger," *Spectrum* 37, no. 3 (March 2000): 31–39.

118. Phillip Clark, "Decline of the Russian Early Warning System," *Jane's Intelligence Review* (January 2001).

119. Norman Friedman, *Seapower and Space: From the Dawn of the Missile Age to Net-Centric Warfare* (Annapolis, MD: Naval Institute Press, 2000).

120. Asif Siddiqi, "Staring at the Sea—The Soviet RORSAT and EORSAT Programme," *Journal of the British Interplanetary Society* 52 (1999): 397–416.

121. Fritz Muse, "RORSATS: The Veiled Threat," *Journal of the British Interplanetary Society* 57, supplement 1 (2004): 42–49.

Advanced Research Projects Agency (DARPA)–USAF effort to develop a satellite to monitor aircraft flight, is discussed by Day.[122]

There is no comprehensive published history of space surveillance, either American or Russian. Some early histories by Hayes, Thomas, and Engle and Drummond are now quite dated but describe passive satellite tracking in the early 1960s.[123] They also include a substantial amount on satellite command and control as it existed at the time. More recent information can be found in *Jane's Space Directory*.[124] An unpublished independent study project by Evans at the University of North Dakota used these sources and a few others to provide an overview history of the U.S. Space Surveillance Network.[125] Spires's *Beyond Horizons* provides some information on the history of the Space Surveillance Network as well.[126] Powell describes the Ground-Based Electro-Optical Deep Space Surveillance (GEODSS) system.[127] The evolution of space surveillance into asteroid detection after the collision of Shoemaker-Levy 9 with Jupiter in 1994 is narrated by Mesco.[128] The history of the Soviet/Russian system remains undocumented, with only a couple of brief papers in English describing the system and even briefer mentions of its history.[129] An interesting case study of academic participation in space tracking is presented by Wikles and Gleditsch.[130] Another specific case study is the tracking of Cosmos 954, which fell on Canada in 1978.[131]

122. Dwayne A. Day, "Jewel in the Sky: The US Military Satellite that Never Made It," *Spaceflight* 47, no. 4 (2005): 147–154.

123. Eugene Hayes, *The Smithsonian's Satellite Tracking Program: Its History and Organization* (Washington, DC: Smithsonian, 1962); Shirley Thomas, *Satellite Tracking Facilities: Their History and Operation* (New York: Holt, 1963); Eloise Engle and Kenneth H. Drummond, *Sky Rangers: Satellite Tracking Around the World* (New York: John Day Co., 1965).

124. *Jane's Space Directory* (Alexandria, VA: Jane's Information Group, annual).

125. Brad M. Evans, "The History of the Space Surveillance Network and its Capabilities" (unpublished Independent Study Project, Department of Space Studies, University of North Dakota, summer 2003).

126. Spires, *Beyond Horizons*.

127. Joel Powell, "Satellite Tracking with GEODSS," *Spaceflight* 27, no. 3 (1985): 129–130.

128. James C. Mesco, "Watch the Skies," *Quest: The History of Spaceflight Quarterly* 6, no. 4 (1998): 35–40.

129. G. Batyr, S. Veniaminov, V. Dicky, V. Yurasov, A. Menshicov, Z. Khotorovsky, "The Current State of Russian Space Surveillance System and its Capability in Surveying Space Debris," paper no. ESA SD-01 in *Proceedings of the First European Conference on Space Debris* (held in Darmstadt, Germany, 5–7 April 1993, European Space Agency); Z. N. Khutorovsky, "Low-Perigee Satellite Catalog Maintenance: Issues of Methodology" (paper presented at the Second European Congress on Space Debris, Darmstadt, Germany, 17–19 March 1997).

130. Owen Wilkes and Nils Petter Gleditsch, "Optical Satellite Tracking: A Case Study in University Participation in Preparation for Space Warfare," *Journal of Peace Research* 15, no. 3 (1978): 205–225.

131. Leo Heaps, *Operation Morning Light: Terror in Our Skies, The Story of Cosmos 954* (London: Paddington Press, 1978).

COMMAND AND CONTROL

Relaying data to and from space systems and ground centers in order to control these devices and to initiate and control military responses to strategic and tactical events is crucial to both nuclear and conventional warfare. With each generation and type of space vehicle, and in many cases with each specific project, are built operations control centers and mechanisms to integrate and analyze the data and to distribute the data coming from the space systems to appropriate people and groups on the ground. Despite the unquestioned fact that all space systems require ground control, this topic has received, with a few notable exceptions, remarkably little attention from historians or other scholars. Most studies focus on the devices that go into space, to the detriment of what happens on the ground to control them.

There are at least two types of ground control systems. The first type includes systems that directly control the operations of spacecraft. To do this, the engineering and sensor data are sent to the Earth (downlinked) from the spacecraft and distributed to a mission operations team, which then sends commands up (uplinked) to the spacecraft to control its operation. The second type includes systems that take the sensor data from spacecraft and then operate on and distribute those data for other functions. The best U.S. example of the former is the satellite command and control complexes at Schriever AFB near Colorado Springs, Colorado, the Air Force Satellite Control Facility. The best example of the latter are the military command and control facilities of the Cheyenne Mountain Complex, also near Colorado Springs, which receive sensor data from all around the world, combine them into an integrated picture of air and space threats to the North American continent, and then use and send those integrated data to decision-makers who must determine how to respond to any perceived threats.

The stories of the two types of ground control systems appear in different kinds of histories. The histories of ground control systems that operate spacecraft are, to the extent they exist at all, usually tied to the history of the projects and spacecraft for which they were built. Thus, in most cases, one finds the ground control story in the general histories of the projects for which they were created. In some cases, these ground control systems are modified to also control other spacecraft, in which case they take on lives of their own, partially separated from the specific systems they control. Such is the story of the Air Force Satellite Control Facility, which began as the facility that controlled the CORONA satellites but later expanded to control other spacecraft as well.

The histories of classical command and control systems such as those residing in Cheyenne Mountain are usually separate from the specific systems that contribute data because the point of these systems is to combine data from different systems and assemble it into formats usable to decision-makers. Thus

Space Defense Center inside Cheyenne Mountain, June 1984. *(Official USAF photo. Air Force Space Command, Office of History)*

the histories depend on sensor systems and higher level political and operational decisions as well as the specifics of the "combination" of the data.

The origins of the North American command and control system start with the early-warning systems described in the previous section. As various radar systems were developed and deployed around the northern periphery of the continent, the United States developed the first real-time computer to automate the translation of radar data into a "user-friendly" graphical interface that would allow Air Force enlisted personnel to identify incoming Soviet bombers and direct U.S. fighters and missiles to intercept them. This system, called the Semi-Automatic Ground Environment, or SAGE, was a major milestone in the development of computing hardware and software. Developed by the Lincoln Laboratory of the Massachusetts Institute of technology, SAGE led to the creation of the Air Force–funded, nonprofit MITRE Corporation to complete its development, and also the System Development Corporation, which spun off from RAND Corporation to create SAGE's software.

In 1957, Canada and the United States formed North American Air Defense Command, or NORAD, to jointly protect the continent, given that the radar systems needed to detect Soviet bombers were located on both U.S. and Canadian soil. The central command center was established at Ent Air Force Base in Colorado Springs, Colorado, that same year. In 1959, the U.S.

Joint Chiefs of Staff selected Cheyenne Mountain, just southwest of Colorado Springs, to be the location of an underground, nuclear-hardened facility to house NORAD. Into the tunnels of Cheyenne Mountain, which was completed in 1965, went the command facilities for the SAGE air defense network, the Ballistic Missile Early Warning System (BMEWS), and what became the Space Surveillance Network. Tying these three separate systems together into a single command center was the 425L Command Operations Center computing and display system, which used Philco 2000 computers. On 1 January 1966, Air Force Systems Command handed over operations to NORAD, whose commander, by treaty, was always an American, and whose deputy commander was always a Canadian. The NORAD Combat Operations Center became operational in February 1967 when the Space Defense Center system, 496L, was completed. Data from NORAD were fed to the American and Canadian national authorities.

Increases in Soviet threats and in corresponding American detection systems such as phased-array radars led to the Cheyenne Mountain Improvement Program, called 427M. This new system would have to integrate with a global command and control system, known as the World-Wide Military Command and Control System (WWMCCS), which used Honeywell Information System 6060 computers. Philco-Ford won a contract for system integration and testing, and the communications gear, while System Development Corporation won the contract for the Space Computation Center software and displays. The system also eventually included UNIVAC 1100/42 systems for satellite early warning. NORAD itself developed much of the system software. 427M was finally completed in 1979 but suffered some false nuclear attack warnings, which led quickly to studies and investigations as to the cause, which turned out to be faulty computer chips.

The 427M program was a set of largely disjointed "stovepipe" projects, which were combined later into the next major upgrade, which became known as the Cheyenne Mountain Upgrade Program. This program came to include a variety of backup systems, both electronic and physical. The USAF developed backup facilities at Offutt AFB near Omaha, Nebraska (the home of Strategic Air Command), and at Peterson AFB in Colorado Springs, along with an existing NORAD backup facility at Malmstrom AFB near Great Falls, Montana. The various upgrades, like their predecessors, ran into cost overruns and schedule slips that accompanied their technical problems. Again came a variety of investigations, which again pointed to problems with systems integration of the many sensors, computers, and facilities. The Cheyenne Mountain Upgrade program finally reached full operational capability (FOC) in October 1998.

In the 1991 Gulf War, Defense Support Program data on Iraqi ballistic missile launches fed into NORAD and then to military units in the Gulf. From that time forward, the military has taken a variety of measures to improve speed and accuracy of ballistic missile and other data from "strategic" sources such as NORAD to tactical units in wartime. By the early 21st century, another series of upgrades were under way, this time to take advantage of technical improvements in computer workstations and computer networks such as the Internet and World Wide Web.

Information from NORAD feeds into the highest level political and military authorities so as to determine, in the worst case, whether a nuclear counterstrike should be launched or whether any other measures are required. With the advent of ballistic missiles, the time available for the nuclear "go code" decision from detection of the ballistic missiles from space and from ground-based radar shrank from hours down to 15 to 30 minutes. Furthermore, hydrogen bombs in space or the upper atmosphere would disrupt the ionosphere, thereby disrupting most long-range radio communications, and destroy ground-based wire communication systems near nuclear impact points. One space-based solution to this problem in the 1960s and 1970s was the creation of the Emergency Rocket Communications System, which would launch Blue Scout (1963–1967) or Minuteman (after 1967) rockets from Wallops Island, Virginia, to high altitude, from where it would send an Emergency Action Message such as the nuclear go-code by radio, thus bypassing ionospheric disruptions. In the 1980s, the Reagan administration approved creation of the Milstar satellite communications system, which was nuclear-hardened so as to send the Emergency Action Message to American nuclear forces around the world during a nuclear war. The end of the Cold War reduced, but did not eliminate, threats to the U.S. command and control system.

The Soviet Union faced similar problems, compounded by the political control of nuclear weapons by the Soviet secret police, the KGB. By the late 1960s, the Soviets created the Signal system, which could detect an attempt by a crew to perform an unauthorized ballistic missile launch. In the 1970s, the Molniya satellite communications system enhanced Soviet command and control, although these satellites were vulnerable to nuclear attack in space. By the 1980s, the Soviets created an automatic nuclear response system known as Perimetr, much like the hypothetical "doomsday machine" satirized in the early 1960s film *Dr. Strangelove*. This system, deployed in 1985, would automatically authorize nuclear retaliation even if the national authorities were dead. The Soviets also developed their own ballistic-missile-based communications system like the American Emergency Rocket Communications System.

There are two recent works on satellite mission control systems. Mudgway's *Uplink-Downlink* describes the evolution of Jet Propulsion Laboratory's Deep Space Network.[132] This is almost entirely a civilian story, but the military

---

132. Douglas J. Mudgway, *Uplink-Downlink: A History of the Deep Space Network, 1957–1997* (Washington, DC: NASA SP-2001-4227, 2001).

origins of the program are detailed in chapter 2. Arnold's *Spying from Space* is the first major published study of a military satellite control system, the Air Force Satellite Control Facility.[133] Spires's *Beyond Horizons* also has discussions of satellite control in the USAF among its many other topics.[134]

The SAGE system has a small but significant literature in the history of computing. The foremost reference is Redmond and Smith's tome, *From Whirlwind to MITRE*.[135] Jacobs's *The SAGE Air Defense System* gives an anecdotal account of SAGE's development.[136] Edwards's eclectic *The Closed World* put SAGE into a broader Cold War context through a postmodern discourse analysis.[137] In 1983, the *Annals of the History of Computing* published a SAGE special issue that included a collection of articles on various facets of the computer system.[138] Two institutional histories link SAGE to broader issues in command and control: MITRE Corporation's corporate history and Baum's history of System Development Corporation.[139] Hughes's *Rescuing Prometheus* also has a chapter on SAGE.[140] Dyer and Dennis produced a new history of MITRE in 1998.[141]

Larger scale command and control systems and their ties to the national command authorities, such as NORAD and WWMCCS, have a surprisingly limited literature, given the importance of the subject for the survival of the United States in wartime. An early external description of NORAD is in DeVere and Johnson.[142] Chapman provides a full history of NORAD's Cheyenne Mountain Complex up to 1989 in *Legacy of Peace*.[143] The history of WWMCCS is told in Pearson, *The World Wide Military Command and Control System*.[144] Control of British and North Atlantic Treaty Organization (NATO) nuclear forces to the mid-1960s is discussed in Twigge and Scott's *Planning*

133. David Christopher Arnold, *Spying from Space: Constructing America's Satellite Command and Control Systems* (College Station: Texas A&M Press University, 2005).

134. Spires, *Beyond Horizons*.

135. Kent C. Redmond and Thomas A. Smith, *From Whirlwind to MITRE: The R&D Story of the SAGE Air Defense Computer* (Cambridge, MA: MIT Press, 2000).

136. John F. Jacobs, *The SAGE Air Defense System: A Personal History* (Bedford, MA: MITRE Corporation, 1986).

137. Paul N. Edwards, *The Closed World: Computers and the Politics of Discourse in Cold War America* (Cambridge, MA: MIT Press, 1996).

138. *Annals of the History of Computing* 5, no. 4 (October 1983).

139. *MITRE: The First Twenty Years* (Bedford, MA: MITRE Corporation, 1979); Claude Baum, *The System Builders: The Story of SDC* (Santa Monica, CA: System Development Corporation, 1981).

140. Thomas P. Hughes, *Rescuing Prometheus* (New York: Pantheon Books, 1998), chap. 2.

141. Davis Dyer and Michael Aaron Dennis, *Architects of Information Advantage: The MITRE Corporation Since 1958* (Montgomery, AL: Community Communications, 1998).

142. G. T. DeVere and N. L. Johnson, "The NORAD Space Network," *Spaceflight* 27, nos. 7–8 (1985): 306–309.

143. Richard G. Chapman, Jr., *Legacy of Peace: Mountain with a Mission, NORAD's Cheyenne Mountain Combat Operations Center, The Cold War Years: 1946–1989* (Colorado Springs, CO: Mountain Express Printing, 1989).

144. David E. Pearson, *The World Wide Military Command and Control System: Evolution and Effectiveness* (Maxwell AFB, AL: Air University Press, 2000).

Armageddon.[145] Blair's *Strategic Command and Control* from 1985 remains a valuable source on the overall control of nuclear forces,[146] as is Bracken's 1983 *Command and Control of Nuclear Forces.*[147] For the Soviet Union and Russia, Zaloga's *The Kremlin's Nuclear Sword* is the best introduction, with information on Signal, Perimetr, etc.[148]

COMMUNICATIONS

Separate from the issue of warfare are everyday military communications for logistics, as well as tactical communications for conventional force operations. The United States has particular need for worldwide communications due to the distribution of American military forces around the globe during and after the Cold War. The first communications satellite experiment was Project SCORE (Signal Communication by Orbiting Relay Equipment), which used a modified Atlas ICBM to broadcast a taped message from President Eisenhower in 1958. The Army Signal Corps launched the first repeater satellite, Courier, in 1960, while working on a more sophisticated satellite known as Advent. Advent was too ambitious and was canceled in 1962, but in 1964, the Department of Defense created the Initial Defense Communications Satellite Program (IDCSP), managed by the Defense Communications Agency. The Air Force built the satellites, while the Army Satellite Communications Agency handled the ground segment. IDCSP consisted of a constellation of simple Philco satellites in medium-Earth orbit, the first seven of which were launched in 1966. The military, from that time to the present, also leased transponders on commercial communications satellites for less sensitive logistical and other information.

The second generation of military satellites was known as the Defense Satellite Communications System II, or DSCS (pronounced "discus") II. Built by TRW, the first pair of these much more capable satellites were launched in 1971. Whereas IDCSP satellites could each handle 11 tactical-quality voice circuits, DSCS II satellites each had capacity for 1,300 voice channels and could communicate with much smaller antennas on the ground. DSCS III satellites, built by General Electric and first launched in 1982, were even more capable, with antijamming capabilities and spot beams. DSCS III satellites continue to operate into the 21st century.

---

145. Stephen Twigge and Len Scott, *Planning Armageddon: Britain, the United States and the Command of Western Nuclear Forces, 1945–1964* (Amsterdam: Harwood Academic Publishers, 2000).

146. Bruce G. Blair, *Strategic Command and Control: Redefining the Nuclear Threat* (Washington, DC: Brookings Institution Press, 1985).

147. Paul Bracken, *The Command and Control of Nuclear Forces* (New Haven, CT: Yale University Press, 1983).

148. Zaloga, *The Kremlin's Nuclear Sword.*

In the meantime, the Navy wanted its own system for mobile fleet communications. The Lincoln Laboratory of MIT, with funding from all of the services, created a series of experimental satellites to test a variety of frequency ranges and capabilities. The first military satellites operated in Super High Frequency (SHF), which required very large ground antennas. Mobile communications required smaller ground antennas, often using Ultra-High Frequencies (UHF). Lincoln Experimental Satellites 3-6 tested these capabilities, leading to the Hughes-built Tacsat, which conclusively proved the utility of UHF communications for the U.S. Navy in particular. The Navy then funded development of the Fleet Satellite Communications (FLTSATCOM) system in the 1970s, but development delays led to purchase of the so-called "Gapfiller" satellites, also built by Hughes. Gapfiller and FLTSATCOM were both used in 1980s, with two FLTSATCOM satellites, controlled from Point Mugu, California, remaining in operation as of February 2005.

The USAF originally developed the Milstar communications satellites in the 1980s for low-rate, nuclear-hardened communications capabilities to ensure the nuclear "go-code" could be sent in nuclear war. When the Cold War ended, the remaining Milstar satellites were modified for higher-rate communication capabilities for tactical purposes. Since the 1970s, the increasing use of imagery for strategic and tactical purposes has driven the development of satellite communication capabilities towards ever greater speeds. The KH-11 reconnaissance satellites, which were the first to use radio signals to send imagery, required communications satellites such as the Satellite Data System to relay the data. Later systems, such as the Lacrosse radar-based reconnaissance satellite, used the Tracking and Data Relay Satellite System also used by NASA. The Ultra-High Frequency Follow-On system, first launched in 1993, is the replacement for the aging FLTSATCOM design. With ever greater demand for communications bandwidth largely driven by sending digital imagery, the U.S. military began leasing significant amounts of time and transponders from commercial carriers, including its 2000 deal with Iridium Satellite LLC to lease the Iridium global satellite constellation that had gone bankrupt.

The Soviet Union likewise developed military communications systems, starting with the well-known Molniya satellites in 1965. Because of the far northern latitudes of the Soviet Union, the Soviets have predominantly used medium-Earth-orbit systems to ensure coverage over the Poles. Later, the Soviets combined communications with navigational capabilities with the Tsiklon (first launched 1967) and later Tsiklon M system (first launched 1974). The Kristal and Strela satellite constellations were also developed, along with the geosynchronous Raduga communications system.

Military satellite communications have also been crucial to other countries, starting with the United Kingdom for the Royal Navy, which developed and operated its Skynet system starting in 1969, and to NATO, which since the 1970s has had its own series of satellites. China developed its Dong Fang Hong communications satellites starting in 1984. Many other countries have military satellite communication capabilities through their own domestic communications satellites. These satellites are generally mixed military-civilian systems.

No comprehensive history of satellite communications, or of military satellite communications, exists. However, some historical research has begun. The origin of satellite communications is best told in Whalen's *The Origins of Satellite Communications*, including the relationships between the military, NASA, and industry in its formative period in the 1950s and early 1960s.[149] Butrica's edited *Beyond the Ionosphere* contains a collection of historical papers on a variety of communications satellite topics, including military efforts of the USAF, Navy, and MIT's Lincoln Laboratories.[150] Martin's *Communication Satellites*, now in its fourth edition, is an essential reference, providing a brief overview of all communications satellites up to its publication date, including source information on where to get further data.[151] Spires and Sturdevant provide an overview of USAF military satellite communications, which is reproduced in *Beyond the Ionosphere*.[152] Van Trees et al. provide an overview of satellite communications in a 2004 article.[153] Lee's *History of the Defense Satellite Communications System* is one of the few works devoted exclusively to military space communications.[154] Davis described Project SCORE in a 1999 article.[155] Richelson describes the Satellite Data System (SDS) in a 1982 article in the *Journal of the British Interplanetary Society*.[156] Day's 1999 *Spaceflight* article discusses SDS and its three launches from the Space Shuttle.[157] The U.S. and Soviet navies' use of communications satellites is well told in Friedman's

---

149. David J. Whalen, *The Origins of Satellite Communications, 1945–1965* (Washington, DC: Smithsonian Institution Press, 2002).

150. Andrew J. Butrica, ed., *Beyond the Ionosphere: Fifty Years of Satellite Communications* (Washington, DC: NASA SP-4217, 1997).

151. Donald H. Martin, *Communications Satellites*, 4th ed. (Reston, VA: AIAA, 2000).

152. David N. Spires and Rick W. Sturdevant, "From Advent to Milstar: The U.S. Air Force and the Challenges of Military Satellite Communications," *Journal of the British Interplanetary Society* 50, no. 6 (1997): 207–214.

153. Harry L. Van Trees, Harry D. Raduege, Rick W. Sturdevant, and Ronald E. Thompson, "Military Satellite Communications: From Concept to Reality," in *The Limitless Sky: Air Force Science and Technology Contributions to the Nation*, ed. Alexander H. Levis (Washington, DC: Air Force History and Museums Program, 2004), pp. 175–209.

154. Major Robert E. Lee, *History of the Defense Satellite Communications System (1964–1986)*, Air Command and Staff College Report No. 87-1545 (Maxwell AFB, AL: Air University Press, 1987).

155. Deane Davis, "The Talking Satellite: A Reminiscence of Project SCORE," *Journal of the British Interplanetary Society* 52 (1999): 239–258.

156. Jeffrey Richelson, "The Satellite Data System," *Journal of the British Interplanetary Society* 37, no. 5 (1984): 226–228.

157. Dwayne A. Day, "Out of the Shadows: The Shuttle's Secret Payloads," *Spaceflight* 41, no. 2 (February 1999): 78–84.

Seapower and Space.[158] Getting describes early military communications programs in his autobiography.[159] Recent issues and options for leasing commercial systems are discussed in a RAND study by Bonds et al.[160]

Harvey's *Russia in Space* has an overview of Soviet and Russian communications systems.[161] Hendrickx describes the early Molniya program.[162] The Chinese program, including its communications satellites, is discussed in Clark's overview in the *Journal of the British Interplanetary Society.*[163] Harvey also gives some attention to the Dong Fang Hong satellites in his *The Chinese Space Programme.*[164] Harris describes the British Skynet program.[165]

BALLISTIC MISSILE DEFENSE

Unlike most other areas of military space, defense against intercontinental ballistic missiles (ICBMs) is a subject that has spawned great public interest in the United States, with high-profile political debates highlighting the subject from its inception in the 1960s, and particularly in the mid-1980s with the initiation of Ronald Reagan's Strategic Defense Initiative (SDI), which critics called "Star Wars" after the 1977 film of that name. In turn, these political debates have led to a minor industry of polemical works both for and against ballistic missile defense and its alleged impact on international political and military stability. Amazingly, despite the thousands of pages and dozens of works on the subject, there is no comprehensive history of the actual ballistic missile defense systems and programs. In fact, there are no comprehensive public histories of *any* of the ballistic missile defense systems that have actually been deployed, the SDI program itself, or its Soviet counterparts.

From the moment that Nazi Germany began firing V-2s at London, British and American soldiers, scientists, and engineers began searching for ways to counter these apparently unstoppable weapons. During World War II, the only counter was to attack launch sites and logistics for the V-2. Once in flight, there was nothing that could stop them, due to their extremely high speed. After

---

158. Norman Friedman, *Seapower and Space: From the Dawn of the Missile Age to Net-Centric Warfare* (Annapolis, MD: Naval Institute Press, 2000).

159. Ivan A. Getting, *All in a Lifetime: Science in the Defense of Democracy* (New York: Vantage Press, 1989).

160. Tim Bonds, Michale G. Mattock, Thomas Hamilton, Carl Rhodes, Michael Scheiern, Philip M. Feldman, David R. Frelinger, and Robert Uy, *Employing Commercial Satellite Communications: Wideband Investment Options for the Department of Defense* (Santa Monica, CA: RAND, 2000).

161. Harvey, *Russia in Space: The Failed Frontier?*

162. Bart Hendrickx, "The Early Years of the Molniya Program," *Quest: The History of Spaceflight Quarterly* 6, no. 3 (1998): 28–36.

163. Phillip Clark, "Review of the Chinese Space Programme," *Journal of the British Interplanetary Society* 52 no. 9/10 (1999): 350–376.

164. Harvey, *The Chinese Space Programme: From Conception to Future Capabilities.*

165. R. L. "Dick" Harris, "Military Satellite Communication in the UK," *Spaceflight* 37, no. 10 (1995): 348–352.

the war, the U.S. Army developed its Nike-Ajax surface-to-air missiles, and the Army Air Forces contracted Project THUMPER with General Electric and the University of Michigan for Project WIZARD to investigate using missiles to destroy incoming ballistic missiles. In 1955, the Army contracted with Western Electric to create an antiballistic missile system, which led ultimately to the Nike-Zeus antiballistic missile. In 1958, the Air Force's Project WIZARD was reduced to research on radar and command and control, and the Army gained control of the antiballistic missile program. The Advanced Research Projects Agency (ARPA) developed an idea in July 1960 for a space-based system called Ballistic Missile Boost Intercept, or BAMBI. Nike-Zeus successfully intercepted an Atlas ICBM in 1962 but remained in research and development. Instead, the system's capabilities were developed further to the Nike-X, which used an upgraded Nike-Zeus missile known as Spartan.

In 1967, President Lyndon Johnson approved development and deployment of the SENTINEL system, which was to be a national ballistic missile defense system with 18 missile sites. However, with the growth of the antiwar movement resulting from the Vietnam War, support for SENTINEL shrank, and it was scaled back to the smaller SAFEGUARD system, which was barely approved in 1969. Congress funded only 2 of the 12 proposed sites, which soon shrank to only 1 site north of Grand Forks, North Dakota, to protect a Minuteman ICBM field. President Richard Nixon used the antiballistic missile (ABM) system as a bargaining chip with the Soviet Union, leading to the signing of the ABM Treaty in 1972, which with a further protocol in 1974 allowed the United States and Soviet Union one missile site each. The system itself, which used new phased-array radars, deployed the long-range Spartan and the short-range Sprint missiles, each tipped with nuclear warheads. In September 1975, the system became fully operational, but the next month, Congress terminated its funding. The next year, the Army began deactivation, and by 1977, the site was in "caretaker status," with only its Perimeter Acquisition Radar remaining functional.

The Soviet Union also began development of its own ABM systems in the late 1950s. Initial testing occurred at Sary Shagan in 1956 and led to the creation of the Anti-Missile Defense Forces in 1958. The first successful ballistic missile interception occurred in 1960, with the actual destruction of a test missile in 1961 using conventional explosives. Nuclear testing followed shortly thereafter. After an abortive attempt to deploy a system around Leningrad in the early 1960s, the Soviets deployed their first system, the A-35, around Moscow beginning in 1967. A series of upgrades followed both with the radar and missile systems. The upgraded system, the A-135, became fully operational only in the mid-1990s, with its new missiles, the SH-08 Gazelle and the SH-11 Gorgon, functioning like the American Sprint and Spartan for a layered defense. Thus the Soviet Union, unlike the United States, has kept an operational ABM system in place continuously since the late 1960s.

Even though the United States dismantled its ABM system in the mid-1970s, research and development continued on the relevant technologies. A revival came in March 1983 when President Ronald Reagan announced the Strategic Defense Initiative. After his landslide reelection in 1984, Reagan pushed major funding increases for strategic defense and created the Strategic Defense Initiative Organization (SDIO). SDIO investigated a variety of approaches to ballistic missile defense, including space-based lasers and kinetic kill vehicles, along with a variety of Earth-based approaches. With the end of the Reagan administration, SDI did not die, but it was scaled back, refocused on research, and renamed several times. The possibility of antiballistic missile systems got a boost during the Gulf War of 1991 when Patriot batteries intercepted some Iraqi Scud missiles over Israel and Saudi Arabia. When Pakistan and Iran tested medium-range ballistic missiles in 1998 and North Korea attempted to put a satellite in orbit, the debate over ABM systems heated up again. Accelerated development followed but did not lead to a deployed system, partly due to technical issues. Through 2004, testing of ABM technologies continued with mixed success.

Chun's 2003 articles in *Quest* are a good starting point for the history of Nike-Zeus.[166] These articles rely on the Army's *Missiles Handbook*, published annually in the late 1950s and early 1960s.[167] Lonnquest and Winkler's *Defend and Deter* provides an overview of Cold War missile systems, including Nike-Zeus and SAFEGUARD.[168] Bowen's 2005 *Quest* article provides a short overview of SAFEGUARD,[169] drawing significantly from three internal Army sources.[170] Walker et al. provide a historical site assessment.[171] Bruce-Briggs provides an overview of ABM systems through the early SDI program.[172]

166. Clayton K. S. Chun, "Defending Against Hitler's Vengeance: The U.S. Army and the V-2," *Quest: The History of Spaceflight Quarterly* 10, no. 2 (2003): 45–52; Clayton K. S. Chun, "Nike-Zeus' Thunder and Lightning: From Antiballistic Missile to Antisatellite Interceptor," *Quest: The History of Spaceflight Quarterly* 10, no. 4 (2003): 40–47.

167. *Office Directorate of Progress and Statistical Reporting U.S. Army Missiles Handbook (U)* (Washington, DC: Department of the Army, 1959–61). These documents, from 1959 to 1961, have been declassified.

168. John C. Lonnquest and David F. Winkler, *To Defend and Deter*, USACERL Special Report 97/01 (Champaign, IL: U.S. Army Construction Engineering Research Laboratories, 1996).

169. Gregory S. Bowen, "SAFEGUARD: North Dakota's Front Line in the Cold War," *Quest: The History of Spaceflight Quarterly* 12, no. 1 (2005): 38–50.

170. Bell Laboratories, *ABM Project History* (Whippany, NJ: U.S. Army Ballistic Missile Defense Command, 1975); James H. Kitchens III, *A History of the Huntsville Division: US Army Corps of Engineers: 1967–1976* (Huntsville, AL: U.S. Army Corps of Engineers, 1978); James A. Walker, Frances Martin, and Sharon S. Watkins, *Strategic Defense: Four Decades of Progress* (Washington, DC: Historical Office, U.S. Army Space and Strategic Defense Command, 1995).

171. James A. Walker et al., compilers, *Historic American Engineering Record Documentation for the Stanley R. Mickelson Safeguard Complex*, vol. 1, *Historical Context*, and vol. 2, *Architectural Data & Photographs*, HAER #ND-9 (Huntsville, AL: U.S. Army Space and Strategic Defense Command, September 1996).

172. B. Bruce-Briggs, *The Shield of Faith* (New York: Simon & Schuster, 1988).

The history of the Soviet Union's ABM systems are described in Zaloga's *The Kremlin's Nuclear Sword*, and also in Whitmore's *Quest* articles in 2002–2003.[173] Mathers discusses Soviet ballistic missile defense (BMD) during the Khrushchev era.[174] The Federation of American Scientists also provides good material on Soviet ABM systems.[175] Siddiqi's 1998 *Spaceflight* article describes the Soviet ground- and space-based laser programs FON and Polyus.[176] Newhouse's *Cold Dawn* is the classic introduction to the history of SALT negotiations.[177] Hays provides a good overview of the Strategic Arms Reduction Treaties (START I and START II).[178]

The best starting point to understand SDI's beginnings is Baucom's *The Origins of SDI*.[179] Baucom also provides an overview of SDI's organization, as does Mary FitzGerald.[180] To date, there are no published overview technical histories of SDI and its descendants. However, Frances Fitzgerald provides an overview of SDI politics during the Reagan administration, and Graham does the same for the later Clinton and early G. W. Bush administrations.[181] Simmons and Bythrow describe Delta Star, an SDI Organization experiment to track launchers from space.[182] Lagrasse and Farmin narrate the TSX-5 experiment for the Ballistic Missile Defense Organization.[183]

173. Zaloga, *The Kremlin's Nuclear Sword*; Paul Whitmore, "Red Bear on the Prowl: Space-related Strategic Defense in the Soviet Union, Part I," *Quest: The History of Spaceflight Quarterly* 9, no. 4 (2002): 22–30.

174. Jennifer G. Mathers, "A Fly in Outer Space: Soviet Ballistic Missile Defence During the Khrushchev Period," *Journal of Strategic Studies* 21, no. 2 (1998): 31–59.

175. A. Karpenko, "ABM and Space Defense," Federation of American Scientists Web site, 1999, *http://www.fas.org/spp/starwars/program/soviet/990600-bmd-rus.htm*.

176. Asif A. Siddiqi, "Cold War in Space: A Look Back at the Soviet Union," *Spaceflight* 40, no. 2 (February 1998): 63–68.

177. John Newhouse, *Cold Dawn: The Story of SALT* (New York: Holt, Rinehart, and Winston, 1973).

178. Peter L. Hays, *United States Military Space: Into the Twenty-First Century*, INSS Occasional Paper 42 (USAF Academy, CO, and Maxwell AFB, AL: USAF Institute for National Security Studies and Air University Press, 2002).

179. Donald Baucom, *The Origins of SDI, 1944–1983* (Lawrence: University Press of Kansas, 1993).

180. Donald R. Baucom, "Developing a Management Structure for the Strategic Defense Initiative," in *Organizing for the Use of Space: Historical Perspectives on a Persistent Issue*, ed. Roger D. Launius, vol. 18 (San Diego, CA: Univelt, 1995), pp. 187–215; Mary C. FitzGerald, *The New Revolution in Russian Military Affairs*, Whitehall paper series, no. 26 (London: Royal United Services Institute for Defence Studies, 1994).

181. Frances Fitzgerald, *Way Out There in the Blue: Reagan, Star Wars and the End of the Cold War* (New York: Touchstone Books, 2001); Bradley Graham, *Hit to Kill: The New Battle Over Shielding America from Missile Attack* (New York: PublicAffairs, 2001).

182. Frederick Simmons and Peter Bythrow, "Delta Star: an SDIO Space Experiment," *Crosslink* 2, no. 2 (summer 2001): 23–29.

183. Michael L. La Grassa and James R. Farmin, "TSX-5: Another Step Forward for Space-Based Research," *Crosslink* 2, no. 2 (summer 2001): 30–37.

SPACE INTELLIGENCE AND RECONNAISSANCE

Using space systems to divine the intentions and capabilities of other nations is a crucial aspect of military space, with a significant and growing historical literature. The use of satellites for reconnaissance was presented in RAND's initial study of artificial satellites in 1946. The U.S. government was desperate for information about secretive Soviet efforts, particularly with respect to nuclear and ballistic missile capabilities. In the 1950s, the United States, with cooperation from Great Britain and others, used a variety of means to gather both photographic and electronic intelligence information, including balloon and aircraft overflights. These culminated with the U-2 program, which had its first mission over the Soviet Union in 1956. American officials realized that sooner or later, the Soviets would develop an antiaircraft missile that could shoot down U-2s, an event that transpired in 1960. In the meantime, the United States began development of a satellite that could replace the U-2. Reconnaissance satellites became a top priority of the military and intelligence communities at this time and have remained so to the present day. A major priority for the Eisenhower and Kennedy administrations was the establishment of the principle of "freedom of space," so as to allow American reconnaissance satellites to gather intelligence of the Communist bloc.

The U.S. reconnaissance satellite effort began as the USAF's Project WS-117L in the mid-1950s. It led to the CORONA and Samos programs for reconnaissance and MIDAS for early warning. The USAF-funded Samos program intended to provide real-time intelligence data by sending images from on-board film readout to the ground by radio. Unfortunately, the technology to acquire high-resolution digital imagery was not yet mature, and after 11 test flights with mixed results, the program was canceled. In the meantime, the CIA, with the Eisenhower administration's encouragement, developed the CORONA film-return system. Under the public name of Discoverer, which was proclaimed to launch life science and engineering technology experiments, the CIA began test flights. After 12 consecutive failures, in August 1960 the first CORONA capsule returned successfully from space. The next flight, Discoverer 14, put a camera in orbit and photographed more of the Soviet Union than all previous air overflights combined.

The CORONA program operated until 1972, by which time it orbited a variety of cameras, improving ground resolution from about 40 feet to 6 feet. Various CORONA missions also incorporated stereo cameras, two film buckets to increase mission length, and mapping cameras for military targeting. Some also carried subsatellites that separated from the main satellite once in orbit, generally for electronics and signals intelligence gathering. Shortly after the first successful flight in 1960, the Eisenhower and Kennedy administrations created, in secret, the National Reconnaissance Office (NRO) to manage CORONA and other space intelligence assets. To handle the massive flow of imagery, the U.S. government created the National Photographic Interpretation Center.

CORONA and its successors were crucial to maintaining peace during the Cold War, as first the U.S. and shortly thereafter the Soviet Union monitored each other's nuclear capabilities. This mutual ability and its high value to each side made it possible to sign treaties banning weapons of mass destruction from space, to limit ballistic missile defenses, and to allow the signing of verifiable arms control treaties starting in the 1970s. CORONA proved in the early 1960s that American fears that the Soviets were ahead in the development and deployment of ICBMs were unfounded. In fact, the "missile gap" was massively in favor of the United States. This information allowed the Kennedy and later administrations to scale back nuclear missile deployments and to stand firm against Soviet threats.

A variety of successor systems for optical reconnaissance followed CORONA, starting with the KH-9 Hexagon in the early 1970s and the KH-11 Kennan real-time optical reconnaissance system. While the KH-9 provided higher resolution using film-return methods, the KH-11 fulfilled the USAF's dream of a real-time optical reconnaissance system, which allowed much faster return of data than the slow film-bucket capability. In parallel, the United States also developed a variety of signals and electronic intelligence systems, under a variety of code names such as Rhyolite, Canyon, and Magnum, and eventually an active radar-imaging satellite known as Lacrosse that allowed spy satellites to "see" through clouds and at night. The Advanced KH-11, Lacrosse, and a variety of signals and electronics intelligence satellites continue to operate today.

The Soviet Union initially objected to U.S. reconnaissance systems, but only until it orbited its own systems, at which point Soviet leaders quietly dropped their objections to these highly useful devices. Korolev's OKB-1 developed the first Soviet reconnaissance system, known as Zenit, from the Vostok capsule used to orbit humans, by replacing the human gear with camera systems. Like the United States, the Soviets then developed a variety of improved optical systems, along with their own electronics and signals intelligence satellites. Improved optical satellites, under the name Yantar, first flew in 1974, with the real-time digital Yantar Terilen system first flying in 1982. New systems, known as Orlets and Arkon, are also currently flying.

China, France, Israel, and Japan have also developed space photoreconnaissance capabilities. China's Fanhui Shi Yao Gang Weixing satellites, first successfully launched in 1975, are recoverable optical imaging satellites, probably at least in part for military purposes. France, Italy, and Spain collaborated to develop the Helios reconnaissance satellites, first launched in 1995. A second Helios was launched in 1999, and the second-generation Helios 2A was placed in orbit in December 2004. Israel's Ofeq series of military imaging satellites, first launched in 1988, are now up to Ofeq-5. Japan launched its first pair of Information Gathering Satellites in March 2003 in response to North Korea's attempt to put a satellite in orbit with its Taepodong rocket launch in 1998. A variety of other systems are in development in a number of nations.

The 1990s saw a boom in histories of space intelligence, mainly due to the declassifications and the opening of some former Soviet archives. The NRO's existence was revealed in 1992, in the first Bush administration.[184] In May 1995, a public conference heralded the declassification of CORONA materials, while in August 2002, the National Imagery and Mapping Agency declassified imagery from the KH-7 and KH-9 Mapping Camera.[185] Prior to 1992, Cold War–era attempts to tell the story of space reconnaissance and intelligence systems were necessarily based on many obscure clues with little direct hard evidence. Klass, Kenden, Borrowman, Richelson, Peebles, and Burrows each attempted this prodigious task, with varying degrees of success.[186] Their efforts for CORONA are now outdated but remain valuable for electronics intelligence (ELINT) and signals intelligence (SIGINT) and for optical reconnaissance after CORONA. For signals intelligence, Bamford's recent book on the National Security Agency is a good place to start, although it focuses mainly on nonsatellite programs.[187] McDowell gives an overview of U.S. spy satellite programs, with each satellite's launch date.[188] While significant progress has been made to untangle these programs, many issues and facts will no doubt remain unresolved for decades to come until the relevant sources are declassified.

---

184. Bill Gertz, "The Secret Mission of the NRO," *Air Force Magazine* 76 (June 1993): 60–63.

185. Dwayne A. Day, "US Government Declassifies Reconnaissance Satellites Information," *Spaceflight* 45, no. 3 (2003): 116–117.

186. Philip J. Klass, *Secret Sentries in Space* (New York: Random House, 1971); Anthony Kenden, "U.S. Reconnaissance Satellite Programme," *Spaceflight* (July 1978); Anthony Kenden, "Recent Developments in U.S. Reconnaissance Satellite Programmes," *Journal of the British Interplanetary Society* 35, no. 1 (1982): 31–44; Anthony Kenden, "A New Military Space Mission," *Journal of the British Interplanetary Society* 35, no. 10 (1982): 441–444; Gerald L. Borrowman, "Recent Trends in Orbital Reconnaissance," *Spaceflight* 24, no. 1 (1982): 10–13; Jeffrey Richelson, *United States Strategic Reconnaissance*, ACIS Working Paper (Los Angeles: Center for International and Strategic Affairs, University of California, Los Angeles, 1983); Jeffrey T. Richelson, *America's Secret Eyes in Space* (New York: Harper & Row, 1990); Curtis Peebles, *Guardians—Secret Reconnaissance Satellites* (Novato, CA: Presidio Press, 1987); William E. Burrows, *Deep Black: Space Espionage and National Security* (New York: Berkley Books, 1988).

187. James Bamford, *Body of Secrets: Anatomy of the Ultra-Secret National Security Agency* (New York: Anchor Books, 2001).

188. Jonathan McDowell, "US Reconnaissance Satellite Program Part I: Imaging Satellites," *Quest: The History of Spaceflight Magazine* 4, no. 2 (1995): 22–33; Jonathan McDowell, "U.S. Reconnaissance Satellite Programs Part 2: Beyond Imaging," *Quest: The History of Spaceflight Magazine* 4, no. 4 (1995): 40–45.

In 1995, the rush of works on the CORONA project based on declassified sources started with a public conference whose proceedings resulted in an edited work by Ruffner.[189] That same conference led also to Day et al., *Eye in the Sky*, which provides a number of excellent articles by historians and participants on CORONA.[190] Day also wrote an early, concise overview of CORONA in two *Quest* issues.[191] Day also followed with articles on other articles on various aspects of the CORONA program and its various camera systems,[192] as well as a variety of other reconnaissance and intelligence programs.[193] McDonald also wrote an early work on CORONA.[194] Not surprisingly, those best able to take advantage of the now-opened archives included those who had written on the subjects before. Peebles soon published an overview history of CORONA.[195] Richelson used these new sources, along with others, to publish a work on the Central Intelligence Agency's Directorate of Science and Technology.[196] Burrows's *This New Ocean*, which attempted a comprehen-

189. Kevin C. Ruffner, ed., *Corona: America's First Satellite Program* (Washington, DC: CIA History Staff, Center for the Study of Intelligence, 1995).

190. Dwayne A. Day, John M. Logsdon, and Brian Latell, eds., *Eye in the Sky: The Story of the Corona Spy Satellites* (Washington, DC: Smithsonian Institution Press, 1998).

191. Dwayne A. Day, "CORONA: America's First Spy Satellite Program," *Quest: The History of Spaceflight Magazine* 4, no. 2 (1995): 4–21; Dwayne A. Day, "CORONA: America's First Spy Satellite Program Part II," *Quest: The History of Spaceflight Magazine* 4, no. 3 (1995): 28–36.

192. Dwayne A. Day, "A Failed Phoenix: The KH-6 LANYARD Reconnaissance Satellite," *Spaceflight* 39, no. 5 (May 1997): 170–174; Dwayne A. Day, "Mapping the Dark Side of the World Part 1: The KH-5 ARGON Geodetic Satellite," *Spaceflight* 40, no. 7 (July 1998): 264–269; Dwayne A. Day, "Mapping the Dark Side of the World—Part 2: Secret Geodetic Programmes after ARGON," *Spaceflight* 40, no. 8 (August 1998): 303–310; Dwayne A. Day, "Falling Star," *Spaceflight* 40, no. 11 (1998): 442–445; Dwayne A. Day, "Lucky Number 13: The First Success of the CORONA Reconnaissance Satellite Program," *Spaceflight* 46, no. 4 (2004): 165–169; Dwayne A. Day, "First Light: The First Reconnaissance Satellite," *Spaceflight* 46, no. 8 (2004): 327–331.

193. Dwayne A. Day, "Recon for the Rising Sun," *Spaceflight* 41, no. 10 (1999): 420–423; Dwayne A. Day, "Medium Metal—The NRO's Smaller Satellites," *Spaceflight* 42, no. 1 (2000): 32–40; Dwayne A. Day, "Early American Ferret and Radar Satellites," *Spaceflight* 43, no. 7 (2001): 288–293; Dwayne A. Day, "Single Orbit Darts and Mercury Eyeballs: Early Unbuilt Strategic Reconnaissance Platforms," *Spaceflight* 43, no. 11 (2001): 468–470; Dwayne A. Day, "The Army–Air Force Space Race," *Spaceflight* 44, no. 7 (2002): 300–306; Dwayne A. Day, "Ferrets of the High Frontier: U.S. Air Force Ferret and Heavy Ferret Satellites of the Cold War," *Spaceflight* 46, no. 2 (2004): 74–81; Dwayne A. Day, "Pushing Iron: On-Orbit Support for Heavy Intelligence Satellites," *Spaceflight* 46, no. 7 (2004): 289–293.

194. Robert A. McDonald, *Corona Between the Sun and the Earth: The First NRO Eye in Space* (Annapolis Junction, MD: American Society for Photogrammetry and Remote Sensing, 1997); Robert A. McDonald, "CORONA: A Success for Space Reconnaissance, a Look into the Cold War, and a Revolution for Intelligence," *Photogrammetric Engineering and Remote Sensing* 51, no. 6 (1995): 689–720.

195. Curtis Peebles, *The CORONA Project: America's First Spy Satellites* (Annapolis, MD: Naval Institute Press, 1997).

196. Jeffrey T. Richelson, *The Wizards of Langley: Inside the CIA's Directorate of Science and Technology* (Boulder, CO: Westview Press, 2001).

sive history of the "First Space Age," used the new CORONA materials as well.[197] Taubman's *Secret Empire* is a more recent take on Eisenhower's support of CORONA and its predecessors.[198] Arnold's *Spying from Space* focuses on the command and control (C2) system set up for CORONA and deals with much of CORONA's early history as a result.[199] Temple's 2004 book *Shades of Grey* is another solid contribution to space reconnaissance history.[200] Day has a series of articles about the Samos program.[201] Hall describes the transfer of its camera technology to NASA's Lunar Orbiter, as does Day.[202]

RAND's part in the development of satellite reconnaissance is described in Davies and Harris, *RAND's Role in the Evolution of Balloon and Satellite Observations Systems and Related U.S. Space Technology*.[203] Peebles wrote about the balloon projects in *The Moby Dick Project*.[204] Hall sets the stage for satellite reconnaissance with a history of aerial overflights of the Soviet bloc.[205] U.S. Air Force Project 117L, which gave rise to CORONA as well as MIDAS, is discussed in Coolbaugh's 1998 article and in Perry's, as well as in Bowen's overviews of the genesis of military space efforts.[206] Other CORONA-related works include McDonald's edited *CORONA: Between the Sun and the Earth*,

197. Burrows, *This New Ocean*.

198. Philip Taubman, *Secret Empire: Eisenhower, the CIA, and the Hidden Story of America's Space Espionage* (New York: Simon & Schuster, 2003).

199. David Christopher Arnold, *Spying From Space: Constructing America's Satellite Command and Control Networks* (College Station: Texas A&M University Press, 2005).

200. L. Parker Temple, *Shades of Grey: National Security and the Evolution of Space Reconnaissance* (Reston, VA: AIAA, 2004).

201. Dwayne A. Day, "A Sheep in Wolf's Clothing: The Samos E-5 Recoverable Satellite—Part One," *Spaceflight* 44, no. 10 (2002): 424–431; Dwayne A. Day, "A Square Peg in a Cone-Shaped Hole: The Samos E-5 Recoverable Satellite—Part Two," *Spaceflight* 45, no. 2 (2003): 71–79; Dwayne A. Day, "From Cameras to Monkeys to Men: The Samos E-5 Recoverable Satellite—Part Three," *Spaceflight* 45, no. 9 (2003): 380–389.

202. R. Cargill Hall, *SAMOS to the Moon: The Clandestine Transfer of Reconnaissance Technology Between Federal Agencies* (Chantilly, VA: NRO History Office, October 2001); Dwayne A. Day, "From Above the Iron Curtain to Around the Moon: Lunar Orbiter and the Samos Spy Satellite," *Spaceflight* 47, no. 2 (2005): 66–71.

203. Merton E. Davies and William R. Harris, *RAND's Role in the Evolution of Balloon and Satellite Observation Systems and Related U.S. Space Technology* (Santa Monica, CA: RAND, 1988).

204. Curtis Peebles, *The Moby Dick Project: Reconnaissance Balloons Over Russia* (Washington, DC: Smithsonian Institution Press, 1991).

205. R. Cargill Hall, "The Truth about Overflights: Military Reconnaissance over Russia before the U-2, One of the Cold War's Best-Kept Secrets," *MHQ: The Quarterly Journal of Military History* 9, no. 3 (1997): 25–39.

206. J. S. Coolbaugh, "Genesis of the USAF's First Satellite Programme," *Journal of the British Interplanetary Society* 51, no. 8 (1998): 283–300; R. L. Perry, *Origins of the USAF Space Program, 1945–1956*, vol. 5, *History of DCAS, 1961*, Air Force Systems Command Historical Publications Series 62-24-10 (Los Angeles, CA: Air Force Systems Command, Space Systems Division, 1961); Lee Bowen, *The Threshold of Space: The Air Force in the National Space Program, 1945–1959* (Wright-Patterson AFB, OH: USAF Historical Division Liaison Office, 1960).

Oder et al.'s *The CORONA Story*, and Lindgren's *Trust but Verify*.[207] There have been concerns about errors in Lindgren's work.[208] Institutional works on the NRO I discuss later in this essay.

The politics of the freedom of space has been the focus of a number of historians. Stephen Ambrose, in his research on Dwight Eisenhower, was among the first to note the importance of the issue in 1981.[209] Rostow analyzed the Open Skies policy one year later.[210] McDougall's . . . *The Heavens and the Earth* provided the first full-length analysis of the issues involved.[211] Hall, with deeper archival research and materials available, revisited the topic in 1995.[212] Day followed with his assessment in 1998.[213] Neufeld revisited the issue in 2000.[214] The most recent assessment is by Bille and Lishock in 2004.[215]

Other relevant materials include McElheny's biography of Eastman Kodak's influential Edwin Land, as well as autobiographies of Richard Bissell and George Kistiakowsky.[216] Ranelagh's overview of the CIA, *The Agency*, contains some information on spy satellite programs.[217] The GRAB SIGINT satellite is described by a 1997 Naval Research Laboratory publication and

---

207. Robert A. McDonald, ed., *CORONA: Between the Sun and the Earth: The First NRO Reconnaissance Eye in Space* (Bethesda, MD: American Society for Photogrammetry and Remote Sensing, 1997); Frederic C. E. Oder, James C. Fitzpatrick, and Paul E. Worthman, *The CORONA Story* (Washington, DC: National Reconnaissance Office, 1997); David T. Lindgren, *Trust but Verify: Imagery Analysis in the Cold War* (Annapolis, MD: Naval Institute Press, 2000).

208. Dwayne A. Day, "*Trust but Verify: Imagery Analysis in the Cold War*, Review," *Technology and Culture* 42, no. 4 (2001): 822–823.

209. Stephen E. Ambrose, *Ike's Spies: Eisenhower and the Espionage Establishment* (Garden City, NY: Doubleday & Co., 1981).

210. W. W. Rostow, *Open Skies: Eisenhower's Proposal of July 21, 1955* (Austin: University of Texas Press, 1982).

211. McDougall, . . . *The Heavens and the Earth*.

212. R. Cargill Hall, "Origins of U.S. Space Policy: Eisenhower, Open Skies, and Freedom of Space," in *Exploring the Unknown: Selected Documents in the History of the U.S. Civil Space Program,* ed. John M. Logsdon, vol. 1, *Organizing for Exploration* (Washington, DC: NASA SP-4218, 1995), pp. 213–229; R. Cargill Hall, "The Eisenhower Administration and the Cold War: Framing American Astronautics to Serve National Security," *Prologue* 27, no. 1 (spring 1995): 59–72.

213. Dwayne A. Day, "A Strategy for Reconnaissance: Dwight D. Eisenhower and Freedom of Space," in *Eye in the Sky*, ed. Day, Logsdon, and Latell, pp. 119–142.

214. Michael J. Neufeld, "Orbiter, Overflight, and the First Satellite: New Light on the Vanguard Decision," in *Reconsidering Sputnik: Forty Years Since the Soviet Satellite*, ed. Roger D. Launius, John M. Logsdon, and Robert W. Smith (Amsterdam: Harwood Academic Publishers, 2000).

215. Matt Bille and Erika Lishock, *The First Space Race: Launching the World's First Satellites* (College Station: Texas A&M Press, 2004), chap. 4.

216. Victor K. McElheny, *Insisting on the Impossible: The Life of Edwin Land, Inventor of Instant Photography* (Reading, MA: Perseus Books, 1998); Richard M. Bissell with Jonathan E. Lewis and Francis T. Pudlo, *Reflections of a Cold Warrior: From Yalta to the Bay of Pigs* (New Haven, CT: Yale University Press, 1996); George Kistiakowsky, *A Scientist at the White House: The Private Diary of President Eisenhower's Special Assistant for Science and Technology* (Cambridge, MA: Harvard University Press, 1976).

217. John Ranelagh, *The Agency: The Rise and Decline of the CIA, from Wild Bill Donovan to William Casey* (New York: Simon & Schuster, 1986).

in Day's "Listening from Above."[218] Ball's *Pine Gap* provides information on U.S. signals intelligence, as do Pike's "CANYON, RHYOLITE, and AQUACADE" and Day's "Ferrets Above."[219] Bamford's 1982 *The Puzzle Palace*, 2002 *Body of Secrets*, and Lindsey's popular book *The Falcon and the Snowman* also provide information on spy satellites, in particular from the Boyce and Lee spy case.[220] An unusual and insightful look at a company's role is found in Lewis's *Spy Capitalism*, which discusses Itek Corporation.[221] Day provided a recent overview of the intelligence space program in 2002.[222]

Non-U.S. reconnaissance systems have significantly less literature. What exists is mostly concerned with the Soviet Union and Russia. Harvey's *Russia in Space* provides an overview.[223] Gorin describes Soviet and Russian optical reconnaissance systems articles in the *Journal of the British Interplanetary Society*, as does Clark.[224] Clark also describes Chinese recoverable satellites, which are probably partly military in nature, in a 1998 *Quest* article.[225] Zorn has a short article on the development of the Israeli satellite intelligence program.[226] A flavor of the interactions between military and civilian systems can be seen in Baker et al., Steinberg, and Dehqanzada and Florini.[227] There are no histories

---

218. *GRAB: Galactic Radiation and Background* (Washington, DC: Naval Research Laboratory, 1997); Dwayne A. Day, "Listening from Above: The First Signals Intelligence Satellite," *Spaceflight* 41, no. 8 (August 1999): 338–346.

219. Desmond Ball, *Pine Gap: Australia and the U.S. Geostationary Signals Intelligence Satellite Program* (Sydney: Allen & Unwin, 1988); Christopher Anson Pike, "CANYON, RHYOLITE, and AQUACADE," *Spaceflight* 37, no. 11 (November 1995): 381–383; Dwayne A. Day, "Ferrets Above: American Signals Intelligence Satellites During the 1960s," *International Journal of Intelligence and CounterIntelligence* 17, no. 3 (2004): 449–467.

220. James Bamford, *The Puzzle Palace: A Report on NSA, America's Most Secret Agency* (Boston: Houghton Mifflin, 1982); James Bamford, *Body of Secrets: Anatomy of the Ultra-Secret National Security Agency* (New York: Anchor, 2002); Robert Lindsey, *The Falcon and the Snowman: A True Story of Friendship and Espionage* (New York: Simon & Schuster, 1979).

221. Jonathan E. Lewis, *Spy Capitalism: ITEK and the CIA* (New Haven, CT: Yale University Press, 2002).

222. Day, "Intelligence Space Program," pp. 371–388.

223. Harvey, *Russia in Space: The Failed Frontier?*

224. Peter Gorin, "Zenit—the First Soviet Photoreconnaissance Satellite," *Journal of the British Interplanetary Society* 50, no. 11 (1997): 441–448; Peter Gorin, "Black Amber—Russian Yantar-Class Optical Reconnaissance Satellites," *Journal of the British Interplanetary Society* 51 (1998): 309–320; P. S. Clark, "Russian Fifth Generation Photoreconnaissance Satellites," *Journal of the British Interplanetary Society* 52 (1999): 133–150.

225. Phillip S. Clark, "Development of China's Recoverable Satellites," *Quest: The History of Spaceflight Quarterly* 6, no. 2 (1998): 36–43.

226. E. L. Zorn, "Israel's Quest for Satellite Intelligence," *Studies in Intelligence* 10 (winter–spring 2001): 33–38.

227. John C. Baker, Kevin M. O'Connell, and Ray A. Williamson, eds., *Commercial Observation Satellites: At the Leading Edge of Global Transparency* (Santa Monica, CA: RAND and ASPRS, 2001); Gerald M. Steinberg, *Commercial Observation Satellites in the Middle East and Persian Gulf* (Santa Monica, CA: RAND, 2001); Yahya A. Dehqanzada and Ann M. Florini, *Secrets for Sale: How Commercial Satellite Imagery Will Change the World* (Washington, DC: Carnegie Endowment for International Peace, 2000).

yet of European, Japanese, or other military space reconnaissance systems, but some information on these can be found at the Federation of American Scientists Web site[228] and Internet searches of newspapers and blogs.

Finally, an area garnering recent attention is the use of satellite reconnaissance data for a variety of intelligence purposes. This is shown by a recent spate of work on American assessment (largely based on satellite imagery) of the Soviet manned lunar program in the 1960s. The best research on this so far is a two-part series, "The Moon in the Crosshairs," by Day and Siddiqi in 2003 and 2004.[229] Day has followed this with several other articles.[230] Pesavento and Vick have also ventured into this territory, although some of their claims have been challenged.[231]

MILITARY HUMAN SPACEFLIGHT

The American and Soviet military services have been involved with human spaceflight programs from the late 1950s to the present, starting with supplying astronauts and cosmonauts to the fledgling human flight programs, moving on to studies and designs for piloted space reconnaissance and bombing vehicles, and then designing and operating manned military space stations. While most people realize that many astronauts and cosmonauts have been military pilots, few have pondered why the military lent many of its top personnel to civilian spaceflight programs. Even fewer people realize that the U.S. and USSR have had manned military space programs and that the Soviets even operated manned military space stations in the 1970s.

Eugen Sänger developed the idea of a manned space bomber in the 1940s and studied the concept in World War II Nazi Germany. This "Silver Bird" vehicle would drop a bomb on New York, skip off the atmosphere, and return to Germany. Walter Dornberger, who headed the German Army's ballistic missile efforts in World War II, brought the idea to the Bell Aircraft Corporation

---

228. See *http://www.fas.org.*

229. Dwayne A. Day and Asif Siddiqi, "The Moon in the Crosshairs: CIA Intelligence on the Soviet Manned Lunar Programme, Part 1—Launch Complex J," *Spaceflight* 45, no. 11 (2003): 466–475; Dwayne A. Day and Asif Siddiqi, "The Moon in the Crosshairs: CIA Intelligence on the Soviet Manned Lunar Programme, Part 2—The J Vehicle," *Spaceflight* 46, no. 3 (2004): 10–11, 114–125.

230. Dwayne A. Day, "The Secret of Complex J," *Air Force* 87, no. 7 (July 2004): 72–76; Dwayne A. Day, "In the Shadows of the Moon Race," *Spaceflight* 46, no. 11 (2004): 436–440; Dwayne A. Day, "From the Shadows to the Stars: James Webb's Use of Intelligence Data in the Race to the Moon," *Air Power History* 51, no. 4 (winter 2004).

231. Peter Pesavento and Charles P. Vick, "The Moon Race 'End Game': A New Assessment of Soviet Crewed Lunar Aspirations—Part 1," *Quest: The History of Spaceflight Quarterly* 11, no. 1 (2004): 6–30; Peter Pesavento and Charles P. Vick, "The Moon Race 'End Game': A New Assessment of Soviet Crewed Lunar Aspirations—Part 2," *Quest: The History of Spaceflight Quarterly* 11, no. 2 (2004): 6–57.

in the United States, which in 1952 proposed to study the concept further with USAF funding. The Bell study, along with the USAF's preference for manned bombers over missile systems, resulted in the USAF issuing requirements for a hypersonic strategic bombardment system in 1955. Several feasibility studies were consolidated in October 1957 into the Dyna-Soar program, which would initially design a hypersonic manned research vehicle. By late 1961, with the mass of Dyna-Soar growing and Soviet competition increasing with Gagarin's flight, the USAF dropped suborbital tests and approved the development of the powerful Titan III launcher to put Dyna-Soar into space. However, the success of CORONA and the Soviet Zenit systems ensured that priority for both nations' military space efforts went to reconnaissance satellites. By 1963, each side was willing to tolerate each other's reconnaissance satellites, and threats to this toleration such as potential antisatellite systems like Dyna-Soar were unwelcome. Secretary of Defense Robert McNamara, who was skeptical of its mission, canceled it in December 1963.

However, McNamara agreed that piloted reconnaissance platforms had military potential, so at the same time that he canceled Dyna-Soar, he approved the Manned Orbiting Laboratory (MOL) program to investigate. MOL's immediate lineage included ideas to modify the Gemini capsule—the so-called "Blue Gemini" program—as part of a military space station program called the "Manned Orbital Development System." When the DOD began to consider taking over Gemini, NASA objected vociferously, and the DOD backed down. Ultimately, the USAF decided to modify the Gemini capsule to transport astronauts to the MOL, which would be carried behind the capsule on a Titan III launcher. As MOL's schedule slipped and its cost grew, NASA pushed its Apollo Applications Program (soon to become *Skylab*) and the Vietnam War intensified, increasing pressure to cancel MOL. The success of CORONA and the need for funds to develop its successor robotic reconnaissance craft (the KH-9 Hexagon) led to MOL's cancellation in June 1969.

Human military spaceflight did not end with MOL, as the military considered its participation in NASA's Space Shuttle program. The military's requirements significantly influenced the Shuttle's design, and in the late 1970s, the USAF prepared to fly Shuttle missions by building its own operations center and launch facility, as well as training military astronauts for classified missions. In the 1980s, U.S. military men flew a number of classified missions on the Space Shuttle, the details of which generally remain hidden from the public. One of the missions is known to have deployed two Defense Satellite Communications System III satellites. Others were most likely National Reconnaissance Office missions to deploy various reconnaissance systems. However, the *Challenger* accident of 1986 and the resulting new priorities for the Shuttle soon ended military Shuttle missions.

Similar aspirations for human military missions also spurred the Soviets to develop programs. A "Raketoplan" explored concepts similar to Dyna-Soar. The Soviets also undertook a military space station program. Officially called Salyut, the second, third, and fifth were all Almaz military stations, launched in 1973, 1974, and 1976. Soyuz missions 14, 15, 21, 23, and 24 were all military missions to the *Salyut 3* and *5* stations, performing a variety of military tasks, mostly to determine the value of using cosmonauts for reconnaissance. After these missions, the Soviets concluded that automated satellites were more effective than humans in space, as the humans had limited amounts of time available for observations, as they had to eat, sleep, and maintain the station. This, combined with the much higher costs of human flights, ended human military missions.

Both American and Soviet armed forces also lent military pilots to their respective civilian space programs. From World War II to the early 1960s, military test pilots aimed to go higher and faster, and their efforts, along with the medical experiments, observations, and flight suits made along the way, paved the way for civilian space missions. In the 1950s and 1960s, the relatively high prestige of spaceflight and the potential for human military missions in space made this a reasonable proposition for the armed forces. Before NASA's creation, the military controlled the space program by default. The Army and Air Force competed in early studies and proposals to put humans in space, including the Army's Project Adam and Project Horizon and the Air Force's "Man-In-Space-Soonest," which had one of the worst acronyms possible, MISS.

With NASA's creation, the military's role changed from one of leadership and control to one of support. Over time, as human-piloted missions and crewed space stations faded from military viability in the 1970s and 1980s, the number of military personnel becoming astronauts and cosmonauts has decreased somewhat. The military rationales for the continued movement of military pilots into civilian space programs have become less clear and, to date, have not been investigated by historians. Also, the military continues to support human flight programs with launch range support and a variety of other capabilities. These have declined over time as the civilian programs have frequently developed their own capabilities for astronaut testing, etc.

Myhra describes Sänger's early orbital bomber program in Nazi Germany.[232] Killebrew gives a history of the USAF's efforts to find a role for military men in space.[233] A short history of Dyna-Soar can be found in *Quest* issue 3, num-

---

232. David Myhra, *Sänger: Germany's Orbital Rocket Bomber in World War II* (Atglen, PA: Schiffer, 2002).

233. Major Timothy D. Killebrew, *Military Man in Space: A History of Air Force Efforts to Find a Manned Space Mission*, Air Command and Staff College Report No. 87-1425 (Maxwell AFB, AL: Air University Press, 1987).

ber 4, which has a number of articles on the program, particularly those by Houchin and by Smith.[234] Houchin's 1997 *Journal of the British Interplanetary Society* article is also insightful.[235] Godwin's recent book on Dyna-Soar is a compilation of original documents.[236] MOL's history is also relatively obscure. Both Peebles[237] and Pealer[238] created three-part series on the project. Houchin's 1995 article investigates the question of NASA's relationship to MOL.[239] Spires's *Beyond Horizons* also describes these programs, along with earlier efforts, such as MISS.[240] Strom provides a brief history of MOL.[241] Jenkins's *Space Shuttle* describes its first hundred missions, a number of which were classified military missions.[242] Day provides an overview of NASA-DOD relations in an overview article in *Exploring the Unknown*.[243] Powell and Day describe military Shuttle missions.[244]

Siddiqi covers the 1960s development of the Soviet Raketoplan and Spiral, along with the 1960s development of Almaz, in *Challenge to Apollo*.[245] He also describes the military Almaz program and consequent Soyuz flights to the military stations in two articles in the *Journal of the British Interplanetary*

234. Roy F. Houchin II, "Why the Air Force Proposed the Dyna-Soar X-20 Program," *Quest: The History of Spaceflight Magazine* 3, no. 4 (1994): 5–12; Terry Smith, "The Dyna-Soar X-20: A Historical Overview," *Quest: The History of Spaceflight Magazine* 3, no. 4 (1994): 13–18; Roy F. Houchin II, "Why the Dyna-Soar X-20 Program Was Cancelled," *Quest: The History of Spaceflight Magazine* 3, no. 4 (1994): 35–37.

235. Roy F. Houchin II, "Air Force-Office of the Secretary of Defense Rivalry."

236. Robert, Godwin, ed., *Dyna-Soar Hypersonic Strategic Weapons System* (Burlington, Ontario: Apogee, 2003).

237. Curtis Peebles, "The Manned Orbiting Laboratory—Part 1," *Spaceflight* 22, no. 4 (1980): 155–160; Curtis Peebles, "The Manned Orbiting Laboratory—Part 2," *Spaceflight* 22, nos. 7–8 (1980): 270–272; Curtis Peebles, "The Manned Orbiting Laboratory—Part 3," *Spaceflight* 24, no. 6 (1982): 274–277.

238. Donald Pealer, "Manned Orbiting Laboratory (MOL) Part I," *Quest: The History of Spaceflight Magazine* 4, no. 3 (1995): 4–17; Donald Pealer, "Manned Orbiting Laboratory (MOL) Part II," *Quest: The History of Spaceflight Magazine* 4, no. 4 (1995): 28–35; Donald Pealer, "Manned Orbiting Laboratory (MOL) Part 3," *Quest: The Magazine of Spaceflight* 5, no. 2 (1996): 16–23.

239. Roy F. Houchin II, "Interagency Rivalry, NASA, The Air Force, and MOL," *Quest: The History of Spaceflight Quarterly* 4, no. 4 (1995): 36–39.

240. Spires, *Beyond Horizons*.

241. Steven R. Strom, "The Best Laid Plans: A History of the Manned Orbiting Laboratory," *Crosslink* 5, no. 2 (summer 2004): 11–15.

242. Dennis R. Jenkins, *Space Shuttle: The History of the National Space Transportation System: The First 100 Missions* (Stillwater, MN: Voyageur Press, 2001).

243. Dwayne A. Day, "Invitation to Struggle: The History of Civilian-Military Relations in Space," in *Exploring the Unknown: Selected Documents in the History of the U.S. Civilian Space Program*, ed. John M. Logsdon, vol. 2, *External Relationships* (Washington, DC: NASA SP-4407, 1996), pp. 233–270.

244. Joel W. Powell, "'Secret' Shuttle Payloads Revealed," *Spaceflight* 35, no. 5 (1993): 152–154; Dwayne A. Day, "Secret Shuttle Mission Revealed," *Spaceflight* 40, no. 7 (1998): 256–257.

245. Siddiqi, *Challenge to Apollo*.

Society.[246] Lantratov describes the early Soyuz manned reconnaissance designs.[247] Zimmerman's recent history of space stations also briefly discusses the Soviet military missions.[248] Pesavento[249] describes the Russian shuttle projects, as does Garber.[250]

The military's ballooning experiments at extreme altitudes are described in Ryan's *The Pre-Astronauts*, as well as DeVorkin's *Race to the Stratosphere.*[251] Gantz provides a late-1950s view of USAF astronaut training, and Erickson's dissertation looks at this as one aspect of a larger NASA-DOD relationship.[252] Military involvement with the development of spacesuits is described in Harris's *The Origins and Technology of the Advanced Extra-Vehicular Spacesuit.*[253] Mallan, De Monchaux, and Kozloski also have monographs on the history of spacesuits, including their military origins.[254] There are no published overview histories of military test-pilot training, aerospace medicine, creation of launch facilities and range support, etc. On aerospace medicine, the best source so far is Mackowski's 2002 dissertation.[255] Important early sources include Armstrong's *Aerospace Medicine* and Campbell's *Earthman/Spaceman/*

246. Asif A. Siddiqi, "The Almaz Space Station Complex: A History, 1964–1992, Part 1: 1964–1976," *Journal of the British Interplanetary Society* 52, no. 11/12 (2001): 389–416; Asif A. Siddiqi, "The Almaz Space Station Complex: A History, 1964–1992, Part 2: 1976–1992," *Journal of the British Interplanetary Society* 55, no. 1/2 (2002): 35–67.

247. Konstantin Lantratov, "Soyuz-Based Manned Reconnaissance Spacecraft," trans. Bart Hendrickx, *Quest: The History of Spaceflight Quarterly* 6, no. 1 (1998): 5–21.

248. Robert Zimmerman, *Leaving Earth: Space Stations, Rival Superpowers, and the Quest for Interplanetary Travel* (Washington, DC: Joseph Henry Press, 2003).

249. Peter Pesavento, "Russian Space Shuttle Projects, 1957–1994, Part 1," *Spaceflight* 37, no. 5 (1995): 158–164; Peter Pesavento, "Russian Space Shuttle Projects, 1957–1994, Part 2," *Spaceflight* 37, no. 6 (1995): 192–199; Peter Pesavento, "Russian Space Shuttle Projects, 1957–1994, Part 3," *Spaceflight* 37, no. 7 (1995): 226–233; Peter Pesavento, "Russian Space Shuttle Projects, 1957–1994, Part 4," *Spaceflight* 37, no. 8 (1995): 264–266.

250. Steve Garber, "A Cold Snow Falls: The Soviet Buran Space Shuttle," *Quest: The History of Spaceflight Quarterly* 9, no. 5 (2002): 42–51.

251. Craig Ryan, *The Pre-Astronauts: Manned Ballooning on the Edge of Space* (Annapolis, MD: Naval Institute Press, 1995); David H. DeVorkin, *Race to the Stratosphere: Manned Scientific Ballooning in America* (New York: Springer-Verlag, 1989).

252. Kenneth F. Gantz, ed., *Men in Space: The United States Air Force Program for Developing the Spacecraft Crew* (New York: Duell, 1959); Mark A. Erickson, "The Evolution of the NASA-DoD Relationship from Sputnik to the Lunar Landing" (Ph.D. diss., The George Washington University, 1997).

253. Gary L. Harris, *The Origins and Technology of the Advanced Extra-Vehicular Spacesuit*, AAS History Series, no. 24 (San Diego: Univelt, 2001).

254. Lloyd Mallan, *Suiting Up for Space: The Evolution of the Space Suit* (New York: John Day Company, 1971); Nicholas De Monchaux, *Space Suit* (New York: Springer Verlag, 2002); Lillian D. Kozloski, *U.S. Space Gear: Outfitting the Astronaut* (Washington, DC: Smithsonian Institution Press, 2000).

255. Maura Phillips Mackowski, "Human Factors: Aerospace Medicine and the Origins of Manned Space Flight in the United States" (Ph.D. diss., Arizona State University, 2002).

UNIVERSAL MAN.[256] Early studies of USAF experiments related to human spaceflight can be found in Mallan and Meeter.[257] Information on military astronauts and their training can be found indirectly through numerous astronaut biographies and autobiographies, which I will not list here. Also, Swenson et al.'s early history of Mercury, *This New Ocean*, discusses some of the early military-based astronaut training and selection.[258] Siddiqi's *Challenge to Apollo* describes similar military origins for cosmonauts.[259]

In 1959, Singer discussed the potential of military Moon bases.[260] Springer describes the U.S. Army's Project Adam in a 1994 *Quest* article and the Army's Project Horizon Moon base study in his 1999 "Securing the High Ground."[261] Burrows and Richelson also discuss military Moon base efforts.[262] Stoff describes plans for a military version of the Apollo Lunar Module.[263]

WEATHER AND SCIENCE

The military has funded and developed a variety of experiments and systems to understand space and atmospheric environments and to support space operations. This intersects with literature in the history of science in the development of space science and meteorology. Prior to NASA's existence, space science was almost exclusively funded by the military. The military has had scientific advisers ever since World War II to help guide its technology and scientific programs. The Office of Naval Research became a "proto–National Science Foundation" in the late 1940s and 1950s, funding a variety of research, while the USAF established a Scientific Advisory Board that periodically provided studies and advice, as well as a Chief Scientist's Office to coordinate with academic advisers. The military as a whole used the Research

---

256. Harry G. Armstrong, *Aerospace Medicine* (Baltimore, MD: Williams & Wilkins, 1961); Paul A. Campbell, *Earthman/Spaceman/UNIVERSAL MAN?* (New York: Pageant Press, 1965).

257. Lloyd Mallan, *Men, Rockets, and Space Rats* (New York: Messner, 1961); George F. Meeter, *The Holloman Story: Eyewitness Accounts of Space Age Research* (Albuquerque: University of New Mexico Press, 1967).

258. Loyd S. Swenson, Jr., James M. Grimwood, and Charles C. Alexander, *This New Ocean: A History of Project Mercury* (Washington, DC: NASA SP-4201, 1966).

259. Siddiqi, *Challenge to Apollo.*

260. Lieutenant Colonel S. E. Singer, "The Military Potential of the Moon," *Air University Quarterly Review* 11 (summer 1959): 31–53.

261. Anthony M. Springer, "PROJECT ADAM, The Army's Man in Space Program," *Quest: The History of Spaceflight Magazine* 3, nos. 2–3 (1994): 46–47; Anthony M. Springer, "Securing the High Ground: The Army's Quest for the Moon," *Quest: The History of Spaceflight Quarterly* 7, no. 2 (1999): 34–39.

262. William E. Burrows, "Securing the High Ground," *Air & Space Smithsonian* 8 (December 1993/January 1994): 64–69; Jeffrey T. Richelson, "Shootin' for the Moon," *Bulletin of the Atomic Scientists* (September/October 2000).

263. Joshua Stoff, "The Lunar Module's Evil Twin," *Air and Space* (October/November 2000).

Technicians check out DMSP Block 5D-3 satellite, late 1990s. *(Official USAF photo. Air Force Space Command, Office of History)*

and Development Board, which was to help coordinate academic efforts for science and technology development after World War II and into the 1950s.

Science experiments aboard American V-2 rocket firings in the late 1940s and early 1950s were coordinated by the Naval Research Laboratory. These military-supported experiments, along with a variety of ground-based studies of the upper atmosphere, were the training ground for many of NASA's early space scientists. Similarly, all of the early space science experiments placed on board pre-NASA Explorer and Pioneer missions were military-funded.

The U.S. Army developed the initial Television and Infrared Observation System (TIROS) weather satellite program, which it turned over to NASA in 1958. The military continued funding certain aspects of space science even after NASA's arrival on the scene in late 1958 and created its own operational programs to monitor Earth and space weather due to their impact on a variety of military operations. The National Reconnaissance Office modified the TIROS design to create the Defense Meteorological Satellite Program (DMSP), which was to ensure that CORONA photography over the Soviet Union took pictures of the ground instead of cloud tops. DMSP continued under USAF control until 1998, supporting a variety of tactical as well as strategic uses. In May 1998, operational responsibility for DMSP transferred to the National Oceanic and Atmospheric Administration (NOAA). Interestingly, the National Weather Service used the DMSP as the basis for its operational satellites in the 1960s instead of NASA's Nimbus. In the early 21st century, military and civilian needs are to be met with the National Polar-Orbiting Environmental Satellite System (NPOESS).

As the impact of solar storms on radio communication became increasingly apparent, both civilian and military groups established groups to monitor space weather and issue warnings and advisories to satellite operators. In the 1980s and 1990s, the military's desire to test ABM technologies in space without violating the ABM Treaty led to the Clementine program, which found surprising evidence for water on the Moon. In the Soviet Union, the Meteor weather satellite program was a military-civilian system from the start, with military specifications provided by the Third Directorate of the Chief Directorate of Reactive Armaments (GURVO) and the design handled by the All-Union Scientific Research Institute of Electromechanics (VNIIEM).

Another major scientific and application initiative was the development of geodesy. This was crucial for military operations planning, both for airborne and ballistic missile strikes from the U.S. to the USSR and vice versa. In the 1950s, knowledge of the exact size and shape of the Earth was insufficient for ballistic missile targeting, as the uncertainty in the distance from North American to Asia was in error between 20 to 30 miles. In addition, the Earth's shape influences the gravity field, which affects ballistic missile trajectories. Thorough mapping of the Earth's surface was essential and was advocated by

Amrom Katz of RAND Corporation in the late 1950s. Development work began on mapping cameras for the USAF Samos program. However, mapping from space began in earnest with the U.S. Army in 1959, when it started the Argon program, which put the KH-5 mapping camera on board CORONA spacecraft. Other mapping cameras were also developed and integrated with the CORONA program.

The other aspect to geodesy was the study of the Earth's gravitational field through experimental satellites. Scientists developed several techniques. One was to measure a satellite's position in orbit through visual sightings at different points on the Earth, such as occurred with the 1960s American Echo 1 and PAGEOS (Passive Geodetic Earth Orbiting Satellite) satellites. Another method was to have a satellite send two radio signals at differing wavelengths and then observe the Doppler-effect frequency shifts from the ground. The U.S. Transit system, as well as the French Diapason and Diademe satellites of the 1960s, operated with this principle. Passive satellites with mirrors that can reflect laser beams from Earth have also been launched, such as the French Starlette. Military geodetic satellites have generally predated civilian systems, and civilian geodetic experiments have been among the first satellites of nations with ballistic missiles, such as France and China. The U.S. military began its Anna 1A and 1B optical ranging satellites in 1962, followed quickly by the Gravity Gradient Stabilization Experiment satellites, the Sequential Collation of Range satellites, and the Geodetic Earth Orbiting Satellite. The Soviets started their geodesy experiments with the Sfera series in 1968, followed by the Musson series beginning in 1980. The U.S. Global Positioning System is also used for geodetic purposes.

Sapolsky's history of the Office of Naval Research is a good introduction to the role of ONR.[264] Van Keuren narrates the scientific cover for intelligence gathering by the Naval Research Laboratory, while McDowell provides an overview of its satellites.[265] Leslie's *The Cold War and American Science* describes military interactions with MIT and Stanford, including some related to space.[266] The role of Johns Hopkins University's Applied Physics Laboratory is told by Klingaman.[267] Sturm describes the creation and evolution of the USAF's

264. Harvey M. Sapolsky, *Science and the Navy: The History of the Office of Naval Research* (Princeton: Princeton University Press, 1990).

265. D. K. Van Keuren, "Cold War Science in Black and White: US Intelligence Gathering and Its Scientific Cover at the Naval Research Laboratory, 1948–62," *Social Studies of Science* 31, no. 2 (2001): 207–229; Jonathan McDowell, "Naval Research Laboratory Satellites 60-89," *Journal of the British Interplanetary Society* 50, no. 11 (1997): 427–432.

266. Stuart W. Leslie, *The Cold War and American Science: The Military-Industrial-Academic Complex at MIT and Stanford* (New York: Columbia University Press, 1993).

267. William K. Klingaman, *APL—Fifty Years of Service to a Nation: A History of the Johns Hopkins University Applied Physics Laboratory* (Laurel, MD: The Johns Hopkins University Applied Physics Laboratory, 1993).

Scientific Advisory Board (SAB) up to 1964.[268] Gorn's *Harnessing the Genie* also discusses the SAB in its relation to technology forecasting.[269] Komons describes the history of the USAF Office of Scientific Research up to the early 1960s.[270] Day's *Lightning Rod* narrates the history of the USAF Office of Chief Scientist.[271] Liebowitz's chronology provides information on the Cambridge Field Station and its evolution to the Air Force Geophysics Laboratory.[272]

Dick describes the long history of the U.S. Naval Observatory and its relationship to astronomy and space science.[273] Doel's general history of pre–Space Age planetary science contains important information about the military's role in its creation.[274] The history of the American V-2 experiments is told in DeVorkin's *Science with a Vengeance*.[275] Bille and Lishock's *The First Space Race* describes the military's role in launching the first satellites, including scientific aspects.[276] Needell's *Science, Cold War and the American State* portrays military-science relationships through the life of Lloyd Berkner, a leader of early Cold War atmospheric and space science.[277] Newell's *Beyond the Atmosphere* and Butrica's *To See the Unseen* both begin with descriptions of military-funded or -approved space science prior to the founding of NASA.[278] Vanguard, along with its Navy origins and science, is described in Green's early NASA history.[279] Paulikas and Strom describe The Aerospace Corporation's early efforts in understanding the space environment.[280] Hendrickx narrates

268. Thomas A. Sturm, *The USAF Scientific Advisory Board: Its First Twenty Years, 1944–1964* (Washington, DC: Office of Air Force History, 1986).

269. Michael H. Gorn, *Harnessing the Genie: Science and Technology Forecasting for the Air Force, 1944–1986* (Washington, DC: Office of Air Force History, 1988).

270. Nick A. Komons, *Science and the Air Force: A History of the Air Force Office of Scientific Research* (Arlington, VA: Historical Division, Office of Information, Office of Aerospace Research, 1966).

271. Dwayne A. Day, *Lightning Rod: A History of the Air Force Chief Scientist Office* (Washington, DC: JSAF Chief Scientist's Office, 2000).

272. Ruth P. Liebowitz, *Chronology: From the Cambridge Field Station to the Air Force Geophysics Laboratory, 1945–1985* (Hanscom AFB, MA: AF Geophysics Laboratory, 1985).

273. Steven J. Dick, *Sky and Ocean Joined: The U.S. Naval Observatory, 1830–2000* (New York: Cambridge University Press, 2003).

274. Ronald E. Doel, *Solar System Astronomy in America: Communities, Patronage, and Interdisciplinary Research, 1920–1960* (Cambridge: Cambridge University Press, 1996).

275. David H. DeVorkin, *Science with a Vengeance: How the Military Created the US Space Sciences after World War II* (New York: Springer-Verlag, 1992).

276. Bille and Lishock, *The First Space Race.*

277. Allan A. Needell, *Science, Cold War and the American State: Lloyd V. Berkner and the Balance of Professional Ideals* (Amsterdam: Harwood Academic Publishers, 2000).

278. Homer E. Newell, *Beyond the Atmosphere: Early Years of Space Science* (Washington, DC: NASA SP-4211, 1980); Andrew J. Butrica, *To See the Unseen: A History of Planetary Radar Astronomy* (Washington, DC: NASA SP-4218, 1996).

279. Green and Lomask, *Vanguard: A History.*

280. George A. Paulikas and Steven R. Strom, "A Decade of Space Observations: The Early Years of the Space Physics Laboratory," *Crosslink* 4, no. 2 (2003): 6–9.

the story of the Soviet Elektron program, which was a scientific response to U.S. discoveries with Explorer.[281]

Day describes the Argon system and other mapping programs linked to CORONA.[282] Geodesy and its links to military space have become topics for recent research, particularly a series of articles by Warner[283] and another series by Cloud.[284] *The Cambridge Encyclopedia of Space* has a good introduction to geodesy that describes the various geodesy missions.[285] Doel has recently ventured into the military's influence on Earth science as well.[286] Cloud looks at the links between the intelligence and civilian remote sensing programs.[287]

The best overview of the origins of the Defense Meteorological Satellite Program is Hall's recently declassified article.[288] This same 2002 *Quest* issue also contains an informative interview with the program's first manager, Thomas Haig.[289] Abel gives a history of DMSP up to 1982; Brandli shows how DMSP was used in Southeast Asia in the 1960s and 1970s; and Day provides a short history on the origins of the program.[290] Bates and Fuller give a gen-

281. Bart Hendrickx, "Elektron: The Soviet Response to Explorer," *Quest: The History of Spaceflight Quarterly* 8, no. 1 (2000): 37–45.

282. Day, "Mapping the Dark Side of the World Part 1," pp. 264–269; Day, "Mapping the Dark Side of the World Part 2," pp. 303–310.

283. Deborah Jean Warner, "Political Geodesy: The Army, the Air Force, and the World Geodetic System of 1960," *Annals of Science* 59, no. 4 (2002): 363–389; Deborah Jean Warner, "From Tallahassee to Timbuktu: Cold War Efforts to Measure Intercontinental Distances," *Historical Studies in the Physical and Biological Sciences* 30, no. 2 (2000): 393–415.

284. John Cloud, "Crossing the Olentangy River: The Figure of the Earth and the Military-Industrial-Academic Complex, 1947–1972," *Studies in the History and Philosophy of Modern Physics* 31B, no. 3 (2000): 371–404; John Cloud, "Imaging the World in a Barrel: CORONA and the Clandestine Convergence of the Earth Sciences," *Social Studies of Science* 31, no. 2 (2001): 231–251; John Cloud, "Hidden in Plain Sight: The CORONA Reconnaissance Satellite Programme and Clandestine Cold War Science," *Annals of Science* 58, no. 2 (2001): 203–209; John Cloud, "Reviewing the Earth: Remote Sensing and Cold War Clandestine Knowledge Production," *Quest: The History of Spaceflight Quarterly* 8, no. 3 (2001): 4–16.

285. Fernand Verger, Isabelle Sourbès-Verger, and Reymond Ghirardi, with contributions by Xavier Pasco, *The Cambridge Encyclopedia of Space: Missions, Applications, and Exploration* (Cambridge: Cambridge University Press, 2003).

286. Ronald E. Doel, "Constituting the Postwar Earth Sciences: The Military's Influence on the Environmental Sciences in the USA after 1945," *Social Studies of Science* 33, no. 5 (2003): 635–666.

287. Cloud, "Re-Viewing the Earth."

288. R. Cargill Hall, "A History of the Military Polar Orbiting Meteorological Satellite Program," *Quest: The History of Spaceflight Quarterly* 9, no. 2 (2002): 4–25.

289. David Arnold, "An Interview with Colonel Thomas O. Haig," *Quest: The History of Spaceflight Quarterly* 9, no. 2 (2002): 53–61.

290. Major Michael D. Abel, *History of the Defense Meteorological Satellite Program: Origin Through 1982*, Air Command and Staff College Report No. 87-0020 (Maxwell AFB, AL: Air University Press, 1987); Henry W. Brandli, "The Use of Meteorological Satellites in Southeast Asia Operations," *Aerospace Historian* 29, no. 3 (September 1982): 172–175; Dwayne A. Day, "Dark Clouds: The Classified Origins of the Defense Meteorological Satellite Program," *Spaceflight* 43, no. 9 (2001): 382–385.

eral history of military weather forecasting, while Nebeker provides a general history of which the military is a part.[291] Gavaghan discusses the military origins of TIROS and early weather satellites in *Something New Under the Sun*, based largely on interviews with Verner Suomi.[292] Hendrickx's 2004 history of Meteor is the best source for the Soviet and Russian weather satellites.[293]

Space weather and its relationship to the Sun have received little historical attention. Hufbauer's *Exploring the Sun* describes some USAF efforts in solar and space weather observations. Myers wrote a study of space weather operations.[294]

NAVIGATION

Developed initially for nuclear warfare, space-based navigation has become a worldwide commercial and civilian utility, as well as a major contributor to conventional warfare. Space-based navigation developed from ideas generated from tracking the first satellites from Earth. Scientists worked out the nuances of determining precise satellite positions and orbital trajectories. Once they determined the orbital positions and parameters with precision, scientists at Johns Hopkins University Applied Physics Laboratory realized that it was possible to reverse the procedure. Knowing precise positions in orbit, one can use satellites to determine precise positions on Earth. This would be extremely useful for ships, which had to calculate their positions on featureless oceans. Thus was born the Transit program, which used the Doppler effect from satellite radio signals to determine ship and submarine positions. The U.S. Navy was particularly interested, because it needed precise position measurements for its Polaris submarines to determine the initial firing positions of submarine-launched ballistic missiles.

Transit worked well for ships but was inadequate for aircraft, because its signals were useful in only two dimensions and there were not enough Transit satellites to ensure that there were enough signals to triangulate positions at all times. The U.S. Army, Navy, and Air Force all experimented in the 1960s with technologies to improve upon Transit, but each had different capabilities. In 1973, the Secretary of Defense ordered the combination of the various programs and technologies into the Navstar Global Positioning System (GPS)

---

291. Charles C. Bates and John F. Fuller, *America's Weather Warriors: 1814–1985* (College Station: Texas A&M University Press, 1986); Frederik Nebeker, *Calculating the Weather: Meteorology in the 20th Century* (New York: Academic Press, 1995).

292. Helen Gavaghan, *Something New Under the Sun: Satellites and the Beginning of the Space Age* (New York: Springer-Verlag, 1998).

293. Bart Hendrickx, "A History of Soviet/Russian Meteorological Satellites," *Journal of the British Interplanetary Society* 57, supplement 1, *Space Chronicle* (2004): 56–102.

294. Master Sergeant Gary P. Myers, "A Portrait of the 4000th Satellite Operations Group" (internal report, 1983).

program. The first test satellites were put in orbit in 1978, but not until 1993 was a full constellation of 24 satellites in place. GPS proved its worth in the 1991 Gulf War as it helped guide Army units over the faceless desert, Navy ships around Iraqi minefields, Air Force aircraft to precise target points, and precision weaponry fired from Navy and Air Force units. Since that time, the U.S. military has converted more and more of its munitions to GPS-based precision munitions, since these proved vastly more effective than conventional ordnance. The use of GPS is now tightly woven with virtually all U.S. military operations. In addition, GPS has spawned a vast commercial market, which greatly exceeds the military's use in terms of receivers sold. GPS has become a global utility, which complicates U.S. military plans. Politically, it can no longer simply shut down civilian access to high-precision signals, even though it had originally intended to do so in wartime.

The Soviet Union was not far behind in the development of its own navigational systems. The Soviets first tested the Tsiklon communications and navigation satellite in 1967, and it became formally operational in 1971. Like Transit, it was used primarily for naval navigation. An improved version, Parus, was first tested in 1974 and operational in 1977. The Soviets next fielded an all-service geodetic and navigational system known as Kristal, which was tested for the Soviet Navy in 1971, and the all-service version in 1984. The Global Navigation Satellite System (GLONASS), the equivalent to GPS, first flew in 1982, but since the fall of the USSR, Russia has been unable to maintain the full constellation.

After 2000, China and Japan flew their first navigational satellites, and Europe, in partnership with China, India, and other nations, is beginning its Galileo program, which will sell its services to military as well as civilian users.

Historical information on navigational satellites remains surprisingly limited. For a longer view of U.S. navigation since the 19th century up to GPS, and also because the U.S. Naval Observatory provides the time for GPS, see Dick's *Sky and Ocean Joined*.[295] Gavaghan discusses the early work of John Hopkins University's Applied Physics Laboratory in the creation of Transit,[296] as do Danchik[297] and Guier and Weiffenbach.[298] Qualkinbush gives an overview of Transit.[299] Friedman provides details of the U.S. and Soviet navigational systems in terms of their utility for naval operations, including Transit,

295. Steven J. Dick, *Sky and Ocean Joined*.

296. Gavaghan, *Something New Under the Sun*.

297. Robert J. Danchik, "An Overview of Transit Development," *Johns Hopkins APL Technical Digest* 19, no. 1 (1998): 18–26.

298. William H. Guier and George C. Weiffenbach, "Genesis of Satellite Navigation," *Johns Hopkins APL Technical Digest* 19, no. 1 (1998): 14–17.

299. Robert Qualkinbush, "Transit: The US Navy Pioneer Navigation Satellite," *Journal of the British Interplanetary Society* 50, no. 11 (1997): 403–426.

GPS, Tsiklon, Parus, and Kristal.[300] The GPS story is extremely important but as yet has no full history. Alford provides a history up to 1985.[301] Bradley has a few papers on the subject.[302] Two articles in *Quest* 11, number 3, provide good overviews of the development of GPS: a historical overview by Banther and an interview of Bradford Parkinson, one of the program's founders.[303] Parkinson has written three historical articles on GPS.[304] Chapter 28 in Getting's *All in a Lifetime* discusses his role in early navigation at The Aerospace Corporation, as does his short paper in *IEEE Spectrum*.[305] Rip and Hasik's recent book, *The Precision Revolution*, is an outstanding look at the impact of space-based navigation on war-fighting.[306] Harvey provides a brief overview of Russian navigational satellites.[307] Forden analyzes the functions of China's Beidou regional navigational satellite system.[308]

ANTISATELLITES AND SPACE WARFARE

Both the United States and Russia have had the capability to destroy each other's satellites from the 1960s, with both sides deploying systems. In the United States, antisatellite weapons have been politically sensitive. Because the United States placed such high value on its space reconnaissance capabilities, political leaders have been wary about creating provocative antisatellite

300. Friedman, *Seapower and Space*.

301. Major Dennis L. Alford, *History of the NAVSTAR Global Positioning System (1963–1985)*, Air Command and Staff College Report No. 86-0050 (Maxwell AFB, AL: Air University Press, 1986).

302. George W. Bradley III, "Historical Origins of the Global Positioning System" (prepared for the History of Technology Conference, Andrews AFB, MD, 24 October 1995); George W. Bradley III, "NAVSTAR Global Positioning System Decision" (prepared for the CAMP Military History Symposium, Rapid City, SD, 11 May 2001); George W. Bradley III, "Origins of the Global Positioning System," in *Technology and the Air Force: A Retrospective Assessment*, ed. Jacob Neufeld, George M. Watson, Jr., and David Chenoweth (Washington, DC: Air Force History and Museums Program, 1997), pp. 245–254.

303. Chris Banther, "A Look into the History of American Satellite Navigation," *Quest: The History of Spaceflight Quarterly* 11, no. 3 (2004): 40–48; Steven R. Strom, "An Interview with Dr. Bradford Parkinson," *Quest: The History of Spaceflight Quarterly* 11, no. 3 (2004): 49–59.

304. Bradford W. Parkinson et al., "A History of Satellite Navigation," *Navigation: Journal of the Institute of Navigation* 42, no. 1, special issue (1995): 109–164; Bradford W. Parkinson and Stephen W. Gilbert, "NAVSTAR Global Positioning System—Ten Years Later," *Proceedings of the IEEE* 71, no. 1 (October 1983): 1177–1186; Bradford W. Parkinson, "Introduction and Heritage of NAVSTAR, the Global Positioning System," in *Global Positioning System: Theory and Applications*, ed. Bradford W. Parkinson and James J. Spiker, Jr., vol. 1 (Washington, DC: AIAA, 1996).

305. Ivan A. Getting, *All in a Lifetime: Science in the Defense of Democracy* (New York: Vantage Press, 1989); Ivan A. Getting, "The Global Positioning System," *IEEE Spectrum* (December 1993): 36–38, 43–47.

306. Michael Russell Rip and James M. Hasik, *The Precision Revolution: GPS and the Future of Aerial Warfare* (Annapolis, MD: Naval Institute Press, 2002).

307. Harvey, *Russia in Space: The Failed Frontier?*

308. Geoffrey Forden, "China's Satellite-Based Navigation System: Implications for Conventional and Strategic Forces," *Breakthroughs* 13, no. 1 (spring 2004): 19–28.

(ASAT) weapons, for fear of provoking the Soviet Union into developing the capability. Despite (or regardless of) American fears or sensitivities, the Soviets developed their own ASAT systems.

American antisatellite capabilities were generally direct spinoffs from other technologies and systems. Dyna-Soar, discussed earlier, was to have a satellite inspection and destruction capability. Ballistic missile defense systems, whether Earth- or space-based, were easily modified to attack satellites as well as missiles, at least in low-Earth orbit. Finally, ballistic missiles provided the orbital boost capabilities to launch antisatellite weapons. All that was really needed was to wait for the satellite to get within range of the booster and then fire it with precise timing.

Early ASAT weapons depended on whether nuclear detonations in space could disable satellites. The first American in-space nuclear test occurred with Project Argus, which was launched in August 1958 and detonated a 2-kiloton weapon, while the Explorer IV satellite measured the resulting change in radiation. Further tests, culminating in the much larger 1962 Starfish Prime nuclear tests in space over Johnston Island in the Pacific Ocean, confirmed that in-space nuclear explosions created radiation intensities that were deadly to both friendly and enemy satellites, as well as knocking out electrical power in the Hawaiian Islands hundreds of miles distant. The data from these tests confirmed that nuclear weapons could destroy satellites, but also that they were indiscriminate in their effects, which led shortly thereafter to the U.S. and USSR agreeing to ban nuclear tests in space.

American ASAT testing began seriously in October 1959, when the USAF's project Bold Orion used a B-47 bomber to air-launch a Martin Corporation missile, which came within 4 miles of the Explorer VI satellite. The Navy explored ship- and air-launched ASAT systems, culminating in two air-launched tests in 1962. In the meantime, the USAF was developing the larger scale SAINT, or Satellite Inspector for Space Defense, which started with a General Operational Requirement to develop a satellite defense system in June 1958. The USAF-managed program was contracted to Radio Corporation of America, which designed a rendezvous-capable vehicle with on-board radar to be launched with an Atlas-Agena. As it became clear that SAINT could not intercept some targets of interest, such as Fractional Orbit Bombardment systems, the USAF canceled it, and its mission migrated to Dyna-Soar.

In parallel, the U.S. Army was extending the capability of its Nike-Zeus ballistic missile defense system to have low-Earth-orbit ASAT functions. This became Program 505 Mudflap, which was the first U.S. operational ASAT system, deployed at Kwajalein Atoll from 1963 to 1967. Replacing it was the USAF's Program 437, which used Thor launchers with nuclear warheads launched from Johnston Island to intercept Soviet satellites. It was operational from 1964 to 1970, when it went on standby status before being terminated in 1975.

The Chelomey design bureau, OKB-52, designed the Soviet Istrebitel Sputnikov (IS) co-orbital ASAT satellite, which first flew in November 1963. A series of tests of the system continued through 1971, including operational tests in 1968 in which the IS satellite successfully exploded near its target satellite. After halting for a few years, the Soviets restarted ASAT tests in 1976, which spurred the Ford administration to restart an American ASAT program, the Miniature Homing Vehicle, an air-launched system that used the fourth stage from a Scout launch vehicle to boost it to space. The United States also funded particle beam and laser beam research programs for potential ASAT and BMD applications, as did the Soviet Union. Since the mid-1980s, U.S. ASAT research, if it continues, appears to have been folded into the Strategic Defense Initiative, and later the Ballistic Missile Defense and National Missile Defense programs. Russian ASAT research remains cloaked, but no space tests appear to have occurred since the demise of the Soviet Union. Nonetheless, both nations, as well as China, have the capability to build ASATs.

Although published in 1985, Stares's *The Militarization of Space* remains a good starting reference for antisatellite systems, describing the politics and basic programs of both U.S. and Soviet systems.[309] Manno provides similar information.[310] Kilgo's 2004 *Quest* article provides an overview of U.S. ASAT programs.[311] Chun has written a number of recent articles on the history of U.S. ASAT systems. He describes SAINT in his "A Falling Star."[312] In a later article, "Nike-Zeus' Thunder and Lightning," he narrates the genesis of the Army's Program 505.[313] The story of the USAF's Program 437 is told in *Shooting Down a "Star."*[314] This work draws from Austerman's *Program 437.*[315] The Miniature Homing Vehicle program is described in Stares's book, in Day's "Arming the High Frontier," and in Spires's *Beyond Horizons.*[316] Siddiqi

309. Paul B. Stares, *The Militarization of Space: U.S. Policy, 1945–1984* (Ithaca, NY: Cornell University Press, 1985).

310. Jack Manno, *Arming the Heavens: The Hidden Military Agenda for Space, 1945–1995* (New York: Dodd, Mead, & Co., 1984).

311. Robert Kilgo, "The History of the United States Anti-Satellite Program and the Evolution to Space Control and Offensive and Defensive Counterspace," *Quest: The History of Spaceflight Quarterly* 11, no. 3 (2004): 30–39.

312. Clayton K. S. Chun, "A Falling Star: SAINT, America's First Antisatellite System," *Quest: The History of Spaceflight Quarterly* 6, no. 2 (1998): 44–48.

313. Clayton K. S. Chun, "Nike-Zeus' Thunder and Lightning: From Antiballistic Missile to Antisatellite Interceptor," *Quest: The History of Spaceflight Quarterly* 10, no. 4 (2003): 40–47.

314. Clayton K. S. Chun, *Shooting Down a "Star": Program 437, the US Nuclear ASAT System and Present-Day Copycat Killers*, CADRE Paper No. 6 (Maxwell AFB, AL: Air University Press, 2000).

315. Wayne R. Austerman, *Program 437: The Air Force's First Antisatellite System* (Peterson AFB, CO: Office of History, 1991).

316. Dwayne A. Day, "Arming the High Frontier: A Brief History of the F-15 Anti-Satellite Weapon," *Spaceflight* 46, no. 12 (2004): 467–471; Spires, *Beyond Horizons.* Siddiqi

narrates the history of the Chelomey ASAT system in an article in *Journal of the British Interplanetary Society.*[317] Onkst describes CIA and NRO responses to Soviet antisatellite systems between 1962 and 1971.[318]

ORGANIZATION, MANAGEMENT, AND ACQUISITION

The history of human activities in space is intimately tied to the development of sophisticated technologies. In military terms, the research and development leading to the creation of these technologies is called "acquisition." The unique characteristics of the space environment drove the creation of new managerial methods for military technology acquisition called "systems management." Space systems are also operated differently from most Earth-based systems, leading to unique operational processes as well. These developmental and operational differences have also led to the creation of new organizations within the services that handle these unique acquisition and operations processes.

In the late 1940s through the early 1960s, the military services competed for "roles and missions" related to nuclear weapons, ballistic missiles, and finally space systems. The novelty of nuclear weapons and of the space environment meant that none of the services had a clear-cut, unchallengeable claim to these technologies or to space. The Army saw ballistic missiles as extensions to its classical artillery. The USAF saw space as a natural extension of flying. The Navy believed space had unique characteristics crucial for its mission on and in the oceans and did not want either the Army or the Air Force to monopolize space.

Army Ordnance handled the bulk of the Army's missile efforts, controlling von Braun's Army Ballistic Missile Agency and funding Jet Propulsion Laboratory (JPL) to develop the Corporal ballistic missile. Early Air Force missile efforts were managed by Air Research and Development Command and Air Materiel Command, which themselves battled over who controlled what portions of the development process. The Navy's efforts were concentrated in the Naval Research Laboratory, with some programs in the Office of Naval Research.

Sputnik highlighted American space deficiencies, leading to a variety of changes. The Advanced Research Projects Agency (ARPA) was formed to coordinate military space activities. However, it was unsuccessful in this role,

317. Siddiqi, *Challenge to Apollo*; Asif A. Siddiqi, "The Soviet Co-Orbital Antisatellite System: A Synopsis," *Journal of the British Interplanetary Society* 50, no. 6 (1997): 225–240.

318. D. H. Onkst, "Check and Counter-Check: The CIA's and NRO's Response to Soviet Anti-Satellite Systems, 1962–1971," *Journal of the British Interplanetary Society* 51, no. 8 (August 1998): 301–308.

and the services pushed ARPA aside to instead focus on advanced research in which the services were not immediately interested. Space was too important to be left to a separate agency. The DOD also created the Deputy Secretary of Defense for Research and Engineering (DDR&E) to coordinate and control military research, while the Secretary of Defense was given more budget authority, which Robert McNamara in the 1960s used to exert control over the services. By the end of the 1950s, the Army had mostly lost the battle for space, relinquishing JPL and ABMA to NASA. However, it retained programs in ballistic missile defense, playing the leading role for BMD and for the Program 505 Mudflap antisatellite system. The Navy successfully prevented an Air Force monopoly, retaining operational control of satellites intended for naval support such as Transit.

The Air Force won the majority of the turf battles, partially assisted by its concept of "aerospace," the "indivisible medium" of air and space that the Air Force claimed could not be separated and was the natural single medium for operations above the Earth's surface. In 1961, the USAF reorganized its research and development activities, creating Air Force Systems Command for the acquisition of all major programs. Since all space programs were, in the early days, development programs, this centralized the management of many NS space systems. McNamara rewarded the USAF by officially awarding it the bulk of the "space mission."

However, this was only a partial bureaucratic victory, because other organizations gained or retained influence over certain aspects of NS space. This included the National Reconnaissance Office, which forced the USAF to share responsibility for reconnaissance satellites with the Central Intelligence Agency, and the Defense Communications Agency, which exerted control over various aspects of communications satellites and ground systems. Other organizations that remained involved with military space included Lincoln Laboratory, which was funded by all three services, and the National Security Agency, which operated ground stations that received and interpreted signals intelligence data.

The next major changes to the organization of American NS space occurred in the early 1980s, due to two major spurs: the Space Shuttle and Reagan's Strategic Defense Initiative. By the late 1970s, the USAF was building new facilities to handle Space Shuttle military operations, including a launch pad at Vandenberg AFB, a new control facility in Colorado Springs, as well as classified facilities at NASA's Johnson Space Center near Houston. The question of what organization would handle Shuttle operations, as well as Reagan administration concerns about the USAF's fractured space operations, led the USAF to centralize its satellite operations into a new major command, USAF Space Command, based in Colorado Springs. The Army and Navy followed suit, creating Army Space Command and Naval Space

Command, respectively. The next step was to create a single unified command, called United States Space Command, to centralize operational control of all military space assets. Space Command eventually wrested control of launch operations from Systems Command, and Systems Command itself was soon deactivated, with its functions handed to a newly created Air Materiel Command, which brought USAF organizational changes full circle, almost identical to its late-1940s form. In the early 2000s, after the 11 September 2001 terrorist attacks on the World Trade Center and the Pentagon, U.S. Space Command was deactivated and its functions split between Strategic Command and a new Northern Command that concentrated on defense of the North American continent.

In the Soviet Union, ballistic missile and space forces evolved differently. Initially, ballistic missiles and the early space programs were coordinated among several research institutes and design bureaus but organized by Sergei Korolev's Special Design Bureau-1 (OKB-1) in Kaliningrad near Moscow. The Soviet leadership soon fomented internal competition for ballistic missiles by giving responsibility for some of these systems to Mikhail Yangel's OKB-586 in Dnepropetrovsk, which soon moved into spacecraft design as well. A third design bureau, Vladimir Chelomey's OKB-52 in Reutov, gained strength during Nikita Khrushchev's reign, influenced by the fact that Chelomey hired Khrushchev's son, Sergei. Chelomey developed ballistic missiles, as well as antisatellite systems and the Almaz manned reconnaissance orbital station. While these "big three" design bureaus were the most prominent, many others were involved with specialized aspects of Soviet military space programs, from subsystems to specific satellite types, such as Mikoyan's OKB-155 that worked on Spiral, Kozlov's OKB-1 Branch 3 that focused on reconnaissance, Savin's OKB-41 that worked on EORSAT and RORSAT, etc.

Most design bureaus reported to the Ministry of Armaments (MV) until 1965, when they were transferred to the Ministry of Machine Building (MOM) under Dmitry Ustinov. Some design bureaus, such as Mikoyan's, reported to the Ministry of Aviation Industry. In the 1950s and 1960s, the Soviets kept design, accomplished in the design bureaus, separate from production, handled in a variety of factories and plants. In the mid-1970s, the Soviets combined design bureaus and associated factories into Scientific-Production Associations, or NPOs. Thus OKB-1 and various bureaus combined into NPO Energia, while OKB-52 became NPO Mashinostroyenia and OKB-586 became NPO Yuzhnoye.

System operations were handled through the Ministry of Defense, which controlled the Army, Navy, and Air Force. Nikita Khrushchev, wanting to emphasize the importance of ballistic missiles, created the Strategic Missile Forces (or Missile Forces of Strategic Designation—RVSN), which from 1959 to 1981 operated ballistic missile and space systems. Air defense systems, which evolved into the ballistic missile defense and warning systems, were operated by the Forces of Anti-Missile Defense (V-PRO), formed in 1958. Soviet military space programs were centralized in 1964 in the Central Directorate of Space Systems (TsUKOS) of the RVSN and, in 1970, called the Chief Directorate of the Space Systems (GUKOS). In 1981, GUKOS was separated from the RVSN and placed directly under the Ministry of Defense. Renamed the Directorate of the Space Systems Commander (UNKS) in 1986, space systems were formed into a separate military service in 1992, the Military Space Forces (VKS). Between 1997 and 2001, the military space forces were once again subordinated to the RVSN but, in 2001, were once again made an independent force, the Space Forces (KVR). In 2000, when the National Air Defense service was disbanded, its strategic defense functions were transferred to the Space Forces.

China's military space program began when Tsien Hsue-Shen, a brilliant rocket theorist working for the California Institute of Technology and a founding member of JPL, returned to Communist China from the United States in 1955. In January 1956, the government founded the Institute of Mechanics in Beijing with Tsien in charge. By October, the government heeded Tsien's proposal to develop rockets, creating the Fifth Academy of the Ministry of National Defense, with Tsien at its head. The Fifth Academy acquired Soviet R1 and R2 missiles, along with Soviet technicians and blueprints. The Chinese satellite program began on a small scale when engineers from the Shanghai Institute of Machine and Electrical Design went to Beijing to work with Tsien. They returned to Shanghai and started to work, but not until 1965 did the Shanghai institute, under the authority of the Seventh Ministry of Machine Building (the Fifth Academy's new designation) and with assistance from the Chinese Academy of Sciences, get authorization to work with local factories to build satellites. The Shanghai group eventually became the Shanghai Academy of Spaceflight Technology. In 1982, the Seventh Ministry became the Ministry of Space Industry (MASI), which had several academies under it developing various systems and subsystems. Information on other military space organizations exists through primary sources, but there has been little historical work published in open literature.

The evolving organizational structures reflect a deeper set of evolving managerial and engineering processes that were also created along with space systems. Ballistic missile and space systems both require levels of reliability significantly higher than most typical Earth-bound technologies. Neither ballistic missiles nor space systems (with a few exceptions like the Shuttle orbiter) return once placed in space; therefore, components, except for software, cannot be replaced. Rocket engines are extremely dangerous and have extreme temperatures and pressures. The space environment also has extremes of temperature along with radiation, while the lack of air confounds conventional heating and cooling methods. Finally, ballistic missiles and space systems are composed of a multiplicity of individually complex technologies, connected in complex ways.

The combination of these factors led designers to create systems engineering, which is the set of methods to coordinate the organizational communication and complexity of space systems. These methods, which include environmental and systems testing, quality control, change control, design reviews, and configuration control, came to symbolize the extremes of preplanning, controlled manufacturing, and rigorous testing that characterized the space industry. They went hand in hand with managerial innovations such as project management, configuration boards, matrix management, network scheduling tools, and program control rooms. Starting with ballistic missile programs of the U.S. Army, Navy, and Air Force, these methods formed through the mutual interactions of government, industry, and academia and led also to the creation of nonprofit organizations such as RAND Corporation, The Aerospace Corporation, and MITRE Corporation to help the government analyze and coordinate complex technological systems. By the mid-1960s, the bulk of these processes and institutions were in place, as the DOD instituted systems management across all of the services. Since that time, a variety of managerial reforms have been attempted, which somewhat modify these techniques or allow flexibility for program managers to select from a menu of the systems management tools. However, at the start of the 21st century, the core of these methods remained in place for space systems and ballistic missiles.

Virtually all military organizations have institutional histories, and thus there are a host of internal studies that either have been or someday will be declassified. These generally provide a solid base for institutional and managerial histories. I will not attempt to describe them all here. The best procedure for historians is to consult the military organization (or its successor) in which they are interested and request access to the appropriate institutional histories, as well as starting with the regular publications described below.

Spires's *Beyond Horizons* is the best starting point for the USAF's space organization and executive management. Neufeld's *Ballistic Missiles in the USAF* provides a similar basis for ballistic missiles,[319] as does Schaffel's *The Emerging Shield* for continental defense. Waldron provides an overview of the Space and Missile Systems Center.[320] No such overview works exist for the U.S. Army's space efforts, or for the U.S. Navy, ARPA, or the Strategic Defense Initiative Organization and its successors. A few lower-level monographs and articles exist. Neufeld's *Research and Development in the United States*

319. Neufeld, *Ballistic Missiles in the United States Air Force.*
320. Waldron, *Historical Overview of the Space and Missile Systems Center.*

Air Force is an interview with key actors: Bernard Schriever, James Doolittle, Samuel Phillips, Robert Marsh, and Ivan Getting.[321] Tunyavongs describes the politics of the foundation of Air Force Space Command.[322] Sapolsky's *Science and the Navy* narrates the history of ONR, while McDowell describes a variety of Naval Research Laboratory satellite projects.[323] Sigethy's 1980 dissertation on the organization of USAF basic research is a good stating point for that area.[324] Lambeth's short 2004 article in *Air Force Magazine* describes some of the politics of military space.[325]

Institutional histories of the intelligence space organizations exist, but most remain classified. However, some of these histories have become available over time. The National Security Archive at George Washington University has a variety of original documents, many of which are posted online, regarding the intelligence space programs, in particular those of the CIA, NRO, and NSA.[326] Richelson's "Undercover in Outer Space" provides an overview of the NRO.[327] Perry's declassified history, *Management of the National Reconnaissance Program, 1960–1965*, is an outstanding early work on the organizational problems of reconnaissance.[328] Laurie reviews the relationship of the NRO and Congress.[329] Other points of view of the NRO include the CIA's *Office of Special Projects, 1965–1970* and *CORONA Program History*.[330] Day describes the relationships between some of these various histories in his 2000 "Rashomon in Space."[331]

U.S. military–funded nonprofits and academically managed organizations have received their share of historical work, both from the nonprofits

321. Jacob Neufeld, ed., *Research and Development in the United States Air Force* (Washington, DC: Center for Air Force History, 1993).

322. T. Tony Tunyavongs, "A Political History of the Establishment of Air Force Space Command," *Quest: The History of Spaceflight Quarterly* 9, no. 1 (2001): 31–43.

323. Sapolsky, *Science and the Navy*; McDowell, "Naval Research Laboratory Satellites 60-89," pp. 427-432.

324. Robert Sigethy, "The Air Force Organization for Basic Research, 1945–1970: A Study in Change" (Ph.D. diss., The American University, 1980).

325. Benjamin S. Lambeth, "A Short History of Military Space," *Air Force Magazine* 87, no. 12 (2004): 60–64.

326. George Washington University, *The National Security Archive*, http://www.gwu.edu/~nsarchiv/index.html.

327. Jeffrey T. Richelson, "Undercover in Outer Space: The Creation and Evolution of the NRO," *International Journal of Intelligence and CounterIntelligence* 13, no. 3 (fall 2000): 301–344.

328. Robert Perry, *A History of Satellite Reconnaissance*, vol. 5, *Management of the National Reconnaissance Program, 1960–1965* (Washington, DC: NRO, 1969).

329. Clayton D. Laurie, *Congress and the National Reconnaissance Office* (Washington, DC: Office of the Historian, NRO, June 2001).

330. *Office of Special Projects, 1965–1970*, vol. 1 (Washington, DC: CIA, 1973), chaps. I–II; Directorate of Science and Technology, CIA, "CORONA Program History, vol. 2, Governmental Activities" (internal document, 19 May 1976).

331. Dwayne A. Day, "Rashomon in Space: A Short Review of Official Spy Satellite Histories," *Quest: The History of Spaceflight Quarterly* 8, no. 2 (2000): 45–53.

themselves and from scholars. Mark and Levine provide an overview of these institutions.[332] RAND Corporation is the most famous of these organizations, whose history is described in an early book by Smith, in Jardini's dissertation, and, most recently, by Collins.[333] Baum describes the RAND spinoff for air defense, System Development Corporation.[334] Freeman describes MIT's Lincoln Laboratory, also initially established for air defense, as was the MITRE Corporation, which wrote its own internal history, with a more recent history by Dyer and Dennis.[335] The Aerospace Corporation did its own internal histories up to 1980 and had a couple of other student thesis histories written about it in the early 1970s.[336] Koppes provides an excellent history of JPL through 1980, including its military roots.[337]

The history of the U.S. aerospace industry from the standpoint of businesses, which are contracted by the military, is best overviewed in Bilstein's *The American Aerospace Industry*.[338] Markusen et al. perform a series of local economic impact studies of military contracting and influences, which include the space sector, in *The Rise of the Gunbelt*.[339] Similar studies for Colorado are Sturdevant and Spires's "Mile-High Ventures" and Spires's "Walter Orr Roberts."[340] Baker and Baker provide a similar story for the foundation of the

332. Hans Mark and Arnold Levine, *The Management of Research Institutions: A Look at Government Laboratories* (Washington, DC: NASA SP-481, 1984).

333. Bruce L. R. Smith, *The RAND Corporation* (Cambridge, MA: Harvard University Press, 1966); David Jardini, "Out of the Blue Yonder: The RAND Corporation's Diversification into Social Welfare Research, 1946–1968" (Ph.D. diss., Carnegie Mellon University, 1996); Martin J. Collins, *Cold War Laboratory: RAND, the Air Force, and the American State, 1945–1950* (Washington, DC: Smithsonian Institution Press, 2002).

334. Claude Baum, *The System Builders: The Story of SDC* (Santa Monica, CA: System Development Corporation, 1981).

335. Eva C. Freeman, ed., *MIT Lincoln Laboratory: Technology in the National Interest* (Lexington, MA: Lincoln Laboratory, MIT, 1995); MITRE Corporation, *MITRE: The First Twenty Years, A History of MITRE Corporation, 1958–1978* (Bedford, MA: The MITRE Corporation, 1979); Dyer and Dennis, *Architects of Information Advantage*.

336. *The Aerospace Corporation: Its Work, 1960–1980* (El Segundo, CA: The Aerospace Corporation, 1980); *The Aerospace Corporation: Its People, 1980* (El Segundo, CA: The Aerospace Corporation, 1980); *The Aerospace Corporation, 1960–1970, Serving America* (El Segundo, CA: The Aerospace Corporation, 1970); James Franklin Wheeler, "The Aerospace Corporation, Past, Present, and Future" (master's thesis, Air Force Institute of Technology, Air University, 1973); Harold P. Wheeler, "The Aerospace Corporation, Then and Now," Air War College Research Report no. 4474 (Maxwell AFB, AL: Air University, 1971).

337. Koppes, *JPL and the American Space Program*.

338. Roger E. Bilstein, *The American Aerospace Industry: From Workshop to Global Enterprise* (New York: Twayne Publishers, 1996).

339. Ann Markusen, Scott Campbell, Peter Hall, and Sabina Deitrick, *The Rise of the Gunbelt: The Military Remapping of Industrial America* (Oxford: Oxford University Press, 1991).

340. Rick W. Sturdevant and David N. Spires, "Mile-High Ventures: Highlights from Colorado Aerospace History, 1923–1997," *Journal of the West* 36, no. 3 (July 1997): 67–77; David N. Spires, "Walter Orr Roberts and the Development of Boulder's Aerospace Community," *Quest: The History of Spaceflight Quarterly* 6, no. 4 (winter 1998): 5–14.

space community in Utah.[341] Commercial space systems have had an increasing impact on military space. An overview of these issues is found in Klotz and in Logsdon and Acker.[342]

There are a number of works about various aerospace companies, including their contracts and relations with the military. These include Aerojet,[343] Boeing,[344] Convair,[345] General Dynamics,[346] General Electric's Aerospace Group,[347] Itek,[348] Lockheed,[349] McDonnell Douglas,[350] Martin Marietta,[351] Reaction Motors,[352] Rocketdyne,[353] Thiokol,[354] and TRW.[355]

Siddiqi's *Challenge to Apollo* is the best starting point for the institutional history of the Soviet ballistic missile and space programs,[356] along with his 1997 *Spaceflight* article that he later put into an appendix in *Challenge to Apollo*. The other essential reference is Zaloga's *The Kremlin's Nuclear Sword*.[357] A simple introduction to the organizational evolution of the Soviet and Russian

---

341. Doran J. Baker and Kay D. Baker, "Outer Space Exploration from Utah: Leon Linford and Rocket Science," *Quest: The History of Spaceflight Quarterly* 12, no. 3 (2005): 6–15.

342. Frank G. Klotz, *Space, Commerce, and National Security* (New York: Council on Foreign Relations Press, 1998); John M. Logsdon and Russell J. Acker, eds., *Merchants and Guardians: Balancing U.S. Interests in Global Space Commerce* (Washington, DC: Space Policy Institute, George Washington University, May 1999).

343. *The Aerojet: The Creative Company* (Los Angeles: Stewart F. Cooper Company, 1995).

344. Eugene E. Bauer, *Boeing in Peace and War* (Enumclaw, WA: Taba Publications, 1991); Guy Norris and Mark Wagner, *Boeing* (Osceola, WI: MBI Publishing, 1998); T. M. Sell, *Wings of Power: Boeing and the Politics of Growth in the Northwest* (Seattle: University of Washington Press, 2001); Robert J. Serling, *Legend and Legacy: The Story of Boeing and Its People* (New York: St. Martin's Press, 1992).

345. Bill Yenne, *Into the Sunset: The Convair Story* (Lyme, CT: Greenwich Pub. Group, 1995).

346. Roger Franklin, *The Defender: The Story of General Dynamics* (New York: Harper & Row, 1986); Jacob Goodwin, *Brotherhood of Arms: General Dynamics and the Business of Defending America* (New York: Times Books, 1985).

347. Major A. Johnson, *Progress in Defense and Space: A History of the Aerospace Group of the General Electric Company* (Major A. Johnson, 1993).

348. Jonathan Lewis, *Spy Capitalism: ITEK and the CIA* (New Haven, CT: Yale University Press, 2002).

349. Walter Boyne, *Beyond the Horizons, The Lockheed Story* (New York: St. Martin's Press, 1998).

350. Douglas J. Ingells, *The McDonnell Douglas Story* (Fallbrook, CA: Aero Publishers, 1979); Bill Yenne, *McDonnell Douglas: A Tale of Two Giants* (London: Arms and Armour, 1985).

351. William B. Harwood, *Raise Heaven and Earth: The Story of Martin Marietta People and Their Pioneering Accomplishments* (New York: Simon & Schuster, 1993).

352. Frederick I. Ordway III and Frank H. Winter, "Reaction Motors Inc.: A Corporate History," AIAA Paper 82-277, 1982.

353. *Thirty Years of Rocketdyne* (Canoga Park, CA: Rocketdyne Division, Rockwell International Corporation, 1985).

354. E. S. Sutton, "From Polymers to Propellants to Rockets—A History of Thiokol," AIAA Paper 99-2929 (35th AIAA/American Society of Mechanical Engineers/Society of Automotive Engineers/American Society for Engineering Education Joint Propulsion Conference and Exhibit, Los Angeles, CA, June 1999).

355. Davis Dyer, *TRW: Pioneering Technology and Innovation Since 1900* (Boston: Harvard Business School Press, 1998).

356. Siddiqi, *Challenge to Apollo*.

357. Zaloga, *The Kremlin's Nuclear Sword*.

military space forces is provided in Gorin's "Russian Space Forces" article in the forthcoming ABC-CLIO space history encyclopedia, *Space Exploration and Humanity*.[358] Berkowitz provided an early look at the organization of the USSR's space units.[359] For a first-person view of the early organization of Soviet rocketry, see Chertok's recently translated memoir.[360] Clark provides an overview history of Yangel's design bureau, now Yuzhnoye.[361]

For China, Chang's biography of Tsien Hsue-Shen, *Thread of the Silkworm*, is the best starting point.[362] Chapter 4 of Johnson-Freese's *The Chinese Space Program* provides a basic organizational overview and history, as does Harvey's *China's Space Program*.[363]

On acquisition and management, Lonnquest's 1996 dissertation, "The Face of Atlas," is an outstanding study of Bernard Schriever's role in the creation of the Atlas ballistic missile. Johnson's *The United States Air Force and the Culture of Innovation* investigates the development of management and systems engineering of USAF ballistic missile and air defense programs in the 1950s, while *The Secret of Apollo* contains a shorter version of the ballistic missile story but adds JPL, the NASA manned space program, and the early European space programs.[364] Hughes also tackles these topics in *Rescuing Prometheus*.[365] A short overview of USAF acquisition is provided by Benson.[366] All of these works draw from Gorn's outstanding study, *Vulcan's Forge*.[367] York's 1970 book explains his role in the organization of NS space in *Race to Oblivion*.[368] A critical but historical assessment of USAF acquisition by a key early participant

---

358. Peter A. Gorin, "Russian Space Forces," in *Space Exploration and Humanity: A Historical Encyclopedia*, ed. Stephen B. Johnson et al. (Santa Barbara, CA: ABC-CLIO, forthcoming, expected publication 2007).

359. M. J. Berkowitz, "To Lift the Veil of Secrecy: USSR Ministry of Defence Space Units," *Journal of the British Interplanetary Society* 46, no. 5 (1993): 191–198.

360. Boris Chertok, *Rockets and People*, vol. 1 (Washington, DC: NASA SP-2005-4110, 2005).

361. Phillip S. Clark, "The History and Projects of the Yuzhnoye Design Bureau," *Journal of the British Interplanetary Society* 49, no. 7 (1996): 267–276.

362. Chang, *Thread of the Silkworm*.

363. John Johnson-Freese, *The Chinese Space Program: A Mystery Within a Maze* (Malabar, FL: Krieger Publishing Company, 1998); Brian Harvey, *China's Space Program: From Conception to Manned Spaceflight* (Chichester, U.K.: Springer-Praxis, 2004).

364. Stephen B. Johnson, *The United States Air Force and the Culture of Innovation, 1945–1965* (Washington, DC: USAF History and Museums Program, 2002); Stephen B. Johnson, *The Secret of Apollo: Systems Management in American and European Space Programs* (Baltimore, MD: Johns Hopkins, 2002).

365. Thomas P. Hughes, *Rescuing Prometheus* (New York: Pantheon, 1998).

366. Lawrence R. Benson, *Acquisition Management in the United States Air Force and Its Predecessors* (Washington, DC: Air Force History and Museums Program, 1997).

367. Michael H. Gorn, *Vulcan's Forge: The Making of an Air Force Command for Weapon Acquisition (1950–1985)*, vol. 1, *Narrative* (Andrews AFB, MD: History Office, HQ Air Force Systems Command, 1985).

368. Herbert F. York, *Race to Oblivion: A Participant's View of the Arms Race* (New York: Simon & Schuster, 1970).

can be found in Hall's *The Art of Destructive Management*.[369] Finally, there is currently ongoing a project by the Department of Defense called the Defense Acquisition History Project, which is to produce a six-volume series on the subject in 2007 and 2008.

Space Power Theory

To date, there is no dedicated monograph on the history of military space doctrine and space power theory, perhaps because there is no single work that commands doctrinal allegiance. Over the centuries, but particularly since the Napoleonic era, military commanders and thinkers have developed a variety of theories and doctrines on the nature of war. As warfare expanded from the land to the sea and to the air, major thinkers for each, which include Sun Tzu, Jomini, and Clausewitz for land warfare; Mahan and Corbett for naval warfare; and Douhet, Mitchell, and Warden for air warfare, developed theories and doctrines that have become the basis for understanding conflict in these domains ever since. To date, no such comprehensive, fundamental theory has been developed for space.

The first attempts to understand the implications of space were reactions to the Nazi V-2 project, such as the 1946 RAND study, which discussed the potential for space assets to enhance certain military activities, such as reconnaissance and weather prediction. RAND also noted the potential political prestige effects of launching the first artificial satellite. In the 1950s, Strategic Air Command's ability to deliver nuclear weapons in a devastating strategic bombing campaign was at the forefront of doctrine, and ballistic missiles were seen as an alternative means to deliver nuclear weapons. Defense-oriented activities, such as early-warning systems, were of distinctly lesser importance.

With the launch of Sputnik in 1957 and the consequent reaction in the United States to launch satellites and to organizationally control space activities, the USAF ultimately won the lion's share of military space programs. General Thomas White defined and propagated the term "aerospace" in 1959 to press the USAF's claim that air and space were a continuous medium with no definite boundary, and hence that it was natural for the Air Force to control operations in this single environment. This claim is debatable at best, but it aided the USAF's bureaucratic cause, as the Kennedy administration in 1961 awarded the USAF the largest share of military space projects and functions.

The next major spur to space power theorizing came in the 1980s, as a theoretical counterpart to the formation of USAF, Army, Navy, and U.S. Space Commands and Reagan's Strategic Defense Initiative. By the late 1980s,

---

369. Edward N. Hall, *The Art of Destructive Management: What Hath Man Wrought?* (New York: Vantage Press, 1984).

Lupton formulated his four-part conceptual division of space doctrines: sanctuary, survivability, control, and high ground. At the same time, the USAF created a four-part division of its activities, which remain its major means of categorizing its activities: space support, force enhancement, space control, and force application. These two conceptualizations remain the basic frameworks for discussion in the early 21st century, although others have been postulated, the most significant of which is probably the extrapolation from Warden's theory of airpower to postulate space as an economic center of gravity.

Serious theorizing continued into the 1990s and into the first decade of the 21st century, but as yet, no comprehensive theory of space warfare has emerged. A number of recent authors, including Dolman, Hays, Lambakis, Preston, Watts, Gray, Sheldon, and others, have continued the debate.

Specific histories of space power and doctrine are few. Futrell's authoritative *Ideas, Concepts, Doctrine* volumes are the starting point for understanding the history of USAF theories and doctrine, including the intrusion of space into the service.[370] Equally authoritative on the political aspects of the military and some of the debates is McDougall's . . . *The Heavens and the Earth*.[371] Hays's dissertation investigates the relationship between space programs and attempts to create a military space doctrine.[372]

The term "aerospace," along with its evolution and influence, has caught some attention. In two articles, Terry narrates the formulation of the aerospace doctrine in the late 1950s, during the formative years of the space program.[373] Jennings focuses on the conflict over the term "aerospace" itself and its use in doctrine.[374] Rothstein investigates the evolution of the concept from airpower theory.[375] Houchin reviews the impact of hypersonic technologies on aerospace doctrine.[376]

Given the relative paucity of historical work, historians will need to read the major proponents directly. Lupton's *On Space Warfare* is often consid-

---

370. Robert Frank Futrell, *Ideas, Concepts, Doctrine: Basic Thinking in the United States Air Force, 1907–1960*, vol. 1 (Maxwell AFB, AL: Air University Press, 1989); Robert Frank Futrell, *Ideas, Concepts, Doctrine: Basic Thinking in the United States Air Force, 1961–1984*, vol. 2 (Maxwell AFB, AL: Air University Press, 1989).

371. McDougall, . . . *The Heavens and the Earth*.

372. Peter Lang Hays, "Struggling Towards Space Doctrine: U.S. Military Space Plans, Programs, and Perspectives During the Cold War" (Ph.D. diss., Tufts University, 1994).

373. Michael R. Terry, "Formulation of Aerospace Doctrine from 1955–1959," *Air Power History* 38, no. 1 (1991): 47–54; Michael Terry, "The Icarus Paradox: Air Force Doctrine and Space Technology," *Quest: The History of Spaceflight Quarterly* 6, no. 3 (1998): 37–43.

374. Frank W. Jennings, "Doctrinal Conflict Over the Word Aerospace," *Airpower Journal* 4 (1990): 46–58.

375. Stephen M. Rothstein, "Dead on Arrival? The Development of the Aerospace Concept, 1944–1958" (master's thesis, School of Advanced Airpower Studies, Maxwell AFB, AL: Air University Press, 2001).

376. Roy F. Houchin II, "Hypersonic Technology and Aerospace Doctrine," *Air Power History* 46, no. 3 (1999): 4–17.

ered the starting point for space power theory.[377] Mantz developed his own theory of space combat in *The New Sword*.[378] Dolman's *Astropolitik* provides another important view on the political aspects of space power.[379] Lambakis's *On the Edge of Earth* is a good overview of current ideas.[380] Preston et al.'s *Space Weapons, Earth Wars* focuses on the political and technical issues of space weapons.[381] Oberg provides an overview of the USAF's official doctrine at the end of the 20th century.[382] Watts provides an informed analysis of trends relevant for military space.[383] Shaw attempts to mirror Alfred Thayer Mahan's influence on history.[384] Two other important recent works on space power theory are by Smith[385] and Lambeth.[386]

Hays et al.'s *Spacepower for a New Millennium* is a compilation of recent papers on U.S. military space, a number of which relate to theoretical aspects.[387] DeBlois's 1999 *Beyond the Paths of Heaven* is a compendium of papers on space power thought.[388] Lambright's collection on space policy contains some theoretical papers.[389] *Air & Space Power Journal* (and its predecessor, *Aerospace Power Journal*, which went by other names earlier) often has papers on military space doctrinal issues.

Although typical for other military functions, there are few works that focus on space systems in combat, for the simple reason that only recently have they been in combat. The First Persian Gulf War of 1991 was the first war in which space systems played an important role, which is documented by Kutyna, Campen, and Berkowitz.[390]

---

377. David E. Lupton, *On Space Warfare: A Space Power Doctrine* (Maxwell AFB, AL: Air University Press, 1988).

378. Michael R. Mantz, *The New Sword: A Theory of Space Combat Power* (Maxwell AFB, AL: Air University Press, 1995).

379. Everett C. Dolman, *Astropolitik: Classical Geopolitics in the Space Age* (London: Frank Cass, 2002).

380. Steven Lambakis, *On the Edge of Earth: The Future of American Space Power* (Lexington: University Press of Kentucky, 2001).

381. Robert Preston, Dana J. Johnson, Sean Edwards, and Michael Miller, *Space Weapons, Earth Wars*, MR-1209-AF (Santa Monica, CA: RAND Corporation, 2002).

382. James Oberg, *Space Power Theory* (Washington, DC: GPO, 1999).

383. Barry D. Watts, *The Military Use of Space: A Diagnostic Assessment* (Washington, DC: Center for Strategic and Budgetary Assessments, February 2001).

384. John E. Shaw, "The Influence of Space Power Upon History, 1944–1998," *Air Power History* 46, no. 4 (1999): 20–29.

385. M. V. Smith, *Ten Propositions Regarding Spacepower* (Maxwell AFB, AL: Air University Press, 2002).

386. Benjamin S. Lambeth, *Mastering the Ultimate High Ground: Next Steps in the Military Uses of Space* (Santa Monica, CA: RAND, 2003).

387. Peter L. Hays, James M. Smith, Alan R. Van Tassel, and Guy M. Walsh, eds., *Spacepower for a New Millennium: Space and U.S. National Security* (New York: McGraw-Hill, 2000).

388. Bruce M. DeBlois, ed., *Beyond the Paths of Heaven: The Emergence of Space Power Thought* (Maxwell AFB, AL: Air University Press, 1999).

389. W. Henry Lambright, ed., *Space Policy in the 21st Century* (Baltimore, MD: Johns Hopkins, 2003).

390. Donald J. Kutyna, "Indispensable: Space Systems in the Persian Gulf War," *Air Power History* 46, no. 1 (1999): 28–43; Alan D. Campen, ed., *The First Information War: The Story of Communications,*

continued on the next page

Conclusion—Holes in the Literature

What can we observe from the rather lengthy treatise on sources provided above? First and foremost, there is no area of military space that has a comprehensive treatment with both in-depth analysis and crosscutting synthesis. Some sectors, such as launcher and ballistic missiles, as well as robotic intelligence and reconnaissance, have an extensive literature. Others, such as command and control, communications, navigation, and space power theory, have received very little historical attention. The remainder have had some historical research done but remain significantly underdeveloped: early-warning and space surveillance; ballistic missile defense; human flight; weather and science; antisatellite systems; and organization, management, and acquisition. Needless to say, this leaves the overall state of military space history as significantly underdeveloped, with a few pockets of significant work and a few areas almost completely blank.

Even in areas that have extensive literature, there remain gaping holes. In those sectors with virtually no historical research, almost the entire sector is a historical blank slate. I give my thumbnail assessment of missing research for each sector below.

Holes in the Research

- Ballistic missiles and launch vehicles: synthetic overview, U.S. ballistic missiles after 1965, ballistic missiles outside the United States and Russia/USSR, nuclear warfare strategies after 1960s, effect of the end of the Cold War.

- Early warning and space surveillance: synthetic overview, U.S. overview, space surveillance, Cold War radar systems history.

- Command and control: synthetic overview, U.S./Canada relationship with NORAD, system-of-systems history, conventional versus nuclear command and control, C2 computing after SAGE, C2 and human factors research.

- Ballistic missile defense: synthetic overview, U.S. overview, project histories, SDI and later programs, unbiased political and arms control studies, strategic versus theater missile defense, technical history of BMD.

continued from the previous page

Computers, and Intelligence Systems in the Persian Gulf War (Fairfax, VA: Armed Forces Communications and Electronics Association International Press, 1992); Bruce D. Berkowitz, *The New Face of War: How War Will Be Fought in the 21st Century* (New York: Free Press, 2003).

- Robotic intelligence and reconnaissance: synthetic overview, non-U.S. reconnaissance, post-CORONA reconnaissance, politics of commercial remote sensing, uses of satellite intelligence, economics of sector.

- Military human flight: synthetic overview, military-civilian relationships with astronauts, aerospace medicine, hypersonic technologies overview, Space Shuttle and Buran military aspects, Raketoplan/Spiral, technical program histories, MOL versus Almaz.

- Weather and science: synthetic overview, Clementine, military-civilian weather political interactions, project histories, institutions and institutional relationships.

- Navigation: synthetic overview, full project histories, non-U.S. navigation systems, strategic to tactical and commercial applications, politics and economics of navigation.

- Antisatellite systems and space warfare: synthetic overview, full-length project studies (both U.S. and USSR), relationship to BMD and space warfare, new political history (beyond Stares).

- Organization, management, and acquisition: synthetic overview; Army, Navy, DOD, Missile Defense Agency space institutional histories; 1970s–present acquisition; comparative studies to other types of systems (aircraft, C2, naval, etc.).

- Space power theory: synthetic overview; relationships of theory to doctrine and practice; studies of theorists and their theories; relation to other military theories; connections to political, technical, and institutional changes.

There would be great value to the militaries of spacefaring nations, governmental leaders and managers, and the general public to have histories of the many areas that remain underdeveloped. Given that the existence of military space activities is no longer classified, and given the changing world since the demise of the Soviet Union and the rise of global terrorism, broader and deeper knowledge of the actual uses of space will be of great benefit. More research, both from the military itself and from external scholars, will be necessary to make the history of national security space as informed and thorough as the great and growing importance of these activities deserves.

CHAPTER 16

CRITICAL THEORY AS A TOOLBOX: SUGGESTIONS FOR SPACE HISTORY'S RELATIONSHIP TO THE HISTORY SUBDISCIPLINES

Margaret A. Weitekamp

After the loss of the Space Shuttle *Columbia* in February 2003, I spoke on a number of radio programs. In the days after the accident, I had written a newspaper editorial reflecting on my fellowship year at the National Aeronautics and Space Administration (NASA) Headquarters History Office in 1997–1998. As a result, the small upstate New York college where I was teaching put my name on its Web site as a local space expert. Busy with classes, I accepted the invitations that fit most easily into my schedule. All but one went smoothly. Too late to cancel, I realized that I had agreed to be the guest for a Las Vegas radio personality whose regional following loved him for his right-wing political opinions and his penchant for controversy. Halfway through the hour-long program, a loyal listener began his question with an apology. He had missed my introduction at the beginning of the hour: "I'm sorry," he asked me, "I didn't hear . . . . Are you a NASA critic or a NASA apologist?"

His question took me aback. I did not consider myself to be either. As an historian of 20th-century America, I studied space history because it allowed me to investigate the intersections of many different themes—politics, society, culture, science, technology, gender, and race—all in one subject. Although historians' conclusions certainly support or criticize particular policy decisions, I saw doing space history as investigating what spaceflight efforts could reveal about a particular time and place: how specific historical contexts shaped which projects were pursued, why historical actors made particular decisions, and how spaceflight technologies have been embedded in their cultural contexts. Regrouping, I tried to explain the role of the professional historian to the listener.

For many years, the caller's assessment of space experts as entrenched in one camp or the other—as either boosters/apologists or critics/exposers—would not have been wrong. In a 2000 *Space Policy* article, Roger D. Launius, then the NASA Chief Historian, argued that space history could be categorized into three parts, including two categories that were more sophisticated but not altogether different than the caller's binary options. The first, the "historiography of expectation" (my caller's "apologists"), is, according to Launius, "unabashedly celebratory and includes not only the so-called 'Huntsville School' of writing but also those fascinated with the machinery and those who use space history to promulgate the space exploration agenda for the future." The second group, the exposés, used space history to question the validity of space exploration efforts at all. Finally, Launius outlined a third category of scholarship that he called the New Aerospace History: "professionally-trained scholars of differing ideologies and prerogatives who concentrate on questions other than whether or not space exploration is justifiable."[1]

Launius's choice of name for this school of historiography, the "New Aerospace History," self-consciously positioned the newest space history scholarship as descended from the New Social History advanced beginning in the 1960s and 1970s. By doing so, he emphasized the active engagement of the New Aerospace History with recent scholarship in the broader field of history. At the same time, he marked the place of space history as a growing subdiscipline within a field still shaped by the New Social History. Indeed, the very subject of this paper—a study of the relationship of space history to the history subdisciplines—reflects the proliferation of subject areas created when historians wrestling with questions of race, class, ethnicity, and gender challenged the artificial nature of the consensus school's master narrative. As a result, mapping the 50 years of space history's expansion means surveying it against the shifting background of a complex and changing discipline.

Such a survey requires two different approaches. First, this analysis reviews and outlines space history's evolution since the beginning of the Space Age. Because the aim of this piece is to survey the field, the bibliography included in the notes offers a sample of relevant works but not a complete accounting of any subdivision of the field.[2] Second, the paper offers some perspective on space history's current relationship to the rest of the discipline of history as practiced in the United States. When examined in these two ways, space history exists both in "relation to" other history subdisciplines (a terminology which implies separation from the other subfields and an internal cohesion within space history, two points that deserve questioning in their own right) and in a continually evolving "relationship with" the rest of the discipline. As this essay maps those dynamics, it also offers some suggestions.

Although the New Aerospace History developed in dialogue with current historical scholarship, the insights of the New Social History have still been only incompletely incorporated into space history. This deficit is not

---

1. Roger D. Launius, "The Historical Dimension of Space Exploration: Reflections and Possibilities," *Space Policy* 16 (2000): 23–38.

2. Asif Siddiqi's chapter in this volume offers a more complete current historiography.

attributable to a lack of source material, but rather to a limited perspective on what it would mean to integrate the study of race, class, ethnicity, and gender into space history more fully. Bringing the insights of the New Social History to space history is not a call for more compensatory histories of the still-understudied women in the space field or for separate histories of each minority group or ethnicity working in any particular segments of space exploration. (Although compensating for past omissions remains a useful contribution to the field, it is just the first step in historical analysis.) If the New Social History has taught historians anything, it is that gender, race, ethnicity, and class exist in every history—for both privileged and marginalized groups. Gender identity shapes the historical experience of both women and men. Racial identity affects the lives of White people just as much as it does for people of color. Bringing this perspective into analyses of technologies or politics requires a new set of tools.

New developments in the humanities—specifically critical theory—offer a toolbox of concepts and methods that will allow space history to delve further into questions of identity, power, and point of view. If the tools of critical theory can be adapted without straying too far from the narrative tradition of historical scholarship (that is, by adopting its principles and insights without overreliance on theoretical terminology, which can become opaque jargon), the result will bring space history into more fruitful dialogue with the rest of the scholarly community while bringing the insights of recent scholarship to a wider readership.

A BRIEF HISTORY OF SPACE HISTORY

The active study of space history began with the very first successful orbital flights in the late 1950s. After the flights of Soviet artificial satellites Sputniks I and II in 1957, spaceflight efforts in the United States generated awareness by both participants and observers that these events were historic; the participants were "making history." Because American lawmakers were also cognizant of the history-making potential of U.S. space efforts—and of the need to publicize American achievements to the rest of the world—the 1958 National Aeronautics and Space Act included, alongside the directives for the creation of a civilian space agency, the mandate that NASA "provide for the widest practicable and appropriate dissemination of information concerning its activities and the results thereof."[3] In practical terms, this directive provided the basis for the creation and maintenance of NASA's history offices, archives,

---

3. "National Aeronautics and Space Act of 1958," Public Law 85-568, in *Exploring the Unknown: Selected Documents in the History of the U.S. Civil Space Program*, vol. 1, *Organizing for Exploration*, ed. John M. Logsdon (Washington, DC: NASA SP-4407, 1995), p. 337.

The 1959 NASA Seal. *(NASA photo no. GPN-2002-000195)*

and libraries. The space agency even began a fine arts program, sponsoring a still-ongoing effort to commission artists to record NASA's achievements through sketches, paintings, and other art forms.[4]

The story of how NASA came to interpret its mandate to include a history program began, at least in part, with Melvin Kranzberg, one of the fathers of the history of technology and a key figure in the creation of the NASA History Office. Kranzberg was a faculty member at the Case University of Technology in Cleveland, Ohio, when Case's president, T. Keith Glennan, was asked by President Dwight D. Eisenhower to become the founding Administrator of NASA. In 1958, Kranzberg persuaded Glennan to create a history office at the new civilian space agency in the tradition of the successful history offices working in the armed forces and in other federal agencies. The

---

4. For history and individual artists in the NASA Art Program, see Anne Collins Goodyear, "The Relationship of Art to Science and Technology in the United States: Five Case Studies,
continued on the next page

founding of the NASA History Office and the beginning of space history as a field occurred at the same time that the broader discipline of history began to see the development of distinct subfields organized by topic and approach.[5]

Around the same time that his discussions with Glennan were inspiring the new NASA History Office, Kranzberg also helped to found the Society for the History of Technology (SHOT). Kranzberg saw the history of technology as the latest development in the study of the past: the newest link in a chain of histories that offered fresh topics of study and modes of analysis to the expanding field. In May 1962, he published an article in *Science* magazine titled "The Newest History: Science and Technology." In it, he compared the history of technology to James Harvey Robinson's *The New History* (1912), published exactly 50 years earlier. As Kranzberg noted, at the same time that Robinson was developing his New History, another historian, George Sarton, was also offering the field a groundbreaking new subject for consideration: a new history of science. In all three cases, changing world events, social movements, and academic developments inspired historians to rethink their conceptions and interpretations of the past.[6]

The development of innovative historical approaches—and thus of new historical subfields—drove the central argument of Kranzberg's *Science* article. For the history of technology, Kranzberg argued, the launch of Sputnik I on 4 October 1957 marked the beginning of a new era. In response, the United States needed not only a technological response in the form of a space program, but also a study of "technology and science as essential components of our culture, affected by and affecting every other aspect of society." Building on the tradition of change and growth in the historical field, Kranzberg saw new histories as extending and expanding a vital and changing discipline. In his words, "Just as the 'new' history triumphed over the 'old' but never succeeded in dislodging it completely, so today the 'new' history is itself being supplemented

---

continued from the previous page

1957–1971" (Ph.D. diss., The University of Texas at Austin, 2002); Anne Collins Goodyear, "NASA and the Political Economy of Art, 1962–1974," in *The Political Economy of Art: Creating the Modern Nation of Culture*, ed. Julie Codell (Newark: University of Delaware Press, forthcoming); Anne Collins, "Art, Technology, and the American Space Program, 1962–1972," *Intertexts* 3, no. 2 (fall 1999): 124–146; Anne Collins Goodyear, "On the Threshold of Space: Norman Rockwell's Longest Step," *Architecture and Design for Space: Vision and Reality* exhibit catalog (New York: Harry N. Abrams, Inc., 2001), pp. 102–107 (exhibit shown at the Art Institute of Chicago, 24 March–21 October 2001); "Robert Rauschenberg's Space-Age Allegory, 1959–1970," in *1998 National Aerospace Conference Proceedings* (Dayton, OH: Wright State University, 1999): 82–91.

5. For Kranzberg's influence on the creation of NASA's History Office, see Roger D. Launius, "NASA History and the Challenge of Keeping the Contemporary Past," *Public Historian* 21 (summer 1999): 63–81.

6. Margaret Rossiter, ed., *Catching Up With the Visions: Essays on the Occasion of the 75th Anniversary of the Founding of the History of Science Society* (Chicago: University of Chicago Book for the History of Science Society, 1999), a supplement to *Isis* 90.

by the 'newest' history."[7] Kranzberg's *Science* article is particularly instructive for a discussion of how today's space history has evolved because his analysis of American historiography up to 1962 offers a useful model for thinking about how new histories expand the discipline of history. In addition, it points out the close link between space history and the history of technology, which continues to be a vital and important subfield for space history.

If the NASA History Office's existence can be traced to Glennan and Kranzberg, its reputation for scholarly rigor began with the first NASA Historian, Eugene "Gene" M. Emme. From the beginning of its life, the NASA History Office worked to balance two major charges: collecting and archiving the history of U.S. civil space exploration efforts for use by historians, scholars, and the press, and interpreting that material to advise the space agency on ongoing decisions. In addition to managing these tasks, Emme put the program on the path to real scholarly publishing. He instituted the practice of peer review for historical manuscripts published by the NASA History Office, a process that parallels the one used by academic presses and one which has allowed NASA's history program to develop into a respected site for both research and publishing. As the first in a series of interpretive volumes recording the details of historic space achievements within a narrative structure, Swenson, Grimwood, and Alexander's *This New Ocean: A History of Project Mercury* set the tone for NASA's authoritative recording of space history. Within its first two decades, NASA's project histories also included books on Gemini, Vanguard, and Apollo.[8] Within the structures of the U.S. space agency, the NASA History Office focused on American space efforts, emphases that also characterized the field of space history generally.

The NASA History Office also began the ongoing relationship between space history and oral history. As a research technique, the tape-recorded interview came into its own in the 1940s and became a useful tool for recording histories both "from the bottom up" and "from the top down."[9] By 1966, the Oral History Association provided a professional organization for oral historians to share their work while developing and refining the ethical and practical guidelines for productive oral histories. For an endeavor like spaceflight,

---

7. Melvin Kranzberg, "The Newest History: Science and Technology," *Science* 136, no. 3515 (11 May 1962): 463–468.

8. Loyd S. Swenson, Jr., James M. Grimwood, and Charles C. Alexander, *This New Ocean: A History of Project Mercury* (Washington, DC: NASA SP-4201, 1966); Constance McLaughlin Green and Milton Lomask, *Vanguard: A History* (Washington, DC: NASA SP-4202, 1970); Barton C. Hacker and James M. Grimwood, *On the Shoulders of Titans: A History of Project Gemini* (Washington, DC: NASA SP-4203, 1977). See also Launius, "NASA History," pp. 63–81.

9. Paul Thompson, *The Voice of the Past: Oral History*, 3rd ed. (Oxford: Oxford University Press, 2000); Edward D. Ives, *The Tape-Recorded Interview: A Manual for Fieldworkers in Folklore and Oral History* (Knoxville: University of Tennessee Press, 1995). The best practical handbook is Donald A. Ritchie, *Doing Oral History: A Practical Guide*, 2nd ed. (Oxford: Oxford University Press, 2003).

which required the work of so many different managers, engineers, scientists, and pilots, oral history became a key means of recording the full history of various space programs, NASA Centers, and historical actors. NASA continues to use oral history as a major tool for collecting, preserving, and disseminating space history.[10]

If the early years of space history (and its relationships with the history subdisciplines) can largely be traced through a history of the NASA History Office, once the field developed into some maturity in the 1980s, the story got much more complex. From where it began in the early 1980s, space history underwent dramatic growth and transformation. Because a full analysis of that historiography would be too long and involved for this piece (and has already been done extraordinarily well elsewhere, as noted above),[11] an outline serves better as a way of noting the relationships between the growing subfield and the changes happening in the discipline of history as a whole. Three events mark key points in the evolution of space history: a 1981 Smithsonian proseminar, Walter McDougall's Pulitzer Prize–winning 1985 book, and Asif Siddiqi's 2000 history of the Soviet space program, *Challenge to Apollo.*

In 1981, a Smithsonian Institution proseminar in space history hosted at the National Air and Space Museum marked the emergence of space history as a recognized field. David DeVorkin and Pamela Mack of the then–Department of Space Science and Exploration called the meeting to bring together scholars working on space history in order to assess the progress made over the previous 15 years. The report of the meeting in *Isis* recorded a successful and growing subdiscipline, noting that "the field is already marked with a respectable number of books, monographs, dissertations, and works-in-progress." The questions being asked at this meeting offer a sense of the state of development of the field. Three issues dominated discussion: first, "Is space history best considered part of the history of science or of the history of technology?"; second, "Can space science be considered a coherent discipline?"; and finally, "How should space historians confront the peculiar state of sources in this field?"[12]

In debating the first question, historians of science and historians of technology who worked on space topics found themselves in active discussion about the commonalities and differences between their home subfields. The discussion of space history's place quickly made it clear just how much space history required the insights of both subdisciplines. Requiring space history to be either one or the other would be insufficient. (The divisions between these

---

10. See Roger D. Launius, "We Can Lick Gravity But Sometimes the Paperwork Is Overwhelming: NASA, Oral History, and the Contemporary Past," *Oral History Review* 30, no. 2 (summer/fall 2003): 111–128.

11. Launius, "Historical Dimension," *Space Policy*, pp. 23–38.

12. Richard F. Hirsh, "Proseminar on Space History, 22 May 1981," *Isis* 73, no. 266 (1982): 96–97.

two subdisciplines and the professional organizations that represent them are only just beginning to be healed. The November 2005 joint meeting between the Society for the History of Technology and the History of Science Society in Minneapolis, Minnesota, marked a renewed attempt to bridge this gap).[13] As a subject centered on the relationships among science, technology, and the state, the history of spaceflight pushed historians to address science and technology as social and political activities.

Space historians at the 1981 Smithsonian proseminar also shared a common set of anxieties about sources. Many faced significant problems getting full access to documentation that was still considered sensitive during the renewed Cold War tensions of the early Reagan administration. At the same time, massive space projects generated so much paperwork that they became difficult to interpret. In the opinions of those attending the Smithsonian event, government records from active or recently active programs were "abundant but poorly organized." Again, this recorded discussion provides a useful benchmark for assessing space history. Given how much space history would expand by the early 1990s, when the end of the Cold War led to an explosion of newly available materials, the question of sources provides a striking point of comparison.[14]

One of the solutions offered for dealing with incomplete or sensitive records was oral history. The proseminar's organizers quickly took up that charge. Between 1981 and 1990, the Department of Space History at the Smithsonian Institution's National Air and Space Museum organized several oral history projects. These included the Space Astronomy Oral History Project, the Space Telescope History Project, the Glennan-Webb-Seamans Project for Research in Space History, and the RAND History Project. In all, the interviews conducted reflected the principal investigators' interests in space science, as well as in management and political themes in space history. In the final catalog of these oral histories, the organizers acknowledge that their understanding of the interactions between science, technology, and the state changed considerably over the course of the oral history projects. This insight reflects the scholars' own intellectual growth during the course of the project through the 1980s, but it also reflects the state of the field. In the midst of their work, space history underwent an evolutionary leap.[15]

---

13. The organizations had unsuccessful joint meetings in Pittsburgh in 1986 and in Madison, WI, in 1991. See Terry S. Reynolds, "From the President's Desk: 'Time to Try Again?'" *SHOT Newsletter* (April 2000), available online at *http://shot.press.jhu.edu/Newsletters/archive/2000_April/presdesk.htm* (accessed 21 April 2005).

14. Hirsh, "Proseminar," pp. 96–97.

15. Martin J. Collins with Jo Ann Bailey and Patricia Fredericks, "Oral History on Space, Science, and Technology: A Catalogue of the Collection of the Department of Space History, National Air and Space Museum" (Washington, DC: Smithsonian Institution, 1993), pp. i–v.

Walter McDougall began his Pulitzer Prize–winning analysis of space history with a metaphor of evolution: the image of the first fish-turned-amphibian. In that moment, he suggested, biological adaptation jumped forward, not in a slow, incremental progression, but in a saltation, an evolutionary leap. According to McDougall, this metaphor also described the transformed relationship between the state and research and development (R&D) in the years after the Second World War. In many ways, . . . *The Heavens and the Earth* was also a saltation for space history. McDougall's work was a watershed book for its comprehensive consideration of space history as a part of political history.[16]

Twenty years later, McDougall's work remains a required first reference on many topics for most space historians (both popular and academic). At a 1997 40th-anniversary conference commemorating the launch of Sputnik, many historians began their analyses with a reference to McDougall's work.[17] In considering how space history exists both in relation to (that is, standing separately) and in active relationship with particular historical subdisciplines, however, McDougall's work solidified a link between space history and political history that remains strong. Few would consider writing a space history without some serious consideration of party politics, national legislators, or foreign and domestic policy. More so, political historians welcome discussion of space history as an avenue into broader topics.

Just as McDougall's example required space historians to place space history in its political context, so also by the mid-1980s, new developments in the history of technology required historians to reconsider how technologies existed as embedded in their social contexts. As a result of the ongoing relationship between space historians and historians of technology (who are often one and the same), space history and the history of technology grew and broadened in similar ways over the years. In a 1986 *Technology and Culture* article, Kranzberg published his famous "six laws of technology," guiding principles that emphasized the role of technology as an inherently human endeavor, embedded in culture. Likewise, space history has deepened its understanding of space technologies—and indeed, of space programs—as embedded in particular social, political, and cultural contexts. Within the Cold War context of the early space race, however, for the first 20 years of space history, most U.S. authors focused on American space efforts, in part because these stories resonated with the public and in part because the ongo-

---

16. Walter McDougall, . . . *The Heavens and the Earth: A Political History of the Space Age* (New York: Basic Books, 1985), p. 3. Because of its length, . . . *The Heavens and the Earth* is not easily assigned in a classroom setting. A digestible history of space exploration that encompasses the political and social contexts is still needed.

17. "Reconsidering Sputnik: 40 Years Since the Soviet Satellite Symposium" (held in Washington, DC, 30 September–1 October 1997).

ing diplomatic stalemate with the Soviet Union made information about the Soviet side of the story all but impossible to access.[18]

Another saltation for space history happened at the end of the Cold War, when the fall of the Berlin Wall in 1989 presaged the disintegration of the Soviet Union in 1991. Not only did these geopolitical changes have major impacts on the way that spaceflight would be conducted from that point onward (thus requiring historians to rethink how space history would be written from then on), but these changes also created a boom in possibilities for space history. New sources emerged, both through the declassification of military or other classified space projects in the United States and through the release of previously secret sources from the former Soviet Union.

New sources yielded new histories. One that compares to Walter McDougall's in scope and impact is Asif Siddiqi's *Challenge to Apollo: The Soviet Union and the Space Race, 1945–1974*. Working in the Russian-language documents made newly available by the release of uncensored records after 1988, Siddiqi reconstructed the history of the Soviet space program from the early 1930s Group for the Investigation of Reactive Engines and Reactive Flight (GIRD) to the end of the N1L3 program in 1974. Comprehensive, detailed, and yet still very readable, his narrative offers new dimensions and backstories to known events, revealing details about the people and the decision-making processes that created the Soviet space program. In doing so, the book presents a clear look at the history of Soviet space efforts, the outlines of which had previously only been gleaned from censored records or American intelligence. The result, Siddiqi suggests, sheds new light on human space exploration as a whole: "What may be possible now is to take a second look not only at the Soviet space program, but also the U.S. space program—that is, to reconsider again humanity's first attempts to take leave of this planet."[19] In the United States, the end of the Cold War also opened new topics for space researchers, permitting histories of previously classified programs (for example, the CORONA spy satellites).[20]

Indeed, the number of topics that constitute space history has multiplied in recent years. As it now stands, space history encompasses the history of human spaceflight, including reevaluations of programs, centers, technologies,

---

18. Melvin Kranzberg, "Technology and History: 'Kranzberg's Laws,'" *Technology and Culture* 27 (1986): 544–560.

19. Asif A. Siddiqi, *Challenge to Apollo: The Soviet Union and the Space Race, 1945–1974* (Washington, DC: NASA SP-2000-4408, 2000), p. x. Also republished as a two-volume set: Asif A. Siddiqi, *The Soviet Space Race with Apollo* (Gainesville: University Press of Florida, 2003), and Asif A. Siddiqi, *Sputnik and the Soviet Space Challenge* (Gainesville: University Press of Florida, 2003).

20. Dwayne A. Day, John M. Logsdon, and Brian Latell, eds., *Eye in the Sky: The Story of the Corona Spy Satellites*, Smithsonian History of Aviation Series (Washington, DC: Smithsonian Institution Press, 1998).

events, and people, including both military and civilian spaceflight projects and technologies.[21] The recent addition of commercial space ventures and a nascent space tourism industry should soon join these topics. Human spaceflight makes up only a part of the picture, however. Space history must also include satellite programs, launch vehicles, and planetary exploration. The history of space science and of astronomy is also a part of space history.[22] Although most of what is written focuses on stories of success, accounts of incomplete, failed, or abandoned projects also illuminate the forces that shape space exploration. And space history is most decidedly international. As the number of countries participating in space efforts has increased, space history reflects an expansion beyond the previous U.S.-Soviet/Russian focus. In part, this breadth of topic and diversity of approach define the New Aerospace History.[23]

THE NEW AEROSPACE HISTORY

More so, however, the New Aerospace History developed in the 1990s as a result of the increasing professionalization of space history. Like other related subdisciplines, space history evolved from histories written by participants and practitioners into a field being advanced by professionally trained historians.[24] Roger D. Launius, the NASA Chief Historian in the 1990s, also led the push for space history to engage the cutting-edge scholarship in the wider discipline. During his tenure leading the NASA Headquarters History Office from 1990 through 2002, Launius worked to develop the Agency's publishing efforts as a way of creating opportunities for a rigorous practice of space history. For instance, in addition to commissioning new volumes for the exist-

21. See, for example, Andrew Chaiken, *A Man on the Moon: The Voyages of the Apollo Astronauts* (New York: Viking Press, 1994); Roger D. Launius, "NASA and the Decision to Build the Space Shuttle, 1969–72," *The Historian* 57 (autumn 1994): 17–34; Robert A. Divine, *The Sputnik Challenge: Eisenhower's Response to the Soviet Satellite* (New York: Oxford University Press, 1993); Roger D. Launius and Howard E. McCurdy, eds., *Spaceflight and the Myth of Presidential Leadership* (Urbana: University of Illinois Press, 1997); W. Henry Lambright, *Powering Apollo: James E. Webb of NASA* (Baltimore: Johns Hopkins, 1995); James J. Harford, *Korolev: How One Man Masterminded the Soviet Drive to Beat America to the Moon* (New York: John Wiley & Sons, Inc., 1997).

22. See, for instance, Pamela Mack, *Viewing the Earth: The Social Construction of Landsat* (Cambridge, MA: The MIT Press, 1990); "Developing U.S. Launch Capability: The Role of Civil-Military Cooperation" (paper presented at the American Association for the Advancement of Science conference, Washington, DC, 5 November 1999); David DeVorkin, *Science with a Vengeance: How the Military Created the US Space Sciences After World War II* (New York: Springer, 1993).

23. See, for instance, Margaret A. Weitekamp, *Right Stuff, Wrong Sex: American's First Women in Space Program* (Baltimore: Johns Hopkins, 2004); John Krige and Arturo Russo, "Europe in Space, 1960–1973," European Space Agency SP-1172 (Noordwijk, Netherlands: ESA Publications Division, 1994).

24. Similar trends exist in the history of technology. At the 13 January 2005 meeting of the Historical Seminar in Contemporary Science and Technology at the Smithsonian Institution's National Air and Space Museum, a spirited debate arose between those celebrating the prevalence of professional historians in the field and those lamenting the absence of trained engineers.

ing Special Publications series, Launius also began the NASA Monographs in Aerospace History, a series of slim paperback volumes focused on specific topics. Throughout his efforts, Launius aimed to bring NASA's publishing to a new level of scholarly excellence, an effort that was recognized by the larger history community when the Agency's history books began to win prizes from professional organizations. Through the development of a professionalized history, space history forged new connections with other subdisciplines at the same time that it also became a somewhat more coherent subfield.[25]

As with so many things, the status and standing of space history as a subdiscipline can be measured through its funding and visibility. Several significant fellowships exist for emerging and established scholars. The American Historical Association (AHA) and NASA have offered a joint full-year predoctoral or postdoctoral aerospace history fellowship each year since 1986. And several different fellowships for graduate students (at the master's, predoctoral, and postdoctoral levels) and senior scholars exist at the Smithsonian Institution's National Air and Space Museum. Space history is also a consistent presence at major scholarly conferences including the AHA, the Society for the History of Technology (SHOT), the Organization of American Historians (OAH), and the American Studies Association (ASA).

Space history also has a tradition of gathering scholars and participants to celebrate and commemorate major anniversaries in the history of the field. Beginning with events and symposia held to mark the first 25 years of the Space Age, such conferences have recorded the state of the field at various points in its existence. This very volume follows in that tradition. As the proceedings of the NASA History Division's "Critical Issues in the History of Spaceflight" symposium, the articles contained here offer a current indicator of the subject's breadth and diversity—and of participants' sense of the field as a coherent enough one to warrant such a meeting.[26]

As much as space history has become a more internally coherent field, however, its employment opportunities, graduate study, and publishing trends reflect its roots in many different subdisciplines. Although dedicated space history jobs can be found at NASA (at Headquarters or the Centers), the Smithsonian's National Air and Space Museum, or the Space Policy Institute

---

25. For instance, the Organization of American Historians (OAH) awarded its 1998 Richard W. Leopold Prize to Andrew Butrica's *To See the Unseen: A History of Planetary Radar Astronomy* (Washington, DC: NASA SP-4218, 1996).

26. Allan Needell, ed., *The First 25 Years in Space: A Symposium* (Washington, DC: Smithsonian Institution Press, 1983); Alex Roland, ed., *A Spacefaring People: Perspectives on Early Spaceflight* (Washington, DC: NASA SP-4405, 1985); Martin J. Collins and Sylvia D. Fries, eds., *A Spacefaring Nation: Perspectives on American Space History and Policy* (Washington, DC: Smithsonian Institution Press, 1991); Stephen J. Garber, ed., *Looking Forward, Looking Backward: Forty Years of U.S. Human Spaceflight Symposium* (Washington, DC: NASA SP-2002-4107, 2002).

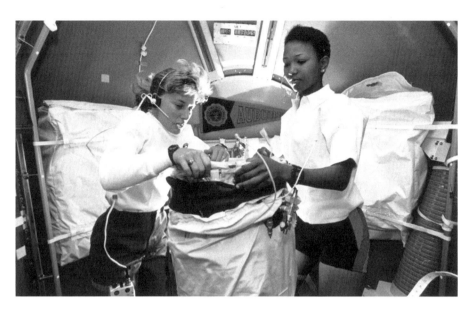

Jan Davis and Mae Jemison on STS-47. *(NASA photo no. GPN-2004-00023)*

at George Washington University, most space history experts continue to find homes in non-space-specific academic jobs in history or political science. (In a rare occurrence, the University of Central Florida offered and filled a full-time, tenure-track space history position in 2005.) The many intersections of space history with the other history subdisciplines offer employment opportunities that are at least as ample as any academic field's opportunities are. Likewise, junior scholars engaged in graduate work have focused on space topics while earning degrees in history and political science as well as fields as diverse as geography and communications.[27] Opportunities for publishing peer-reviewed articles also reflect the roots of space history as a topic studied by many different types of historians. Except for *Space Policy*, few professional journals have space topics as a central focus.

The inherently interdisciplinary nature of space history can be seen in some of its best new works. For instance, Howard McCurdy's *Space and the American Imagination* combines social and cultural history with public policy analysis to show how popular culture influenced policy-making. McCurdy analyzes how "space boosters" in the 1950s and 1960s used magazines, television shows, and movies to create the groundswell of support needed to loose the massive amounts of public funding required to carry out space exploration initiatives. McCurdy's detailed analysis persuasively links comics and

27. Kathy Keltner, for example, is writing a communications Ph.D. dissertation at Ohio University.

Congress. What might have seemed like an unlikely junction between unrelated fields is now a connection being followed by other scholars.[28]

Some likely connections are only just being explored. Despite what might seem like natural areas of overlap, very few scholars have actively pursued work at the juncture between environmental history and space history. As areas of history that both study the intersections of science, technology, and culture, space history and environmental history have much to say to each other. In a field that is building on its histories of national parks and natural spaces, environmental history investigates the intersections between nature, technology, and public policy. Environmental historians have taken on roads, cars, and urban/suburban sprawl as topics but have stopped short of dealing with outer space. As much as many environmental historians have not considered outer space as "nature" or even as a natural place, neither have space historians looked to environmental history for ways to think about space as an environment. Environmental history might also offer models for thinking about the Earth and low-Earth orbit as "natural." New work by scholars such as Neil Maher demonstrates the extent to which exploring space is less about finding nature in outer space than it is about obtaining new perspectives on nature on Earth. In the environmental historian's triad of investigating the intersections between nature, technology, and culture, space historians often ignore nature. The need for intersection between these subfields is a development being echoed by historians of science and technology. Both the History of Science Society (HSS) and SHOT now have environmental history special interest groups (called the "Earth and Environment Forum" and "Envirotech," respectively). Despite these forays into interdisciplinarity, space history has often lagged behind the evolution of the discipline as practiced in the United States.[29]

By the 1980s, the New Social History had fundamentally transformed the discipline's practice, becoming formalized through established journals, academic appointments, and professional organizations. The rejection of the consensus school led to renewed attention to the lives of ordinary people and a new set of narratives that challenged the accepted periodization of U.S. and world history. Although critics complained that the field of history was becoming fractured or that a common American identity was being lost,[30] advocates

---

28. Howard McCurdy, *Space and the American Imagination* (Washington, DC: Smithsonian Institution Press, 1997).

29. Two examples are Neil Maher, "On Shooting the Moon," Gallery in *Environmental History* 9 (July 2004): 526–531, and Erik M. Conway, "The World According to GARP: Scientific Internationalism and the Construction of Global Meteorology, 1961–1980" (paper presented at the International Commission on History of Meteorology, Polling, Germany, 5–9 July 2004). "Envirotech" was founded at the August 2000 SHOT meeting in Munich, Germany.

30. Arthur M. Schlesinger, Jr., *The Disuniting of America: Reflections on a Multicultural Society* (New York: W. W. Norton & Company, 1998).

for the New Social History argued that particular attention to women, laborers, people of color, the poor, or people with disabilities revealed aspects of the past that had been systematically ignored by the previous, more unified narrative. Growing scholarship demonstrated how exclusionary and limited the master narrative had needed to be in order to maintain its cohesiveness.

Through the 1970s and the 1980s, scholars developed subfields with new modes of analysis that focused on questions of difference and power. In 1990, when Eric Foner edited a new collection of essays for the AHA called *The New American History*, in addition to essays on various periods of U.S. history, the volume included attention to six "major themes in the American experience." These included "Social History," "U.S. Women's History," "African-American History," "American Labor History," "Ethnicity and Immigration," and diplomatic history. If these topics can be considered a rudimentary breakdown of the established subfields in American history and of the concerns of the New Social History, then an examination of these areas offers insight into how well space history has engaged each of them. In the parlance of many historians, this longer list is often simplified to class, race, ethnicity, and gender.[31]

Political scientists working on space topics have addressed questions of class or labor history in space history through their analyses of NASA as a complex organization and NASA's management culture. Sadly, these subjects became all too relevant after the losses of two Space Shuttles, *Challenger* in 1986 and *Columbia* in 2003. Both the Rogers Commission and the Columbia Accident Investigation Board diagnosed organizational cultures that had become inured to risk. In addition, they found communication and project management problems that contributed directly to the loss of the two Shuttle crews. As a result, scholars have paid particular attention to NASA's decision-making culture. Many other aspects of NASA as a labor force remain unexamined, however. Although the individual stories of astronauts, flight controllers, and rocket scientists have been recorded, the collective stories of the thousands of people who made particular space projects work offer many opportunities for thinking about the space agency as a workplace.[32]

Labor practices and environments, including the relationship of the space agency with contract work, a key characteristic of NASA's labor structure—and of the larger aerospace industry—remain an underdeveloped topic. For instance, the Grumman Corporation, the engineering company that won the

---

31. Eric Foner, ed., *The New American History* (Philadelphia: Temple University Press, 1990), p. vi.

32. Howard McCurdy, *Inside NASA: High Technology and Organizational Change in the U.S. Space Program* (Baltimore: Johns Hopkins, 1993). See also Diane Vaughan, *The Challenger Launch Decision: Risky Technology, Culture, and Deviance at NASA* (Chicago: University of Chicago Press, 1996); Joseph J. Trento, *Prescription for Disaster: From the Glory of Apollo to the Betrayal of the Shuttle* (New York: Crown Publishers, Inc., 1987); Greg Klerkx, *Lost in Space: The Fall of NASA and the Dream of a New Space Age* (New York: Pantheon Books, 2004).

NASA contract to design and manufacture the Lunar Modules for the Apollo Program, never unionized because Grumman self-consciously promoted a sense of community at its facilities while discouraging labor organizing. In a very different example, engineers working at space work sites like the Jet Propulsion Laboratory came to understand that layoffs were a part of the business plan. Aerospace companies hired highly skilled workers when contracts began, only to dismiss them when contracts ended. These two stories are small pieces of a larger story about how shifting relationships between NASA, aerospace contractors, and the larger aerospace industry shaped and reshaped what it meant to do space work from the beginning of the Space Age through the end of the Cold War.[33] Finally, the labor history of the U.S. space program should also include the entire communities that grew up around NASA Centers, when long-term projects like Mercury, Gemini, or Apollo required entire families to relocate. The transformations of places like Huntsville, Alabama, or Cape Canaveral, Florida, or Tysons Corner, Virginia, illustrate how the work of science and technology industries transformed landscapes, creating new communities and cultures.[34]

If the labor history of space has only just begun to be explored, questions of race and ethnicity have been almost entirely ignored. Only one book has dealt with race or ethnicity as a primary topic. J. Alfred Phelps's collective biography, *They Had a Dream: The Story of African-American Astronauts*, offers chapter-length biographies of African American astronauts as basic compensatory history (adding omitted names and events to the historical record without a broader analysis of their social, political, or cultural contexts).[35] Such work is a necessary beginning, but much more remains to be done. Given the sophistication of the analysis in African American history, Asian American history, and Native American history, and the emergence of interest in whiteness as a constructed racial category, space history's lack of analysis of race betrays an unspoken but distinct discomfort. The aspect of the New Social History that has received the most attention in space history has been women's contributions. In recent years, there has been a sudden flurry of attention to women in space. In 1996, when I began my dissertation research on Randy Lovelace's Woman in Space Program, a short-lived and privately funded proj-

---

33. M. G. Lord, *Astro Turf: The Private Life of Rocket Science* (New York: Walker & Company, 2005). See also Joan Lisa Bromberg, *NASA and the Space Industry* (Baltimore: Johns Hopkins, 1999). Lord's memoir of her father's work at the Jet Propulsion Laboratory (JPL) offers useful insights into JPL as a workplace. Bromberg addresses NASA's relationship with contracting companies as a business history while calling for future scholars to return to this subject through primary research.

34. Paul Ceruzzi, *From Tysons Corner to Internet Alley: High Technology in Northern Virginia, 1945–2001* (New Brunswick, NJ: Rutgers University Press, forthcoming in 2006).

35. J. Alfred Phelps, *They Had a Dream: The Story of African-American Astronauts* (Novato, CA: Presidio, 1994).

ect that tested women pilots for astronaut fitness in the early 1960s, only two short pieces and a book chapter had been written about the subject.[36] By the time my book was published in 2004, however, it counted as the fourth major treatment of that specific program in six years.[37] In addition, three new books have recently been published documenting women's successes as astronauts and cosmonauts.[38] In all, there are seven new books published since 2002 about women and space.[39] Another dissertation about NASA's first women astronauts connects the question of women astronauts to the literature in the history of science and technology.[40]

This attention reflects the increased visibility of women in the astronaut corps, the most visible face of NASA's programs. Yet, despite the attention to the subject, space history can still only be considered as working in relation to women's history but not in any real dialogue with women's history or women's studies. Most of the new accounts amount to compensatory history, adding women to the historical account with little attempt to contextualize the histories by using them to make a broader critique or reassessment of the time in which they are set. And little to no work has offered a critical analysis of the role of gender (both femininity and masculinity) in a particular time or place. Investigating the treatment of women can expand what is known about the complex, intersecting, social, cultural, and political contexts of the U.S. space program.

A partial solution for development in the neglected areas may lie in a subfield that has a long relationship with space history: oral history. Oral history continues to be a useful tool, technique, and intersecting subfield for space

---

36. Joseph D. Atkinson and Jay M. Shafritz, "The First Efforts of Women and Minorities to Become Astronauts," chap. 5 in *The Real Stuff: A History of NASA's Astronaut Recruitment Program* (New York: Praeger, 1985); Sheryll Goecke Powers, *Women in Flight Research at NASA Dryden Flight Research Center from 1946 to 1995*, Monographs in Aerospace History, no. 6 (Washington, DC: NASA, 1997); Sylvia D. Fries, "The History of Women in NASA," NASA TM-108100, Women's Equality Day talk, Marshall Space Flight Center, 23 August 1991.

37. Leslie Haynesworth and David Toomey, *Amelia Earhart's Daughters: The Wild and Glorious Story of American Women Aviators from World War II to the Dawn of the Space* Age (New York: William Morrow & Co., 1998); Stephanie Nolen, *Promised the Moon: The Untold Story of the First Women in the Space Race* (New York: Four Walls Eight Windows, 2002); Martha Ackmann, *The Mercury 13: The Untold Story of Thirteen American Women and the Dream of Space Flight* (New York: Random House, 2003); Weitekamp, *Right Stuff, Wrong Sex*.

38. Pamela Freni, *Space for Women: A History of Women with the Right Stuff* (Santa Ana, CA: Seven Locks Press, 2002); Laura S. Woodmansee, *Women Astronauts* (Burlington, Ontario: Apogee Books, 2002); Bettyann Holtzmann Kevles, *Almost Heaven: The Story of Women in Space* (New York: Basic Books, 2003).

39. In addition to those listed above, see also Laura S. Woodmansee, *Women of Space: Cool Careers on the Final Frontier*, Apogee Books Space Series 38 (Burlington, Ontario: Collector's Guide Publishing Inc., 2003).

40. Amy Foster, "Sex in Space: The Politics and Logistics of Sexually Integrating NASA's Astronaut Corps" (Ph.D. diss., Auburn University, 2005).

historians. In 1996, NASA's Johnson Space Center History Office initiated an oral history project to interview NASA employees and contractors from the Mercury, Gemini, Apollo, and *Skylab* programs, as well as to convert decaying oral history reel-to-reel tapes to more stable media. An analysis and reflection on NASA's history and continuing work with oral history can be found in Roger Launius's 2003 article in a special issue of the *Oral History Review* about oral history in the federal government.[41] As we continue to lose the original participants in early space efforts, the need to preserve space history in comprehensive, well-researched, -documented, and -preserved interviews is becoming all the more important. Furthermore, the current scholarship in oral history demands consideration of what recorded interviews reveal about race, class, gender, status, and power. Perhaps a closer relationship between oral history and space history, two subdisciplines that have been closely linked for some time, could provide one avenue for the New Aerospace History to develop in its integration of the insights of the New Social History.

In 2000, Roger Launius identified a New Aerospace History that seeks to engage with the scholarship and insights of the New Social History. And, as just outlined, much remains to be done. But in many ways, the scholarly world has already moved beyond the ideas of the New Social History. If space history is going to engage with the insights provided by the explosion of historical scholarship in the last 20 years, space historians must begin to grapple with the influences of critical theory.

CRITICAL THEORY AS A TOOLBOX

Critical theory is an umbrella term that encompasses the diverse and often divergent theoretical schools of structuralist, poststructuralist, feminist, Marxist, postmodern, and psychoanalytic theory that emerged since the 1970s in literary and anthropological analysis. Critical theory concerns itself with the differences between representations and reality and, in particular, the ways in which language constructs what is perceived. One part of this analysis is the complex social construction of various identities (race, class, gender, sexuality, etc.). Critical theory looks at how cultures and institutions construct some identities as privileged while marginalizing or denying others. (A similar dynamic also occurs on a national or international level, underlying colonialism and postcolonial relationships between states and peoples.) Critical theory questions the seeming obviousness of these categories, pointing out how assumptions about naturalness are part of the construction of privilege (and thus also of marginalization). The postmodern component of critical theory addresses globalization, consumerism, and the fragmentation

---

41. Launius, "We Can Lick Gravity."

of authority. Such scholarship often pursues discourse analysis, a study of how the way that a topic is discussed shapes its reality. Epistemological questions of how meaning is made and how we know what we know also drive this analysis. Critical theory thrives on juxtaposing texts (which include not only literal, written texts, but also any cultural form that can be read for meaning, including images, music, movies, or television). It embraces contradictions, often frustrating those who want definitive characterizations. In recent years, the exploration of these questions using critical theory has proven to be so fruitful that entire new research fields now exist, including cultural studies, queer theory, and critical race theory.

Historians began to engage literary theory in the late 1970s. In fact, by the time I entered graduate school in the early 1990s, there was a perceptible divide in the history department where I studied at Cornell University. On the one side, Dominic LaCapra led the School of Criticism and Theory, a summer institute begun in 1976 that brought together faculty and graduate students for an intensive six-week theory "boot camp" premised on the idea that an understanding of theory is fundamental to humanistic studies. On the other side, empiricists, including my adviser, taught the intensive study of primary documents—not as texts to be juxtaposed at will, but as evidence of the reality of the past.

The theorists argued that overarching concepts of hegemony, power, and privilege unlocked the central debates raised by the histories they analyzed. They embraced Foucault's suggestion that all history is really about the present, not the past, and that the "real" or "true" past was unknowable. They wrote comfortably for a scholarly audience, preferring analysis to narrative (which is all constructed anyway). The empiricists lamented the impenetrability of theoretical jargon and the ahistorical problems of bringing the postmodern European theory of Foucault to bear on czarist Russia, colonial Latin America, or premodern China. They believed that sufficient research could reveal a past that might not be objectively perceived but that was nonetheless real. They believed in the power of history as a tale well told, in the tradition of the scholar-writer. As I did with the radio caller mentioned at the beginning of this piece, I find that I resist fitting neatly into one category or the other. Although I completed my Ph.D. as a broadly trained Americanist rooted in empirical research, my first job—teaching women's studies, a very theory-centered field—became an informal three-year postdoc in critical theory.

Space history, of course, fits both camps. On the one hand, the history of spaceflight can easily be told as a modernist narrative of progress achieved through rationality and hierarchy. For that matter, space history also fits well into American exceptionalism, the model of U.S. history as an example for the world. On the other hand, critical theory also applies. National and international space efforts cannot be understood without consideration of the mass media, mass consumption, and the mass production that feeds it. Globalization is also a crucial context for space history.

Indeed, the very topic of this essay, an analysis of the historiography of space history and its relationship with the other history subdisciplines, follows an epistemological line of inquiry. It seeks to illuminate critical issues in the history of spaceflight through an analysis of how the field of space history has been constructed and what other fields have been influencing the questions asked—at base, investigating how we do what we do, to the end of understanding how we know what we know. Over the last 10 years, critical theory has become an entrenched part of scholarly discourse, enabling useful critiques of power and difference that bridge national and international studies and bring race, gender, and class into the center of political and social analyses.

For those interested in space history, analyzing the broader cultural settings provides a new way to understand how space efforts resonated. Two examples help make the point. In her 1998 book *Aliens in America*, Jodi Dean analyzed the pre-Y2K fascination with aliens and UFOs as a part of the 1990s trend of interest in space-themed things. Dean suggests that Ron Howard's 1995 film *Apollo 13* transformed the story of a 1970 space accident into a tale that reflected 1990s American preoccupations with a safe return to home that is witnessed through television. Likewise, British scholar Debra Shaw analyzed the spacesuit as cultural icon in the context of broader American popular culture. In both cases, the authors used space as part of their analyses, but neither author is particularly interested in actual spaceflight. A wonderful opportunity exists here for a scholar to work on the cultural imagery of space while also taking spaceflight seriously as something real, not merely as a convenient text.[42]

One of the best examples of a scholar executing sophisticated theoretical analyses in plain language while taking spaceflight seriously is Constance Penley's analysis of NASA in the first half of her book *NASA/TREK*. Written in the wake of the Space Shuttle *Challenger*'s January 1986 explosion, media studies scholar and cultural critic Penley addressed the public's fixation on Christa McAuliffe, the "ordinary citizen"/teacher whose inclusion on the flight accounted for the intense media coverage of the much-postponed launch. Her analysis revealed how widely circulated sick jokes about the public deaths of the Shuttle astronauts betrayed cultural discomfort with women's presence in the highly technological Space Shuttle. Penley's arguments are carefully made and easy to read even as they draw on a vast literature in feminist theory. Penley moves beyond a simple accounting of women's or men's roles

---

42. Jodi Dean, *Aliens in America: Conspiracy Cultures From Outerspace to Cyberspace* (Ithaca, NY: Cornell University Press, 1998); Debra Benita Shaw, "Bodies Out of This World: The Space Suit as Cultural Icon," *Science as Culture* 13, no. 1 (March 2004): 123–144.

to consider how ideas about gender are embedded in customs, organizational structures, and social practices.[43]

The construction of masculinity is just as important as the construction of femininity. In *Astro Turf*, her memoir of her father, a 1960s Jet Propulsion Laboratory engineer, M. G. Lord's deeply personal story also offers a model for a nuanced analysis of the constructions of gender at NASA Centers. Lord explores the rocket engineer as an archetype of 1960s masculinity, a stereotype which she acknowledges "no human person can ever fully embody. The buzz-cut cowboys of Mission Control, homogenous as a Rockette kick-line, were a cold-war fiction, along the lines of other cold-war fictions—the notion, for instance, that hard-drinking, womanizing test pilots, when selected to be astronauts, metamorphosed into temperate family men." Lord's reflections demonstrate that a monolithic masculinity did not exist. Rather, different archetypes of masculinity existed in flight control, or planetary probe engineering, or the astronaut corps: constructions of masculinity that were specific not only to a particular time and place, but also to different jobs. More so, she illustrates in easily comprehensible prose how abstract constructions of masculinity had real effects even though individual men did not conform to the stereotypes.[44]

Analyses of masculinity are also being developed in histories of the images of astronauts. Roger Launius's ongoing reevaluation of the Apollo astronauts in myth and memory offers an insightful analysis of the men's personal backgrounds. With only one exception, NASA's Apollo astronauts were working-class or middle-class men who benefited from military service and the GI Bill—a story that mirrored the postwar American dream, the ideal of the best that America had to offer. The cultural story told by Apollo's models of masculinity provides a marked contrast with the characterizations observed when the nation mourned the *Columbia* astronauts. In that case, the reaction to the *Columbia* tragedy represented a little-noticed but significant shift in the way that astronauts have been depicted. More than just the absence of the previously disproportionate attention to the female members of the crew (as Penley noted after the *Challenger* disaster), the aftermath of the *Columbia* loss included a noticeable focus on the male astronauts as husbands and fathers. The *Columbia* coverage revealed a new conceptualization of men as active, nurturing parents, not just as "family men" (a term that describes a kind of dependability that serves as a workplace asset but which said little about a man's real role as an integral part of his family's life). In both examples, the images of the astronauts reflect the cultural context in which they lived.[45]

---

43. Constance Penley, *NASA/TREK: Popular Science and Sex in America* (New York: Verso, 1997).

44. M. G. Lord, *Astro Turf*, p. 16.

45. Roger D. Launius, "Heroes in a Vacuum: The Apollo Astronaut as Cultural Icon" (presented at the Organization of American Historians 2005 Annual Meeting, San Jose, CA, 3 April 2005);
continued on the next page

A practical model for this kind of wide-ranging gender analysis can also be found in some recent work in diplomatic history. Frank Costigliola's close reading of George Kennan's famous long telegram advocating containment noticed that Kennan cast the Soviet Union and the United States in gender-laden metaphors. Costigliola argues that Kennan's appeal to cultural ideas about proper gender roles reinforced his arguments about necessary U.S. action. Likewise, Robert Dean offers a very useful analysis of the particular brand of upper-class, White masculinity that defined and drove John F. Kennedy and his New Frontiersmen. Examining White House decision-makers throughout the 1960s, Dean points out how gendered metaphors of strength and weakness underlay foreign policy-makers' understanding of international situations, specifically the Cold War. Dean shows how the gendered metaphors used to understand foreign policy led to real Cold War decisions, bringing ideas about gender into crucial national actions. In both cases, gender does not mean "women" but rather the social construction of both masculinity and femininity.[46]

In much the same way, critical race theory has demonstrated that race also requires a more complex treatment than the oversimplified American preoccupation with rigid Black/White racial categories. Critical race theory demonstrates that race is mutable, not biologically determined, and yet nonetheless real. Because race categories have been historically constructed and carried (and still carry) real consequences for people of all colors, the construction of those categories and what they meant at a particular place and time provide the best way to analyze their historical influence and multiple meanings.

The best examples of this kind of work are being carried out in cultural studies. In *Astrofuturism: Science, Race, and Visions of Utopia in Space*, De Witt Douglas Kilgore employs well-grounded race analysis as a part of his examination of the connections between space science fiction and utopian visions of the future set in space. Another author analyzing race in space-themed popular culture is Daniel Bernardi, whose work on *Star Trek* investigates how America's obsession with race played out in the multiple incarnations of Gene Roddenberry's cult hit television show and its many spin-offs. For Bernardi, "'race' refers to a multifaceted, omnipresent but utterly historical category of meanings." How these meanings are constructed in particular times and places

---

continued from the previous page
Margaret A. Weitekamp, "Mourning Men and Women: Gender in the Coverage of the Space Shuttle Columbia Accident and Other Space Tragedies" (presented at the Organization of American Historians 2005 Annual Meeting, San Jose, CA, 3 April 2005).

46. Frank Costigliola, "'Unceasing Pressure for Penetration': Gender, Pathology, and Emotion in George Kennan's Formation of the Cold War," *Journal of American History* 83, no. 4 (March 1997); Robert D. Dean, *Imperial Brotherhood: Gender and the Making of Cold War Foreign Policy* (Amherst: University of Massachusetts Press, 2003).

informs his work, allowing his analysis to account for changes in race relations over time. As a result, Bernardi's work avoids reinforcing racial categories.[47]

Having more complex, theoretically grounded conceptions of race also allows scholars to examine the social and historical construction of whiteness. In addition to the historians documenting the contested construction of White racial identity in the United States, other scholars have been exploring the impact of White privilege: the unearned and usually unnoticed advantages that accompany being White in America. For space history, an awareness of whiteness as a contested identity, which carried real meaning for people's day-to-day lives, opens new topics for investigation. For instance, it would be very interesting to examine a place like Huntsville, Alabama, where whiteness took on several different historical meanings. By the 1950s, the Army Ballistic Missile Agency in Huntsville welcomed German rocket scientists, who had been brought into the U.S. through Project Paperclip. These men found themselves living and working in a state just beginning to wrestle with the fundamental questions raised by the Civil Rights movement. Little race history presents itself to be written when the focus remains narrowed to documenting the historical presence of African American workers. But if one considers the multiple and varied meanings of whiteness, this history offers intriguing possibilities for reinvestigating a formative site for space history.[48]

One of the reasons that space history has not always embraced all of the aspects of the New Social History is that many scholars dismiss the focus on race, class, ethnicity, and gender as forced or unnecessary due to the lack of women or minorities in a field. The previously ignored women's stories have been largely uncovered and already told, the argument goes. Having few people of color working in various space programs means that little race history presents itself to be written. Few labor problems beg for a class history analysis. But when considering critical theory, the question becomes, not how does one write an appropriately attentive history of each race or ethnicity, but rather, how did the space program deal with race or ethnicity? Not where are the women, but how did the space program deal with gender for both men and women? Not where are the gays, but why is the space program so relentlessly straight (and, for that matter, so reluctant to broach the topic of sexuality at all)?

Such questions are relevant even if the identities being analyzed were not noticed or commented upon at the time. Indeed, one of the defining

---

47. De Witt Douglas Kilgore, *Astrofuturism: Science, Race, and Visions of Utopia in Space* (Philadelphia: University of Pennsylvania Press, 2003); Daniel Leonard Bernardi, Star Trek *and History: Race-ing Toward a White Future* (New Brunswick, NJ: Rutgers University Press, 1998), p. 15.

48. Matthew Frye Jacobson, *Whiteness of a Different Color: European Immigrants and the Alchemy of Race* (Cambridge, MA: Harvard University Press, 1999); Peggy McIntosh, "White Privilege: Unpacking the Invisible Knapsack," in *Women: Images & Realities, A Multicultural Anthology*, ed. Amy Kesselman, Lily D. McNair, and Nancy Schniedewind, 3rd ed. (New York: McGraw-Hill, 2003), pp. 424–427.

Dr. Wernher von Braun greeting a crowd at the Gulf South State Fair in Picayune, Mississippi, in October 1963. *(NASA photo no. GPN-2000-000538)*

characteristics of privilege is obliviousness. White privilege, for instance, includes the assumption of whiteness as the norm, a condition that does not need to be named (in contrast to the way that Blackness, for instance, does not go unnamed). Even though participants did not comment on the impact of whiteness or masculinity in the historical moment, the contemporary social construction of those identities continued to shape historical actors' experiences. The insight that all history contains gender, race, ethnicity, and class opens up new possibilities for integrating these elements into the ongoing discussions of technologies and politics in any space history.

One of the admitted drawbacks of critical theory is the jargon that accompanies it. As one teaching Web site suggests, "The hardest part of understanding and working with critical theory is grasping and using the new vocabulary, but, as with all languages, the new vocabulary will empower you and enhance your exposition of already existing thoughts and ideas."[49] I disagree. The concepts and insights of critical theory empower scholars. The vocabulary can be cumbersome and obfuscating. The examples offered above, however, demonstrate that critical theory can be employed in the service of an historical analysis while still using plain language. Keeping in mind the importance of narrative and craft in the writing of history will allow space historians to integrate these insights into readable histories. Critical theory does not offer all of the answers for the development of space history, but sampling from this toolbox can move the field forward.

---

49. Dino Felluga, "General Introduction to the Site," *Introductory Guide to Critical Theory*, updated 28 November 2003, *http://www.purdue.edu/guidetotheory/introduction/* (accessed 16 February 2005).

CHAPTER 17

SPACE ARTIFACTS: ARE THEY HISTORICAL EVIDENCE?

David A. DeVorkin

Museum collections . . . show you not what there was but what was collected.

> —Jim Bennett, "Scientific Instruments,"
> in *Research Methods Guide*, Department
> of History and Philosophy of Science,
> University of Cambridge

Anyone sensitive to the immense costs involved in collecting and preserving the material legacy of modern culture must question such expenditures at one time or another. Can the needs of history, for instance, justify the effort and expense it takes to identify, acquire, transport, preserve, inventory, evaluate, and possibly even to exhibit some object of note? An 11th-century astrolabe, a Galilean telescope, or the fabulously mysterious and insightful Antikythera mechanism all, no doubt, have provided valuable insight into historic events, capabilities, unwritten norms of practice, and cultural imperatives. But what of the modern stuff, essentially the past 50 years of the Space Age? What does the *Freedom 7* capsule tell us? Or what can Apollo 11, or Armstrong's chronograph, or the backup mirror to the Hubble Space Telescope tell us that other forms of documentation cannot reveal? Why collect and preserve material artifacts of the Space Age when there is, indeed, a mountain of documentation readily accessible that can tell us everything we might possibly want to know or can answer every question we can imagine to ask?

The act of collecting and properly preserving objects that somehow represent or inform the history of the exploration of outer space is one of the most expensive and labor-intensive ways of preserving the record of space history. As an historical activity, it is far more expensive, requiring a broad range of talents and expertise and an infrastructure at least an order of magnitude greater than that required for any library or archival facility devoted to space history, and it is many orders of magnitude greater than what is required by an individual scholar to pursue publishable space history. Why, then, do institutions and historians engage in such activity? Are the payoffs and returns proportionally worth the effort and expense? Can the payoffs be measured on scales that compare to the professional payoffs resulting from other forms of historical inquiry and outreach, or are the payoffs of a wholly different character, so removed or distinct from familiar intellectual processes and modes of communication that they demand a distinct scale for evaluation separate from, or complementary to, those in place within academe? This essay will raise these questions and explore them.

WHY ARTIFACTS ARE MARGINALIZED
AS HISTORICAL EVIDENCE

In a 1962 essay in *Science*, filled with the exuberance of establishing a new discipline, Mel Kranzberg argued that there were ample reasons to support the history of science and technology disciplines as the "'newest' history."[1] Speaking more about technology than science, the newest history, he argued, offered promise of reconnecting the two cultures, as if to counter C. P. Snow's allegations. As Kranzberg wished to describe it, "It is about human work [in science and technology] . . . . Indeed, the search for truth and order and beauty in science is comparable to the same striving in literature, art, poetry." It is a very human activity to search for truth, order, and beauty, and the nature of the search reflects changing intellectual climates, human inventiveness and imagination, and human values and social systems. Technology plays an intimate part in all of this because its significance "lies in what it does." Again, following Kranzberg, "the significance of technology is in its use by human beings."[2]

In his 1962 essay, he explored the significance of the study of the history of science and technology, and its possible applications, and identified typical questions modern practitioners of the "newest history" ask and how the exploration of their answers might benefit society. Above all, Kranzberg placed humans at the center of attention as well as the institutions they build and the nations they defend. He used the telephone to describe what is important about technology: At one level, the telephone is merely a system of wires, circuits, and switches, transmitters and receivers of electrical signals. Issues historians have addressed have included who invented it and why they did it, motivations, resources available, assumptions, "but the human meaning of the telephone lies in its transmission of sound for long distances between persons." The telephone has changed the way people live their lives and communicate

---

1. Melvin Kranzberg, "The Newest History: Science and Technology," *Science* 136, no. 3515 (11 May 1962): 465.
2. Ibid., p. 466.

with others. Using the telephone and other examples, Kranzberg's message is "that science and technology have social consequences."[3]

Kranzberg's article, at one level, reflects modern practice. It leaves the strong impression that the "things" of technology do not constitute the knowledge base, but that they do represent history in some amorphous way. Indeed, in his campaign to increase attention by historians and scientists to the value of the history of science and technology, he emphasized its social application and minimized issues relating to what one might call a "material culture" focus. "Things" do appear prominently, and Kranzberg is clearly sensitive to the ills of Neoplatonic aristocratic dualism, the emphasis of brain over hand. But for the sake of his argument in 1962, things merely symbolize human goals and aspirations and adorn the titles, texts, and images of the literature of the history of science and technology.

Thus, in the 40-odd years since Kranzberg's essay in *Science*, an unintended consequence of his campaign, and of those of his generation, was a certain neglect of the things of science and technology, the material artifacts, as sources of information themselves. They could be sacralized and celebrated and even revered, but they did not, in and of themselves, provide a knowledge base. And as a new literature emerged in the history and technology of space exploration, a consequence of the increased interest in the field overall since Sputnik, it also reflected the same priorities of the newest history and did not include things in its formal knowledge base.

Things do matter in the geological and biological sciences, as well as in the broader ranges of natural history including anthropology, archaeology, and paleontology. Collections do constitute primary knowledge. After all, these disciplines largely grew up around collections that had to be organized and preserved somehow, and the present structure of these museums and their collections still represents the organized data the scientist needs.[4] But for the disciplines engaged in space history, where we might find historians of technology and science, or social and cultural history, military history, business history, American history, American studies, along with a smattering of sociologists, economists, policy specialists, and psychologists, to say nothing of those who came from backgrounds in aerospace itself, none of these areas of inquiry grew up around a practice of collecting artifacts, organizing and classifying them, and searching for new knowledge in the effort, through empirical analysis or some form of rational argument. As a result, although those engaged in curatorial functions most definitely think about their collections and treat them to all the standards required of their codes of ethics and institutional capabilities, few of them actually have utilized these collections as

---

3. Ibid., p. 466.
4. Bernard S. Finn, "The Science Museum Today," *Technology and Culture* 6, no. 1 (1965): 78–79.

primary evidence in their historical research and writings. Many have written about their collections and the objects in them, of course, ordering them by age, manufacturer, speed, function, and capability, because they are fascinated by or are somehow attracted to objects, but the data they employ are of the more traditional kind: the written and spoken word, images, pictorial representations, and the like.[5]

To make this last observation, Joseph Corn surveyed a decade's worth of articles in *Technology and Culture*, the quarterly publication of the Society for the History of Technology. He found that less than 15 percent of the authors "employed any material evidence" and, of these, most wrote on ancient or early modern technologies. "Rhetoric to the contrary, then, the history of technology as a field is not deeply committed to learning from things."[6] Corn takes this farther to identify factors that detract from the use of things as evidence and also argues that because of these social factors limiting how historians communicate processes and the influences upon them, in fact, the survival of the real thing (the true artifact and even the facsimile) is more important than one might appreciate from the published record alone.

THINGS AS "CONGEALED CULTURE"

After all, things do exist, have existed, and are constantly on the minds of at least some historians, especially those who find themselves working in museums or training those who might see museums as a career goal. Things constitute the "corpse" of much of what we call science and technology, and so they have been regarded by some as holding out potential as a source of diagnostic or even forensic knowledge offering insights unavailable otherwise. Given the emphasis on people and institutions fostered by Kranzberg and almost all subsequent workers, this potential has remained largely locked up in the things themselves, which has led at least one prominent historian of technology, Thomas Parke Hughes, to refer to them as "congealed culture."[7]

Hughes's rhetorical concept has been applied by scholars to various and sundry objects, institutions, and individuals, mainly to describe a static relic or an art object, "a kind of tomb for the creative spirit" that has somehow been transported into a context wholly unlike that of its creation: the art gallery, living room, museum, or historic site. The term has also been used to

---

5. Joseph J. Corn, "Object Lessons/Object Myths? What Historians of Technology Learn from Things," in *Learning from Things, Method and Theory of Material Culture Studies*, ed. W. David Kingery (Washington, DC: Smithsonian Institution Press, 1996), pp. 35–54.

6. Ibid., p. 37.

7. Thomas Hughes, commentary in Pamela Mack and David DeVorkin, "Proseminar in Space History," *Technology and Culture* 23 (1982): 202–206.

encapsulate entrenched personalities, hopelessly outdated or resistive bureau-cracies, and static libraries and the books they contain.[8] Hughes, however, had no such negative thoughts in mind when he used the term at a May 1981 "Proseminar in Space History" at the National Air and Space Museum. There, he was expressing his feeling that it was the best we could hope for in material culture, but to utilize it we had to learn how to obtain the proper tools to capture the essence of an artifact and to understand how it represents an amalgam of interests, motivations, ideas, questions, and techniques that are representative of the culture that conceived of it, paid for it, built it, and used it. At least, that is what some participants took away from his commentary.[9] Hughes's remark embodied the perennial challenge facing curators of objects, or things, to find ways to unpack all the forces and drives that brought that artifact into existence and played a part in its lifetime of use. Curators trained as historians have certainly done much of this. The literature of space his-tory is rich in the study of the technologies and the objects representing them that made space travel possible. But the question in my mind then and now is, where is the survival of the artifact itself in all this effort? And what is its role in history: as historical evidence leading to new knowledge, or as a commodity, an ornament that somehow illustrates or celebrates, but does not necessarily inform the past?

Kranzberg's assertion that the history of technology focuses on human actions did not prevent almost half of the articles in *Technology and Culture* scanned by Corn from dealing somehow with devices: tools, weapons, instru-ments, objects with a function. But historical studies of things are subject to a wide variety of perspective: "What's nuts and bolts to one historian is 'congealed culture' to another," Larry Owens once observed, implying that things can be described in terms of their "brute facts" of existence, to excru-ciating detail, but they also "embody conceptual schemes and logical strate-gies for dealing with the world." The historian's task, ideally, is to employ interpretive and descriptive tools that present an integrated portrait of the machine/object/thing and the ideas and aspirations it embodies. Owens's very definition of a good historian [of technology] was someone with sensitivity to "socioeconomic and institutional environments."[10]

---

8. John S. Duffield, "Political Culture and State Behavior: Why Germany Confounds Neorealism," *International Organization* 53 (1999): 765–803, noting Jepperson and Swidler describ-ing institutions as "congealed" culture; John D. Kelly, "Nature, Natives, and Nations: Glorification and Asymmetries in Museum Representation, Fiji and Hawaii," *Ethnos* 65, no. 2 (1 July 2000): 195–216; Shaun Gray, quoted in "Aesthetics of Computer Graphics," *pixxelpoint,* http://www. pixxelpoint.org/2001/article-01.html.

9. Discussions with Pamela Mack over the years.

10. Larry Owens, "Book Review," *Isis* 78 (1987): 625–626 (review of Michael R. Williams, *A History of Computing Technology* [Englewood Cliffs, NJ: Prentice-Hall, 1985]).

A machine can certainly embody ideas and assumptions. First, implicit in its design are ideas about the way nature works, as well as assumptions of the ways humans work, as well as assumptions about how a particular human goal can be met. Take the telescope: it definitely embodies basic assumptions about how nature works. Although invented before systematic rules in geometrical optics provided guidance, empirical or experimental exploration soon showed how to build telescopes with greater magnification, resolution, and light-gathering power. Following the development of astronomical telescope technology, then, how it changed over time, has the potential of revealing how technical limitations, intellectual drives, and social issues influenced the development of each of these powers or inhibited their growth for one reason or another. Yet, with but few exceptions, histories of telescope technology in the past tended not to be organized this way and instead were chronological and periodized, or centered on observatory development. And with even fewer exceptions, mainly the work of Albert van Helden and others noted below, histories of telescope technologies have not required the survival of the telescopes themselves. Yet telescopes are lovingly preserved and beautifully displayed throughout European culture as an enduring legacy of human achievement and curiosity. Faced with this situation, any curator of things must, at some point in life, pause and ask, "Why?"

This essay, then, is an exploration of these questions: Is the existence of an artifact useful to history, or does its value reside elsewhere? Is there a sensible difference, in researching and writing history, having the actual artifact involved in that history at hand or not? We will begin by looking at institutional rationales for collecting, then at individual arguments, and finally we will sum up by suggesting some alternative ways to justify the effort.

ADDRESSING THE ISSUE: RATIONALES FOR COLLECTING

It is surprising that there doesn't appear to be a literature critical of the act of formal collecting. There is a literature defending and rationalizing collecting and a smaller literature looking into the psychological motives that stimulate collecting on both individual and collective bases, but there appears not to be one questioning the value or importance of collecting. Of course, I raise this as an observation in the hopes that a reader who has read more widely than I have at this point will offer a correction and direct me to what I have missed. Until that happens, however, I will labor under the assumption that collecting is a core act of human culture, bound up some way in a search for identity and even for power and transcendence.[11] But I will also

---

11. Werner Muensterberger, *Collecting: An Unruly Passion* (Princeton, NJ: Princeton University Press, 1994).

accept the possibility that formal collecting, by institutions and nations, is a self-conscious act that in and of itself is artificial enough to warrant rationalization. Therefore, we should begin by looking at the rationalizations people and organizations have given for collecting.

Institutions and organizations are, first and foremost, composed of individuals, and these individuals act singly and collectively out of both personal and professional motivations. Personal motivations to collect derive from a wide variety of impulses and drives: collecting can provide a sense of identity, personal exploration, security and validation, self-worth, transcendence, and power. All manner of people collect all imaginable things, from stamps, coins, and baseball cards to M&M items, cars, telescopes, and phonograph records.[12] It is one of our more basic instincts and seems to be shared among many cultures. Styles vary, of course, from astute collectors to indiscriminate hoarders. Individuals rarely rationalize why they collect, nor do they need to. But institutions, especially public ones or those existing on private or corporate philanthropy, typically try to, because of the costs involved.

Historians, museum professionals, anthropologists, geologists, biologists, collectors of all types, and their institutions have presented numerous and varied arguments for preservation. In the cultural arena, possibly the most pervasive effort was established by the National Park Service emerging from the Historic Sites Act of 1935: "To preserve places of national significance that retain exceptional value as commemorating or illustrating the history of the United States for the inspiration and benefit of the people."[13] The 1946 enabling legislation that ultimately gave life to the National Air and Space Museum in 1976, which we always cite in the various editions of the introduction to our "Collections Rationale," calls upon us to "memorialize the national development of aviation and space flight." Our charge is to "serve as the repository for, preserve, and display aeronautical and space flight equipment and data of historical interest and significance to the progress of aviation and space flight, and provide educational material for the historical study of aviation and space flight and their technologies."[14]

In order to carry out its designated task, the Park Service has mounted numerous "theme studies" and has created a standardized "National Register

12. Ibid.; Igor Kopytoff, "The Cultural Biography of Things: Commoditization as Process," in *The Social Life of Things: Commodities in Cultural Perspective*, ed. Arjun Appadurai (Cambridge: Cambridge University Press, 1986), pp. 64–91; Frederick Kunkle, "A Heart Melts at Sight of All Things M&M's," *Washington Post* (10 February 2005): Montgomery Extra, pp. 16–17.

13. National Historic Landmarks Survey, "Surveying American History," June 2003, *http:// www.cr.nps.gov/nhl/*, p. 1 (accessed 10 February 2005).

14. Public Law 79-722, chap. 955, 70th Cong., 2nd sess., 12 August 1946. "Initially the legislation did not mention 'space,' but this was added and now serves as basis for the Museum's Mission Statement, as promulgated July 29, 1996," according to the Division of Space History, "Collections Rationale," 2005, NASM Curatorial Files, Washington, DC.

Nomination Form" that contains room for not only describing the candidate, but for including a narrative statement of historical, cultural, and architectural significance and how these characteristics meet a set of criteria maintained by the NPS. Reproduced in full, it reads:

> The quality of significance in American history, architecture, archeology, engineering, and culture is present in districts, sites, buildings, structures, and objects that possess integrity of location, design, setting, materials, workmanship, feeling, and association, and:
>
> A. That are associated with events that have made a significant contribution to the broad patterns of our history; or
>
> B. That are associated with the lives of significant persons in or past; or
>
> C. That embody the distinctive characteristics of a type, period, or method of construction, or that represent the work of a master, or that possess high artistic values, or that represent a significant and distinguishable entity whose components may lack individual distinction; or
>
> D. That have yielded or may be likely to yield, [sic] information important in history or prehistory.[15]

Commemorating, validating, and illuminating historical events, lives of note, or objects of construction or manufacture within their original environments is thus the domain of an agency concerned with such diverse issues as land use and national identity. An entity of the Department of the Interior, it promotes programs in public recreation and education, with preservation at its core: more than half of the parks represent land management "set aside as symbols and evidence of our history and prehistory."[16]

The process followed by curators at the National Air and Space Museum is somewhat different than the National Park Service, though many of its criteria map onto those of the NPS. Symbolism and national identity pervade the collection. Although collecting activity ranges over the whole of the 20th century, collecting in space history itself was heavily augmented

---

15. National Register Bulletin, "How to Apply the National Register Criteria for Evaluation," *National Register Publications*, *http://www.cr.nps.gov/nr/publications/bulletins/nrb15/nrb15_2.htm* (accessed 10 February 2005).

16. "History," National Park Service Web site, *http://www.cr.nps.gov/history/hisnps/* (accessed 10 February 2005).

by an agreement between NASA and the Smithsonian set out in 1967 and modified on numerous occasions. This special agreement was set up because NASA realized that it was rapidly becoming responsible for "a growing number of artifacts, many with great historical value and others with great value for educational, exhibition, and other purposes, relating to the development, demonstration, and application of aeronautical and astronautical science and technology of flight."[17]

NASA decided that the Smithsonian was a more appropriate place to take on this responsibility since NASA did not really want to be in the business of managing a large collection of iconic objects that attracted wide public and political attention. Further, it sought out both a political buffer and a means of historical validation. Left unsaid but implicit in the act of agreement was the fact that in making this arrangement, the Smithsonian was also tacitly agreeing to a formal method of removing objects from the commodity sphere (commercial trading and speculation in space artifacts) and placing them into a singularized and sacralized sphere, to adopt (for the moment) the notions and rhetoric of the economic anthropologist. If one views the NASA/NASM Transfer Agreement as a cultural act from this perspective, one can see it as an example of culture counteracting commoditization (in fact, curators in the department have made this point repeatedly)—since the essence of culture is discrimination, and societies typically set aside or set apart certain objects they deem to be sacred. Anthropology teaches us that culture demands that certain things be singular, unexchangeable, and "publicly precluded from being commoditized."[18] Typically, such constraints are imposed by the state, seeking to create a symbolic inventory akin to the crown jewels of monarchies and reflecting the power of the state itself. National museums, then, can be likened to agencies of the state and mechanisms through which the state imposes its eminent domain to sacralize particular objects. To my knowledge, however, no other federal agency has this form of continuing formal agreement with the Smithsonian. Therefore, the existence of the act itself defines NASA as a unique cultural entity, and it would be useful if, sometime in the future, someone examined the agreement in that light.

This agreement, however, does not compel the Smithsonian to collect a NASA object but gives it first right of refusal. In addition, this arrangement does not limit the Smithsonian's interest to collecting NASA artifacts, since significant programs exist elsewhere within our culture and our focus is space history, not NASA history. Our department has thus identified issues of

17. "Agreement Between the National Aeronautics and Space Administration and the Smithsonian Institution Concerning the Custody and Management of NASA Historical Artifacts," signed 10 March 1967, in the introduction to Division of Space History's "Collections Rationale," 2005.

18. Kopytoff, "Cultural Biography of Things," p. 73.

concern when evaluating any object for inclusion in the national collection, independent of national origin or the part of the government, academe, or industry responsible for it. These are placed within a context that we hope and expect will somehow illuminate and inform space history generally. Choices are made based upon

1. the unique qualities of [the object]

2. the relationship of flown items to engineering prototypes, backups, and models

3. the place for ground support equipment such as simulators, operational consoles, test stands, and the like, and

4. the different metrics of culture, history, and technology that come into play when assessing the historical value of a space artifact.[19]

Within the agreement set forth by the two agencies, one also finds rhetoric describing what should be collected, again offering some guidance on how and why, and overall it attests to NASA's view that these objects possess cultural and educational, as well as technical, value. We maintain no other agreement with any other agency or institution in this country or with any nation. However, although there is a tacit understanding that the criteria we utilize to collect any object remain independent of the originating institution, our special agreement with NASA creates an institutional bias that we cannot and should not ever forget or ignore. The quotation from Jim Bennett at the outset of this paper should always be kept in mind: that collections represent choices made and therefore should not be construed as history but as part of history.

Thus far, looking at the rhetoric of these two very different collecting agencies, NASM and the Park Service, one finds consistent appeal to the need to memorialize, display, educate, or stimulate. These goals are presumed by museum professionals and, again, are the results of choices, both individual and collective. Even though these choices are socially conditioned, one can easily find in the rhetoric of museology a presumption of warrant: the International Council of Museums offers, for instance, a "Code of Ethics" for museums that identifies their collective purpose and their unique responsibilities.[20] Excerpting relevant elements, we find that according to ICOM, "Museums preserve, interpret and promote the natural and cultural inheri-

---

19. "Preface," Division of Space History, "Collections Rationale," 2005.

20. ICOM, "ICOM Code of Ethics for Museums," 2004, *http://icom.museum/ethics.html* (accessed 20 February 2005).

tance of humanity" and hold their collections "in trust for the benefit of society and its development." Museums are, in effect, social institutions that exist to "acquire, preserve and promote their collections as a contribution to safeguarding the natural, cultural and scientific heritage." ICOM sees these collections as a "significant public inheritance" that must be protected by law and international legislation. Throughout its ethics statement, there is a strong and explicit sense of stewardship "that includes rightful ownership, permanence, documentation, accessibility and responsible disposal."

Central to ICOM's warrant is that "museums hold primary evidence for establishing and furthering knowledge." Professional staff within museums are responsible not only for collections care and public accessibility, but for the interpretation of the collection as "primary evidence." Indeed, the notion of "primary evidence" stands at the very core of ICOM's ethics statement. ICOM, which represents all types of museums, including art, technology, and the natural sciences, asserts without example or citation that what museums collect constitutes primary evidence. It recognizes that the designation of primary evidence should not be "governed by current intellectual trends or museum usage" and offers out hope that primary evidence will be used to make a "contribution to knowledge that it would be in the public interest to preserve." Thus, according to ICOM, museums should regard collections as both a present and a future potential resource for knowledge production. The overall policy of the Smithsonian Institution reflects this sensibility, reaffirmed by its Board of Regents in 2001: "Collections serve as an intellectual base for scholarship, discovery, exhibition, and education."[21]

From the standpoint of the collecting institution, then, whose statements are largely bureaucratic and organizational, to say nothing of being self-serving, one finds arguments that still presume the value of collecting, rather than demonstrate value. Once again, it would be easy to reinterpret ICOM's assertions using the perspective of the economic anthropologist: "Power often asserts itself symbolically precisely by insisting on its right to singularize an object, or a set or class of objects."[22] Taken together with ICOM's view, these two interpretations offer copious evidence for rationalizing why we collect.

Each assumes that collections will be useful to "memorialize" or to "educate" and "inform" and even to "inspire." Each also assumes that collections constitute "primary evidence" for historical and scientific inquiry. Indeed, the economic anthropologist goes to considerable and quite convincing lengths to

---

21. Board of Regents, "Smithsonian Collections Management Guidelines," SD-600, 26 October 2001, p. 37.

22. Kopytoff, "Cultural Biography of Things," p. 73. Sometimes that power is tested. When a National Park Service theme study promised to designate a number of observatories as potential candidates for landmark status, observatory directors objected, fearing that such a designation would limit their power to modify their equipment and buildings. Landmark status was not conferred.

argue how a biography of a thing reveals new knowledge about culture. One can learn a lot, for instance, about inheritance rules and practices, as well as family structure, by following how a particular object moved through a family down through the generations. The biography of a thing, therefore, is not only contained in its production, but in its use and treatment as a commodity, and if that thing is somehow removed from the world of commerce and deified as a sacred object, its biography needs somehow to be preserved and made accessible in order for it to illuminate the culture involved.[23] Historians acting as curators might see this as a new way to appreciate the importance of the "provenance" of an object, the history of who owned the object and the conditions of transfer from one hand to another. But few, to my knowledge, have knowingly explored how provenance informs us about the overall culture—its values, priorities, and stability—within which the object moved. Economic anthropologists have long used these techniques to map out change among generations. Historians might take a cue from this and look for ways to apply provenance.

Why Preserve Objects?
The Views and Actions of Individuals

In his survey of a decade's worth of articles in the journal *Technology and Culture*, Joseph Corn also identified ways that a few historians used objects as primary evidence, showing that indeed there is potential knowledge if the right questions are asked. He points to five different ways scholars have used objects as primary source material:

1. Looking at the object in use or (if a machine) in motion can reveal information about the tacit shop practices and techniques of the culture that produced it.

2. Performing a technical analysis of a manufactured object can reveal the process of manufacture, through contemporary accounts as well as retrospective accounts by producers and users.

3. Simulating an object can test behavior and evaluate design expertise through models.

4. Testing actual objects through use can reveal norms of precision.

5. Microscopic analysis of surface markings and looking for consistency in dimension and weight may be evidence of skill and motive.

---

23. Kopytoff, "Cultural Biography of Things," p. 66.

Corn identified each of these methods in specific case studies, mainly of objects from periods where other forms of documentation were not plentiful. Using a case study of a pin-making machine by Steve Lubar to illustrate the first modality, Corn argues that documentary sources (patent records) showed that there were many ways to make a pin machine, but the way the sample was made indicates choices based upon "specific beliefs and practices" because it mimicked manual assembly-line practices.[24] This was only apparent when Lubar experienced the machine functioning, which underscores an argument recently made by Deborah Jean Warner that objects—scientific and technological artifacts and instruments—are interesting because they are functional and therefore should be interpreted in terms of their "performance characteristics." Performance characteristics include all aspects of the building and use of the objects—the skills of design and manufacture involved, the ways to operate them, repair them, and finally, how they are disposed of after their production life.[25]

None of Corn's methodologies apply across the board, and there is significant overlap between some of them. Still and all, it is a useful exercise in articulating how objects have been found to increase historical knowledge and understanding. One finds examples from the history of astronomy that fit one or more of these methods. For instance, there is the famous case of the Antikythera mechanism that significantly improved understanding of the complexity obtained by the Greeks in gearing and clockwork.[26] Modern interferometric studies of optical elements of 17th-century telescope makers like Torricelli, Divini, and Campani revealed the level of their optical polishing technologies and improved understanding of the limits of telescopic knowledge of that time.[27] However, once we get beyond the 17th and into the 18th and 19th centuries, it is typically archival investigation that yields the most telling information about technological capabilities, as in Robert Smith and Richard Baum's excellent study of William Lassell's reflectors, whose optical imperfections led him to believe that he had detected a ring around the planet Neptune even though he was aware of those imperfections.[28] But examples are harder to find when one moves into the contemporary era. This trend is

---

24. Corn, "Object Lessons/Object Myths?" p. 37.

25. Deborah Jean Warner, "A Matter of Gravity, with reflections on the differences between Gizmos and Works of Art" (unpublished manuscript; text kindly provided by Warner in advance of presentation, March 2005).

26. Derek De Solla Price, *Gears from the Greeks: The Antikythera Mechanism—A Calendar Computer from ca. 80 B.C.* (New York: Science History Publications, 1975).

27. Mara Miniati, Albert Van Helden, Vincenzo Greco, and Giuseppe Molesini, "Seventeenth-century Telescope Optics of Torricelli, Divini, and Campani," *Applied Optics* 41 (February 2002): 644–647.

28. Robert W. Smith and Richard Baum, "William Lassell and the Ring of Neptune: A Case Study in Instrumental Failure," *Journal for the History of Astronomy* 15 (1984): 6–15.

likely similar for all types of collecting. The history of the technical museum in Western culture reflects this trend.

Originally collections of antique instruments, machines, patent models, and industrial products, in the 20th century, technical museums became venues to commemorate "native scientific and technological genius" as well as to supplement the academic program attendant to a technical education: if there was a trend, it was toward commemoration and pedagogy. Thus the technical museum became what Robert Multhauf has described as "a laboratory course extended in space rather than in time, arranged in some historical sequence to exploit the value of historic apparatus." These museums were also initially regarded as repositories of knowledge and inspiration insofar as they acted as places where inventors, designers, and engineers could go to get new ideas or to solve specific problems in design and manufacture. This application, however, closest to Corn's ideal methodologies, proved to be transitory; it was merely a passing interest through the early 20th century. And so the trend moved on once again, when technical museums returned to promote industrial products and act as places for the "preservation of our cultural heritage and to the inspiration of young people with an interest in science and technology."[29] Multhauf's goal in this 1958 essay was to highlight the limitations of perspective: "Unlike the engineer of the last century," he pointed out, "we begin our training, and rest our work, upon a basis of knowledge much of which is outside our own experience."[30] Therefore, for Multhauf, technical museums were the best places where one could explore, through utilizing all available primary and secondary source material, the many ways that discovery and invention happen, the very human artificial element in the inventive process.

Like Kranzberg, Multhauf did not actually regard the thing itself as embodying knowledge, but rather as a locus for the gathering in of knowledge in all forms and with increasing and changing perspective over time. His allusion to how the experience of the engineer of the last century differs from our experience offers testimony to how one needs to read an artifact: a worker who experienced the development of a technology before it was successful and before the principles upon which it was based were fully worked out would see that artifact very differently than someone looking at it years later, after it had proven itself and the principles it embodied. All the doubt, uncertainty, and promise congealed within the artifact can only be assumed, unless one has at hand numerous accounts of attempts made in that day to solve the same functional problem or goal, like attaining the facility of traveling in space and then having to learning how to work in that new theater, or how to build a

---

29. Robert P. Multhauf, "The Function of the Technical Museum in Engineering Education," *Journal of Engineering Education* 49 (December 1958): 200.

30. Ibid., p. 200.

pin machine that would be acceptable to piece workers, or a rifle that could be assembled, disassembled, and made reliable in the field.

To a certain extent, episodes in the recent history of the National Air and Space Museum's space history collection bear out this transitory phase, but they also show that it lingers even today and no doubt will be present in the future. The NASA/NASM Transfer Agreement explicitly states that if NASA decides that an object it had transferred to the national collection somehow reacquired its usefulness to the space program, it would be recalled. Sometimes this works, sometimes not. When the Viking 1 lander failed to call home from Mars in November 1982, NASA engineers came to the Museum to inspect the computer inside the engineering model we display in the Milestones of Flight gallery, hoping that their inspection might help them figure out how to regain communication with the lander. Unfortunately, the box holding the on-board computer in our example, although real, was empty of its contents. Our *Skylab* orbital workshop, originally built for flight, has been on display since 1976, though modified to allow visitors to walk through the living quarters. In the early 1980s, Marshall Space Flight Center engineers requested the return of a set of circulating air fans and, a few years later, came to inspect the toilet systems, since surviving documentation was apparently unobtainable when they were looking for ways to adapt these designs for new human space initiatives. And on occasion, engineers and scientists have expressed interest in everything from our Saturn F-1 engines to the backup Hubble Space Telescope mirror now on display. In the case of the engines, the engineers sought out the technical documentation we held in our archives rather than the object itself.

Multhauf's views on the use of objects in pedagogy were reflected in at least one of Corn's methodologies, as well as by some of the presentations at a 1975 conference at the Winterthur Museum held to explore how material objects are useful to the study of American history. Historians, archaeologists, ethnologists, American studies specialists, and even a molecular chemist spoke from their perspectives and experiences. James V. Kavanaugh suggested how a course in American studies could be augmented by using anthropological techniques upon "accumulated material evidence" to more fully explore the culture of invention in American life.[31] Cary Carson, Saint Mary's City Commission, echoing Corn's later observations, argued that artifacts have not contributed at all to "developing the main themes of American history" but have, in their design and arrangement, especially in the buildings of surviving early communities, certainly helped to fill in the details and provided new

---

31. James V. Kavanaugh, "The Artifact in American Culture: The Development of an Undergraduate Program in American Studies," in *Material Culture and the Study of American Life*, ed. Ian M. G. Quimby (New York: W. W. Norton, 1978), pp. 65–74.

insights. Facing the allegation that things "have seldom been a source of ideas for historians," he argued that by looking differently at objects, the mind is certainly capable of thinking up questions that they can answer or contentions they might prove or disprove. Embracing Kranzberg's "New History," Carson argues that "bottom up" history can often best be reconstructed by looking at the details of living environments, and thereby it can pose new questions. The experience of life, of "society as a working organism," can best be appreciated by somehow encountering the material vestiges of that experience. Although he applied his methodology to 17th- and 18th-century life on the Eastern Shore of Maryland, showing how "architecture became the instrument of segregation" and other insights, one might map these concepts into an exploration of the contemporary dwellings of scientific instruments and space operations.[32] Building upon a recent comment by Pam Mack, it is one thing to examine graphic profiles or even photographs of the interior of a Mercury capsule. But it is quite another to actually experience that tiny space, looking from the outside, of course.[33] Possibly someday someone might ask the crowded and complex chamber specific questions relating to the actual role of the astronaut in the Mercury, Gemini, and Apollo eras that cannot be answered as completely or as poignantly by other forms of indirect documentation. One might also find such reminiscences in debriefing documents after the flights. Definitely riding in a machine and being part of its operation is a most valuable experience. Many historians have expressed how important it has been for them to fly in an aircraft they have studied; Ron Davies at NASM recently commented that it was an essential experience, even though his primary data came from airline timetables.[34]

Probably the most eloquent argument for the value of experience at the 1975 Winterthur Conference was Brooke Hindle, who was the lead speaker. Hindle was then Director of the National Museum of American History, and he took the occasion to explore the essence of material culture in his now-classic "How Much Is a Piece of the True Cross Worth?" Hindle identified the factors that led him to what we today might call "priceless." Pondering Lenin's body, Dolley Madison's gown, Ben Franklin's printing press, he first stated that artifacts provide "direct, three-dimensional evidence of individuals who otherwise exist only as abstractions in words, paintings, or monuments."[35] In order to utilize them properly, however, one has to know how

---

32. Cary Carson, "Doing History with Material Culture," in *Material Culture and the Study of American Life*, ed. Quimby, pp. 42–50.

33. Telephone conversations with Pamela Mack, February 2005.

34. Ron Davies, personal communication in response to informal questionnaire sent to NASM curators, February 2005.

35. Brooke Hindle, "How Much Is a Piece of the True Cross Worth?" in *Material Culture and the Study of American Life*, ed. Quimby, p. 6.

to apply "linguistic models to the nonverbal, three-dimensional world." This, however, was not a simple matter for Hindle, who felt that language "floats on top of the material world" and so remains separate from it. One must walk the battlefields, cruise the oceans, make landfall as explorers did, to find the words appropriate to the experience. Only in this experiential way, Hindle felt, "the abstractions of language are penetrated by direct knowledge of life's complex multidimensional and instantaneous character."[36]

Hindle's concept of the importance of experiential reality underscores what is, in fact, both a compelling but essentially still abstract circumstance. He did not describe any one set of analytical tools one must bring to the experience in order to sense it and then reduce it to language. He provided examples, as all writers of this genre tend to do, and many of those are compelling, such as Eugene Ferguson's attempt to reconstruct the methods of artisans by showing how they thought in pictures, suggesting that one might do the same for the builders of machinery. His strongest suit, of course, is how the techniques of industrial archaeology have radically changed our view of Eli Whitney's role in the development of interchangeable parts. This was indeed a wonderful example of how, in a manner suggested by Carson and others, asking the right questions of a set of artifacts yielded new knowledge about their history and provided a correction to the broader history of industrial technology.

The success of the interchangeability study naturally raises the question of what is important about today's space technology, especially what is important that might be studied by examining artifacts in the ways Corn and others suggest. Is the ability to exchange parts important in the technology of space history, does it define modern capabilities and practice? Does it typify an era? The answer is probably no, at least not in the way rifles illuminated manufacturing techniques of their day. However, a modern counterpart might be the ability to ensure consistency and reliability across a very widely spread-out system or infrastructure. How sure is an instrument developer, for instance, that his instrument will work within the environment of a satellite housing that has been launched into space? What steps does that developer take to design his instrument to be as forgiving and robust as possible—resistive to vast swings of temperature, pressure, and acceleration, yet sensitive enough to get the job done effectively? This is only one of many questions about "integration" that has been an issue ever since scientific instruments were flown on vessels that were not under the direct control of the instrument maker or scientist.[37] The need to integrate a scientific instrument into a system used either remotely or by surrogates changes the way science is done and certainly

---

36. Ibid., pp. 9–11.

37. David DeVorkin, *Race to the Stratosphere: Manned Scientific Ballooning in America* (New York: Springer-Verlag, 1986).

changes the experience of the scientist, much as the telephone changed how we communicate. A more obvious approach might be to compare designs of instruments flown on different vehicles, looking for changes or shifts that are only understood in terms of the capabilities of the vehicle. These and other questions can be asked by historians of space artifacts, whether they be launch vehicles, manned or unmanned craft, subsystems, or instruments.

Historians of this contemporary scene might be more interested in issues such as how nations achieve new levels of capability or performance (as with Campani's lenses), how design variations reveal compromise, or how adaptations were made to existing technologies to make them work in the space environment or to survive launch or landing. But unlike the study of Campani's lenses, it is doubtful that the space historian will ask these questions of the artifacts themselves.

Indeed, one usually finds questions directed to the nature of the individuals or organizations that produced the technology. Among historians contributing to the *Osiris* volume "Instruments" in the early 1990s, Robert Smith and Joseph Tatarewicz represented space history, showing how the technical complexity of the Hubble Space Telescope not only symbolizes the complexity of the institutions and motivations involved in creating the thing, but also revealing how these motivations were often in conflict. It is clear from their study of how the largely untested charge-coupled device (CCD) became the detector of choice for the critical Wide Field/Planetary Camera that one can only hope to understand the ultimate technological artifact through the interactions of conflicting institutional priorities between science, the military, and NASA, each possessing different goals, different resources, and different agendas.[38]

This study of the CCD and the complexity of HST gets about as close to the artifact as I have seen in the literature of space history. It is typical of a small but hopefully growing literature that uses some characteristic of the hardware to inform a larger story. But the majority of the literature of space history is still rather far from this sort of treatment. Major characteristics include early practitioner histories, going into great detail describing examples of early rocketry and speculative space vehicles but asking few, if any, questions about them that informed broader historical interests. The NASA-sponsored histories of the 1970s, '80s, and '90s focused, correctly, on the elucidation of missions and the application of broad technologies, rarely focusing on specific examples of the technology and questions about their origin and application. Among the synthetic reviews and disciplinary histories, one often finds descriptions of objects, who built them and why, and what they did, but

---

38. Robert W. Smith and Joseph N. Tatarewicz, "Counting on Invention: Devices and Black Boxes in Very Big Science," *Osiris* 9 (1994): 101–123.

rarely, if ever, is an artifact in a collection at the center of attention or used in any explicit way in the analysis.

One can find this attitude explicitly stated in some of the papers from the XIX Scientific Instrument Symposium in September 2000, held in Wadham College, Oxford, where a session was devoted to "Instruments in the 20th Century," organized by Paolo Brenni. Speakers said the usual things, like how instruments might provide useful information when other documentation is lacking, but gave no hint in their abstracts of the kind of information one might extract from an instrument other than suggesting that one look at an instrument or actually use it in performing an experiment. The most refreshing remarks about the value of collecting were made by Roland Wittje, who pointed out that any collection of 20th-century instrumentation was for purposes of exhibition and not for the study of history.[39] In other venues, historians have said much the same thing. Marvin Bolt of the Adler Planetarium and Astronomical Instrument Collection, echoing a strong and persistent theme among educators, presented demonstrable evidence for how historical replicas can reveal physical and chemical processes more simply than modern devices. Others concentrated on how, reflecting Hindle, an encounter with an historical object can stimulate greater interest in the subject matter surrounding the actions of that device and the efforts of their human creators and users.

We have touched on Hindle's experiential argument before. It continues to appear in a wide range of studies. An excellent example is Paul Forman's recent study of three mechanical wave guides from I. I. Rabi's early research program that were part of the museum's accession of his materials after his death and his office was cleaned out at Columbia. Paul was already interested in Rabi, of course, but, stimulated by the existence of these relics, he realized that their survival after all these years confirmed that Rabi regarded these early experiments very dearly and saved the devices as a result, even though they were completely overshadowed by his later work that won him the Nobel Prize. This encouraged Paul to search out the nature of his early work, and he found it to be more significant than hitherto realized. These wave guides also confirmed designs previously known only from publications.[40]

At the same 1999 Artefact Conference where Paul Forman reported on Rabi's devices, Paul Ceruzzi recalled an incident where someone examining a circuit board recognized that it was probably designed by the legend-

39. Roland Wittje, "How Can Scientific Instruments Teach the Historian about 20th Century Physics?" in *Session VII A: Instruments in the 20th Century*, session abstract, *http://www.sic.iuhps. org/conf2000/ox_s07a.htm* (accessed 18 January 2005).

40. Paul Forman, "Researching Rabi's Relics: Using the Electron to Determine Nuclear Moments Before Magnetic Resonance," in *Exposing Electronics*, ed. Bernard Finn, Robert Bud, and Helmuth Trischler (Netherlands: Harwood, 2000), pp. 161–174.

ary Seymour Cray because it had specific design earmarks that Cray had pioneered, specifically his "cordwood packaging" technique that achieved greater densities than hitherto attained. There were no markings on the board other than the known fact that it was part of a military mainframe called the Naval Tactical Data System, or NTDS, built by Sperry. This was a highly specialized machine known only within military circles, and nothing was known about its design. It was also not generally known that Cray worked for Sperry, although he left Sperry before the NTDS was delivered. There is little in the published record linking Cray to the NTDS—no reference in the technical manuals or other contemporary descriptions. In presenting this analysis of a design style and using it to discern design origins in other computing devices, Paul examined a CDC 3800 acquired by the National Air and Space Museum, finding the same packaging design, even though no documentation has yet been found identifying it as a Cray design. Paul describes this as a "reading of the text of the machine itself" and is using it as a guide to search for traditional documentation.[41]

"Reading the text of the machine itself" includes many other areas beyond the survival, existence, or design style of a device, but quite frankly, it is a circumstance that is not as common as one might like. However, there are ways to increase the chances that a reading of an artifact will result in new, useful knowledge. Here I offer two examples from my personal experience: one involves documentation efforts, and the other involves exhibit preparation. Both, by their nature, required the survival, existence, and availability of artifacts.

The first example deals with the use of video to document objects. In the late 1980s, the Smithsonian decided to experiment with the use of video recording to better document its collections. This program, sponsored by the Sloan Foundation, brought together artifacts with their makers and users. As part of this effort, between 1988 and 1990, I interviewed sets of scientists and technicians who had been involved in space research at the Naval Research Laboratory from the 1940s through the 1980s. During the course of these interviews, sessions were devoted to voice-overs of a series of slow pans through laboratories and workspaces, followed by on-camera "enactments" and, following that, by direct examination of artifacts, mainly x-ray and ultraviolet detectors, collimators, and other elements of flight systems. I could fill many pages with examples of how this experience produced evidence that documented the interface between an instrument and its builder, as well as the interaction between the instrument and the laboratory environment within which it was designed and tested in prototype fashion. We documented design

---

41. Paul Ceruzzi, "The Mind's Eye and the Computers of Seymour Cray," in *Exposing Electronics*, ed. Finn, Bud, and Trischler, pp. 151–160.

Typical x-ray ionization chamber designed, built, and used by the Naval Research Laboratory team on sounding rocket flights in the 1950s and early satellite systems. *(File no. A1988-0012000, NASM Curatorial Files)*

choices, instrumental styles, experimental procedures, and testing methodologies, not merely through reminiscences, but through recording the tactical connection between instrument and builder. On one occasion, one scientist demonstrated the methods used to fill halogen Geiger counters with gas and then test them for sensitivity. He used a contemporary filling station as a backdrop, but his hands twisted invisible dials and stopcocks as if he was using one from the 1950s. They were literally imprinted in his tactile memory. These explorations of working environments gave body to other sessions where the people who built these detectors talked about them while they handled them. Edward T. Byram was faced with many detectors he had built, laid out on a table in front of him. He rarely took his eyes off the detectors during the interview, and when asked if his efforts making these devices work properly were frustrating, he replied: "I was never frustrated. I enjoyed fighting them. It wasn't a frustration, it was a challenge. It was mind over Geiger tube."[42] His behavior matched his rhetoric—throughout the interview, Byram's gaze

---

42. E. T. Byram, quoted in David H. DeVorkin, "Preserving a Tool-Building Culture: Videohistory and Scientific Rocketry," in *A Practical Introduction to Videohistory*, ed. Terri A. Schorzman (Malabar, FL: Krieger Publishing, 1993), pp. 125–137.

Early halogen counter with an entrance window of mica, capable of sensing ionizing ultraviolet radiation. Note the suspended anode just behind the mica window. This is a tube similar to the one Kreplin tapped during his video-history interviews. *(File no. A1988-0010000, author digital file, NASM Curatorial Files)*

remained on the tubes. Obviously, he was still very attached to them, attached to devising ways to adhere exotic radiation entrance windows onto their shells and ways to ensure that the halogen gas mixtures he was filling them with did not leak or cause the seals to deteriorate. And finally, one of Byram's colleagues, Robert Kreplin, was also asked to talk about the tubes he built. He held an early example while he talked, and in the review, I noticed that as he discussed ways to test the mechanical integrity of these detectors, which had to survive the launch of a rocket, he instinctively tapped the side of the tube and peered through the mica window at a small protruding wire anode. His tapping was reminiscent of the group's concern for the survival of the anode, which in later models was supported at both ends.[43]

Although my basic goal for these interviews was to produce a collective profile of what I deemed to be a tool-building culture at the Naval Research Laboratory and to explore aspects of that culture, I also came away with a better appreciation for how these people organized themselves, raised issues and

43. Image of Kreplin holding a tube, in DeVorkin, "Preserving a Tool-Building Culture," p. 134.

problems, and dealt with outside entities first in the Navy and then at NASA. In a very definite, though not explicit, way, I feel that the surviving artifacts that we interrogated, and which are now in the collection, stimulated memories and physical responses and led to discussions between team members that rekindled behaviors that I could actually discern. From this vicarious experience, I feel I gained a fuller portrait of this tool-building culture.

As my second example of the stimulus generated by a surviving artifact, I turn to recent activities preparing for NASM's new Udvar-Hazy Center. Curators had an unprecedented opportunity to examine a significant portion of the collections in a process that included improving documentation, preservation techniques, and methods of monitoring them, since from now on they will be on permanent display/storage. In the past, various factors have limited our access to these objects. They were stored off-site, sealed and boxed up, and required manpower and coordination for examination. One of the dozens of objects I had never had the chance to fully inspect was a model of the Explorer VII satellite identified as a "full scale replica." It had been acquired on paper in 1976, inventoried several years later, but never actually examined at the Alabama Space and Rocket Center, where it was presumably on display. It finally was shipped to the Garber Facility in 1989. It was quickly inspected, but the box was never actually opened, nor were the insides of the object inspected. As a replica, it was, frankly, not of great interest. As to documentation, we were left with hardly more than a shipping document.

In the years leading up to preparation for the Udvar-Hazy facility, our department's sensitivity for the critical importance of adequate documentation vastly improved. Udvar-Hazy afforded me a chance to acquire intimate knowledge of a set of early satellites and the scientific instruments they hopefully contained, so I opted to examine Explorer VII as part of a suite of first-generation geophysical satellites.

Typically, anything marked as a mock-up or replica or even reconstructed satellite is not going to contain actual flight hardware, so I was really not expecting much. However, many of those objects hauled out and destined for Udvar-Hazy labeled replica or model have turned out to be very real. Based upon my experience with the videotaping of NRL detectors, I quickly realized that the detectors in the skin of Explorer VII were, in fact, real. One detector had a clear entrance window revealing a small chamber that had a single wire on the cylindrical axis, just like the one Kreplin was tapping. Explorer VII may well have been a flight backup, which means that everything about it is real. Documents in our technical files in the NASM library confirmed that the detectors were indeed built by the NRL group, and other elements of the craft closely matched the descriptions in an extensive Technical Note.

None of this effort would ever have been made if I had not been compelled to answer detailed questions raised by an intimate inspection of an

Explorer VII before cleaning and evaluation. Note that the artifact inventory tags were tied to a damaged x-ray detector similar to those examined at the Naval Research Laboratory and recorded during video-history sessions. See the image on page 593 for an intact example. *(File no. A1978-1109000, author digital file, NASM Curatorial Files)*

artifact. Explorer VII is interesting as a representative of the state of tech-nology available for multifaceted studies of solar radiation and the nature of the low–Earth-orbit environment in the late 1950s. As with any early flight, there were some technical "firsts" and at least one first for science: the detec-tion of micrometeorite impacts. But whether or not the remnants of the craft itself reveal anything beyond what is still available from our technical files, at NASA, in our archives and oral histories (with people like Herbert Friedman and James Van Allen, another instrument principal investigator on Explorer VII), or from the published literature, it remains a fact that in the process of inspection and evaluation, more documentation was gathered and consistently filed away than was available before, and hence is likely to be retrievable in the future. Scattered documentation was collected, recorded, and filed away, hopefully someday to be of use in some unpredictable way, stimulated by motivations that we cannot predict. My contention is that the motivation would come either from the recognition someday that this was a watershed flight in space history (the first application of passive techniques of thermal stabilization) or that an artifact that has survived in a major collection calls out, by its very existence, for attention to the fine structure of nuts-and-bolts history, for only through such efforts is a full picture of the nature of the first years of true space research likely to emerge.

So, Why Collect?

As I prepared my remarks for the "Critical Issues" conference, I queried colleagues at NASM, asking them questions stimulated in part by Corn's findings but also by my inability thus far to find unequivocal evidence of how an object relating to space history has actually been used as a source of historical knowledge. I also queried aeronautics curators as a cursory check on a collecting area where documentation tends to be not as rich or institutionally based. In general, the responses confirmed the impressions I was getting from the literature and from experience. Curators (John Anderson, Michael Neufeld, Ron Davies, Tom Crouch, and Jeremy Kinney) typically felt that direct and personal experience with an artifact stimulated them to make historical inquiries. Neufeld, in particular, felt that an encounter with an historical object can stimulate intellectual interests and makes the past seem more real, less dry and distant even for academic historians, but how much they drive any historiography is questionable. Others, like Tom Crouch, felt that they learned from these inspections and gained important intellectual insights. For Crouch, "interpretation . . . was in large measure based on a combination of examining the objects and knowing the documentary record." Jeremy Kinney reported that what he learned from his detailed inspection of variable pitch propellers in the collection is reflected in his publications in significant ways, but that his physical inspections largely confirmed textual descriptions in primary sources. All felt more or less strongly that the survival of artifacts could be a stimulus to researching and writing history. Artifacts provide information on design and shop practices that run hand in hand with the intellectual methods of aeronautical engineering. As for the limits on collecting and the importance of the survival of the "real thing," Tom Crouch added that it is impossible to preserve all the details of a machine (the written and visual records are approximations); close examination always reveals more detail—small mechanical details. For Crouch, one of the museum's failings is the lack of attention to machine tools and production machinery—transitions from one medium (wood) to another (metal) and from metal to modern composite materials are always constrained by fundamental changes in tooling and production machinery. Reflecting issues raised by Warner and others, he also sees a problem with collecting "black boxes" if it is not possible to "turn them on" and examine their behavior. Finally, reflecting Jim Bennett's qualification cited under the title to this essay, he suggests that we all have to consider carefully what we collect for exhibit and what we collect for research—these are not necessarily the same class of object, and selection rules may tend to be very different.[44]

From the arguments so far reviewed from the literature, from the responses of my Museum colleagues here, and from my own experiences, what conclusions do I draw as to the value of collection and preservation? Here is a brief summary of my impressions. Objects can provide the following:

1. Validation—material proof that something happened in space history (Hindle). This requires solid information on provenance, however, and requires as well that the object that is experienced by the visitor was actually the very same one involved in the historical episode it preserves. Collections in space history are rather peculiar in that, as often is the case, the actual historical object that performed the act or the function deemed worthy of note is not accessible—it has been used up or lost in the process of conducting its business, or, simply put, it is still "up there" where we put it, and we have no known means or the wherewithal of retrieving it. There are very notable exceptions, of course: vehicles that have returned to Earth as part of their mission or, even rarer, have been returned to Earth through some conscious act unrelated to the historical event or process that made it noteworthy. For all the rest, we are left with some form of surrogate: an exact flight backup, just like the flight model in every way except that it, in and of itself, did not experience the final act of making history but was still very much a part of that history. It had a role in that process but definitely comes in second place. Third place are various levels of engineering models and mock-ups, reconstructed replicas using parts that were fabricated out of the same computer program, melt, or block. And a distant fourth is all sorts of replicas or reproductions. Are these approximations merely surrogates for the "True Cross," or does each and every one of them tell a particular story that is available no where else in quite the same way? What does their existence, and their survival today, reveal about the culture in which they were made?

2. Celebration—sense of transcendence promoted by physically encountering an object that made history. Accompanies commemorative

or memorializing events, lends visibility and weight to these efforts (NASM legislation).

3. Inspiration—evidence of challenges met or exceeded, handicaps overcome, struggles vindicated. Promotes insight into ways to illustrate basic principles of science and technology (Multhauf, Corn, Bolt).

4. Illumination—preserves something about an historical event, era, or trend that, when means of interpretation are devised, provides additional knowledge that otherwise would not be available. Objects can survive for specific reasons, and searching out those reasons illuminates history (Corn, Lubar, Ceruzzi, Forman).

5. Stimulation—the preservation of an object stimulates interest in it and efforts to learn about it and the history it symbolizes or represents. It also obligates those responsible for its curation to ensure that adequate documentation is collected and preserved to understand it in the future (Explorer VII, Forman, curatorial questionnaire).

Of course, neither celebration nor inspiration actually requires the survival of an artifact, though it would clearly help. Even illumination and stimulation are possible without the real thing, though impact would be even more restrictive. Nothing but the actual object, however, can provide validation—no facsimile, replica, reproduction, or description will ever suffice, although the survival of any of these items still stands testimony at some level.

AFTERWORD

If the survival of an artifact is useful to history in any of the five categories listed above, one still has to look beyond history to the institutions that house and somehow represent it to ask how they react to the suggestion that collections are important to their own survival. In a recent Smithsonian survey cited as significant by the *Washington Post*, 60 percent of the respondents claimed that they were visiting the Mall museums to see "the real thing," whether it is Dorothy's red shoes or the Apollo 11 capsule.[45] The *Post* itself was concerned with what motivates programming at the Smithsonian in its efforts to overcome the tourist slump after 11 September 2001. Ironically, the part of

---

45. "Smithsonian Institution Office of Planning and Analysis Report" (internal document, 2004), quoted in J. Trescott, "The Smithsonian's Concession to the Bottom Line," *Washington Post* (13 April 2005): A1, A8.

the Smithsonian being covered by the *Post* reporter and as reported by her, its Business Ventures arm, responded as if this fact gave it a "mandate" to push IMAX films, simulators, jazz concerts, and anything else it could imagine would raise revenues. The irony was, unfortunately, lost on the *Post* reporter. Yet the fact remains, the public, when asked in this instance, reified "the real thing" just as Hindle argued it should. This runs counter to opinions voiced by museum watchers and critics in studies over the past several years, who have claimed that, in the face of theme parks and Disneylands, public tastes have shifted "to immersion in an environment, to an appeal to all the senses, to action and interactivity, to excitement, and beyond that to aliveness." And in response to this shift, many modern museums have "shifted their allegiance from real objects to real experience."[46] Oddly, these are just the sorts of experiences that, at least in the case of Smithsonian Business Ventures, a museum can charge money for. No one knows if it is a viable strategy for long-term survival of these institutions as collecting agencies.

---

46. Randolph Starn, "A Historian's Brief Guide to New Museum Studies," *American Historical Review* (February 2005): 92 (citing statements by David Lowenthal and Hilde S. Hein).

Postscript

AFTERWORD:
COMMUNITY AND EXPLANATION IN SPACE HISTORY (?)

Martin Collins

The deep significance of certain problems for the advance of mathematical science in general and the important role which they play in the work of the individual investigator are not to be denied. As long as a branch of science offers an abundance of problems, so long is it alive; a lack of problems foreshadows extinction or the cessation of independent development. Just as every human undertaking pursues certain objects, so also mathematical research requires its problems. It is by the solution of problems that the investigator tests the temper of his steel; he finds new methods and new outlooks, and gains a wider and freer horizon.
— David Hilbert, "Mathematical Problems," 1900

Hilbert, a pivotal figure in mathematics in the late 19th and early 20th centuries, addressed these thoughts in 1900 to the International Congress of Mathematicians in Paris—a quadrennial, premier gathering of practitioners.[1] In this year, poised at the crossing between centuries, Hilbert and his colleagues self-consciously took stock as to professional ethos, standards, and research. Holding center stage, Hilbert presented (in what was recognized then and since as) one of the most significant templates for research in mathematics, providing a conceptual outlook for his discipline and a list of 23 outstanding problems that engaged, in different ways, the foundations of the field. These problems, for Hilbert, served as vital links among practice, theory, and tacit notions of professional community—and, more broadly, connected all of these to European culture.

Hilbert's moment, of course, is not ours. The early modernist temperament, confidently ascendant, shines through—the application of reason seems to confront few limits. Too, the "profession" as social and intellectual instrument

---

1. David Hilbert, "Mathematical Problems: Lecture Delivered before the International Congress of Mathematicians at Paris in 1900" (trans. by Dr. Maby Winton Newson), *Bulletin of the American Mathematical Society* 8 (1902): 437–479.

(with scientific disciplines as model) seems to promise a progressive, well-oiled engine to extend and secure reason's reach—indefinitely. Hilbert's moment reminds us of the arc (choose your geometric tracing) from then to now and of the undermining of that world's foundational assumptions. Late modern and postmodern perspectives have made problematic the relationship among researcher, subject, and knowledge claims—the ontology and epistemology of the "true" and the "objective"—as well as the idea of the profession as self-contained and self-regulated in its pursuits.

Hilbert's thoughts and their context provide one of many possible points of departure for reflecting on notions of discipline and profession in space history. The aim here is not an extended comparison with that earlier example, but a simple reminder of the history of these inventions and the complex of issues embedded within them. Nor is this afterword a critique of the seminar's papers, separately or collectively, nor is it a meta-literature review. Rather, it is an attempt to foreground assumptions that lingered backstage in many of the papers, assumptions that speak to our sense of what space history is and does. The "(?)" in the title above signals one part of this examination: How and why we self-identify as "space history" rather than as space "something else"—and what such a choice implies.

All of these bits of critique, of course, are right out of the big playbook of analytic moves—a (pseudo-)knowing voice taking us from background to foreground, dark to light, fuzzy to sharp—yet necessary and useful as part of thinking about "critical issues." Reviews of this type are the rare occasions when "we" and "I"—as much as our subject matter—are preeminently our concern.

This discussion will touch on two intertwined themes central to notions of discipline and profession (hereafter, for concision, I will use "discipline" to speak for both), yet largely submerged in this "Critical Issues" conference: community and explanation (our equivalent of Hilbert's problems). The former, with a nod to Robert Merton, refers to norms and shared practices, and the latter to how we conceptually establish ourselves in relation to history and related disciplines and how we frame the aims of inquiry. In outline, these themes almost are banally familiar; in application to space history, they may illuminate "we" choices that should be explicit rather than implicit.

COMMUNITY

Is there something missing in our sense of community, in the way we approach our research domain?

Such a question may seem off. The very fact of this seminar, of a thoughtful focus on critical issues, signaled health and intellectual robustness. It reflected a shared interest (at least to the level of subject matter) among scholars from a range of disciplines. In narrative and argument, the individual papers were "right on," embodying a knowledgeable engagement with the best methodological practice. And as a topical map of the field, the seminar provided a thorough survey.

Yet there were two prominent lacunae: a *collective* self-awareness and openness on the importance of methodology—of presenting, testing, and critiquing the conceptual tools that define a field as a domain of inquiry—*and* of the core aims of research. In short, the seminar shied away from "discipline"—of explicit correlation between methods and aims, a communal orientation and set of commitments that distinguish a discipline from a subject area. Open, systematic exchange, rumination, and struggle on points of method and explanatory focus (and attaching value to such) are integral to community practice within a discipline. To suggest the relative absence of such a mindset at the seminar is not to cast aspersion, but to bring to the fore an important choice: discipline (or more accurately, subdiscipline) or subject specialty? This choice was a largely unspoken tension within the seminar: we assumed discipline but performed as subject specialty.

Does this distinction or choice matter? Space history as subject specialty can be commodious, welcoming a range of scholarship under a big tent—the seminar was a showcase. Under this mode of practice, we can learn from each other, then head back to our respective subfields—a kind of intellectual tourist model of research. And that may be enough. Space history best may be pursued as loose confederation sans discipline. But does the notion of discipline offer us opportunities worth considering? I'll highlight a few issues associated with method first, and then with aims and explanation.

Method, of course, is not an end in itself. In recent years, its reputation has edged, on occasion, into a preoccupation with dreary, constant self-reference and culture-war comedy (recall Alan Sokal's parody of postmodernism and deconstructionism in *Social Text*, a journal edited by the noted Stanley Fish).[2] And, of course, to refer to method abstractly, without accounting for the many variants that have effloresced since the early 1960s, is to risk reduction and simplicity (Margaret Weitekamp's essay, nonsimplistically, gives one overview of this landscape). But the point for us is primarily attitudinal—to note that a primary motivation for the boon in critique was a fundamental (and from a disciplinary perspective, necessary) question: How do we know and justify what we know (about the world)?[3] Taking that question seriously is not to turn historians (or other disciplines represented at the seminar) into phi-

---

2. Alan Sokal, "Transgressing the Boundaries: Towards a Transformative Hermeneutics of Quantum Gravity," *Social Text* 46/47 (1996): 217–252.

3. For one recent overview of the centrality of epistemology and of the genealogy of critique in relation to the modern and postmodern, see Bruno Latour, "Has Critique Run Out of Steam: From Matters of Fact to Matters of Concern," *Critical Inquiry* 30 (2004): 225–248.

losophers, but to foreground the importance of method in any discipline's practice—in establishing the constructs and categories through which we do research. Such a concern, for example, was at the heart of the formation of science and technology studies. There, method served two functions: to elevate epistemological issues as just noted and, equally important, as a focus for creating a sense of community among a congeries of disciplines. Method and explicit talk about method provided a means for drawing disciplines together and connecting and building on each community's research perspectives. This orientation and the norm which is its corollary led to the designation of "studies" to characterize this research enterprise.

With this "Critical Issues" seminar, space history, I think, faces, à la Yogi Berra, the simple recognition that there is indeed a fork in the road.[4] The seminar was notable in that this norm regarding methodology (making how we do research as much a part of the discussion as the subject of research) was largely absent. How, for example, do we fruitfully interconnect history, political science, sociology, and policy studies (the four major disciplines represented at the seminar)? What are the issues arising from—the goals for—such interconnection? Are we after merely some vague expansion of "context"? Or can explanatory aims be integrated? Are there research projects that emerge from such interconnection that confront basic issues of historical change and explanation? To entertain these (or similar) questions is to probe the possibilities of discipline and space history as multidisciplinary crossroad.

The supercilious "(?)" I keep pinning on the rump of space history is a suggestion that self-description matters—as it did with science and technology studies, women's studies, and other domains of research—that it is a marker of our notion of community (subject specialty or [sub]discipline); goals; and organized, professional self-reflection. If we wish to emulate these examples, then we might rethink how we designate what we do—to self-identify in a way that embraces a range of disciplines in a common enterprise. Space history, hereafter in this essay, will stand, awkwardly, for history as well as a constellation of other disciplines interested in the cultural meaning and manifestations of space activity.

Discipline is more than branding—as with that bygone image of the teacher rapping knuckles with a ruler, it imposes restraints, or, at the very least, "stop and think" road markers. One of those is a commitment to clarity in what methodological tools are in play and how they interrelate. In the seminar, this issue most often came to the fore in the several papers that sought to join a policy voice (concerned with the prescriptive and/or normative) and either a historical, political science, or sociological voice (concerned, primarily, with description and explanation). These two voices involve a different use of time and

---

4. For those unfamiliar with Yogi Berra wisdom, the aphorism is: "If you come to a fork in the road, take it."

tense, approach to conditionals and normative claims, and modes of argument. They have, succinctly, different narrative standards. To shift casually between these narrative strategies is to muddy what is at stake analytically in space history as subject matter—and to blur the distinction between the methods and aims of explanation and the elucidation of options in service of policy.

These concerns raise a deeper consideration regarding notions of discipline and space history: the stance of practitioners to subject matter and, especially, to participants in that subject matter. As noted by Roger Launius, many writers on space history conflate their research with a "fer it or agin it" normative stance on space exploration's social or political value.[5] The prevalence of this stance, of the conflation of the normative (which presumes at least a partial overlap between analytic practitioners and participants) with narrative explanation, is profoundly antidisciplinary—that is, undermines discipline's formative characteristic: independence in intellectual standards.[6] In an important but not often cited essay, "Independence, not Transcendence, for the Historian of Science," Paul Forman addressed a closely similar set of relations in history of science.[7] His argument was that for history of science to be *history* of science it had to frame questions, methods, and aims distinct from those of science. History of science, if it was to be a profession (sporting more than the social trappings of self-regulation), had to separate itself from providing social and intellectual justification for (or dismantling of) science and scientists' self-image. It had to be thoroughly, completely historical. The thick veins of a similar predicament run through space history. If space history leans toward discipline rather than subject specialty, then issues of practice and boundary setting will need to come to the fore.

EXPLANATION

Let's take as given that understanding and explaining change—socially, culturally, historically—is a fundamental task of our effort, that description *and* causality loom large. To think in terms of discipline, then, is to ask in what ways space history might define its research domain and to focus an

---

5. Roger D. Launius, "The Historical Dimension of Space Exploration: Reflections and Possibilities," *Space Policy* 16 (2000): 23–38.

6. This point is not to say that history cannot be "useful" or be applied to a wide range of cultural concerns. The notion of utility, too, has been a central methodological issue for sociology and political science since their formation as disciplines in the 19th century. The issue is whether "usefulness" provides the right intellectual basis for a discipline that aims toward explanation. For one of the most prominent statements on history and its application, see Richard Neustadt and Ernest May, *Thinking in Time: The Uses of History for Decision-Makers* (New York: Free Press; London: Collier Macmillan, 1986). Importantly, for Neustadt, "usefulness" is a byproduct of disciplinarity, not an aim.

7. Paul Forman, "Independence, not Transcendence, for the Historian of Science," *Isis* 82 (1991): 71–86.

explanatory enterprise.[8] In contrast to space history as subject specialty, this slight shift of the kaleidoscope has ramifications.

Consider Walter McDougall's seminal . . . *The Heavens and the Earth.*[9] This work, I think, is most often viewed as an exemplar of contextualization, of rescuing space history from the modalities of program and institutional history and binding the subject to the perspectives of diplomatic history and political science. For McDougall, the emergence of spaceflight was not a subplot, but a major narrative thread of the all-encompassing drama of the Cold War. Less noted, but more germane here, is that McDougall's work was not merely context-expanding and thesis-driven, but broadly explanatory—an identifiable set of causes (preeminently ideology, new social/technical tools associated with systems thinking, and an uptick in the concentration of power in the federal government and willingness to apply it—in combination, technocracy) structured historical description and shaped historical change on a broad cultural scale. It was an explanatory framework that, with national variations, provided a common way of seeing events in the U.S., USSR, and Europe and that yielded good, scholarly fruit: a coherent (albeit contestable) notion of period.[10]

It is a revealing feature of space history scholarship that this call to explanation generated little to no resonance in the community.[11] The work was absorbed into mantras for "context" (an improvement over "internalism," but still a half-rigorous notion for a research program) and into quasi-policy research agendas (Why did spaceflight diminish in the U.S.'s political agenda?). Vestigial trails of this absorption were evident throughout the seminar. Embedded in this stance is a stultifying (and unexamined) research proposition: that with the undoing of the 1960s framework for spaceflight, space history itself has no deep relation to explanatory accounts of change in the world after 1970 and, thus, provides no distinctive vantage onto recent history. Or, stated somewhat differently, the field has no overarching explanatory outlook.

---

8. For adherents to history-as-special-kind-of-narrative, explanation as an aim of inquiry is not a given. The description versus explanation stands in history have, well, a tangled history. The best account, seen through the lens of the "objective," is Peter Novick, *That Noble Dream: The "Objectivity Question" and the American Historical Profession* (Cambridge: Cambridge University Press, 1988). For an enlightening but philosophically muddled defense of description and its relation to explanation, see Allan Megill, "Recounting the Past: 'Description,' Explanation and Narrative in Historiography," *American Historical Review* 94 (1989): 627–653.

9. Walter A. McDougall, . . . *The Heavens and the Earth: A Political History of the Space Age* (New York: Basic Books, 1985).

10. These views were more succinctly argued in an earlier essay: Walter McDougall, "Technocracy and Statecraft in the Space Age: Toward the History of a Saltation," *American Historical Review* 87 (1982): 1010–1040.

11. To be clear: My claim is that the space history literature seems to have had little interest in the explanatory element of McDougall's work. Science and technology studies, Cold War studies, and diplomatic history did engage and integrate the issues of explanation posed by McDougall into the problematics of their respective fields.

If this is true, then space history is history at the margins—or, more generously, serves as a source of case studies for other disciplinary projects (for example, for sociology in understanding organizations that manage risk or pursue complex problems, or for political science in unpacking decision-making in technology-infused modern polities) or as a site for exploring important but largely disaggregated themes: exploration, innovation, and so on. This may be the proper research stance—but it should be arrived at through open examination, not happenstance. If space history is to have a self-identity, to be discipline-oriented, then a concern with explanation, with the relation of space history to macrostructures of change, seems a crucial starting point. This is not to say that the only "good" history is grandly explanatory, that theory reigns over the hard work of research—but rather that models of explanation, and debate about them, are central to discipline.[12]

Perhaps, historiographically, the Cold War lulled us into anticipating an explanatory framework (in that case, the centrality of state action joined to an overweening interest in science and technology) that was conceptually straightforward, yet capable of informing and being informed by a variety of "local" histories. Is there an explanatory framework, post–*Heavens and the Earth*, which offers us the start of a (contingent) synthetic view of space history and currents of change in recent decades? Significantly, McDougall hinted at the necessity of new interpretive frames nearly contemporaneously with publication of *Heavens and the Earth* in his article "Space-Age Europe: Gaullism, Euro-Gaullism, and the American Dilemma."[13] Here he argued that to comprehend the European approach to space, the market as well as the state needed to be incorporated into post-Apollo and broadly geopolitical analyses.

But a deeper, sustained response to the problem of post–World War II periodization arose in other scholarly quarters—in sociology, literary criticism, and philosophy—as part of the tangled reflections on demarcating the modern from the postmodern. For discussion and to retain a focus on explanation in space history, I will pull only a few threads from this skein. Postmodernism (by no means monolithic) is best known for its epistemological and ontological claims, particularly Francois Lyotard's dictum that grand narratives (transcendent certainties about human nature, culture, knowledge

---

12. Exemplars of such history abound in which explanation, problems, research, cases, and debate create a disciplinary ecology. A beautiful recent snapshot of such scholarship is captured in Leonard Rosenband's retrospective on David Landes's seminal *The Unbound Prometheus*: Leonard Rosenband, "Never Just Business: David Landes, *The Unbound Prometheus*," *Technology and Culture* 46 (2005): 168–176. Seminar contributor Phil Scranton has been at the center of a related research ecology for most of his career through his scholarship on innovation, business, and U.S. industrialization.

13. Walter A. McDougall, "Space-Age Europe: Gaullism, Euro-Gaullism, and the American Dilemma," *Technology and Culture* 26 (1985): 179–203.

of the "real") no longer could withstand scrutiny.[14] But, more important for our purpose here, it advances a set of causes that provide tentative reference points for description and explanation of contemporary culture (generally, and not coincidentally for space history, taking the 1970s as watershed). And these causes that shape the postmodern condition ring familiar: that, relatively, economic and political power have shifted from states to markets; that communications technologies and systems (and that large-scale technological systems of many types) have been integral to this shift; and that, in combination, these factors have remapped processes of cultural production, changing the ways in which states, corporations, communities, ethnic groups, and individuals exercise power and create identities. The acceleration of actions and reactions across national borders, of the collapse of geographic distance, of the sense that everyone and every place potentially seem proximate become, in this frame, distinctive features of a global cultural landscape. This new condition mirrors, with steroids injected, McLuhan's infamous "global village," a notion partially rooted in the beginnings of the Space Age. That postmodernism has attempted to specify a template of material causes for such phenomena is not too surprising. The leading exponents draw heavily on the Marxist intellectual tradition that links base to superstructure, economics to culture.[15] Too, in considering causes and explanation, postmodernism must be seen in conjunction with its close cousin, globalism. The two roughly map onto each other—but with one important difference in emphasis: globalism addresses more directly the functioning of political and economic power on the transnational stage, particularly as regards the U.S. dominant position in the aftermath of the Cold War.

Should these turns of thinking be of interest to space history? Readers already will have their pointy objects handy to prick the above notions of any seeming juice;—at minimum, we might skeptically ask if such causal assertions have empirical weight. But as a set of ideas, as a heuristic, I think this framework is provocative for space history, suggesting ways to explore notions of causation and change that (re)integrate the field into broader structures of meaning and reconceptualize our sense of the problem map of the field.[16]

---

14. See Jean-Francois Lyotard, *The Postmodern Condition: A Report on Knowledge* (Minneapolis: University of Minnesota, 1984 [originally published in 1979]), and Frederic Jameson, *Postmodernism, or, The Cultural Logic of Late Capitalism* (Durham: Duke University Press, 1991).

15. The most lucid and extended account of the interconnections among epistemology, causes, and cultural effects in the postmodern is David Harvey, *The Condition of Postmodernity: An Enquiry into the Origins of Cultural Change* (Oxford, England: Basil Blackwell, 1989).

16. One case study that attempts to connect space history with the interpretive frames of postmodernism and globalism is Peter Redfield, "The Half-Life of Empire in Outer-Space," *Social Studies of Science* 32 (2002): 791–825. See also Martin Collins, "One World . . . One Telephone: Iridium, One Look at the Making of a Global Age," *History and Technology* 21 (2005): 301–324.

At a simple level of correspondence, elements of space activity seem central to postmodern/global analysis: a variety of space-based transnational technical systems—communications, navigation, and surveillance and monitoring—are implicated in their material and causal account of the how and why of recent cultural change. Notably, these applied space systems—with origins in markets, civilian government, and military and intelligence agencies—have not been, relatively, the center of policy debate (at least in the U.S.), nor the perceived strategic focus of scholarship. Through the lens of postmodernism/globalism, though, one is inclined to start from the proposition that these undertakings, individually and collectively, are crucial sites for research, for understanding the very composition of a post-1960s world order. This emphasis on the applied, on systems of culture and technology that intersect with the global "everyday" (including constructions of power and identity), if taken seriously, de-centers NASA and human space exploration as the signature markers of geopolitics—an orientation that aligns with the halting course of decisions and funding in post-Moon-landing space history—yet provides a rich, alternative framework for making the "international" a central area of investigation.

Conceptually, postmodernism and globalism, too, reemphasize, in comparison to the focus on state-centered accounts in Cold War space history, the centrality of markets and culture (in particular, their distinctive interrelation in the postmodern) as loci of change. To say this is almost to be "history 101" obvious. But the persistent legacy of Cold War literature seems to have pushed the obvious to the margins. Regarding markets, postmodernism/globalism highlights the issue of developing better accounts of state-market configurations in the post-Moon-landing years as a means to situate all space activities, including human space exploration. Thus considered, NASA's travails in the 1970s and after might be examined as part of a broader framework of change, rather than as (often, normatively, bemoaned) exemplars of half-measure policy-making and institutional diminishment. This, I think, put somewhat differently, is Asif Siddiqi's point on the "problem" of Apollo in space history scholarship.

Regarding culture, postmodernism points us to a fundamental (and not yet fully articulated) question in the field: How do we analytically frame the interconnections between space activity and culture? Is this primarily a dynamic in which cultural tropes broadly circulate to be plucked, adopted, and made instrumental by a range of groups (individuals, civil associations, markets, government)? Is such a process marked by a mutual perfusion of the real and fictional—with broad consequences for creators and consumers of space-related cultural productions? Or, more narrowly, is space culture merely an overlay, a gloss, a distraction from the meaty acts of political and business decision-making? In considering such questions, postmodernism advances an important claim: that culture, especially in the post–World War II years, serves not primarily as a set of restraints but becomes an active instrument, a resource to be deployed by the powerful and, in the information age, by the less powerful. Culture, thus, is not only precepts, assumptions, ways of doing, but is as protean and purpose-driven as capital—and is as central to macrostructures of change.[17] Seen from this angle, culture is not a side motif in space history, but a central explanatory problem.

These reflections are not meant to suggest that we swallow the bait and barb of postmodernism and globalism claims without care. Rather it is, first and foremost, to point to the value of generating our own conscientious engagement with scholarly discussions of historical change—with a suggestion that postmodernism and globalism make provocative foils. Second, it is to acknowledge the special place of historical (and other) research in situating and grounding explanatory claims in the empirical. This latter task is not insignificant. In the case of postmodernism/globalism, an outlook primarily developed by disciplines other than history, it theorizes about but often fails to engage the tough actualities of how technology, markets, states, and cultures

---

17. A little more needs to be said here. Consider two prominent definitions of culture.

First, Clifford Geertz: "The concept of culture I espouse . . . is essentially a semiotic one. Believing, with Max Weber, that man is an animal suspended in webs of significance he himself has spun, I take culture to be those webs, and the analysis of it to be therefore not an experimental science in search of law but an interpretative one in search of meaning. It is explication I am after . . ." (Clifford Geertz, *The Interpretation of Cultures: Selected Essays* [New York: Basic Books, 1973], pp. 4–5).

Second, Raymond Williams: "Culture is ordinary: that is the first fact. Every human society has its own shape, its own purposes, its own meanings. Every human society expresses these, in institutions, and in arts and learning. The making of a society is the finding of common meanings and directions, and its growth is an active debate and amendment under the pressures of experience, contact, and discovery, writing themselves into the land. The growing society is there, yet it is also made and remade in every individual mind. The making of a mind is, first, the slow learning of shapes, purposes, and meanings, so that work, observation and communication are possible. Then, second, but equal in importance, is the testing of these in experience, the making of new observations, comparisons, and meanings. A culture has two aspects: the known meanings and directions, which its members are trained to; the new observations and meanings, which are offered and tested. These are the ordinary processes of human societies and human minds, and we see through them the nature of a culture: that it is always both traditional and creative; that it is both the most ordinary common meanings and the finest individual meanings" (Raymond Williams, *Culture and Society* [New York: Columbia University Press, 1958]).

These notions of culture were used implicitly and effectively in a number of the essays, especially in those that focused on institutions and communities (including those by Diane Vaughan, Todd La Porte, Phil Scranton, and Alexander Brown). Postmodernism/globalism does not upend these analytic frameworks but does claim that "capital" and new communications technologies (to be unwarrantedly deterministic) have assumed an enhanced role in lifting and disconnecting cultural products from their local settings, resulting in two strata of culture: that associated with capital and global information flows and that associated with the "traditional" and the local. This dynamic is the core idea in a range of global "clash" literature, including work by Samuel Huntington, Bernard Barber, and Thomas Friedman.

interrelate, of how technological innovation occurs and integrates with culture, and of how the local and global are variously co-related.

But those are tepid aspirations and disciplinary goals. More, I think, post-modernism and globalism highlight the ways in which space and associated undertakings are central to the modern condition, and provide crucial sites of research for an explanatory enterprise. Three longstanding, deep cultural-technical themes of space history—exploration, control, and systems—also are integral to unpacking and giving substance to postmodern and globalism accounts. Space history, too, raises issues that these literatures largely side-step: for example (and crucially), analytic engagement with the meaning and import of military and intelligence space activities (especially in relation to the U.S.). The scale of these activities (in terms of funding, geospatial reach, and range of technologies) and, for a significant fraction, their relative invisibility behind barriers of security classification, pose important issues for models that seek to interconnect states, markets, and global technical systems. As Stephen Johnson's essay ably and starkly details, great chunks of military and intelligence space activity are shadowed. The emphasis in the postmodern has been on the *public* transnational modalities of interconnection. But what are we to make of the largely secret global systems of surveillance and control, of their interconnections with markets, of their integration into models of explanation and change? Given classification barriers, this now may be an unresearchable question, but as a discipline, it is of the first consequence to pose it.[18]

I have offered here at least two layers of idiosyncrasy: my own reading of the present state of the field and of possible responses and questions that ask us to consider the implications of subject specialty versus discipline. My goal here was not to make the case for a particular formulation of community, but to argue that we face a substantive choice—and that space history possesses the intellectual heft to make that choice important. We have come to a fork in the road. Take it.

---

18. An important reconnaissance of this problem in the Cold War and after (especially post-9/11) is Peter Galison, "Removing Knowledge," *Critical Inquiry* 31 (2004): 229–243. On the comparative scope of unclassified and classified information in the United States, the money quotation is: "In fact, the classified universe, as it is sometimes called, is certainly not smaller and very probably much larger than this unclassified one" (p. 229). The "unclassified one" that Galison uses as a point of reference is the total estimated page count of *all* material in the Library of Congress: 7.5 billion pages. The significance of classified activity in understanding the Cold War or "postmodern" eras is yet to be fully mapped.

ABOUT THE AUTHORS

Alexander Brown is a graduate student in MIT's Program in Science, Technology and Society. His research examines changing engineering practice from the 1960s to the 1990s. Using accidents and their investigations as a lens, he examines the changing cultures of engineering within NASA. He is tracking changes in engineering practices from Apollo 1 to *Challenger* to *Columbia*.

Andrew J. Butrica earned his Ph.D. in history of technology and science from Iowa State University in 1986. He then worked at the Thomas A. Edison Papers Project and the Center for Research in the History of Science and Technology in Paris. Subsequently, as a contract historian, he has researched and written *Out of Thin Air*, a history of Air Products and Chemicals, Inc., a Fortune 500 firm; *Beyond the Ionosphere*, a history of satellite communications; *To See the Unseen*, a history of planetary radar astronomy, which won the Leopold Prize of the Organization of American Historians; and *Single Stage To Orbit: Politics, Space Technology, and the Quest for Reusable Rocketry*, which was recently published by Johns Hopkins University Press and won the 2005 Michael C. Robinson Prize of the National Council on Public History. In addition, he was the historian on NASA's X-33 Program and currently is a historian on the Defense Acquisition History Project responsible for researching and writing the fourth volume of *From the Reagan Buildup to the End of the Cold War, 1981–1990*.

Martin J. Collins is a curator in the Smithsonian National Air and Space Museum. He received his Ph.D. from the University of Maryland in history of science and technology. Book publications include *Cold War Laboratory: RAND, the Air Force, and the American State* (2002) and *Showcasing Space*, edited with Douglas Millard, volume 6 of the Artefacts series: Studies in the History of Science and Technology (2005). He is currently working on a history of the Iridium satellite telephone venture.

David H. DeVorkin is presently the curator of the history of astronomy and the space sciences within the Department of Space History at the National Air and Space Museum (NASM). He was brought to NASM in January 1981 as an associate curator, and in the interim, he has held the positions of both acting chairman of the Space Science and Exploration Department (1983–84) and chairman of the department (1984–86). Since 1985, he has held the concurrent position of chairman of the Advisory Committee to the Smithsonian Videohistory Program, and he spent the summer and fall of 1991 as a visiting member of the Institute for Advanced Study in Princeton. His major research interests are in the origins and development of modern astrophysics during the 20th century and the origins of the space sciences in the V-2 and early Aerobee eras through the International Geophysical Year (IGY). He specializes as well in the history of space astronomy and in the government patronage of science in the postwar era. He recently has published a major biography of the Princeton astronomer Henry Norris Russell, which has been critically acclaimed. DeVorkin has authored over 100 scholarly papers and has authored, edited, or compiled 9 monographs in the history of, and education in, astronomy and the space sciences. His works have appeared in the *Journal for the History of Astronomy*, *Sky & Telescope*, *Isis*, *Scientific American*, *Minerva*, *Science*, *Historical Studies in the Physical Sciences*, *Physics Today*, and elsewhere. He holds a Ph.D. in the history of science from the University of Leicester (1978), a master of philosophy in astronomy from Yale (1970), an M.S. in astronomy from San Diego State College (1968), and a B.S. in astronomy/physics from the University of California, Los Angeles (UCLA) (1966). His present external and research activities are supported by an ongoing National Science Foundation (NSF) grant and a newly approved scholar's grant from NSF (with Patrick McCray). An educational outreach grant from NASA is pending. His present major contract is with National Geographic for a book on the Hubble Space Telescope.

Steven J. Dick is the Chief Historian for NASA. He obtained his B.S. in astrophysics (1971) and his M.A. and Ph.D. (1977) in history and philosophy of science from Indiana University. He worked as an astronomer and historian of science at the U.S. Naval Observatory in Washington, DC, for 24 years before coming to NASA Headquarters in 2003. Among his books are *Plurality of Worlds: The Origins of the Extraterrestrial Life Debate from Democritus to Kant* (1982), *The Biological Universe: The Twentieth Century Extraterrestrial Life Debate and the Limits of Science* (1996), and *Life on Other Worlds* (1998). The last book has been translated into Chinese, Italian, Czech, and Polish. His most recent books are *The Living Universe: NASA and the Development of Astrobiology* (2004) and a comprehensive history of the U.S. Naval Observatory, *Sky and Ocean Joined: The U.S. Naval Observatory, 1830–2000* (2003). The latter received the Pendleton Prize of the Society for History in the Federal Government. He also is editor of *Many Worlds: The New Universe, Extraterrestrial Life and the Theological Implications* (2000) and (with Keith Cowing) of the proceedings of the NASA Administrator's symposium *Risk and Exploration: Earth, Sea and the Stars* (2005). He is the recipient of the Navy Meritorious

Civilian Service Medal. He received the NASA Group Achievement Award for his role in NASA's multidisciplinary program in astrobiology. He has served as chairman of the Historical Astronomy Division of the American Astronomical Society and as president of the History of Astronomy Commission of the International Astronomical Union, and he is the immediate past president of the Philosophical Society of Washington. He is also a member of the International Academy of Astronautics.

Slava Gerovitch received two doctorates in the history of science and technology from the Russian Academy of Sciences (1992) and from MIT (1999). His book *From Newspeak to Cyberspeak: A History of Soviet Cybernetics* (2002) received an honorable mention from the American Association for the Advancement of Slavic Studies for an outstanding monograph in Russian, Eurasian, or East European studies. He published articles on the history and historiography of Soviet science and technology in the journals *Technology and Culture*, *Social Studies of Science*, and the *Russian Review*, and in the collections *Science and Ideology*, *Cultures of Control*, and *Universities and Empire*. Currently, he is a research associate at the Dibner Institute for the History of Science and Technology at MIT, serving as the principal investigator on the NSF-funded project "Trusting the Machine: Onboard Computing, Automation, and Human Control in the Soviet Space Program."

Peter L. Hays is a policy analyst supporting the planning integration division of the National Security Space Office. A retired lieutenant colonel with 25 years of service in the U.S. Air Force (USAF), his previous positions include executive editor of *Joint Force Quarterly*, professor at the School of Advanced Airpower Studies (SAAS), associate professor of political science and division chief for the international relations and defense policy curriculum, and director of the USAF Institute for National Security Studies. He holds Ph.D. and M.A.L.D. degrees in international relations from the Fletcher School of Law and Diplomacy at Tufts University and an M.A. in defense and strategic studies from the University of Southern California. A 1979 honor graduate of the USAF Academy, Hays was a command pilot with over 3,200 hours of flying time, primarily in the C-141 Starlifter. He has focused his studies and research on U.S. national security space policy by developing space policy courses at the USAF Academy, SAAS, National Defense University, and George Washington University; serving as a research assistant at the White House Office of Science and Technology Policy and the National Space Council; and writing a dissertation on the U.S. military space doctrine. Hays is author of *United States Military Space* (2002) and is a coeditor of *Space Power for a New Millennium* (2000), *Countering the Proliferation and Use of Weapons of Mass Destruction* (1998), and the seventh edition of *American Defense Policy* (1997).

Stephen Johnson is an associate professor in the Space Studies Department of the University of North Dakota (UND) teaching military space, space history, and management and economics of space endeavors. He is the author of *The United States Air Force and the Culture of Innovation, 1945–1965* (2002) and *The Secret of Apollo: Systems Management in American and European Space Programs* (2002). He is also the editor of *Quest: The History of Spaceflight Quarterly*. His current research involves the development of cognitive psychology and artificial intelligence, space industry management and economics, dependable system design, and the history of space science and technology. Prior to coming to UND, he worked in the aerospace industry for 15 years, managing computer simulation laboratories, designing space probes, and developing engineering processes. He received his doctorate in 1997 in the history of science and technology from the University of Minnesota, where he was also the associate director of the Babbage Institute for the history of computing.

John Krige is the Kranzberg Professor in the School of History, Technology, and Society at the Georgia Institute of Technology, Atlanta, Georgia; for academic year 2004–05, he was the Charles A. Lindbergh Professor at the National Air and Space Museum in Washington, DC. His historical interest is in the relations between science, technology, and foreign policy in the first two decades of the Cold War. He has coauthored a three-volume history of CERN, the European Organization for Nuclear Research, and a two-volume history of ESA, the European Space Agency. Krige serves on the editorial board of several international journals and is the executive editor of *History and Technology*, published by Routledge. His most recent book, *American Hegemony and the Postwar Reconstruction of Science in Europe*, is scheduled for publication by MIT Press in 2006. He currently is working on U.S.-European relations in the field of rocketry.

Todd R. La Porte has been a professor of political science at the University of California, Berkeley, since 1965, where he was also associate director of the Institute of Governmental Studies (1973–88). He received his B.A. from the University of Dubuque (1953) and his M.A. and Ph.D. from Stanford University (1962). He also held faculty posts at the University of Southern California and Stanford University. He teaches and publishes in the areas of organization theory; technology and politics (and assessment); and the organizational and decision-making dynamics of large, complex, technologically intensive organizations, as well as public attitudes toward advanced technologies and the problems of governance in a technological society. He was a principal of the Berkeley High Reliability Organization Project, a multidisciplinary team that studied the organizational aspects of safety-critical systems such as nuclear power, air traffic control, and nuclear aircraft carriers. Currently, his research concerns the evolution of large-scale organizations operating technologies demanding a very high level of reliable (nearly failure-free) performance and the relationship of large-scale technical systems to political legitimacy, especially in the nuclear domain. This took him to Los Alamos National Laboratory (1998–2003) to examine the institutional challenges of multigeneration nuclear missions. He was elected to the National Academy of Public Administration (1985) and was a Fellow at the Woodrow Wilson International Center for Scholars, Smithsonian Institution, and Research Fellow at the Wissenschaftszentrum (Sciences Center), Berlin, and the Max Planck Institute for Social Research, Cologne. He has been a member of the Board on Radioactive Waste Management and served on panels of the Committee on Human Factors and Transportation Research Board, National Academy of Sciences. He served on the Secretary of Energy Advisory Board, Department of Energy, and chaired its Task Force on Radioactive Waste Management, examining questions of institutional trustworthiness; he also was on the Technical Review Committee, Nuclear Materials Technology Division, Los Alamos National Laboratory. Additionally, he has served as a member of the Committee on Long-Term Institutional Management of Department of Energy (DOE) Legacy Waste Sites: Phase Two and the Committee on Principles and Operational Strategies for Staged Repository Systems, both of the Board on Radioactive Waste Management, National Academy of Sciences (2001–03). He is currently a faculty affiliate, Decision Science Division, Los Alamos National Laboratory.

Roger D. Launius is Chair of the Division of Space History at the Smithsonian Institution's National Air and Space Museum in Washington, DC. Between 1990 and 2002, he served as Chief Historian of NASA. A graduate of Graceland College in Lamoni, Iowa, he received his Ph.D. from Louisiana State University, Baton Rouge, in 1982. He has written or edited more than 20 books on aerospace history, including *Space: A Journey to Our Future* (2004); *Space Stations: Base Camps to the Stars* (2003), which received the American Institute of Aeronautics and Astronautics' (AIAA's) history manuscript prize; *Flight: A Celebration of 100 Years in Art and Literature* (2003); *Reconsidering a Century of Flight* (2003); *To Reach the High Frontier: A History of U.S. Launch Vehicles* (2002); *Imagining Space: Achievements, Possibilities, Projections, 1950–2050* (2001); *Reconsidering Sputnik: Forty Years Since the Soviet Satellite* (2000); *Innovation and the Development of Flight* (1999); *Frontiers of Space Exploration* (1998; rev. ed., 2004); *Spaceflight and the Myth of Presidential Leadership* (1997); and *NASA: A History of the U.S. Civil Space Program* (1994; rev. ed., 2001). He is frequently consulted by the electronic and print media for his views on space issues. His research interests encompass all areas of U.S. and space history and policy history, especially cultural aspects of the subject and the role of executive decision-makers and their efforts to define space exploration.

John M. Logsdon is the director of the Space Policy Institute at George Washington University's Elliott School of International Affairs, where he is also a professor of political science and international affairs. He holds a B.S. in physics from Xavier University (1960) and a Ph.D. in political science from New York University (1970). Dr. Logsdon's research interests focus on the policy and historical aspects of U.S. and international space activities. Dr. Logsdon is the author of *The Decision to Go to the Moon: Project Apollo and the National Interest* and is general editor of the eight-volume series *Exploring the Unknown: Selected Documents in the History of the U.S. Civil Space Program*. He has written numerous articles and reports on space policy and history. He is frequently consulted by the electronic and print media for his views on space issues. Dr. Logsdon recently served as a member of the Columbia Accident Investigation Board. He is a former member of the NASA Advisory Council and a current member of the Commercial Space Transportation Advisory Committee of the Department of Transportation. He is a recipient of the NASA Distinguished Public Service and Public Service Medals and a Fellow of the American Institute of Aeronautics and Astronautics and of the American Association for the Advancement of Science.

Howard McCurdy is a professor of public affairs and chair of the Public Administration Department at American University in Washington, DC. An expert on space policy, he recently authored *Faster, Better, Cheaper*, a critical analysis of cost-cutting initiatives in the U.S. space program. An earlier study of NASA's organizational culture, *Inside NASA*, won the 1994 Henry Adams prize for that year's best history on the federal government. He also has written *Space and the American Imagination* and coedited *Spaceflight and the Myth of Presidential Leadership*. His work appears in scholarly journals such as *Public Administration Review* and *Space Policy*. He is often consulted by the media on public policy issues and has appeared on national news outlets such as the *NewsHour with Jim Lehrer*, National Public Radio, and *NBC Nightly News*. Professor McCurdy received his bachelor's and master's degrees from the University of Washington and his doctorate from Cornell University.

David A. Mindell is Dibner Professor of the History of Engineering and Manufacturing and Professor of Engineering Systems at MIT. He is the founder and director of MIT's "DeepArch" research group in technology, archaeology, and the deep sea. His research interests include technology policy (historical and current), the history of automation in the military, the history of electronics and computing, new theories of engineering systems, deep-ocean robotic archaeology, and the history of space exploration. His book *War, Technology, and Experience aboard the USS Monitor* was published in April 2000 by Johns Hopkins University Press. His second book, *Between Human and Machine: Feedback, Control, and Computing before Cybernetics*, was published in the spring of 2002.

Mindell is currently researching a book on the history of the Apollo Guidance Computer that guided the astronauts to the Moon. From 1992 to 1995, Mindell was a National Science Foundation Graduate Fellow, and from 1995 to 1996, he was a Fellow at the Dibner Institute for the History of Science and Technology at MIT. He is an adjunct researcher at the Institute for Exploration. Before coming to MIT, Mindell worked as a research engineer in the Deep Submergence Laboratory of the Woods Hole Oceanographic Institution, where he is currently a visiting investigator. Mindell has consulted on engineering and policy for a number of industrial and research organizations, including the National Academy of Sciences. He has degrees in literature and electrical engineering from Yale University and a doctorate in the history of technology from MIT.

Stephen Pyne is a professor with the human dimensions faculty in the School of Life Sciences at Arizona State University, where he teaches courses on environmental history; the history of exploration; and the history, ecology, and management of fire. He is the author of 16 books, which mostly concern the history of fire on Earth. Of the others, three develop themes in the history of science and exploration (*Grove Karl Gilbert*, *The Ice: A Journey to Antarctica*, and *How the Canyon Became Grand*), as have several essays on space exploration. He is currently writing a fire history of Canada.

Philip Scranton is Board of Governors Professor in history of industry and technology at Rutgers University and director of the Center for the History of Business, Technology, and Society at the Hagley Museum and Library. He edits a monograph series, *Studies in Industry and Society*, for the Johns Hopkins University Press and coedits a collected-essays series, *Hagley Perspectives in Business and Culture*, for the University of Pennsylvania Press. He is author, coauthor, or editor of nine books; serves as a consultant for historical museums and public programming; and, most recently, held the Charles Lindbergh Chair (2003–04) at the National Air and Space Museum. His current project focuses on Cold War innovation and specialty production, especially regarding jet propulsion, NASA space capsules, instrumentation, and the new materials "explosion" in and following the Second World War.

Asif Siddiqi received his Ph.D. in history from Carnegie Mellon University in 2004. He specializes in the history of science and technology with a focus on the history of astronautics. He is the author of *Challenge to Apollo: The Soviet Union and the Space Race, 1945–1974* (2000) and the series editor for *Rockets and People*, the four-volume memoirs of Russian space engineer Boris Chertok. He is currently a Visiting Fellow in the American Academy of Arts & Sciences in Cambridge, Massachusetts, and took up a position as assistant professor of history at Fordham University in New York in the fall of 2004.

Diane Vaughan received her Ph.D. in sociology from Ohio State University in 1979 and began teaching at Boston College in 1982, where she is now a professor of sociology. She has been awarded fellowships at Yale (1979–82), Centre for Socio-Legal Studies, Oxford (1986–87), the American Bar Foundation (1988–89), the Institute for Advanced Study, Princeton (1996–97), and the John Simon Guggenheim Memorial Foundation (2003–04). Her areas of specialization are the sociology of organizations; sociology of culture; deviance and social control; field methods; research design; and science, knowledge, and technology. Much of her research has examined the "dark side" of organizations: mistake, misconduct, and disaster. In 2003, she worked with the Columbia Accident Investigation Board on their analysis and report on the loss of the Space Shuttle *Columbia*. Her books include *Controlling Unlawful Organizational Behavior*, *Uncoupling*, and *The Challenger Launch Decision*, which was awarded the Rachel Carson Prize, the Robert K. Merton Award, and Honorable Mention for Distinguished Contribution to Scholarship of the American Sociological Association, and was nominated for the National Book Award and Pulitzer Prize. Her current projects are *Theorizing: Analogy, Cases, and Comparative Social Organization* and *Dead Reckoning: Air Traffic Control in the Early 21st Century*.

Margaret A. Weitekamp earned her B.A. summa cum laude from the University of Pittsburgh and her Ph.D. in history at Cornell University in May 2001. Currently, she is a curator in the Division of Space History at the National Air and Space Museum, Smithsonian Institution, in Washington, DC. There, she oversees the "Social and Cultural Dimensions of Spaceflight" collection, almost 3,500 individual pieces of space memorabilia and space science fiction objects. These social and cultural products of the Space Age— everything from toys and games to clothing and stamps, medals and awards, buttons and pins, and even comics and trading cards—complete the story about spaceflight told by the museum's collection of space technologies. Her first book, *Right Stuff, Wrong Sex: America's First Women in Space Program*, was published by the Johns Hopkins University Press as a part of the Gender Relations in the American Experience series in November 2004. The book investigates shifting ideas about gender in the early 1960s through the history of a privately funded project that tested women pilots for astronaut fitness at the beginning of the Space Age. In addition, Weitekamp also has contributed to an edited anthology entitled *Impossible to Hold: Women and Culture in the 1960s* (2005). She won the Smithsonian Institution's National Air and Space Museum Aviation/Space Writers Award in 2002. She spent academic year 1997–98 in residence at the NASA Headquarters History Office in Washington, DC, as the American Historical Association/NASA Aerospace History Fellow. She was a 1993 Mellon Fellow in the Humanities.

ACRONYMS AND ABBREVIATIONS

AAAS	American Association for the Advancement of Science
AAS	American Astronautical Society
ABC	American Broadcasting Company
ABM	antiballistic missile
ABMA	Army Ballistic Missile Agency
ACES	Air Collection and Enrichment System
AFB	Air Force Base
AGARD	Advisory Group for Aerospace Research and Development
AHA	American Historical Association
AIAA	American Institute of Aeronautics and Astronautics
AmRoc	American Rocket Company
ARDC	Air Research and Development Command
ARPA	Advanced Research Projects Agency
ASA	American Studies Association
ASAT	antisatellite
ASEE	American Society for Electrical Engineering; American Society for Engineering Education
ASME	American Society of Mechanical Engineers
ASSET	Aerothermodynamic/elastic Structural Systems Environmental Tests
AT&T	American Telephone and Telegraph Company
ATV	Advanced Technology Vehicle
BAMBI	Ballistic Missile Boost Intercept
BMD	ballistic missile defense
BMDO	Ballistic Missile Defense Organization
BMEWS	Ballistic Missile Early Warning System
BOR	Unpiloted Orbital Rocketplane (in Russian)

C2	command and control
C³I	command, control, communications, and intelligence
CAIB	Columbia Accident Investigation Board
Caltech	California Institute of Technology
CAPCOM	capsule communicator
CBI	Chicago Bridge and Iron
CCD	charge-coupled device
CCP	Contract Change Proposal
CDC	Centers for Disease Control
CELV	Complementary Expendable Launch Vehicle
CEO	chief executive officer
CERN	European Organization for Nuclear Research
CEV	Crew Exploration Vehicle
CIA	Central Intelligence Agency
CMLC	Civilian-Military Liaison Committee (of the House)
comsat	commercial satellite; communications satellite
COMSAT	Communications Satellite Corporation
COMSTAC	Commercial Space Transportation Advisory Committee
CRV	Crew Recovery Vehicle
C/SGT	Continental/SemiGlobal Transport
CSM	Command and Service Module (Apollo)
CSOC	Consolidated Space Operations Center
DARPA	Defense Advanced Research Projects Agency
DCAS	Defense Contract Administration Services
DC-X	Delta Clipper-Experimental
DC-XA	Delta Clipper-Experimental Advanced
DDR&E	Director of Defense Research and Engineering; Deputy Secretary of Defense for Research and Engineering; designing, developing, researching, and engineering
DEW	Distant Early Warning
DMSP	Defense Meteorological Satellite Program
DNA	deoxyribonucleic acid
DOD	Department of Defense

DOE	Department of Energy
DSCS II	Defense Satellite Communications System II
DSP	Defense Support Program
Dyna-Soar	Dynamic Soaring
ECR	Engineering Change Request
EEC	European Economic Community
EELV	Evolved Expendable Launch Vehicle
ELDO	European Launcher Development Organisation
ELINT	electronics intelligence
ELV	expendable launch vehicle
EOP	Executive Office of the President
EORSAT	Electronic Intelligence Ocean Reconnaissance Satellite
EOS	Earth Observing System
EPA	Environmental Protection Agency
ESA	European Space Agency
ESRO	European Space Research Organisation
EST	eastern standard time
ET	External Tank (of the Space Shuttle)
EVA	extravehicular activity
FCC	Federal Communications Commission
FEMA	Federal Emergency Management Agency
FLATs	First Lady Astronaut Trainees
FLTSATCOM	Fleet Satellite Communications
FOBS	Fractional Orbiting Bombardment System
FOC	full operational capability
FOIA	Freedom of Information Act
FRC	Flight Research Center
FRR	Flight Readiness Review
FY	fiscal year
GALCIT	Guggenheim Aeronautical Laboratory at the California Institute of Technology
GAO	General Accounting Office
GE	General Electric

GEODSS	Ground-Based Electro-Optical Deep Space Surveillance
GIRD	Group for the Investigation of Reactive Engines and Reactive Flight
GLONASS	Global Navigation Satellite System
GOES	Geostationary Operational Environmental Satellite
GPO	Gemini Program Office; Government Printing Office
GPS	Global Positioning System
GRAB	Galactic Radiation and Background
GUKOS	Chief Directorate of the Space Systems (Soviet)
GURVO	Chief Directorate of Reactive Armaments (Soviet)
HAER	Historic American Engineering Record
HL	Horizontal Lander
HRO	high-reliability organization
HSS	History of Science Society
HST	Hubble Space Telescope
Hum Vee, Hummer	High Mobility Multipurpose Wheeled Vehicle
IAU	International Astronomical Union
ICBM	intercontinental ballistic missile
ICO	Intermediate Circular Orbit
ICOM	International Council of Museums
IDCSP	Initial Defense Communications Satellite Program
IEEE	Institute of Electrical and Electronics Engineers
IGY	International Geophysical Year
ILV	Industrial Launch Vehicle
IMINT	overhead photoreconnaissance
IMP	Imager for Mars Pathfinder
INPO	Institute for Nuclear Power Operators
INTELSAT	International Telecommunications Satellite Consortium
IRBM	intermediate-range ballistic missile
IS	Istrebitel Sputnikov
ISS	International Space Station
ISTS	International Symposium on Space Technology and Science

IUS	Inertial Upper Stage
JBIS	*Journal of the British Interplanetary Society*
JCS	Joint Chiefs of Staff
JDFN	Joint Defence Facility Nurrungar
JPL	Jet Propulsion Laboratory
JSC	Johnson Space Center
KGB	Soviet secret police organization
KSC	Kennedy Space Center
KVR	Space Forces (Soviet)
L-5	Lagrange Point 5
LACES	Liquid Air Collection Engine System
LANL	Los Alamos National Laboratory
LBJ	Lyndon Baines Johnson
LEM	Lunar Excursion Module
LLRV	Lunar Landing Research Vehicle
LLTV	Lunar Landing Training Vehicle
LM	Lunar Module (Apollo)
LOR	lunar orbit rendezvous
MASI	Ministry of Space Industry (Chinese)
MASINT	measurement and signature intelligence
MASTIF	Multiple Axis Space Test Inertia Facility
MCC	Mission Control Center
MDA	Missile Defense Agency
MIDAS	Missile Defense Alarm System
MISS	Manned Military Space System; Man-In-Space-Soonest
MIT	Massachusetts Institute of Technology
MMT	Mission Management Team
MODS	Military Orbital Development System
MOL	Manned Orbiting Laboratory
MOM	Ministry of Machine Building (Soviet)
MOU	Memorandum of Understanding
MSC	Manned Spacecraft Center (later renamed Johnson Space Center)

MSE	Manned Spaceflight Engineer
MV	Ministry of Armaments (Soviet)
NA	North American
NAA	North American Aviation
NAC	NASA Advisory Council
NACA	National Advisory Committee for Aeronautics
NARA	National Archives and Records Administration
NASA	National Aeronautics and Space Administration
NASC	National Aeronautics and Space Council
NASCAR	National Association for Stock Car Auto Racing
NASM	National Air and Space Museum
NASP	National Aero-Space Plane
NATO	North Atlantic Treaty Organization
NEO	Near-Earth Object
NESC	NASA Engineering and Safety Center
NLS	National Launch System
NM	nautical mile
NOAA	National Oceanic and Atmospheric Administration
NORAD	North American Air Defense Command
NOTS	Naval Ordnance Test Station
NOTSNIK	designation combining "Naval Ordnance Test Station" and "Sputnik"
NPO	Scientific-Production Association (Soviet)
NPOESS	National Polar-Orbiting Environmental Satellite System
NPS	National Park Service
NRC	Nuclear Regulatory Commission
NRL	Naval Research Laboratory
NRO	National Reconnaissance Office
NS	national security
NSAM	National Security Action Memorandum
NSC	National Security Council
NSD	National Security Directive
NSDD	National Security Decision Directive

NSF	National Science Foundation
NSI	*NASA and the Space Industry*
NSIAD	National Security and International Affairs Division (within GAO)
NSPD	National Space Policy Directive
NSSD	National Security Study Directive
NSTC	National Science and Technology Council
NTDS	Naval Tactical Data System
OAH	Organization of American Historians
OCST	Office of Commercial Space Transportation
OKB	designation for a Soviet design bureau (such as OKB-1 or OKB-52)
OMB	Office of Management and Budget
ONR	Office of Naval Research
OSD	Office of the Secretary of Defense
OSHA	Occupational Safety and Health Administration
OSP	Orbital Space Plane
OSTP	Office of Science and Technology Policy
PAGEOS	Passive Geodetic Earth Orbiting Satellite
PAWS	Phased Array Warning System
PD	Presidential Directive
PSAC	President's Science Advisory Committee
psi	pounds per square inch
PUC	Public Utility Commissions
QA	quality assurance
R&D	research and development
RAF	Royal Air Force (United Kingdom)
RASV	Reusable Aerodynamic Space Vehicle
RCA	Radio Corporation of America
RFP	Request for Proposals
RGAE	Russian State Archive of the Economy
RLV	reusable launch vehicle
ROLS	Recoverable Orbital Launch System

RORSAT	Radar Ocean Reconnaissance Satellite
R.U.R.	*Rossum's Universal Robots*
RVSN	Missile Forces of Strategic Designation (Soviet)
SAAS	School of Advanced Airpower Studies
SAB	Scientific Advisory Board (USAF)
SAC	Strategic Air Command
SAE	Society of Automotive Engineers
SAEF	Spacecraft Assembly and Encapsulation Facility
SAGE	Semi-Automatic Ground Environment
SAINT	Satellite Inspector for Space Defense
SALT	Strategic Arms Limitations Talks
SATCOM	RCA Americom Satellite "SATCOM" series
SBIRS	Space-Based Infrared System
SCORE	Signal Communication by Orbiting Relay Equipment
SCOT	social construction of technology
scramjet	supersonic combustion ramjet
SDC	System Development Corporation
SDI	Strategic Defense Initiative
SDIO	Strategic Defense Initiative Organization
SDS	Satellite Data System
SEI	Space Exploration Initiative
SETI	Search for Extraterrestrial Intelligence
SETP	Society for Experimental Test Pilots
SHF	Super High Frequency
SHOT	Society for the History of Technology
SIG	Senior Interagency Group
SIGINT	signals intelligence
SIR	Shuttle Imaging Radar
SLC	space launch complex
SLV	Standardized Launch Vehicle
SMV	Space Maneuver Vehicle
SOPC	Shuttle Operations and Planning Complex

SR&QA	Office of Safety, Reliability and Quality Assurance (NASA)
SRB	Solid Rocket Booster
SRM	Solid Rocket Motor
SRTC	Solid Rocket Technical Committee
SSI	Space Services, Inc.
SSME	Space Shuttle Main Engine
SSTO	single stage to orbit
STAIF	Space Technology & Applications International Forum
START	Strategic Arms Reduction Treaties
STG	Space Task Group
STP	space test program
STS	Space Transportation System
TAOS	Thrust-Assisted Orbiter Shuttle
TAV	TransAtmospheric Vehicle
TCP	Technological Capabilities Panel
TDRSS	Tracking and Data Relay Satellite System
TIROS	Television and Infrared Observation System
TQ	transition quarter
TRW	Thompson Ramo Wooldridge Corporation
TsUKOS	Central Directorate of Space Systems (Soviet)
UCLA	University of California, Los Angeles
UHF	Ultra-High Frequencies
UND	University of North Dakota
UNKS	Directorate of the Space Systems Commander (Soviet)
unk-unks	unknown unknowns
U.S.	United States
US-A	active radar (Soviet)
USA	U.S. Army
USACERL	U.S. Army Construction Engineering Research Laboratories
USAF	U.S. Air Force
US-P	passive radar (Soviet)
USSR	Union of Soviet Socialist Republics

VKS	Military Space Forces (Soviet)
VNIIEM	All-Union Scientific Research Institute of Electromechanics (Soviet)
V-PRO	Forces of Anti-Missile Defense (Soviet)
WDD	Western Development Division (USAF)
WS	Weapons System (USAF)
WWMCCS	World-Wide Military Command and Control System

THE NASA HISTORY SERIES

REFERENCE WORKS, NASA SP-4000:

Grimwood, James M. *Project Mercury: A Chronology.* NASA SP-4001, 1963.

Grimwood, James M., and C. Barton Hacker, with Peter J. Vorzimmer. *Project Gemini Technology and Operations: A Chronology.* NASA SP-4002, 1969.

Link, Mae Mills. *Space Medicine in Project Mercury.* NASA SP-4003, 1965.

Astronautics and Aeronautics, 1963: *Chronology of Science, Technology, and Policy.* NASA SP-4004, 1964.

Astronautics and Aeronautics, 1964: *Chronology of Science, Technology, and Policy.* NASA SP-4005, 1965.

Astronautics and Aeronautics, 1965: *Chronology of Science, Technology, and Policy.* NASA SP-4006, 1966.

Astronautics and Aeronautics, 1966: *Chronology of Science, Technology, and Policy.* NASA SP-4007, 1967.

Astronautics and Aeronautics, 1967: *Chronology of Science, Technology, and Policy.* NASA SP-4008, 1968.

Ertel, Ivan D., and Mary Louise Morse. *The Apollo Spacecraft: A Chronology, Volume I, Through November 7, 1962.* NASA SP-4009, 1969.

Morse, Mary Louise, and Jean Kernahan Bays. *The Apollo Spacecraft: A Chronology, Volume II, November 8, 1962–September 30, 1964.* NASA SP-4009, 1973.

Brooks, Courtney G., and Ivan D. Ertel. *The Apollo Spacecraft: A Chronology, Volume III, October 1, 1964–January 20, 1966.* NASA SP-4009, 1973.

Ertel, Ivan D., and Roland W. Newkirk, with Courtney G. Brooks. *The Apollo Spacecraft: A Chronology, Volume IV, January 21, 1966–July 13, 1974.* NASA SP-4009, 1978.

Astronautics and Aeronautics, 1968: Chronology of Science, Technology, and Policy. NASA SP-4010, 1969.

Newkirk, Roland W., and Ivan D. Ertel, with Courtney G. Brooks. *Skylab: A Chronology.* NASA SP-4011, 1977.

Van Nimmen, Jane, and Leonard C. Bruno, with Robert L. Rosholt. *NASA Historical Data Book, Volume I: NASA Resources, 1958–1968*. NASA SP-4012, 1976; rep. ed. 1988.

Ezell, Linda Neuman. *NASA Historical Data Book, Volume II: Programs and Projects, 1958–1968*. NASA SP-4012, 1988.

Ezell, Linda Neuman. *NASA Historical Data Book, Volume III: Programs and Projects, 1969–1978*. NASA SP-4012, 1988.

Gawdiak, Ihor Y., with Helen Fedor, compilers. *NASA Historical Data Book, Volume IV: NASA Resources, 1969–1978*. NASA SP-4012, 1994.

Rumerman, Judy A., compiler. *NASA Historical Data Book, 1979–1988: Volume V, NASA Launch Systems, Space Transportation, Human Spaceflight, and Space Science*. NASA SP-4012, 1999.

Rumerman, Judy A., compiler. *NASA Historical Data Book, Volume VI: NASA Space Applications, Aeronautics and Space Research and Technology, Tracking and Data Acquisition/Space Operations, Commercial Programs, and Resources, 1979–1988*. NASA SP-2000-4012, 2000.

Astronautics and Aeronautics, 1969: Chronology of Science, Technology, and Policy. NASA SP-4014, 1970.

Astronautics and Aeronautics, 1970: Chronology of Science, Technology, and Policy. NASA SP-4015, 1972.

Astronautics and Aeronautics, 1971: Chronology of Science, Technology, and Policy. NASA SP-4016, 1972.

Astronautics and Aeronautics, 1972: Chronology of Science, Technology, and Policy. NASA SP-4017, 1974.

Astronautics and Aeronautics, 1973: Chronology of Science, Technology, and Policy. NASA SP-4018, 1975.

Astronautics and Aeronautics, 1974: Chronology of Science, Technology, and Policy. NASA SP-4019, 1977.

Astronautics and Aeronautics, 1975: Chronology of Science, Technology, and Policy. NASA SP-4020, 1979.

Astronautics and Aeronautics, 1976: Chronology of Science, Technology, and Policy. NASA SP-4021, 1984.

Astronautics and Aeronautics, 1977: Chronology of Science, Technology, and Policy. NASA SP-4022, 1986.

Astronautics and Aeronautics, 1978: Chronology of Science, Technology, and Policy. NASA SP-4023, 1986.

Astronautics and Aeronautics, 1979–1984: Chronology of Science, Technology, and Policy. NASA SP-4024, 1988.

Astronautics and Aeronautics, 1985: Chronology of Science, Technology, and Policy. NASA SP-4025, 1990.

Noordung, Hermann. *The Problem of Space Travel: The Rocket Motor.* Edited by Ernst Stuhlinger and J. D. Hunley, with Jennifer Garland. NASA SP-4026, 1995.

Astronautics and Aeronautics, 1986–1990: A Chronology. NASA SP-4027, 1997.

Astronautics and Aeronautics, 1990–1995: A Chronology. NASA SP-2000-4028, 2000.

MANAGEMENT HISTORIES, NASA SP-4100:

Rosholt, Robert L. *An Administrative History of NASA, 1958–1963.* NASA SP-4101, 1966.

Levine, Arnold S. *Managing NASA in the Apollo Era.* NASA SP-4102, 1982.

Roland, Alex. *Model Research: The National Advisory Committee for Aeronautics, 1915–1958.* NASA SP-4103, 1985.

Fries, Sylvia D. NASA *Engineers and the Age of Apollo.* NASA SP-4104, 1992.

Glennan, T. Keith. *The Birth of NASA: The Diary of T. Keith Glennan.* J. D. Hunley, editor. NASA SP-4105, 1993.

Seamans, Robert C., Jr. *Aiming at Targets: The Autobiography of Robert C. Seamans, Jr.* NASA SP-4106, 1996.

Garber, Stephen J., editor. *Looking Backward, Looking Forward: Forty Years of U.S. Human Spaceflight Symposium.* NASA SP-2002-4107, 2002.

Chertok, Boris. *Rockets and People, Volume I.* NASA SP-2005-4110, 2005.

Laufer, Dr. Alexander, Todd Post, and Dr. Edward J. Hoffman. *Shared Voyage: Learning and Unlearning from Remarkable Projects.* NASA SP-2005-4111, 2005.

Dawson, Virginia P., and Mark D. Bowles, editors. *Realizing the Dream of Flight.* NASA SP-2005-4112, 2005.

PROJECT HISTORIES, NASA SP-4200:

Swenson, Loyd S., Jr., James M. Grimwood, and Charles C. Alexander. *This New Ocean: A History of Project Mercury.* NASA SP-4201, 1966; rep. ed. 1998.

Green, Constance McLaughlin, and Milton Lomask. *Vanguard: A History.* NASA SP-4202, 1970; rep. ed. Smithsonian Institution Press, 1971.

Hacker, Barton C., and James M. Grimwood. *On the Shoulders of Titans: A History of Project Gemini.* NASA SP-4203, 1977.

Benson, Charles D., and William Barnaby Faherty. *Moonport: A History of Apollo Launch Facilities and Operations.* NASA SP-4204, 1978.

Brooks, Courtney G., James M. Grimwood, and Loyd S. Swenson, Jr. *Chariots for Apollo: A History of Manned Lunar Spacecraft.* NASA SP-4205, 1979.

Bilstein, Roger E. *Stages to Saturn: A Technological History of the Apollo/Saturn Launch Vehicles.* NASA SP-4206, 1980, rep. ed. 1997.

SP-4207 not published.

Compton, W. David, and Charles D. Benson. *Living and Working in Space: A History of Skylab.* NASA SP-4208, 1983.

Ezell, Edward Clinton, and Linda Neuman Ezell. *The Partnership: A History of the Apollo-Soyuz Test Project.* NASA SP-4209, 1978.

Hall, R. Cargill. *Lunar Impact: A History of Project Ranger.* NASA SP-4210, 1977.

Newell, Homer E. *Beyond the Atmosphere: Early Years of Space Science.* NASA SP-4211, 1980.

Ezell, Edward Clinton, and Linda Neuman Ezell. *On Mars: Exploration of the Red Planet, 1958–1978.* NASA SP-4212, 1984.

Pitts, John A. *The Human Factor: Biomedicine in the Manned Space Program to 1980.* NASA SP-4213, 1985.

Compton, W. David. *Where No Man Has Gone Before: A History of Apollo Lunar Exploration Missions.* NASA SP-4214, 1989.

Naugle, John E. *First Among Equals: The Selection of NASA Space Science Experiments.* NASA SP-4215, 1991.

Wallace, Lane E. *Airborne Trailblazer: Two Decades with NASA Langley's Boeing 737 Flying Laboratory.* NASA SP-4216, 1994.

Butrica, Andrew J., editor. *Beyond the Ionosphere: Fifty Years of Satellite Communication.* NASA SP-4217, 1997.

Butrica, Andrew J. *To See the Unseen: A History of Planetary Radar Astronomy.* NASA SP-4218, 1996.

Mack, Pamela E., editor. *From Engineering Science to Big Science: The NACA and NASA Collier Trophy Research Project Winners.* NASA SP-4219, 1998.

Reed, R. Dale, with Darlene Lister. *Wingless Flight: The Lifting Body Story.* NASA SP-4220, 1997.

Heppenheimer, T. A. *The Space Shuttle Decision: NASA's Search for a Reusable Space Vehicle.* NASA SP-4221, 1999.

Hunley, J. D., editor. *Toward Mach 2: The Douglas D-558 Program.* NASA SP-4222, 1999.

Swanson, Glen E., editor. *"Before this Decade Is Out . . .": Personal Reflections on the Apollo Program.* NASA SP-4223, 1999.

Tomayko, James E. *Computers Take Flight: A History of NASA's Pioneering Digital Fly-by-Wire Project.* NASA SP-2000-4224, 2000.

Morgan, Clay. *Shuttle-Mir: The U.S. and Russia Share History's Highest Stage.* NASA SP-2001-4225, 2001.

Leary, William M. *"We Freeze to Please": A History of NASA's Icing Research Tunnel and the Quest for Flight Safety.* NASA SP-2002-4226, 2002.

Mudgway, Douglas J. *Uplink-Downlink: A History of the Deep Space Network 1957–1997.* NASA SP-2001-4227, 2001.

Dawson, Virginia P., and Mark D. Bowles. *Taming Liquid Hydrogen: The Centaur Upper Stage Rocket, 1958–2002.* NASA SP-2004-4230, 2004.

CENTER HISTORIES, NASA SP-4300:

Rosenthal, Alfred. *Venture into Space: Early Years of Goddard Space Flight Center.* NASA SP-4301, 1985.

Hartman, Edwin P. *Adventures in Research: A History of Ames Research Center, 1940–1965.* NASA SP-4302, 1970.

Hallion, Richard P. *On the Frontier: Flight Research at Dryden, 1946–1981.* NASA SP-4303, 1984.

Muenger, Elizabeth A. *Searching the Horizon: A History of Ames Research Center, 1940–1976.* NASA SP-4304, 1985.

Hansen, James R. *Engineer in Charge: A History of the Langley Aeronautical Laboratory, 1917–1958.* NASA SP-4305, 1987.

Dawson, Virginia P. *Engines and Innovation: Lewis Laboratory and American Propulsion Technology.* NASA SP-4306, 1991.

Dethloff, Henry C. *"Suddenly Tomorrow Came . . .": A History of the Johnson Space Center.* NASA SP-4307, 1993.

Hansen, James R. *Spaceflight Revolution: NASA Langley Research Center from Sputnik to Apollo.* NASA SP-4308, 1995.

Wallace, Lane E. *Flights of Discovery: 50 Years at the NASA Dryden Flight Research Center.* NASA SP-4309, 1996.

Herring, Mack R. *Way Station to Space: A History of the John C. Stennis Space Center.* NASA SP-4310, 1997.

Wallace, Harold D., Jr. *Wallops Station and the Creation of the American Space Program.* NASA SP-4311, 1997.

Wallace, Lane E. *Dreams, Hopes, Realities: NASA's Goddard Space Flight Center, The First Forty Years.* NASA SP-4312, 1999.

Dunar, Andrew J., and Stephen P. Waring. *Power to Explore: A History of the Marshall Space Flight Center.* NASA SP-4313, 1999.

Bugos, Glenn E. *Atmosphere of Freedom: Sixty Years at the NASA Ames Research Center.* NASA SP-2000-4314, 2000.

GENERAL HISTORIES, NASA SP-4400:

Corliss, William R. *NASA Sounding Rockets, 1958–1968: A Historical Summary.* NASA SP-4401, 1971.

Wells, Helen T., Susan H. Whiteley, and Carrie Karegeannes. *Origins of NASA Names.* NASA SP-4402, 1976.

Anderson, Frank W., Jr. *Orders of Magnitude: A History of NACA and NASA, 1915–1980.* NASA SP-4403, 1981.

Sloop, John L. *Liquid Hydrogen as a Propulsion Fuel, 1945–1959.* NASA SP-4404, 1978.

Roland, Alex. *A Spacefaring People: Perspectives on Early Spaceflight.* NASA SP-4405, 1985.

Bilstein, Roger E. *Orders of Magnitude: A History of the NACA and NASA, 1915–1990.* NASA SP-4406, 1989.

Logsdon, John M., editor, with Linda J. Lear, Jannelle Warren-Findley, Ray A. Williamson, and Dwayne A. Day. *Exploring the Unknown: Selected Documents in the History of the U.S. Civil Space Program, Volume I, Organizing for Exploration.* NASA SP-4407, 1995.

Logsdon, John M., editor, with Dwayne A. Day and Roger D. Launius. *Exploring the Unknown: Selected Documents in the History of the U.S. Civil Space Program, Volume II, External Relationships.* NASA SP-4407, 1996.

Logsdon, John M., editor, with Roger D. Launius, David H. Onkst, and Stephen J. Garber. *Exploring the Unknown: Selected Documents in the History of the U.S. Civil Space Program, Volume III, Using Space.* NASA SP-4407, 1998.

Logsdon, John M., general editor, with Ray A. Williamson, Roger D. Launius, Russell J. Acker, Stephen J. Garber, and Jonathan L. Friedman. *Exploring the Unknown: Selected Documents in the History of the U.S. Civil Space Program, Volume IV, Accessing Space.* NASA SP-4407, 1999.

Logsdon, John M., general editor, with Amy Paige Snyder, Roger D. Launius, Stephen J. Garber, and Regan Anne Newport. *Exploring the Unknown: Selected Documents in the History of the U.S. Civil Space Program, Volume V, Exploring the Cosmos.* NASA SP-2001-4407, 2001.

Siddiqi, Asif A. *Challenge to Apollo: The Soviet Union and the Space Race, 1945–1974.* NASA SP-2000-4408, 2000.

MONOGRAPHS IN AEROSPACE HISTORY, NASA SP-4500:

Launius, Roger D., and Aaron K. Gillette, compilers. *Toward a History of the Space Shuttle: An Annotated Bibliography.* Monographs in Aerospace History, No. 1, 1992.

Launius, Roger D., and J. D. Hunley, compilers. *An Annotated Bibliography of the Apollo Program.* Monographs in Aerospace History, No. 2, 1994.

Launius, Roger D. *Apollo: A Retrospective Analysis.* Monographs in Aerospace History, No. 3, 1994.

Hansen, James R. *Enchanted Rendezvous: John C. Houbolt and the Genesis of the Lunar-Orbit Rendezvous Concept.* Monographs in Aerospace History, No. 4, 1995.

Gorn, Michael H. *Hugh L. Dryden's Career in Aviation and Space.* Monographs in Aerospace History, No. 5, 1996.

Powers, Sheryll Goecke. *Women in Flight Research at NASA Dryden Flight Research Center, from 1946 to 1995.* Monographs in Aerospace History, No. 6, 1997.

Portree, David S. F., and Robert C. Trevino. *Walking to Olympus: An EVA Chronology.* Monographs in Aerospace History, No. 7, 1997.

Logsdon, John M., moderator. *Legislative Origins of the National Aeronautics and Space Act of 1958: Proceedings of an Oral History Workshop.* Monographs in Aerospace History, No. 8, 1998.

Rumerman, Judy A., compiler. *U.S. Human Spaceflight, A Record of Achievement, 1961–1998.* Monographs in Aerospace History, No. 9, 1998.

Portree, David S. F. *NASA's Origins and the Dawn of the Space Age.* Monographs in Aerospace History, No. 10, 1998.

Logsdon, John M. *Together in Orbit: The Origins of International Cooperation in the Space Station.* Monographs in Aerospace History, No. 11, 1998.

Phillips, W. Hewitt. *Journey in Aeronautical Research: A Career at NASA Langley Research Center.* Monographs in Aerospace History, No. 12, 1998.

Braslow, Albert L. *A History of Suction-Type Laminar-Flow Control with Emphasis on Flight Research.* Monographs in Aerospace History, No. 13, 1999.

Logsdon, John M., moderator. *Managing the Moon Program: Lessons Learned From Apollo.* Monographs in Aerospace History, No. 14, 1999.

Perminov, V. G. *The Difficult Road to Mars: A Brief History of Mars Exploration in the Soviet Union.* Monographs in Aerospace History, No. 15, 1999.

Tucker, Tom. *Touchdown: The Development of Propulsion Controlled Aircraft at NASA Dryden.* Monographs in Aerospace History, No. 16, 1999.

Maisel, Martin D., Demo J. Giulianetti, and Daniel C. Dugan. *The History of the XV-15 Tilt Rotor Research Aircraft: From Concept to Flight.* Monographs in Aerospace History, No. 17, 2000.

Jenkins, Dennis R. *Hypersonics Before the Shuttle: A Concise History of the X-15 Research Airplane.* Monographs in Aerospace History, No. 18, 2000.

Chambers, Joseph R. *Partners in Freedom: Contributions of the Langley Research Center to U.S. Military Aircraft in the 1990s.* Monographs in Aerospace History, No. 19, 2000.

Waltman, Gene L. *Black Magic and Gremlins: Analog Flight Simulations at NASA's Flight Research Center.* Monographs in Aerospace History, No. 20, 2000.

Portree, David S. F. *Humans to Mars: Fifty Years of Mission Planning, 1950–2000.* Monographs in Aerospace History, No. 21, 2001.

Thompson, Milton O., with J. D. Hunley. *Flight Research: Problems Encountered and What They Should Teach Us.* Monographs in Aerospace History, No. 22, 2000.

Tucker, Tom. *The Eclipse Project.* Monographs in Aerospace History, No. 23, 2000.

Siddiqi, Asif A. *Deep Space Chronicle: A Chronology of Deep Space and Planetary Probes, 1958–2000.* Monographs in Aerospace History, No. 24, 2002.

Merlin, Peter W. *Mach 3+: NASA/USAF YF-12 Flight Research, 1969–1979.* Monographs in Aerospace History, No. 25, 2001.

Anderson, Seth B. *Memoirs of an Aeronautical Engineer—Flight Tests at Ames Research Center: 1940–1970.* Monographs in Aerospace History, No. 26, 2002.

Renstrom, Arthur G. *Wilbur and Orville Wright: A Bibliography Commemorating the One-Hundredth Anniversary of the First Powered Flight on December 17, 1903.* Monographs in Aerospace History, No. 27, 2002.

No monograph 28.

Chambers, Joseph R. *Concept to Reality: Contributions of the NASA Langley Research Center to U.S. Civil Aircraft of the 1990s.* Monographs in Aerospace History, No. 29, 2003.

Peebles, Curtis, editor. *The Spoken Word: Recollections of Dryden History, The Early Years.* Monographs in Aerospace History, No. 30, 2003.

Jenkins, Dennis R., Tony Landis, and Jay Miller. *American X-Vehicles: An Inventory—X-1 to X-50.* Monographs in Aerospace History, No. 31, 2003.

Renstrom, Arthur G. *Wilbur and Orville Wright Chronology.* Monographs in Aerospace History, No. 32, 2003.

Bowles, Mark D., and Robert S. Arrighi. *NASA's Nuclear Frontier: The Plum Brook Reactor Facility, 1941–2002.* Monographs in Aerospace History, No. 33, 2004.

McCurdy, Howard E. *Low-Cost Innovation in Spaceflight.* Monographs in Aerospace History, No. 36, 2005.

Seamans, Robert C. *Project Apollo: The Tough Decisions.* Monographs in Aerospace History, No. 37, 2005.

Lambright, Henry W. *NASA and the Environment: The Case of Ozone Depletion.* Monographs in Aerospace History, No. 38, 2005.

Phillips, W. Hewitt. *Journey Into Space Research: Continuation of a Career at NASA Langley Research Center.* Monographs in Aerospace History, No. 40, 2005.

ELECTRONIC MEDIA, NASA SP-4600:

Remembering Apollo 11: The 30th Anniversary Data Archive CD-ROM. NASA SP-4601, 1999.

The Mission Transcript Collection: U.S. Human Spaceflight Missions from Mercury Redstone 3 to Apollo 17. NASA SP-2000-4602, 2001.

Shuttle-Mir: The United States and Russia Share History's Highest Stage. NASA SP-2001-4603, 2002.

U.S. Centennial of Flight Commission Presents Born of Dreams—Inspired by Freedom. NASA SP-2004-4604, 2004.

Of Ashes and Atoms: A Documentary on the NASA Plum Brook Reactor Facility. NASA SP-2005-4605, 2005.

Taming Liquid Hydrogen: The Centaur Upper Stage Rocket Interactive CD-ROM. NASA SP-2004-4606, 2004.

Fueling Space Exploration: The History of NASA's Rocket Engine Test Facility DVD. NASA SP-2005-4607, 2005.

CONFERENCE PROCEEDINGS, NASA SP-4700:

Dick, Steven J., and Keith L. Cowing, editors. *Risk and Exploration: Earth, Sea and the Stars.* NASA SP-2005-4701, 2005.

INDEX